T0190255

Lecture Notes in Computer Science 12306

More information about this series at http://www.springer.com/series/7412

Yuxin Peng · Qingshan Liu ·
Huchuan Lu · Zhenan Sun ·
Chenglin Liu · Xilin Chen ·
Hongbin Zha · Jian Yang (Eds.)

Pattern Recognition and Computer Vision

Third Chinese Conference, PRCV 2020
Nanjing, China, October 16–18, 2020
Proceedings, Part II

 Springer

Editors
Yuxin Peng
Peking University
Beijing, China

Huchuan Lu
Dalian University of Technology
Dalian, China

Chenglin Liu
Chinese Academy of Sciences
Beijing, China

Hongbin Zha
Peking University
Beijing, China

Qingshan Liu
Nanjing University of Information Science
and Technology
Nanjing, China

Zhenan Sun
Chinese Academy of Sciences
Beijing, China

Xilin Chen
Institute of Computing Technology
Chinese Academy of Sciences
Beijing, China

Jian Yang
Nanjing University of Science
and Technology
Nanjing, China

ISSN 0302-9743   ISSN 1611-3349 (electronic)
Lecture Notes in Computer Science
ISBN 978-3-030-60638-1   ISBN 978-3-030-60639-8 (eBook)
https://doi.org/10.1007/978-3-030-60639-8

LNCS Sublibrary: SL6 – Image Processing, Computer Vision, Pattern Recognition, and Graphics

This Springer imprint is published by the registered company Springer Nature Switzerland AG
The registered company address is: Gewerbestrasse 11, 6330 Cham, Switzerland

Preface

Welcome to the proceedings of the Third Chinese Conference on Pattern Recognition and Computer Vision (PRCV 2020) held in Nanjing, China.

PRCV is the merger of Chinese Conference on Pattern Recognition (CCPR) and Chinese Conference on Computer Vision (CCCV), which are both the most influential Chinese conferences on pattern recognition and computer vision, respectively. Pattern recognition and computer vision are closely interrelated and the two communities are largely overlapping. The goal of merging CCPR and CCCV into PRCV is to further boost the impact of the Chinese community in these two core areas of artificial intelligence and further improve the quality of academic communication. Accordingly, PRCV is co-sponsored by four major academic societies of China: the Chinese Association for Artificial Intelligence (CAAI), the China Computer Federation (CCF), the Chinese Association of Automation (CAA), and the China Society of Image and Graphics (CSIG).

PRCV aims at providing an interactive communication platform for researchers from academia and industry. It promotes not only academic exchange, but also communication between academia and industry. In order to keep at the frontier of academic trends and share the latest research achievements, innovative ideas, and scientific methods in the fields of pattern recognition and computer vision, international and local leading experts and professors are invited to deliver keynote speeches, introducing the latest advances in theories and methods in the fields of pattern recognition and computer vision.

PRCV 2020 was hosted by Nanjing University of Science and Technology and was co-hosted by Nanjing University of Information Science and Technology, Southeast University, and JiangSu Association of Artificial Intelligence. We received 402 full submissions. Each submission was reviewed by at least three reviewers selected from the Program Committee and other qualified researchers. Based on the reviewers' reports, 158 papers were finally accepted for presentation at the conference, including 30 orals, 60 spotlights, and 68 posters. The acceptance rate is 39%. The proceedings of PRCV 2020 are published by Springer.

We are grateful to the keynote speakers, Prof. Nanning Zheng from Xi'an Jiaotong University, China, Prof. Jean Ponce from PSL University, France, Prof. Mubarak Shah from University of Central Florida, USA, and Prof. Dacheng Tao from The University of Sydney, Australia.

We give sincere thanks to the authors of all submitted papers, the Program Committee members and the reviewers, and the Organizing Committee. Without their contributions, this conference would not be a success. Special thanks also go to all of the sponsors and the organizers of the special forums; their support made the conference a success. We are also grateful to Springer for publishing the proceedings and especially to Ms. Celine (Lanlan) Chang of Springer Asia for her efforts in coordinating the publication.

We hope you find the proceedings enjoyable and fruitful.

September 2020

Yuxin Peng
Qingshan Liu
Huchuan Lu
Zhenan Sun
Chenglin Liu
Xilin Chen
Hongbin Zha
Jian Yang

Organization

Steering Committee Chair

Tieniu Tan — Institute of Automation, Chinese Academy of Sciences, China

Steering Committee

Xilin Chen	Institute of Computing Technology, Chinese Academy of Sciences, China
Chenglin Liu	Institute of Automation, Chinese Academy of Sciences, China
Yong Rui	Lenovo, China
Hongbin Zha	Peking University, China
Nanning Zheng	Xi'an Jiaotong University, China
Jie Zhou	Tsinghua University, China

Steering Committee Secretariat

Liang Wang — Institute of Automation, Chinese Academy of Sciences, China

General Chairs

Chenglin Liu	Institute of Automation, Chinese Academy of Sciences, China
Xilin Chen	Institute of Computing Technology, Chinese Academy of Sciences, China
Hongbin Zha	Peking University, China
Jian Yang	Nanjing University of Science and Technology, China

Program Chairs

Yuxin Peng	Peking University, China
Qingshan Liu	Nanjing University of Information Science and Technology, China
Huchuan Lu	Dalian University of Technology, China
Zhenan Sun	Institute of Automation, Chinese Academy of Sciences, China

Organizing Chairs

Xin Geng   Southeast University, China
Jianfeng Lu   Nanjing University of Science and Technology, China
Liang Xiao   Nanjing University of Science and Technology, China
Jinshan Pan   Nanjing University of Science and Technology, China

Publicity Chairs

Zhaoxiang Zhang   Institute of Automation, Chinese Academy of Sciences,
                              China
Jiaying Liu   Peking University, China
Wankou Yang   Southeast University, China
Lianfa Bai   Nanjing University of Science and Technology, China

International Liaison Chairs

Jingyi Yu   ShanghaiTech University, China
Shiguang Shan   Institute of Computing Technology, Chinese Academy
                              of Sciences, China

Local Coordination Chairs

Wei Fang   JiangSu Association of Artificial Intelligence, China
Jinhui Tang   Nanjing University of Science and Technology, China

Publication Chairs

Risheng Liu   Dalian University of Technology, China
Zhen Cui   Nanjing University of Science and Technology, China

Tutorial Chairs

Gang Pan   Zhejiang University, China
Xiaotong Yuan   Nanjing University of Information Science
                              and Technology, China

Workshop Chairs

Xiang Bai   Huazhong University of Science and Technology,
                              China
Shanshan Zhang   Nanjing University of Science and Technology, China

Special Issue Chairs

Jiwen Lu	Tsinghua University, China
Weishi Zheng	Sun Yat-sen University, China

Sponsorship Chairs

Lianwen Jin	South China University of Technology, China
Jinfeng Yang	Civil Aviation University of China, China
Ming-Ming Cheng	Nankai University, China
Chen Gong	Nanjing University of Science and Technology, China

Demo Chairs

Zechao Li	Nanjing University of Science and Technology, China
Jun Li	Nanjing University of Science and Technology, China

Competition Chairs

Wangmeng Zuo	Harbin Institute of Technology, China
Jin Xie	Nanjing University of Science and Technology, China
Wei Jia	Hefei University of Technology, China

PhD Forum Chairs

Tianzhu Zhang	University of Science and Technology of China, China
Guangcan Liu	Nanjing University of Information Science and Technology, China

Web Chair

Zhichao Lian	Nanjing University of Science and Technology, China

Finance Chair

Jianjun Qian	Nanjing University of Science and Technology, China

Registration Chairs

Guangyu Li	Nanjing University of Science and Technology, China
Weili Guo	Nanjing University of Science and Technology, China

Area Chairs

Zhen Cui	Nanjing University of Science and Technology, China
Yuming Fang	Jiangxi University of Finance and Economics, China

Contents – Part II

Pattern Recognition and Application

Pattern Recognition and Application

Assessing Action Quality via Attentive Spatio-Temporal Convolutional Networks

Jiahao Wang, Zhengyin Du, Annan Li$^{(\boxtimes)}$, and Yunhong Wang

Beijing Advanced Innovation Center for Big Data and Brain Computing,
Beihang University, Beijing, China
{jhwang,duzhy,liannan,yhwang}@buaa.edu.cn

Abstract. Action quality assessment, which aims at evaluating the performance of specific actions, has drawn more and more attention due to its extensive demand in sports, health care, etc. Unlike action recognition, in which a few typical frames are sufficient for classification, action quality assessment requires analysis at a fine temporal granularity to discover the subtle motion difference. In this paper, we propose a novel spatio-temporal framework for action quality assessment at full-frame-rate (25fps), which consists of two steps: i.e. spatio-temporal feature extraction and temporal feature fusion, respectively. In the first step, to generate representative spatio-temporal dynamics, we utilize a spatial convolutional network (SCN) together with specially designed temporal convolutional networks (TCNs) and train them by a two-stage strategy. In the second step, we introduce an attention mechanism to fuse features in the temporal dimension according to their impact on the overall performance. Compared with existing three dimensional convolutional neural networks (3D-CNN) based methods, our model is capable of capturing more action quality relevant details. As a by-product, our model can also attend to the highlight moments in sports videos, which gives a better interpretation of the score. Extensive experiments on three public benchmarks demonstrate that the proposed method has distinct advantage in action quality assessment and achieves improvement over the state-of-the-art.

Keywords: Action quality assessment · Temporal convolution · Attentive fusion

1 Introduction

Automatically assessing action quality can facilitate a wide range of real-world applications, such as automated sports refereeing, surgical skill scoring, exercise therapy guidance and so on. Automated scoring systems can play an important role in various sports events when the scores given by human referees are

This work was supported by the National Key Research and Development Plan of China (Grant No. 2016YFB1001002) and the Foundation for Innovative Research Groups through the National Natural Science Foundation of China (Grant No. 61421003).

Y. Peng et al. (Eds.): PRCV 2020, LNCS 12306, pp. 3–16, 2020.
https://doi.org/10.1007/978-3-030-60639-8_1

t_1 $\quad\quad$ t_2 $\quad\quad$ t_3 $\quad\quad$ t_4 $\quad\quad$ t_5 $\quad\quad$ t_6

Fig. 1. In sports events, the critical moments are usually heterogeneously distributed. (a) The falling fault in figure skating (t_3, t_4). (b) The splash fault in diving (t_6).

questioned due to partiality or divergence [20,21]. Consequently, there is an increasing demand for developing action quality assessment systems that can generate accurate and objective results for sports videos.

Action recognition [3,4,22,24,26,30,31] is usually considered to be the most relevant computer vision problem to action quality assessment. Although both of the tasks need modeling human body movement in a spatio-temporal manner, two reasons make action quality assessment different from action recognition. Firstly, the quality score is usually determined by subtle difference among actions [17,20,21] while in action recognition the class-differences are often significant [30]. Secondly, the temporal distribution of frame-level action impact on the overall result is more heterogeneous in action quality assessment due to the uncertainty in sports events. For example, in figure skating events, as illustrated in Fig. 1(a), the skater falls down in frame t_3 and t_4 but quickly adjusts the posture afterward. In diving competitions, as shown in Fig. 1(b), the diver performs well in the air from frame t_1 to t_5 but makes a large splash in frame t_6. These critical moments are usually randomly distributed throughout the entire sports video.

Previous works mainly use discrete cosine transform (DCT) [21] or 3D-CNNs [17,20,29] to generate spatio-temporal action representations. The aforementioned subtleness and elusiveness in sports videos are overlooked in such approaches. For 3D-CNN based methods, due to the constrain of computing capability and memory consumption, these methods take sparsely sampled video frames (usually 16 frames for each video clip) as the input for extracting clip-level features, which will be simply averaged or concatenated to generate the final video representations. Such down-sampling inevitably causes loss in subtle details. Besides that, the averaging operation brings smoothing effects, salient moments shown in Fig. 1 will be smeared by adjacent frames.

In this paper, to overcome the limitations of existing methods, we revisit the spatio-temporal modeling in action quality assessment and formulate it as a two-step problem, i.e. spatio-temporal feature extraction and temporal feature fusion, respectively. The overall framework of our approach is illustrated in Fig. 2. In the first step, to capture fine-grained visual cues, we process videos at

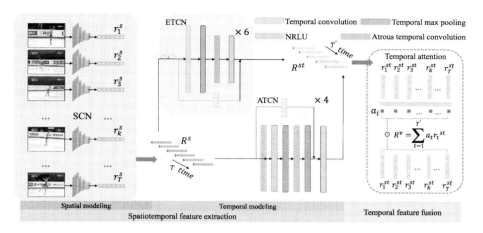

Fig. 2. Overall framework of the proposed method. Given the input video, the SCN firstly captures frame-level spatial representation R^s, on which the TCNs further extracts spatio-temporal features R^{st}. Finally, we employ attentive fusion on R^{st} to generate video representation R^v for score prediction.

full-frame-rate (25fps). Specifically, we employ a spatial convolutional network (SCN) [25] to capture spatial features and then utilize specially designed temporal convolutional networks (TCNs) [2, 16] to generate spatio-temporal representations. In the second step, to better exploit the randomly distributed key frames, we propose attentive temporal fusion to fuse spatio-temporal features for score prediction. The introduced temporal attention mechanism performs notably better than averaging or concatenating schemes especially for long videos. Extensive experiments on three public benchmarks, i.e. UNLV-Skating, UNLV-Vault and UNLV-Dive [20], demonstrate our superiority over the state-of-the-art.

In summary, our contributions are three-fold:

– We propose a novel approach for action quality assessment which employs a SCN together with specially designed TCNs for spatio-temporal modeling. Our method is capable of capturing detailed action quality cues by utilizing video frames at full-frame-rate.
– To exploit the randomly distributed key frames in sports videos, we introduce the attentive temporal fusion method which adaptively attends to critical actions.
– Our method achieves improvement over the state-of-the-art on three public action quality assessment benchmarks.

2 Related Work

2.1 Action Recognition

Action recognition aims to classify videos containing different human actions. Early researchers use hand-crafted features to express action information, such as space-time interest points (STIP) [13], histogram of optical flow (HOF) [14]. Recently, inspired by the success of deep architectures [8,12,25] in image classification [23], plenty of works adopt deep neural networks to pursuit more accurate recognition results. 3D-CNNs like C3D [26] and I3D [4] directly take video clips as input and model spatio-temporal information of videos in an end-to-end manner. In comparison, two-stream CNNs [24,31] use static video frames and optical flow features to respectively model the spatial and temporal information. As the difference of visual patterns between action classes are often significant, there are also some works that attempt to recognize actions more efficiently with single image [3] or few-shot learning [30].

2.2   Action Quality Assessment

The purpose of action quality assessment is to assess how well a certain kind of action is performed. As aforementioned, it is a nontrivial problem as it requires detailed action cues to determine the overall quality of action performance. Moreover, the influence of different moments within the whole action evolution varies drastically, which makes it challenging to appropriately fuse frame-level action representations. Over the past few years, several works [17,18,20,21,27,29] have been dedicated to explore automated action quality assessment in sports events (e.g. diving, figure skating, etc.), of which early methods tend to use human pose [21,27] features while the recent ones [17,20,29] employ 3D-CNNs [22,26] to directly extract features from video frames. Pirsiavash et al. [20] apply the discrete cosine transform (DCT) on pose features extracted by a pose estimator [19] and use support vector regression (SVR) to predict the action scores. In [20] and [17], C3D network [26] is used to extract clip-level spatio-temporal features. Xiang et al. [29] use P3D network [22] for stage-wise feature extraction. Due to the constrain of computational cost and memory demand, these 3D-CNN based methods usually take sparsely sampled video clips as input, which is likely to loss important action quality details. To fuse clip-level features, long short-term memory (LSTM) [9], average pooling and feature concatenation are adopted in [20], [29] and [17], respectively. Nevertheless, none of these methods learns to pay attention to the critical actions in videos.

3   Approach

3.1   Problem Formulation

Suppose we have a video $V\{v_t\}_{t=1}^{T}$, where T is the frame number and v_t represents the t-th frame, action quality assessment can be substantially formulated as a many-to-one problem, where all frames in V are mapped into a single score s to describe how well the action in V is performed. As mentioned in Sect. 1, we decompose the whole problem into two steps, i.e. spatio-temporal feature extraction and temporal feature fusion, which can be denoted as follows:

$$R^{st}\{r_t^{st}\}_{t=1}^{T'} = \mathcal{ST}(V),\tag{1}$$

$$s = \mathcal{F}(R^{st}). \tag{2}$$

Equation (1) represents the process of spatio-temporal feature extraction where we employ a certain method $\mathcal{ST}(\cdot)$ to extract spatio-temporal features $\{r_t^{st}\}_{t=1}^{T'}$ of temporal length T' from input video V. We denote the generated spatio-temporal representations as R^{st}. Next, as formulated in Eq. (2), we use the temporal fusion method $\mathcal{F}(\cdot)$ to fuse R^{st} and generate the final action quality score s.

3.2   Two-Stage Spatio-Temporal Feature Extraction

As action quality assessment needs to pay attention to detailed action cues, we desire a system that operates at full-frame-rate to retain motion details. However, current 3D-CNN methods are difficult to achieve that due to the limitation of computation and memory resources. Fortunately, recent research shows the temporal convolutional network (TCN) is a powerful and efficient tool for sequence modeling tasks including temporal action segmentation [16], machine translation [2] and so on. Also, experiments in [2] show TCNs have a series of merits including lower training difficulty and memory demand. Inspired by that, we propose a two-stage spatio-temporal framework, which consists of a SCN for spatial modeling and a TCN for temporal modeling, respectively.

Spatial Convolutional Network. In the first stage, the SCN is applied on every frame of the input video which we define as

$$R^s\{r_t^s\}_{t=1}^{T} = \mathcal{S}(V). \tag{3}$$

For every input video V of frame number T, we use the SCN (denoted as $\mathcal{S}(\cdot)$) to extract the same number of spatial representations $\{r_t^s\}_{t=1}^{T}$, which we denote as R^s. Specifically, we adopt the Inception-v3 network [25] pretained on the ImageNet dataset [6] as the backbone model for SCN. Since the task of ImageNet classification has a considerable gap with action quality assessment, we finetune the model on corresponding action quality assessment datasets. The detailed training process is introduced in Sect. 4.1.

Temporal Convolutional Network. In the second stage, we further extract spatio-temporal features with motion cues and contextual information on the basis of R^s. Most sports events are composed of a series of motions, e.g., a complete figure skating performance consists of actions of jumping, spinning, etc. Therefore, we must capture spatio-temporal features with rich temporal dynamics to model these movements. We employ the TCN [2,10,16] to generate spatio-temporal representations which can be formulated as

$$R^{st}\{r_t^{st}\}_{t=1}^{T'} = \mathcal{T}(R^s), \tag{4}$$

where R^s is the spatial features. We utilize a TCN model $\mathcal{T}(\cdot)$ to produce spatio-temporal representations R^{st} of temporal length T'. Note that the temporal length of R^{st} and R^s, i.e. T' and T can be different. We define TCNs with $T' < T$ as encoding TCN and $T' = T$ as plain TCN. We comprehensively investigate both architectures. For encoding TCN, we devise the encoding temporal convolutional network (ETCN). For plain TCN, we introduce the atrous temporal convolutional network (ATCN).

Encoding Temporal Convolutional Network. As shown in Fig. 3(a), encoding temporal convolutional network (ETCN) utilizes temporal convolutions [10] which convolve features in the temporal dimension. ETCN is designed under the paradigm of encoding networks, which decrease the output size after each convolutional layer. As illustrated in Fig. 3(c), ETCN is composed of encoding convolution blocks, which is formed by a sequence of temporal convolution, spatial dropout [16] and temporal max pooling. Inspired by the residual learning in [8], we also use residual connections in every block as it facilitates the training of deep networks. For each encoding block, we simply add up the input and output of the residual function if they have the same depth, otherwise we employ an 1×1 convolution to alter the depth of identity mapping. After the element-wise addition, we use a normalized rectified linear unit (NRLU) [16] layer to get the output of the block. We employ a ETCN with 6 convolution blocks, which have output dimensionalities of $\{256, 256, 256, 512, 512, 512\}$. Max pooling is only used in the first 4 convolution blocks with the stride of 2, which makes T'/T equals to $1/16$.

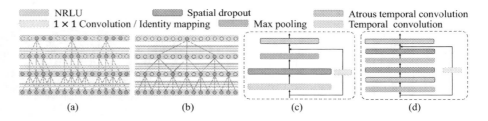

Fig. 3. Temporal convolutional networks. (a) An example of temporal convolutions in ETCN with kernel size 3. (b) An example of atrous convolutions in ATCN with kernel size 3 and exponentially increasing atrous rates. (c) Structure of encoding convolution block. (d) Structure of atrous convolution block.

Atrous Temporal Convolutional Network. As the temporal resolution reduction may cause information loss in output features, we propose the atrous temporal convolutional network (ATCN) which has an output feature length T' equals to the input length T. To achieve a field-of-view comparable to the ETCN without pooling layers, we utilize the atrous temporal convolution [10,28]. An illustration of the atrous convolution is shown in Fig. 3(b). For the l-th convolutional block, we apply an atrous rate of $2^{(l-1)}$ which effectively enlarges the field-of-view of the neurons. Our ATCN is composed of atrous convolution blocks, which is presented in Fig. 3(d). An atrous convolution block consists of two consecutive sequences of atrous temporal convolution, NRLU and spatial dropout. We also use the same residual connections as ETCN in atrous convolution blocks. In practice, we employ an ATCN with 4 atrous convolution blocks, which have output dimentionalities of $\{256, 256, 512, 512\}$.

3.3   Attentive Temporal Feature Fusion

After obtaining the spatio-temporal representations of the input video, we perform feature fusion in the temporal dimension so as to generate the final prediction. Obviously, during the execution of a sports event, there are some moments that play key roles in the final scoring. As mentioned in Sect. 1, a diver can be punished in final score for the splash at the last second of the video. Meanwhile, the attention mechanism have been demonstrated to be an effective solution for non-uniformly distributed temporal problems including machine translation [1], image generation [7]. Motivated by that, we propose our attentive temporal fusion strategy.

Given the spatio-temporal representations $R^{st}\{r_t^{st}\}_{t=1}^{T'}$ from TCNs, our attention weights are generated via a temporal convolution layer with kernel size 3 followed by a softmax layer, which can be formulated as

$$a_t = \frac{e^{p_t}}{\sum_{i=1}^{T'} e^{p_i}}, \tag{5}$$

where p_i denotes the convolution output at the i-th time step and a_t is the corresponding attention weight of r_t^{st}. With these attention weights, we calculate weighted average of R^{st} in the temporal dimension to produce video representation R^v as

$$R^v = \sum_{t=1}^{T'} a_t r_t^{st}. \tag{6}$$

Note that the fusion strategy is cascaded with the TCNs so as to be optimized jointly when training. Finally, we employ two fully connected (FC) layers to produce the final quality score prediction. For comparison, we also investigate the temporal average pooling scheme where every a_t equals to $1/T'$.

3.4   Loss Function

For traditional regression problems, the mean square error (MSE) is usually adopted as the loss function. We take MSE loss as the first term in our loss function, which is denoted as

$$L_{mse} = \frac{1}{2N} \sum_{i=1}^{N} (p_i - g_i)^2. \tag{7}$$

Given a training batch of size N, we use p_i and g_i to denote the prediction and ground truth scores of the i-th sample. As presented in Eq. (7), the MSE term minimizes the gap of score value between prediction and ground truth scores for each training sample. In action quality assessment tasks, it is important to retain the correct score ranking correlations between different samples. In other words, we want the predictions to reflect appropriate ranking order of individuals. Nevertheless, merely minimizing score value distance of each individual with the MSE loss often leads to inaccurate ranking order. Hence, we propose a correlation loss term to constrain the ranking correlations between different training samples, which can be formulated as

$$L_{corr} = \frac{1}{N(N-1)} \sum_{i=1}^{N} \sum_{j=1, j>i}^{N} ((p_j - p_i) - (g_j - g_i))^2, \tag{8}$$

where p and g represent the same values as in Eq. (7). For each pair of two predicted scores $\{p_i, p_j\}$, the correlation loss term computes the difference of correlation distance, i.e. $p_i - p_j$, between the prediction and the ground truth. The overall loss function is then formed as

$$L = \alpha L_{mse} + \beta L_{corr}, \tag{9}$$

where α and β are the weights for corresponding loss terms.

4   Experiments

4.1   Training and Implementation Details

SCN Training. As mentioned in Sect. 3.2, to adapt the Inception-v3 network to the action quality assessment task, for each dataset, we finetune the SCN with 10% of training data. In practice, we finetune the last two inception blocks of the SCN with action score supervision. Two additional FC layers, which respectively have output dimensionalities of 1024 and 1 are added on top of the last inception block to produce a single score. We simply take the mean square error (MSE) loss as loss function. Later when extracting spatio-temporal features with TCNs, we directly use output features from the last average pooling layer of the SCN.

TCN Training. Our TCNs take the spatial features captured by SCN as input and produce a single score for each video after attentive temporal fusion. As the input features have a dimensionality of 2048, both TCNs employ a 1×1 temporal convolution with output size 256 for dimensionality reduction. Also, we set the kernel size and convolution stride in TCNs to 7 and 1, respectively. The integrated loss introduced in Sect. 3.4 is adopted for TCN training.

Implementation Details. Our models are implemented with Keras [5]. For both SCN and TCN training, we use the ADAM [11] optimizer with the initial learning rate of 0.0001. For TCNs, we employ exponential learning rate decay every 50 training steps with an decay rate of 0.9. The spatial dropout probability is set to 0.1. We use batch size of 32 for SCN training and 8 for TCN training. To facilitate batch training for TCNs, we use zero padding at the end of each video to ensure the sequence lengths are equal. We also use horizontal flipping, random rotation and cropping for data augmentation when training the SCN.

4.2   Datasets and Metrics

Our framework is evaluated on three public action quality assessment datasets [20], which respectively contain three different Olympic events, i.e. figure skating, gymnastic vault and platform diving. All video frames are captured at 25fps for detailed visual cues. We present video number, average and max video length (in frames) of these datasets in Table 1. For UNLV-Dive and UNLV-Vault, the training/testing splits are the same as those in previous works, which are respectively 300/70 and 120/56. For UNLV-Skating, we repeat the experiments 200 times with random splits as in [20,21].

Table 1. General information of datasets.

Dataset	Sports event	#Videos	Avg. length	Max length
UNLV-Skating	Figure skating	171	4500	5823
UNLV-Vault	Gymnastic vault	176	75	100
UNLV-Dive	10 m platform diving	370	110	150

Note that the length of videos in UNLV-Skating dataset is nearly 50 times longer than the ones in other datasets, which means it is more challenging to aggregate subtle action information across time on this dataset. This feature hinders the performance of 3D-CNN models, because they have high temporal down-sampling rate without any additional temporal fusion method. We take the Spearman's rank correlation (SRC) used in previous works [17,20,21,27,29] as evaluation metrics. SRC assesses how well the relationship between two variables can be described using a monotonic function, which means predictions achieving higher SRC have a stronger ranking correlation with the ground-truth scores and vice versa. The range of SRC is between -1.0 and 1.0.

Table 2. Comparison with the state-of-the-art methods.

Method	UNLV-Skating	UNLV-Vault	UNLV-Dive
Pose+DCT [21]	0.35	0.10	0.53
ConvISA [15]	0.45	–	–
C3D+LSTM [20]	–	0.05	0.27
C3D+SVR [20]	0.53	0.66	0.78
C3D+CNN [17]	0.58	0.70	0.80
ScoringNet [18]	–	0.70	0.84
S3D [29]	–	–	**0.86**
Ours (SCN+ATCN)	**0.71**	**0.76**	0.85
Ours (SCN+ETCN)	0.59	0.73	0.81

4.3 Comparison with the State-of-the-Art

We compare the performance of our approach on three benchmarks with other
state-of-the-art methods [15,17,20,21,29]. Table 2 shows the results measured
in SRC. It can be observed that both of our SCN-TCN approaches outperform
the state-of-the-art methods on UNLV-Skating and UNLV-Vault datasets and
achieve a competitive performance on UNLV-Dive dataset. This result demonstrates the distinct advantage of our SCN-TCN framework in action quality
assessment. Furthermore, ATCN performs much better than ETCN on longer
videos (UNLV-Skating), indicating that the temporal length reduction degrades
the spatio-temporal representations of TCN in action quality assessment tasks.

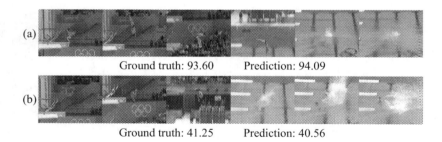

(a) Ground truth: 93.60   Prediction: 94.09

(b) Ground truth: 41.25   Prediction: 40.56

Fig. 4. Examples of two predictions from UNLV-Dive dataset. (a) An athlete with
good performance (video #179). (b) An athlete with bad performance (video #291).
Our model makes accurate prediction in both cases.

Notably, our method boost the performance on UNLV-Skating dataset by
a large margin (22%). As analyzed in [17,21], down-sampled input hinders the
performance of 3D-CNNs on long videos. This result further demonstrates the
capability of our model in capturing detailed action cues. Figure 4 shows two

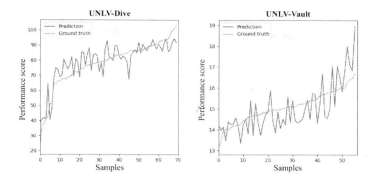

Fig. 5. Ground truth and predicted scores of UNLV-Dive and UNLV-Vault. To better visualize the ranking correlations, we rearrange the samples in an ascending order of ground truth scores.

qualitative examples of the ATCN model on UNLV-Dive dataset. It shows that our model can make accurate predictions for athletes of both good and bad performance, indicating that our framework can capture discriminative spatio-temporal representations for different performance levels. The predicted scores of UNLV-Dive and UNLV-Vault datasets generated by our ATCN model together with the ground truth scores are presented in Fig. 5. As the figure shows, for UNLV-Dive dataset, our model preserves most of the original ranking corre-lations while making predicted scores close to the ground-truth values. For UNLV-Vault, the predictions fluctuate more intensively around the ground truth, which is probably caused by the insufficiency of training samples and the blur of motions in gymnastic vault videos.

Although S3D [29] achieves a better result on UNLV-Dive dataset, they use additional manually annotated segmentation labels for temporal segmentation. Besides, annotating segmentation labels for sports videos is a labor-intensive work, which will take enormous efforts on datasets with larger scale or longer video length.

4.4   Effectiveness of Temporal Attention

In order to demonstrate the effectiveness of the proposed temporal attention mechanism, we compare it with simple temporal average pooling method and present the result in Table 3. The two evaluated models, which both employs ATCN, are exactly the same except the temporal fusion layers. Obviously, the one with temporal attention performs better on all three datasets. This result indicates that our temporal attention mechanism is capable of learning better fusion weights than sample average fusion. Note that for longer videos like figure skating, the attention mechanism gains more improvement. This demonstrates that the longer the sports event is, the more temporal attention is needed to determine the critical actions.

Table 3. Comparison of temporal fusion methods.

Method	UNLV-Skating	UNLV-Vault	UNLV-Dive
SCN+ATCN (average)	0.65	0.72	0.82
SCN+ATCN (attention)	**0.71**	**0.76**	**0.85**

An qualitative example of the attention weights derived by our ATCN model is visualized in Fig. 6, in which three frames with high attention values and two frames with low values are illustrated. Note that in the first high-value frame, the skater accidentally tumbles, while in the other two frames the skater performs twist and one-foot glide, respectively. These frames are worth paying attention to because each of them has significant influence on the overall performance, whether negatively or positively. In contrast, in the frames with low attention values, the skater remains in basic skating postures, which has little impact on the final score. This result demonstrates that our temporal attention mechanism is capable of capturing actions that contribute the most to the overall performance. As a by-product, the derived attention weights can provide a better understanding of the predicted scores and facilitate other video analysis tasks, such as sports highlight generation and video retrieval.

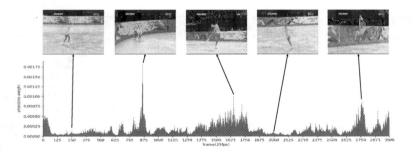

Fig. 6. Visualization of attention weights for video #18 from UNLV-Skating. Note that the three peak-value frames have significant influence on the overall performance, especially for the first one, where the skater accidentally tumbles. For the frames with low attention values, the skater remains in basic skating postures. Our model learns to focus on critical actions with this attention mechanism.

5   Conclusion

In this paper, we revisit the spatio-temporal modeling for action quality assessment and formulate it as a two-step problem consisting of spatio-temporal feature extraction and temporal feature fusion. In the first step, we introduce an effective SCN-TCN framework, in which fine-grained spatial features are captured by processing videos at full-frame-rate. For temporal modeling, two kinds of specially designed TCNs are evaluated. In the second step, we propose attentive temporal fusion which learns to focus on the critical sports actions. Experimental results demonstrate the effectiveness of the methodology and the superiority over the state-of-the-art.

References

1. Bahdanau, D., Cho, K., Bengio, Y.: Neural machine translation by jointly learning to align and translate. arXiv preprint arXiv:1409.0473 (2014)
2. Bai, S., Kolter, J.Z., Koltun, V.: An empirical evaluation of generic convolutional and recurrent networks for sequence modeling. arXiv preprint arXiv:1803.01271 (2018)
3. Bilen, H., Fernando, B., Gavves, E., Vedaldi, A., Gould, S.: Dynamic image networks for action recognition. In: CVPR, pp. 3034–3042 (2016)
4. Carreira, J., Zisserman, A.: Quo vadis, action recognition? A new model and the kinetics dataset. In: CVPR, pp. 6299–6308 (2017)
5. Chollet, F., et al.: Keras: the python deep learning library. Astrophysics Source Code Library (2018)
6. Deng, J., Dong, W., Socher, R., Li, L.J., Li, K., Fei-Fei, L.: Imagenet: a large-scale hierarchical image database. In: CVPR, pp. 248–255 (2009)
7. Gregor, K., Danihelka, I., Graves, A., Rezende, D.J., Wierstra, D.: Draw: a recurrent neural network for image generation. arXiv preprint arXiv:1502.04623 (2015)
8. He, K., Zhang, X., Ren, S., Sun, J.: Deep residual learning for image recognition. In: CVPR, pp. 770–778 (2016)
9. Hochreiter, S., Schmidhuber, J.: Long short-term memory. Neural Comput. **9**(8), 1735–1780 (1997)
10. Holschneider, M., Kronland-Martinet, R., Morlet, J., Tchamitchian, P.: A real-time algorithm for signal analysis with the help of the wavelet transform. In: Combes, J.M., Grossmann, A., Tchamitchian, P. (eds.) Wavelets, pp. 286–297. Springer, Heidelberg (1990). https://doi.org/10.1007/978-3-642-75988-8_28
11. Kingma, D.P., Ba, J.: Adam: a method for stochastic optimization. arXiv preprint arXiv:1412.6980 (2014)
12. Krizhevsky, A., Sutskever, I., Hinton, G.E.: Imagenet classification with deep convolutional neural networks. In: NeurIPS, pp. 1097–1105 (2012)
13. Laptev, I.: On space-time interest points. Int. J. Comput. Vision **64**(2–3), 107–123 (2005)
14. Laptev, I., Marszalek, M., Schmid, C., Rozenfeld, B.: Learning realistic human actions from movies. In: CVPR, pp. 1–8 (2008)
15. Le, Q.V., Zou, W.Y., Yeung, S.Y., Ng, A.Y.: Learning hierarchical invariant spatiotemporal features for action recognition with independent subspace analysis. In: CVPR, pp. 3361–3368 (2011)
16. Lea, C., Flynn, M.D., Vidal, R., Reiter, A., Hager, G.D.: Temporal convolutional networks for action segmentation and detection. In: CVPR, pp. 1003–1012 (2017)
17. Li, Y., Chai, X., Chen, X.: End-to-end learning for action quality assessment. In: PCM, pp. 125–134 (2018)
18. Li, Y., Chai, X., Chen, X.: ScoringNet: learning key fragment for action quality assessment with ranking loss in skilled sports. In: Jawahar, C.V., Li, H., Mori, G., Schindler, K. (eds.) ACCV 2018. LNCS, vol. 11366, pp. 149–164. Springer, Cham (2019). https://doi.org/10.1007/978-3-030-20876-9_10

19. Park, D., Ramanan, D.: N-best maximal decoders for part models. In: ICCV, pp. 2627–2634 (2011)
20. Parmar, P., Morris, B.T.: Learning to score olympic events. In: CVPRW, pp. 76–84 (2017)
21. Pirsiavash, H., Vondrick, C., Torralba, A.: Assessing the quality of actions. In: Fleet, D., Pajdla, T., Schiele, B., Tuytelaars, T. (eds.) ECCV 2014. LNCS, vol. 8694, pp. 556–571. Springer, Cham (2014). https://doi.org/10.1007/978-3-319-10599-4_36
22. Qiu, Z., Yao, T., Mei, T.: Learning spatio-temporal representation with pseudo-3D residual networks. In: ICCV, pp. 5534–5542 (2017)
23. Russakovsky, O., et al.: Imagenet large scale visual recognition challenge. Int. J. Comput. Vision **115**(3), 211–252 (2015)
24. Simonyan, K., Zisserman, A.: Two-stream convolutional networks for action recognition in videos. In: NeurIPS, pp. 568–576 (2014)
25. Szegedy, C., Vanhoucke, V., Ioffe, S., Shlens, J., Wojna, Z.: Rethinking the inception architecture for computer vision. In: CVPR, pp. 2818–2826 (2016)
26. Tran, D., Bourdev, L., Fergus, R., Torresani, L., Paluri, M.: Learning spatiotemporal features with 3D convolutional networks. In: ICCV, pp. 4489–4497 (2015)
27. Venkataraman, V., Vlachos, I., Turaga, P.K.: Dynamical regularity for action analysis. In: BMVC, pp. 67–1 (2015)
28. Wang, J., Du, Z., Li, A., Wang, Y.: Atrous temporal convolutional network for video action segmentation. In: ICIP, pp. 1585–1589 (2019)
29. Xiang, X., Tian, Y., Reiter, A., Hager, G.D., Tran, T.D.: S3D: stacking segmental P3D for action quality assessment. In: ICIP, pp. 928–932 (2018)
30. Xu, B., Ye, H., Zheng, Y., Wang, H., Luwang, T., Jiang, Y.G.: Dense dilated network for few shot action recognition. In: ICMR, pp. 379–387 (2018)
31. Zhu, Y., Lan, Z., Newsam, S., Hauptmann, A.G.: Hidden two-stream convolutional networks for action recognition. arXiv preprint arXiv:1704.00389 (2017)

Axial Data Modeling with Collapsed Nonparametric Watson Mixture Models and Its Application to Depth Image Analysis

Lin Yang[1], Yuhang Liu[1], and Wentao Fan[1,2,3]✉ iD

[1] Department of Computer Science and Technology,
Huaqiao University, Xiamen, Fujian, China
{19014083028,1625161012}@stu.hqu.edu.cn, fwt@hqu.edu.cn
[2] Xiamen Key Laboratory of Computer Vision and Pattern Recognition,
Huaqiao University, Xiamen, China
[3] Key Laboratory of Computer Vision and Machine Learning (Huaqiao University),
Fujian Province University, Xiamen, China

Abstract. Recently, axial data (i.e. the observations are axes of direction) have been involved with various fields ranging from blind speech separation to gene expression data clustering. In this paper, axial data modeling is performed by proposing a nonparametric infinite Watson mixture model which is constructed in a collapsed space (denoted by Co-InWMM) where the mixing coefficients are integrated out. Then, an effective collapsed variational Bayes (CVB) inference method is theoretically developed to learn the Co-InWMM with closed-from solutions. The proposed Co-InWMM with CVB inference for modeling axial data is validated through both synthetical data sets and a challenging application regarding depth image analysis.

Keywords: Axial data modeling · Mixture model · Watson distributions · Collapsed variational Bayes · Depth image analysis

1 Introduction

In recent years, directional data (i.e. the "direction" of the data is more important than their magnitude) analysis has drawn significant attention in various fields [12,14]. Typical directional data are the data that are normalized to have unit norm, which lie on the surface of the unit sphere. Since directional data

The first author is a student. The completion of this work was supported by the National Natural Science Foundation of China (61876068), the Natural Science Foundation of Fujian Province (2018J01094), the Promotion Program for Young and Middle-aged Teacher in Science and Technology Research of Huaqiao University (ZQNPY510) and Provincial Key Laboratory of Computer Vision and Machine Learning of Educational Department of Fujian Province (201902).

© Springer Nature Switzerland AG 2020
Y. Peng et al. (Eds.): PRCV 2020, LNCS 12306, pp. 17–28, 2020.
https://doi.org/10.1007/978-3-030-60639-8_2

are better represented on a manifold, the nonlinear nature of manifolds implies that common distributions such as the multivariate Gaussian distribution can not be used to model and analyze directional data. Alternatively, distributions that are defined on the unit hypersphere are more appropriate and effective to model directional data.

One of the most basic directional distributions is the von Mises-Fisher (vMF) distribution, which is defined on the unit hyperspahere ($\mathbb{S}^{D-1}$) and has similar characteristics to those of the multivariate Gaussian distribution defined in the Euclidean space $\mathbb{R}^D$. Although vMF distributions were widely involved with directional data modeling, it is not a universal solution to all types directional data. For instance, resent reach works have demonstrated that *axial data* where the observations are axes of direction (i.e. the unit vectors $\pm\boldsymbol{X}$ are indistinguishable) are better modeled with Watson distributions rather than with vMF [1]. As a special type of directional data, axial data have found their applications in various applications, such as blind speech separation [21], speech clustering in distributed microphone arrays [17], differentiation between normal and schizophrenic brains [13], gene expression data clustering analysis [7], etc.

Different methods have been proposed to learn Watson distributions or its natural extension the Watson mixture model (WMM). The major difficulty of learning Watson-based models lies on the fact that no analytically solution to the inference of the concentration parameters of Watson distributions can be found. Thus, approximation methods were proposed to solve this problem. A simple approximation method for large concentrations has been proposed in [13] to learn Watson distributions with the maximum likelihood (ML) estimates. This learning method, however, can not deal axial data with higher dimensions. In [1], an approximation to ML estimates has been proposed within an expectation maximization (EM) framework to learn WMMs. However, this method is prone to the problem of over-fitting. A better alternative method to ML estimates is the variational Bayes (VB) [4,9], a method that approximates posterior distributions through optimization. In [18], a VB inference method was proposed to learn WMMs and demonstrated better performance than the ML estimates. Although closed-form solutions can be obtained by this method, the evaluation of the model complexity (i.e. the number of mixture components model that best fit the data) requires extra effort. Specifically, the VB inference method in [18] treats the mixing coefficients of the WMM as random variables which are assigned with a Dirichlet prior. Then, model selection was performed by removing the components with small responsibilities. A more elegant solution to the model selection problem in modeling WMMs was proposed in [7], where a nonparametric framework known as the *Dirichlet process mixture model* [3,10] was adopted to define the WMM with an infinite number of components. By applying VB inference method to learn the infinite WMM (In-WMM), the number of mixture component can be freely initialized and will be adjusted automatically as the data set increases [7].

Although both VB inference methods ([18] and [7]) are effective to learn WMMs, to ensure closed-form solutions, the VB inference has to adopt the mean-field assumption [2] where parameters are assumed to be independent. This assumption, however, is not realistic in the WMM or In-WMM in which the mixing coefficients and the latent indicator variables are obviously closely related. This issue can be addressed by applying VB inference in a collapsed space where parameters are marginalized out, which leads to the so-called collapsed VB (CVB) inference framework [20]. As described in [11,20], the mean-field assumption is more satisfied with CVB without the concern of dependencies between parameters. Thus, in this work we focus on developing an effective CVB inference method to learn the In-WMM in a collapsed space where the mixing coefficients are integrated out.

We summarize the contributions of this work as follows. Firstly, a collapsed infinite WMM (Co-InWMM) is proposed for modeling axial data by marginalizing out the mixing coefficients. Secondly, an effective CVB inference method is theoretically developed to learn Co-InWMM with closed-from solutions. Lastly, the proposed Co-InWMM with CVB inference is validated through both synthetical data sets and a challenging application about depth image analysis.

2   The Collapsed Infinite WMM

2.1   Infinite Watson Mixture Models

Given a data set $\mathcal{X} = \{\boldsymbol{x}_i\}_{i=1}^N$ which contains N axial random vectors (i.e. $\boldsymbol{x}$ and $-\boldsymbol{x}$ are equivalent), each D-dimensional data vector can be represented as a unit vector (i.e. $\|\boldsymbol{x}\|_2 = 1$) defined on a $(D-1)$-dimensional unit hypersphere $\mathbb{S}^{D-1}$. If each vector $\boldsymbol{x}$ is a drawn from a mixture of an infinite number of Watson distributions, then the probability density function of this infinite Watson mixture model (InWMM) is given by

$$p(\boldsymbol{x}|\boldsymbol{\pi},\boldsymbol{\mu},\boldsymbol{\gamma}) = \sum_{k=1}^{\infty} \pi_k \mathcal{W}(\boldsymbol{x}|\boldsymbol{\mu}_k,\gamma_k) \tag{1}$$

where $\boldsymbol{\pi} = \{\pi_k\}_{k=1}^{\infty}$ represent the mixing coefficients that should be nonnegative and sum to 1; $\boldsymbol{\mu} \in \mathbb{S}^{D-1}$ denotes the *mean direction* with $\|\boldsymbol{\mu}\|_2 = 1$, and $\gamma \in \mathbb{R}$ represents the *concentration*. $\mathcal{W}(\boldsymbol{x}_i|\boldsymbol{\mu}_k,\gamma_k)$ indicates the Watson distribution associated with the kth component of the mixture model and is defined by

$$\mathcal{W}(\boldsymbol{x}|\boldsymbol{\mu}_k,\gamma_k) = \frac{\Gamma(D/2)}{2\pi^{D/2}M(\frac{1}{2},\frac{D}{2},\gamma_k)} \exp[\gamma_k(\boldsymbol{\mu}_k^T\boldsymbol{x})^2] \tag{2}$$

where $M(a,b,\cdot)$ represents the Kummer function (also known as the confluent hypergeometric function) which is given by

$$M(a,b,\gamma) = \sum_{n=0}^{\infty} \frac{\Gamma(a+n)\Gamma(b)}{\Gamma(a)\Gamma(b+n)} \frac{\gamma^n}{n!} \tag{3}$$

where $\Gamma(\cdot)$ denotes the Gamma function.

Next, each vector $\boldsymbol{x}_i$ is assigned with a latent indicator variable z_i which is used to indicate the component from which $\boldsymbol{x}_i$ is drawn. For the data set $\mathcal{X}$, the distribution of indicator variables $\boldsymbol{z} = \{z_i\}_{i=1}^N$ can be represented by

$$p(\boldsymbol{z}|\boldsymbol{\pi}) = \prod_{i=1}^{N} \prod_{k=1}^{\infty} \pi_k^{1[z_i=k]} \tag{4}$$

where $\mathbf{1}[\cdot]$ denotes the indicator function which equals 1 when $z_i = k$, otherwise it equals 0.

2.2   Prior Distributions

The InWMM is constructed using a Bayesian framework, in which each unknown variable is assigned with a prior distribution. A nonparametric prior namely Dirichlet process [10] is considered for the mixing coefficients $\boldsymbol{\pi}$, and is defined in terms of a stick-breaking representation [3] as

$$\pi_k = \pi_k' \prod_{s=1}^{k-1}(1 - \pi_s'), \quad \pi_k' \sim \text{Beta}(1, \varpi_k), \quad G = \sum_{k=1}^{\infty} \pi_k \delta_{\theta_k}, \quad \theta_k \sim H \tag{5}$$

where G is a drawn from the Dirichlet process $G \sim DP(\varpi, H)$ with the base distribution H and scaling parameter ϖ, where δ_{θ_k} is an atom at θ_k.

Following [7,18], a Watson-Gamma prior is selected for parameters $\boldsymbol{\mu}$ and $\boldsymbol{\gamma}$ as

$$p(\boldsymbol{\mu}, \boldsymbol{\gamma}) = \prod_{k=1}^{\infty} \mathcal{W}(\boldsymbol{\mu}_k|\boldsymbol{m}_k, \beta_k\gamma_k)\mathcal{G}(\gamma_k|a_k, b_k) \tag{6}$$

where $\mathcal{G}(\cdot)$ indicates the Gamma distribution.

2.3   Collapsed Infinite Watson Mixture Models

According to several recent works in the literature of mixture modeling [5,6], better performance often would be obtained when model learning was conducted in a collapsed space where some or all of the parameters are marginalized out. In our case, inspired from [5,6,11], we re-formulate a collapsed version of InWMM (i.e. the Co-InWMM) by marginalizing out the mixing coefficients $\boldsymbol{\pi}$. Consequently, the latent variable $\boldsymbol{z}$ does not depend on the mixing coefficients $\boldsymbol{\pi}$ anymore and is distributed as

$$p(\boldsymbol{z}) = \prod_{k=1}^{\infty} \frac{\Gamma(1 + n_k)\Gamma(\varpi_k + n_{>k})}{\Gamma(1 + \varpi_k + n_{\geq k})} \tag{7}$$

where $n_k = \sum_{i=1}^N \mathbf{1}[z_i = k]$ indicates the number of data instances from the kth component, $n_{>k} = \sum_{i=1}^N \mathbf{1}[z_i > k]$, and $n_{\geq k} = n_k + n_{>k}$.

The conditional distribution of $z_i = k$ given the current state of all except one variable z_i is

$$p(z_i = k|\boldsymbol{z}^{\neg i}) \propto (1 + n_k^{\neg i})(\varpi_k + n_{>k}^{\neg i})(1 + \varpi_k + n_{\geq k}^{\neg i})^{-1} \tag{8}$$

where the superscript $\neg i$ indicates the associated ith term is removed.

The joint distribution of all latent and random variables in the Co-InWMM is given by

$$p(\mathcal{X}, \boldsymbol{z}, \boldsymbol{\mu}, \boldsymbol{\gamma}) = \prod_{i=1}^{N} p(\boldsymbol{x}_i | \boldsymbol{\mu}_{z_i}, \gamma_{z_i}) p(z_i) \prod_{k=1}^{\infty} p(\boldsymbol{\mu}_k, \gamma_k) \qquad (9)$$

In contrast with the InWMM as described in Eq. (1), the Co-InWMM has two major advantages: 1) the explicit dependency between latent variables $\boldsymbol{z}$ and mixing coefficients $\boldsymbol{\pi}$ is broken, which will be in favor of the mean-filed variational Bayes model learning method as developed in the following section; 2) a smaller number of parameters are obtained by integrating out $\boldsymbol{\pi}$, which leads to a faster inference process with better performance.

3   Model Learning

In this section, based on the VB inference methods that were respectively proposed in [7,18] for learning finite WMM and InWMM, we develop an effective method based on collapsed variational Bayes (CVB) [11,20] to learn the proposed Co-InWMM with closed-form solutions.

3.1   Mean-Field Collapsed Variational Inference

VB inference is an effective method for approximating posterior dentistries in Bayesian models. In our case, VB is adopted to approximate the true posterior $p(\Theta | \mathcal{X})$ with an approximated posterior $q(\Theta)$ (also referred to as *variational posterior*), where $\Theta = \{\boldsymbol{z}, \boldsymbol{\mu}, \boldsymbol{\gamma}\}$ denotes the set of all latent and random variables of the Co-InWMM. VB inference solves the problem of approximation though optimization, by minimizing the Kullback-Leibler (KL) divergence between $q(\Theta)$ and $p(\Theta | \mathcal{X})$, which is equivalent to maximizing the lower bound of $\ln p(\mathcal{X})$ that is defined by

$$\mathcal{L}(q) = \int q(\Theta) \ln[p(\mathcal{X}, \Theta)/q(\Theta)] d\Theta \qquad (10)$$

To perform VB inference for learning Co-InWMM which contains an infinite number of mixture components, a common technique is to truncate the stick-breaking representation of Co-InWMM at a finite value K as

$$\pi'_K = 1, \quad \sum_{k=1}^{K} \pi_k = 1, \quad \pi_k = 0 \ \text{ when } \ k > K \qquad (11)$$

where K can be freely initialized and would be inferred automatically through VB inference.

To obtain closed-from solutions, *mean-field* assumption [2] is often adopted in VB inference to factorize the variational posterior as the product of independent factors, where each factor represents variational posterior of the corresponding variable. In [7], the variational posterior of InWMM with truncation was factorized as

$$q(\Theta) = q(\boldsymbol{\pi})q(\boldsymbol{z})q(\boldsymbol{\mu}, \boldsymbol{\gamma}) \tag{12}$$

This factorization assumption, however, clearly violates the fact that latent variables $\boldsymbol{z}$ and mixing coefficients $\boldsymbol{\pi}$ are closely related with strong dependency as demonstrated in Eq. (4). The mean-field assumption is more satisfied in Co-InWMM where $\boldsymbol{\pi}$ are marginalized out as

$$q(\Theta) = \prod_{i=1}^{N} \left[q(z_i) \right] \prod_{k=1}^{K} \left[q(\boldsymbol{\mu}_k, \gamma_k) \right] \tag{13}$$

Then, we can obtain the following update equations by maximizing the lower bound $\mathcal{L}(q)$ with respect to each variational posterior

$$q(\boldsymbol{z}) = \prod_{i=1}^{N} \prod_{k=1}^{K} r_{ik}^{\mathbf{1}[z_i=k]} \tag{14}$$

$$q(\boldsymbol{\mu}, \boldsymbol{\gamma}) = \prod_{k=1}^{K} \mathcal{W}(\boldsymbol{\mu}_k | \boldsymbol{m}_k^*, \beta_k^* \gamma_k) \mathcal{G}(\gamma_k | a_k^*, b_k^*) \tag{15}$$

where the hyperparameters in the above variational posteriors are calculated by

$$r_{ik} = \frac{\widetilde{r}_{ik}}{\sum_{s=1}^{K} \widetilde{r}_{is}}, \tag{16}$$

$$
\begin{aligned}
\widetilde{r}_{ik} =& \ln \Gamma(\frac{D}{2}) - \frac{D}{2} \ln 2\pi + \frac{D}{2} \langle \ln \gamma_k \rangle - \ln[\bar{\gamma}_k^{\frac{D}{2}} M(\frac{1}{2}, \frac{D}{2}, \bar{\gamma}_k)] \\
&- \frac{\partial}{\partial \bar{\gamma}_k} \left[\ln \bar{\gamma}_k^{\frac{D}{2}} M(\frac{1}{2}, \frac{D}{2}, \bar{\gamma}_k) \right] (\langle \gamma_k \rangle - \bar{\gamma}_k) \\
&+ \bar{\gamma}_k \vartheta(\beta_k^* \bar{\gamma}_k) + \{ \bar{\gamma}_k [\vartheta(\beta_k^* \bar{\gamma}_k) + \beta_k^* \bar{\gamma}_k \vartheta'(\beta_k^* \bar{\gamma}_k)] \\
&\times (\langle \ln \gamma_k \rangle + \ln \beta_k^* - \ln \beta_k^* \bar{\gamma}_k) \} (\boldsymbol{m}_k^{*T} \boldsymbol{X}_i)^2 \\
&+ \langle \ln(1 + n_k^{-i}) \rangle - \langle \ln(1 + \varpi_k + n_{\geq k}^{-i}) \rangle \\
&+ \sum_{j<k} \left[\langle \ln(\varpi_j + n_{>j}^{-i}) \rangle - \langle \ln(1 + \varpi_j + n_{\geq j}^{-i}) \rangle \right]
\end{aligned} \tag{17}
$$

$$a_k^* = a_k + \frac{D}{2} (1 + \sum_{i=1}^{N} \langle z_{i=k} \rangle) + \beta_k^* \bar{\gamma}_k \frac{\partial}{\partial \beta_k^* \bar{\gamma}_k} \ln M(\frac{1}{2}, \frac{D}{2}, \beta_k^* \bar{\gamma}_k) \tag{18}$$

$$
\begin{aligned}
b_k^* =& b_k + \sum_{i=1}^{N} \langle z_{i=k} \rangle \frac{\partial}{\partial \bar{\gamma}_k} \left[\ln \bar{\gamma}_k^{\frac{D}{2}} M(\frac{1}{2}, \frac{D}{2}, \bar{\gamma}_k) \right] \\
&+ \beta_k \frac{\partial}{\partial \beta_k \bar{\gamma}_k} \left[\ln(\beta_k \bar{\gamma}_k)^{\frac{D}{2}} M(\frac{1}{2}, \frac{D}{2}, \beta_k \bar{\gamma}_k) \right]
\end{aligned} \tag{19}
$$

$$A = \beta_k \boldsymbol{m}_k \boldsymbol{m}_k^T + \sum_{i=1}^{N} \langle z_{i=k} \rangle \boldsymbol{x}_i \boldsymbol{x}_i^T \tag{20}$$

where $\vartheta(x) = \frac{\partial}{\partial x} \ln M\left(\frac{1}{2}, \frac{D}{2}, x\right)$, β_k^* is the largest eigenvalue of A, $\boldsymbol{m}_k^*$ represents the corresponding eigenvector to β_k^*. The expected values in above equations are given by

$$\langle z_{i=k} \rangle = r_{ik}, \qquad \bar{\gamma}_k = a_k^*/b_k^*, \qquad \langle \ln \gamma_k \rangle = \psi(a_k^*) - \ln b_k^* \tag{21}$$

$$\langle \ln(1 + n_k^{\neg i}) \rangle \approx \ln(1 + \langle n_k^{\neg i} \rangle), \tag{22}$$

$$\langle \ln(\varpi_k + n_{>k}^{\neg i}) \rangle \approx \ln(\varpi_k + \langle n_{>k}^{\neg i} \rangle) \tag{23}$$

$$\langle \ln(1 + \varpi_k + n_{\geq k}^{\neg i}) \rangle \approx \ln(1 + \varpi_k + \langle n_{\geq k}^{\neg i} \rangle) \tag{24}$$

$$\langle n_k^{\neg i} \rangle = \sum_{i' \neq i} r_{i'k}, \qquad \langle n_{>k}^{\neg i} \rangle = \sum_{i' \neq i} \sum_{s=k+1}^{K} r_{i's}, \qquad \langle n_{\geq k}^{\neg i} \rangle = \langle n_k^{\neg i} \rangle + \langle n_{>k}^{\neg i} \rangle \tag{25}$$

where the expected values of $\ln(1 + n_k^{\neg i})$, $\ln(\varpi_k + n_{>k}^{\neg i})$, and $\ln(1 + \varpi_k + n_{\geq k}^{\neg i})$ were acquired according to Gaussian approximations [20] with 0th-order Taylor approximation [15]. Our CVB inference method for learning the Co-InWMM is analogous to the maximum likelihood expectation maximization (EM) algorithm, which is summarized in Algorithm 1.

Algorithm 1. CVB Inference of the Co-InWMM.

1: Initialize the truncation level K.
2: Initialize the hyper-parameters a_k, b_k, ϖ_k, and β_k.
3: Apply K-Means algorithm to initialize r_{ik}.
4: **repeat**
5:      *The variational E-step*:
6:      Estimate the expected values in (21)~(25), use the current distributions over
        the model parameters.
7:      *The variational M-step*:
8:      Update the variational posteriors with (14) and (15) based on the estimated
        expected values.
9: **until** The convergence criterion is satisfied.

---

4   Experimental Results

The proposed Co-InWMM with CVB inference is evaluated through two experiments involved with both simulated data and a application about depth image analysis. In our experiments, the truncation level K is initialized to 10, ϖ_k and β_k are set to 1, a_k and b_k are initialized to 1 and 0.01, respectively. These initial values were found through cross validation.

4.1   Synthetic Data

The principal purpose of conducting experiments on synthetic axial data is to validate the "correctness" of the proposed CVB inference algorithm in learning the proposed Co-InWMM. This is fulfilled by verifying the discrepancy between computed values of the parameters and their true values. A synthetic data set was generated to conduct the experiments. This data set contains 900 3-dimensional data instances which are drawn from 3 Watson distributions (as demonstrated in Fig. 1).

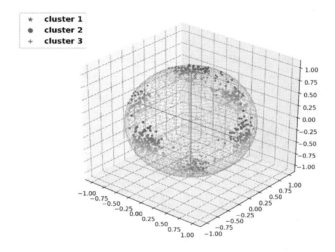

Fig. 1. The synthetic data set.

The true parameters that were used to generate the data set and the estimated parameters by CVB inference method are shown in Table 1. According to this table, the proposed learning algorithm is able to effectively learn the Co-InWMM with estimated values of parameters that are vary close to the true ones.

4.2   Depth Image Analysis

In this experiment, we apply the proposed Co-InWMM to a challenging application namely depth image analysis. We use the NYU-V2 depth data set [16] to conduct our experiments. This data set includes 1449 rgb-d images collected from three different cities in the United States, consisting of 464 indoor different scenes across 26 scene classes in commercial buildings and residences. Following [8], we compute surface normals of depth images and then apply Co-InWMM for clustering the normals. It is worth noting that the axially symmetric property of WMM can naturally overcome the ambiguity signals caused by the normal vector which calculated by plane fitting method.

Table 1. Parameters estimation of the synthetic data set.

N_k	k	μ_{k1}	μ_{k2}	μ_{k3}	κ_k	$\hat{\mu}_{k1}$	$\hat{\mu}_{k2}$	$\hat{\mu}_{k3}$	$\hat{\kappa}_k$
300	1	0	0	1	15	0.01	0.02	0.99	15.46
300	2	0	1	0	22	0.01	0.99	−0.01	22.77
300	3	1	0	0	17	0.99	0.00	−0.01	17.19

Figure 2 shows the number of estimated clusters for all NYU-V2 depth data set obtained by finite WMM with the Integrated Completed Likelihood (ICL) criteria [8] and the proposed Co-InWMM. As can we can see from the figure, most of the images contain 3–4 clusters. It is note worthy that, the WMM method in [8] has to calculate the ICL criteria with different number of clusters in order to determine the optimal number. In contrast, our model can detect the number of clusters automatically with a single run.

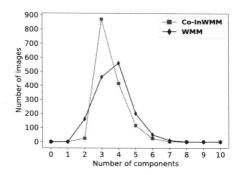

Fig. 2. Estimated number of clusters for NYU-V2 depth data set.

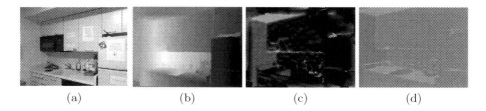

| (a) | (b) | (c) | (d) |

Fig. 3. Cluster example in the NYU-V2 depth data set. (a) rgb image; (b) depth image; (c) normals; (d) results by Co-InWMM.

Figure 3 shows the example of depth image analysis. From the results we observe that, different clusters represent different image regions and also represent the segment plane associated with the scene with a specific axis. Other

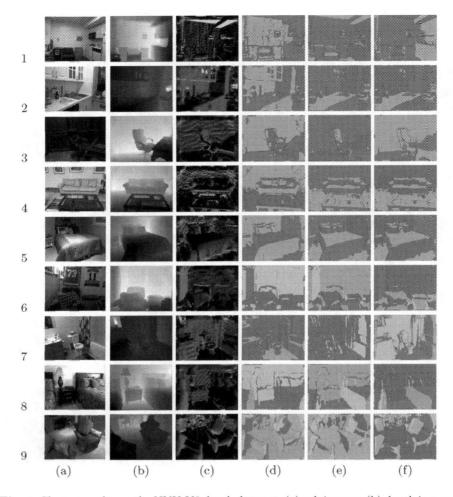

Fig. 4. Cluster results on the NYU-V2 depth data set. (a) rgb images; (b) depth images; (c) normals; (d) results by WMM; (e) results by In-WMM; (f) results by Co-InWMM.

results can be seen in Fig. 4. Through the results, we can see that some classes represent some nonplanar objects (see case-7 and case-9 of Fig. 4), which means that our method can find nonplanar objects. From case-3 and case-5, we can see a lot of noise on the normal vector, but our method can still identify plane and nonplaner objects well. In addition, similar to [8], we also find that the data with lower prior probability will be divided into fewer clusters. In order to solve this problem, a reasonable solution is to highlight each cluster by preprocessing the normal vector to make the clustering more accurate.

In order to show the superiority of our model, we compare it with Kmeans, finite vMFMM [19], finite WMM [8] and the In-WMM proposed in [7] in terms of clustering performance on normals and computational runtime. It should be

Table 2. Results obtained by different methods in terms of MI and computational runtime (in min.)

Algorithm	MI	Time
Kmeans	0.293	241.39
vMFMM [19]	0.329	273.19
WMM [8]	0.335	270.00
In-WMM [7]	0.347	164.22
Co-InWMM	**0.355**	**156.97**

noted that the first three algorithms use ICL criteria to calculate the optimal cluster number. We use mutual information (MI) to evaluate the performance of clustering. The specific results are shown in Table 2. Based on the results shown in this table, it is obvious that the Co-InWMM is able to provide better clustering performance in terms of the highest MI value. Moreover, the Co-InWMM is more computational efficient than other tested methods in terms of the shortest computational runtime. This result demonstrates the advantages of constructing the nonparametric infinite WMM in a collapsed space, where mixing coefficients are integrated out and thus leads to a smaller number of parameters that have to be estimated.

5   Conclusion

In this paper, we proposed a collapsed infinite Watson mixture model for modeling axial data where the mixing coefficients are integrated out. We developed an effective collapsed variational Bayes inference method to learn the proposed model with closed-from solutions. The effectiveness of the proposed Co-InWMM with CVB inference for modeling axial data was verified through experiments that were conducted on both synthetical data sets and a challenging application regarding depth image analysis.

References

1. Bijral, A.S., Breitenbach, M., Grudic, G.Z.: Mixture of watson distributions: a generative model for hyperspherical embeddings. In: Proceedings of the Eleventh International Conference on Artificial Intelligence and Statistics, pp. 35–42 (2007)
2. Bishop, C.M.: Pattern Recognition and Machine Learning. Springer, New York (2006)
3. Blei, D.M., Jordan, M.I.: Variational inference for Dirichlet process mixtures. Bayesian Anal. **1**, 121–144 (2005)
4. Blei, D.M., Kucukelbir, A., Mcauliffe, J.: Variational inference: a review for statisticians. J. Am. Stat. Assoc. **112**(518), 859–877 (2017)
5. Fan, W., Bouguila, N.: Modeling and clustering positive vectors via nonparametric mixture models of Liouville distributions. IEEE Trans. Neural Netw. Learn. Syst. 1–11 (2019). https://doi.org/10.1109/TNNLS.2019.2938830

6. Fan, W., Bouguila, N.: Simultaneous clustering and feature selection via nonparametric Pitman-Yor process mixture models. Int. J. Mach. Learn. Cybernet. **10**(10), 2753–2766 (2019)
7. Fan, W., Bouguila, N., Du, J., Liu, X.: Axially symmetric data clustering through Dirichlet process mixture models of Watson distributions. IEEE Trans. Neural Netw. Learn. Syst. **30**(6), 1683–1694 (2019)
8. Hasnat, M.A., Alata, O., Trémeau, A.: Unsupervised clustering of depth images using watson mixture model. In: 2014 22nd International Conference on Pattern Recognition, pp. 214–219 (2014)
9. Jordan, M.I., Ghahramani, Z., Jaakkola, T.S., Saul, L.K.: An introduction to variational methods for graphical models. Mach. Learn. **37**(2), 183–233 (1999)
10. Korwar, R.M., Hollander, M.: Contributions to the theory of Dirichlet processes. Ann. Probab. **1**, 705–711 (1973)
11. Kurihara, K., Welling, M., Teh, Y.W.: Collapsed variational Dirichlet process mixture models. In: Proceedings of International Joint Conference on Artificial Intelligence (IJCAI), pp. 2796–2801 (2007)
12. Ley, C., Verdebout, T.: Applied Directional Statistics: Modern Methods and Case Studies. Chapman and Hall/CRC, Boca Raton (2018)
13. Mardia, K.V., Dryden, I.L.: The complex Watson distribution and shape analysis. J. Roy. Stat. Soc.: Ser. B (Stat. Methodol.) **61**(4), 913–926 (1999)
14. Mardia, K.V., Jupp, P.E.: Directional Statistics. Wiley, Hoboken (2000)
15. Sato, I., Nakagawa, H.: Rethinking collapsed variational bayes inference for LDA. In: Proceedings of the 29th International Conference on Machine Learning, ICML 2012 (2012)
16. Silberman, N., Hoiem, D., Kohli, P., Fergus, R.: Indoor segmentation and support inference from RGBD images. In: Fitzgibbon, A., Lazebnik, S., Perona, P., Sato, Y., Schmid, C. (eds.) ECCV 2012. LNCS, vol. 7576, pp. 746–760. Springer, Heidelberg (2012). https://doi.org/10.1007/978-3-642-33715-4_54
17. Souden, M., Kinoshita, K., Nakatani, T.: An integration of source location cues for speech clustering in distributed microphone arrays. In: 2013 IEEE International Conference on Acoustics, Speech and Signal Processing, pp. 111–115 (2013)
18. Taghia, J., Leijon, A.: Variational inference for Watson mixture model. IEEE Trans. Pattern Anal. Mach. Intell. **38**(9), 1886–1900 (2016)
19. Taghia, J., Ma, Z., Leijon, A.: Bayesian estimation of the von-Mises fisher mixture model with variational inference. IEEE Trans. Pattern Anal. Mach. Intell. **36**(9), 1701–1715 (2014)
20. Teh, Y.W., Newman, D., Welling, M.: A collapsed variational Bayesian inference algorithm for latent Dirichlet allocation. In: Proceedings of Advances in Neural Information Processing Systems (NIPS), pp. 1353–1360 (2007)
21. Vu, D.H.T., Haeb-Umbach, R.: Blind speech separation employing directional statistics in an expectation maximization framework. In: 2010 IEEE International Conference on Acoustics, Speech and Signal Processing, pp. 241–244 (2010)

Diagonal Symmetric Pattern Based Illumination Invariant Measure for Severe Illumination Variations

Changhui Hu[1,3]($\boxtimes$), Mengjun Ye[2], Yang Zhang[1], and Xiaobo Lu[1]

[1] School of Automation, Southeast University, Nanjing 210096, China
hchseu@qq.com
[2] College of Mechatronics and Control Engineering, Hubei Normal University,
Huangshi 435002, China
[3] College of Automation and College of Artificial Intelligence,
Nanjing University of Posts and Telecommunications, Nanjing 210023, China

Abstract. This paper proposes a diagonal symmetric pattern (DSP) to develop the illumination invariant measure for severe illumination variations. Firstly, the subtraction of two diagonal symmetric pixels is defined as the DSP unit in the face local region, which may be positive or negative. The DSP model is obtained by combining the positive and negative DSP units. Then, the DSP model can be used to generate several DSP images based on the 4×4 block region by controlling the proportions of positive and negative DSP units, which results in the DSP image. The single DSP image with the arctangent function can develop the DSP-face. Multi DSP images employ the extended sparse representation classification (ESRC) as the classifier that can form the DSP images based classification (DSPC). Further, the DSP model is integrated with the pre-trained deep learning (PDL) model to construct the DSP-PDL model. Finally, the experimental results on the Extended Yale B, CMU PIE and VGGFace2 test face databases indicate that the proposed methods are efficient to tackle severe illumination variations.

Keywords: Diagonal symmetric pattern · Illumination invariant measure · Severe illumination variations · Single sample face recognition

1 Introduction

The illumination variation problem is tough and inevitable in face recognition, even the deep learning feature performed unsatisfactorily [1]. The severe illumination variation is considered as one of the tough issues for the face image in the outdoor environment, such as the driver face image in the intelligent transportation systems [2]. Hence, it is significance to address illumination variations in face recognition, especially for severe illumination variations. As numerous approaches have been proposed to tackle severe illumination variations, some significant works are selected to review here.

The illumination holding based approach and the illumination eliminating based approach are two categories of methods to address illumination variations, where the

Y. Peng et al. (Eds.): PRCV 2020, LNCS 12306, pp. 29–40, 2020.
https://doi.org/10.1007/978-3-030-60639-8_3

illumination eliminating based approach is more efficient to tackle severe illumination variations. Most of illumination eliminating based approaches were developed based on the lambertian reflectance model [3]. The face reflectance [4], the face high-frequency facial features [1, 5], and the face illumination invariant measures [2, 5–7] are very efficient to tackle severe illumination variations.

The face reflectance approach [4] employed the lambertian reflectance model [3] to estimate the reflectance and the illumination from the illumination contaminated face image simultaneously. The high-frequency single value decomposition face (HFSVD-face) [5] firstly used the frequency interpretation of the single value decomposition algorithm to extract the high-frequency facial feature of the illumination contaminated face image, Recently, HFSVD-face was extended to the orthogonal triangular with column pivoting (QRCP) decomposition algorithm, which resulted in that the QRCP decomposition was first used to construct the QRCP-face [1].

The illumination invariant measure [2, 5–7] constructed the reflectance based pattern by eliminating the illumination of the face image. The Weber-face [6] proposed a simple reflectance based pattern that the difference of the center pixel and its neighbor pixel was divided by the center pixel in the 3×3 block region, which can eliminate the illumination of the face image, since the face illumination invariant measure assumes that illumination intensities of neighborhood pixels are approximately equal in the face local region. Then, the Weber-face was extended to the logarithm domain, and several illumination invariant measures were proposed such as [2, 5], and [7], since the illumination invariant measure of the logarithm domain was proved to have better tolerance to illumination variations than that of the pixel domain in mathematics [5]. The multiscale logarithm difference edgemaps (MSLDE) [7] was obtained from multi local edge regions of the logarithm face. The local near neighbor face (LNN-face) [5] was attained from multi local block regions of the logarithm face. In [7] and [5], different weights were assigned to different local edge or block regions, whereas the edge region based generalized illumination robust face (EGIR-face) and the block region based generalized illumination robust face (EGIR-face) [2] equally treated different local regions, and removed the weights associated with the edge and block regions. EGIR-face and BGIR-face were obtained from local edge and block regions of the logarithm face image.

The local binary pattern (LBP) based approach [8, 9] was an efficient hand-crafted facial descriptor, and robust to various facial variations. The centre symmetric pattern (CSP) was widely used in the LBP based facial feature. The centre symmetric local binary pattern (CSLBP) [8] employed the symmetric pixel pairs around the centre pixel in the 3×3 block region to code the facial feature. Recently, the centre symmetric quadruple pattern (CSQP) [9] extended the CSP to the quadruple space. The quadruple space was based on the 4×4 block region, which meant that CSQP coded the LBP based facial feature in the face local region with the size of 4×4 pixels.

Nowadays, the deep learning feature is the best for face recognition, which requires massive available face images to train. VGG [10] was trained by 2.6M internet face images (2622 persons and 1000 images per person). ArcFace [11] was trained by 85742 persons and 5.8M internet face images. As large-scale face images for training the deep learning model are collected via internet, the deep learning feature performed very well on internet face images. However, the internet face images are not with severe illumination variations, thus the deep learning feature performed unsatisfactorily under severe illumination variations [1, 2].

2 Diagonal Symmetric Pattern Based Illumination Invariant Measure

2.1 The Centre Symmetric Quadruple Pattern

The centre symmetric pattern was widely used in the LBP based approach, and the recent one is the CSQP [9] that extended the centre symmetric pattern to quadruple space. The quadruple space is based on a 4×4 block region, which means that the CSQP codes the LBP based facial feature in a face local region with the size of 4×4 pixels. The CSQP divided the local kernel of the size 4×4 into 4 sub-blocks of the size 2×2. Figure 1 shows the centre symmetric quadruple pattern. Suppose $m \geq n$, the pixel image I is with m rows and n columns. In Fig. 1, $I(i, j)$ denotes the pixel intensity at the location (i, j), where (i, j) denotes the location of the image point of the i-th row and the j-th column.

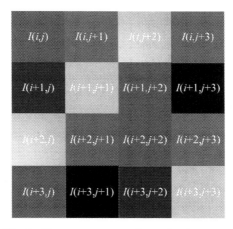

Fig. 1. The centre symmetric quadruple pattern.

The pixel-blocks with the same color in Fig. 2 are compared to generate the CSQP [9] as below

$$
\begin{aligned}
A(i,j) = \\
2^7 \times C(I(i,j), I(i+2,j+2)) + 2^6 \times C(I(i,j+1), I(i+2,j+3)) \\
+2^5 \times C(I(i+1,j), I(i+3,j+2)) + 2^4 \times C(I(i+1,j+1), I(i+3,j+3)) \quad (1) \\
+2^3 \times C(I(i,j+2), I(i+2,j)) + 2^2 \times C(I(i,j+3), I(i+2,j+1)) \\
+2^1 \times C\big(I(i+1,j+2), I(i+3,j) + 2^0 \times C(I(i+1,j+3), I(i+3,j+1)))
\end{aligned}
$$

$$
C(I_1, I_2) = \begin{cases} 1, & if \quad I_1 > I_2 \\ 0, & if \quad I_1 \leq I_2 \end{cases} \quad (2)
$$

Where I_1 and I_2 are intensities (or grayscales) of two pixels in the CSQP. From formulas (1) and (2), the CSQP based LBP feature $A(i, j)$ is a decimal number.

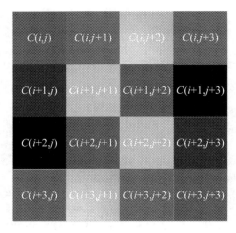

Fig. 2. The diagonal symmetric pattern.

2.2 The Diagonal Symmetric Pattern

The CSQP employs the centre symmetric sub-blocks to develop the facial feature in the face local region. The centre of the CSQP is the intersection of the horizontal and vertical axes that divide the 4×4 face local block region into 4 sub-blocks [9]. In this paper, we propose the pixel-wise diagonal symmetric pattern (DSP) model, whose symmetric centre is also the intersection of the horizontal and vertical axes, and the pixel pairs of the DSP are strictly diagonal symmetry.

Figure 2 shows the proposed pixel-wise DSP model, which is a 4×4 block of the logarithm image C, where C is the logarithm version of the pixel image I in Fig. 1. The proposed DSP model incorporates 16 pixel-blocks, and the pixel-blocks with the same color such as two yellow blocks, are diagonal symmetrical pixels in Fig. 2. The DSP based illumination invariant measures are defined as below.

$$P_1 = C(i, j) - C(i + 3, j + 3) = \ln(R(i, j)) - \ln(R(i + 3, j + 3)) \tag{3}$$

$$P_2 = C(i, j + 1) - C(i + 3, j + 2) = \ln(R(i, j + 1)) - \ln(R(i + 3, j + 2)) \tag{4}$$

$$P_3 = C(i + 1, j) - C(i + 2, j + 3) = \ln(R(i + 1, j)) - \ln(R(i + 2, j + 3)) \tag{5}$$

$$P_4 = C(i + 1, j + 1) - C(i + 2, j + 2) = \ln(R(i + 1, j + 1)) - \ln(R(i + 2, j + 2)) \tag{6}$$

$$P_5 = C(i, j + 2) - C(i + 3, j + 1) = \ln(R(i, j + 2)) - \ln(R(i + 3, j + 1)) \tag{7}$$

$$P_6 = C(i, j + 3) - C(i + 3, j) = \ln(R(i, j + 3)) - \ln(R(i + 3, j)) \tag{8}$$

$$P_7 = C(i + 1, j + 2) - C(i + 2, j + 1) = \ln(R(i + 1, j + 2)) - \ln(R(i + 2, j + 1)) \tag{9}$$

$$P_8 = C(i+1, j+3) - C(i+2, j) = \ln(R(i+1, j+3)) - \ln(R(i+2, j)) \qquad (10)$$

From the lambertian reflectance model [3], the logarithm image C can be presented as $C(i, j) = lnI(i, j) = lnR(i, j) + lnL(i, j)$, where R and L are the reflectance and the illumination respectively. As illumination intensities are approximately equal in the DSP, P_k $(k = 1, 2, ..., 8)$ in Fig. 2 are independent of the illumination accordingly. In point view of numerical sign, P_k $(k = 1, 2, ..., 8)$ may be positive or negative. We divide P_k $(k = 1, 2, ..., 8)$ into positive DSP units (DSP+) and negative DSP units (DSP-), where $P_k^+ > 0$ and $P_k^- < 0$ denote the positive DSP unit and the negative DSP unit respectively. The summation of all DSP units can be attained as

$$\sum\nolimits_{k=1}^{8} P_k = \underbrace{\sum P_k^+}_{\text{DSP +}} + \underbrace{\sum P_k^-}_{\text{DSP -}}, k = 1, 2, \cdots, 8 \qquad (11)$$

According to formula (11), the illumination invariant measure under the DSP with 4×4 block region can be presented as

$$\text{DSP}(i, j) = \alpha \underbrace{\sum I_k^+}_{\text{DSP +}} + (2-\alpha) \underbrace{\sum I_k^-}_{\text{DSP -}} \qquad (12)$$

Formula (12) is the termed as the DSP image in this paper. The DSP face (DSP-face) is obtained by the DSP image with the arctangent function, which is presented as below

$$\text{DSP - face}(i, j) = arctan\left(4\left(\alpha \underbrace{\sum I_k^+}_{\text{DSP +}} + (2-\alpha) \underbrace{\sum I_k^-}_{\text{DSP -}} \right) \right) \qquad (13)$$

Some DSP images and DSP-faces are shown in Fig. 3. Compared with [2, 5] and [7], the DSP image and the DSP-face are quite different from previous illumination invariant measures.

3 Classification

3.1 Single DSP Image Classification

From [2, 5, 6] and [7], the illumination invariant measure with the saturation function (i.e. DSP-face) is more efficient than the illumination invariant measure without the saturation function (i.e. DSP image) for the single image recognition under the nearest neighbor classifier, DSP-face in formula (13) is directly used to tackle the single DSP image recognition under the nearest neighbor classifier. The parameter $\alpha = 0.4$ in formula (5) is adopted, which is the same as [2] recommended.

3.2 Multi DSP Image Classification

Formula (4) is used to generate multi training DSP images of the single training image by different parameter α. Multi DSP images employ the noise robust ESRC [12] to tackle severe illumination variation face recognition with the single sample problem, which is

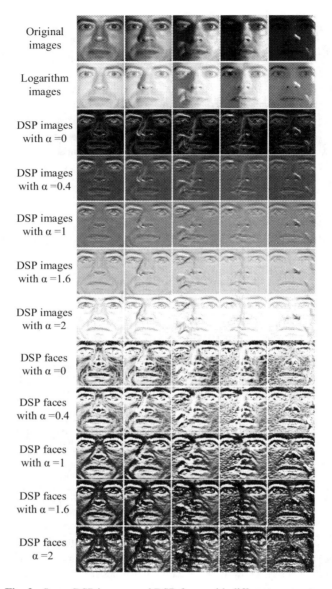

Fig. 3. Some DSP images and DSP-faces with different parameters.

the same as [2, Formulas (8) and (9)]. In this paper, we selected three DSP images with α = 0.4, 1, and 1.6 to form multi training DSP images of each single training image, which is the same as recommended by [2]. Accordingly, the DSP image of the test image is generated by α = 1. The DSP image of each generic image is generated by α = 1. Here, ESRC with multi DSP images is termed as multi DSP images based classification (DSPC).

3.3 Multi DSP Images and the Pre-trained Deep Learning Model Based Classification

Similar with [2, Formulas (10) and (11)], the proposed DSP model can be integrated with the pre-trained deep learning model. The representation residual of DSPC can be integrated with the representation residual of ESRC of the deep learning feature to conduct classification, which is termed as multi DSP images and the pre-trained deep learning model based classification (DSP-PDL). In this paper, the pre-trained deep learning models VGG [10] and ArcFace [11] are adopted. Multi DSP images and VGG (or ArcFace) based classification is briefly termed as DSP-VGG (or DSP-ArcFace) in this paper.

4 Experiments

This paper proposes the DSP model to address single sample face recognition under severe illumination variations. The performances of the proposed methods are validated on the Extended Yale B [13], CMU PIE [14] and VGGFace2 test [15] face databases. In this paper, all cropped face images and the experimental setting are the same as [2]. The recognition rates of Tables 1, 2 and 3 are from [2, Tables 3, 4 and 7] except for the proposed methods. For the proposed DSPC and DSP-VGG/ArcFace, three DSP images (i.e. $\alpha = 0.4$, 1, and 1.6) are generated.

4.1 Experiments on the Extended Yale B Face Database

The Extended Yale B face database [13] contains grayscale images of 28 persons. 64 frontal face images of each person are divided into subsets 1–5 with illumination variations from slight to severe. Subsets 1–5 consist of 7, 12, 12, 14 and 19 images per person respectively. Figure 4 shows some images of one person in the Extended Yale B face database.

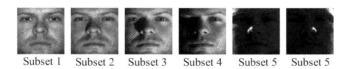

Subset 1   Subset 2   Subset 3   Subset 4   Subset 5   Subset 5

Fig. 4. Some images of one person in the Extended Yale B face database.

The Extended Yale B face database is with extremely challenging illumination variations. Subsets 1–2 face images are with slight illumination variations, and subset 3 face images are with small scale cast shadows. Subset 4 face images are with moderate scale cast shadows, and subset 5 face images are with large scale cast shadows (or severe holistic illumination variations).

From Table 1, DSP-face outperforms EGIR-face, BGIR-face, MSLDE and LNN-face on all Extended Yale B datasets, except that DSP-face lags behind BGIR-face on subset 4. The reason can be explained as that moderate scale cast shadows (i.e. subset

4 images) incorporate more edges of cast shadows than large scale cast shadows (i.e. subset 5 images) as shown in Fig. 4, and edges of cast shadows of face images violate the assumption of the illumination invariant measure that illumination intensities are approximately equal in the face local block region.

DSPC outperforms DSP-face. There may be two reasons, one is that multi DSP images incorporate more intra class variation information than the single DSP image, and the other one is that ESRC is more robust than the nearest neighbor classifier under illumination variations. DSPC performs better than EGIRC and BGIRC under severe illumination variations such as on subsets 1–5.

Table 1. The average recognition rates of the compared methods on the Extended Yale B face database.

Approach	Subset 4	Subset 5	Subsets 1–5
Weber-face [6]	58.66	92.52	74.21
MSLDE [7]	53.35	81.45	60.27
LNN-face [5]	61.59	92.02	70.32
CSQP [9]	59.04	87.67	65.53
EGIR-face [2]	61.69	88.12	66.74
BGIR-face [2]	70.15	93.27	72.75
DSP-face	**62.97**	**94.84**	**77.01**
VGG [10]	47.14	27.67	45.32
ArcFace [11]	53.28	30.93	49.71
EGIRC [2]	75.62	96.31	83.59
BGIRC [2]	78.53	97.30	86.69
DSPC	**78.31**	**97.30**	**88.66**
EGIR-VGG [2]	81.79	84.30	82.28
BGIR-VGG [2]	82.45	82.84	83.53
DSP-VGG	81.58	88.52	87.49
EGIR-ArcFace [2]	79.49	83.13	78.24
BGIR-ArcFace [2]	79.19	81.73	78.97
DSP-ArcFace	**79.71**	**88.95**	**84.53**

VGG/ArcFace was trained by large scale light internet face images, without considering severe illumination variations, which performs unsatisfactorily under severe illumination variations such as on all Extended Yale B datasets. DSP-VGG is superior to EGIR/BGIR-VGG on subset 5 and subsets 1–5. DSP-ArcFace outperforms EGIR/BGIR-ArcFace on all Extended Yale B datasets. Despite DSP-VGG/ArcFace is not able to attain the highest recognition rates on all Extended Yale B datasets, DSP-VGG/ArcFace achieves very high recognition rates on all these datasets. Hence, the proposed DSP-PDL model is able to have the advantages of both the DSP model and the pre-trained deep learning model to tackle face recognition.

4.2   Experiments on the CMU PIE Face Database

The CMU PIE [14] face database contains color images of 68 persons. 21 images of each person from each of C27 (frontal camera), C29 (horizontal 22.5° camera) and C09 (above camera) in CMU PIE illumination set are selected. CMU PIE face images are with slight/moderate/severe illumination variations. From [14], pose variation of C29 is larger than that of C09. Figure 5 shows some images of one person in the CMU PIE face database.

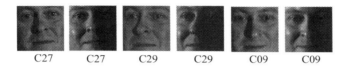

C27   C27   C29   C29   C09   C09

Fig. 5. Some images of one person in the CMU PIE face database.

Table 2. The average recognition rates of the compared methods on the CMU PIE face database.

Approach	C27	C29	C09
Weber-face [6]	89.17	84.00	89.17
MSLDE [7]	81.01	77.57	80.04
LNN-face [5]	89.26	84.67	88.29
CSQP [9]	86.36	82.46	83.21
EGIR-face [2]	82.12	83.50	83.33
BGIR-face [2]	89.30	89.25	89.72
DSP-face	**98.02**	**97.46**	**97.76**
VGG [10]	87.33	76.91	86.67
ArcFace [11]	91.90	78.02	97.51
EGIR-VGG [2]	98.88	95.48	98.52
BGIR-VGG [2]	99.08	95.91	98.88
DSP-VGG	**99.48**	**98.13**	**99.36**
EGIR-ArcFace [2]	98.40	93.38	99.07
BGIR-ArcFace [2]	98.66	93.92	99.37
DSP-ArcFace	**99.45**	**97.60**	**99.79**

Some CMU PIE face images are bright (i.e. slight illumination variations), and others are with partial dark (i.e. moderate/severe illumination variations). Illumination variations of CMU PIE are not as extreme as those of Extended Yale B. From Fig. 5, images in each of C27, C29 and C09 are with the same pose (i.e. frontal, 22.5° profile and downward respectively).

From Table 2, DSP-face achieves very high recognition rates on C27, C29 and C09, and performs much better than EGIR-face, BGIR-face, MSLDE and LNN-face on all CMU PIE datasets. It can be seen that DSP-face is very robust to severe illumination variations under fixed pose.

DSP-VGG/ArcFace performs very well on all CMU PIE datasets. As DSP-face is superior to EGIR-face and BGIR-face under severe illumination variations on C27, C29 and C09, DSP-VGG/ArcFace is better than EGIR-VGG/ArcFace and BGIE-VGG/ArcFace.

4.3   Experiments on the VGGFace2 Test Face Database

The VGGFace2 test face database [15] incorporates color images of 500 persons, which are with large variations in pose, age, illumination, ethnicity and profession. The VGGFace2 test is a large scale face database with more than 160,000 images, which consists of large scale bright internet face images with large pose/expression variations, and illumination variations of VGGFace2 are not as severe as those of Extended Yale B and CMU PIE. Figure 6 shows some images of one person in the VGGFace2 test face database.

Fig. 6. Some images of one person in the VGGFace2 test face database.

Table 3. The recognition rates (%) of some compared methods on the VGGFace2 test face database.

EGIR-face [2]	BGIR-face [2]	**DSP-face**	VGG [10]	ArcFace [11]	ArcFace + ESRC	EGIRC-VGG [2]	BGIRC-VGG [2]	**DSP-VGG**
2.93	2.55	**3.04**	28.80	34.84	35.67	41.98	44.19	**43.76**

From Table 3, the shallow illumination invariant approaches achieve very low recognition rates compared with the deep learning feature VGG/AracFace. DSP-face is superior to EGIR-face and BGIR-face. DSP-VGG achieve high recognition rates, since VGG is the dominant feature of the DSP-PDL model on VGGFace2 test.

4.4   Discussions

In comparison with the data-driven based deep learning methods VGG [10] and ArcFace [11] that required a training processing, the proposed DSP-face are the model-driven based illumination invariant measures, which do not depend on large scale face images training, such as previous illumination invariant measures EGIR-face [2], BGIR-face [2], LNN-face [5] and MSLDE [7]. Moreover, DSP-face in (13) employs only one parameter α.

From experimental results of Extended Yale B, CMU PIE and VGGFace2 test datasets, DSP-face achieves higher recognition rates than EGIR-face and BGIR-face except on subset 4 of Extended Yale B, which illustrates that DSP-face is superior to EGIR-face and BGIR-face under severe illumination variations. DSP-VGG/ArcFace achieves very high recognition rates on all datasets of Extended Yale B and CMU PIE, except on Extended Yale B subset 5 with extremely severe illumination variations, since the pre-trained deep learning model is restricted to frontal face images with severe illumination variations, whereas this is insufficient to deny that DSP-VGG/ArcFace are the best approaches to tackle severe illumination variations. Hence, the DSP-PDL model is able to have the advantages of both the DSP model (i.e. robust to severe illumination variations) and the pre-trained deep learning model (i.e. robust to slight illumination variations and large pose variations) to tackle face recognition. Moreover, DSP-face outperforms EGIR-face and BGIR-face on VGGFace2 test, which indicates that DSP-face is superior to EGIR-face and BGIR-face under slight/moderate illumination variations.

5 Conclusion

This paper proposes the DSP model to address single sample face recognition under severe illumination variations. DSP-face achieves higher recognition rates compared with previous illumination invariant approaches EGIR-face, BGIR-face, LNN-face and MSLDE under severe illumination variations. DSPC is efficient to severe illumination variations, due to the fact that multi DSP images cover more discriminative information of the face image. Further, the proposed DSP model is integrated with the pre-trained deep learning model to have the advantages of both the DSP model and the pre-trained deep learning model.

Although the proposed DSP model can efficiently tackle severe illumination variations, the authors realized that further work must be done to improve the proposed model under pose variations and well illumination.

Acknowledgments. This work was supported in part by National Natural Science Foundation of China (No. 61802203), in part by National Science Foundation of Jiangsu Province (No. BK20180761), in part by China Postdoctoral Science Foundation (No. 2019M651653), in part by Postdoctoral Research Funding Program of Jiangsu Province (No. 2019K124), and in part by NUPTSF (No. NY218119).

References

1. Hu, C., Lu, X., Liu, P., Jing, X., Yue, D.: Single sample face recognition under varying illumination via QRCP decomposition. IEEE Trans. Image Process. **28**(5), 2624–2638 (2019)

2. Hu, C., Zhang, Y., Wu, F., Lu, X., Liu, P., Jing, X.: Toward driver face recognition in the intelligent traffic monitoring systems. IEEE Trans. Intell. Transp. Syst. (to be published). https://doi.org/10.1109/tits.2019.2945923

3. Horn, B.: Robot Vision. MIT Press, Cambridge (1997)

4. Fu, X., Zeng, D., Huang, Y., Zhang, X., Ding, X.: A weighted variational model for simultaneous reflectance and illumination estimation. In: Proceedings of the IEEE Conference on Computer Vision and Pattern Recognition, Las Vegas, pp. 2782–2790. IEEE (2016)

5. Hu, C., Lu, X., Ye, M., Zeng, W.: Singular value decomposition and local near neighbors for face recognition under varying illumination. Pattern Recogn. **64**, 60–83 (2017)

6. Wang, B., Li, W., Yang, W., Liao, Q.: Illumination normalization based on weber's law with application to face recognition. IEEE Signal Process. Lett. **18**(8), 462–465 (2011)

7. Lai, Z., Dai, D., Ren, C., Huang, K.: Multiscale logarithm difference edgemaps for face recognition against varying lighting conditions. IEEE Trans. Image Process. **24**(6), 1735–1747 (2015)

8. Heikkilä, M., Pietikäinen, M., Schmid, C.: Description of interest regions with local binary patterns. Pattern Recogn. **42**(3), 425–436 (2009)

9. Chakraborty, S., Singh, S., Chakraborty, P.: Centre symmetric quadruple pattern: a novel descriptor for facial image recognition and retrieval. Pattern Recogn. Lett. **115**, 50–58 (2018)

10. Parkhi, O., Vedaldi, A., Zisserman, A.: Deep face recognition. In: Proceedings of the British Machine Vision Conference, Swansea, pp. 1–12, BMVA Press (2015)

11. Deng, J., Guo, J., Xue, N., Zafeiriou, S.: Arcface: additive angular margin loss for deep face recognition. In: Proceedings of the IEEE Conference on Computer Vision and Pattern Recognition, Long Beach, pp. 4690–4699. IEEE (2019)

12. Deng, W., Hu, J., Guo, J.: Extended SRC: undersampled face recognition via intraclass variant dictionary. IEEE Trans. Pattern Anal. Mach. Intell. **34**(9), 1864–1870 (2012)

13. Georghiades, A., Belhumeur, P., Kriegman, D.: From few to many: illumination cone models for face recognition under variable lighting and pose. IEEE Trans. Pattern Anal. Mach. Intell. **23**(6), 643–660 (2001)

14. Sim, T., Baker, S., Bsat, M.: The CMU pose, illumination, and expression database. IEEE Trans. Pattern Anal. Mach. Intell. **25**(12), 504–507 (2003)

15. Cao, Q., Shen, L., Xie, W., Parkhi, O.M., Zisserman, A.: VGGFace2: a dataset for recognising faces across pose and age. In: Proceedings of the IEEE Conference on Automatic Face and Gesture Recognition, Xi'an, pp. 67–74, IEEE (2018)

Multi-level Temporal Pyramid Network for Action Detection

Xiang Wang[1], Changxin Gao[1], Shiwei Zhang[2], and Nong Sang[1(✉)]

[1] Key Laboratory of Image Processing and Intelligent Control,
School of Artificial Intelligence and Automation,
Huazhong University of Science and Technology, Wuhan, China
{u201613707,cgao,nsang}@hust.edu.cn
[2] DAMO Academy, Alibaba Group, Hangzhou, China
zhangjin.zsw@alibaba-inc.com

Abstract. Currently, one-stage frameworks have been widely applied for temporal action detection, but they still suffer from the challenge that the action instances span a wide range of time. The reason is that these one-stage detectors, e.g., Single Shot Multi-Box Detector (SSD), extract temporal features only applying a single-level layer for each head, which is not discriminative enough to perform classification and regression. In this paper, we propose a Multi-Level Temporal Pyramid Network (MLTPN) to improve the discrimination of the features. Specially, we first fuse the features from multiple layers with different temporal resolutions, to encode multi-layer temporal information. We then apply a multi-level feature pyramid architecture on the features to enhance their discriminative abilities. Finally, we design a simple yet effective feature fusion module to fuse the multi-level multi-scale features. By this means, the proposed MLTPN can learn rich and discriminative features for different action instances with different durations. We evaluate MLTPN on two challenging datasets: THUMOS'14 and Activitynet v1.3, and the experimental results show that MLTPN obtains competitive performance on Activitynet v1.3 and outperforms the state-of-the-art approaches on THUMOS'14 significantly.

Keywords: Action detection · One-stage · Feature

1 Introduction

The purpose of temporal action detection in long untrimmed videos is to temporally localize intervals where actions occur and simultaneously predict the action categories. It serves as a key technology in video retrieval, anomaly detection and human-machine interaction, hence it has been receiving an increasing

The first author of this paper is a graduate student.
This work is supported by the National Natural Science Foundation of China under grant 61871435 and the Fundamental Research Funds for the Central Universities no. 2019kfyXKJC024.

Y. Peng et al. (Eds.): PRCV 2020, LNCS 12306, pp. 41–54, 2020.
https://doi.org/10.1007/978-3-030-60639-8_4

mount of attention from both academia and industry. Recently, temporal action detection has achieved great achievements on some public datasets [1,29]. However, this task is still challenging because the duration of action varies dramatically by ranging from fractions of a second to several minutes. Recent existing approaches can be divided into three categories: multi-stage approach [5,9,27], two-stage approach [7,8] and one-stage approach [10]. Among these methods, one-stage methods, which are mainly inspired by SSD [14], are more efficient and practical in many direct and indirect applications. Typically, as with Single Shot Action Detector (SSAD) network [10], these methods apply a single-level layer to detect actions for each head. Because the action instances span a wide range of time, the SSD-like methods tend to apply the different layers to detect action with different durations, e.g., the shallow layers detect short actions and deep layers for long actions. However, a single layer may be not discriminative enough to conduct action classification and localization, which can be proved in Fig. 1 to some extent. Therefore, current performance of the methods are not satisfactory actually. We believe that improving the discrimination of each layer can effectively improve action detection performance. Based on the above observations, we proposed a Multi-Level Temporal Pyramid Network (MLTPN) to improve the performance of one-stage action detection, as is shown in Fig. 2. In our proposed MLTPN, we first encode multi-layer temporal information by fusing multiple layers with different temporal resolutions. Second, inspired by the successful applications of multi-scale feature pyramid network [23] in object detection, we propose to embed several feature pyramid networks on the encoded features to learn multi-level and discriminative features with different scales. Third, we design a simple yet effective feature fusion module to fuse the multi-level multi-scale features. By this means, the features encoded by the proposed MLTPN are more discriminative, and the undergoing multi-scale property is suitable for detection different action instances with different durations. Moreover, to further improve the performance, we apply GIoU loss to regress the temporal boundaries. To evaluate the effectiveness of the proposed MLTPN, we conduct experiments on THUMOS'14 and Activitynet v1.3 datasets, and the results show that MLTPN achieves a significant improvement on both datasets. Particularly, the result of MLTPN outperforms the state-of-the-art approaches on THUMOS'14 significantly.

In summary, we make the following three contributions:

- We propose a multi-level temporal feature pyramid network to improve the discrimination of each layer for one-stage frameworks;
- We design a simple but effective module to fuse multi-level multi-scale features;
- We extensively evaluate MLTPN on the challenging THUMOS'14 dataset and achieve state-of-the-art performance.

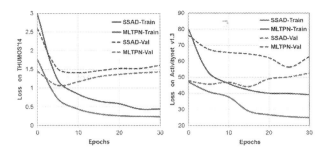

Fig. 1. Losses of SSAD and MLTPN on the challenging THUMOS'14 (left) and Activitynet v1.3 (right) benchmarks. Universally, SSAD losses are much higher than our MLTPN during both the training and validation phase. The results demonstrate that MLTPN can encode more temporal information, and the encoded features are more discriminative to a certain degree. Better viewed in original color pdf. (Color figure online)

2   Relation Work

For temporal action detection task, recent existing approaches can be intuitively divided into three categories: multi-stage approach, two-stage approach and one-stage approach.

Multi-stage approach, particularly, Boundary Sensitive Network (BSN) [9] first locates temporal boundaries (start and end), then directly combines these boundaries as proposals. Next BSN retrieves proposals by evaluating the confidence of whether a proposal contains an action within its region. Finally, using the proposals to localize action. Boundary-Matching Network (BMN) [27] is an improvement of BSN, first BMN predicts start and end boundary probabilities by a sub-network. Then the boundary probabilities are applied to extensively enumerate the proposals, which is followed by a boundary-matching confidence map to densely evaluate confidence of all proposals. Based on the proposals, then refine the boundaries and predict the corresponding categories. Multi-granularity Generator (MGG) [24] first uses a bilinear matching model to exploit the rich local information within the video. Then two components, namely segment proposal producer and frame actionness producer, are combined to perform the task of temporal at two distinct granularities. Finally, using the proposals to localize action. These methods achieve impressive performance, but it is inefficient because of its long pipeline actually. P-GCN [26] exploits the proposal-proposal relations using Graph Convolutional Networks, first based on the already obtained proposals, P-GCN constructs an action proposal graph, where each proposal is represented as a node and their relations between two proposals as an edge. Finally P-GCN applies the Graph Convolutional Networks over the graph to model the relations among different proposals and learn representations for the action classification and localization.

As for two-stage approach, Region Convolutional 3D Network (R-C3D) [7], first encodes the video streams using a 3D fully convolutional network, then

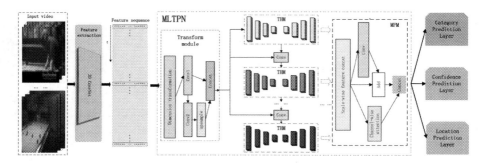

Fig. 2. An overview of our MLTPN. We first utilize a 3D ConvNet to extract the features of the input video. Then the obtained feature sequence is input into Transform module to encode multi-layer temporal information. Afterwards each THM generates a group of multi-scale features, and then the cascaded multiple alternating joint Temporal H-shaped Modules (THMs) and Conv general multi-level pyramid features. Finally, Merge Feature Module (MFM) fuses the multi-level pyramid features for detection.

generates candidate temporal regions containing activities, and finally classifies selected regions into specific activities. TAL-Net [8] is an improvement of R-C3D, compared to R-C3D,TAL-net improves receptive field alignment, better exploits the temporal context of actions for both proposal generation and action classification by appropriately extending receptive fields. Although, these methods have great improved on temporal action detection, The two-stage methods are Faster RCNN-like procedure, and suffer from another drawback that they are limited by fixed length inputs. They need to down sample the frames to fit the GPU memory, *e.g.*, 3 FPS is applied in [7]. Therefore, they lose some temporal information, which may result in a sub-optimal solution.

In contrast, one-stage methods are mainly spirited by Single Shot MultiBox Detector (SSD), classification and localization at the same time, hence they are more efficient. SSAD [10] based on 1D temporal convolutional layers to skip the proposal generation step via directly detecting action instances in untrimmed videos. However, SSAD extracts temporal features only applying a single-level layer for each head, which is not discriminative enough to perform classification and localization. Therefore, we propose a Multi-Level Temporal Pyramid Network (MLTPN) to improve the discrimination of the features. In particular, SSAD is our baseline.

3   The Proposed Method

3.1   Overview

For the input video, we first use a 3D convolution network to extract features, and concatenate these features. Then the features are input into Transform module, to do the dimensional transformation and fuse the features from multiple layers. Afterwards, We apply a multi-level feature pyramid architecture on the features.

Finally, the multi-level multi-scale features are input into the Merge Feature Module (MFM). MFM performs a certain fusion of the features, making our features more robust and more conducive. Finally, the features are used for action classification and localization.

3.2   Feature Extracting

To detect action instances in temporal dimension is the ultimate target of action localization. Given the video frame sequence, We uniformly sample the sequence of video frame into several small consecutive snippets and then extract visual features within each snippet. In particular, a sequence of snippets level feature $\{f_i\}_{i=0}^{T-1}$ are extracted, where T is temporal length. We further feed the features into two 1D convolutional layers (with temporal kernel size 3, stride 2) to increase the temporal size of receptive fields.

3.3   MLTPN

Our MLTPN consists of three parts, namely transform module, Temporal H-Shaped Module (THM), Merge Feature Module (MFM).

In the transform module, we use $t \times c$ to represent the obtained feature map, where t is the temporal length and c is the dimension of the representation. In particular, same as [5], we first use a dimensional transformation, features change from $t \times c$ to $1 \times t \times c$, and two convolutions to improve the dimension of features, so that our features contain multi-layer information and are more conducive to classification. 'Conv1' with $k1$ kernels, kernel size (3, 1), stride (1, 1). 'Conv2' with $k2$ kernels, kernel size (3, 1), stride (2, 1), and 'Conv2' also followed by one up-sample operation (Nearest neighbor interpolation or Linear interpolation) map back to the same size as output of 'Conv1' and we concatenate the outputs of 'Conv1' and 'Conv2'. The final output is $k \times t \times c$, where k is equal to $k1 + k2$. Here, transform module can extract the multi-scare features from backbone.

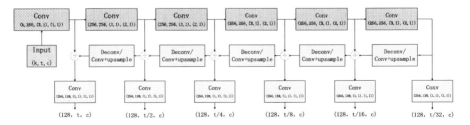

Fig. 3. Details of THM. The numbers in brackets of Conv block: input channels, output channels, conv kernel size, stride size.

As shown in the Fig. 3, inspired by m2det [23], we design the THM. The encoder is a series of convolution layers (with kernel size (3, 1), stride (2, 1)).

The output of these layers is reference set of feature map. At the same time, in order to make our feature representation more robust and more informative, we cascade multiple THMs. Finally the resulting multi-level multi-scale features, which are more conducive to subsequent tasks. The difference between cascaded multiple THMs and FPN (Feature Pyramid Network) includes several following aspects: 1) In FPN, each feature map in the pyramid is mainly or even solely constructed from single-level layers of the backbone. However, in our proposed cascaded multiple THMs, we embed several feature pyramid networks on the encoded features to learn multi-level and discriminative features with different scales; 2) The THM decoder takes the outputs of encoder layers as its reference set of feature maps, while the original FPN chooses the output of the last layer of each stage. Also the difference between cascaded multiple THMs and m2det is: In m2det, as the network deepens, the features of the model are reduced in both width and height, which is more suitable for object detection in the picture. However, our cascaded multiple THMs are reduced in one time dimension and the other is unchanged, which can well retain more semantic information suitable for classification.

Conv between two adjacent THM consists of two convolution (with temporal kernel size 1, stride 1) for dimensionality reduction and a concat operation.

The function of MFM is to fuse the features of different THM layers. First we concatenate the same scale features obtained by multiple THMs into $(l *$ $128) \times t' \times c$, where l is the number of THM. The value of t' is $t/4$, $t/8$, ... etc. Then we use a small residual module and a channel-wise attention module to fuse the multi-level features and the convolution in the small residual module shares parameters with different resolution features from the THM modules.

3.4   Network Optimization

For classification, we reshape the output of the category branch in each cell to a $(C + 1) \times N$ matrix, $C + 1$ represents the total action categories plus one background category. N represents the number of anchor. The probability of being in the ith category is

$$P(C_i) = \frac{e^{o_i}}{\sum_{j=1}^{C} e^{o_j}}, i = 0, 1, 2, ..., C, \tag{1}$$

where o_i is the output of the network, which corresponds to the ith class. We utilize the standard softmax loss, which can be formulated as

$$L_{cls} = - \sum_{n=0}^{C} I_{n=c} \log(p(C_n)) \tag{2}$$

where $I_{n=c}$ is an indicator function, equals to 1 if n is the ground truth class label c, otherwise 0.

We use the outputs of confidence branch as predictions of the IoU (Intersection over Union) values between the proposals and the ground truth. Note that the IoU here represents the temporal IoU, and there is only one temporal dimension. At the same time, for the sake of confusion, we will always use IoU to represent the temporal IoU. The IoU overlap loss is Smooth L1 loss (S_{L1}), so the loss is formulated as

$$L_{conf} = S_{L1}(p_{iou} - g_{iou}) \qquad (3)$$

where g_{iou} is the ground truth IoU value between the proposal and its closest ground truth.

For location prediction, the localization loss is only applied to positive samples. The same as [35]:

$$Iou_i = \frac{|p_i \cap g_i|}{|p_i \cup g_i|} \qquad (4)$$

$$GIou_i = Iou_i - \frac{|B_i \backslash (p_i \cup g_i)|}{|B_i|} \qquad (5)$$

$$L_{reg} = 1 - GIou_i \qquad (6)$$

where g_i is the closest ground truth of the proposal p_i. B_i is the shortest continuous time interval including ground truth g_i and the proposal p_i.

We jointly train to optimize the above three loss functions. And the final loss is formulated as

$$L_{cos\,t} = \alpha_1 \frac{1}{N_{cls}} \sum_i L_{cls}(p_i, p_i^*) \\ + \alpha_2 \frac{1}{N_{conf}} \sum_i L_{conf}(Iou_i, Iou^*) + \alpha_3 \frac{1}{N_{reg}} \sum_i L_{reg}(GIou_i) \qquad (7)$$

where α_1, α_2 and α_3 are weight coefficients (empirically set to 1, 10 and 0.3). N_{cls}, N_{conf} and N_{reg} are the number of samples. p_i is the predicted probability and p_i^* is the ground truth label. Iou_i is the predicted IoU between the sample and the closest ground truth, and Iou^* is the real IoU. $GIou_i$ is the $GIoU$ between the sample and the closest ground truth.

4 Experiments

We conduct experiments on two challenging datasets, THUMOS'14 and Activitynet v1.3 respectively. The impact of different experimental settings of MLTPN is investigated by ablation studies. Comparison of MLTPN and other state-of-the-art methods is also reported.

4.1 Dataset and Experimental Settings

Dataset. THUMOS'14 has 200 and 213 videos with temporal annotations in validation and testing set from 20 classes. The training set of THUMOS'14 is trimmed videos and cannot be used for training of temporal action detection for untrimmed videos. Therefore, like SSAD, we use its validation set to training our model and report results on its testing set. The Activitynet v1.3 dataset contains 19994 videos in 200 classes. The numbers of videos for training, validation and testing are 10024, 4926, and 5044. The labels of testing set are not publicly available and we report results on its validation set.

Evaluation Metrics. We follow the official evaluation metrics in each dataset for action detection task. On THUMOS'14, the mean average precision (mAP) with IoU thresholds 0.3, 0.4, 0.5, 0.6, 0.7 are adopted. On Activitynet v1.3, the mAP with IoU thresholds between 0.5 and 0.95 (inclusive) with a step size 0.05 are exploited for comparison.

Features. In order to extract snippet-level video feature, we first sample frames from each video at its own original frame rate on THUMOS'14 and 5fps on Activitynet v1.3. Then we apply a TV-L1 [41] algorithm to get the optical flow of each frame. On THUMOS'14, we use open source I3D model [37] pretrained on Kinetics to extract the I3D features, also we use the TSN models pretrained on Activitynet v1.3 to extract two-stream features. On activitynet v1.3, the pretrained two TSN models with Senet152 [42] are used to extract two-stream features.

Implementation Details. For Thumos'14, In the training phase, we train the model using Stochastic Gradient Descent (SGD) with momentum of 0.9, weight decay of 0.0001 and the batch-size is set to 16. We set the initial learning rate at 0.001 and reduce once with a ratio of 0.1 after 15 epochs. For activitynet v1.3, we train the model using adam [43] method. We set the initial learning rate at 0.0001 and reduce once with a ratio of 0.1 after 15 epochs. The model is trained from scatch. Also we prevent model overfitting by using an early-stop strategy. In the testing phase, since there are little overlap between action instances of same category in temporal action detection task, we take a strict threshold in NMS, which is set to 0.2.

Table 1. Study of different feature settings on THUMOS'14 in terms of mAP@IoU(%)

IoU	0.3	0.4	0.5	0.6	0.7
TSN RGB	39.2	33.4	26.0	16.2	9.3
TSN FLOW	46.6	43.1	35.0	24.5	13.4
TSN Fusion	**57.2**	**51.7**	**42.5**	**29.2**	**16.2**
I3D RGB	56.4	51.1	41.9	29.6	17.3
I3D FLOW	62.6	58.0	49.4	34.9	20.0
I3D Fusion	**66.0**	**62.6**	**53.3**	**37.0**	**21.2**

Table 2. Study of different anchor settings on THUMOS'14 in terms of mAP@IoU(%)

IoU	0.3	0.4	0.5	0.6	0.7
5 anchors	**66.3**	61.3	50.0	35.1	18.9
6 anchors	63.8	59.1	50.0	33.2	19.1
7 anchors	66.0	**62.6**	**53.3**	**37.0**	**21.2**
8 anchors	65.9	61.6	50.1	33.3	19.2
9 anchors	65.5	61.6	51.6	35.0	19.7

Table 3. Study of different THM settings on THUMOS'14 in terms of mAP@IoU(%)

IoU	0.3	0.4	0.5	0.6	0.7
4 THM	65.1	59.8	50.4	35.8	20.4
5 THM	64.8	60.6	51.0	35.8	**21.9**
6 THM	**66.0**	**62.6**	**53.3**	**37.0**	21.2
7 THM	63.8	59.2	48.8	34.2	19.3
8 THM	63.3	57.3	47.8	35.6	**21.9**

4.2 Ablation Studies

Feature Settings. We use the features generated on the TSN and I3D models for comparison experiments, as shown in Table 1. It can be seen that the I3D feature performance is much better than TSN features, and there is a large gap. Moreover, it is common that FLOW features perform better than RGB features. As expected, the fusion features perform better than using only the RGB or FLOW features, indicating that in this task, the RGB features and the FLOW features have a certain complementary effect.

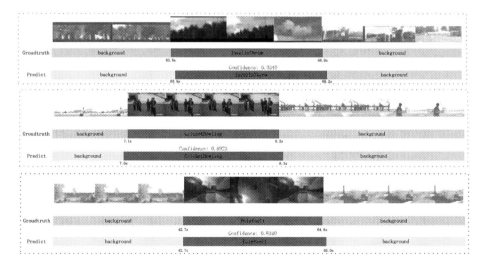

Fig. 4. Visualization of predicted action instances by our MLTPN on THUMOS'14.

Anchor Settings. We study the effect of different number of anchor segments on our performance. As shown in Table 2, the setting of 7 anchors achieved the best results with the IoU thresholds of 0.4, 0.5, 0.6, 0.7. The best results are obtained with the setting of 5 anchors with the IoU threshold of 0.3. We can say that the number of anchors cannot be too much, because many redundant detections are involved, and it also cannot be too few. Too few, there are too few examples of training, which is not conducive to training a good result.

Table 4. Performance comparisons of temporal action detection on THUMOS'14, measured by mAP at different IoU thresholds. The contents of the brackets represent the feature extractor used.

THUMOS'14, mAP@IoU(%)					
Approach	0.3	0.4	0.5	0.6	0.7
Multi-stage & two-stage action localization					
Wang et al. [2]	14.0	11.7	8.3	–	–
FTP [33]	–	–	13.5	–	–
DAP [34]	–	–	13.9	–	–
Oneata et al. [36]	27.0	20.8	14.4	–	–
Yuan et al. [38]	33.6	26.1	18.8	–	–
SCNN [6]	36.3	28.7	19.0	10.3	5.3
SST [15]	41.2	31.5	20.0	10.9	4.7
CDC [21]	40.1	29.4	23.3	13.1	7.9
TURN [16]	46.3	35.5	24.5	14.1	6.3
TCN [18]	–	33.3	25.6	15.9	9.0
R-C3D [7]	44.8	35.6	28.9	19.1	9.3
SSN [19]	51.9	41.0	29.8	19.6	10.7
CBR [22]	50.1	41.3	31.0	19.1	9.9
BSN [9]	53.5	45.0	36.9	28.4	20.0
MGG [24]	53.9	46.8	37.4	29.5	**21.3**
BMN [27]	56.0	47.4	38.8	29.7	20.5
TAL-Net [8]	53.2	48.5	42.8	33.8	20.8
P-GCN [26]	63.6	57.8	49.1	–	–
One-stage action localization					
Richard et al. [32]	30.0	23.2	15.2	–	–
Yeung et al. [11]	36.0	26.4	17.1	–	–
SMS [28]	36.5	27.8	17.8	–	–
SS-TAD [12]	45.7	–	29.2	–	9.6
SSAD [10] (TSN)	52.6	46.2	36.6	22.8	12.0
SSAD [10] (I3D)	62.53	55.14	42.1	27.43	13.5
MLTPN (TSN)	57.2	51.7	42.5	29.2	16.2
MLTPN (I3D)	**66.0**	**62.6**	**53.3**	**37.0**	21.2

Number of THM. We study the effect of the different number of THM modules in our MLTPN. The results are shown in the Table 3. We find that when the number of THM modules is 6, on the THUMOS'14 dataset, the performance is best with the IoU thresholds of 0.3, 0.4, 0.5, 0.6, and when the number of THM modules is 5 and 8, the performance is best with the IoU threshold of 0.7.

Table 5. Performance comparisons of temporal action detection on Activitynet v1.3. (∗) indicates the method that uses the external video labels from [44]

Activitynet v1.3@IoU(%)				
Approach	Validation			
	0.5	0.75	0.95	Average
Multi-stage & two-stage action localization				
Wang et al. [39]	45.11	4.11	0.05	16.41
CDC [21]	45.30	26.00	0.20	23.80
TAG-D [31]	39.12	23.48	5.49	23.98
TAL-Net [8]	38.23	18.30	1.30	20.22
BSN∗ [9]	46.45	29.96	8.02	30.03
BMN∗ [27]	50.07	34.78	8.29	33.85
P-GCN [26]	42.90	28.14	2.47	26.99
P-GCN∗ [26]	48.26	33.16	3.27	31.11
One-stage action localization				
Singh et al. [25]	26.01	15.22	2.61	14.62
SSAD [10]	33.91	23.16	3.74	22.26
MLTPN	**44.86**	**28.96**	**4.30**	**28.27**

4.3 Comparison with the State-of-the-Arts

THUMOS'14. We compare our MLTPN with some current state-of-the-art methods on THUMOS'14 and summarize the results in Table 4. We can see that our method has a great improvement over the current methods. Especially at IoU 0.3, 0.4, 0.5, 0.6, the performance is even better than the multi-stage methods and two-stage methods, achieving the current best results and it is very close to the best results at IoU 0.7. It is worth noting that the performance of MLTPN is much better than that of SSAD when MLTPN and SSAD use the same feature extractor (e.g. TSN, I3D). This fully proves the effectiveness of our method. And Fig. 4 shows the visualization of prediction results of three action categories respectively.

Activitynet v1.3. On Activitynet v1.3 dataset, we compare with some current state-of-the-art methods. As shown in the Table 5, regarding the average mAP, MLTPN outperforms SSAD by 6.01% with the same features and the same experimental Settings for fair comparison. Please refer to Sect. 4.1 for experimental Settings. In fact, MLTPN has better performance than many current multi-stage & two-stage methods (i.g., CDC, TAL-Net) and our method can make direct inference to the detection results without using any other labels, while many other multi-stage methods require step-by-step inference, which is very troublesome and time-consuming. For example, in the BMN approach, the classification results of the proposals must be obtained before they can be combined with the final detection results. And under the same experimental environment configuration, the calculation speed of BMN is 0.363 s per video (excluding classification), and MLTPN is 0.284 s per video (including classification).

5 Conclusion

In this paper, we proposed a Multi-Level Temporal Pyramid Network (MLTPN) for action detection. Our MLTPN drops the proposal generation step and can directly predict action instances in untrimmed videos. Specially, we first fuse the features from multiple layers with different temporal resolutions, to encode multi-layer temporal information. We then apply a multi-level feature pyramid architecture on the feature to enhance their discriminative abilities. Finally, we design a simple yet effective feature fusion module to fuse the multi-level multi-scale feature. Our MLTPN can learn rich and discriminative features for different action instances with different durations. As the experimental results show, MLTPN obtains competitive performance on Activitynet v1.3 and outperforms the state-of-the-art approaches on THUMOS'14 significantly.

References

1. Jiang, Y.-G., et al.: THUMOS challenge: action recognition with a large number of classes (2014)
2. Wang, L., Qiao, Y., Tang, X.: Action recognition and detection by combining motion and appearance features. THUMOS14 Action Recognit. Challenge **1**(2), 2 (2014)
3. Simonyan, K., Zisserman, A.: Two-stream convolutional networks for action recognition in videos. In: Advances in Neural Information Processing Systems (2014)
4. Long, F., et al.: Gaussian temporal awareness networks for action localization. In: Proceedings of the IEEE Conference on Computer Vision and Pattern Recognition (2019)
5. Li, X., et al.: Deep Concept-wise Temporal Convolutional Networks for Action Localization. arXiv preprint arXiv:1908.09442 (2019)
6. Shou, Z., Wang, D., Chang, S.-F.: Temporal action localization in untrimmed videos via multi-stage CNNs. In: Proceedings of the IEEE Conference on Computer Vision and Pattern Recognition (2016)
7. Xu, H., Das, A., Saenko, K.: R-C3D: region convolutional 3D network for temporal activity detection. In: Proceedings of the IEEE International Conference on Computer Vision (2017)
8. Chao, Y.-W., et al.: Rethinking the faster R-CNN architecture for temporal action localization. In: Proceedings of the IEEE Conference on Computer Vision and Pattern Recognition (2018)
9. Lin, T., et al.: BSN: boundary sensitive network for temporal action proposal generation. In: Proceedings of the European Conference on Computer Vision (ECCV) (2018)
10. Lin, T., Zhao, X., Shou, Z.: Single shot temporal action detection. In: Proceedings of the 25th ACM International Conference on Multimedia (2017)
11. Yeung, S., et al.: End-to-end learning of action detection from frame glimpses in videos. In: Proceedings of the IEEE Conference on Computer Vision and Pattern Recognition (2016)

12. Buch, S., et al.: End-to-end, single-stream temporal action detection in untrimmed videos. In: BMVC, vol. 2 (2017)
13. Ren, S., et al.: Faster R-CNN: towards real-time object detection with region proposal networks. In: Advances in Neural Information Processing Systems (2015)
14. Liu, W., et al.: SSD: single shot MultiBox detector. In: Leibe, B., Matas, J., Sebe, N., Welling, M. (eds.) ECCV 2016. LNCS, vol. 9905, pp. 21–37. Springer, Cham (2016). https://doi.org/10.1007/978-3-319-46448-0_2
15. Buch, S., et al.: SST: single-stream temporal action proposals. In: Proceedings of the IEEE Conference on Computer Vision and Pattern Recognition (2017)
16. Gao, J., et al.: Turn tap: temporal unit regression network for temporal action proposals. In: Proceedings of the IEEE International Conference on Computer Vision (2017)
17. Gao, J., Chen, K., Nevatia, R.: CTAP: complementary temporal action proposal generation. In: Proceedings of the European Conference on Computer Vision (ECCV) (2018)
18. Dai, X., et al.: Temporal context network for activity localization in videos. In: Proceedings of the IEEE International Conference on Computer Vision (2017)
19. Zhao, Y., et al.: Temporal action detection with structured segment networks. In: Proceedings of the IEEE International Conference on Computer Vision (2017)
20. Kay, W., et al.: The kinetics human action video dataset. arXiv preprint arXiv:1705.06950 (2017)
21. Shou, Z., et al.: CDC: convolutional-de-convolutional networks for precise temporal action localization in untrimmed videos. In: Proceedings of the IEEE Conference on Computer Vision and Pattern Recognition (2017)
22. Gao, J., Yang, Z., Nevatia, R.: Cascaded boundary regression for temporal action detection. arXiv preprint arXiv:1705.01180 (2017)
23. Zhao, Q., et al.: M2Det: a single-shot object detector based on multi-level feature pyramid network. In: Proceedings of the AAAI Conference on Artificial Intelligence, vol. 33 (2019)
24. Liu, Y., et al.: Multi-granularity generator for temporal action proposal. In: Proceedings of the IEEE Conference on Computer Vision and Pattern Recognition (2019)
25. Singh, B., et al.: A multi-stream bi-directional recurrent neural network for fine-grained action detection. In: Proceedings of the IEEE Conference on Computer Vision and Pattern Recognition (2016)
26. Zeng, R., et al.: Graph convolutional networks for temporal action localization. In: Proceedings of the IEEE International Conference on Computer Vision (2019)
27. Lin, T., et al.: BMN: boundary-matching network for temporal action proposal generation. In: Proceedings of the IEEE International Conference on Computer Vision (2019)
28. Yuan, Z., et al.: Temporal action localization by structured maximal sums. In: Proceedings of the IEEE Conference on Computer Vision and Pattern Recognition (2017)
29. Caba Heilbron, F., et al.: Activitynet: a large-scale video benchmark for human activity understanding. In: Proceedings of the IEEE Conference on Computer Vision and Pattern Recognition (2015)
30. Singh, G., Cuzzolin, F.: Untrimmed video classification for activity detection: submission to activitynet challenge. arXiv preprint arXiv:1607.01979 (2016)
31. Xiong, Y., et al.: A pursuit of temporal accuracy in general activity detection. arXiv preprint arXiv:1703.02716 (2017)

32. Richard, A., Gall, J.: Temporal action detection using a statistical language model. In: Proceedings of the IEEE Conference on Computer Vision and Pattern Recognition (2016)
33. Caba Heilbron, F., Carlos Niebles, J., Ghanem, B.: Fast temporal activity proposals for efficient detection of human actions in untrimmed videos. In: Proceedings of the IEEE Conference on Computer Vision and Pattern Recognition (2016)
34. Escorcia, V., Caba Heilbron, F., Niebles, J.C., Ghanem, B.: DAPs: deep action proposals for action understanding. In: Leibe, B., Matas, J., Sebe, N., Welling, M. (eds.) ECCV 2016. LNCS, vol. 9907, pp. 768–784. Springer, Cham (2016). https://doi.org/10.1007/978-3-319-46487-9_47
35. Rezatofighi, H., et al.: Generalized intersection over union: a metric and a loss for bounding box regression. In: Proceedings of the IEEE Conference on Computer Vision and Pattern Recognition (2019)
36. Oneata, D., Verbeek, J., Schmid, C.: The lear submission at thumos 2014 (2014)
37. Carreira, J., Zisserman, A.: Quo vadis, action recognition? A new model and the kinetics dataset. In: Proceedings of the IEEE Conference on Computer Vision and Pattern Recognition (2017)
38. Yuan, J., et al.: Temporal action localization with pyramid of score distribution features. In: Proceedings of the IEEE Conference on Computer Vision and Pattern Recognition (2016)
39. Wang, R., Tao, D.: UTS at activitynet 2016. AcitivityNet Large Scale Activity Recognition Challenge 8, 2016 (2016)
40. Lin, T., Zhao, X., Shou, Z.: Temporal convolution based action proposal: submission to activitynet 2017. arXiv preprint arXiv:1707.06750 (2017)
41. Zach, C., Pock, T., Bischof, H.: A duality based approach for realtime TV-L^1 optical flow. In: Hamprecht, F.A., Schnörr, C., Jähne, B. (eds.) DAGM 2007. LNCS, vol. 4713, pp. 214–223. Springer, Heidelberg (2007). https://doi.org/10.1007/978-3-540-74936-3_22
42. Hu, J., Shen, L., Sun, G.: Squeeze-and-excitation networks. In: Proceedings of the IEEE Conference on Computer Vision and Pattern Recognition (2018)
43. Kingma, D.P., Ba, J.: Adam: a method for stochastic optimization. arXiv preprint arXiv:1412.6980 (2014)
44. Zhao, Y., et al.: Cuhk & ethz & siat submission to activitynet challenge 2017. arXiv preprint arXiv:1710.08011 8 (2017)

Anchor-Free One-Stage Online Multi-object Tracking

Zongwei Zhou[1,2(✉)], Yangxi Li[3], Jin Gao[1,2], Junliang Xing[1,2], Liang Li[4],
and Weiming Hu[5,6]

[1] University of Chinese Academy of Sciences, Beijing, China
{zongwei.zhou,jin.gao.jlxing}@nlpr.ia.ac.cn,
liyangxi@outlook.com
[2] Institute of Automation, Chinese Academy of Sciences, Beijing, China
[3] National Computer Network Emergency Response Technical Team/Coordination
Center of China, Beijing, China
[4] The Brain Science Center, Beijing Institute of Basic Medical Sciences,
Beijing, China
liang.li.brain@aliyun.com
[5] CAS Center for Excellence in Brian Science and Intelligence Technology,
National Laboratory of Pattern Recognition, CASIA, Beijing, China
wmhu@nlpr.ia.ac.cn
[6] School of Artificial Intelligence, University of Chinese Academy of Sciences,
Beijing, China

Abstract. Current multi-object tracking (MOT) algorithms are dominated by the tracking-by-detection paradigm, which divides MOT into three independent sub-tasks of target detection, appearance embedding, and data association. To improve the efficiency of this tracking paradigm, this paper presents an anchor-free one-stage learning framework to perform target detection and appearance embedding in a unified network, which learns for each point in the feature pyramid of the input image an object detection prediction and a feature representation. Two effective training strategies are proposed to reduce missed detections in dense pedestrian scenes. Moreover, an improved non-maximum suppression procedure is introduced to obtain more accurate box detections and appearance embeddings by taking the box spatial and appearance similarities into account simultaneously. Experiments show that our MOT algorithm achieves real-time tracking speed while obtaining comparable tracking performance to state-of-the-art MOT trackers. Code will be released to facilitate further studies of this problem.

Keywords: Anchor-free · One-stage · Multi-object tracking

Student paper.
This work is supported by the national key R&D program of China (No. 2018AA-A0102802, No. 2018AAA0102803, No. 2018AAA0102800), the NSFC-general technology collaborative Fund for basic research (Grant No. U1636218, U1736106), the NSFC (Grant No. 61751212, 61721004), Beijing Natural Science Foundation (Grant No. L172051), the Key Research Program of Frontier Sciences, CAS, Grant No. QYZDJ-SSW-JSC040, and the NNSF of Guangdong (No. 2018B030311046).

© Springer Nature Switzerland AG 2020
Y. Peng et al. (Eds.): PRCV 2020, LNCS 12306, pp. 55–68, 2020.
https://doi.org/10.1007/978-3-030-60639-8_5

56   Z. Zhou et al.

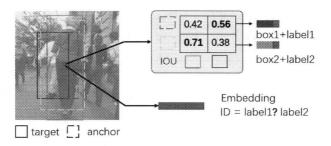

Fig. 1. The label ambiguity of features in an anchor-based MOT tracker.

1 Introduction

Multi-Object Tracking (MOT), *a.k.a* Multi-Target Tracking (MTT), is critical in video analysis systems ranging from video surveillance to autonomous driving. The objective of MOT is to determine the trajectories of multiple objects simultaneously by localizing and associating targets with the same identity across multiple frames. It is a very difficult task due to challenging factors like large variations in intra-target appearance and frequent inter-target interactions [13].

Tracking-by-detection is the main paradigm for the current multi-object tracking algorithms. It usually includes three steps: object detection in each frame, appearance embedding of each object, and data association across frames. Integrating these steps in one algorithm is usually difficult, especially if real time performance is required. For a MOT framework using a common simple association strategy (*e.g.* Hungarian algorithm), its computing resources are mainly consumed in separated object detection and appearance embedding steps. These two steps can share low-level features to improve the tracking speed. This suggests unifying object detection and appearance embedding in one step.

At present, there are two main schemes for joint detection and embedding learning. One is a two-stage framework similar to Faster-RCNN [17], and the other is a one-stage framework similar to SSD [11]. In the two-stage framework [25], the first stage uses a Region Proposal Network [17] to detect targets, and the second stage uses metric learning supervision to replace classification supervision in Faster-RCNN to learn target embedding. Although it saves some computation by sharing the low-level features, the two-stage design still limits its tracking speed. Moreover, generating a large number of region proposals improves accuracy but reduces efficiency. The solutions in one-stage framework are not well studied yet. The existing methods, such as AJDE [23], learn a joint detector and embedding model based on an anchor-based network, which relies on some predefined proposals named anchor boxes. The framework achieves near real-time tracking speed, but still has two disadvantages. As shown in Fig. 1, according to the Intersection-Over-Union (IOU) values, different anchor boxes (dotted boxes) at the same location are responsible for different targets (solid boxes), but only one feature vector is obtained, making the labels of features ambiguous. The other disadvantage stems from anchor-based structures, such as the manual configuration of hyper-parameters to define anchors and the complex architecture of detection subnets based on the predefined anchors.

To improve the tracking speed and avoid the disadvantages of the anchor-based structure, an anchor-free one-stage network is proposed in this work, where the bounding boxes and their corresponding appearance features are simultaneously extracted from the locations on feature maps directly, rather than predefined anchor boxes. We notice that the idea in [26] is similar to ours, but the method in [26] is more focused on the design of backbone, while our method focuses on the processing of joint object detection and embedding. We name the locations on feature maps as samples in the following. Unlike in general object detection tasks, the targets in multi-target tracking, especially multi-pedestrian tracking, tend to have similar scales and large occlusions. Thus, general anchor-free detectors (such as FCOS [21]) have a large number of missed detections in MOT due to attention bias and feature selection. Attention bias means that objects with good views tend to draw more attention from the detector making the partially occluded objects being easily missed. The feature selection issue arises because each target is scaled to a single pyramid level. This causes that multiple targets with similar scales may be assigned to same locations, especially if one target occludes another. The embedded features of the targets sharing the same location are ambiguous in that case. Therefore, the proposed model includes two strategies to reduce missed detection while incorporating embedding into the detector. First, the samples used for detection and embedding are re-weighted in the contribution to the network loss based on their distance to the object center. Second, the box regression ranges overlap in adjacent pyramid levels. A multi-task loss is introduced to train the model end-to-end.

Our precise embedding facilitates an improved Non-Maximum Suppression (NMS). The traditional NMS operator only considers Intersection-Over-Union (IOU) values between detections. The appearance information is ignored. As a result, many true targets are suppressed in crowded scenes. The improved NMS suppresses proposals, using both overlaps between the detections and the similarity of the appearances within the detected boxes for reducing over-suppression. The main contributions of this work are in three-fold:

- An anchor-free one-stage joint detection and embedding learning network is presented for online multi-target tracking. The model achieves real-time tracking speed while obtaining state-of-the-art tracking performance.
- Two effective training strategies are proposed to detect targets with similar scales in crowded scenes. The strategies are regression range overlapping and samples re-weighting.
- An improved NMS operator is designed to incorporate both the box spatial and appearance similarity to reduce false negatives in crowded scenes.

We develop a high performance online multi-object tracking system by incorporating the proposed network into a hierarchical data association pipeline. Extensive experimental analyses and evaluations on the MOT benchmark demonstrate the effectiveness and the efficiency of the proposed approach.

2   Related Work

Separate Detection and Embedding for MOT. These methods are dominant in the tracking-by-detection paradigm. Some of these methods build embedding networks upon the detections provided by the MOT benchmark to associate detections across frames, such as DeepSort [24], MOTDT [12], and DAN [19]. Other methods design both detectors and feature extractors to track targets. For example, POI [27] proposes a pedestrian detector based on Faster R-CNN, and Tracktor [1] uses the previous tracking results as proposals to detect the new bounding boxes of the targets for tracking. The single object tracking-based trackers [32] can also be regarded as detectors based on template matching. All these methods need an additional extractor after the detector to handle long-time occlusions. The overall inference time for these methods is approximately equal to the sum of the times for detection and extraction. This makes real-time operation difficult to achieve.

Joint Detection and Embedding for MOT. These methods reduce the tracking time calculations by combining the detection and the embedding into one step. MOTS [22], STAM [3] and D&T [6] integrate the embedding into a detector in a two-stage network, while AJDE [23] is a one-stage model. In a two-stage model, the detection and the embedding share the low-level features. The embedding is then extracted from the Region-of-Interest (ROI) after the detection. Due to the sequential nature of detection and embedding, the two-stage structure still has a limited tracking speed. Besides, since each target is processed separately in the second stage, the runtime of embedding is proportional to the number of targets. The one-stage model, AJDE, adds an embedding branch to the detection header of the SSD framework to carry out detection and embedding in parallel. This speeds up the tracking while maintaining tracking performance. But it suffers from the anchor-based structures, such as the manual configuration of anchor hyper-parameters and the complex architecture of detection header. Besides, the corresponding relationship between embedding and anchor boxes at the same location is not always one-to-one correspondent (Fig. 1). The proposed model is an anchor-free one-stage network, which overcomes the disadvantages of anchor-based structure and further improves the tracking speed.

3   Our Approach

3.1   Anchor-Free Joint Detection and Embedding

Network Architecture. As shown in Fig. 2, the network consists of a backbone, a feature pyramid and one prediction header per pyramid level, in a fully convolutional style. The backbone can include commonly used convolutional networks, such as ResNet50 [7]. The feature pyramid is adopted to deal effectively with large scale variations between targets. A pyramid level is represented as P_m where m denotes the level number. The level has $1/s_m$ resolution of the input

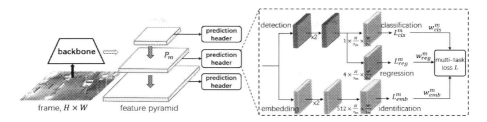

Fig. 2. Architecture of the anchor-free joint detection and embedding model.

frame size, where s_m is the stride of down-sampling. A prediction header contains two task-specific subnets, *i.e.* detection and embedding. The embedding subnet has three 3×3 convolutional layers and the output layer extracts a 512-dimensional discriminative feature from each location on the feature map. The detection subnet contains two 3×3 convolutional layers followed by two branches for classification and bounding box regression. The classification branch outputs the probability that each location is a positive sample. The regression branch predicts the distances from each sample to the boundaries of a corresponding target if the sample is positive.

Supervision Targets. A target in a frame $I \in R^{3 \times W \times H}$ is denoted as $B = (x, y, w, h, c)$ where (x, y) is the center position, w, h are the box width and height respectively. The $c \in Z^k$ is the partially annotated identity label, where -1 indicates a target without an identity label. Given a target, we first assign it to one pyramid level according to its scale. Specifically, the target is assigned to the mth pyramid level P_m if $\max(w, h) \in [a_m, b_m]$, where $[a_m, b_m]$ is the predefined regression range of bounding box in P_m. We overlap the predefined regression ranges in adjacent pyramid levels to improve the recall by providing more proposals from different granularity, especially for close and similar-scaled targets. Next we define the positive samples in the mth pyramid level. Each sample p_{mij} with $i = 1, 2, \ldots, W/s_m$ and $j = 1, 2, \ldots, H/s_m$ on P_m has a corresponding image spatial location (X_{mij}, Y_{mij}) where $X_{mij} = s_m(i - 0.5)$ and $Y_{mij} = s_m(j - 0.5)$. The sample is set as positive if its centerness to any one target B_k assigned to P_m is larger than a threshold τ_c. The centerness is the same as that defined in FCOS [21], *i.e.*,

$$\mathrm{CT}(p_{mij}, B_k) = \sqrt{\frac{\min(l_{mij}^k, r_{mij}^k)}{\max(l_{mij}^k, r_{mij}^k)} \frac{\min(t_{mij}^k, b_{mij}^k)}{\max(t_{mij}^k, b_{mij}^k)}}, \tag{1}$$

where $(l_{mij}^k, r_{mij}^k, t_{mij}^k, b_{mij}^k)$ denotes the distances between (X_{mij}, Y_{mij}) and the left, right, top and bottom boundaries of target B_k.

If the centernesses of a positive sample p_{mij} to multiple targets are all larger than the threshold τ_c, the sample is regarded as ambiguous. The target B_{k^*} with maximal centerness is chosen as the responsible object of the ambiguous sample, where $k^* = \text{argmax}_k\{\text{CT}(p_{mij}, B_k)|k = 1, 2, \ldots, K\}$, and K is the number of targets assigned in mth level. The centerness map on P_m is defined as:

$$M_{mij} = \max_{k=1,2,\ldots,K} \text{CT}(p_{mij}, B_k). \tag{2}$$

Multi-task Loss Function. The loss function of the proposed anchor-free joint detection and embedding model consists of three components for different tasks, the point classification, the box regression, and discriminative feature extraction.

In the detection subnet, we use the IOU loss L_{reg} as in FCOS to regress bounding boxes B_{k^*} from positive samples. For the point classification, the hard-designation of positives and negatives brings more difficulties for training. To reduce the ambiguity of the samples between hard positives and negatives, we apply the centerness map M to re-weight the contributions of ambiguous samples. The focal weight [10] on hard examples are also adopted to combat the extreme class imbalance between positive and negative samples. Let ρ_{mij} be the network's estimated probability indicating whether the sample p_{mij} is positive, and γ be the focusing hyper-parameter. Then, the classification loss in mth pyramid level can be formulated as:

$$L_{cls}^m = -\frac{1}{K} \sum_{i=1}^{W/s_m} \sum_{j=1}^{H/s_m} \alpha_{mij}(1 - \hat{\rho}_{mij})^\gamma \log(\hat{\rho}_{mij}), \tag{3}$$

where

$$\hat{\rho}_{mij} = \begin{cases} \rho_{mij}, & \text{if } M_{mij} > \tau_c \\ 1 - \rho_{mij}, & \text{otherwise} \end{cases}, \quad \alpha_{mij} = \begin{cases} 1, & \text{if } M_{mij} > \tau_c \\ (1 - M_{mij})^\beta, & \text{otherwise} \end{cases}. \tag{4}$$

The focusing hyper-parameter γ is experimentally set to 2 as suggested in the Focal Loss [10], and the hyper-parameter β controls the penalty on the ambiguous samples to reduce their contributions to the total loss.

The objective of the embedding subnet is to learn an embedding space where observations of the same target are close to each other, while observations of different targets are far apart. We transform the metric learning problem into the classification problem like many re-identification (ReID) models [20,30]. Then the cross-entropy loss is used to extract discriminative features. Let $\mathbf{f}_{mij} \in R^{512}$ be the output feature in p_{mij} and c_k be the class label of B_k regressed in p_{mij}. Let $W \in R^{512 \times N}$ be the learnable parameters of the last classifier layer, where N is the number of targets. Then, the embedding loss is defined as follows,

$$L_{emb}^m = -\sum_{ij:M_{mij} > \tau_c} \log \frac{e^{(\mathbf{W}^T \mathbf{f}_{mij})_{c_k}}}{\sum_q e^{(\mathbf{W}^T \mathbf{f}_{mij})_q}}. \tag{5}$$

The automatic learning scheme for loss weights proposed in [8] is adopted to combine these three losses. The total multi-task loss with automatic loss balancing is formulated as,

$$L = \sum_{m,T \in \{cls,reg,emb\}} \frac{1}{e^{w_T^m}} L_T^m + w_T^m, \tag{6}$$

where $w_T^m, T \in \{cls, reg, emb\}$ is the learnable weight parameters.

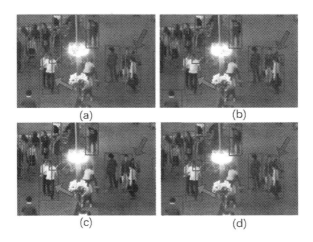

Fig. 3. An exemplar of ENMS. (a), (b), (c), (d) show the detections without NMS, with NMS, with NMS using appearance similarity and with ENMS respectively.

3.2 Appearance Enhanced NMS (ENMS)

NMS is an integral part of the object detection pipeline. The detected boxes are first sorted according to scores. The box with the highest score is then selected. All the other boxes that have a significant overlap with it are suppressed. This process is applied recursively to the remaining boxes until the final detection result is obtained. Though NMS is efficient in suppressing false positives, it also over-suppresses in dense scenes as it does not take any appearances into account. As shown in Fig. 3, the raw proposals are given in Fig. 3(a) and the detections processed by NMS are given in Fig. 3(b). The arrows in Fig. 3(b) point to targets that are wrongly suppressed by NMS.

Benefiting from the joint model introduced in the last subsection (Sect. 3.1), which provides detection and embedding simultaneously, we can use the discriminative feature to enhance the NMS operator. Formally, given the raw proposals $\mathcal{B} = \{(B_k, \rho_k, \mathbf{f}_k)k = 1, 2, \ldots, N\}$ and an empty set $\mathcal{B}_f = \emptyset$, where $B_k, \rho_k, \mathbf{f}_k$ denotes the regressed boxes, predicted scores and features respectively, the most reliable proposals $(B_{k^*}, \rho_{k^*}, \mathbf{f}_{k^*})$, $k^* = \text{argmax}_k \rho_k$ are selected firstly. Then get the false proposals of B_{K^*} based on the box overlap and the appearance similarity, $i.e.$,

$$\mathcal{B}_s = \{(B_k, \rho_k, \mathbf{f}_k)|\text{IOU}(B_k, B_{k^*}) > \tau_i \cap \mathbf{f}_{k^*}^T \mathbf{f}_k > \tau_e\}, \tag{7}$$

where τ_i, τ_e are predefined thresholds for IOU and appearance similarity respectively. Update the set $\mathcal{B}_f = \mathcal{B}_f \cup \{B_{k^*}\}$ and apply the above process recursively in $\mathcal{B} = \mathcal{B} \setminus (\mathcal{B}_s \cup \{B_{k^*}\})$ until $\mathcal{B} = \emptyset$. The set $\mathcal{B}_f$ contains the final detections. The detections obtained after suppressing false positives using only appearance similarity are shown in Fig. 3(c), while that obtained using the proposed ENMS are shown in Fig. 3(d). It shows that the ENMS reduces false suppressions in dense scenes.

3.3 Tracking Pipeline

The proposed anchor-free joint detection and embedding model and the appearance enhanced NMS operator are combined with the hierarchical association strategy in MOTDT [12] to form the tracking pipeline in our tracking algorithm.

- Step 1. Given a new frame, obtain the proposals and corresponding features using the proposed anchor-free joint detection and embedding model.
- Step 2. Filter the proposals using the enhanced NMS.
- Step 3. Assign the filtered detections to existing tracklets using feature similarities with a threshold ε_d for the minimum similarity. The similarity is also limited by the distance between the detection and prediction of the tracklet in order to meet the constraint of spatial continuity. That is, the target motion offset in consecutive frames is small. The tracklet feature is online updated as,

$$\mathbf{f}_t = \eta\mathbf{f}_{t-1} + (1 - \eta)\mathbf{f}_k, \qquad (8)$$

where η is the momentum cofficient and set as 0.9 as in AJDE [23], $\mathbf{f}_k$ is the feature of associated detection and $\mathbf{f}_t$ denotes the track feature at time t.
- Step 4. Associate the remaining candidates with unassociated tracklets based on the IOU values between candidates and predictions with a threshold ε_{iou}.
- Step 5. Mark any untracked track as lost. Initialize a new trajectory with any unmatched detection with a confidence higher than ε_p. Terminate any trajectory that remains lost for over ε_n successive frames or exits the field of view. Additionally, any new tracks will be deleted if they are lost within the first two frames. This is to suppress false trajectories.
- Step 6. Repeat above steps for the next frame until no more frames arrive.

Table 1. Statistics of the training set.

Dataset	ETH	CP	CT	M16	CS	PRW	Total
#Img	2K	3K	27K	5.3K	11K	6K	54.3K
#Box	17K	21K	46K	112K	55K	18K	270K
#ID	–	–	0.6K	0.5K	7K	0.5K	8.7K

Table 2. Quantitative analysis of two training strategies.

OR	RW	MOTA	Pre	Rec	IDS	IDF1	mAP	TFR$_{0.1}$
×	×	63.4	80.3	75.7	366	66.3	81.2	88.3
×	✓	66.7	85.7	81.1	98	67.1	82.7	89.0
✓	×	70.2	**88.9**	81.9	103	66.1	82.2	90.8
✓	✓	**71.9**	88.4	**83.4**	**78**	**69.8**	**82.8**	**91.7**

4 Experiments

4.1 Experimental Settings

Datasets. Since we transform metric learning into a classification problem, datasets for pedestrian detection, pedestrian ReID and multi-pedestrian tracking are all used to train the anchor-free joint detection and embedding model. The statistics of the training sets are shown in Table 1. ETH dataset [4] and CityPersons (CP) dataset [28] are used for person detection. We mark their targets ID as −1 in training as they have no identity annotations. PRW dataset [29] and CUHK-SYSU (CS) dataset are derived from the ReID task. CalTech (CT) dataset [25] and MOT16 (MT) dataset [15] are collected from the MOT task. The sequences in the ETH dataset that overlap with the MOT16 test set are excluded for fair evaluation. The model is first analyzed on the MOT15 dataset [9] after excluding the sequences appeared in the training, then its performance is compared with the SOTA methods on the MOT16 test set.

Evaluation Metrics. The CLEAR MOT metrics [2] are used to analyze the tracking performance. They include multiple object tracking accuracy (MOTA, ↑), the number of mostly tracked targets (MT, >80% covered, ↑), the number of mostly lost targets (ML, <20% covered, ↓), false positive (FP, ↓), false negative (FN, ↓) precision (Pre, ↑), recall (Rec, ↑), and identity switches(IDs, ↓). Additionally, ID F1 score (IDF1, ↑) [18] is also employed to measure the identity-preserving ability of trackers. IDF1 denotes the ratio of correctly identified detections over the average number of ground-truth and computed detections. To evaluate the detection accuracy and the appearance embedding, we also use the metrics defined in [23], *i.e.*, the average precision (AP, ↑) at IOU threshold of 0.5 over the Caltech validation set and the true positive rate at false accept rate 0.1 ($\text{TFR}_{0.1}$, ↑) over the CUHK-SYSU and PRW validation sets. Here ↑ means higher is better, and ↓ means lower is better.

Implementation Details. We employ DarkNet-53 [16] as the backbone network. The network is trained for 60 epochs with Stochastic Gradient Descent (SGD) optimizer and the batchsize is set as 16. The learning rate is initialized as 10^{-2} and is decreased by 0.1 at the 30th and 50th epoch. The input resolution is 1088×608 if not specified and the data augmentation techniques, such as random rotation, random scale and color jittering, are applied to reduce over-fitting. The predefined regression ranges $[a_m, b_m], m = 1, 2, 3$ are set as $[0, 160], [64, 320]$ and $[256, 608]$ respectively. The parameters τ_i and τ_e used in the improved NMS are set to 0.5 and 0.2 respectively according to the experiment analysis shown in Fig. 4. The parameter τ_c used for selecting positive anchor points and the parameter β used in Eq. (5) are analyzed in next subsection. For data association, we set $\varepsilon_d = 0.5, \varepsilon_{iou} = 0.5, \varepsilon_p = 0.6$ and ε_n as the frame rate of the sequence.

4.2 Ablation Study

Analysis of the Hyper-parameters. Table 4 analyzes the effects of the hyper-parameters τ_c and β, where τ_c is the centerness threshold used to select positive samples and β is the exponent in Eq. (4) used to penalize ambiguous samples. When $\beta = 1.0$, the higher τ_c obtains a better Pre as the positives are more concentrated, but concentrated positives mean more ambiguity in negatives which decrease the Rec. When $\tau_c = 0.8$, the performances of $\beta > 0$ are all, except Pre, better than the performances with $\beta = 0$. This means that the weighted focal loss is more effective than the original focal loss in MOT. The higher Pre is obtained at $\beta = 0$ because samples around positives are marked as hard negative samples. This enhances the certainty of positives, but also introduces the ambiguity of negatives leading to a lower Rec. The proposed model has the best performance at $\tau_c = 0.8$ and $\beta = 1.0$, so we use these settings when we evaluate the proposed method on MOT benchmark.

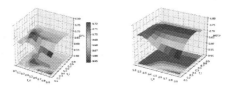

Fig. 4. The analysis of IOU threshold t_i and appearance similarity threshold t_s on the performance of the appearance enhanced NMS on MOT15 train dataset. (a) and (b) are IDF1 and MOTA respectively.

Table 3. Analysis of the appearance enhanced NMS (ENMS). × and ✓ indicate whether the ENMS module is used.

Method	ENMS	MOTA	Pre	Rec	IDs	IDF1	FPS
AJDE	×	67.3	84.0	85.1	203	66.4	29.4
	✓	62.0	78.3	89.6	366	67.4	28.7
AFOS[†]	×	67.7	87.0	80.4	99	66.2	31.3
	✓	68.9	86.8	82.1	106	66.6	30.8
AFOS	×	71.9	88.4	83.4	78	69.8	31.6
	✓	72.8	86.3	87.2	80	70.5	30.8

Analysis of Two Training Strategies. Two training strategies, *i.e.* overlapping regression ranges (OR) and samples re-weighting (RW), are designed to deal with crowded scenes. The quantitative analyses of the strategies with the enhanced NMS are shown in Table 2. Both Pre and Rec are improved by overlapping regression ranges because targets with similar scales in crowded scenes are assigned to different pyramid levels to reduce their interaction. This also makes it possible to extract more discriminative embedding to reduce the ID switching in MOT. Samples re-weighting further enhances tracking performance by improving Rec, as the contribution of ambiguous samples among hard positives and negatives is reduced. Both strategies improve detection and tracking performance by enhancing the discrimination of targets in crowded scenes.

Analysis of the Appearance Enhanced NMS. Enhanced NMS (ENMS) introduces feature similarity to conventional NMS to reduce over-suppression. As can be seen from Table 3, AFOS and AFOS[†], which represent results with and without overlapping regression ranges respectively, all benefit from ENMS.

Table 4. Analysis of the hyper parameters on MOT15-train.

τ_c	β	AP	$TFR_{0.1}$	MOTA	Pre	Rec	IDs	IDF1
0.70	1.0	83.3	90.9	70.7	87.2	83.8	96	69.0
0.75	1.0	83.3	89.2	71.2	87.4	**84.2**	109	67.5
0.80	1.0	82.8	**91.7**	**71.9**	88.4	83.4	**78**	**69.8**
0.85	1.0	83.7	90.4	71.7	88.4	83.4	106	68.7
0.80	0.0	82.2	90.9	70.2	**88.9**	81.9	103	66.1
0.80	0.05	**83.9**	91.7	71.8	88.2	82.9	97	66.8
0.80	2.0	82.6	91.5	71.2	87.5	84.0	93	69.8

Table 5. Failure analysis of each subset of MOT16-test.

Sets	Density	FP	FN
MOT16-01	14.2	284 (2.0%)	2675 (5.5%)
MOT16-03	69.7	**10928 (75.3%)**	**15068 (31.2%)**
MOT16-06	9.7	651 (4.5%)	4013 (8.3%)
MOT16-07	32.6	967 (6.7%)	5321 (11.0%)
MOT16-08	26.8	667 (4.6%)	8702 (18.0%)
MOT16-12	9.2	420 (2.9%)	2874 (6.0%)
MOT16-14	24.6	594 (4.1%)	9635 (20.0%)
Total	30.8	14511	48288

Table 6. Comparison with the state-of-the-art online MOT trackers under the private detectors on the MOT16 benchmark. In each column of the one-stage and two-stage methods, the best result is in **bold**.

#Stage	Tracker	Det	Emb	#Box	#Id	MOTA	IDF1	MT	ML	FP	FN	IDs	FPS
Two-stage	DeepSORT_2	FRCNN	WRN	429K	1.2K	61.4	62.2	32.8	**18.2**	12852	56668	781	<8.1
	RAR16wVGG	FRCNN	Inception	429K	—	63.0	63.8	**39.9**	22.1	13663	53248	**482**	<1.5
	TAP	FRCNN	MRCNN	429K	—	64.8	**73.5**	38.5	21.6	12980	**50635**	571	<8.2
	CNNMTT	FRCNN	5-layer	429K	0.2K	65.2	62.2	32.4	21.3	6578	55896	946	<6.4
	POI	FRCNN	QAN	429K	16K	**66.1**	65.1	34.0	20.8	**5061**	55194	805	<6.0
One-stage	AJDE_864	Anchor-box	JDE	270K	8.7K	62.1	56.9	34.4	**16.7**	—	—	1608	32.1
	AJDE_1088	Anchor-box	JDE	270K	8.7K	64.4	55.8	**35.4**	20.0	**9172**	54160	1544	25.4
	AFOS_864 (ours)	Anchor-free	JDE	270K	8.7K	63.2	59.0	33.6	22.9	13268	52277	1485	**34.3**
	AFOS_1088 (ours)	Anchor-free	JDE	270K	8.7K	**64.8**	**63.1**	35.0	22.9	14511	**48288**	**1300**	26.5

Although ENMS slightly decreases the Pre values, it improves the recall rates (Rec) by reducing false suppressions. This improves the MOTA. By reducing false suppression, the models also achieve the higher IDF1, which measures the continuity of the trajectory. In addition, the slightly slower speed of the model using ENMS than that of the model using conventional NMS is because ENMS calculates the appearance similarity. For AJDE, we find the performance with ENMS is worse than that with conventional NMS. The reason is that the label ambiguity of embeddings in the training process of AJDE leads to a confusing of the targets in the crowded scenes. This reduces the performance of the ENMS. On the contrary, our model overcomes the label ambiguity, which facilitates ENMS to further improve tracking performance.

4.3 Evaluation on MOT Benchmark

The proposed method is compared with several state-of-the-art trackers under private detectors, such as DeepSORT_2 [24], RAR16wVGG [5], TAP [31], CNNMTT[14], POI [27] and AJDE [23], on the test sets of MOT16. Their configurations and performances are summarized in Table 6. It can be seen from Table 6 that the joint models (AJDE and the proposed AFOS) run at least $3\times$ faster than existing methods while achieving comparable overall tracking accuracy, e.g., as measuted by the MOTA metric.

Compared with AJDE, the proposed method AFOS obtains better IDs and IDF1 as it extracts more discriminative features and avoids the label ambiguity. With the enhanced NMS, AFOS also suppress more false negatives. Note we didn't compare APOS with the performance of AJDE with ENMS because the feature ambiguous in AJDE reduces the performance of ENMS which analyzed in Sect. 4.2. As AFOS is an anchor-free model while AJDE is an anchor-based model, AFOS is faster than AJDE. AFOS reaches a real-time speed, *i.e.*, 26.5 FPS for images of size 1088×608. When the image resolution is down-sampled to 864×480, the speed of AFOS can be further increased to 34.3 FPS with only a minor performance drop ($\Delta = -1.6\%$ MOTA). All the experiments are performed on an NVIDIA Tesla V100 GPU.

Analysis of Tracking Failures. One may notice that AFOS has a much better FN but a worse FP compared to other methods. We analyze the performance of each subset in Table 5 and find that the FP and FN mainly come from MOT16-03 (75.3% and 31.2% respectively). This is because the targets in MOT16-03 are densely distributed with severe occlusions. Many targets are assigned to the same pyramid level, making them difficult to distinguish.

5   Conclusion and Future Work

In this paper, we have proposed a new MOT tracker named AFOS, which allows target detection and appearance embedding to be learned in an anchor-free joint model. AFOS achieves real-time tracking speed with a tracking performance comparable to that of state-of-the-art MOT trackers. Moreover, in order to benefit from the anchor-free joint detection and embedding model, we introduce an appearance enhanced NMS, which combines the appearance similarity with the conventional NMS to prevent over-suppression. We analyze the tracking failures in the proposed model, and plan to perform occlusion model in AFOS to further improve its performance in densely crowded scenes in future work.

References

1. Bergmann, P., Meinhardt, T., Leal-Taixe, L.: Tracking without bells and whistles. arXiv preprint arXiv:1903.05625 (2019)
2. Bernardin, K., Stiefelhagen, R.: Evaluating multiple object tracking performance: the CLEAR MOT metrics. JIVP **1**, 1–10 (2008)
3. Chu, Q., Wanli, O., Li, H., Wang, X., Liu, B., Yu, N.: Online multi-object tracking using CNN-based single object tracker with spatial-temporal attention mechanism. In: ICCV, pp. 4836–4845 (2017)
4. Ess, A., Leibe, B., Schindler, K., Van Gool, L.: A mobile vision system for robust multi-person tracking. In: CVPR, pp. 1–8 (2008)
5. Fang, K., Xiang, Y., Li, X., Savarese, S.: Recurrent autoregressive networks for online multi-object tracking. In: WACV, pp. 466–475 (2018)

6. Feichtenhofer, C., Pinz, A., Zisserman, A.: Detect to track and track to detect. In: ICCV, pp. 3038–3046 (2017)
7. He, K., Zhang, X., Ren, S., Sun, J.: Deep residual learning for image recognition. In: CVPR, pp. 770–778 (2016)
8. Kendall, A., Gal, Y., Cipolla, R.: Multi-task learning using uncertainty to weigh losses for scene geometry and semantics. In: CVPR, pp. 7482–7491 (2018)
9. Leal-Taixé, L., Milan, A., Reid, I., Roth, S., Schindler, K.: MOTchallenge 2015: towards a benchmark for multi-target tracking. arXiv preprint arXiv:1504.01942 (2015)
10. Lin, T.Y., Goyal, P., Girshick, R., He, K., Dollár, P.: Focal loss for dense object detection. In: ICCV, pp. 2980–2988 (2017)
11. Liu, W., et al.: SSD: single shot multibox detector. In: Leibe, B., Matas, J., Sebe, N., Welling, M. (eds.) ECCV 2016. LNCS, vol. 9905, pp. 21–37. Springer, Cham (2016). https://doi.org/10.1007/978-3-319-46448-0_2
12. Long, C., Haizhou, A., Zijie, Z., Chong, S.: Real-time multiple people tracking with deeply learned candidate selection and person re-identification. In: ICME, pp. 1–6 (2018)
13. Luo, W., Xing, J., Zhang, X., Zhao, X., Kim, T.K.: Multiple object tracking: a literature review. arXiv preprint arXiv:1409.7618 (2014)
14. Mahmoudi, N., Ahadi, S.M., Rahmati, M.: Multi-target tracking using CNN-based features: CNNMTT. Multimed. Tools. Appl. **78**(6), 7077–7096 (2019)
15. Milan, A., Leal-Taixé, L., Reid, I., Roth, S., Schindler, K.: MOT16: a benchmark for multi-object tracking. arXiv preprint arXiv:1603.00831 (2016)
16. Redmon, J., Divvala, S., Girshick, R., Farhadi, A.: You only look once: unified, real-time object detection. In: CVPR, pp. 779–788 (2016)
17. Ren, S., He, K., Girshick, R., Sun, J.: Faster R-CNN: towards real-time object detection with region proposal networks. In: NIPS, pp. 91–99 (2015)
18. Ristani, E., Solera, F., Zou, R., Cucchiara, R., Tomasi, C.: Performance measures and a data set for multi-target, multi-camera tracking. In: Hua, G., Jégou, H. (eds.) ECCV 2016. LNCS, vol. 9914, pp. 17–35. Springer, Cham (2016). https://doi.org/10.1007/978-3-319-48881-3_2
19. Sun, S., Akhtar, N., Song, H., Mian, A.S., Shah, M.: Deep affinity network for multiple object tracking. arXiv preprint arXiv:1810.11780 (2019)
20. Sun, Y., Zheng, L., Yang, Y., Tian, Q., Wang, S.: Beyond part models: person retrieval with refined part pooling (and a strong convolutional baseline). In: Ferrari, V., Hebert, M., Sminchisescu, C., Weiss, Y. (eds.) ECCV 2018. LNCS, vol. 11208, pp. 501–518. Springer, Cham (2018). https://doi.org/10.1007/978-3-030-01225-0_30
21. Tian, Z., Shen, C., Chen, H., He, T.: FCOS: Fully convolutional one-stage object detection. arXiv preprint arXiv:1904.01355 (2019)
22. Voigtlaender, P., et al.: MOTS: multi-object tracking and segmentation. In: CVPR, pp. 7942–7951 (2019)
23. Wang, Z., Zheng, L., Liu, Y., Wang, S.: Towards real-time multi-object tracking. arXiv preprint arXiv:1909.12605 (2019)
24. Wojke, N., Bewley, A., Paulus, D.: Simple online and realtime tracking with a deep association metric. In: ICIP, pp. 3645–3649 (2017)
25. Xiao, T., Li, S., Wang, B., Lin, L., Wang, X.: Joint detection and identification feature learning for person search. In: CVPR, pp. 3415–3424 (2017)
26. Zhan, Y., Wang, C., Wang, X., Zeng, W., Liu, W.: A simple baseline for multi-object tracking. arXiv preprint arXiv:2004.01888v2 (2020)

27. Yu, F., Li, W., Li, Q., Liu, Yu., Shi, X., Yan, J.: POI: multiple object tracking with high performance detection and appearance feature. In: Hua, G., Jégou, H. (eds.) ECCV 2016. LNCS, vol. 9914, pp. 36–42. Springer, Cham (2016). https://doi.org/10.1007/978-3-319-48881-3_3
28. Zhang, S., Benenson, R., Schiele, B.: CityPersons: a diverse dataset for pedestrian detection. In: CVPR, pp. 3213–3221 (2017)
29. Zheng, L., Zhang, H., Sun, S., Chandraker, M., Yang, Y., Tian, Q.: Person re-identification in the wild. In: CVPR, pp. 1367–1376 (2017)
30. Zheng, Z., Zheng, L., Yang, Y.: A discriminatively learned CNN embedding for person reidentification. ACM TOMM **14**(1), 1–20 (2017)
31. Zhou, Z., Xing, J., Zhang, M., Hu, W.: Online multi-target tracking with tensor-based high-order graph matching. In: ICPR, pp. 1809–1814 (2018)
32. Zhu, J., Yang, H., Liu, N., Kim, M., Zhang, W., Yang, M.-H.: Online multi-object tracking with dual matching attention networks. In: Ferrari, V., Hebert, M., Sminchisescu, C., Weiss, Y. (eds.) ECCV 2018. LNCS, vol. 11209, pp. 379–396. Springer, Cham (2018). https://doi.org/10.1007/978-3-030-01228-1_23

Multi-view and Multi-label Method with Three-Way Decision-Based Clustering

Changming Zhu[1][(✉)] ⓘD, Lin Ma[1], Panhong Wang[1], and Duoqian Miao[2]

[1] Shanghai Maritime University, Shanghai 201306, China
cmzhu@shmtu.edu.cn
[2] Tongji University, Shanghai 201804, China

Abstract. In real-world multi-view multi-label clustering and some classification tasks, instances and the corresponding clusters has at least three kinds of relationships, belong-to definitely, not belong-to definitely, and uncertain. Some learning machines consider only two of them, for example, belong-to definitely and not belong-to definitely. Moreover, three-way decision-based clustering (TDC) strategy is a good method to make the belongingness of instances to a cluster depend on the probabilities of uncertain instances belonging to core regions. Thus in our work, we take the notion of classical multi-view multi-label learning machines as the basic and introduce TDC so as to develop a multi-view and multi-label method with three-way decision-based clustering (MVML-TDC) and consider the relationships between instances and clusters. Experimental results validate that MVML-TDC achieves a better average performance and an acceptable running time.

Keywords: Three-way decision-based clustering · Multi-view learning · Multi-label learning

1 Introduction

1.1 Background and Problems

Since multi-view multi-label data sets exist in real-world applications widely, thus some related tasks are put forward. Among them, two tasks are of a general nature. One is clustering and the other is classification. Classical multi-view multi-label clustering methods include connectivity constrained clustering algorithm for traffic segmentation [1] and classical multi-view multi-label classification methods include latent semantic aware multi-view multi-label learning (LSA-MML) [2], latent multi-view subspace clustering (LMSC) [3]. Although there are many learning methods developed to process these tasks, there is a key problem should to be solved. As we know, in real-world applications, instances and the clusters (or sub classes) might have gradual relationships. Namely, there are three relationships between an instance and a cluster, namely, belong-to definitely, not belong-to definitely, and uncertain. In most of the existing studies,

Y. Peng et al. (Eds.): PRCV 2020, LNCS 12306, pp. 69–80, 2020.
https://doi.org/10.1007/978-3-030-60639-8_6

a cluster is represented by a single set. Any set naturally divides the space into two regions. Instances belong to the cluster if they are elements of the set, otherwise they do not. Here, only two relationships are considered, no matter in hard clustering or in soft clustering. They are typically based on two-way (i.e., binary) decisions. Then for the third relationship, which means the instance may or may not belong to the cluster, one cannot make decisions based on the presently obtained knowledge, i.e, two-way decisions, but they can make further decisions once more information becomes available. This method is referred to as three-way decisions. According to the present learning methods, they don't adopt three-way decisions. Thus, the improvement of performance is limited.

1.2 Objectives

In order to solve the above this problem, we develop a multi-view and multi-label method with three-way decision-based clustering (MVML-TDC). First, we adopt an three-way decision-based clustering (TDC) which is one kind of three-way decisions developed by Yu et al. [4] to improve clustering performance and produce the group partition of the data set. Second, we construct a multi-view and multi-label model with the introduction of TDC.

1.3 Novelty and Contributions

The novelty of MVML-TDC is that in the field of multi-view multi-label learning, it is the first attempt for the combination of multi-view multi-label learning and three-way decisions. The proposed method can improve the classification and clustering performances simultaneously.

The contributions of MVML-TDC are (1) it has a better ability to process multi-view multi-label data sets; (2) it won't add too much running time and moves forward research of multi-view multi-label learning.

2   Three-Way Decision-Based Clustering [4]

Suppose there is a data set $X = \{x_1, ..., x_i, ..., x_n\}$ and it should be clustered with g clusters, namely, $C = \{C_1, ..., C_m, ..., C_g\}$ and for each cluster C_m, it can be represented by $C_m = (Co(C_m), Fr(C_m))$ further. Here, $Co(C_m) = CoreRegion(C_m)$ and $x \in CoreRegion(C_m)$ represents that x belongs to cluster C_m definitely. Meanwhile, $Fr(C_m) = FringeRegion(C_m)$ and $x \in FringeRegion(C_m)$ represents that x might belong to cluster C_m. Then, $Tr(C_m) = X - Co(C_m) - Fr(C_m) = TrivialRegion(C_m)$ and $x \in TrivialRegion(C_m)$ represents that x does not belong to cluster C_m definitely. Moreover, for each cluster, we have $X = Co(C_m) \bigcup Fr(C_m) \bigcup Tr(C_m)$, $Co(C_m) \bigcap Fr(C_m) = \varnothing$, $Co(C_m) \bigcap Tr(C_m) = \varnothing$, $Tr(C_m) \bigcap Fr(C_m) = \varnothing$. Thus, $C = \{(Co(C_1), Fr(C_1)), ..., (Co(C_g), Fr(C_g))\}$. What's more, we let $Y = \{y_1, ..., y_i, ..., y_n\}$ indicates the label matrix and y_i represents the label of x_i. Then TDC is carried out as a sequence of the following steps.

2.1   Pairwise Constraints

According to [4], pairwise constraints offer typical prior information for semi-supervised clustering. In their work, they introduce must-link (positive association) and cannot-link (negative association) to reflect the constraint relations between the data points, i.e., instances. Must-link constraint requires that the two instances must belong to the same cluster, and this relation is denoted by $ML = \{(x_i, x_j)|y_i = y_j, \quad for \quad i \neq j, \quad x_i, x_j \in X, \quad y_i, y_j \in Y\}$. Cannot-link constraint requires that the two objects must belong to different clusters, and this relation is denoted by $CL = \{(x_p, x_q)|y_p = y_q, \quad for \quad p \neq q, \quad x_p, x_q \in X, \quad y_p, y_q \in Y\}$. Then for instances $x_i, x_j, x_g \in X$, [5] said that must-link constraint shows the following transitivity properties for instances.

$$
\begin{aligned}
& (x_i, x_j) \in ML \quad \& \quad (x_j, x_g) \in ML \\
& \Rightarrow (x_i, x_g) \in ML \\
& (x_i, x_j) \in ML \quad \& \quad (x_j, x_g) \in CL \\
& \Rightarrow (x_i, x_g) \in CL
\end{aligned}
\tag{1}
$$

Then in [4], we set a matrix $R \in \mathbb{R}^{n \times n}$ to store the constraint pairs and the update of R is given below. First, R is initialized as $\varnothing$. Then, when a pair of instances are must-link constraint relation, namely $(x_i, x_j) \in ML$, the corresponding value of element in R is updated to 1; when a pair of instances are cannot-link constraint relation, namely $(x_i, x_j) \in CL$, the corresponding value of element in R is updated to 0. At the end of each iteration, we update R according to the response of expert and the transitivity properties of Eq. (1).

Next, one updates the consensus similarity matrix $W^\star$ in the following manner. Here, $W^\star = (Z^\star + (Z^\star)^T)/2$ where $Z^\star$ is a consensus low-rank matrix which is derived from X. The method to get $Z^\star$ can be refer to [4].

$$
\begin{aligned}
& if \quad (x_i, x_j) \in ML, \quad then \quad w_{ij} = w_{ji} = 1 \\
& if \quad (x_i, x_j) \in CL, \quad then \quad w_{ij} = w_{ji} = 0
\end{aligned}
\tag{2}
$$

2.2   Initialize Core Regions Construction

When we obtain the pairwise constraints of X, they should initialize core regions construction. The method to realize the initialization is farthest-first traversal scheme and this scheme aims to select the core instances. Core instances indicate the ones which locate on the fringe of a cluster and contain more information than instances in the center of a cluster. The basic idea of farthest-first traversal of a set of instances is to find K instances such that they are far from each other. Details are given below.

First, we let the original X be the *CandidateSet* and l be the count of the number of constructed core regions. Initial value of l is 0. Second, for each cluster

$(Co(C_m), Fr(C_m))$, it is initialized as $\varnothing$, namely, $Co(C_m) = \varnothing$ and $Fr(C_m) = \varnothing$. Third, we first select a starting instance x from X at random and put x into the $Co(C_1)$. Fourth, we choose the next instance to be farthest from the untraversed set $CandidateSet$ by Eq. (3). Being specific, $AllCo$ is the set of all core instances, namely, $AllCo = \bigcup_{p=1}^{l} Co(C_p)$. According to min-max criterion, the distance between x and $AllCo$ is $d(x, AllCo(p)) = \min\limits_{y \in AllCo(p)} ||x - y||$. Then, the farthest one is determined as follows.

$$x \leftarrow arg \max\limits_{x \in CandidateSet} d(x, AllCo) \tag{3}$$
$$= arg \max\limits_{x \in CandidateSet, y \in AllCo(p)} (\min ||x - y||)$$

After that, we should decide whether x and an instance $x_i \in Co(C_p)$ ($1 \leq p \leq l$) are in the same cluster. They make pair-wise queries through the form as: do instances x and x_i belong to the same cluster? If the ML constraint is satisfied, x is assigned to $Co(C_m)$ and it should be removed from $CandidateSet$. If no one ML constraint is satisfied after traversing all core regions, a new core region $Co(C_{l+1})$ is constructed and assign x to the new core region $Co(C_{l+1})$. With the above procedure, we can divide the $CandidateSet$ into g clusters.

2.3 Extend Core Regions and Construct Fringe Regions

Once we initialize the core regions construction, they should extend these core regions and construct fringe regions. Concretely speaking, let $N(x)$ be a set of q neighbor instances of x. Then we extend the core regions by observing the relationship between the x (which is an unlabeled instance) and x_i (which is a labeled random instance from $Co(C_m)$) with the following three-way decision rules. Namely, if x is the neighbor of x_i and x_i is also the neighbor of x, then we can say that x is much similar with x_i and they should both belong to the core region. If x is the neighbor of x_i, but x_i is not the neighbor of x, we say they are not similar and x should belong to the fringe region. Otherwise, if x is not the neighbor of x_i, x should belong to the $Tr(C_m)$.

$$if \quad (x \in N(x_i)) \wedge (x_i \in N(x)), \tag{4}$$
$$then \quad Co(C_m) = Co(C_m) \cup \{x\}$$
$$if \quad (x \in N(x_i)) \wedge (x_i \notin N(x)),$$
$$then \quad Fr(C_m) = Fr(C_m) \cup \{x\}$$
$$if \quad (x \notin N(x_i)),$$
$$then \quad Tr(C_m) = Tr(C_m) \cup \{x\}$$

2.4 Select the Most Informative Instance $x^\star$ from Fringe Regions

Then we adopt the active learning strategy to improve the performance of clustering. The objective of this strategy is to select the most informative instance $x^\star$

from fringe regions. Concretely speaking, we measure the uncertainty of instance based on the similarity firstly. Namely, on the base of consensus similarity matrix $W^\star$, let $w_{.j}$ denote the similarity between x and x_j and $U = \bigcup_{m=1}^{g} Fr(C_m)$ denote all uncertain instances currently, then they adopt the following formula to estimate the probability of an uncertain instance x belonging to a core region $Co(C_m)$ where $|Co(C_m)|$ is the number of instances in the core region $Co(C_m)$.

$$P(x \in Co(C_m)) = \frac{\frac{1}{|Co(C_m)|} \sum_{x_j \in Co(C_m)} w_{.j}}{\sum_{s=1}^{g} \frac{1}{|Co(C_s)|} \sum_{x_j \in Co(C_s)} w_{.j}} \tag{5}$$

Second, we use Eq. (6) to measure the uncertainty of an instance by the entropy.

$$H(x) = -\frac{1}{g} \sum_{t=1}^{g} (P(x \in Co(C_m)) log_2 P(x \in Co(C_m))) \tag{6}$$

where $x \in U$.

Then, the most informative instance $x^\star$ is selected by Eq. (7).

$$x^\star = arg \max_{x \in U} H(x) \tag{7}$$

2.5 Construct Pairwise Query

Finally, we construct pairwise query with the following method. First, they sort the clusters by $P(x^\star \in Co(C_m))$ in descending order where $1 \leq m \leq g$. Second, for each cluster, they select one instance x_i from $Co(C_m)$ in random and query the constraint relationship between $x^\star$ and x_i. If $(x^\star, x_i) \in ML$, then $Co(C_m) = Co(C_m) \cup \{x^\star\}$. At last, they adopt Eq. (1) to update the matrix R and Eq. (2) to update the matrix $W^\star$.

With the above five steps, we can cluster X into g clusters. Details of TDC can be found in [4].

3 Multi-view and Multi-label Method with Three-Way Decision-Based Clustering

3.1 Data Preparation

Suppose there is a data set $X = \{x_1, ..., x_i, ..., x_n\}$ with v views and $X^1, ..., X^j, ..., X^v$ is a data matrix of each view. Here, $i \in [1, n]$ and $j \in [1, v]$. For jth view, i.e., $X^j \in \mathbb{R}^{d_j \times n}$ where d_j is the feature dimension of jth view, it consists of information from n instances, namely, $X^j = \{x_1^j, ..., x_i^j, ..., x_n^j\}$ and $x_i^j = \{x_{i1}^j, ..., x_{ip}^j, ..., x_{id_j}^j\} \in \mathbb{R}^{d_j \times 1}$ represents the information of jth view for ith instance and x_{ip}^j represents pth feature of ith instance in the jth view where $p \in [1, d_j]$. For ith instance, i.e., $x_i \in \mathbb{R}^{d \times 1}$ can be represented by $x_i = \{x_i^{1^T}, ..., x_i^{j^T}, ..., x_i^{v^T}\}^T$ where $d = \sum_{j=1}^{v} d_j$. Thus X can also be rewritten as $X = \begin{pmatrix} X^1 \\ \vdots \\ X^j \\ \vdots \\ X^v \end{pmatrix} \in \mathbb{R}^{d \times n}$. Furthermore, X is also a multi-label data set and

in different views, an instance always possesses different labels, thus suppose $y_i^j \in \mathbb{R}^{l_j \times 1}$ is a label vector of ith instance in the jth view and each component of y_i^j indicates the label of x_i^j for the corresponding class. l_j represents that at jth view, instances have l_j classes. If the rth component of y_i^j, namely, $y_{ir}^j = 1$, it means x_i^j belongs to rth class definitely. If $y_{ir}^j = -1$, this indicates that x_i^j does not belong to rth class definitely. If $y_{ir}^j = 0$, this means whether x_i^j belongs to rth class or not is not available. Then $y_i = \{y_i^{1^T}, ..., y_i^{j^T}, ..., y_i^{v^T}\}^T$ represents the label of ith instance, $Y^j = \{y_1^j, ..., y_i^j, ..., y_n^j\} \in \mathbb{R}^{l_j \times n}$ represents the label

matrix of jth view, and we let $Y = \begin{pmatrix} Y^1 \\ \vdots \\ Y^j \\ \vdots \\ Y^v \end{pmatrix} \in \mathbb{R}^{l \times n}$ indicates the label matrix

for X where $l = \sum_{j=1}^{v} l_j$. Furthermore, since in many cases, instances have two kinds of labels, one is predicted labels and the other is real labels. So here, we let Y, Y^j, y_i^j, y_i represent the predicted ones and $\widetilde{Y}$, $\widetilde{Y^j}$, $\widetilde{y_i^j}$, $\widetilde{y_i}$ represent the real ones. Definitions of $\widetilde{\star}$ is similar with the $\star$.

3.2   Framework of MVML-TDC

On the base of TDC, our proposed MVML-TDC is carried out with the following method.

According to the above contents, we know if $\widetilde{Y}$ is low-rank, it can be written as the low-rank decomposition, i.e., $\widetilde{Y} = UV$, where $U \in \mathbb{R}^{l \times k}$, $V \in \mathbb{R}^{k \times n}$, and $rank(\widetilde{Y}) = k < l$. U has a function to project the original labels to the latent label space while V can be treated as the latent labels that are more compact and more semantically abstract than the original labels. For $\widetilde{Y^j}$, it can also be decomposed by $\widetilde{Y^j} = U^j V^j$ where $U^j \in \mathbb{R}^{l \times k_j}$, $V^j \in \mathbb{R}^{k_j \times n}$, and $rank(\widetilde{Y^j}) = k_j < l$. Since in real-world applications, due to the lack of manpower or making equipment failure, labels are only partially observed and some instances maybe lost a few labels in some views, so we always want to minimize the reconstruction error on the observed labels, i.e., $\min_{U,V,U^j,V^j} ||\Pi_\Omega(Y - UV)||_F^2 + \sum_{j=1}^{v} ||\Pi_{\Omega^j}(Y^j - U^j V^j)||_F^2$. Here, $||\star||_F^2$ represents the square of Frobenius norm for $\star$, Ω (Ω^j) consists of indices of the observed labels in Y (Y^j), $[||\Pi_\Omega(A)|||]_{ij} = A_{ij}$ if $(i,j) \in \Omega$, and 0 otherwise (similar to Ω^j case).

After that, we adopt a linear mapping $W \in \mathbb{R}^{d \times k}$ ($W^j \in \mathbb{R}^{d_j \times k_j}$) to map instances to the latent labels and W (W^j) is learned by

$$\min_{W,V,W^j,V^j} \left\|V - W^T X\right\|_F^2 + \sum_{j=1}^v \left\|V^j - {W^j}^T X^j\right\|_F^2.$$

Moreover, in order to introduce the local label correlation, we divide the data set into several groups with TDC. Namely, for X, it is partitioned into g groups, i.e., $X = \{X_1, X_2, ..., X_g\}$ and each part $X_m \in \mathbb{R}^{d \times n_m}$ where n_m is the number of instances in X_m. Then under j-th view, X^j is also divided into g^j groups, i.e., $X^j = \{X_1^j, X_2^j, ..., X_{g^j}^j\}$ and m-th group of X^j is $X_m^j \in \mathbb{R}^{d_j \times n_m^j}$. Then since the prediction on instance x_i is $sign(f(x_i))$ where $f(x_i) = UW^T x_i \in \mathbb{R}^{l \times 1}$, so $F_0 = [f(x_1), f(x_2), ..., f(x_n)] = UW^T X$ represents the classifier output matrix of X. Similarly, $F_0^j = U^j {W^j}^T X^j$, $F_m = UW^T X_m$, $F_m^j = U^j {W^j}^T X_m^j$ represent the classifier output matrices of X^j, X_m, X_m^j respectively. The dimensions of F_0, F_0^j, F_m, F_m^j are $l \times n$, $l \times n$, $l \times n_m$, $l \times n_m^j$ respectively.

Then, on the base of X, X_m, X^j, X_m^j and their corresponding observed label matrices, we compute the label correlation matrices. Take X as instance, $S_0 = \{[S_0]_{pq}\}$ denotes global label correlation matrix and $[S_0]_{pq} = \frac{y_{p,:} y_{q,:}^T}{||y_{p,:}|| ||y_{q,:}||}$ represents the global label correlation of p-th label with respect to q-th label and $y_{p,:}$ is the p-th row of Y. Then we let L_0 be the Laplacian matrix of S_0. Similarly, for X_m, $S_m = \{[S_m]_{pq}\}$ is the corresponding local label correlation matrix and L_m is its Laplacian matrix. Then, under j-th view, for X^j and X_m^j $S_0^j = \{[S_0^j]_{pq}\}$ and $S_m^j = \{[S_m^j]_{pq}\}$ are the corresponding global label correlation matrix and local label correlation matrix, L_0^j and L_m^j are their corresponding Laplacian matrices. Dimensions of S_0, L_0, S_m, L_m, S_0^j, L_0^j, S_m^j, L_m^j are both $l \times l$.

According to the above definitions, we want the classifier outputs can be closer if two labels are more positively correlated and then, we should minimize

$$tr(F_0^T L_0 F_0) + \sum_{m=1}^g tr(F_m^T L_m F_m) + \sum_{j=1}^v (tr({F_0^j}^T L_0^j F_0^j) + \sum_{m=1}^{g^j} tr({F_m^j}^T L_m^j F_m^j))$$ where

$tr(A)$ represents the trace of A.

Furthermore, refer to LSA-MML and LMSC [2,3] which introduce a consensus multi-view representation to encode the complementary information from different views, we adopt the same way. Suppose P is a latent representation matrix (i.e., consensus multi-view representation), B^j is the basic matrix corresponding to j-th view, then $\sum_{j=1}^v \left\|X^j - B^j P\right\|_F^2$ searches a comprehensive multi-view representation and $\sum_{j \neq t} IND(B^j, B^t)$ is used to measure the independence between different views where $IND(B^j, B^t) = -HSIC(B^j, B^t)$ and $HSIC$ is a Hilbert-Schmidt independence criterion estimator [2].

So according to the above contents, our goal is to solve the following optimization problem.

$$\min_{U,W,V,U^j,W^j,V^j} ||\amalg_\Omega(Y - UV)||_F^2 \tag{8}$$

$$+ \lambda_0 ||V - W^T X||_F^2 + \lambda_1 \Re(U, V, W, U^j, V^j, W^j, P, B^j)$$

$$+ \sum_{j=1}^{v} (\lambda_2 ||\amalg_{\Omega^j}(Y^j - U^j V^j)||_F^2 + \lambda_3 ||V^j - W^{j^T} X^j||_F^2)$$

$$+ \lambda_4 tr(F_0^T L_0 F_0) + \lambda_5 \sum_{m=1}^{g} tr(F_m^T L_m F_m)$$

$$+ \sum_{j=1}^{v} (\lambda_6^j tr(F_0^{j^T} L_0^j F_0^j) + \lambda_7^j \sum_{m=1}^{g^j} tr(F_m^{j^T} L_m^j F_m^j))$$

$$+ \lambda_8 \sum_{j=1}^{v} ||X^j - B^j P||_F^2 + \lambda_9 \sum_{j \neq t} IND(B^j, B^t)$$

where λs are tradeoff parameters, λ^js are tradeoff parameters corresponding to j-th views, $\Re(U, V, W, U^j, V^j, W^j, P, B^j) = ||U||_F^2 + ||V||_F^2 + |||W|_F^2 + ||U^j||_F^2 + ||V^j||_F^2 + ||W^j||_F^2 + ||P||_F^2 + ||B^j||_F^2$ is the regularizer.

3.3 Solution

In order to solve Eq. (8), we adopt alternating optimization. Namely, in each iteration, we update one of the variables in $\{Z_m, U, V, W, Z_m^j, U^j, V^j, W^j, P, B^j\}$ with gradient descent and leave the others fixed. After we get the ∇_A where $A \in \{Z_m, U, V, W, Z_m^j, U^j, V^j, W^j, P, B^j\}$, we can use $A := A - \eta \nabla_A$ to update A where η is the step size. Once we get the optimal results of these parameters, we can use $UW^T X$ to compute the classifier outputs for X. For X^j and the group partitions C_m^j and C_m, the outputs can be produced with the corresponding optimal matrices including U^js, W^js and others.

3.4 Computational Complexity

As in [4], the computational complexity of TDC is $O(ng^2) + O(Frng) + max(O(Fr), O(g))$ where Fr is the number of instances in all fringe regions. Then according to [6], the computational complexity of a classical multi-view multi-label learning machine is $O(Gn^2)$ where G is a constant. Thus, the computational complexity should be $O(ng^2) + O(Frng) + max(O(Fr), O(g)) + O(Gn^2) = O(Sn^2)$ where S is another constant.

4   Experiments

In order to demonstrate the performance of the developed MVML-TDC, we consider some benchmark data sets for experiments and related experimental results are shown in Subsect. 4.2.

4.1   Experimental Setting

Data Set. In our experiments, three kinds of data sets are adopted. The first kind is 3 multi-view data sets including Pascal VOC 2007 (VOC)[1], MIR-Flickr (MIR)[2], and 3Source[3]. The second kind is 2 multi-label data sets including Arts and Business which are also adopted in [7–9]. The third kind is a multi-view multi-label data set, i.e., NUS-WIDE [10,11].

Compared Method. Since three kinds of data sets are adopted in our experiments, thus for the fair comparison, we also adopt three kinds of learning methods for comparisons. They are multi-view learning methods, multi-label ones, and multi-view multi-label ones. For the multi-view ones, multiple-view multiple-learner semi-supervised learning method (MVMLSS) [12], LMSC [3], multi-view low-rank dictionary learning (MLDL) [13] are adopted. For the multi-label ones, label-specific features and local pairwise label correlation based multi-label learning (LF-LPLC) [14], multi-label classification machine with hierarchical embedding (MLCHE) [15], multi-label learning with global and local label correlation (GLOCAL) [7] are adopted. For the multi-view multi-label ones, multi-view based multi-label propagation (MVMLP) [11], semi-supervised dimension reduction for multi-label and multi-view learning (SSDR-MML) [16], LSA-MML [2] are adopted.

Parameter Setting. For the compared methods, the parameter settings of them can be found in the respective references. Then for the proposed MVML-TDC, the setting of TDC can refer to [4]. For others, we can refer to [6]. Although it seems we should adjust too many parameters, but [6] has validated that only ones which corresponds to global and local label correlations have a larger influence on the performance of the learning machine. Thus, we won't show the influence of these parameters and the ones corresponding to global and local label correlations are set to be 10^{-3}.

4.2   Experimental Results

AUC and Precision. We adopt AUC and precision to show the effectiveness of MVML-TDC for the classification tasks. In general, a higher AUC or a higher precision brings a better classification performance. Table 1 and Table 2 give the average AUC and precision respectively for the test sets for each data set. In these tables, ●/○ indicates that MVML-TDC is significantly better/worse than the corresponding method (pairwise t-tests at 95% significance level). The best average AUC or precision for each data set is shown in bold. / represents no result since the related method cannot process the corresponding data set. From these tables, it is found that in most cases, MVML-TDC has a better performance

---

[1] http://host.robots.ox.ac.uk/pascal/VOC/.

[2] http://press.liacs.nl/mirflickr/.

[3] http://mlg.ucd.ie/datasets/3sources.html.

Table 1. Average AUC (mean ± std.) of MVML-TDC with compared methods for test instances.

Data sets	MVML-TDC	LMSC	MVMLSS	MLDL	LF-LPLC
VOC	**0.729 ± 0.009**	0.687 ± 0.016 ●	0.702 ± 0.035 ●	0.686 ± 0.035 ●	/
MIR	**0.522 ± 0.017**	0.488 ± 0.008 ●	0.516 ± 0.036 ●	0.520 ± 0.046	/
3Source	**0.731 ± 0.001**	0.711 ± 0.037 ●	0.693 ± 0.045 ●	0.725 ± 0.014	/
Arts	**0.890 ± 0.007**	/	/	/	0.820 ± 0.005 ●
Business	0.958 ± 0.003	/	/	/	0.926 ± 0.003 ●
NUS-WIDE	**0.850 ± 0.027**	/	/	/	/
win/tie/loss		3 / 0 / 0	3 / 0 / 0	1 / 2 / 0	2 / 0 / 0
Data sets	MLCHE	GLOCAL	MVMLP	SSDR-MML	LSA-MML
VOC	/	/	0.701 ± 0.003 ●	0.626 ± 0.001 ●	0.666 ± 0.049 ●
MIR	/	/	0.512 ± 0.037 ●	0.514 ± 0.025 ●	0.475 ± 0.002 ●
3Source	/	/	0.647 ± 0.003 ●	0.704 ± 0.029 ●	0.688 ± 0.023 ●
Arts	0.782 ± 0.005 ●	0.887 ± 0.005	0.788 ± 0.005 ●	0.814 ± 0.005 ●	0.771 ± 0.005 ●
Business	0.961 ± 0.003	0.950 ± 0.003	0.808 ± 0.003 ●	0.839 ± 0.003 ●	**0.962 ± 0.003**
NUS-WIDE	/	/	0.822 ± 0.031 ●	0.796 ± 0.056 ●	0.801 ± 0.056 ●
win/tie/loss	1 / 1 / 0	0 / 2 / 0	6 / 0 / 0	6 / 0 / 0	5 / 1 / 0

Table 2. Average precision (mean ± std.) of MVML-TDC with compared methods for test instances.

Data sets	MVML-TDC	LMSC	MVMLSS	MLDL	LF-LPLC
VOC	**0.701 ± 0.004**	0.617 ± 0.006 ●	0.648 ± 0.047 ●	0.683 ± 0.028 ●	/
MIR	**0.521 ± 0.002**	0.471 ± 0.042 ●	0.440 ± 0.004 ●	0.496 ± 0.004 ●	/
3Source	**0.703 ± 0.014**	0.630 ± 0.050 ●	0.614 ± 0.034 ●	0.655 ± 0.025 ●	/
Arts	**0.655 ± 0.008**	/	/	/	0.634 ± 0.005 ●
Business	**0.916 ± 0.004**	/	/	/	0.902 ± 0.004 ●
NUS-WIDE	**0.871 ± 0.011**	/	/	/	/
win/tie/loss		3 / 0 / 0	3 / 0 / 0	3 / 0 / 0	2 / 0 / 0
Data sets	MLCHE	GLOCAL	MVMLP	SSDR-MML	LSA-MML
VOC	/	/	0.594 ± 0.011 ●	0.682 ± 0.018 ●	0.637 ± 0.034 ●
MIR	/	/	0.462 ± 0.002 ●	0.464 ± 0.045 ●	0.486 ± 0.006 ●
3Source	/	/	0.624 ± 0.005 ●	0.692 ± 0.035 ●	0.680 ± 0.002 ●
Arts	0.580 ± 0.005 ●	0.608 ± 0.005 ●	0.603 ± 0.005 ●	0.627 ± 0.004 ●	0.639 ± 0.004 ●
Business	0.839 ± 0.004 ●	0.803 ± 0.004 ●	0.909 ± 0.004	0.905 ± 0.004 ●	0.895 ± 0.004 ●
NUS-WIDE	/	/	0.782 ± 0.015 ●	0.876 ± 0.011	0.853 ± 0.011 ●
win/tie/loss	2 / 0 / 0	2 / 0 / 0	5 / 1 / 0	5 / 1 / 0	6 / 0 / 0

and according to the win/tie/loss counts, the proposed MVML-TDC is clearly superior to other compared learning methods, as it wins for most times and less loses.

Running Time. In Sect. 3.4, we said that the computational complexity of MVML-TDC can be reduced to $O(Sn^2)$ where S is a constant. This computational complexity is always smaller than $O(n^3)$ which is the computational complexity of many traditional methods, it still larger than some linear learning methods. Thus, we show the running time of these compared methods and observe the difference. Table 3 shows the related experimental results. From this

Table 3. Running time (in seconds) of MVML-TDC with compared methods.

Data sets	MVML-TDC	LMSC	MVMLSS	MLDL	LF-LPLC
VOC	80.37	73.03	81.61	78.39	/
MIR	377.49	376.59	379.69	377.98	/
3Source	0.23	0.23	0.21	0.23	/
Arts	58.31	/	/	/	53.47
Business	52.72	/	/	/	49.09
NUS-WIDE	42.09	/	/	/	/
Data sets	MLCHE	GLOCAL	MVMLP	SSDR-MML	LSA-MML
VOC	/	/	77.72	80.75	80.43
MIR	/	/	360.02	360.96	368.03
3Source	/	/	0.22	0.22	0.22
Arts	58.13	56.16	54.40	57.75	55.69
Business	49.72	52.49	53.43	53.76	49.30
NUS-WIDE	/	/	35.98	37.12	39.81

table, we find that our proposed method cost a little more running time which is also accepted by us.

5   Conclusions and Future Studies

In real-world applications, multi-view multi-label data sets are widely used and traditional learning methods always produce worse performances when these data sets exhibit complicate topologies. One main reason is that they cannot reveal the uncertain relationship between instances and the corresponding clusters. In this work, we develop a multi-view multi-label learning method with three-way decision-based clustering (MVML-TDC) to overcome such a problem. In MVML-TDC, it makes the belonging of instances to a cluster depend on the probability with three-way decision-based clustering strategy. Experimental results validate that (1) MVML-TDC achieves a better average AUC and precision in statistical; (2) the running time of MVML-TDC won't add too much.

Acknowledgment. This work is supported by 'Chenguang Program' supported by Shanghai Education Development Foundation and Shanghai Municipal Education Commission under grant number 18CG54. Furthermore, this work is also sponsored by Project funded by China Postdoctoral Science Foundation under grant number 2019M651576, National Natural Science Foundation of China (CN) under grant number 61602296, Natural Science Foundation of Shanghai (CN) under grant number 16ZR1414500. The authors would like to thank their supports.

References

1. Fu, Y.J., et al.: Service usage analysis in mobile messaging apps: a multi-label multi-view perspective. In: IEEE 16th International Conference on Data Mining (ICDM) (2016). https://doi.org/10.1109/ICDM.2016.0106
2. Zhang, C.Q., Yu, Z.W., Hu, Q.H., Zhu, P.F., Liu, X.W., Wang, X.B.: Latent semantic aware multi-view multi-label classification. In: Thirty-Second AAAI Conference on Artificial Intelligence, pp. 4414–4421 (2018)
3. Zhang, C.Q., et al.: Generalized latent multi-view subspace clustering. IEEE Trans. Pattern Anal. Mach. Intell. (2018). https://doi.org/10.1109/TPAMI.2018.2877660
4. Yu, H., Wang, X.C., Wang, G.Y., Zeng, X.H.: An active three-way clustering method via low-rank matrices for multi-view data. Inf. Sci. (2018). https://doi.org/10.1016/j.ins.2018.03.009
5. Klein, D., Kamvar, S.D., Manning, C.D.: From instance-level constraints to space-level constraints: making the most of prior knowledge in data clustering. Technical report, Stanford (2002)
6. Zhu, C.M., Miao, D.Q., Wang, Z., Zhou, R.G., Wei, L., Zhang, X.F.: Global and local multi-view multi-label learning. Neurocomputing 371, 67–77 (2020)
7. Zhu, Y., Kwok, J.T., Zhou, Z.H.: Multi-label learning with global and local label correlation. IEEE Trans. Knowl. Data Eng. 99, 1–24 (2017)
8. Zhang, J., et al.: Multi-label learning with label-specific features by resolving label correlations. Knowl. Based Syst. 159, 148–157 (2018)
9. Huang, J., et al.: Improving multi-label classification with missing labels by learning label-specific features. Inf. Sci. 492, 124–146 (2019)
10. Chua, T.S., Tang, J., Hong, R., Li, H., Luo, Z., Zheng, Y.: NUS-WIDE: a real-world web image database from national university of Singapore. In: Proceedings of the ACM International Conference on Image and Video Retrieval, p. 48 (2009)
11. He, Z.Y., Chen, C., Bu, J.J., Li, P., Cai, D.: Multi-view based multi-label propagation for image annotation. Neurocomputing 168, 853–860 (2015)
12. Sun, S.L., Zhang, Q.J.: Multiple-view Multiple-learner Semi-supervised Learning. Neural Process. Lett. 34, 229–240 (2011)
13. Wu, F., Jing, X.Y., You, X.G., Yue, D., Hu, R.M., Yang, J.Y.: Multi-view low-rank dictionary learning for image classification. Pattern Recogn. 50, 143–154 (2016)
14. Weng, W., Lin, Y.J., Wu, S.X., Li, Y.W., Kang, Y.: Multi-label learning based on label-specific features and local pairwise label correlation. Neurocomputing 273, 385–394 (2018)
15. Kumar, V., Pujari, A.K., Padmanabhan, V., Sahu, S.K., Kagita, V.R.: Multi-label classification using hierarchical embedding. Expert Syst. Appl. 91, 263–269 (2018)
16. Qian, B.Y., Wang, X., Ye, J.P., Davidson, I.: A reconstruction error based framework for multi-label and multi-view learning. IEEE Trans. Knowl. Data Eng. 27(3), 594–607 (2015)

Grouping and Recurrent Feature Encoding Based Multi-task Learning for Pedestrian Attribute Recognition

Shaofei Zheng, Bangjie Tang, Huadong Pan[✉], Xingming Zhang, and Jun Yin

Zhejiang Dahua Technology Co., Ltd., Hangzhou, China
shaofeizheng@foxmail.com, tang_bangjie@dahuatech.com,
pan_huadong@dahuatech.com, zhang_xingming@dahuatech.com,
yin_jun@dahuatech.com

Abstract. Pedestrian attribute recognition (PAR) in surveillance is to predict pedestrian visual features (somatotype, wearing style, etc.). Existing methods usually elaborately design complex multi-label deep neural networks to solve it, which is hard to take advantage of attribute correlations and prone to suffering from the negative transfer problem. In this paper, we proposed a grouping and recurrent feature encoding based multi-task learning method to solve these problems. We group attributes adaptively based on attribute learning state and use Bi-direction recurrent neural network (Bi-RNN) to acquire the encodings of different groups to build a auxiliary learning task. We optimize group learning and feature encoding simultaneously in an end-to-end multi-task learning (MTL) manner. Furthermore, we establish dynamic loss module to enable the model learn the weight automatically for different tasks in a closed-loop way. Finally, after finishing training, the proposed method allow us to remove auxiliary module and merge all group into one to get a concise yet effective model without weakening the performance. Extensive experimental results in two public datasets, PA-100K and RAP has demonstrated the performance superiority of our method.

Keywords: PAR · Correlation · Grouping · MTL · Bi-RNN · Dynamic loss

1 Introduction

Different from low-level features, pedestrian attributes (gender, haircut and etc.) can be seen as high-level semantic information which is more robust to viewing condition diversity and can be described and retrieved. Hence, many visual tasks integrate attributes into their algorithms to improve performance, such as person re-identification [1,2] and person retrieval [3]. PAR, which is shown in Fig. 1, has always been one of the most important visual tasks in video structural analysis because of its great role.

© Springer Nature Switzerland AG 2020
Y. Peng et al. (Eds.): PRCV 2020, LNCS 12306, pp. 81–94, 2020.
https://doi.org/10.1007/978-3-030-60639-8_7

Fig. 1. The general framework for pedestrian attribute recognition. The model parses pedestrian images and outputs visual attributes.

Traditional PAR methods usually focus on developing feature representations [4,5], however, hand-crafted features aren't robust enough to various pedestrian appearance. Recently years, the rise of deep learning (DL) has made computer vision an impressive performance, researchers now prefer to use DL to solve PAR. Earlier methods regard PAR as a multi-label classification problem and establish *fork* network (shallow layers extract public representations, while the deepers establish specific branches for each attribute [6,7]) for attribute prediction. Part-based methods are also used by most people. They usually fuse local and global information to acquire more powerful features [8–12]. However, these methods either involve too much pre-processing or contain multiple modules for image region extracting, which make them overly-complicated. Drawn inspiration from real visual system, attention based works have been proposed [13–16]. By focusing on different image regions, attribute features can be enhanced. Similarly, weakly-supervised methods concentrate on locating attribute regions through image-level labels [17]. However, some attributes are abstract concepts and don't correspond to certain regions, such as gender and age, which limits the development of these methods. Some researchers note that there are correlations among attributes, for example, whether a person has long hair can be easily inferred if female is recognized, so they propose to explore the attribute

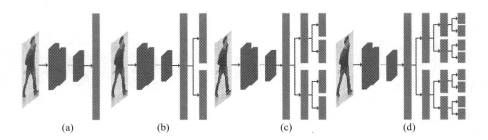

Fig. 2. The process of grouping attributes. (a) indicates all attributes are trained jointly and the average accuracy and loss of each attribute are recorded. (b) shows attributes are divided into two groups, those with higher average accuracy and lower average loss are divided into one group, and the others are divided into another group. (c) and (d) repeat the process and leaf nodes is the attribute groups.

correlations [18–20], but these methods depends heavily on the prediction order of attributes.

We argue that a superior PAR model should have three advantages, 1) simple data processing, 2) powerful feature extraction ability, 3) concise model. Taking these into account, we propose a group and recurrent feature encoding based MTL method. First, we don't do any special pre- and post-processing for images except for re-scaling. Second, the region sizes and semantic levels of different attributes are different, so we only take the *inception* network which excels in capturing different receptive fields as backbone to fusion features from multiple levels. We observe that straightly learning all attributes jointly is prone to suffering from the negative transfer because of different attribute learning difficulties, so we adaptively group attributes according to their learning state, which is shown in Fig. 2, to alleviate this phenomenon. Finally, it is arduous for a simple model to extract robust features, so we use Bi-RNN to design a auxiliary module (recurrent feature encoding) to explorer attribute correlations for improving the feature representation of the model. We can simultaneously optimize group learning and feature encoding in an end-to-end way. In order to balance the learning of two tasks, we establish dynamic loss module to enable the model learn the weight automatically for two tasks in a closed-loop way.

To summary, the main contributions of this work are four-fold. (1) An adaptive attribute grouping method is proposed to group attributes according to their learning state rather than experience to alleviate the negative transfer among multiple attributes. (2) we use Bi-LSTM to establish the auxiliary module to integrate the attribute correlations into feature learning in the way of MTL. (3) We enable the model learn the weights for different tasks adaptively instead of setting a fixed weight by experience to balance the learning between different tasks. (4) After training, the proposed method allow us to remove the auxiliary module and merge all groups into one without weakening performance to get a concise yet effective model.

2   Related Work

2.1   Pedestrian Attributes Recognition

PAR is not a new concept, earlier works rely on hand-crafted features and traditional machine learning methods [4,5] to solve it, however, they ignore the attribute correlations. Some graph model methods are proposed to model the attribute correlations [21–23], but they calculated the relationships of attribute pairs, too much computation overhead when predicting more attributes.

Following the DL renaissance, researches usually apply ConvNets to PAR. Sudowe *et al.* [6] and Li *et al.* [7] input pedestrian images into CNN and establish multiple network branches to predict attributes. Wang *et al.* [18] encoded the feature of image regions by RNN and then weighted fusion was performed on the encoded features, finally decoded the fused features to predict the attributes sequentially. Zhao *et al.* [19,20] grouped attribute and used RNN to predict the attributes of different groups sequentially. Fabbri *et al.* [10] divided the images into four parts and fused part features for attribute prediction. Li *et al.* [12] proposed to extract the part regions according to pedestrian keypoints and fused local region features and global information for attribute recognition. Liu *et al.* [13] fused the multi-level attention feature map for attribute prediction. Sarfraz *et al.* [14] proposed to take the view cues into consideration to better estimate attributes.

Despite of the promising progress recently, however, they either fail to take into account the negative knowledge transfer or lack of considering the correlation among attributes. In this work, we group attributes adaptively to allevite negative knowledge transfer. Furthermore, we encode group features and establish MTL architecture to take attribute correlations into attribute inference. Note that the feature encoding module can be removed after finishing training, which allows us to obtain a simpler model.

2.2   Multi-task Learning

MTL has been widely used in computer vision, especially when some tasks are correlated or under-sampled [24]. MTL can optimize related tasks simultaneously and share knowledge in multiple tasks. It has been demonstrated that feature sharing can boost the performance of some or sometimes all of the tasks [24].

Many researchers use deep multi-task learning to solve corresponding problems. Zhang *et al.* [25] proposed a deep cascaded multi-task framework which exploits the inherent correlation between face detection and alignment. Zhang *et al.* [26] proposed to optimize the facial landmark localization with the help of other heterogeneous but subtly correlated tasks (such as gender and appearance attributes). Jou *et al.* [27] proposed a multi-task cross-residual network for knowledge transfer. Abdulnabi *et al.* [28] proposed a multi-task CNN model to allow sharing of visual knowledge between tasks to learn facial attributes. Lu *et al.* [29] proposed a bottom-up approach that starts with a thin network and dynamically widens it greedily during training using a criterion that promotes grouping of similar tasks.

We build multi-task architecture draws inspiration from the above methods. We use Bi-LSTM to acquire the feature encoding of different groups, based on encoding, the same attribute classification task and optimization objective as group learning is established. In order to balance two task, we enable the model learn the weight automatically to control the loss between them in a closed-loop way.

3   Approach

Before introducing our method, we give the definition of the problem. Given N training images $\{I_1, I_2, ..., I_N\}$ and each sample has k attribute tags for training. Each tag is in set $T = \{t_1, t_2, ..., t_k\}$. $G = \{g_1, g_2, ..., g_m\}$ is attribute groups, where $g_i \cap g_j = \emptyset$ ($i \neq j$ and $i, j \leq m$) and $\cup_{i=1}^{m}(g_i) = T$. For sample I_i, there is a label vector $V_i = \{v_i^1, v_i^2, ..., v_i^k\}$, where $v_i^j = 1$ if I_i has attribute t_i and $v_i^j = 0$

otherwise. We aim to learn visual attribute recognition models $R_I : I \to \{0,1\}$ to recognize the attributes of sample I.

3.1 Attribute Grouping

We group attributes according to the attribute learning state rather than prior distribution of attribute locations [19]. Attributes which are harder to learn and easier to learn should be divided into different groups to weaken negative knowledge transfer.

During training, we regard the average loss and average accuracy as attribute learning state. The whole grouping process is shown in Fig. 2. We build FC layer at the rear of the backbone, followed by a sigmoid layer which forwards prediction results. The sigmoid cross entropy loss, which is defined in (1), is introduced in multi-attribute learning.

$$L = -\frac{1}{k\varrho} \sum_{i=1}^{\varrho} \sum_{j=1}^{k} (y_{ij}\ln p_{ij} + (1 - y_{ij})\ln(1 - p_{ij})) \tag{1}$$

$$p_{ij} = -\frac{1}{1 + exp(-x_{ij})} \tag{2}$$

where y_{ij} is j'th attribute of example I_i. x is the output of the last FC layer and p_{ij} is the output probability. We record the loss and accuracy during training and acquire the average loss and average accuracy of each attribute according to (3) and (4).

$$C = \left\{ c_j = \frac{1}{\vartheta} \sum_{i=1}^{\vartheta} \varsigma_{ij} | j = 1, ..., k \right\} \tag{3}$$

$$\Gamma = \left\{ \tau_j = \frac{1}{\vartheta} \sum_{i=1}^{\vartheta} \gamma_{ij} | j = 1, ..., k \right\} \tag{4}$$

where ς_{ij} and γ_{ij} respectively is the loss and accuracy of attribute j in the i'th iteration. ϑ is the iterations number. We set two thresholds μ and ν for loss and accuracy respectively. Attributes with average loss less than μ and average accuracy greater than ν are divided into one group, and the others are regarded as another group, which is defined as (5) and (6).

$$g_1 = \{t_j | c_j < \mu, \tau_j > \nu, j = 1, ..., k\} \tag{5}$$

$$g_2 = \{t_j | t_j \notin g_1, j = 1, ..., k\} \tag{6}$$

After that, we get the attribute groups shown in (b) of Fig. 2. We recursively execute this procedure on different attribute groups, a network structure similar to binary tree is obtained in turn shown in (c) and (d) of Fig. 2. Different leaf nodes represent different attribute groups. After grouping, we establish branches for each attribute group at the rear of the backbone network, which is shown in grouping learning module of Fig. 3. Attributes are grouped in this adaptive way rather than in terms of subjective awareness, we can ensure that the learning state of all attributes in each group are similar to each other. The learning difficulty of different attribute groups is different and each group is respectively responsible for its own attribute, in this way, negative knowledge transfer among attributes is weakened effectively. We compared the effect of attribute grouping with Zhao *et al.* [19], which divides attributes into different groups according to semantic relationship and attribute location, the comparison results. which is shown in Table 3 of Sect. 4.6, demonstrates that our grouping method outperform that of Zhao *et al.* [19].

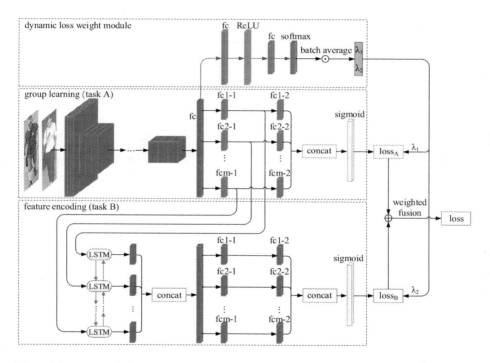

Fig. 3. Overview of the proposed method. According to the grouping results, we build *m* branchs for different attribute groups. The first FC layer of each group serves as the input of the Bi-LSTM. We fuse the output of Bi-LSTM as the feature encodings of attribute groups. Based on the encodings, we design another learning task whose structure is consistent with attribute grouping structure. Group learning task and feature encoding task are optimized simultaneously in an end-to-end way. To balance multiple tasks, we enable the network learn the weights for different tasks through dynamic loss weights module instead of setting the weights manually.

3.2   Group Feature Encoding Based Multi-task Learning

After the above processes, we establish a branch with two FC layers for each attribute group. In order to integrate the correlations among different groups into attribute reasoning, we design an auxiliary module which use Bi-LSTM to encode the features extracted from the first FC layer of each attribute group sequentially, which is shown in Fig. 3. We fuse the output of each time step of Bi-LSTM to acquire attribute group feature encodings to establish auxiliary optimization task whose structure is the same as that of attribute grouping. Group learning (task A) extracts the unique features of each attribute group, while the auxiliary task (task B) encodes the features of different attribute groups, the former provides feature input for the latter, and the latter explorers the attribute correlations, so that different groups integrate the correlation information into feature learning. The two tasks work together to improve attribute features in a complementary way and can be optimized in an end-to-end way simultaneously.

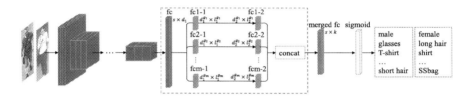

Fig. 4. After finishing training, we remove the feature encoding module and merge all attribute groups and the shared fc layer (the dotted part) into one followed by a sigmoid layer which output attribute prediction results.

After training, we remove the auxiliary module and simplify FC layer parameters of different attribute groups. Note that we deliberately don't connect any nonlinear layer after the first FC layer of each grouping branches to ensure the parameters of the two FC layers can be merged. We regard $F_j^{g_i} \in R^{d_j^{g_i} \times l_j^{g_i}}$ as the parameters of the j'th FC layers of the i'th attribute group, where $j \in \{1, 2\}$ and $l_2^{g_1} + \cdots + l_2^{g_m} = k$. We observe that $d_1^{g_1} = \cdots = d_1^{g_m}$ because the first FC layer of each group is connected behind the backbone network, we abbreviate them as d_1 for convenience. After finishing training, the FC layers parameters of the i'th group can be merged into $F^{g_i} \in R^{d_1 \times l_2^{g_i}}$. We further combine the FC layers parameters of each group into $F^g \in R^{d_1 \times (l_2^{g_1} + \cdots + l_2^{g_m})}$ by concatenating the parameters of all groups on the second dimension. Because there is no nonlinear layer connected behind the shared fc layer, we can merge F^g and the parameters of the shared fc into $F \in R^{s \times k}$. Finally, all attribute groups are merged into one FC layer, which is shown in Fig. 4, followed by a sigmoid layer, which is responsible for predicting the existence or not of all attributes. In this way, we reduce the volume of the network and get a concise but efficient inference model.

3.3 Loss Function

We establish two loss components for each optimization task, sigmoid weighted cross entropy loss and probability difference loss, which are defined in (7) and (8).

$$L_1 = -\frac{1}{k\varrho} \sum_{i=1}^{\varrho} \sum_{j=1}^{k} (w_j' y_{ij} \ln p_{ij} + w_j''(1 - y_{ij}) \ln(1 - p_{ij})) \tag{7}$$

$$L_2 = 1 - \frac{1}{k\varrho} \sum_{i=1}^{\varrho} \sum_{j=1}^{k} (\sigma_j' y_{ij}(p_{ij})^{\varphi'} + \sigma_j''(1 - y_{ij})((1 - p_{ij})^{\varphi''})) \tag{8}$$

where $w_j' = \exp(1 - r_j)$ and $w_j'' = \exp(r_j)$ is the weight of positive and negative samples of the j'th attribute and r_j is the positive sample ratio of the j'th attribute. σ_j', σ_j'', φ' and φ'' are hyper-parameters and set to 1 experiencely. For group learning and feature encoding, which are shown in Fig. 3, their optimization objectives are built in the same way, as shown in (9)

$$L_{A(B)} = L_1 + L_2 \tag{9}$$

We optimize two tasks simultaneously in a multi-task learning manner and the final optimization objectives is defined as (10)

$$Loss = K^T \begin{pmatrix} L_A \\ L_B \end{pmatrix} \tag{10}$$

Where $K = (\lambda_1, \lambda_2)^T$ is generated from dynamic loss weight module which is shown in the top of Fig. 3. It is calculated by (11).

$$K = \frac{1}{\varrho} \sum_{i=1}^{\varrho} f_i \tag{11}$$

where f_i is the output of sigmoid for the i'th sample. We enable the model to learn the loss weights for different optimization objectives adaptively rather than manually setting them to improve the balance between different tasks.

4 Experiment

4.1 Dataset

We use two publicly available pedestrian attribute datasets, RAP [9] and PA-100K [13] for evaluations. The former consists of 33268 training images and 8317 validation images, which collected from 26 indoor surveillance cameras. Each image is labelled with 72 attributes and we select 51 binary attributes for evaluation following the official protocol. The PA-100K dataset is captured from 598 real outdoor surveillance cameras and includes 100000 images which are randomly split into training, validation and test sets with a ratio of 8:1:1. Every image is labelled by 26 attributes, the label is either 0 or 1, indicating the presence or absence of corresponding attributes respectively.

4.2   Performance Metrics

We adapt five metrics to evaluate the performance, one of which is label-based evaluation criteria proposed by Deng *et al.* [4], which is defined as (12).

$$mA = \frac{1}{2k} \sum_{i=1}^{k} (\frac{TP_i}{P_i} + \frac{TN_i}{N_i}) \tag{12}$$

where TP_i and TN_i are the number of correctly predicted positive and negative samples of i'th attribute respectively. P_i and N_i are the number of positive and negative samples of i'th attribute respectively. Example-based evaluation criteria includes four indicators, as defined below.

$$Acc = \frac{1}{N} \sum_{i=1}^{N} \frac{|Y_i \cap f(I_i)|}{|Y_i \cup f(I_i)|} \qquad Prec = \frac{1}{N} \sum_{i=1}^{N} \frac{|Y_i \cap f(I_i)|}{|f(I_i)|}$$
$$Rec = \frac{1}{N} \sum_{i=1}^{N} \frac{|Y_i \cap f(I_i)|}{|Y_i|} \qquad F1 = \frac{2 \times Prec \times Rec}{Prec + Rec} \tag{13}$$

where Y_i is the ground truth labels of the i'th sample, $f(I_i)$ is the predicted positive labels for i'th sample. $|\cdot|$ means the set cardinality.

4.3   Competitors

Our method is compared against 12 state-of-the-art methods, five of which are CNN based methods, the others are CNN-RNN based methods. **ACN** [6] and **DeepMAR** [7] trains a CNN model to predict all attributes jointly, shallow layers share weights and transfer knowledge among attributes while the deepers learn specific features for each attributes. **HPNet** [13] is an attention model which integrates local and global information for better attribute recognition. **WPAL** [17] fuses the deep features from different layers fed them into FSPP module to locate attribute regions and predict attribute categories. **PDGM** [12] uses the person keypoints to locate the part regions and fuses the region features and image information for attribute recognition. **VSPAR** [14] takes the view cues into consideration to estimate the view weights which is used for learning specialized view-specific multi-label attribute predictions. **JRL** [18] divides the pedestrian image into several parts and uses an encoder-decoder architecture to mine the relationship among parts and attributes, finally, attributes are decoded sequencely. **RCRA** [19] uses Convolutional-LSTM to explore the correlations among attributes and introduce visual attention module to highlight the region of interest on the feature map. **GRL** [20] is formulated to recognize human attributes by group step by step to pay attention to both intra-group and inter-group relationships. **LGNet** [30] assigns attribute-specific weights to local features based on the affinity among pre-extracted proposals and attribute locations. **CCR** [31] is a CNN-RNN based sequential prediction model designed to encode the scene context and inter-person social relations for modeling multiple people in an image. **SCR** [32] is a state-of-the-art multi-label image classification model that exploits the ground-truth attribute labels for strongly supervised deep learning and richer image embedding.

4.4   Implementation Details

We use tensorflow to train model and in order to verify the effectiveness of the method, we choose two *inception* structures, *inception*-v2 and *inception*-v4, which pretrained from ImageNet image classification task as the backbone network. The optimization method is SGD. The initial learning rate of multi-task training is 0.1 and reduced to 0.00001 by a factor of 0.1 at last. We set the thresholds μ and ν to the mean of the average loss and average accuracy of all attributes in the current group.

Table 1. Evaluation on PA-100K and RAP with bold best result. We use inception-v4 to extract image features, and then use group and recurrent feature encoding based multi-task learning method to train the network in an end-to-end way. After training, we remove the recurrent encoding module and combine the FC layers to get an optimized simple model, which is used for the attribute prediction. The bold fonts indicate the best results.

	RAP(%)					PA-100K(%)				
	mA	*Acc*	*Prec*	*Rec*	*F*1	*mA*	*Acc*	*Prec*	*Rec*	*F*1
ACN	69.66	62.61	80.12	72.26	75.98	–	–	–	–	–
DeepMAR	73.79	62.02	74.94	76.21	75.56	72.70	70.39	82.24	80.42	81.32
HP-Net	76.12	65.39	77.33	78.79	78.05	74.21	72.19	82.97	82.09	82.53
WPAL	79.48	53.30	60.82	78.80	68.65	–	–	–	–	–
PGDM	74.31	64.57	78.86	75.90	77.35	74.95	73.08	84.36	82.24	83.29
VSPAR	77.70	67.35	79.51	79.67	79.59	76.32	73.00	84.99	81.49	83.20
JRL	77.81	–	78.11	78.98	78.58	–	–	–	–	–
RCRA	81.16	–	79.45	79.23	79.34	–	–	–	–	–
GRL	**81.20**	–	77.70	80.90	79.29	–	–	–	–	–
LGNet	78.68	68.00	**80.36**	79.82	80.09	76.96	75.55	86.99	83.17	85.04
CCR	70.13	–	71.03	71.20	70.23	–	–	–	–	–
SCR	74.21	–	75.11	76.52	75.83	–	–	–	–	–
Ours	78.04	**68.11**	79.19	**81.37**	**80.26**	**80.07**	**78.95**	**87.83**	**86.82**	**87.32**

4.5   Result

The experiment results of our method and competitors are shown in Table 1. Although competitors are all the recent state-of-the-art methods, the experiments show the performance superiority of our method even though we don't use any data augmentation. For RAP, among all the comparators, our method is superior in three evaluation metrics (*Acc*, *Rec* and *F*1) compared with any one.

Table 2. Effect analysis of recurrent feature encoding (RFE) module. We take two different networking, *inception*-v2 and -v4 as backbones and verify the effect improvement of RFE on two datasets. The bold fonts indicate the better results.

Dataset	Method			mA	Acc	$Prec$	Rec	$F1$
	v2	v4	RFE					
RAP (%)	✓			74.60	65.69	77.95	79.07	78.51
	✓		✓	**76.00**	**66.66**	**78.31**	**80.13**	**79.21**
		✓		76.60	66.76	78.52	80.07	79.28
		✓	✓	**78.04**	**68.11**	**79.19**	**81.37**	**80.26**
PA-100K (%)	✓			77.65	76.63	86.79	84.92	85.85
	✓		✓	**78.73**	**77.77**	**87.22**	**85.84**	**86.53**
		✓		79.49	77.82	86.95	86.40	86.68
		✓	✓	**80.07**	**78.95**	**87.83**	**86.82**	**87.32**

Table 3. We take *inception*-v2 as backbone and compare the attribute grouping methods with Zhao *et al.* [19] which is a representative of attribute division with prior location information as the competitor. The bold fonts indicate the better results.

Dataset	Method	mA	Acc	$Prec$	Rec	$F1$
RAP (%)	Zhao *et al.*	73.92	64.53	77.93	77.66	77.38
	Our method	**74.60**	**65.69**	**77.95**	**79.07**	**78.51**
PA-100K (%)	Zhao *et al.*	77.26	75.80	86.33	84.23	85.27
	Our method	**77.65**	**76.63**	**86.79**	**84.92**	**85.85**

Note that $F1$ is the comprehensive performance of $Prec$ and Rec, our method outperforms all competitors in $F1$ and mA demonstrates the superior performance. Specifically, compared with **DeepMAR**, **HPNet**, **PGDM**, **JRL**, **CCR** and **SCR**, our method is superior to them and achieves the highest value on all evaluation metrics. **GRL**, **RCRA**, **WPAL** and **LGNet** respectively are the first, second, third and forth best method in mA (improving 3.16%, 3.12%, 1.44% and 0.64% than that of our method), but our method outperforms **GRL** in $Prec$, Rec and $F1$ (improving 1.49%, 0.47% and 0.97%), outperforms **RCRA** in Rec and $F1$ (improving 2.14% and 0.92%), outperforms **WPAL** in Acc, $Prec$, Rec and $F1$ (improving 14.81%, 18.37%, 2.57% and 11.61%), outperforms **LGNet** in Acc, $Prec$ and $F1$ (improving 0.11%, 1.55% and 0.17%). **LGNet**, **ACN**, **VSPAR** and **RCRA** all outperforms our method in $Prec$, but our method is far superior in other metrics compare with **ACN** and **VSPAR**. Respectively improving (8.38%, 5.59%, 9.11%, 4.28%) and (0.34%, 0.76%, 1.7%, 0.67%). For **LGNet** and **RCRA**, according to the comparison results mentioned above, our method outperforms **LGNet** and **RCRA** in (Acc, Rec and F) and (Rec and F). For PA-100K, it can be clearly observed that our method is far superior to all comparators in five evaluation criterias. Specifically, our method outperforms

DeepMAR, **HPNet**, **PGDM**, **VSPAR** and **LGNet** in mA, Acc, $Prec$, Rec and $F1$, improving (7.37%, 8.56%, 5.59%, 6.4%, 6%), (5.86%, 6.76%, 4.86%, 4.73%, 4.79%), (5.12%, 5.87%, 3.47%, 4.58%, 4.03%), (3.75%, 5.95%, 2.84%, 5.33%, 4.12%) and (3.11%, 3.4%, 0.84%, 3.65%, 2.28%) respectively. The experiment result shows clearly the benefit of the proposed group and recurrent feature encoding based multi-task learning method in PAR. This is mainly due to the capacity of recurrent learning in mining both the intra-group and inter-group correlations.

Table 4. We take *inception*-v2 as backbone and compare the dynamic loss with a set of fixed weights (0.5, 0.5). The bold fonts indicate the better results.

Dataset	Method	mA	Acc	$Prec$	Rec	$F1$
RAP (%)	Fixed weights	75.15	66.58	**78.44**	79.88	79.15
	Dynamic weight	**76.00**	**66.66**	78.31	**80.13**	**79.21**
PA-100K (%)	Fixed weights	77.72	77.16	**87.44**	84.79	86.10
	Dynamic weight	**78.73**	**77.77**	87.22	**85.84**	**86.53**

4.6 Ablation Studies

(1) attribute grouping. We compare our attribute grouping with that of Zhao *et al.* [19], which is shown in Table 3. The effectiveness of our attribute grouping method can be clearly observed because of its better performance in five evaluation criterias compared with Zhao *et al.*'s method.

(2) recurrent feature encoding. To demonstrate the superiority of the recurrent feature encoding, we compare the results of adding recurrent learning module and not adding it on two network structures (*inception*-v2 and -v4), which is shown in Table 2. For these two datasets and two backbones, it can be clearly observed that our proposed recurrent encoding task can improve the performance in all evaluation criterias.

(3) dynamic loss weight. Because of its infinity, we can't enumerate all loss weights for grouping learning task and recurrent encoding task in our multi-task learning. Our dynamic loss module is only compared with a set of fixed weights (0.5, 0.5). The result is shown in Table 4. The dynamic loss module can significantly improve all criterias except *Prec* because of its unique ability to balance two tasks.

5   Conclusion

In this work, we proposed a novel group and recurrent feature encoding based multi-task learning method for pedestrian attribute recognition. First, we adaptively divide attributes into several groups according to the learning state of attributes to weaken negative knowledge transfer. Then establish recurrent feature encoding module to model the correlations among attributes. We simultaneously train attribute grouping learning task and feature encoding task in an end-to-end way. We establish optimization objectives for the two tasks respectively and generate dynamic weights for the two optimization objectives through the dynamic loss weight module. Extensive experiments demonstrate the advantages of our method.

References

1. Wang, J., Zhu, X., Gong, S., Li, W.: Transferable joint attribute-identity deep learning for unsupervised person re-identification. In: Proceedings of the IEEE Conference on Computer Vision and Pattern Recognition, pp. 2275–2284 (2018)
2. Lin, Y., et al.: Improving person re-identification by attribute and identity learning. Pattern Recogn. **95**, 151–161 (2019)
3. Feris, R., Bobbitt, R., Brown, L., Pankanti, S.: Attribute-based people search: lessons learnt from a practical surveillance system. In: Proceedings of International Conference on Multimedia Retrieval, pp. 153–160 (2014)
4. Deng, Y., Luo, P., Loy, C.C., Tang, X.: Pedestrian attribute recognition at far distance. In: Proceedings of the 22nd ACM International Conference on Multimedia, pp. 789–792 (2014)
5. Layne, R., Hospedales, T.M., Gong, S.: Attributes-based re-identification. In: Gong, S., Cristani, M., Yan, S., Loy, C.C. (eds.) Person Re-Identification. ACVPR, pp. 93–117. Springer, London (2014). https://doi.org/10.1007/978-1-4471-6296-4_5
6. Sudowe, P., Spitzer, H., Leibe, B.: Person attribute recognition with a jointly-trained holistic CNN model. In: Proceedings of the IEEE International Conference on Computer Vision Workshops, pp. 87–95 (2015)
7. Li, D., Chen, X., Huang, K.: Multi-attribute learning for pedestrian attribute recognition in surveillance scenarios. In: 2015 3rd IAPR Asian Conference on Pattern Recognition (ACPR), pp. 111–115. IEEE (2015)
8. Gkioxari, G., Girshick, R., Malik, J.: Contextual action recognition with R* CNN. In: Proceedings of the IEEE International Conference on Computer Vision, pp. 1080–1088 (2015)
9. Li, D., Zhang, Z., Chen, X., Ling, H., Huang, K.: A richly annotated dataset for pedestrian attribute recognition. arXiv preprint arXiv:1603.07054 (2016)
10. Fabbri, M., Calderara, S., Cucchiara, R.: Generative adversarial models for people attribute recognition in surveillance. In: 2017 14th IEEE International Conference on Advanced Video and Signal Based Surveillance (AVSS), pp. 1–6. IEEE (2017)
11. Yang, L., Zhu, L., Wei, Y., Liang, S., Tan, P.: Attribute recognition from adaptive parts. arXiv preprint arXiv:1607.01437 (2016)
12. Li, D., Chen, X., Zhang, Z., Huang, K.: Pose guided deep model for pedestrian attribute recognition in surveillance scenarios. In: 2018 IEEE International Conference on Multimedia and Expo (ICME), pp. 1–6. IEEE (2018)
13. Liu, X., Zhao, H., Tian, M., Sheng, L., Shao, J., Yi, S., Yan, J., Wang, X.: Hydraplus-net: attentive deep features for pedestrian analysis. In: Proceedings of the IEEE International Conference on Computer Vision, pp. 350–359 (2017)
14. Sarfraz, M.S., Schumann, A., Wang, Y., Stiefelhagen, R.: Deep view-sensitive pedestrian attribute inference in an end-to-end model. arXiv preprint arXiv:1707.06089 (2017)

15. Sarafianos, N., Xu, X., Kakadiaris, I.A.: Deep imbalanced attribute classification using visual attention aggregation. In: Ferrari, V., Hebert, M., Sminchisescu, C., Weiss, Y. (eds.) ECCV 2018. LNCS, vol. 11215, pp. 708–725. Springer, Cham (2018). https://doi.org/10.1007/978-3-030-01252-6_42
16. Guo, H., Fan, X., Wang, S.: Human attribute recognition by refining attention heat map. Pattern Recogn. Lett. **94**, 38–45 (2017)
17. Yu, K., Leng, B., Zhang, Z., Li, D., Huang, K.: Weakly-supervised learning of mid-level features for pedestrian attribute recognition and localization. arXiv preprint arXiv:1611.05603 (2016)
18. Wang, J., Zhu, X., Gong, S., Li, W.: Attribute recognition by joint recurrent learning of context and correlation. In: Proceedings of the IEEE International Conference on Computer Vision, pp. 531–540 (2017)
19. Zhao, X., Sang, L., Ding, G., Guo, Y., Jin, X.: Grouping attribute recognition for pedestrian with joint recurrent learning. In: IJCAI, pp. 3177–3183 (2018)
20. Zhao, X., Sang, L., Ding, G., Han, J., Di, N., Yan, C.: Recurrent attention model for pedestrian attribute recognition. In: Proceedings of the AAAI Conference on Artificial Intelligence, vol. 33, pp. 9275–9282 (2019)
21. Chen, H., Gallagher, A., Girod, B.: Describing clothing by semantic attributes. In: Fitzgibbon, A., Lazebnik, S., Perona, P., Sato, Y., Schmid, C. (eds.) ECCV 2012. LNCS, vol. 7574, pp. 609–623. Springer, Heidelberg (2012). https://doi.org/10.1007/978-3-642-33712-3_44
22. Deng, Y., Luo, P., Loy, C.C., Tang, X.: Learning to recognize pedestrian attribute. arXiv preprint arXiv:1501.00901 (2015)
23. Shi, Z., Hospedales, T.M., Xiang, T.: Transferring a semantic representation for person re-identification and search. In: Proceedings of the IEEE Conference on Computer Vision and Pattern Recognition, pp. 4184–4193 (2015)
24. Argyriou, A., Evgeniou, T., Pontil, M.: Multi-task feature learning. In: Advances in Neural Information Processing Systems, pp. 41–48 (2007)
25. Zhang, K., Zhang, Z., Li, Z., Qiao, Y.: Joint face detection and alignment using multitask cascaded convolutional networks. IEEE Signal Process. Lett. **23**(10), 1499–1503 (2016)
26. Zhang, Z., Luo, P., Loy, C.C., Tang, X.: Learning deep representation for face alignment with auxiliary attributes. IEEE Trans. Pattern Anal. Mach. Intell. **38**(5), 918–930 (2015)
27. Jou, B., Chang, S.F.: Deep cross residual learning for multitask visual recognition. In: Proceedings of the 24th ACM International Conference on Multimedia, pp. 998–1007 (2016)
28. Abdulnabi, A.H., Wang, G., Lu, J., Jia, K.: Multi-task CNN model for attribute prediction. IEEE Trans. Multimed. **17**(11), 1949–1959 (2015)
29. Lu, Y., Kumar, A., Zhai, S., Cheng, Y., Javidi, T., Feris, R.: Fully-adaptive feature sharing in multi-task networks with applications in person attribute classification. In: Proceedings of the IEEE Conference on Computer Vision and Pattern Recognition, pp. 5334–5343 (2017)
30. Liu, P., Liu, X., Yan, J., Shao, J.: Localization guided learning for pedestrian attribute recognition. arXiv preprint arXiv:1808.09102 (2018)
31. Li, Y., Lin, G., Zhuang, B., Liu, L., Shen, C., van den Hengel, A.: Sequential person recognition in photo albums with a recurrent network. In: Proceedings of the IEEE Conference on Computer Vision and Pattern Recognition, pp. 1338–1346 (2017)
32. Liu, F., Xiang, T., Hospedales, T.M., Yang, W., Sun, C.: Semantic regularisation for recurrent image annotation. In: Proceedings of the IEEE Conference on Computer Vision and Pattern Recognition, pp. 2872–2880 (2017)

Collaborative Classification for Woodland Data Using Similar Multi-concentrated Network

Yixuan Zhu[1], Mengmeng Zhang[2], Wei Li[2(✉)], Ran Tao[2], and Qiong Ran[1]

[1] Beijing University of Chemical Technology, Beijing 100029, China
[2] Beijing Institute of Technology, Beijing 100081, China
liwei089@ieee.org

Abstract. With the increasing of the forest area and complexity of tree species, collaborative classification using multi-source remote sensing data has been drawn increasing attention. Fusion of hyperspectral and LiDAR data can improve to acquire a comprehensive information which is conductive to the forest land classification. In this work, a similar multi-concentrate network focusing on the fine classification of tree species, denoted as SMCN, is proposed for woodland data. More specific, a preprocessing stage named pixel screening for data intensity critical control is firstly designed. Then, a similar multi-concentrate network is developed to capture spectral and spatial features from hyperspectral and LiDAR data and make specific connections, respectively. Experimental results validated on Belgian data have favorably demonstrated that the proposed SMCN outperforms other state-of-the-art methods.

Keywords: Multi-source remote sensing data · Collaborative classification · Convolutional neural network · Woodland classification

1 Introduction

With the development of geospatial science and sensor technology, classification technologies of remote sensing image faced to forest land information have made great progress [13]. Collaborative classification of hyperspectral image (HSI) and light detection and ranging (LiDAR) data takes advantages of the complementary information from multi-source data [4]. For example, hyperspectral image provides abundant biophysical and chemical canopy properties information which is convenient to discriminate various materials of interest target [4,10]. And LiDAR data provides elevation information which can be acquired free from the limit of time and weather conditions, it is more suited to assess the horizontal and vertical canopy structure of forest area [4].

Many studies conclude that combining multi-sensor data could achieve better classification accuracy than using either data set individually. Collaborative

This work was supported by the National Natural Science Foundation of China under Grants NSFC-91638201, 61922013.

Y. Peng et al. (Eds.): PRCV 2020, LNCS 12306, pp. 95–101, 2020.
https://doi.org/10.1007/978-3-030-60639-8_8

classification is beneficial to synthesize diverse forest information to more accurate forest data classification performance [8,12]. Liao et al. proposed a new deep fusion framework to integrate the complementary information from multi-sensor data [6]. Recently-proposed dense Convolutional Network [1] and UNET network [7] demonstrated that they can be used as an effective method for tree species classification. However, these deep learning architectures might not perform better for tree species mapping in complex and closed forest canopies.

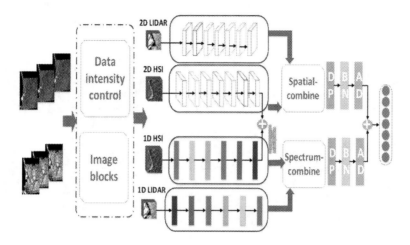

Fig. 1. The proposed collaborative classification framework for forest area.

Based on difficult characteristics of complex tree species, a preprocessing method is proposed for data intensity control which reduces the impact of excessive pixel differences on network training. A similar multi-concentrated network, denoted as SMCN, is further designed for focusing on reducing the mutual interference between spectral and spatial signatures which can effectively combine the respective feature. The similar and a little different structure guarantees the consistency of the features. At the same time, the specific information supplement mode for the spectral features and spatial features makes the network more flexible. A real remote sensing scene has been employed to validate the effectiveness of the proposed SMCN.

2   Proposed SMCN Classification

The proposed SMCN framework is designed to comprehensively learn and reasonably distinguish the difference of multi-sources data in spectral and spatial features. The overall structure is illustrated in Fig. 1.

Firstly, a screening process for original data is designed to ensure the critical control of data intensity. When the network is trained, if pixel range of some channels is much larger than other channels, it may affect the network only extract features of a large pixel range and lose useful pixel information of small channels. Through comparing the pixel range of the popular remote sensing data sets, a 10-fold difference between the spectral pixel values of hyperspectral image in Belgium data may affect classification. After origin data normalized band-by-band, it has improved visually (as shown in Fig. 2). Because it is a separate normalization operation for each band, difference in the spectrum is also retained while reduces the effect of excessive pixel range at the same time.

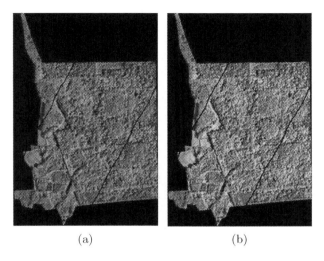

(a)   (b)

Fig. 2. Example of spectral pixel inspection: (a) the original image, and (b) the image after normalized.

Most previous work only pay attention to the study of spatial information, while proposed network considers the multi-branch to learn different features. Proposed SMCN divided the same location of data into 2D and 1D image block, respectively. The former focuses on spatial features and the latter is concentrating on spectral features. A one-dimensional processing channel for spectral features, including two 1-D convolution layers, batch normalization [3], two activation layers, a max-pooling layer, and the flatten layer. It focuses on the center pixel p_c, through batch normalization to set a high learning rate for accelerating convergence in each training mini-batch. The leaky rectified linear unit (ReLU) [9] is used as activation and the convolutional and max-pooling layer are adopted to solve features simultaneously. To facilitate subsequent processing, the output spectral features $F_{p(ij)}^{spec}$ is solved by flatten layer.

To ensure that the spectral and spatial characteristics of data can be well combined, the structure of spatial branch is as similar as one-dimensional processing channel. It only changed the links and parameter of network. The input data is image block with radius r around the center pixel p_c. After the flatten layer, the spatial and spectral features were concatenated into the full-connection layer. The output can be further expressed as,

$$\mathbf{L}_{out} = f\left(\mathbf{W} \cdot \left(\mathbf{F}_{p(ij)}^{spc} \| \mathbf{F}_{p(ij)}^{spa}\right) + \mathbf{b}\right) \tag{1}$$

where $\mathbf{W}$ and $\mathbf{b}$ are the weights and bias of the full connection, $\|$ denotes the simple superposition method of concatenating the spatial and spectral feature vectors.

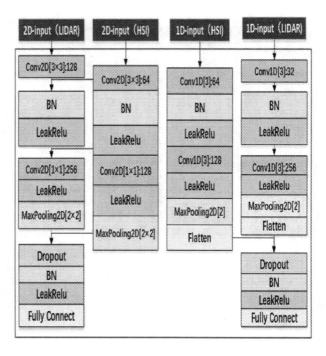

Fig. 3. The parameter of deep-mining module.

During training, use HSI image to train the two branch CNN at beginning. After fixing the weight of trained branch, introduce LiDAR data to fine-tune the network. The extraction and analysis of spectral characteristics is focusing on the central pixel of image block, which is independent of each other and have no corresponding domain information. Therefore, only a simple superposition is used. In the branch of spatial features, it not only focuses on the center pixel, but also considers the spatial features $\mathbf{F}_{p(ij)}^{spa}$ from the surrounding domain of center target. Therefore, LiDAR features are passed to HSI branch in stages continuously to correct the learning of forest information. Finally, perform superposition between different source feature map and passing the fusion features to subsequent layers. The final layer usually has the nodes of classification category, it is denoted as P_{ij}^n which is a discrete probability distribution values for each category,

Table 1. The classification performance (%) of different window sizes.

Belgium data	3×3	5×5	7×7
OA	86.28	87.68	87.37
AA	84.28	84.82	84.95
Kappa	82.42	84.17	83.79

$$P_{ij}^n = \frac{\exp\left(\theta_n | \mathbf{L}_{out}\right)}{\sum_{n=1}^{N} \exp\left(\theta_n | \mathbf{L}_{out}\right)}. \tag{2}$$

Because the initial random weights are far away from the optimal value, training the specific dual-concentrate network of HSI with a large learning rate in the first stage. When the training of HSI network is completed and the weights of the HSI branches are fixed, LiDAR features are transmitted phase by phase and fine-tuned the network with a small learning rate. The learning rate of the dual-concentrate structure of HSI is set to 0.01, and the network does fine-tune at the learning rate of 0.0001 during the adding of LiDAR data, optimizer is Adam. Figure 3 shows the parameter information of the proposed network in details.

3   Experimental Results and Analysis

TensorFlow is an open source library that can employ Keras as an application interface for machine intelligence. Based on the personal computer equipped with Ubuntu 14.04 and Nvidia GTX 1080, Tensorflow1.3.0 and Keras2.1.2 construct the integral network. Most programs are implemented using Python language, some simple processing use MATLAB language.

Belgium data is used to validate the performance of the proposed network. It represents a forest area reserved at the western part of Belgium. A total of 1450 trees were labeled for the seven species. Tree distribution in the upper canopy was common beech (27.6%), copper beech (5.5%), pedunculate oak (20.6%), common ash (4.6%), larch (8.2%), poplar (28.6%) and sweet chestnut (4.6%). Around 20% samples are used for training, the remaining samples are used for testing. We only use a multi-band image of 11 PH bands (i.e., full-waveform LiDAR data) and 286 band hyperspectral data. It covering the visible and short wave infrared wavelength (372-2498nm). The specific category information can be acquired in Table 2.

Use overall accuracy (OA), average accuracy (AA), and Kappa coefficients as evaluation indicator. Table 1 lists the classification performance of the patch with different sizes. It demonstrates that the size of image block has impact on the classification performance of different data sets, the best size of Belgium data is 5×5.

Table 2. Comparison of the classification accuracy (OA%) among the proposed SMCN.

No.	Class(training/testing)	Classification performance				
		SVM	ELM	Two branch CNN	Contex CNN	Proposed SMCN
1	Beech(88/321)	79.13	75.39	76.64	63.55	**84.74**
2	Ash(13/54)	14.81	55.56	64.81	38.89	**77.78**
3	Larch(23/93)	79.57	82.80	91.4	73.12	86.02
4	Poplar(83/333)	97	93.39	97	92.19	**96.7**
5	Copper beech (16/64)	100	96.88	93.75	93.75	**100**
6	Chestnut(13/54)	33.33	51.85	48.15	22.22	**64.81**
7	Oak (60/243)	75.62	75.62	79.75	75.21	**85.12**
OA(%)		79.59	80.36	83.38	73.56	**87.94**
AA(%)		68.49	75.93	78.78	65.56	**85.02**
Kappa		73.45	74.72	78.70	65.84	**84.51**
Training time (in *Seconds*)			943.92		140.32	**227.68**

To demonstrate the performance of the proposed SMCN framework for multi-source remote sensing data classification, some traditional and state-of-the-art methods are compared, such as SVM, ELM [2], Two-Branch CNN [11], Contex CNN [5], paper [6]. Experimental results listed in Table 2 prove that the proposed SMCN performs better than aforementioned methods, all kinds of classification results are excellent.

The distribution of training and testing samples for all the comparison methods is the same as [6], nearly 20% samples are used for training. At the same time, based on the different proportion between train and test samples, Table 3 indicates that the proposed network still has good classification performance on fine classification of tree species. As the number of training samples increases, the classification of the network becomes more accurate.

Table 3. The classification performance (OA%) on the different proportion between training and testing samples.

No.	Compared methods	OA with different proportions		
		1:9	2:8	3:7
1	SVM	75.86	79.59	84.14
2	ELM	78.62	80.36	86.34
3	Contex CNN	72.95	73.56	85.72
4	Two Branch CNN	81.72	83.38	89.31
5	Proposed SMCN	**87.86**	**87.94**	**91.86**

4 Conclusion

A collaborative classification method based on the proposed SMCN using HSI and LiDAR data has been studied for forest area. In the proposed method, each center pixel of the image block was combined with the spatial information of the image for deep analysis after learning the relevant information between the bands. For consensus between the different source information, the structure of each branch was similar and different. Compared with 3D convolution, the proposed SMCN has faster speed and better flexibility without taking up too much memory. Experimental results confirmed that the proposed SMCN was more effective.

References

1. Hartling, S., Sagan, V., Sidike, P., Maimaitijiang, M., Carron, J.: Urban tree species classification using a worldview-2/3 and lidar data fusion approach and deep learning. Sensors **19**(6), 1424–8220 (2019)
2. Huang, G.B., Zhu, Q.Y., Siew, C.K.: Extreme learning machine: theory and applications. Neurocomputing **70**(1), 489–501 (2006)
3. Ioffe, S., Szegedy, C.: Batch normalization: Accelerating deep network training by reducing internal covariate shift. CoRR (2015)
4. Koetz, B., et al.: Fusion of imaging spectrometer and LIDAR data over combined radiative transfer models for forest canopy characterization. Remot. Sens. Environ. **106**(4), 449–459 (2007)
5. Lee, H., Kwon, H.: Going deeper with contextual CNN for hyperspectral image classification. IEEE Trans. Image Process. **26**(10), 4843–4855 (2017)
6. Liao, W., Coillie, F.V., Gao, L., Li, L., Chanussot, J.: Deep learning for fusion of APEX hyperspectral and full-waveform LiDAR remote sensing data for tree species mapping. IEEE Access **6**, 68716–68729 (2018)
7. Liu, J., Wang, X., Wang, T.: Classification of tree species and stock volume estimation in ground forest images using deep learning. Sensors **166**, 0168–1699 (2019)
8. Luo, S., et al.: Fusion of airborne LIDAR data and hyperspectral imagery for aboveground and belowground forest biomass estimation. Ecol. Indicator. **73**, 378–387 (2017)
9. Maas, A.L., Hannun, A.Y., Ng, A.Y.: Rectifier nonlinearities improve neural network acoustic models. In: 30th ICML, vol. 30, no. 1 (2013)
10. Tao, R., Zhao, X., Li, W., Li, H.C., Du, Q.: Hyperspectral anomaly detection by fractional Fourier entropy. IEEE J. Sel. Top. Appl. Earth Observat. Remot. Sens. **12**(12), 4920–4929 (2019)
11. Xu, X., Li, W., Ran, Q., Du, Q., Gao, L., Zhang, B.: Multisource remote sensing data classification based on convolutional neural network. IEEE Trans. Geosci. Remot. Sens. **56**(2), 937–949 (2018)
12. Zhao, X., et al.: Joint classification of hyperspectral and LiDAR data using hierarchical random walk and deep CNN architecture. IEEE Trans. Geosci. Remot. Sens. **58**, 7355–7370 (2020)
13. Yokoya, N., Grohnfeldt, C., Chanussot, J.: Hyperspectral and multispectral data fusion: a comparative review of the recent literature. IEEE Geosci. Remot. Sens. Mag. **5**(2), 29–56 (2017)

Multi-classifier Guided Discriminative Siamese Tracking Network

Yi Zhu and Baojie Fan[✉]

College of Automation and College of Artificial Intelligence, Nanjing University
of Posts and Telecommunications, Nanjing 210023, China
jobzhuyi@gmail.com, jobfbj@gmail.com

Abstract. Recent siamese trackers regard the tracking task as a similarity learning problem. However, these trackers ignore the background
information and result in lacking the discriminability. Besides, they can
not adapt to some challenges owing to the fixed template during tracking.
To address these problems, we design a multi-classifier guided discriminative siamese tracking network, which consists of three parts: siamese base
tracker, multi-classifier module and an online template update mechanism with bounding box selection strategy. Specifically, the siamese base
tracker is used to generate classification scores and regression offsets.
Then, we utilize the ResNet50 and SeNet50 to extract the features for
training the classifiers online. Besides, we replace the original loss function in online module with a re-weighted loss to balance the sample
weights. After that, we fuse their classification scores maps, combine
with the scores map from siamese base tracker to improve its discrimi-
natibility and boost regression accuracy. Moreover, the fused classification scores map not only can improve the discriminability but also can
guide the template update. Finally, we add a bounding box selection
strategy based on the template update mechanism to get more accurate
results. The extensive experiments on OTB2015, VOT2018, GOT-10k
and VOT2019 demonstrate the competitive performance of our tracker
against the state-of-the-art trackers.

Keywords: Visual tracking · Siamese networks · One-shot learning

1 Introduction

Visual tracking aims to estimate the target state in subsequence, given the first
frame with annotated object. It is a fundamental topic in computer vision field.
Besides, it has wide applications like surveillance [1], security, human-machine
interaction [2] and autonomous vehicles [3], etc. Remarkable endeavors have been
made by researchers in the past few years, but visual tracking task still has much

This work is supported by National Natural Science Foundation of China (No.
61876092), State Key Laboratory of Robotics (No. 2019-O07) and State key Laboratory of Integrated Service Network (ISN20-08).

Y. Peng et al. (Eds.): PRCV 2020, LNCS 12306, pp. 102–113, 2020.
https://doi.org/10.1007/978-3-030-60639-8_9

difficulties to deal with, such as out of view, low resolution, motion, illumination variations, background clutter, etc. [4].

Generally speaking, visual trackers can be divided into two categories: discriminative trackers and the generative ones. Discriminative trackers often train and update a classifier online to distinguish the target from background. While the generative trackers often calculate the similarity between target template and candidate regions, then predict the state of the target in subsequences.

Recently, siamese tracking architectures [5–9] receive great attention due to their real-time speed and high accuracy. They are trained offline. In the tracking phase, they calculate the similarity between target template and search region to locate the target position. Although these trackers achieve great success, they are still limited by their disadvantages. Above all, siamese trackers only utilize the target appearances and ignore the background information, lead to lacking the discriminability. Moreover, these methods can not adapt the challenges like rotation, deformation, etc.

Different from the offline methods, discriminative trackers train a high-quality discriminative classifier online to handle with the deformation and rotation. However, these trackers lack generalization capabilities.

To resolve the mentioned problems, we propose a multi-classifier guided discriminative siamese tracking network with template update module. Foremost, we use the siamese base tracker to generate the regression offsets and classification scores. At the same time, the ResNet [10] and SeNet [11] are used to extract features to train the classifiers online, then we fuse the classification scores maps from online module and offline siamese base tracker to enhance the discriminability of our method. Secondly, we search the target location with highest scores on scores map and calculate its bounding box according to the regression offsets. Moreover, a template update and bounding box selection strategy is added to handle the target deformation, rotation, etc. Benefit from the ResNet [10]and SeNet [11] our tracker can learn a richer feature representation. The main contributions of this paper can be summarized as follows:

1. We propose a novel multi-classifier guided discriminative siamese tracking architecture, which consists of siamese base tracker, online multi-classifier module and a template update mechanism with bounding box selection strategy to fully exploit the background information and improve its discriminability. In addition, we introduce the re-weighted loss function to replace the original loss in ATOM, aiming to balance the weights of training samples in online classification module.
2. Our proposed multi-classifier module not only can enhance the discriminability, but also can guide template update to handle the rotation and deformation, etc. Besides, we propose a bounding box selection strategy to improve the robustness of our method.

2   Related Work

Siamese Tracking Network. Recently, siamese trackers receive great attract due to their high speed and efficiency. SiamFC [5] as the pioneering method, utilizes the fully convolutional siamese network to train offline on the ILSVRC dataset. Owing to its shallow backbone and large offline dataset, SiamFC can run over 80fps with a relatively good performance. Many of the works that follow are based on it. SiamImp-tri [12] improves the SiamFC by introducing a novel triplet loss function. SiamRPN [6] introduces the region proposal network into the tracking, which replaces the target scale estimate with the bounding box regression and runs at the real-time. Its network is offline-trained on the ILSVRC and YouTuBe-BB. DaSiamRPN [13] expands the datasets by introducing the COCO dataset to handle the challenge of occlusion, distractors and out of boundary, which achieves the outstanding performance. C-RPN [8] utilizes the cascaded RPN with feature transfer block to fuse the deep and shallow features and balance the positive with negative samples. SiamRPN++ [7] explores the reason that current tracking architectures contain no more than five layers, and boosts the performance by using a deeper network like ResNet50 [10]. Besides, it replaces the RPN head with a depth-wise cross-correlation to balance the parameters and accuracy. SiamDW [14] presents the two key principles for designing siamese backbone, which makes the network can go deeper and wider. SiamMask [9] combines the task of segmentation and tracking by adding an extra branch, achieving the high scores on VOT2018 [15] benchmark. GradNet [16] proposes a gradient guided network to exploit the discrimiantive information by using gradients. SiamFC++ [17] introduces 4 guidelines of target state estimation to improve the generalization ability of the tracker. SiamCAR [18] designs a novel anchor and proposal free tracker, which simplifies the parameters tuning. SiamBAN [19] introduces a no-prior box siamese tracker with a novel sample label assignment strategy, and achieves competitive results without the sacrifice above the speed.

Discriminative Tracking Approach. The correlation based trackers have successfully been applied to the visual tracking tasks. These methods often distinguish the target by computing reliable confidence scores in a 2-D grid, while using brute force to estimate the target bounding box, leading to huge computation cost. In order to benefit from end-to-end training on tracking data, a few recent works aim to combine the discriminative tracker with the neural network. CFNet [20] introduces the correlation filter into the siamese network, but providing little gains compared with SiamFC [5]. Inspired by the DCF and IoU-Net [21], ATOM [22] designs a novel offline trained IoU net for target regression and updates the classifier online for a better discriminative, which reduces the computing costs and yields a significant improvement on accuracy. SPSTracker [23] improves the performance of ATOM by introducing the Peak Response Pooling and a Boundary Response Truncation to align discriminative features and reduces the variance of feature response. DIMP [24] proposes a novel discriminative

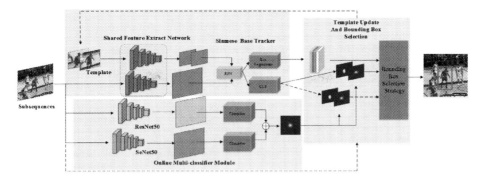

Fig. 1. Overview of our method. It consists of three parts: the siamese base tracker, online multi-classifier module and a template update mechanism with bounding box selection strategy. The siamese base tracker generates the regression offsets and classification scores map, while the online multi-classifier module outputs the fused discriminative scores map. The dashed line denotes the template update, we update each T frame to enhance its robustness.

learning loss and optimization-based architecture, results show that gets the state-of-the-art on each benchmarks (Fig. 1).

3   Proposed Method

In this work, we design a multi-classifier guided discriminative siamese tracking network. We combine the multi-classifier module with the siamese base tracker, which can fully exploit background information, making the tracker more robust. Besides, we propose a template update mechanism with bounding box selection strategy to handle the target deformation. All that will take the advantage of siamese branch and online multi-classifier branch.

3.1   Siamese Network

Before describing our method, we first review the SiamRPN++. It views tracking task as a similarity learning problem and uses the Eq. 1 to find the most similarity proposal region. It consists of two branches, one is classification branch another is regression. To ensure classification and regression, extra layers are used to adjust the channel of $\phi(z)$, $\phi(x)$ for adapting the output forms, denotes as $[\phi(z)]_{cls}$ $[\phi(x)]_{cls}$, $[\phi(z)]_{reg}$, $[\phi(x)]_{reg}$. Therefore, the classification scores f_{cls} and the regression offsets f_{reg} can be presented as follows:

$$f_{cls} = DP([\phi(z)]_{cls}, [\phi(x)]_{cls})$$
$$f_{reg} = DP([\phi(z)]_{reg}, [\phi(x)]_{reg})$$

(1)

where DP denotes the depth-wise cross-correlation, $[\phi(z)]$ serves as kernel. Tracking requires rich representations to tackle the scenarios like motion blur, deformation, etc. Therefore, a layer-wise aggregation is used to improve inference of classification and regression. Weighted sum is simply added after the RPN outputs. The function is defined as follows:

$$f_{offline}^{cls} = \sum_{i=1}^{3} f_{cls}^{i} \quad f_{offline}^{reg} = \sum_{i=1}^{3} f_{reg}^{i} \tag{2}$$

3.2   Online Multi-classifier Module

Since the SiamRPN++ is trained offline, it lacks the video-specific context information and can not utilize the background information, thus we design an online multi-classifier branch to make the tracker more discriminative. Traditional discriminative trackers work by optimizing the loss function shows as:

$$l = \min \sum_{i=1}^{t} \alpha_i (f(x_i) - y_i)^2 \tag{3}$$

where $f(x_i)$ represents the outputs of the network and y_i represents the labels. Our online module is a 2-layer fully convolutional network with a ReLu or LeakReLu activation in each layer. Different from the ATOM, we use a re-weighted loss function defined as:

$$L(w) = \sum_{i=1}^{2} \sum_{j=1}^{m} \varepsilon_{i,j} \gamma_{i,j} ||f(x_{i,j}; w) - y_i||^2$$
$$+ \sum_{i=1}^{2} \sum_{k} \lambda_{i,k} ||w_{i,k}||^2 \tag{4}$$

where $f(s_{i,j}; w)$ is the scores map from online module, y_i represents the Gaussian label of the $f(x_{i,j}; w)$, and $w_{i,k}$ is the regulation of the function. Each γ_j in Eq. 4 is computed as below:

$$\gamma_j = e^{-\beta * \frac{\eta_j}{\eta_{max}}} \tag{5}$$

$$\eta_j = \arg\max_{j} (||f(x_j; w) - y_i||) + \epsilon \tag{6}$$

where η_j denotes the max weight of each training sample. ϵ, β are constant. We use both SeNet50 [11] and ResNet50 [10] as feature extract network on multi-classifier module, benefit from the two deep network, our tracker can get a better feature representation. The Newton-Gaussian optimization strategy is used the same as ATOM [22] for online training.

After getting the online classification scores map, we use the bilinear interpolation to resize it to the same size as in SiamRPN++, the final response map is calculated as:

$$F^{cls}(x, z, w) = \lambda f_{offline}^{cls}(x, z)$$
$$+ (1 - \lambda) f_{online}^{cls}(x, w) \tag{7}$$

where λ is a hyper-parameter which is used to adjust the weight of online and offline scores map.

Algorithm 1. Tracking Algorithm

Input: Subsequences of the video from 1 to L;
Output: Target state in the following frames;

1: Init training set and filters, let t=2;
2: **while** $t \leq L$ **do**
3:     Crop the search region x^t based on previous state ;
4:     Obtain $F^{cls}(x^t, z^1, w)$ via Eq. 7 based on $f^{cls}_{offline}(x^t, z^1)$, $f^{cls}_{online}(x^t, w)$;
5:     Calculate the b_{z^1} according to the $F^{cls}(x^t, z^1, w)$;
6:     **if** z^t is not None: **then**
7:         Obtain $F^{cls}(x^t, z^t, w)$ via Eq. 7 based on $f^{cls}_{offline}(x^t, z^t)$, $f^{cls}_{online}(x^t, w)$;
8:         Calculate B_{final} via Eq. 9 based on $F^{cls}(x^t, z^t, w)$;
9:     **else**
10:         $B_{final} = b_{z^1}$;
11:     **end if**
12:     **if** hard sample **or** t==0 **then**
13:         Run Newton-Gaussian;
14:     **end if**
15:     **if** t/T==0 **then**
16:         Get z^t use Eq. 8;
17:     **end if**
18: **end while**
19: **return** Target state in the subsequences;

3.3 Template Update Mechanism and Bounding Box Selection Strategy

Note that a good template will improve the performance, thus the selection of template needs to be well designed. We store the search region in short term memory each frame and choose the one whose prediction score is higher than threshold θ_1 as the template of next frame, the formula is given by

$$z^t = \begin{cases} z^{n-1} & \text{otherwise} \\ z^n & \max(f^{cls}_{online}(x^n, w)) > \theta_1 \end{cases} \tag{8}$$

where $\max(f^{cls}_{online}(x^n, w))$ denotes the max classification score of online branch and z^{n-1}, z^n its previous frame selected in the short term memory and current frame. In order to further improve the robustness of the tracker, we use a bounding box selection strategy which is calculated as

$$B_{final} = \begin{cases} b_{z^1} & \text{otherwise} \\ b_w & max(F^{cls}(z^n)) - max(F^{cls}(z^1)) > \theta_2, \\ & Dist(b_{z^n}, b_{z^1}) < \theta_3 \end{cases} \tag{9}$$

where b_w is defined as $\delta b_{z^1} + (1 - \delta)b_{z^t}$ represents the weighted bounding box from the current frame, b_{z^1} is the predicted output bounding box from the first template. $Dist$ denotes the distance between predicted boxes, our algorithm is detailed in Algorithm 1.

Table 1. Comparison on two tracking benchmarks OTB2015 and VOT2018. Our tracker is test on the pysot toolkit.

Method	OTB2015		VOT2018		
	AUC	Pr	EAO	A	R
ECO [25]	0.682	0.903	0.280	0.484	0.276
UPDT [26]	0.702	–	0.378	0.536	0.184
LADCF [27]	0.696	0.906	0.389	0.503	0.159
DIMP [24]	0.684	–	0.440	0.594	0.153
ATOM [22]	0.669	0.882	0.401	0.590	0.204
MDNet [28]	0.678	0.909	–	–	–
SiamRPN [6]	0.637	0.851	0.326	0.569	–
DaSiamRPN [13]	0.658	0.875	0.383	0.586	0.28
SiamRPN++ [7]	0.691	0.915	0.414	0.600	0.234
SiamMask [9]	–	–	0.423	0.615	0.248
DROL-RPN[14]	0.715	0.937	0.481	0.616	–
DROL-Mask [14]	–	–	0.434	0.614	–
D3S [29]	–	–	0.489	0.64	0.150
SiamFC++ [17]	0.683	–	0.426	0.587	0.183
SiamFC [5]	0.582	0.771	0.188	0.503	0.585
Siamman [30]	0.705	0.919	0.462	0.605	0.183
SPSTracker [23]	0.692	0.902	0.434	0.612	0.169
SiamDW-RPN [14]	0.67	0.86	–	–	–
Ours	0.714	0.925	0.517	0.609	0.101

4   Experiments

Our method is implemented in pysot toolkit, we employ SiamRPN++ with the trained model provided by the official as our siamese base tracker, and utilize the modified ResNet50 as our backbone.

Online Tracking. We utilize the dual feature extract network: ResNet50 [10] and SeNet50 [11] on multi-classifier module, during tracking we use the 1×1 convolutional with a ReLu activation in the first layer of each network in order to reduce the computations. In last layer we employ a 4×4 kernel for outputting the classification scores. Features are extracted from the image patches with the size of 255×255 in search region. Like ATOM [22] and DRNet [31] we perform data augmentation in the first frame, yielding 30 training samples and total 250 training patches in the following frames. During the tracking we only update the last layers. Besides, we replace the patch in training set with the higher score predicted from the online classifiers.

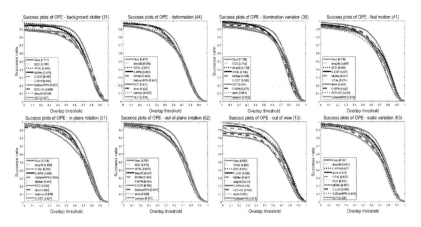

Fig. 2. Overlap success plots with attributes on OTB2015. Our method achieves the best performance when evaluating with the mentioned challenging factors.

Table 2. State-of-the-art comparison on GOT-10k.

GOT-10k	ECO	ATOM	SiamRPN++	SPM	SiamFC	THOR	CCOT	MDNet	Ours
$SR_{.5}$	30.3	63.4	61.8	59.3	40.4	53.8	32.8	30.3	66.5
$SR_{.75}$	9.9	40.2	32.5	35.9	14.4	20.4	10.7	9.9	36.1
AO	29.9	55.6	51.8	51.3	37.4	44.7	32.5	29.9	55.3

Template Update and Bounding Box Selection. To handle the deformation, rotation, etc. We update template each T frame. Notice that, we have dual templates, one is in the first frame, another is chosen during the online template updating, we calculate both bounding boxes predicted from the siamese base tracker and use the Eq. 9 to choose the most suitable one.

Hyper-parameters Setting. We set λ to 0.5 for fusing the online and offline classification scores, θ_1, θ_2, θ_3, δ to 1.5, 0.1, 5 and 0.25 for template update and bounding box selection. And the constant ϵ, β are 0.01, -1 respectively (Fig. 2).

4.1   Comparison with the State-of-the-Art

In this section, we will discuss the performance on four benchmarks: OTB2015, VOT2018, VOT2019 and GOT-10k.

OTB2015: We evaluate our tracker on OTB2015 dataset, which consists of 100 videos. Generally speaking, siamese trackers are less competitive than traditional correlation-based or DCF-based trackers on this benchmark in that most siamese trackers can not handle well with the occlusion, deformation, out of plane, rotation, etc. Results in Table 1 demonstrates that our tracker can get a competitive performance (Fig. 3).

ATOM   DROL   Ours   GT

Fig. 3. Comparison with the state-of-th-art trackers on OTB2015 and VOT2018.

Table 3. State-of-the-art comparison on VOT2019 in EAO and Accuracy.

VOT2019	SiamDW_ST	ATOM	SiamRPN++	SiamMask	ARTCS	SiamCRF_RT	Ours
EAO	0.299	0.292	0.285	0.287	0.294	0.282	0.310
A	0.600	0.603	0.599	0.594	0.602	0.550	0.608
R	0.467	0.411	0.482	0.461	0.456	0.301	0.356

VOT2018: We evaluate our approach on the VOT2018 challenges, which consisting of 60 videos over 7 conditions, they are camera motion, change of illumination, motion and size, occlusion and unassigned. Trackers are evaluated with the accuracy (average overlap over successfully tracked frames), robustness (failure rate) and EAO (excepted average overlap), Our approach outperforms all previous methods by achieving EAO score of 51.7%. From the Table 1, our tracker is the only one whose EAO can exceed 50.0%, and outperforms DROL, DIMP with relative gains of 3.6% and 7.7%.

VOT2019: In VOT2019, 60 sequences in VOT2018 are ranked according to their difficulty, and select the 12 new difficult sequences from the GOT-10k dataset to replace the easy one but maintain the diversity of the dataset. The performance is evaluated in terms of accuracy, robustness and EAO. Our tracker use the default parameters in VOT2018 and improves by 0.8% and 1.5% compared with ATOM and SiamRPN++ respectively. But our tracker still has room for improvement. The difficulties result from the fast motion, occlusion and background clutter (Table 3).

GOT-10k: We evaluate our approach on the GOT-10k [32] test set, which is a large scale with high-diversity benchmark for object tracking in the wild and it consists of 180 videos. The AO represents the average overlap and the $SR_{0.5}$, $SR_{0.75}$ represent the rate of successfully tracked frames whose overlap is exceeds 0.5 and 0.75. The Table 2 shows that our tracker improves the scores by 3.1% and 4.7% for $SR_{0.5}$ compared with ATOM and SiamRPN++.

4.2 Ablation Analyses

In this part, we will perform extra ablation studies to demonstrate the influence of each component in our algorithm. Analyses results include the EAO and accuracy on VOT2018.

Table 4. Ablation study of multi-classifier and template update with bounding box selection mechanism, Number of cls represents the number of classifiers, $Single - cls$ represents the single classifier with ResNet or SeNet. $Interval$, δ denote the update interval of template and weights of bounding boxes. The RL means the re-weighted loss function.

Number of cls	RL	Interval	δ	VOT2018	
				EAO	A
$Single - cls(ResNet)$	✓	T = 8	0.25	0.467	0.605
$Single - cls(SeNet)$	✓	T = 8	0.25	0.449	**0.611**
$Multi - cls$		T = 8	0.25	0.48	**0.611**
$Multi - cls$	✓	T = 8	0.5	0.48	0.607
$Multi - cls$	✓	T = 8	0.333	0.495	0.607
$Multi - cls$	✓	T = 8	0.2	0.493	0.608
$Multi - cls$	✓	T = 8	0.25	**0.517**	0.609
$Multi - cls$	✓	T = 7	0.25	0.512	0.606
$Multi - cls$	✓	T = 6	0.25	0.503	0.610
$Multi - cls$	✓	T = 5	0.25	0.493	0.608
$Multi - cls$	✓	T = 0	0.25	0.468	0.606

Template update is essential in handling with the variation, deformation, etc. We select the different intervals to balance the speed with precision. As shown in Table 4, the EAO criterion increases to 51.7% from 46.8% when the update template mechanism is added and the multi-classifier is more competitive than the single classifier with ResNet or SeNet. A suitable bounding box strategy will improve the performance of our method, thus, we select several weights to find the most suitable parameter in VOT2018. Moreover, compared with the original loss function in online module, our new loss function can boost the performance. Results are showed in Table 4 that the tracker with RL loss can improve the score by 3.7% for EAO.

5   Conclusion

In this work, we propose a multi-classifier guided discriminative siamese tracking network, which consists of three parts: the siamese base tracker, online discriminative multi-classifier, and template update mechanism with bounding box selection strategy. We use the ResNet50 and SeNet50 to extract the features online to get a rich feature representation. Then, we train the online classifiers with re-weighted loss function and fuse their results to get a discriminative score map, combine with the scores map from the siamese base tracker to improve the discriminability of our method. Furthermore, a template update mechanism with bounding box selection strategy is added to improve the robustness of the tracker. Results show that our method can get a competitive performance on the benchmark mentioned above.

References

1. Xing, J.L., Ai, H.Z., Lao, S.H.: Multiple human tracking based on multi-view upper-body detection and discriminative learning. In: 20th International Conference on Pattern Recognition, pp. 1698–1701 (2010)
2. Liu, L.W., Xing, J.L., Ai, H.Z., Ruan, X.: Hand posture recognition using finger geometric feature. In: Proceedings of the 21st International Conference on Pattern Recognition, pp. 565–568 (2012)
3. Lee, K.H., Hwang, J.N.: On-road pedestrian tracking across multiple driving recorders. IEEE Trans. Multimedia **17**(9), 1429–1438 (2015)
4. Wu, Y., Lim, J., Yang, M.H.: Object tracking benchmark. IEEE Trans. Pattern Anal. Mach. Intell. **37**(9), 1834–1848 (2015)
5. Bertinetto, L., Valmadre, J., Henriques, J.F., Vedaldi, A., Torr, P.H.S.: Fully-convolutional Siamese networks for object tracking. In: Hua, G., Jégou, H. (eds.) ECCV 2016. LNCS, vol. 9914, pp. 850–865. Springer, Cham (2016). https://doi.org/10.1007/978-3-319-48881-3_56
6. Li, B., Yan, J., Wu, W., Zhu, Z., Hu, X.: High performance visual tracking with Siamese region proposal network. In: Proceedings of the IEEE Conference on Computer Vision and Pattern Recognition, pp. 8971–8980 (2018)
7. Li, B., Wu, W., Wang, Q., Zhang, F.Y., Xing, J.L., Yan, J.J.: Siamrpn++: evolution of siamese visual tracking with very deep networks. In: Proceedings of the IEEE Conference on Computer Vision and Pattern Recognition, pp. 4282–4291 (2019)
8. Fan, H., Ling, H.B.: Siamese cascaded region proposal networks for real-time visual tracking. In: Proceedings of the IEEE Conference on Computer Vision and Pattern Recognition, pp. 7952–7961 (2019)
9. Wang, Q., Zhang, L., Bertinetto, L., Hu, W.M., Torr, P.H.S.: Fast online object tracking and segmentation: a unifying approach. In: Proceedings of the IEEE Conference on Computer Vision and Pattern Recognition, pp. 1328–1338 (2019)
10. He, K.M., Zhang, X.Y., Ren, S.Q., Sun, J.: Deep residual learning for image recognition. In: Proceedings of the IEEE Conference on Computer Vision and Pattern Recognition, pp. 770–778 (2016)
11. Hu, J., Shen, L., Sun, G.: Squeeze-and-excitation networks. In: Proceedings of the IEEE Conference on Computer Vision and Pattern Recognition, pp. 7132–7141 (2018)
12. Dong, X., Shen, J.: Triplet loss in Siamese network for object tracking. In: Ferrari, V., Hebert, M., Sminchisescu, C., Weiss, Y. (eds.) ECCV 2018. LNCS, vol. 11217, pp. 472–488. Springer, Cham (2018). https://doi.org/10.1007/978-3-030-01261-8_28
13. Zhu, Z., Wang, Q., Li, B., Wu, W., Yan, J., Hu, W.: Distractor-aware Siamese networks for visual object tracking. In: Ferrari, V., Hebert, M., Sminchisescu, C., Weiss, Y. (eds.) ECCV 2018. LNCS, vol. 11213, pp. 103–119. Springer, Cham (2018). https://doi.org/10.1007/978-3-030-01240-3_7
14. Zhang, Z.P., Peng, H.W.: Deeper and wider Siamese networks for real-time visual tracking. In: Proceedings of the IEEE Conference on Computer Vision and Pattern Recognition, pp. 4591–4600 (2019)
15. Kristan, M., et al.: The Sixth Visual Object Tracking VOT2018 Challenge Results. In: Leal-Taixé, L., Roth, S. (eds.) ECCV 2018. LNCS, vol. 11129, pp. 3–53. Springer, Cham (2019). https://doi.org/10.1007/978-3-030-11009-3_1

16. Li, P.X., Chen, B.Y., Ouyang, W.L., Wang, D., Yang, X.Y., Lu, H.C.: GradNet: gradient-guided network for visual object tracking. In: Proceedings of the IEEE International Conference on Computer Vision, pp. 6162–6167 (2019)
17. Xu, Y.D., Wang, Z.Y., Li, Z.X., Ye, Y., Yu, G.: SiamFC++: towards robust and accurate visual tracking with target estimation guidelines. arXiv preprint arXiv:1911.06188 (2019)
18. Guo, D.Y., Wang, J., Cui, Y., Wang, Z.H., Chen, S.Y.: SiamCAR: Siamese fully convolutional classification and regression for visual tracking. arXiv preprint arXiv:1911.06188 (2019)
19. Chen, Z.D., Zhong, B.N., Li, G.R., Zhang, S.P., Ji, R.R.: Siamese box adaptive network for visual tracking. arXiv preprint arXiv:2003.06761 (2020)
20. Valmadre, J., Bertinetto, L., Henriques, J., Vedaldi, A., Torr, P.H.S.: End-to-end representation learning for correlation filter based tracking. In: Proceedings of the IEEE Conference on Computer Vision and Pattern Recognition, pp. 2805–2813 (2017)
21. Jiang, B., Luo, R., Mao, J., Xiao, T., Jiang, Y.: Acquisition of localization confidence for accurate object detection. In: Ferrari, V., Hebert, M., Sminchisescu, C., Weiss, Y. (eds.) Computer Vision – ECCV 2018. LNCS, vol. 11218, pp. 816–832. Springer, Cham (2018). https://doi.org/10.1007/978-3-030-01264-9_48
22. Danelljan, M., Bhat, G., Khan, F.S., Felsberg, M.: Atom: accurate tracking by overlap maximization. In: Proceedings of the IEEE Conference on Computer Vision and Pattern Recognition, pp. 4660–4669 (2019)
23. Hu, Q.T., Zhou, L.J., Wang, X.X., Mao, Y., Zhang, J.L., Ye, Q.X.: SPSTracker: speak suppression of response map for robust object tracking. arXiv preprint arXiv:1912.00597 (2019)
24. Bhat, G., Danelljan, M., Van G.L., Timofte, R.: Learning discriminative model prediction for tracking. arXiv preprint arXiv:1904.07220 (2019)
25. Danelljan, M., Bhat, G., Khan, F.S., Felsberg, M.: ECO: efficient convolution operators for tracking. In: Proceedings of the IEEE Conference on Computer Vision and Pattern Recognition, pp. 6638–6646 (2016)
26. Bhat, G., Johnander, J., Danelljan, M., Khan, F.S., Felsberg, M.: Unveiling the power of deep tracking. In: Ferrari, V., Hebert, M., Sminchisescu, C., Weiss, Y. (eds.) ECCV 2018. LNCS, vol. 11206, pp. 493–509. Springer, Cham (2018). https://doi.org/10.1007/978-3-030-01216-8_30
27. Xu, T.Y., Feng, Z.H., Wu, X.J., Kittler, J.: Learning adaptive discriminative correlation filters via temporal consistency preserving spatial feature selection for robust visual object tracking. IEEE Trans. Image Process. **28**(11), 5596–5609 (2019)
28. Nam, H., Han, B.: Learning multi-domain convolutional neural networks for visual tracking. In: Proceedings of the IEEE Conference on Computer Vision and Pattern Recognition, pp. 4293–4302 (2016)
29. Lukežič, A., Matas, J., Kristan, M.: D3S-A discriminative single shot segmentation tracker. arXiv preprint arXiv:1911.08862 (2019)
30. Zhou, W.Z., Wen, L.Y., Zhang, L.B., Du, D.W., Luo, T.J., Wu, Y.J.: SiamMan: Siamese motion-aware network for visual tracking. arXiv preprint arXiv:1912.05515 (2019)
31. Kristan, M., et al.: The seventh visual object tracking vot2019 challenge results. In: Proceedings of the IEEE International Conference on Computer Vision Workshops, p. 1 (2019)
32. Huang, L.H., Zhao, X., Huang, K.Q.: Got-10k: a large high-diversity benchmark for generic object tracking in the wild. arXiv preprint arXiv:1810.11981 (2018)

Noise Resistant Focal Loss for Object Detection

Zibo Hu, Kun Gao$^{(\boxtimes)}$, Xiaodian Zhang, and Zeyang Dou

Beijing Institute of Technology, Beijing 100081, China
gaokun@bit.edu.cn

Abstract. Noise robustness and hard example mining are two important aspects in object detection. A common view is that the two techniques are contradictory and they cannot be combined. In this paper, we show that there is a possibility to combine the best of two techniques. We find that, even using the hard example mining technique, recent deep neural network-based object detectors themselves have abilities to distinguish correct annotations and wrong annotations during the early stage of training. Based on this observation, we design a simple strategy to separate the wrong annotations from training data, reducing their loss weights and correcting their labels during training. The proposed method is simple, it doesn't add any computational overhead during model inference. Moreover, the proposed method combines the hard example mining and noise resistance property in one model. Experiments on PASCAL VOC and DOTA datasets show that the proposed method not only archieves competitive performances on clean dataset, but also outperforms the baseline by a large margin when data contain severe noise.

Keywords: Noise robustness · Hard example mining · Focal loss · Noise resistant focal loss

1 Introduction

Recently, the object detectors, such as Fast RCNN [16] and RetinaNet [5], using deep neural networks have exhibited impressive performances. One reason for this huge success is the easy accessibility of the high quality large-scale datasets. Since data is labeled by people's subjective judgments, building large-scale datasets inevitably introduce label noise. Due to the high capacity of deep neural networks, the presence of noise often degrade detectors' performance. Therefore, it is important to develop noise-resistent training strategies for noisy datasets.

Hard example mining is a popular technique in object detection. It imporves detectors' performances by outstanding importances of hard examples during training. Focal loss [5] is a representative method in hard example mining. It reduces the relative loss for well-classified examples and put more focus on hard

© Springer Nature Switzerland AG 2020
Y. Peng et al. (Eds.): PRCV 2020, LNCS 12306, pp. 114–125, 2020.
https://doi.org/10.1007/978-3-030-60639-8_10

examples. Since noise can be regarded as the extremely hard examples, directly applying the focal loss to noisy datasets deteriorates the performances of detectors. In fact, there is a common belief that the noise robust loss and the hard example mining are contradictory, they treat hard examples in opposite manners. One question arises: Could we combine the best of both world to develop a method which not only focus on the hard examples, but also is resistant to noise?

In this paper, we give the above question a positive answer. We find that, though the hard example mining technique has been used, correct annotations and wrong annotations can be distinguished by recent deep neural network-based object detectors during the early stage of training. Specifically, the network is capable of predicting correct labels of wrong annotations at the early stage of training. With the progress of training, the wrong annotations interference the model, thereby degrading the model performances. Thus, we design noise resistant focal loss to separate the wrong annotations from training data, reducing their loss weights and correcting their labels during training. The proposed method is simple, it doesn't add any operation during model inference. Moreover, the proposed method combines the hard example mining and noise resistance property in one model. Experiments on PASCAL VOC [6] and DOTA [17] datasets show that the proposed method not only archieves competitive performances on clean dataset, but also outperforms the baseline by a large margin when data contain severe noise.

To summary, the contributions of this paper are three-folds:

- We experimentally demonstrate that the neural network-based detector has self discrimination ability for noise labels.
- We propose noise resistant focal loss to combine the advantages of hard example mining and noise robustness. To the best of our knowledge, this is the first paper to combine these contradictory techniques.
- The proposed method shows competitive results on both clean and noisy datasets.

2   Related Work

2.1   Dataset Noise

Excellent datasets, such as PASCAL VOC [6] and MS-COCO [12], in object detection requires high-quality manual annotations with accurate object labels and precise bounding box coordinates. Unfortunately, because the level of the workers is different, it is inevitably to generate noise annotations which contain label noise (i.e., wrong object classes), bounding box noise (i.e., inaccurate object locations) or the mixture of label noise and bounding box noise. Hard example mining is used for the part of object classification. Thus, it is sensitive to the label noise. Label noise could be divided into category-independent label noise which means that the label noise is irrelevant to the categories and category-related label noise which refers to several types of ground truths that are labeled as other fixed labels. In this paper, we focus on category-independent label noise to study the noise robustness under hard example mining.

2.2   Learning with Label Noise

Though noise robust training for object detection is vitally important for the industrial community, we surprisingly find that there is only a few researches in this field. Recently, there are three main methods (weakly-supervised (WS) [4], semi-supervised (SS) [10,14,15] and co-teaching [8]) for training noise-resistant object detectors. WS object detection aims to learn object detectors with only image-level labels. Zhang et al. [18] propose an adaptive sampling method to impose similarity loss on noisy images through instances with high classification scores. SS object detection aims to obtain unlabeled image labels and bounding boxes through existing labels and bounding boxes. Gao et al. [7] focus on the SS setting to train a detector to use a small amount of bounding box annotations and a large amount of image-level annotation information. Co-teaching aims to train two parallel networks, where each network selects small-loss samples to train the other, to achieve noise robustness and correct the noise. Chadwick et al. [2] proposes improved co-teaching to achieve robustness to noisy bounding box. All of these methods do not combine the hard example mining technique. Besides, they design complex model structures which increase computational overhead. In contrast, our method doesn't add any extra computation during inference step.

2.3   Hard Example Mining

Hard example mining, such as online hard example mining (OHEM) [13] and focal loss [5], has been widely used in object detection training process. Generally, hard example mining guides the object detectors to focus on the hard examples by increasing the relative loss weight of these examples during training. OHEM computes loss for all RoIs, sorting them based on this loss to select hard RoIs and setting the loss of non-hard RoIs to 0. Focal loss outstands the relative loss of hard examples by down-weighting easy examples, such that the hard examples dominate the training process. Since noise can be regarded as hard examples, OHEM and focal loss perform the opposite role of noise robustness: they focuses training on a set of hard examples. Therefore, hard example mining and noise robustness are contradictory, directly using hard example mining will degrade detectors' performances. However, as we will show, this is a possibility to combine these two properties in one detector.

3   Method

We surprisingly find that the neural network-based detector itself could identify annotations with wrong categories under hard example mining technique. To demonstrate our point, we conduct the following experiments. The training dataset is NWPU VHR-10 [3] which contains 10 categories. We randomly picked some ground truth bounding boxes (20% and 40%) from the training dataset and replace their categories with random categories. The training detector is RetinaNet [5] using ResNet-50 [9] as the backbone. Following the common training strategy, we use SGD with learning rate 0.0025 and momentum 0.9. We train the detector 60 epochs and divide the learning rate by 10 at 40th and 55th epoch.

For the sake of clarity, we introduce some notations which will be used throughout the paper. Suppose we have a noisy dataset $D = \{(I, G), I \in \chi, G \in \gamma\}$ with K categories, where χ is the set of input images, and $G = \{(g_i, l_i), i = 1, \cdots, N\}$ is the set of annotations including bounding box g_i and the corresponding category l_i. We denote $G_T = \{(g_{i\_t}, l_{i\_t}), i\_t = 1, \cdots, N_{i\_t}\}$ as the set of bounding boxes with correct labels, and $G_F = \{(g_{i\_f}, l_{i\_f}), i\_f = 1, \cdots, N_{i\_f}\}$ as the set of bounding boxes with wrong labels. $G = G_T \bigcup G_F$. We also use $G'_F = \{(g_{i\_f'}, l'_{i\_f'}), i\_f' = 1, \cdots, N_{i\_f'}\}$ to denote G_F with correct labels.

For g_i, $A_i = \{(a_j, P_j^{cls}), j = 1, \cdots, M_i\}$ denotes the set of positive anchor samples, where $P_j^{cls} = \{s_{ij}^1, \cdots, s_{ij}^l, \cdots, s_{ij}^k\}$ is the classification probability distribution vector for the positive anchor a_j whose IoU with the nearest bounding box is greater than 0.5. s_{ij}^{max} is the maximum predicted score of P_j^{cls}, and s_{ij}^{ref} means the maximum prediction score of P_j^{cls} except the score of annotated category. We use $r_{ij}^{l_i} \in \{1, \cdots, k\}$ to denote the rank of $s_{ij}^{l_i}$ in P_j^{cls}.

Equations 2, 3 and 4 define ACC_{it}, ACC_f and $ACC_{f\text{-}c}$. ACC_{it} represents the accuracy of classification prediction for the positive samples of all labels. In terms of wrong annotated bounding boxes, ACC_f represents the prediction accuracy of wrong labels for the positive samples, and $ACC_{f\text{-}c}$ means the prediction accuracy of right labels for the positive samples.

$$(s_{ij}^{l_i} = s_{ij}^{max}) = \begin{cases} 1 & if\ s_{ij}^{l_i} = s_{ij}^{max} \\ 0 & if\ s_{ij}^{l_i} \neq s_{ij}^{max} \end{cases} \tag{1}$$

$$ACC_{it} = \frac{\sum_{i=1}^{N} \sum_{j=1}^{M_i} (s_{ij}^{l_i} = s_{ij}^{max})}{\sum_{i=1}^{N} M_i}, (g_i, l_i) \in G \tag{2}$$

$$ACC_f = \frac{\sum_{i=1}^{N} \sum_{j=1}^{M_i} (s_{ij}^{l_i} = s_{ij}^{max})}{\sum_{i=1}^{N} M_i}, (g_i, l_i) \in G_F \tag{3}$$

$$ACC_{f\text{-}c} = \frac{\sum_{i=1}^{N_2} \sum_{j=1}^{M_i} (s_{ij}^{l'_i} = s_{ij}^{max})}{\sum_{i=1}^{N} M_i}, (g_i, l_i) \in G'_F \tag{4}$$

Equations 5, 6 and 7 define ΔS_{it}, ΔS_t and ΔS_f. ΔS_{it} is the mean of the absolute value of the difference between $s_{ij}^{l_i}$ and s_{ij}^{ref} in G. For G_T and G_F, we utilize ΔS_t and ΔS_f to represent the average of the difference between $s_{ij}^{l_i}$ and s_{ij}^{ref}, respectively. Equation 8 defines ΔS_{g_i} as the similar measurement for individual bounding box g_i.

$$\Delta S_{it} = \frac{\sum_{i=1}^{N} \sum_{j=1}^{M_i} |s_{ij}^{l_i} - s_{ij}^{ref}|}{\sum_{i=1}^{N} M_i}, (g_i, l_i) \in G \tag{5}$$

$$\Delta S_t = \frac{\sum_{i=1}^{N} \sum_{j=1}^{M_i} (s_{ij}^{l_i} - s_{ij}^{ref})}{\sum_{i=1}^{N} M_i}, (g_i, l_i) \in G_T. \tag{6}$$

$$\Delta S_f = \frac{\sum_{i=1}^{N} \sum_{j=1}^{M_i} (s_{ij}^{l_i} - s_{ij}^{ref})}{\sum_{i=1}^{N} M_i}, (g_i, l_i) \in G_F. \tag{7}$$

$$\Delta S_{g_i} = \frac{\sum_{j=1}^{M_i} s_{ij}^{l_i} - s_{ij}^{ref}}{M_i} \tag{8}$$

In Eqs. 9 and 10, R_t and R_f respectively represents the mean of the relative ranking $r_{ij}^{l_i}/k$ in G_T and G_F. Equation 11 defines R_{g_i} as the similar measurement for individual bounding box g_i.

$$R_t = \frac{\sum_{i=1}^{N} \sum_{j=1}^{M_i} r_{ij}^{l_i}/k}{\sum_{i=1}^{N} M_i}, (g_i, l_i) \in G_T \tag{9}$$

$$R_f = \frac{\sum_{i=1}^{N} \sum_{j=1}^{M_i} r_{ij}^{l_i}/k}{\sum_{i=1}^{N} M_i}, (g_i, l_i) \in G_F \tag{10}$$

$$R_{g_i} = \frac{\sum_{j=1}^{M_i} r_{ij}^{l_i}/k}{M_i} \tag{11}$$

We record ACC_{it}, ACC_f, ACC_{f_c}, ΔS_{it}, ΔS_t, ΔS_f, R_t and R_f every iteration. For the sake of observation, we averaged the above parameters every 10 iterations.

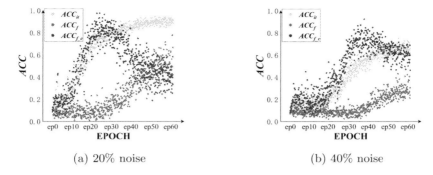

(a) 20% noise   (b) 40% noise

Fig. 1. Results of ACC_{it}(Green), ACC_f(Red), ACC_{f_c}(Blue) at the NWPU VHR-10 dataset with 20% and 40% noise. (Color figure online)

3.1 Analysis of Experimental Results

In Fig. 1, with the training of the network, ACC_{it} and ACC_f keep increasing during the whole training. While ACC_{f_c} increases in the early stage of training and then decreases during the late stage. The above observetion shows the correct labels in G_F can be predicted in the early stage of training. However, in the late of the training, the wrong labels mislead the network, and the network tends to predict the wrong category of g_i in G_F.

Comparing Fig. 1 and Fig. 2, the trend of ΔS_{it}, ΔS_t and ACC_t are similar, but ΔS_f and ACC_{f_c} have opposite trends. It indicates that ΔS_{it} can reflect the accuracy of the classification prediction. From Fig. 2 and Fig. 3, there are significant differences in ΔS_t and ΔS_f as well as in R_t and R_f. These shows that the network can separate G_F from G, moreover the correct labels of g_i in G_F can be predicted in the early stage of training. However, when training real datasets, ΔS_f and R_f are unknown. Fortunately, ΔS_{g_i} and ΔS_f have a similar pattern for $g_i \in G_F$, so does R_{gi} and R_f. Therefore, we can use ΔS_{it}, ΔS_{g_i} and R_{g_i} to separate G_F from G.

3.2 Noise Resistant Focal Loss

According to the above experiment, we propose noise resistant focal loss. Noise resistant focal loss is implemented in two steps. The first step is *judgement* which identify whether g_i is in G_F, and the second step, *treatment*, is used to reduce the relative weight of loss and correct the wrong labels.

Judgement. Based on the above observation, ΔS_{g_i}, ΔS_{it} and R_{g_i} can be used as indicators to judge whether g_i is in G_F. On the one hand, when $\Delta S_{g_i} < -0.15$, ACC_{it} keeps a high value and g_i in G_F can be distinguished intuitively. On the other hand, when $\Delta S_{it} > 0.15$ and $R_{g_i} > 0.3$, g_i in G_F has sufficient confidence. Therefore, we use the two conditions to judge whether g_i is in G_F.

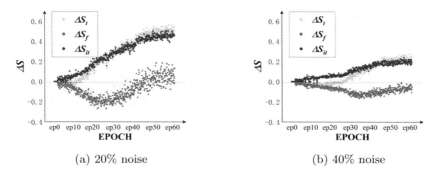

(a) 20% noise   (b) 40% noise

Fig. 2. Results of ΔS_{it}(Blue), ΔS_t(Green) and ΔS_f(Red) at the NWPU VHR-10 dataset with 20% and 40% noise. (Color figure online)

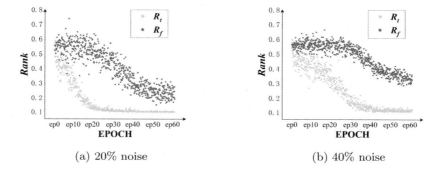

(a) 20% noise   (b) 40% noise

Fig. 3. Results of R_t(Green), R_f(Red) at the NWPU VHR-10 dataset with 20% and 40% noise. (Color figure online)

Treatment. Wrong annotated samples are very hard samples. Due to the hard example mining, the classification loss weights of wrong annotated samples are amplified. Therefore, the classification loss of g_i satisfying *Judgement* is reduced to 0.2. Then, we try to correct the wrong label. As Eq. 12 shows that $\overline{P_i^{cls}}$ denotes the mean vector of all P_j^{cls} in A_i.

$$\overline{P_j^{cls}} = \frac{\sum_{j=1}^{M_i} P_j^{cls}}{M_i} \tag{12}$$

We further define $\overline{s_i^{1st}}$ and $\overline{s_i^{2nd}}$, which represent the maximum score and the second largest score in $\overline{P_i^{cls}}$, respectively. By analyzing ΔS_{g_i} and R_{g_i}, $T_i^{cls} = \overline{s_i^{1st}} - \overline{s_i^{2nd}}$ can judge that the predicted label is the correct label of g_i in G_F. Particularly, if $T_i^{cls} > 0.15$, we use the category of $\overline{s_i^{1st}}$ to correct wrong label besides reducing the weight of loss to 0.2.

Algorithm 1: Noise resistant focal loss

Input: parameters: ΔS_{g_i}, ΔS_{it}, R_{g_i}, T_i^{cls}

1 **for** *each iteration* **do**
2   **if** $\Delta S_{g_i} < -0.15$ *or* $\Delta S_{it} > 0.15$ & $R_{g_i} > 0.3$ **then**
3   *weight* 0.2;
4   **if** $T_i^{cls} > 0.15$ **then**
5   *correct wrong label*;
6   **end**
7   **end**
8 **end**

The whole algorithm of noise resistant focal loss is shown in Algorithm 1. While the above method uses a hard threshold, it is applicable to different datasets.

4 Experiments

4.1 Datasets and Implementation Details

We use PASCAL VOC [6] and DOTA [17] to evaluate the noise resistant focal loss. PASCAL VOC is widely used in object detection with high quality bounding boxes and label annotations. PASCAL VOC contains 20 object categories and more than 20000 images. We use the union set of VOC 2007 trainval and VOC 2012 trainval as our training data, and VOC 2007 test as our test data. We use the mean average precision ($mAP_{0.5}$ and $mAP_{0.75}$) as the evaluation metric. DOTA is the largest dataset for object detection in aerial images with both horizontal and oriented bounding box annotations. The dataset contains 15 categories and 2806 images from different sensors and platforms. For DOTA, we use both the training and validation sets for training, and the testing set for testing. We report the mean average precision ($mAP_{0.5}$) as the evaluation metric.

As for noise, we follow the previous works [1,11] and create random wrong label to simulate human mistakes of different severity. Specifically, we randomly choose $N\%$ of the training samples and change each of their labels to another random label. For PASCAl VOC, N contains 0, 10, 20, 30, 40, and 60, and for DOTA, N contains 0, 20, and 40.

For PASCAL VOC, we adopt the proposed method into the ResNet-50 [9] based RetinaNet [5] with 1 NVIDIA 2080TI GPU. For DOTA, We adopt the proposed method into the RetinaNet-O [5] with 4 NVIDIA 2080TI GPU. We implement the detectors based on the mmdetection codebase. For PASCAL VOC, we use SGD with a learning rate of 0.0025, a momentum of 0.9 and a weight decay of 0.0001. We train detector 12 epochs, and decrease the learning rate at epoch 9. For DOTA, only learning rate of 0.01 and decrease the learning rate at epoch 8 and epoch 10 are different from PASCAL VOC.

Table 1. Comparison with focal loss on PASCAl VOC.

	Loss function	0%	10%	20%	30%	40%	60%
$mAP_{0.5}$	Focal loss	80.1	78.3	76.8	75.5	72.6	57.5
	Noise resistant focal loss	80.1	**79.5**	**78.5**	**78.5**	**77.8**	**68.4**
$mAP_{0.75}$	Focal loss	56.7	54.1	52.6	51.2	49.1	38.9
	Noise resistant focal loss	**57.1**	**56.6**	**56.5**	**56.2**	**55.8**	**48.6**

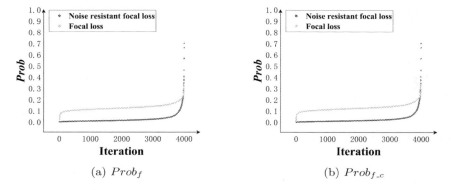

(a) $Prob_f$ (b) $Prob_{f\_c}$

Fig. 4. Results of $Prob_f$(Green) ,$Prob_{f\_c}$(Red) in 12th epoch with 40% noise. (Color figure online)

4.2 Comparison with Focal Loss

Result on PASCAL VOC. Table 1 shows the comparison result on PASCAL VOC dataset, where the training data contains different levels of label noise. Our method significantly outperforms the focal loss. For clean datasets, our method can do not affect the performance in $mAP_{0.5}$ and even achieve 0.4% improvement in $mAP_{0.75}$. For noisy data, our method outperforms the baseline by a large margin. Even for 60% noise, our method can even improve 10.9% in $mAP_{0.5}$ and 9.7% in $mAP_{0.75}$.

In order to better verify the effectiveness of our method, we record the mean of $s_{ij}^{l_i}$ ($Prob_f$) and $s_{ij}^{l_i'}$ ($Prob_{f\_c}$) of g_i in G_F in 12th epoch with 40% noise. For the convenience of observation, we draw the figure after sorting $Prob_f$ and $Prob_{f\_c}$. Figure 4 shows the results. It shows noise resistant focal loss can not only improve the classification prediction probability of the correct labels of noise, but also suppress the classification prediction probability of wrong labels of noise. These prove that our method is robust to noise.

Result on DOTA. As shown in Table 2, our method achieves impressive performance on DOTA. For clean dataset, our method only decreased 0.7% in $mAP_{0.5}$

compared with focal loss, and it has little impact on network performance. For noisy data, our method has a significant improvement over focal loss. Even for 40% noise, our method can improve 6.2% in $mAP_{0.5}$ and performs better than focal loss under 20% noise. These shows that our method is still valid on different datasets.

Table 2. Comparison with focal loss on DOTA.

Method	0%	20%	40%
Focal loss	**66.4**	61.9	56.9
Noise resistant focal loss	65.7	**64.4**	**63.1**

Table 3. Results under different thresholds T.

T	Focal loss	0.01	0.05	0.1	0.15	0.25	0.5	1
$mAP_{0.5}$	72.6	**77.8**	77.4	77.2	77.2	76.6	76.1	75.5
$mAP_{0.75}$	49.1	55.3	**55.5**	55.1	55.2	55.1	53.0	53.3

4.3   Ablation Study and Hyperparameter Analysis

In this subsection, we not only conduct ablation experiments to analyze the effectiveness of each method in *judgement* and *treatment*, but also present hyperparameter analysis. All experiments use PASCAL VOC dataset with 40% noise.

The Sensitivity of Threshold T of T_i^{cls}. We use *judgement* as the indicator of noise. For *treatment*, we only change threshold T of T_i^{cls}. We selected 7 values which are 0.01, 0.05. 0.1, 0.15, 0.25, 0.5 and 1. Table 3 shows the results. When $T \leq 0.15$, the performance of the network is very close. It indicates that when T is small, the network is not sensitive to T. Compared with $T \leq 0.15$, the performance of the network is wrose when $T > 0.15$. Therefore, we can choose any $T \leq 0.15$ as the threshold for training.

The Effectiveness of Each Component of *Judgment*. To study the effectiveness of the components of *judgement*, we omit different components in the *judgement* to investigate the effectiveness of each part, including $\Delta S_{gt} < -0.15$ and $\Delta S_{it} > 0.15 \& R_{gt} > 0.3$. We use *treatment* with threshold 0.01 of T_i^{cls} to deal with noise. Table 4 shows the result. Both $\Delta S_{gt} < -0.15$ and $\Delta S_{it} > 0.15$ & $R_{gt} > 0.3$ can effectively judge the noise. Using $\Delta S_{gt} < -0.15$ or $\Delta S_{it} > 0.15$ & $R_{gt} > 0.3$ can achieve the best performance.

Table 4. The effectiveness of each component of *judgement*.

$\Delta S_{gt} < -0.15$	$\Delta S_{it} > 0.15 \, \& \, R_{gt} > 0.3$	$mAP_{0.5}$	$mAP_{0.75}$
		72.6	49.1
✓		77.2	55.2
	✓	77.0	54.7
✓	✓	**77.8**	**55.2**

Table 5. The effectiveness of each component of *treatment*

Weight 0.2	Correct wrong label	$mAP_{0.5}$	$mAP_{0.75}$
		72.6	49.1
✓		75.5	53.3
	✓	18.6	11.4
✓	✓	**77.2**	**55.2**

The Effectiveness of Each Component of *treatment*. We use *judgement* to judge noise and omit different components in the *treatment* to investigate the effectiveness of each part, including *weight* 0.2 and *correct wrong label* ($T_i^{cls} > 0.01$). Table 5 shows the results. Only reducing the weight of loss can effectively improve the performance, while just correcting the label will damage the network. On the basis of reducing the weight, correcting the noise label can achieve the best performance.

5   Conclusion

In this paper, we found recent deep neural network-based object detectors can distinguish correct annotations and wrong annotations in the early stage of training under hard example mining. We proposed noise resistant focal loss which combined hard example mining and noise resistance property in one model. Experiments on PASCAL VOC and DOTA datasets showed that the proposed method archieved competitive performances. At present, we used hard thresholds to train. In the future, we hope to propose an adaptive method to achieve noise robustness.

References

1. Arazo, E., Ortego, D., Albert, P., O'Connor, N., McGuinness, K.: Unsupervised label noise modeling and loss correction. In: ICML 2019: Thirty-Sixth International Conference on Machine Learning, pp. 312–321 (2019)

2. Chadwick, S., Newman, P.: Training object detectors with noisy data. In: 2019 IEEE Intelligent Vehicles Symposium (IV) (2019)
3. Cheng, G., Han, J., Zhou, P., Guo, L.: Multi-class geospatial object detection and geographic image classification based on collection of part detectors. Isprs J. Photogr. Remote Sens. **98**(98), 119–132 (2014)
4. Dietterich, T.G., Lathrop, R.H., Lozano-Prez, T.: Solving the multiple instance problem with axis-parallel rectangles. Artif. Intell. **89**(1), 31–71 (1997)
5. Dollár, P.: Focal loss for dense object detection. IEEE Trans. Pattern Anal. Mach. Intell. **PP**(99), 2999–3007 (2017)
6. Everingham, M., Gool, L., Williams, C.K., Winn, J., Zisserman, A.: The pascal visual object classes (VOC) challenge. Int. J. Comput. Vision **88**(2), 303–338 (2010)
7. Gao, J., Wang, J., Dai, S., Li, L.J., Nevatia, R.: Note-RCNN: noise tolerant ensemble RCNN for semi-supervised object detection. In: 2019 IEEE/CVF International Conference on Computer Vision (ICCV), pp. 9507–9516 (2019)
8. Han, B., et al.: Co-teaching: robust training of deep neural networks with extremely noisy labels. arXiv preprint arXiv:1804.06872 (2018)
9. He, K., Zhang, X., Ren, S., Sun, J.: Deep residual learning for image recognition. In: 2016 IEEE Conference on Computer Vision and Pattern Recognition (CVPR), pp. 770–778 (2016)
10. Hoffman, J., et al.: LSDA: large scale detection through adaptation. In: Advances in Neural Information Processing Systems, vol. 27, pp. 3536–3544 (2014)
11. Jiang, L., Zhou, Z., Leung, T., Li, L.J., Fei-Fei, L.: Learning data-driven curriculum for very deep neural networks on corrupted labels. In: ICML 2018: Thirty-fifth International Conference on Machine Learning (2018)
12. Lin, T.-Y., et al.: Microsoft COCO: common objects in context. In: Fleet, D., Pajdla, T., Schiele, B., Tuytelaars, T. (eds.) ECCV 2014. LNCS, vol. 8693, pp. 740–755. Springer, Cham (2014). https://doi.org/10.1007/978-3-319-10602-1_48
13. Shrivastava, A., Gupta, A., Girshick, R.: [IEEE 2016 IEEE Conference on Computer Vision and Pattern Recognition (CVPR) - Las Vegas, NV, USA (2016.6.27-2016.6.30)] 2016 IEEE Conference on Computer Vision and Pattern Recognition (CVPR) - training region-based object detectors with online hard example. In: IEEE Conference on Computer Vision & Pattern Recognition (2016)
14. Tang, Y., Wang, J., Gao, B., Dellandrea, E., Gaizauskas, R., Chen, L.: Large scale semi-supervised object detection using visual and semantic knowledge transfer. In: 2016 IEEE Conference on Computer Vision and Pattern Recognition (CVPR), pp. 2119–2128 (2016)
15. Uijlings, J.R.R., Popov, S., Ferrari, V.: Revisiting knowledge transfer for training object class detectors. In: 2018 IEEE/CVF Conference on Computer Vision and Pattern Recognition, pp. 1101–1110 (2018)
16. Wang, X., Shrivastava, A., Gupta, A.: A-fast-RCNN: hard positive generation via adversary for object detection. In: Proceedings of the IEEE Conference on Computer Vision and Pattern Recognition, pp. 2606–2615 (2017)
17. Xia, G.S., et al.: DOTA: a large-scale dataset for object detection in aerial images. In: 2018 IEEE/CVF Conference on Computer Vision and Pattern Recognition, pp. 3974–3983 (2018)
18. Zhang, X., Yang, Y., Feng, J.: Learning to localize objects with noisy labeled instances. AAAI 2019 : Thirty-Third AAAI Conference on Artificial Intelligence, vol. 33, no. 1, pp. 9219–9226 (2019)

Global-Local Mutual Guided Learning
for Person Re-identification

Junheng Chen$^{(\boxtimes)}$, Xiao Luan , and Weisheng Li

College of Computer Science and Technology,
Chongqing University of Posts and Telecommunications, Chongqing, China
junhengchen228@gmail.com, {luanxiao,liws}@cqupt.edu.cn

Abstract. Person Re-Identification (Re-ID) plays a significant role in intelligent surveillance systems. Existing popular methods mainly focus on locating regions with specific pre-defined semantics to learn local representations, where pedestrian part-level features are inefficient to fully utilize the global feature information. Besides that, some methods miss out semantic transition information of human body. In this paper, we propose an end-to-end feature learning strategy to get refined feature representations with global-local mutual guided learning. In order to explore global and local information, we design a Global-Local Mutual Guided Network (GLMG-Net). It contains two branches to learn global feature representations, and local feature representations, respectively. For mutual guided module, global features are combined with each local feature by the add-wise operation. In the training process, this module enables branches to guide each other. Comprehensive experiments conducted on the public datasets of Market-1501 and DukeMTMC-ReID indicate that our method outperforms state-of-the-art approaches in several cases. In particular, mean average precision (mAP) scores of our method on those benchmarks are 89.2% and 79.7%, respectively.

Keywords: Person re-identification · Refined feature learning ·
Dual-branch deep network

1 Introduction

Person re-identication (Re-ID) is a tracking technique used in vision-based smart retail and security surveillance. It aims to retrieve a given person among all gallery pedestrian images captured across multiple non-overlapping security camera views at different locations. It is challenging to learn robust feature representations for each person because of large variations of human attributes like poses, clothes. Since pedestrian images are typically captured by surveillance cameras

---

J. Chen—The postgraduate student.
Supported by the National Natural Science Foundation of China (No. 61801068, 61502067, 61972060, U1713213), and Natural Science Foundation of Chongqing (cstc2015jcyjA40013, cstc2015jcyjA40034).

Y. Peng et al. (Eds.): PRCV 2020, LNCS 12306, pp. 126–137, 2020.
https://doi.org/10.1007/978-3-030-60639-8_11

in open area, the imperfect imaging devices and environment may lead to low resolution and complex background in pedestrian images. This will degrade the performance of person Re-ID methods. Benefit from the superiority of mining automatically high discriminative information, deep learning methods recently become more popular than traditional hand-crafted methods in this community. Some recent deep Re-ID methods [10,12,15,18,19] have achieved promising results with high-level identification rate and mean average precision. There are several approaches [1,6,8] only taking advantage of global features from the whole body on pedestrian images. They are hard to extract discriminative features because of high complexity for images captured in surveillance scenes and being sensitive to the missing key parts of human body.

To relieve this dilemma, most existing part-based approaches focus on learning part-informed representations containing salient information in recent years. They can be basically categorized into three types according to local regions generation procedures. Prior knowledge based approaches [17,20,21] utilized prior knowledge like poses or body landmarks to localize the discriminative regions. Nevertheless, robustness of the pose or landmark estimation models greatly influence the identification result. Attention based approaches [13,16] employed attention mechanism to enable models to focus on salient regions by localizing the high activations in deep feature maps. However, the selected regions lack semantic interpretation. Partition scheme based approaches [18,19] designed specific partition scheme to crop deep feature maps into pre-defined patches or stripes. The performance of above methods highly relies on whether pedestrian images are perfectly aligned.

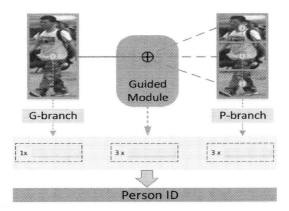

Fig. 1. Illustration of components in our GLMG-Net. We regard whole body of a person as global representation in G-branch in the left column. The right column refers to equally split deep feature maps of a person into three horizontal parts for local information in P-branch. The middle column is Guided-Module for combination of global representation and each part, respectively.

We propose a simple yet effective approach, called Global-Local Mutual Guided Network (GLMG-Net). As shown in Fig. 1, every sub-branch of GLMG-Net has specific contribution to the whole network. By combining global and local information in the mutual guided module, our GLMG-Net aims to extract more refined and discriminative features. The contributions of this paper are as follows:

- GLMG-Net contains two sub-branches, named G-branch and P-branch. Besides, GLMG-Net contains a mutual guided module called Guided-Module. In training process, local salient information in P-branch guide G-branch to learn more centralized and discriminative features, and vice versa.
- We design G-branch for learning global presentations with more useful information. Specifically, we employ ResNet-50 as backbone network of our model, and combine output feature maps of the last and penultimate residual block by manipulating their channel and spatial scale. Experiment results show that most key feature responses concentrate upon foreground of human body. In P-branch, we directly partition last deep feature map into three horizontal stripes for learning local presentations.

2   Related Work

In this part, we review several closely related work including deeply part-based methods and global-local based methods.

Deeply Part-Based Method. Deep learning methods have two main superiorities over hand-crafted. On one hand, deep features generically obtain stronger discriminative ability. On the other hand, some deep learning tools for parsing pedestrians can effectively benefit the part features. For instance, several works [17,21] in Re-ID employed these tools for pedestrian partition and reported encouraging improvement. However, robustness of these tools greatly influence the identification result. Based on strong discriminative ability of deep learning, Sun et al. [18] divided deep feature maps into fixed numbers horizontal stripes for partial features in their models. This simple operation can also bring a good promotion.

Global-Local Based Method. Global features are robust and essential for person Re-ID task. Therefore, several recent works notice that global and local features should be jointly learned. Wang et al. [19] designed a multi-branch network named MGN for learning feature from coarse to fine granularities of pedestrian images. They respectively divide deep feature maps into 1, 2 and 3 stripes in three sub-branches of MGN. Fu et al. [10] introduced a Horizontal Pyramidal Pooling (HPP) for dividing deep feature maps into 1, 2, 4 and 8 stripes. They both tried to retain underlying information between different scale feature stripes.

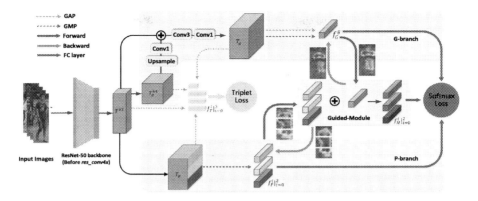

Fig. 2. The architecture of our proposed GLMG-Net for person re-identification.

3   Proposed Method

3.1   Network Architecture

Motivation. Global features are robust to the subtle view changes and internal variations, and are very significant cues for identification. They are always not well exploited together with local features in existing deep part-based models. So, we specially design a branch for refining discriminative global feature in our GLMG-Net. In addition, for current approaches based on global and local feature jointly learning, feature learning processes are independent after dividing deep feature maps, taking limited effect on common convolutional layers of backbone network. We consider to employ a mutual guided module called Guided-Module to realize mutual learning between global features and local features.

The architecture of Global-Local Mutual Guided Network is shown in Fig. 2. The ResNet-50 backbone is split into two branches: G-branch and P-branch. Then Guided-Module combine output features of above branches. For each FC layer, features are reduced from 2048-dim to 256-dim, and then trained for ID prediction with a softmax loss separately. During the testing stage, we concatenate all reduced features of three components to form the final feature representation of a pedestrian image. Notice that fully connected layers for dimension reduction and identity prediction in each component DO NOT share weights with each other. No matter for triplet loss or softmax loss, each feature has an independent supervisory signal. More details will be given in Table 1 and following parts.

Refined Global Feature Extraction. It aims to enrich the global feature information of pedestrian images. We choose the Resnet-50 as the backbone network with some modifications following the previous state-of-the-art [18, 19]. In G-branch, specifically, we remove the average pooling layer and the fully connected layer. The output tensor of last residual block and penultimate residual block are

Table 1. Comparison of the settings for components of GLMG-Net. Notice that the size of input images is set to 384×128.

Component	Feature map	Feature for triplet loss	Feature for softmax loss
Backbone	T^{b3} : $24 \times 8 \times 1024$	f_T^0 : $1 \times 1 \times 1024$	Null
G-branch	T_g^{b4} : $12 \times 4 \times 2048$ T_g : $24 \times 8 \times 2048$	f_T^1 : $1 \times 1 \times 2048$ f_T^2 : $1 \times 1 \times 2048$	f_G^g : $1 \times 1 \times 2048$
P-branch	T_p : $24 \times 8 \times 2048$	f_T^3 : $1 \times 1 \times 2048$	$f_P^i\vert_{i=0}^2$: $1 \times 1 \times 2048$
Guided-Moudle	Null	Null	$f_M^i\vert_{i=0}^2$: $1 \times 1 \times 2048$

named as T_g^{b4} and T^{b3}, respectively. We combine T_g^{b4} and T^{b3} to extract more useful global representation. To simplify the explanation, upsampling, 1×1 convolution and 3×3 convolution with batch normalization [11] and ReLU are named as *Upsample*, *Conv1* and *Conv3*, respectively. First of all, we employ an *Upsample* operation on T_g^{b4} , where its spatial scale is changed from 12×4 to 24×8 and a *Conv1* to reduce 2048-dim features to 1024-dim. Secondly, we directly combine T_g^{b4} and T^{b3} by add-wise operation. Then, we use a *Conv3* on the tensor after combination for regularization and a *Conv1* to increase dimensions from 1024-dim to 2048-dim for convenience of mutual guided learning module. Formally, denote the feature maps extracted by G-branch as T_g.

$$T_g = Conv1(Conv3(T^{b3} + Conv1(Upsample(T_g^{b4})))) \tag{1}$$

Finally, we get global representation f_G^g by global average pooling (GAP) and global max pooling (GMP).

$$f_G^g = GAP(T_g) + GMP(T_g) \tag{2}$$

Mutual Guided Learning Module. It aims to learn more discriminative feature by combining local and global representation. First of all, P-branch shares similar network architecture with the original ResNet-50 but removes the last spatial down-sampling operation to preserve proper areas of reception fields for local features [19]. Therefore, the feature map of last residual block named as T_p which scale is $24 \times 8 \times 2048$. And then, with a global max pooling (GMP), we partition T_p into 3 horizontal stripes and actually they are column vector $f_P^i(i = 0, 1, 2)$ which are represent distinct local information.

$$f_P^i = GMP(T_p), \; i = 0, 1, 2 \tag{3}$$

Secondly, by the use of an add-wise way, every single column vector $f_P^i(i = 0, 1, 2)$ is combined with global representation f_G^g for extracting combined features $f_M^i(i = 0, 1, 2)$.

$$f_M^i = f_G^g + f_P^i, \; i = 0, 1, 2 \tag{4}$$

For testing phases, in order to obtain the most powerful discrimination, we concatenate all the features $\{f_G^g, f_P^i\vert_{i=0}^2, f_M^i\vert_{i=0}^2\}$ as the final learned features.

3.2   Loss Function

We utilize a setting of joint learning with both softmax loss and triplet loss in our proposed method. Following various deep Re-ID methods, we employ softmax loss for classification, and triplet loss for metric learning as the loss functions in training process. Generally, the identification loss is the same as the classification loss. For i-th learned features f_i, softmax loss is formulated as:

$$L_{softmax} = -\sum_{i=1}^{N} \log \frac{e^{W_{y_i}^T f_i}}{\sum_{k=1}^{C} e^{W_k^T f_i}} \tag{5}$$

where W_k corresponds to a weight vector for class k, with the size of mini-batch in training process N and the number of classes in the training dataset C.

4   Experiment

4.1   Datasets and Protocols

This paper uses two mainstream Re-ID datasets for evaluation: Market-1501 [22] and DukeMTMC-reID [14,23]. It is necessary to introduce these datasets and their evaluation protocols before we show our results.

Market-1501. It comprises 32,668 images of 1,501 labeled persons of six camera views. There are 19,732 gallery images and 12,936 training images collected by DPM [9], including 751 identities in the training set and 750 identities in the testing set. Among the testing data, the test probe set has 3,368 images. The test gallery set also includes 2,793 additional distractors, which may have a considerable influence on the retrieval accuracy.

DukeMTMC-ReID. It is a subset of the DukeMTMC [14] dataset specifically collected for person reidentification. It contains 1,812 identities captured by 8 high-resolution cameras. There are 2,228 query images, 16,522 training images and 17,661 gallery images, with 1,404 identities appear in more than two cameras while 408 (distractor) identities appears in only one camera. DukeMTMC-ReID is one of the most challenging Re-ID datasets up to now with common situations in high similarity across persons and large variations within the same identity.

Evaluation Protocol. In our experiment, we use the cumulative matching characteristic (CMC) at Rank-k, $k = 1, 5, 10$, and mean average precision (mAP) to evaluate our approach. CMC represents the accuracy of the person retrieval, and it is accurate when each query only has one ground truth. However, when multiple ground truths exist in the gallery, the goal is to return all right match to user. In this case, CMC may not have enough discriminative ability, but the mAP could reflect the recall. Moreover, for simplicity, all results reported in this paper are under the single-query setting and do not use the re-ranking proposed in [24].

4.2   Implementation

We resize all the image to 384×128. For the backbone network, we use Resnet-50 that initialized with the weights pretrained on ImageNet [7]. During training process, we deploy random horizontal flipping and random erasing to pedestrian images for data augmentation. Each mini-batch is sampled with randomly selected P identities and randomly sampled K images for each identity from the training set to cooperate the requirement of triplet loss. Here we set $P = 8$ and $K = 4$. For the margin parameter for triplet loss, we set to 1.0 in all our experiments. We choose SGD as the optimizer with momentum 0.9. The weight decay factor for L2 regularization is set to 0.0005. We use a warm-up strategy in the first 10 epochs. Learning rate decays to $3e-3$ at epoch 60, and further decays to $3e-4$ at epoch 100. The total training process lasts for 200 epochs. During evaluation, we concatenate the reduced feature vectors together to generate feature representation of query image. Our model is implemented on PyTorch platform and train with one NVIDIA 1080Ti GPU. All datasets share the same experiments setting as above.

Table 2. Comparison results (%) with the state of the art on Market-1501 (4 evaluation metrics: mAP, Rank-1, Rank-5 and Rank-10) and DukeMTMC-reID (2 evaluation metrics: mAP and Rank-1). Numbers in bold indicate the best performance.

Model	Market-1501				DukeMTMC-reID	
	Rank-1	Rank-5	Rank-10	mAP	Rank-1	mAP
MLFN [4]	90.0	–	–	74.3	81.0	62.8
HA-CNN [13]	91.2	–	–	75.7	80.5	63.8
DuATM [16]	91.4	–	–	76.6	81.8	64.6
GP-reid [2]	92.2	97.9	–	81.2	85.2	72.8
Deep-Person [3]	92.3	–	–	79.6	80.9	64.8
DNN_CRF [5]	93.5	–	–	81.6	84.9	69.5
PCB+RPP [18]	93.8	97.5	98.5	81.6	83.3	69.2
HPM [10]	94.2	97.5	98.5	82.7	86.6	74.3
PPS [15]	94.3	97.7	98.7	85.3	88.2	75.9
MGN [19]	95.7	–	–	86.9	88.7	78.4
GLMG-Net (ours)	**95.9**	**98.4**	**99.0**	**89.2**	**89.7**	**79.7**

4.3   Comparison with State-of-the-Art Methods

In this section, we compare our proposed method with current state-of-the-art methods, most of which was proposed in recent two years, on all the candidate datasets. Results in detail are listed as follows:

Results on Market-1501. Person Re-ID results on Market-1501 are given in Table 2. It is worth noting that all results accomplished in the case of single query. The PCB is a convolutional baseline that achieved outstanding result without re-ranking, but we have improved performance by 7.6% and 2.1% on metrics mAP and Rank-1, respectively. MGN considers multiple branches and partitions feature map from coarse to fine granularities which motivates our approach, but we employ a more available strategy that takes advantage of the gradual cues between global and local information. Our method exceeds 2.3% and 0.2% on metrics mAP and Rank-1 comparing to MGN. Our method outperforms other comparative methods, especially showing large margins in terms of metrics mAP.

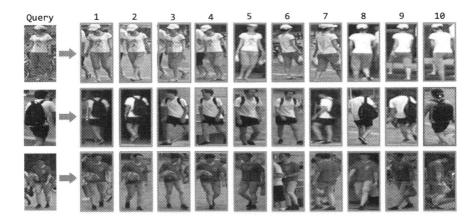

Fig. 3. Top-10 ranking list for some query images on Market-1501 by GLMG-Net. The images with green borders belong to the same identity to the given query image, while the one with red border show the incorrect identity. (Color figure online)

Figure 3 shows top-10 ranking results for some given query pedestrian images. The second pedestrian shows his back carrying a black backpack, but we can obtain his captured images in front view in Rank-3, 4, 5 and 6. The last query image is captured in a low-resolution condition, losing an amount of important information. Nevertheless, from some detailed clues such as the red bag carrying in hand and his red T-shirt with a special logo, all the ranking results are accurate and with high quality. This surprising result dues to our GLMG-Net, which learned features can robustly represent discriminative information of their identities.

Results on DukeMTMC-reID. Results on the challenging DukeMTMC-reID dataset are shown in Table 2. Among the compared methods, MGN is the closest method to our method score, but still below 1.0% Rank-1 and 1.3% mAP score, respectively. GP-reid is a good practice of many useful strategies combined in person Re-ID tasks and achieved the excellent published result. GLMG-Net achieves state-of-the-art result of Rank-1/mAP $= 89.7\%/79.7\%$, outperforming GP-reid by $+4.5\%$ in Rank-1 and $+6.9\%$ in mAP.

4.4   Ablation Study

To verify the effectiveness of each component and setting of GLMG-Net, we design several ablation study with different settings on Market-1501 in single query mode. Notice that other unrelated settings in each comparative experiment are the same as GLMG-Net implementation in Sect. 4.2, and we have carefully tuned all the candidate models and report the best performance with our settings. Table 3 shows the comparison results in different settings related to components of GLMG-Net. We separately analyze each component as follows:

Table 3. Results (%) with different settings on Market-1501. "TP" refers to triplet loss. "G-b, P-b and G-M" refers to "G-branch, P-branch and Guided-Module". "Branch" refers to a sub-branch of GLMG-Net. "Single" refers to a single network with the same setting as the branch with the corresponding name in GLMG-Net. "GLMG-Net (*)" refers to evaluation results only based on trained classifiers of '*' in GLMG-Net.

Model	Rank-1	Rank-5	Rank-10	mAP
ResNet-50+TP	92.0	97.5	98.6	79.6
G-b (Single)	93.5	98.0	98.8	81.6
G-b (Branch)	95.3	98.3	99.0	87.5
P-b (Single)	93.7	97.8	98.7	81.9
P-b (Branch)	95.1	98.2	98.8	86.5
GLMG-Net w/o TP	95.4	98.4	98.9	88.6
GLMG-Net w/o G-b	95.4	98.4	98.9	89.0
GLMG-Net w/o G-M	95.1	**98.5**	99.0	87.9
GLMG-Net (G-b+P-b)	95.4	98.3	98.8	88.3
GLMG-Net (G-M)	95.7	98.4	98.9	88.8
GLMG-Net	**95.9**	98.4	**99.0**	**89.2**

G-branch vs ResNet-50. Noticed that G-branch and ResNet-50 are both used to learning global representations in our paper. Comparing to the baseline ResNet-50 model, from a local view, we can observe G-b (single) makes a significant performance improvement Rank-1/mAP from $92.0\%/79.6\%$ to $93.5\%/81.6\%$ ($+1.5\%/2.0\%$). For our GLMG-Net, from a global view, if we employ ResNet-50 to replace our G-branch, the Rank-1/mAP from $95.9\%/89.2\%$ drop to $95.4\%/89.0\%$ ($-0.5\%/0.2\%$). In addition, we can also observe that feature map activation of G-branch precisely locates at discriminative foreground regions and almost not contains useless background information about identities for pedestrians compared to ResNet-50 in Fig. 4. Results above prove that our

Original   ResNet50-TP G-b(Single) G-b(Branch)   P-b(Single) P-b(Branch)

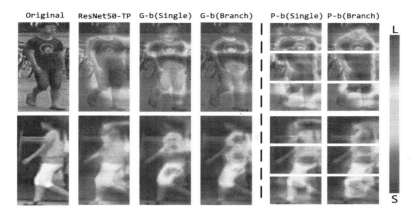

Fig. 4. Visualization of attention maps from ResNet-50+TP, G-b (Single) G-b (Branch), P-b (Single) and P-b (Branch). As shown in the 4th and 6th column, the diverse attention maps from G-b (Branch) and P-b (Branch) accurately located refined and coherent regions of human body.

G-branch is obviously superior to ResNet-50 and our G-branch can learn refined global features.

Guided-Module. We explore the effects of Guided-Module in two aspects. From a global view, the metrics scores of Rank-1/mAP from 95.9%/89.2% drop to 95.1%/87.9% (−0.8%/1.3%) when we remove Guided-Module from GLMG-Net. From a local view, the features from sub-branches also perform better than that from single networks because of Guided-Module. As shown in Fig. 4, Guided-Module guides G-b and P-b to learn ignored information from each other. To be specific, for G-branch, detail information from local features helps it to refine more regions. For P-branch, human sematic information from global features makes feature map activation coherent and well complements transitional information between global and local representations. Comparing to P-b (Single), Rank-1/mAP of P-b (Branch) from 93.7%/81.9% rise to 95.1%/86.5% (+1.4%/4.6%). It shows that the mutual guided cooperation of branches learns more discriminative feature representations than independent networks. As our expected, with the help of Guided-Module, the mutual effects between sub-branches complement the blind spots in their individual learning procedure.

5   Conclusion

In this paper, we construct a global-local mutual guided learning model for person re-identification. Firstly, we design G-branch for learning refined features for global information. Secondly, we employ Guided-Module further enhances both global and local feature representations of pedestrian. It is worth noting that all our results are achieved in a single query setting without using any re-ranking algorithms. Extensive experiments on Market-1501 and DukeMTMC-reID clearly show that the proposed GLMG-Net has achieved state-of-the-art performance on several benchmark datasets.

References

1. Ahmed, E., Jones, M.J., Marks, T.K.: An improved deep learning architecture for person re-identification. In: Proceedings of the IEEE Computer Society Conference on Computer Vision and Pattern Recognition, pp. 3908–3916 (2015)
2. Almazán, J., Gajic, B., Murray, N., Larlus, D.: Re-id done right: towards good practices for person re-identification. CoRR (2018)
3. Bai, X., Yang, M., Huang, T., Dou, Z., Yu, R., Xu, Y.: Deep-person: learning discriminative deep features for person re-identification. Pattern Recogn. **98**, 107036 (2020)
4. Chang, X., Hospedales, T.M., Xiang, T.: Multi-level factorisation net for person re-identification. In: IEEE Conference on Computer Vision and Pattern Recognition, pp. 2109–2118 (2018)
5. Chen, D., Xu, D., Li, H., Sebe, N., Wang, X.: Group consistent similarity learning via deep CRF for person re-identification. In: Proceedings of the IEEE Computer Society Conference on Computer Vision and Pattern Recognition, pp. 8649–8658 (2018)
6. Chen, W., Chen, X., Zhang, J., Huang, K.: A multi-task deep network for person re-identification. In: Proceedings of the Thirty-First AAAI Conference on Artificial Intelligence, pp. 3988–3994 (2017)
7. Deng, J., Dong, W., Socher, R., Li, L., Li, K., Li, F.: ImageNet: a large-scale hierarchical image database. In: IEEE Computer Society Conference on Computer Vision and Pattern Recognition, pp. 248–255 (2009)
8. Ding, S., Lin, L., Wang, G., Chao, H.: Deep feature learning with relative distance comparison for person re-identification. Pattern Recogn. **48**, 2993–3003 (2015)
9. Felzenszwalb, P.F., McAllester, D.A., Ramanan, D.: A discriminatively trained, multiscale, deformable part model. In: 26th IEEE Conference on Computer Society Conference on Computer Vision and Pattern Recognition (2008)
10. Fu, Y., et al.: Horizontal pyramid matching for person re-identification. In: Proceedings of the AAAI Conference on Artificial Intelligence, pp. 8295–8302 (2019)
11. Ioffe, S., Szegedy, C.: Batch normalization: Accelerating deep network training by reducing internal covariate shift. In: Proceedings of the 32nd International Conference on Machine Learning, pp. 448–456 (2015)
12. Li, H., Yang, M., Lai, Z., Zheng, W., Yu, Z.: Pedestrian re-identification based on tree branch network with local and global learning. In: Proceedings - IEEE International Conference on Multimedia and Expo, pp. 694–699 (2019)
13. Li, W., Zhu, X., Gong, S.: Harmonious attention network for person re-identification. In: Proceedings of the IEEE Computer Society Conference on Computer Vision and Pattern Recognition, pp. 2285–2294 (2018)
14. Ristani, E., Solera, F., Zou, R., Cucchiara, R., Tomasi, C.: Performance measures and a data set for multi-target, multi-camera tracking. In: Hua, G., Jégou, H. (eds.) ECCV 2016. LNCS, vol. 9914, pp. 17–35. Springer, Cham (2016). https://doi.org/10.1007/978-3-319-48881-3_2
15. Shen, Y., et al.: A part power set model for scale-free person retrieval. In: Proceedings of the Twenty-Eighth International Joint Conference on Artificial Intelligence, pp. 3397–3403 (2019)

16. Si, J., et al.: Dual attention matching network for context-aware feature sequence based person re-identification. In: Proceedings of the IEEE Computer Society Conference on Computer Vision and Pattern Recognition, pp. 5363–5372 (2018)
17. Su, C., Li, J., Zhang, S., Xing, J., Gao, W., Tian, Q.: Pose-driven deep convolutional model for person re-identification. In: Proceedings of the IEEE International Conference on Computer Vision, pp. 3980–3989 (2017)
18. Sun, Y., Zheng, L., Yang, Y., Tian, Q., Wang, S.: Beyond part models: person retrieval with refined part pooling (and a strong convolutional baseline). In: Ferrari, V., Hebert, M., Sminchisescu, C., Weiss, Y. (eds.) ECCV 2018. LNCS, vol. 11208, pp. 501–518. Springer, Cham (2018). https://doi.org/10.1007/978-3-030-01225-0_30
19. Wang, G., Yuan, Y., Chen, X., Li, J., Zhou, X.: Learning discriminative features with multiple granularities for person re-identification. In: Proceedings of the 2018 ACM Multimedia Conference, pp. 274–282 (2018)
20. Wei, L., Zhang, S., Yao, H., Gao, W., Tian, Q.: GLAD: global-local-alignment descriptor for pedestrian retrieval. In: Proceedings of the 2017 ACM on Multimedia Conference, pp. 420–428 (2017)
21. Zheng, L., Huang, Y., Lu, H., Yang, Y.: Pose-invariant embedding for deep person re-identification. IEEE Trans. Image Process. **28**, 4500–4509 (2019)
22. Zheng, L., Shen, L., Tian, L., Wang, S., Wang, J., Tian, Q.: Scalable person re-identification: a benchmark. In: Proceedings of the IEEE International Conference on Computer Vision, pp. 1116–1124 (2015)
23. Zheng, Z., Zheng, L., Yang, Y.: Unlabeled samples generated by GAN improve the person re-identification baseline in vitro. In: Proceedings of IEEE International Conference on Computer Vision, pp. 3774–3782 (2017)
24. Zhong, Z., Zheng, L., Cao, D., Li, S.: Re-ranking person re-identification with k-reciprocal encoding. In: IEEE Conference on Computer Vision and Pattern Recognition, pp. 3652–3661 (2017)

Handwritten Style Recognition for Chinese Characters on HCL2020 Dataset

Peiyi Hu, Mengqiu Xu, Ming Wu$^{(\boxtimes)}$, Guang Chen, and Chuang Zhang

Beijing University of Posts and Telecommunications, Beijing 100876, China
{hupeiyi,xumengqiu,wuming,chenguang,zhangchuang}@bupt.edu.cn

Abstract. Structural features of Chinese characters provide abundant style information for handwritten style recognition, while prior work on this task has few senses of using structural information. Meanwhile, based on current handwritten Chinese character datasets, it is hard to obtain a good generalization model only by character category and writer information. Therefore, we add the structural information known as morpheme which is the smallest and unique structure in Chinese character into the large handwritten dataset HCL2000 and update it to HCL2020. We also present a deep fusion network (Morpheme-based Handwritten Style Recognition Network, **M-HSRNet**), capturing both overall layout characteristics and detail structural features of characters to recognize handwritten style. The evaluation results of the proposed model on HCL2020 are observed to prove the effectiveness of morpheme. Together with the proposed Morpheme Encoder module, our approach achieves an accuracy of 78.06% in handwritten style recognition, which is 3 points higher than the result without morpheme information.

Keywords: Handwritten Chinese character dataset · Morpheme · Handwritten style recognition · Fusion network

1 Introduction

Handwritten style recognition is one of the most challenging research fields about handwritten character. It aims to identify who this handwritten character belongs to. It has been implemented in historic document analysis fields and anti-crime [3], which requires a high level of domain expertise and heavy work. Handwritten style recognition is divided into page level and character level. On the research of recognition on page level, Jain et al. [11] used K-adjacent segments (KAS) feature to model character contours of 300 writers from IAM

P. Hu and M. Xu—contribute equally to this work.
The first authors are students.

This work was supported in part by MoE-CMCC "Artifical Intelligence" Project No.MCM20190701.

Y. Peng et al. (Eds.): PRCV 2020, LNCS 12306, pp. 138–150, 2020.
https://doi.org/10.1007/978-3-030-60639-8_12

dataset. [24] leverages online writing information and deep CNNs on 187 writers with Chinese page input and on 134 writers with English page input on CASIA Handwriting Database. On the research of recognition on character level, Kamal Parves et al. [12] introduce an approach with the idea of splitting number of small sub-images based on a local feature extracted by writer-specific characteristics. Bensefia Ameur et al. [2] present an original approach based on the analysis of a unique sample of a handwriting word by using the Levenshtein edit distance based on Fisher-Wagner algorithm.

However, few of these methods have considered the influence of the smaller structural units in a character on the style characteristics, which are inconsistent with the experts' handwritten identification methods in real scenes. When recognizing handwritten style in real scene, in addition to comparing the overall layout characteristics of the characters, they have a tendency to pay more attention on some detail structure features which help recognize style information. When human writing, the acting point on each character structure is different, which naturally leads to the difference in the ultimate handwriting style. It means that, more attention to the differences in detail structures can better help us to identify one's handwritten style. For this reason, we extract morpheme information contained in each character and integrating it into models.

The development of handwritten character recognition and handwritten style recognition is accompanied by the support of handwritten character datasets. Nowadays, handwritten character datasets are mainly divided into the following three categories: 1) symbol image datasets [14, 23] with numbers as examples; 2) phonogram character datasets [1, 7] with English and Latin as examples, which express the meanings by the sound; 3) logogram character datasets [5, 16, 18, 22, 26] with Chinese characters and Japanese characters as examples, which express the meanings by shape. Research on handwritten character datasets has also made considerable progress. Taking Chinese character datasets as an example, Zhang H et al. [26] present a large scale offline handwritten Chinese character database HCL2000 to facilitate handwritten Chinese recognition research. Liu et al. [16] release the CASIA handwritten dataset divided into online handwritten database and offline handwritten database, which covers a maximum of 7,185 Chinese characters.

Many of these datasets are hard to obtain a good generation model in handwritten style recognition, which are mainly for the following reasons: 1) some samples in dataset are not accurate, including of erroneous samples or labels caused by data collection and storage. 2) Lacking of in writers' information, make it difficult to be applied in handwritten style recognition. Besides, the annotation information of dataset is incomplete and cannot meet the command of various studies, such as the information of character structures. 3) Storage format of dataset is not universal and cannot be adapted to current deep learning algorithm framework, such as Tensorflow. Therefore, a method of reconstruction and supplementary of annotations should be submitted to make these datasets more complete and universal to promote the development of handwritten style recognition.

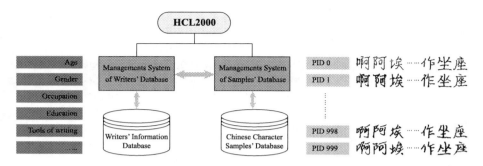

Fig. 1. The system model of HCL2000. The character samples are organized by different writers and stored in PID (Personal Identification) files in Chinese Character Samples' Database. Writers' information includes the gender, age, occupation, education, tools of writing in Writers' Information Database.

In order to solve the shortcomings of models and datasets mentioned above, we re-proposed HCL2020, a new large scale offline handwritten Chinese character dataset with the annotations of character information, writer information, and morpheme information, and utilize a fusion deep network integrating morpheme information to recognize handwritten style. It is useful of the network to focus more on the glyph structure that is easy to carry style information. Experiments on three backbones show that the addition of morpheme information does help identify one's handwritten style.

The contributions of our work can be summarized as follows:

1) we update HCL2000 into a more complete large-scale offline handwritten Chinese character dataset named HCL2020, with annotations of character information, writer information, and morpheme information.
2) We design a deep fusion network named M-HSRnet utilizing the character image information and morpheme information for handwritten style recognition.
3) We demonstrate the impact of the addition of morpheme information via extensive experiments using different backbones.

2   HCL2020

2.1   Introduction of HCL2000

HCL2000 database contains 3,755 frequently used simplified Chinese characters written by 1,000 different writers. Besides the well-preserved sample information stored in handwritten Chinese character samples' sub-database, they also established writers' information sub-database, which contains age, gender, occupation, education and tools of writing information of writers. It uses a system model to control the information that contains not only the Chinese character image, but also the information about writers as shown in Fig. 1. In order to

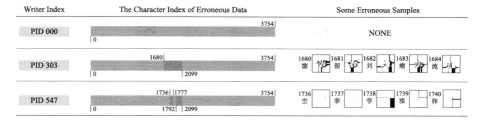

Fig. 2. Some erroneous character samples with writers' index information and charater index information. No erroneous data taking writer (Index:000) as an example; Having some erroneous data taking writer (Index:303,547) as an example.

take advantage of two sub-databases conveniently, the system model builds two management systems. Users can view all Chinese character samples of a writer. Chinese character samples are organized by different writers and stored in PID (Personal Identification) files. The samples written by the same writer are stored by the sort of section code. Each Chinese character sample is presented as 64 × 64 binary pixels.

Although HCL2000 has been the basic dataset for handwritten Chinese character recognition research for nearly 20 years, it has limited its application in deep learning research due to its organizational form and specific storage format. Therefore, we propose a new reconstruction method for HCL2000 database to solve the above problems.

2.2   Reconstruction of HCL2020

In order to make HCL2000 more convenient to employ, we extract the writer's file name as the writer's index through HCL2000 writer information sub-database. We exclude some erroneous character samples, which writer indices are "303, 306, 307, 308, 314, 316, 317, 318, 319, 443, 547", for a total of 4,984 character samples. Some erroneous character samples with writer index information and character index information are shown in Fig. 2. Finally, we obtain a fresh sample dataset written by 1,000 writers, for a total of 3,750,016 samples.

We convert HCL2000 to a more generic data storage format-.npz file format, which can be more convenient used by deep learning algorithms and framework. On the premise of saving storage space and not affecting visual recognition, each character sample is described as 28 × 28 pixel gray-scale image by down-sampling. We compared the effects of different interpolation methods in down-sampling processing of image scaling and finally choose Nearest Interpolation Method [17] for image scaling and obtain the final character dataset.

2.3   Annotations of HCL2020

We obtain handwritten Chinese character samples with character category label and writer number label after the reconstruction of HCL2000. However, there is still a doubt whether those information is sufficient to describe Chinese character because we have not achieved satisfactory generalization results in handwritten style recognition. Therefore, we explore the factors that influence the style recognition of the writer, which contains the information of Chinese character structure [4], the morpheme and strokes. We add above information as attribute labels to the HCL2020.

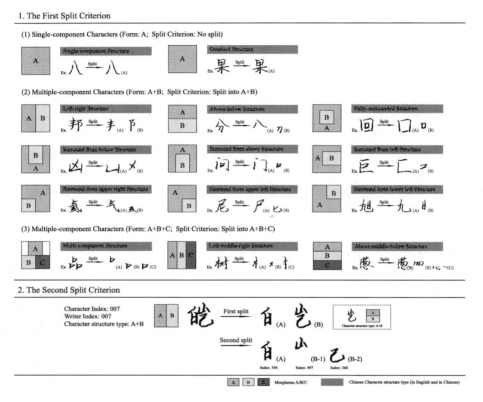

Fig. 3. The criterion of splitting Chinese characters into morphemes based on different character structures and some examples. The first split criterion contains: (1) No split based on single-component characters in the form of 'A'. (2) Split into 'A+B' based on multiple-component characters in the form of 'A, B'. (3) Split into 'A+B+C' based on single-component characters in the form of 'A, B, C'. If morphemes after the first split are still more complicated, a second split will be performed which split criterion is same as first split criterion, taking the Chinese character 皑 as an example. Finally, each morpheme is represented by a morpheme category index.

It is generally considered that there are four levels [6] for Chinese font from whole to the part, namely the Chinese characters, morphemes, strokes, and stroke shapes. Strokes are the most basic element in Chinese characters, which can form morphemes. Complex strokes can be further split into stroke shapes.

Table 1. The difference between HCL2000 dataset and HCL2020 dataset. Compared with HCL2000, HCL2020 has more annotations about character structure information.

	HCL2000 dataset	HCL2020 dataset
Published year	2000	2020
Amount	3,755,000	3,750,016
Size	28 × 28	64 × 64
Description	Binary pixels	Gray-scale pixels
Annotations	Character categories information; Writer information;	Character categories information; Writer information; Character structures information; Morphemes information; Stroke sequence information;

The split from Chinese characters to morphemes is based on different Chinese character structures, which can be subdivided into 14 structures. Among the 3,755 types of Chinese characters, the left-right structures have the most types of Chinese characters. Chinese characters can be disassembled into the form of 'A+B' or 'A+B+C' and can be split to morphemes based on the structural type of Chinese characters, as shown in Fig. 3. If morphemes after the first split are still more complicated, a second split will be performed. We add the morphemes as attribute labels to handwritten Chinese character samples in HCL2020 dataset.

According to statistics, there are 780 types of morphemes in 3,755 types of Chinese characters. We store morpheme information in an array with size of (3755,780). Each row represents the morpheme information contained in a Chinese character. If a morpheme exists in this character, the corresponding index will be set as 1, and the non-appearing morpheme will be set as 0. Thus people can easily get morpheme information according to corresponding character index and use it on other handwritten style recognition datasets.

Building on the above reconstruction and annotations, we re-propose a new large scale offline handwritten Chinese character dataset, named HCL2020. The difference between HCL2000 database and HCL2020 dataset is shown in Table 1. We randomly select 545 handwritten Chinese character samples written by every writer to form HCL2020 test dataset, which contains 545,000 Chinese character samples. The remaining character samples make up the HCL2020 train dataset, which contains 3,205,016 Chinese character samples. The ratio of the train dataset to the test dataset is about 6:1.

3   Proposed Network

With the idea of using a deep neural network to solve the multi-model image fusion problems [8,9,15], we propose a method to fuse the morpheme information and character image to prove the validity of the morpheme information and design a deep network (**M**orpheme-based **H**andwritten **S**tyle **R**ecognition

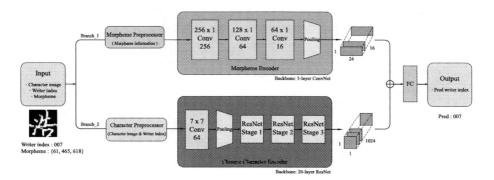

Fig. 4. A brief framework of M-HSRNet. First, we can obtain Chinese character image data, the writer category index and the morpheme information for each Chinese character. Then, we can get an embedding vector to present morpheme information in Morpheme Preprocessor and get an array to present character image in Image Preprocessor. We can extract feature through Morpheme Encoder and Character Encoder by using 3-layer ConvNet and 20-layer ResNet as backbones. Later we concatenate two tensors obtained by encoders and send it to Classifier to recognition and predict the handwritten style. When the M-HSRNet is testing data, it is needed to recognize the category of Chinese characters used in extracting morpheme information.

Network, **M-HSRNet**) to classify handwritten style. A brief framework of M-HSRNet is shown in Fig. 4.

Our model consists of two branches, Character Encoder and Morpheme Encoder. Character Encoder encodes style information contained in a character image, while the Morpheme Encoder extracts the morpheme information of the corresponding words in the image, thereby reminding the network to place emphasis on these structures that are usually carrying style information. Then features extracted by the two branches are concatenated together and passed to the next classification layer. Experiments demonstrate that the network with the addition of the morpheme encoder branch can improve the accuracy points on HCL2020 dataset than without morpheme information. Character Encoder and Morpheme Encoder are introduced as following. More details information about M-HSRnet settings when the network is training or testing is available in Chaprter 4.

Character Encoder. The backbone of Character Encoder employed in our work is a 20-layer ResNet. It has shown remarkable power on image classification tasks in recent years. Features extracted from deeper networks have a tendency to achieve higher performance than that is extracted from shallow networks. However, with the network deepening, gradients may be vanished or explode, and the accuracy may be saturated and then degrade rapidly. Thanks to its exciting property of 'shortcut connection' which maps an identity mapping to push the residual to zero, the network with stacked layers can be trained without considering the above problems. The input of Character Encoder is a three dimensional matrix of a character image with dimension $\mathbf{X} \in R^{d_h \times d_h \times d_w}, d_h = d_w = 28, d_k = 1$.

We perform a linear projection by the shortcut connection in 20-layer ResNet as a backbone to match the dimensions. The residual block is defined as:

$$y = F(x, \{w_i\}) + x. \tag{1}$$

Morpheme Encoder. Morpheme is the smallest and unique structure of the Chinese character. A Chinese character can consist of one or more morphemes. In previous handwritten style recognition experiments, we find that the characters with the highest recognition rate have high commonality in the morpheme. Therefore, we design the morpheme encoder and add the morpheme information to style recognition model. We sort out all the morphemes in 3,755 Chinese characters, a total of 780. Each morpheme has its category index between 0 and 779. The input of this morpheme encoder is a one-hot vector which contains all morpheme in current character with dimension $\mathbf{M} \in R^{d_{mor} \times d_k}, d_{mor} = 780, d_k = 1$. In this vector, the index position corresponding to all the morpheme information contained in the current character is marked as 1 and the rest as 0. Structure of morpheme encoder is a ConvNet stacked by three convolution layers plus a full-connected layer. Each convolution layer is accompanied by a Batch Normalization layer.

H_X is the character embedding feature vector obtained by X through Charater Encoder f_{CE}, and H_M is the morpheme embedding feature vector obtained by M through Morpheme Encoder f_{ME}.

$$H_X = f_{CE}(X, \theta_{ce}) \tag{2}$$

$$H_M = f_{ME}(M, \theta_{me}) \tag{3}$$

We concatenate H_X and H_M as the final feature representation. The probability of the category of handwritten style is defined p_t as following.

$$p(y|x, m) = Softmax(W_i[H_X, H_M] + b) \tag{4}$$

We use cross-entropy loss as the optimization goal to evaluate our network. y is corresponding ground-truth.

$$Loss = -\sum_t y_t \times log(p_t), t \in [0, 100) \tag{5}$$

4   Experiment

4.1   Hanwritten Style Recognition on HCL2020 Dataset

Morpheme Preprocessor. The role of the Morpheme Preprocessor is to embed morpheme index information into a vector which can be easily processed

Table 2. Handwritten Style Recognition results evaluating on HCL2020 dataset using different backbones. The train and test set sizes are 3,210 and 545 per class, for total 100 classes.

Deep leaning algorithms	Parameters	Top-1 .err%
ZFNet [25]	30.53 M	39.07
ZFNet [25]+ Morpheme	32.80 M	**35.71**
VGGNet-16 [20]	34.02 M	34.12
VGGNet-16 [20]+ Morpheme	36.29 M	**27.62**
ResNet-20 [10]	26.32 M	25.69
ResNet-20 [10]+ Morpheme	30.13 M	**21.94**

Table 3. Handwritten Chinese character recognition results evaluating on HCL2020 dataset. The train and test set sizes are 3,205,016 and 545,000.

Deep leaning algorithms	Parameter	Top-1 .err%
ConvNet-3 [19]	39.08 M	8.99
AlexNet [13]	44.80 M	4.64
ZFNet [25]	44.80 M	3.88
VGGNet-16 [20]	347.31 M	2.43
VGGNet-19 [20]	352.61 M	2.34
ResNet-50 [10]	114.93 M	**2.31**

by the Morpheme encoder. Every Chinese character contains at least one morpheme. According to our statistics, 3,755 characters contain a total of 780 morphemes. We labeled these morphemes from 0 to 779. So we set the dimension of the morpheme information of corresponding character as $W_i^M \in R^{d_{mor} \times d_k}$ dimensional embedding. In this morpheme embedding, if a morpheme exists in the character, the corresponding index will be set as 1, and the non-appearing morpheme will be set as 0. Finally we smooth this sparse embedding and then feed it into Morpheme encoder.

OCR Preprocessor. Recognizing handwritten Chinese characters plays an important role because the Chinese character category can link character with its corresponding morpheme information. When M-HSRNet is testing, it needs to recognize the category of Chinese characters used in extracting morpheme information in the morpheme category index set while it can be obtained correctly as label information when training. Also, we evaluate HCL2020 dataset using deep learning algorithms [10, 13, 19–21, 25] in handwritten Chinese character recognition task and use 50-layer ResNet algorithm as a backbone of OCR Preprocessor with data augmentation which achieves the accuracy of 97.69%. All of evaluation results of deep learning algorithms are shown in Table 3.

Evaluation Results. Because there is just one sample for each character of each writer in HCL2020. To ensure the quality of the training set for the handwritten style recognition experiment, we need to select those writers who have obvious personal handwritten style and consistent style between different samples. We finally select 100 writers. We randomly select 545 samples from each writer's 3,755 samples to form the test set, which contains 54,500 samples. The remaining 321,000 samples are used as the train set. The ratio of the train set to the test set was 6:1. We use different backbones in different experimental group to verify the effectiveness of morpheme. Evaluation results are shown in Table 2. We consider for three different backbones: ZFNet [25], VGGNet-16 [20], ResNet-20 [10]. By comparing the performance between vanilla model and our fusion model with morpheme information using various backbones, we hope these results will provide a sense of the relative effectiveness of our morpheme label (Fig. 5).

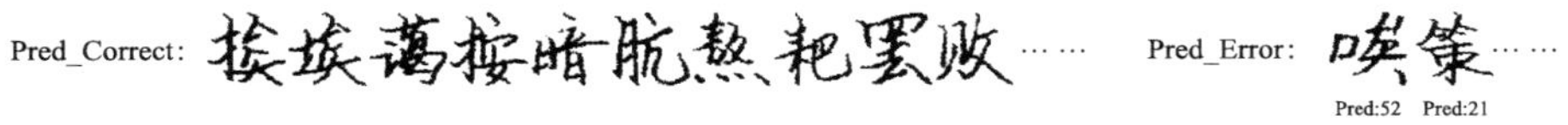

Fig. 5. There are some samples of one writer (Index:007). The left part are the samples which M-HSRNet recognizes correctly, and the right part are the samples can not be recognize.

Furthermore, Chinese character style recognition in real scenes is usually based on multiple Chinese characters. In order to explore the influence of increasing number of Chinese characters as input data, we improve above handwritten style recognition models, of which network allows multiple Chinese characters of same author as input once time. We keep model structure unchanged. As we described above, we utilize a Character Encoder with ResNet-20 as backbone plus a Morpheme Encoder, and then concatenate the hidden layer representations of each input character to get the final style representation of this author. The number of input character increases from 1 to 5. Experiment results are shown in Fig. 6. Red line shows the results of our model with morpheme information when input size of character size increased. We also make contrast between our model and the ResNet-20 backbone. The blue one represents ResNet-20 backbone's performance. Among all sizes, our model performs slightly better than backbone. On the other hand, with the increase of input character size, the accuracy of validation set of the two models both get boosts. When input character size reaches 5, our model can get the accuracy of 95.05%.

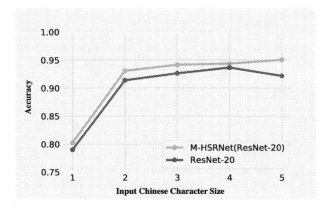

Fig. 6. Model performances on different input character sizes, building on ResNet-20 backbone. The red line shows the results of our model when input size of character size increases, while the blue one represents performance of ResNet-20 without morpheme information. (Color figure online)

5   Conclusions

Based on the study of handwritten identification method presented by professional experts in real scenes, we add morpheme information and update HCL2000 into a more complete large-scale handwritten Chinese character dataset, named HCL2020. We propose M-HSRnet, a deep fusion network, which consists of two network branches to extract morpheme and character image features. We achieve the accuracy of 78.06% in handwritten style recognition task on HCL2020, which is higher than the result of methods without morpheme information. The results show the effectiveness of morpheme information in handwritten style recognition.

For future work, we will explore more deep learning network with morpheme information. Also, we will develop more Chinese character structure features to assist handwritten style recognition task. Besides, we hope handwritten character datasets can be used not only for character recognition and handwritten style recognition, but also for more creative work, such as handwritten style generation, transfer learning, cross-domain retrieval and so on.

References

1. Arvanitopoulos, N., Chevassus, G., Maggetti, D., Süsstrunk, S.: A handwritten French dataset for word spotting: CFRAMUZ. In: Proceedings of the 4th International Workshop on Historical Document Imaging and Processing, pp. 25–30 (2017)
2. Bensefia, A., Paquet, T.: Writer verification based on a single handwriting word samples. EURASIP J. Image Video Process. **2016**(1), 34 (2016)
3. Bulacu, M., Schomaker, L.: Text-independent writer identification and verification using textural and allographic features. IEEE Trans. Pattern Anal. Mach. Intell. **29**(4), 701–717 (2007)

4. Chinese, D.M.: Chinese character structure: a guide to writing characters well. https://www.decodemandarinchinese.com/character-structure/
5. Clanuwat, T., Bober-Irizar, M., Kitamoto, A., Lamb, A., Yamamoto, K., Ha, D.: Deep learning for classical Japanese literature. arXiv preprint arXiv:1812.01718 (2018)
6. Dai, R., Liu, C., Xiao, B.: Chinese character recognition: history, status and prospects. Front. Comput. Sci. China **1**(2), 126–136 (2007)
7. De Campos, T.E., Babu, B.R., Varma, M., et al.: Character recognition in natural images. VISAPP (2), 7 (2009)
8. Deng, X., Dragotti, P.L.: Deep convolutional neural network for multi-modal image restoration and fusion. IEEE Trans. Pattern Anal. Mach. Intell. (2020)
9. Deng, Z., et al.: Deep multi-model fusion for single-image Dehazing. In: Proceedings of the IEEE International Conference on Computer Vision, pp. 2453–2462 (2019)
10. He, K., Zhang, X., Ren, S., Sun, J.: Deep residual learning for image recognition. In: Proceedings of the IEEE Conference on Computer Vision and Pattern Recognition, pp. 770–778 (2016)
11. Jain, R., Doermann, D.: Offline writer identification using k-adjacent segments. In: 2011 International Conference on Document Analysis and Recognition, pp. 769–773. IEEE (2011)
12. Kamal, P., Rahman, F., Mustafiz, S.: A robust authentication system handwritten documents using local features for writer identification. J. Comput. Sci. Eng. **8**(1), 11–16 (2014)
13. Krizhevsky, A., Sutskever, I., Hinton, G.E.: ImageNet classification with deep convolutional neural networks. In: Advances in Neural Information Processing Systems, pp. 1097–1105 (2012)
14. LeCun, Y., Bottou, L., Bengio, Y., Haffner, P.: Gradient-based learning applied to document recognition. Proc. IEEE **86**(11), 2278–2324 (1998)
15. Lee, G., Nho, K., Kang, B., Sohn, K.A., Kim, D.: Predicting Alzheimer's disease progression using multi-modal deep learning approach. Sci. Rep. **9**(1), 1–12 (2019)
16. Liu, C.L., Yin, F., Wang, D.H., Wang, Q.F.: Online and offline handwritten Chinese character recognition: benchmarking on new databases. Pattern Recogn. **46**(1), 155–162 (2013)
17. Olivier, R., Hanqiang, C.: Nearest neighbor value interpolation. Int. J. Adv. Comput. Sci. Appl. **3**(4), 25–30 (2012)
18. Ren, J., Guo, J.: The new edition of HCL2000 and its application (in Chinese). J. Chin. Inf. Process. **19**(5), 99–106 (2005)
19. Sermanet, P., Chintala, S., LeCun, Y.: Convolutional neural networks applied to house numbers digit classification. In: Proceedings of the 21st International Conference on Pattern Recognition (ICPR2012), pp. 3288–3291. IEEE (2012)
20. Simonyan, K., Zisserman, A.: Very deep convolutional networks for large-scale image recognition. arXiv preprint arXiv:1409.1556 (2014)
21. Szegedy, C., et al.: Going deeper with convolutions. In: Proceedings of the IEEE Conference on Computer Vision and Pattern Recognition, pp. 1–9 (2015)
22. Wang, D.H., Liu, C.L., Yu, J.L., Zhou, X.D.: CASIA-OLHWDB1: A database of online handwritten Chinese characters. In: 2009 10th International Conference on Document Analysis and Recognition, pp. 1206–1210. IEEE (2009)
23. Yadav, C., Bottou, L.: Cold case: The lost MNIST digits. In: Advances in Neural Information Processing Systems, pp. 13443–13452 (2019)
24. Yang, W., Jin, L., Liu, M.: DeepWriterID: an end-to-end online text-independent writer identification system. IEEE Intell. Syst. **31**(2), 45–53 (2016)

25. Zeiler, M.D., Fergus, R.: Visualizing and understanding convolutional networks. In: Fleet, D., Pajdla, T., Schiele, B., Tuytelaars, T. (eds.) ECCV 2014. LNCS, vol. 8689, pp. 818–833. Springer, Cham (2014). https://doi.org/10.1007/978-3-319-10590-1_53
26. Zhang, H., Guo, J., Chen, G., Li, C.: HCL 2000-a large-scale handwritten Chinese character database for handwritten character recognition. In: 2009 10th International Conference on Document Analysis and Recognition, pp. 286–290. IEEE (2009)

Multimodal Image Retrieval Based on Eyes Hints and Facial Description Properties

Yuelong Li[1,2], Junyu Bi[1,2], Tongshun Zhang[1,2], and Jianming Wang[1,2(✉)]

[1] Tianjin Key Laboratory of Independent Intelligent Technology and System, Tianjin, People's Republic of China
wangjianming@tiangong.edu.cn
[2] School of Computer Science and Technology, Tiangong University, Tianjin, People's Republic of China

Abstract. Eyes are the most prominent visual components on human face. Obtaining the corresponding face only by the visual hints of eyes is a long time expectation of people. However, since eyes only occupy a small part of the whole face, and they do not contain evident identity recognition features, this is an underdetermined task and hardly to be finished. To cope with the lack of query information, we enroll extra face description properties as a complementary information source, and propose a multimodal image retrieval method based on eyes hints and facial description properties. Furthermore, besides straightforward corresponding facial image retrieval, description properties also provide the capacity of customized retrieval, i.e., through altering description properties, we could obtain various faces with the same given eyes. Our approach is constructed based on deep neural network framework, and here we propose a novel image and property fusion mechanism named Product of Addition and Concatenation (PAC). Here the eyes image and description properties features, respectively acquired by CNN and LSTM, are fused by a carefully designed combination of addition, concatenation, and element-wise product. Through this fusion strategy, both information of distinct categories can be projected into a unified face feature space, and contribute to effective image retrieval. Our method has been experimented and validated on the publicly available CelebA face dataset.

Keywords: Image retrieval · Multimodal information fusion · Eyes hints · Deep neural network · Customized face query

1 Introduction

Image is a very convenient tool to store and demonstrate visual information. Hence, how to query and obtain wanted images from giant image datasets is

J. Wang—Supported by the National Natural Science Foundation of China (No. 61771340), the Tianjin Natural Science Foundation (No. 18JCYBJC15300), the Program for Innovative Research Team in University of Tianjin (No. TD13-5032), and the Tianjin Science and Technology Program (19PTZWHZ00020).

Y. Peng et al. (Eds.): PRCV 2020, LNCS 12306, pp. 151–163, 2020.
https://doi.org/10.1007/978-3-030-60639-8_13

an attractive research topic both academically and industrially [2,19,20]. But since the image belongs to a kind of unstructured information, image retrieval is never an easily conducted task. Furthermore, in the past few decades, accompanied by the rapid development of image capturing and collecting techniques, the number of available images is booming astonishingly. Hence how to effectively acquire wanted images from tremendous candidates is attracting more and more attentions.

Eyes are the most important facial visual features, and image retrieval based on eyes is a long history interesting research topic. It has wide practical value and application significance in the fields of public safety, blind date matching, beauty retouching, and so forth. However, due to their quite limited region area, the unique identity information that could be conveyed by eyes hints is seriously restricted, and hence, by this information source alone, accurate image retrieval is hardly achievable. In order to deal with this problem, in this paper, we designed a multimodal image retrieval method based on eyes hints combined with facial description properties. In our approach, both image and text information are unified as the query source, hence more effective identity information can be utilized to guide the searching procedure. On the other hand, the introduction of description properties can not only directly improve the accuracy of image retrieval, but also introduce more personality and customization. The users can perform customized retrieval by alternating the text facial descriptions, as shown in Fig. 1. Here it can be observed that the image query procedure is customized by the yellow and green background facial description properties respectively.

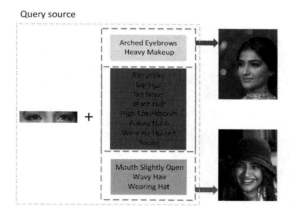

Fig. 1. A demonstration of the image retrieval based on both eyes hints and facial description properties. Here, both images of the rightmost column are the retrieval output, where the top one is based on the far left eyes image and the yellow and blue background facial description properties, while the bottom one is on the same eyes image and the blue and green background properties. (Color figure online)

In recent years, there have been some research findings involving the combination of vision and text information [11]. But most of them focus on visual question answering [4,9], cross-modal retrieval and image captioning [7,27]. The works that are directly related with the multimodal image retrieval specifically discussed in this paper are relatively rare.

In order to solve the mentioned problem, we propose a novel image and property information fusion mechanism. After obtaining the high-level semantic features of the eye images and the description properties respectively through the encoders, we do element wise addition and concatenation between different types of features, then product the results to achieve effective multimodal information fusion. This proposed method is named the Product of Addition and Concatenation (or PAC for short). We will give the detailed introduction in Sect. 3. The approach is experimented on the public available CelebA dataset, and satisfactory performance is achieved.

To summarize, the main contribution of this paper is threefold:

(1) We propose a multimodal image retrieval method based on eyes hints and facial description properties.
(2) A novel cross-modal feature fusion mechanism is introduced which could effective combine both vision and text information.
(3) The multimodal information fusion capacity of deep neural network is beneficially explored.

The rest parts of the paper are organized as follows: Sect. 2 reviews related work; The proposed PAC based image retrieval method will be presented in Sect. 3; Validation experiments are shown in Sect. 4; Sect. 5 concludes the paper.

2 Related Work

Vision Question Answering (VQA). VQA is a typical application that combines both image and text information. Its goal is to automatically generate the natural language answers according to the image and natural language question input. There are generally two ways to combine features in VQA, one of which is the direct combinations, such as, concatenation, element wise multiplication, and element wise addition [14]. Zhou et al. [28] introduce to use the bag-of-words to express the questions, the GoogleNet to extract visual features, and then direct connect the two features. Agrawal et al. [1] worked out to input the product of two feature vectors into a multi-layer perceptron of two hidden layers. Another way to combine features is using bilinear pooling or related schemes in a neural network framework [14]. Fukui et al. [8] proposed to the Multimodal Compact Bilinear (MCB) pooling to combine image features and text features. Due to the high computational cost of MCB, a multimodal low-rank bilinear pooling strategy (MLB) is worked out, where the Hadamard product and linear mapping are used to achieve approximate bilinear pooling [15].

Cross-Modal Retrieval. Cross-modal retrieval uses a certain modal sample to search for other modal samples with similar semantics. The traditional method generally obtains the mapping matrix from the paired symbiotic information of different modal sample pairs, and maps the features of different modalities into a common semantic vector space. Li et al. [17] introduced a cross-modal factor analysis method. Rasiwasia et al. [13,21] designed to apply the canonical correlation analysis (CCA) to the cross-modal retrieval between text and images. In recent years, many researches use deep learning to extract the effective representations of different modalities at the bottom layer, and establish semantic associations of different modalities at the top layer [6,16]. Wei et al. [26] worked out an end-to-end deep canonical correlation analysis method to retrieve text and images. Gu et al. [10] enrolled the Generative Adversarial Networks and Reinforcement Learning for cross-modal retrieval. In their work, the generation process is integrated into the cross-modal feature embedding. Here, not only the global features can be learned but also the local features. Wang et al. [24] believed that the previous methods rarely consider the interrelationship between image and text information during calculating the similarity, so they proposed the Cross-modal Adaptive Message Passing (CAMP) method.

Metric Learning. The goal of metric learning is to maximize the inter-class variations while minimize the intra-class variations, and it is quite common in pattern recognition applications. In neural network based approaches, LeCun et al. [12] designed the contrastive loss to increase inter-class variations. Schroff et al. [22] proposed the triplet loss. Then a large number of subsequent metric learning methods are worked out based on the triplet loss, such as the quadruplet loss [5].

3   Method

As mentioned in the introduction section, our goal is to achieve multimodal image retrieval based on both eyes hints and facial description properties. Here, how to effectively combine the query information coming from distinct categories is the most critical problem. Since both vision and text information are complex and comprehensive, we designed a neural network based information fusing and processing strategy. The main training pipeline is demonstrated in Fig. 2.

Specifically, first, the query eyes image x is encoded by a Light CNN [25]. Light CNN is a light-weight, noise-removable network proposed for face recognition. Here the query eyes image is transformed into 2D spatial feature vector $f_{\text{img}}(x) = \phi_x \in \mathbb{R}^{W \times H \times C}$, where W is the width, H is the height, and $C = 512$ is the number of feature channels. Note that we modify the size of the last fully connected layer of Light CNN from 256 to 512 to make the number of channels of image and text features the same. Second, we encode the facial description properties t with LSTM [28]. We define $f_{\text{text}}(t) = \phi_t \in \mathbb{R}^{L \times S \times d}$ to be the hidden state at the final time step, where L is the sequence length, S is the batch size, and $d = 512$ is the hidden layer size. Finally, both ϕ_x and ϕ_t are combined into

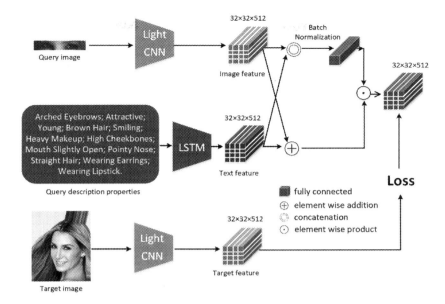

Fig. 2. The training pipeline of our multimodal image retrieval method based on eyes hints and facial description properties.

$\phi_{xt} = f_{\text{combine}}(\phi_x, \phi_t)$ with the proposed PAC method, which will be introduced in Sect. 3.1 in details.

On the other hand, during image retrieval, we calculate the similarity of the fused feature and that of candidate images by cosine distance, and then sort to get the face images that best meets the query conditions.

3.1 Feature Fusion by PAC

In order to effectively achieve multimodal information fusion, we explored a comprehensive combination strategy. Since the element wise addition and catenation are the most common direct way of fusion, our first glance is to contain both operation continuously. But because the dimensions of both information may be different, a co-dimensionalization approach is worked out. In detail, a convolution operation is enrolled to adjust the latitude of the concatenated feature matrix. At the same time, the sigmoid function is introduced to avoid taking too large a value. After the co-dimensionalization, the two types of combination methods are fused again by bitwise multiplication to obtain the final fusion feature. The whole mentioned processing is named the Product of Addition and Concatenation. Specifically,

$$\phi_{xt} = f_{\text{add}}(\phi_x, \phi_t) \odot f_{\text{concat}}(\phi_x, \phi_t), \tag{1}$$

where $\odot$ is element wise product, ϕ_x denotes image feature, ϕ_t is text feature, ϕ_{xt} is the fused feature.

$$f_{\text{add}}(\phi_x, \phi_t) = W_{\text{img}}\phi_x + W_{\text{text}}\phi_t, \tag{2}$$

where $W_{\text{img}}, W_{\text{text}}$ are learnable weights to balance both components.

$$f_{\text{concat}}(\phi_x, \phi_t) = \sigma(W_g \circ [\phi_x, \phi_t]), \tag{3}$$

where $[\phi_x, \phi_t]$ is matrix concatenate, and σ denotes the sigmoid function. We define $W_g\circ$ to be a series of convolution operations: first, the concatenated matrix is normalized in batches, then goes through the ReLU activation function, and finally its the number of channels is reduced from 1024 to 512 through the fully connected layer.

3.2   Loss Function

Clearly, the goal of our training is to make the fused features of faces in the same identity and state closer, while pulling apart the features of distinct images. For this task, we employ a triplet loss, which contains anchors, positive samples, and negative samples. When selecting the triples, we choose negative samples that are the same identity but different states as the anchor in order to enable that the network can distinguish the differences in face states. We define this triple as $T_{\text{state}}(f(x_i^a), f(x_i^p), f(x_i^n))$, where x_i^a is anchor, x_i^p is positive sample, and x_i^n is negative sample. $f(x)$ is embedding constrained to live on the d-dimensional hypersphere [22], where $d = 512$, i.e. $\|f(x)\|_2 = 1$. Similarly, we define $T_{\text{identity}}(f(x_j^a), f(x_j^p), f(x_j^n))$ to denote a triple whose negative sample and anchor have different identities but similar status. We then use the following triplet loss:

$$
\begin{aligned}
L = \sum_i^{N_{\text{state}}} & \left[\|f(x_i^a) - f(x_i^p)\|_2^2 - \|f(x_i^a) - f(x_i^n)\|_2^2 + \alpha \right]_+ \\
+ \sum_j^{N_{\text{identity}}} & \left[\|f(x_j^a) - f(x_j^p)\|_2^2 - \|f(x_j^a) - f(x_j^n)\|_2^2 + \alpha \right]_+,
\end{aligned} \tag{4}
$$

where α is a margin that is enforced between positive and negative pairs. N_{state} is the number of T_{state}, and N_{identity} is the number of T_{identity}. We believe that splitting loss into two parts, status and identity, is beneficial for the network to combine two types of data.

4   Experiments

In this section, the proposed multimodal image retrieval approach based on eyes hints and facial description properties will be experimented both quantitatively and qualitatively.

4.1   Experiment Configurations

Datasets. The experiments are conducted on a widely used face attribute dataset CelebA [18], which contains 202,599 images of 10,177 celebrity identities, and each of its image contains 40 attribute tags. Our experiments utilize 35 of them which do not related with eyes. As shown in Fig. 1, the query eyes images are in a single rectangle shape. We adopt the same subset division introduced in [18], where 40,000 images constitute the testing set, while all the left images form the training set.

Implementation Details. The experiments are realized by PyTorch code. The training is run for 210k iterations with a start learning rate 0.01.

Evaluation Metrics. As to the performance evaluation, we use the most commonly used evaluation metric R@K, which is the abbreviation for Recall at K and is defined as the proportion of correct matchings in top-k retrieved results [24]. Specifically, we adopt the R@1, R@5, R@10, R@50 and R@100 as our evaluation metrics.

4.2   Quantitative Results

In order to objectively evaluate the method performance, several classical information fusion approaches are enrolled as comparison. The MLB is a very classic multimodal fusion method in the field of VQA, which is based on the Hadamard product [15]. The MUTAN is a fusion method based on tensor decomposition applied in the field of VQA [3]. The TIRG method converts multimodal features into two parts called gating and the residual features, where the gating connection uses the input image features as the reference for the output composite features, and the residual connection represents the modifications in the feature space [23].

Table 1. Image retrieval performance on the CelebA dataset.

Methods	R@1	R@5	R@10	R@50	R@100
Image only (light CNN) [25]	3.3	7.6	10.5	20.0	25.7
Text only (LSTM) [28]	1.9	5.6	9.0	24.3	34.8
MLB [15]	3.1	9.1	14.3	32.2	44.7
MUTAN [3]	3.0	9.5	13.2	31.0	42.4
TIRG [23]	3.5	10.4	15.4	33.4	45.3
PAC (ours)	**4.1**	**11.4**	**16.3**	**34.4**	**45.8**

In order to fairly compare their performance with that of ours, during experiments, only the feature fusion part is distinct, while all other components are exactly the same as that of ours.

Table 1 presents the detailed performance. It can be observed that our method is evidently better than other methods on each evaluation indicator. Here, one thing to be mentioned is that, since facial description is not a kind of unique identity information, while the ground truth image used as the evaluation benchmark is unique, the overall performance may not be quite impressive. But this is caused by the nature of this task, rather than the method adopted.

Fig. 3. A few image retrieval outputs on the CelebA dataset. In the green dotted frame of the first column, the eyes image and the description are the query condition. The next five columns are the top five search results obtained by our method, where the ground truth image is surrounded by a solid green border. (Color figure online)

4.3   Qualitative Results

A few of retrieved images are shown in Fig. 3. It can be observed that generally the top five worked out candidates all conform to the query eyes and facial properties. Let's take the first row as an example. It can be easily found that all the five images are "Attractive, Big Nose, Black Hair, Heavy Makeup, High Cheekbones, Mouth Slightly Open, No Beard, Receding Hairline, Smiling, Wearing Earrings, Wearing Lipstick, Wearing Necklace, Young", while their eyes are similar with the query eyes to some extent.

Fig. 4. A few of "unsuccessful" retrieval output on the CelebA dataset. In the green dotted frame of the first column, the eyes image and the description are the query condition. The next five columns are the top five search results obtained according to our method. The images surrounded by solid green border in the rightmost column are the corresponding ground truth images. (Color figure online)

Figure 4 shows some "unsuccessful" retrievals, which means the ground truth image is not in the top five worked out images. It can be observed that, even in those "unsuccessful" retrievals, the obtained images are still compatible with the query eyes and description properties.

On the other hand, Fig. 5 shows some retrieval output according to the same eyes but distinct description properties. It can be seen that the text query condition can directly influence the retrieval output. Hence, it can be claimed that customized image retrieval is achievable by our method.

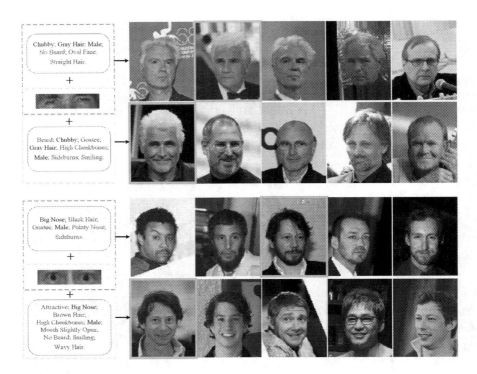

Fig. 5. Image retrieval outputs with the same eye but distinct description properties as the input. The image of the eyes in the dotted frame in the first column is the query image, and the description is the query text. Note that attribute words marked in red are unique. The next five columns are the top five search results obtained according to our method. The images surrounded by solid green border in the rightmost column are the corresponding ground truth images. (Color figure online)

In addition, we calculated the R@1 of each description properties as well, see Fig. 6. Among them, the average is 0.6874. It can be seen that "Male", "No Beard", "Mouth Slightly Open" have the highest recall rate in our model, while "Wearing Necktie", "Blurry", "Bald" have relatively lower recall rate. This phenomenon is identical with our intuition because those latter properties are relatively rare in the candidate dataset.

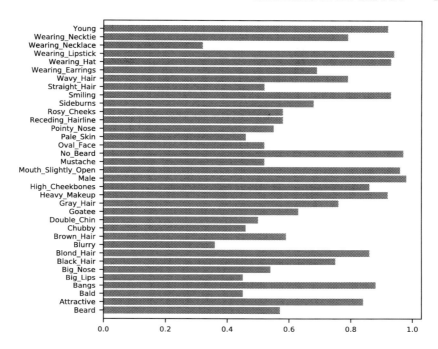

Fig. 6. The accuracy of identifying each properties.

5 Conclusion

In this paper, we propose to use the description properties as supplementary information for eye images in image retrieval tasks. And a novel multimodal information fusion method is worked out for the mission. The effectiveness of the proposed method is verified on a public available dataset, the CelebA. Generally the performance is identical with our expectation. In addition, personalized and customized image retrieval is achievable by the proposed approach. In the near future, we would like to try to extend this method to more general image retrieval problems.

References

1. Antol, S., et al.: VQA: visual question answering. In: International Conference on Computer Vision (2015)
2. Arandjelović, R., Gronat, P., Torii, A., Pajdla, T., Sivic, J.: NetVLAD: CNN architecture for weakly supervised place recognition. In: IEEE Conference on Computer Vision and Pattern Recognition (2016)
3. Ben-Younes, H., Cadene, R., Cord, M., Thome, N.: MUTAN: multimodal tucker fusion for visual question answering. In: IEEE International Conference on Computer Vision (2017)

4. Chen, L., Yan, X., Xiao, J., Zhang, H., Pu, S., Zhuang, Y.: Counterfactual samples synthesizing for robust visual question answering. In: IEEE Conference on Computer Vision and Pattern Recognition (2020)
5. Chen, W., Chen, X., Zhang, J., Huang, K.: Beyond triplet loss: a deep quadruplet network for person re-identification. In: IEEE International Conference on Computer Vision and Pattern Recognition (2017)
6. Eisenschtat, A., Wolf, L.: Linking image and text with 2-way nets. In: IEEE Conference on Computer Vision and Pattern Recognition (2017)
7. Feng, Y., Ma, L., Liu, W., Luo, J.: Unsupervised image captioning. In: IEEE Conference on Computer Vision and Pattern Recognition (2019)
8. Fukui, A., Park, D.H., Yang, D., Rohrbach, A., Darrell, T., Rohrbach, M.: Multimodal compact bilinear pooling for visual question answering and visual grounding. arXiv preprint arXiv:1606.01847 (2016)
9. Gao, D., Li, K., Wang, R., Shan, S., Chen, X.: Multi-modal graph neural network for joint reasoning on vision and scene text. In: IEEE Conference on Computer Vision and Pattern Recognition (2020)
10. Gu, J., Cai, J., Joty, S., Niu, L., Wang, G.: Look, imagine and match: improving textual-visual cross-modal retrieval with generative models. In: IEEE Conference on Computer Vision and Pattern Recognition (2018)
11. Gu, J., Joty, S., Cai, J., Zhao, H., Yang, X., Wang, G.: Unpaired image captioning via scene graph alignments. In: IEEE International Conference on Computer Vision (2019)
12. Hadsell, R., Chopra, S., Lecun, Y.: Dimensionality reduction by learning an invariant mapping. In: IEEE Computer Society Conference on Computer Vision and Pattern Recognition (2006)
13. Hotelling, H.: Relations between two sets of variates (1992)
14. Kafle, K., Kanan, C.: Visual question answering: datasets, algorithms, and future challenges. Comput. Vis. Image Underst. **163**, 3–20 (2017)
15. Kim, J.H., Kim, J., Ha, J.W., Zhang, B.T.: TrimZero: a torch recurrent module for efficient natural language processing. In: Proceedings of KIIS Spring Conference (2016)
16. Kiros, R., Salakhutdinov, R., Zemel, R.: Unifying visual-semantic embeddings with multimodal neural language models. In: International Conference on Machine Learning (2014)
17. Li, D., Dimitrova, N., Li, M., Sethi, I.K.: Multimedia content processing through cross-modal association. In: ACM International Conference on Multimedia (2003)
18. Liu, Z., Luo, P., Wang, X., Tang, X.: Deep learning face attributes in the wild. In: International Conference on Computer Vision (2015)
19. Liu, Z., Ping, L., Shi, Q., Wang, X., Tang, X.: DeepFashion: powering robust clothes recognition and retrieval with rich annotations. In: IEEE Conference on Computer Vision and Pattern Recognition (2016)
20. Ng, Y.H., Yang, F., Davis, L.S.: Exploiting local features from deep networks for image retrieval. In: IEEE Conference on Computer Vision and Pattern Recognition (2015)
21. Rasiwasia, N., et al.: A new approach to cross-modal multimedia retrieval. In: ACM International Conference on Multimedia (2010)
22. Schroff, F., Kalenichenko, D., Philbin, J.: FaceNet: a unified embedding for face recognition and clustering. In: IEEE Conference on Computer Vision and Pattern Recognition (2015)

23. Vo, N., et al.: Composing text and image for image retrieval-an empirical odyssey. In: IEEE International Conference on Computer Vision and Pattern Recognition (2019)
24. Wang, Z., Liu, X., Li, H., Sheng, L., Yan, J., Wang, X., Shao, J.: CAMP: cross-modal adaptive message passing for text-image retrieval. In: International Conference on Computer Vision (2019)
25. Wu, X., He, R., Sun, Z., Tan, T.: A light CNN for deep face representation with noisy labels. IEEE Trans. Inf. Forensics Secur. **13**, 2884–2896 (2018)
26. Yan, F., Mikolajczyk, K.: Deep correlation for matching images and text. In: IEEE Conference on Computer Vision and Pattern Recognition (2015)
27. Yang, X., Tang, K., Zhang, H., Cai, J.: Auto-encoding scene graphs for image captioning. In: IEEE Conference on Computer Vision and Pattern Recognition (2019)
28. Zhou, B., Tian, Y., Sukhbaatar, S., Szlam, A., Fergus, R.: Simple baseline for visual question answering. Comput. Sci. (2015)

Depth-Adaptive Discriminant Projection with Optimal Transport

Peng Wan and Daoqiang Zhang[✉]

College of Computer Science and Technology,
MIIT Key Laboratory of Pattern Analysis and Machine Intelligence,
Nanjing University of Aeronautics and Astronautics, Nanjing, China
dqzhang@nuaa.edu.cn

Abstract. Discriminant projection is a key technique for dimensionality reduction, especially in many classification tasks of high-dimensional, small-sample datasets. In this paper, we propose a Depth-Adaptive Discriminant Projection (DADP) method to improve the discriminability of the projection subspace. First, we propose a novel single-layer discriminant projection by adjusting projected points with nonlinear transformation. Second, the single-layer structure is extended to multiple-layer ones according to the proposed Adaptive Depth Determination Criterion (ADDC), naturally constructing a deep projection model with adaptive depth, which makes DADP allow for the avoidance of over-fitting problem for small-scale datasets. Furthermore, the regularized optimal transport (OT) is introduced to learn the optimal weights for pairwise data, which balances the local and global information incorporated in scatter estimation. Experimental results on several high-dimensional, small-scale datasets show the effectiveness of our algorithm both in terms of visualization and classification prediction.

Keywords: Dimensionlity reduction · Optimal transport · Adaptive depth

1 Introduction

Dimensionality reduction (DR) as a crucial technique for analyzing real-world data, aims at learning a more meaningful and compact low-dimensional representation of high-dimensional data. One major family of DR is discriminant projection, the goal of which is to learn a subspace that preserves class separability for the tasks such as clustering and classification. When label information is available, various supervised discriminant projection techniques [3,4,7,10] have been explored to learn a low-dimensional subspace which both preserve intrinsic structure and discriminative information of input data. The main characteristic of discriminant subspace is that samples from different classes are as separate

---

P. Wan—Currently working toward the Doctor degree in the College of Computer Science and Technology, Nanjing University of Aeronautics and Astronautics.

© Springer Nature Switzerland AG 2020
Y. Peng et al. (Eds.): PRCV 2020, LNCS 12306, pp. 164–174, 2020.
https://doi.org/10.1007/978-3-030-60639-8_14

as possible while samples from the same class are close to each other. Linear Discriminant Analysis (LDA), a classic linear dimensionality reduction method, seeks to find such projection subspace by maximizing the ratio of inter-class and intra-class scatter. However, like most linear models, LDA is hard to uncover the structure of nonlinear data that are very universal in realistic applications. Thus, a large number of nonlinear variants have been developed for addressing this problem.

With the assumption that real-world data could be linearly separable in some space with much higher dimensionality, kernel based techniques, such as Kernel Fisher Discriminant Analysis (KFDA) [5,9] implicitly project observed data into a high-dimensional feature space via a nonlinear mapping determined by kernel function. However, choosing an appropriate kernel for a specific dataset is challenging due to the lack of prior knowledge, which limits the application of kernel based techniques. In contrast, deep architecture based methods directly learn an explicit nonlinear mapping via deep neural network instead of manually selecting a kernel function. GerDA [6] attempts to learn a linearly separable latent representation by fine-tuning a pre-trained restricted Boltzmann machines on a linearly discriminant criterion. For simplicity, DeepLDA [2,8] directly replaces cross entropy loss with Fisher discriminant loss and optimizes several smallest eigenvalues for increasing discriminative variance in all projection directions. However, an optimal nonlinear subspace cannot be learned by the network with an inappropriate architecture. Apart from that, except for the last layer corresponding to a linearly separable subspace, each of the former layers lacks interpretability from the perspective of discriminability. Other problems like the need of large samples to avoid over-fitting and the extreme sensitivity to hyperparameter setting, also severely limit the application of such kinds of methods in real-world problems [11].

Motivated by the limitations of existing methods, we propose a general discriminant subspace learning algorithm called Depth-adaptive Discriminant Projection (DADP), which shares a similar architecture of DNN. Compared with previous works based on DNN, DADP endows the discriminability to each single-layer nonlinear mapping. Moreover, to enhance the discriminative capacity layer-by-layer, the Adaptive Depth Determination Criterion (ADDC) is proposed to determine the depth of our deep architecture. The key idea is to seek a projection space which has a better discriminative performance than that of previous layer. The layer-by-layer projection process would not terminate until there is no more a subspace with better separability. Therefore, the proposed DADP method can adaptively determine the depth for a given dataset.

The main contributions of this paper are summarized in the following three points,

1) Adaptive depth: With the help of ADDC, DADP can adaptively determine the depth for a given real-world datasets. Meanwhile, DADP can avoid the problem of over-fitting for small sample size cases.
2) Optimal projection matrix: DADP selects optimal weighting plans for constructing the inter-class and intra-class scatter matrix respectively via

smoothed optimal transport. Our approach adaptively balances the local and global data correlation incorporated in scatter estimation, which further improves the discriminability of subspace for given datasets.
3) Unrestricted projection dimension: In contrast to LDA, projection dimension of DADP is not bounded by the number of classes. Each layer mapping can freely project data into any lower-dimensional subspace.

2 Preliminary

2.1 Notations

Let $\langle A, B \rangle := \text{tr}\left(A^{\mathrm{T}}B\right)$ denote the Frobenius dot-product of two matrix. Besides, we denote the simplex as $\sum_d := \left\{ x \in \mathbb{R}_+^d : x^T \mathbf{1}_d = 1 \right\}$, where $\mathbf{1}_d$ is the d dimensional vector of ones. For two probability vector r and c in the simplex $\sum_d$, we write $U(r, c)$ be the transport polytope of r and c, namely the polyhedral set of $d \times d$ matrices, which satisfies $U(r, c) := \left\{ \mathrm{T} \in R_+^{d \times d} \,\middle|\, \mathrm{T}\mathbf{1}_d = r, \mathrm{T}^{\mathrm{T}}\mathbf{1}_d = c \right\}$.

2.2 Regularized Optimal Transport

Choosing a suitable metric to compare probability distributions is crucial to many machine learning tasks. Among traditional divergences, the distance defined by optimal transport is capable of capturing geometrical information when the probability space is a metric space. The distance, known as Wasserstein distance or Earth Mover's distance, is solved by linear programming with complexity at least in $O\left(d^3 log\left(d\right)\right)$. To facilitate the applicability of OT in large-scale data, an entropic regularization is introduced to obtain a smoothed version of OT. More importantly, regularized OT can be solved with Sinkhorn's matrix scaling algorithm, which can be computed several orders of magnitude faster [1].

Definition 2. *(Regularized Optimal Transport) Given a $d \times d$ cost matrix M, the cost of mapping r to c using a transport matrix (or a joint probability) T can be quantified as $\langle T, M \rangle$, the regularized optimal transport is defined as,*

$$d_M^\lambda (r, c) := \left\langle T^\lambda, M \right\rangle \tag{1}$$

$$T^\lambda = \underset{T \in U(r,c)}{\arg \min} \langle T, M \rangle - \frac{1}{\lambda} H\left(T\right) \tag{2}$$

where $H\left(T\right) = -\sum_{i=1}^d \sum_{j=1}^d T_{ij} \log T_{ij}$ is the entropy of P, and $\lambda > 0$ is entropic regularization coefficient.

It is noteworthy that the optimal transport T^* is a sparse $d \times d$ matrix with at most $2d - 1$ non-zero elements [1]. In our work, sparsity of T means those sample points merely correlate with their nearest neighbors and global information of projected data has not been sufficiently considered. Hence, smooth regularizers (entropic regularization e.g.) are necessary to take both local and global interactions into account.

3   The Proposed Method

3.1   Problem Statements

In many real-world applications, the volume of available data is ordinarily limited while the dimensionality is very high. Hence, we consider the problem of learning a discriminative subspace for small-scale high-dimensional data. Given a dataset $x_1, x_2, x_3, \cdots x_m = X \in \mathbb{R}^{d \times m}$, which contains m samples from C different classes $c = \{1, 2, \cdots C\}$. We define set $\zeta = \{(c_1, c_2) \,|\, c_1 \neq c_2\}$, denoting the set of all pairs of different categories. The goal is to find a target projected subspace which preserves class discriminability.

3.2   Single-Layer Discriminant Projection

1) *Characterize the inter-class Scatter.* Given that the optimal linear projection matrix is $W \in R^{d \times l}, l \leq d$. Regarding the determination of an optimal projection dimensionality l, we will have a detail discussionality in later section. The distance between pairs of data from different classes is $\left\| W^{\mathrm{T}}x_i - W^{\mathrm{T}}x'_j \right\|$. By tuning the regularization term λ, we attempt to learn the appropriate weights for all pairs from ζ. The weighted sum of the squared distances of point pairs in ζ can be written as:

$$
\begin{aligned}
d_b &= \sum_{i,j \in \zeta} tr \mathrm{T}_{i,j} \left(W^{\mathrm{T}}x_i - W^{\mathrm{T}}x'_j \right) \left(W^{\mathrm{T}}x_i - W^{\mathrm{T}}x'_j \right)^T \\
&= tr W^{\mathrm{T}} \sum_i \sum_j \mathrm{T}_{i,j} \left(x_i - x'_j \right) \left(x_i - x'_j \right)^{\mathrm{T}} W \\
&= tr W^{\mathrm{T}} \mathrm{S}_b W
\end{aligned}
\tag{3}
$$

where $\mathrm{S}_b = \sum\limits_{i,j \in \zeta} \mathrm{T}_{i,j} \left(x_i - x'_j \right) \left(x_i - x'_j \right)^{\mathrm{T}}$ is the inter-class scatter matrix, $\mathrm{T}_{i,j}$ is the transportation weight that is obtained by optimizing a regularized optimal transport problem Eq. (2). The cost matrix M is the pairwise squared Euclidean distance $M_{i,j} = \|x_i - x'_j\| \in R^{n \times m}$, where $x_i \in c_1, x'_j \in c_2$, n and m are the number of samples in class c_1 and c_2 respectively. Note that, x_i and x'_j are feature vectors passed from the last layer. Here, according to Eq. (2), when cost matrix M is fixed, $\mathrm{T}_{i,j} (\lambda_1)$ can be viewed as a function of λ_1. Note that, when λ_1 is large enough, the impact of entropic regularization is so small that the problem boils down to the origin optimal transport. Even in special cases where two classes have equal number of points, the problem has become an optimal matching problem in which each point only is linked to one another. Hence, it is crucial to set a appropriate λ_1 for selecting the effective data pairs.

2) *Characterize the intra-class Scatter.* Similarly, we have weighted sum of squared distances of pairwise points from the same class $c = \{1, 2, \cdots, C\}$,

$$
d_w = tr \left(W^T S_w W \right)
\tag{4}
$$

where $S_w = \sum_{i,j \in c} \mathrm{T}'_{i,j} (\mathrm{x}_i - \mathrm{x}_j)(\mathrm{x}_i - \mathrm{x}_j)^{\mathrm{T}}$ is the intra-class scatter matrix.

$\mathrm{T}'_{i,j}(\lambda_2)$ is also a function of λ_2. Similar to the discussion above, when λ_2 is small enough, regularization term plays the dominant role in the optimization problem. When $\lambda_2 \to 0$, the transport matrix $T := rc^T$ which indicates two distributions are independent to each other and all coupling are used to construct the intra-class scatter. Conversely, when $\lambda_2 \to \infty$, optimal transport cost from an empircial distribution to itself is 0. From the point of the correlation of points, one point only interacts with itself, which is incapable of capturing the global structure of data from the same class. Therefore it is necessary to select a relatively smaller value for λ_2 ($\lambda_2 \ll \lambda_1$) to obtain a smoother transport matrix T'.

Optimal Projection Matrix Selection. We cast optimal projection matrix selection as an optimization problem with the objective of maximizing the trace of inter-class scatter of projected data while minimizing the trace of intra-class scatter of projected data. The objective function is formulated as,

$$J(W) = \max \frac{tr(W^{\mathrm{T}} S_b W)}{tr(W^{\mathrm{T}} S_w W)} \tag{5}$$
$$s.t. W^{\mathrm{T}} W = \mathrm{I}$$

where $W \in R^{d \times l}$, I is $d \times d$ identity matrix. Note that the formulation of $J(W)$ is in form of Raleigh quotients, the solution of optimal W^* can be obtained by solving the general eigenvalue problem, $S_b W = \alpha S_w W$ where $W = (e_1, e_2, \cdots e_n)$, e_i is eigenvector of $S_w^{-1} S_b$ and the corresponding eigenvalue is α_i. To avoid the matrix singularity problem, we adjust the intra-class scatter matrix by adding a multiple of identity matrix I, as follows, $(S_w + \gamma I)^{-1} S_b e = \alpha e$.

Nonlinear Activation. Inspired by the single-layer nonlinear transformation of deep neural network, we adopt similar nonlinear activation approach via $sigmoid(\cdot)$ function. Besides, we adjust linearly projected data by adding an extra bias vector b before nonlinear activation. To obtain a better nonlinear projection subspace, we build the objective function as follows,

$$\min_b \sum_{i=1}^{C} \sum_{y_j \in C_i} \|\delta(y_j + b) - m_i\|^2 \tag{6}$$

where C is the number of total classes, and $\delta(y) = (1 + \exp(y))^{-1}$, $m_i = \frac{1}{K_i} \sum \delta(y_j + b)$ is the mean vector of i-th class. As can be seen from Eq. (6), we hope that the distances of point pairs from the same class are as small as possible after nonlinear mapping. In this paper, we find the bias vector using gradient descent. In addition, we pre-processing the input data or latent representation by whitening before nonlinear mapping at each layer.

3.3   Multiple-Layer Discriminant Projection

It is well-known that powerful data representation learning ability of deep neural network stems from deep architecture, which motivates us to extend single-layer discriminant projection to multiple-layer ones for boosting the discriminability of embedding subspace. Although deep architecture is effective in realistic tasks, it is still intractable to decide how many layers of network are suitable for a given dataset. Hence, we propose a criterion called Adaptive Depth Determination Criterion (ADDC) to adaptively determine the depth of network and select the dimensionality of projection corresponding to each layer.

Algorithm 1. Framework of depth-adaptive discriminant projection.

Input:

   Training set $S = \{(x_1, y_1), (x_2, y_2) \ldots (x_m, y_m)\}$,
   $\lambda_1 \geq 0, \lambda_2 \geq 0, \gamma \geq 0, 0 < \varepsilon < 1$;

Output:

   Adaptive deep discriminant projection model $f(x)$;
 1: Initialize $p_1 = p_0 - 1$, $l = 1$, whitening the training set S, calculate S_w, S_b respectively according to Eq. (3), Eq. (4);
 2: **while** 1 **do**
 3:    **if** $l = 1$ **then**
 4:       calculate W, b according to Eq. (5), Eq. (6).
 5:       **while** $acc(p_1) < \frac{1}{C}$ **do**
 6:          $p_1 = p_1 - 1$
 7:          update W, b
 8:       **end while**
 9:       $l = l + 1$
10:    **end if**
11:    update S_w, S_b, calculate p_l according to Eq. (8).
12:    update W, b
13:    **if** $acc(p_l) > acc(p_{l-1})$ **then**
14:       $l = l + 1$
15:    **else**
16:       **break**
17:    **end if**
18: **end while**

Adaptive Depth Determination Criterion. ADDC determines the projection dimension sequence by two parts, including *initial projection dimension determination* and *subsequent projection dimensions determination*. Here, the quality of the projection subspace is judged by the classification accuracy of a k nearest neighbor algorithm (KNN).

1) *Initial Projection Dimension Determination.* Motivated by the idea of weak classifier in the ensemble learning, we propose a general criterion for determining the initial projection dimension p_1 that raw input data are projected

on, the p_1-dimensional subspace should meet the following conditions: *classification accuracy exceeds random guesses* $\frac{1}{C}$. Simultaneously, to increase the depth of whole network as possible, the first projection dimension p_1 should be as large as possible. Overall, the criterion can be formualted as follows,

$$\max p_1$$
$$s.t. p_1 < p_0 \qquad (7)$$
$$acc(p_1) \geq \frac{1}{C}$$

$acc(p_1)$ is the prediction accuracy of KNN. p_0 is the origin dimension of input data. It is possible that p_1 is very small if we directly search for a lower dimension p_1 that maximizes classification accuracy. Hence, we only constrain the accuracy greater than $\frac{1}{C}$ (accuracy of random guesses).

2) *Subsequent Projection Dimension Determination.* To achieve a depth-adaptive network structure, we determine the subsequent dimensions based on two conditions: (i) the preserve of projection directions with major discriminative variance. (ii) the improvement of the classification accuracy. For the former condition, we preserve those projection directions which correspond to the larger eigenvalues because each eigenvalue α_i^{l-1} quantifies the amount of separation in direction of e_i of the $l-1$ layer Here, eigenvalues are in descending order and criterion can be written as,

$$p_l = \arg\min_j \frac{\sum_{i=1}^{j} \alpha_i^{l-1}}{\sum_{i=1}^{p_{l-1}} \alpha_i^{l-1}} \geq \varepsilon \qquad (8)$$

where ε is the threshold value set around 95%. For the latter condition, if p_i-dimensional subspace does not meet $acc(p_l) > acc(p_{l-1}), l \geq 2$, which means there is no improvement of the performance of the current network and the projection process terminates. Thus, the depth of network can be automatically determined, whcih make DADP applicable to different scaling of datasets.

4   Experiments

As discussed above, shallow models that are linear or nonlinear are hard to uncover essential structure of data with high dimensionality, while deep architecture based models require large volume of data. Hence, in this section, we present the evaluation of the subspace learning capacity of DADP on several small-scale high-dimensional datasets. Experiments are performed on 10 UCI dataset, face datasets, such as AR, ORL and FERET databases and Alzheimer's disease datasets that include 4 types. The performance of discriminative projection of DADP is compared with some classical discriminant analysis methods like LDA, LSDA, some locality preserving methods, such as LFDA, LMNN, and a deep neural network based method DeepLDA. The discriminability of the embedding is judged by the recognition rate of k nearest neighbors classifier with Euclidean distance. Note that we have split the datasets 80%-20% for training set and testing set respectively and repeated the random partition process 20 times. In order to obtain a reliable numercial results, we take the average of 20 independent experimental results as the final recording. Considering the potential impact brought by the dimensionality of projection subspace, we roughly initialized the dimensionality p to $C - 1$ except for DADP, where C is the number of classes.

Table 1. Average test classification accuracy on 10 UCI dataset. Each row represents a dataset. For each dataset, 20 test runs were performed and the average classification accuracy as well as standard deviation are presented. Besides, the highest accuracy for each dataset is in bold.

	LDA	LFDA	LMNN	LSDA	DeepLDA	DADP
Blood transfusion	.5862 ± .0078	.5831 ± .0084	.6033 ± .0088	.6012 ± .0073	.5823 ± .0142	**.7782 ± .0075**
Breast cancer	.9942 ± .0044	.9956 ± .0032	.9972 ± .0045	.9582 ± .0052	.8021 ± .0153	**.9973 ± .0031**
Dermatology	.9522 ± .0042	.9692 ± .0031	**.9842 ± .0048**	.9157 ± .0043	.9122 ± .0084	.9731 ± .0042
Diabetes	.5562 ± .0076	.7592 ± .0074	.7222 ± .0086	.7549 ± .0076	.7533 ± .0094	**.8572 ± .0045**
Ionosphere	.9322 ± .0042	.9476 ± .0056	.9227 ± .0048	.9623 ± .0052	**.9722 ± .0099**	.9426 ± .0025
Isolet	.8058 ± .0097	.9172 ± .0102	.9322 ± .0112	.7592 ± .0087	.9032 ± .0142	**.9422 ± .0074**
Ozone	.9546 ± .0074	.9624 ± .0078	.9726 ± .0048	.9672 ± .0075	**.9886 ± .0098**	.9782 ± .0076
Parkinsons	.5652 ± .0084	.6526 ± .0074	.8017 ± .0072	.6522 ± .0097	.5125 ± .0121	**.8533 ± .0074**
Sonar	.6012 ± .0099	.5912 ± .0083	.7723 ± .0065	.5443 ± .0045	.5712 ± .0154	**.753 ± .0084**
Zoo	.8232 ± .0042	.9642 ± .0051	.9776 ± .0052	.9752 ± .0044	.8122 ± .0087	**.9972 ± .0074**

Table 2. Average test classification accuracy on real-world datasets, including as 4 face datasets and 4 Alzheimer disease (AD) datasets. Experiment results are the average of 20 independent prediction accuracy.

	LDA	LFDA	LMNN	LSDA	DeepLDA	DADP
AR	.8212 ± .0086	.9156 ± .0075	**.9844 ± .0084**	.9786 ± .0059	.8556 ± .0075	.9836 ± .0072
Feret	.4032 ± .0074	.5156 ± .0085	.5032 ± .0102	.7175 ± .0174	.6223 ± .0082	**.7632 ± .0075**
Orl	.8734 ± .0062	.9250 ± .0514	.9652 ± .0047	.9525 ± .0322	.8172 ± .0122	**.9752 ± .0057**
MRI3	.4136 ± .0453	.3556 ± .0268	.5524 ± .1323	.3278 ± .0351	.4524 ± .1422	**.6012 ± .0072**
MRI4	.2732 ± .1023	.2950 ± .0471	.4286 ± .0105	.3200 ± .0949	.4323 ± .1182	**.4332 ± .0961**
PET3	.3130 ± .0962	.3501 ± .0982	.5952 ± .0101	.3389 ± .0315	.6172 ± .1022	**.6207 ± .0418**
PET4	.2621 ± .1049	.2905 ± .1055	**.5238 ± .0840**	.2850 ± .0818	.3448 ± .0977	.4538 ± .0844

4.1   Experiments on UCI Datasets

We carried out experiments on UCI benchmark datasets, including Breast cancer, Diabetes, Isolet, Zoo etc. The goal of our experiment is to validate the ability of our method to learn a separable subspace for high-dimensional data with small number of instances. In our experiments, except for DADP, we empirically fine-tuned the projection dimensionality p and the size of nearest neighborhood k of KNN classifier. Concretely, we initialized the number of neighbors to $k' = 7$ for LFDA, LMNN and the size was adjusted to maximize the classification accuracy by 10-fold cross-validation. In addition, for LMNN, we split 20% of training set as validation, other settings like the maximal iteration times, trade-off parameter between loss and regularizer have been fine-tuned according to the accuracy of each experiment. For LSDA, that needs a graph based embedding learning method, we utilized the algorithm LGE and the required affinity graph matrix and constraint graph matrix are constructed in the same way, where KNN with Euclidean distance was used and the neighborhood size was set to 5, weights between two nodes were calculated in HeatKernel mode. As for DeepLDA, we manually altered network architecture for various UCI datasets and we employed a relatively shadow structure here in order to avoid the overfitting problem. The number of the maximized eigenvalues was set $C - 1$, where C is the number of classes. Last, for our method DADP, we gave a rough setting for $\lambda_1 = 0.01$ and $\lambda_2 = 0.1$ respectively and fine-tuned the regularization term with a much smaller step to assign a more appropriate weights for all data pairs.

As can be observed from the Table 1, DADP is capable of achieving the highest accuracy compared with other methods or sightly lower than the best accuracy, which demonstrates the effectiveness of multiple-layer nonlinear combinations of features. For UCI datasets, DeepLDA could not perform well in most cases and even performed worst compared with other methods for some datasets. Concludely, deep neural network based discriminant analysis methods are not suitable for small-scale datassets. Note that, average accuracy of LMNN and LFDA demonstrates the significance of preserving local structure. Lastly, what we want to emphasize is that for Diabetes dataset, the prediction accuracy of most algorithms are below 80% while DADP achieved the accuracy 85.72%. Similarily, for Blood transfusion, DADP achieves the best classification accuracy at least 15% higher than other methods.

4.2   Experiments on Several Real-World Datasets

The objective of this experiment is to evaluate the capacity of DADP to handle real-world datasets, including three kinds of face datasets and Alzheimer's disease dataset.

Face Datasets. Face datasets include AR, FERET and Orl databases. AR database contains 3120 grayscale images from 120 persons, 26 images for each person. Images from the same class correspond to one person's distinct facial expressions pictured in different light condition. Similarly, FERET and Orl databases include 200×7, 40×10 grayscale images respectively. Additionally, for AR and FERET datasets, we pre-processed original images using PCA and projected images on 200-dimensional feature vectors, for Orl datasets, we cropped and resized the original images to 32×32 and flatten the pixel matrix to a

1024-dimensional feature vector. Besides, we normalized images by scaling pixel values to $[0 \sim 1]$.

As can be seen from Table 2, DADP is able to find a discriminant subspace which preserves separability of different persons. From the perspective of the size of different face datasets, DADP is adaptive to different scaling of datasets. Meanwhile, as expected, DeepLDA still can not perform well owing to the small sample size. Only few hundreds or thousands of pictures are far from sufficient for training deep model in end-to-end fashion.

Alzheimer's Disease Datasets. AD datasets contain 4 types of datasets on Alzheimer's disease, including MRI3, PET3, MRI4, PET4. Each dataset comprises 202 samples that can be categoried into Alzheimer disease patients (AD), Mild Cognitive Impairment (MCI) and NormalControl (NC). In addition, for MCI, patients can be further divided into MCI-C (those MCI patients convert to AD) and MCI-NC (those MCI patients do not convert to AD). Among them, MRI3 represents samples with 3 classes, MRI4 indicates persons with 4 classes, and similar case for PET datasets. Each row is a 93-dimensional features corresponding to different brain regions extracted from MRI images or PET images.

As can be seen from Table 2, DADP is more suitable for classifying AD samples with three classes comparing with other methods. Especially for MRI3, accuracy of DADP is significantly better than other methods. The prediction results of LFDA and LSDA are around the accuracy of random guess, which illustrates that both two kinds of methods are unable to discriminate AD datasets. For all types of AD datasets, The results of LMNN do not make a big difference and the accuracy of LMNN is slightly lower that of DADP for the first three AD diseases and the highest for PET4, which illustrates that nearest neighbors large margin assumption is appropriate for AD datasets to some extent.

5   Conclusion

This paper presents the Depth-Adaptive Discriminant Projection (DADP), a nonlinear discriminant subspace learning model. By virtue of the proposed Adaptive Depth Determination Criterion (ADDC), DADP can adaptively determine the projection dimensionality of each nonlinear layer and the number of non-linear transformation, which makes DADP more flexible than existing discriminant analysis models combined with Deep Neural Network (DNN). Based on the framework of regularized optimal transport (OT), each layer of DADP learns new weights for each pairwise data, leading to the selection of an optimal projection matrix. Unlike most methods in which weighting scheme is designed under some prior assumption (neighborhood, margin), DADP can flexibly balance between the global and local correlations at a category level by adjusting regularization term of Optimal Transport. Experimental results show that DADP is able to efficiently seek a discriminative subspace for small-scale and high-dimensional datasets.

Similar research, such as manifold learning, component analysis and correlation analysis, can be explored in later work. Future work will consider an objective function for bias vector that has a closed solution for the purpose of further reducing the computation cost and enhancing the adaptivity for small-scale dataset with more complex class distribution.

Acknowledge. This work was supported in part by the National Key Research and Development Program of China under Grant 2018YFC2001600 and Grant 2018YFC2001602, in part by the National Natural Science Foundation of China under Grant 61876082, Grant 61861130366 and Grant 61732006 in part by the Royal Society-Academy of Medical Sciences Newton Advanced Fellowship under Grant NAF\R1\180371.

References

1. Cuturi, M.: Sinkhorn distances: lightspeed computation of optimal transport. In: Advances in Neural Information Processing Systems (NIPS 2013), Lake Tahoe, Nevada, USA, pp. 2292–2300 (2013)
2. Dorfer, M., Kelz, R., Widmer, G.: Deep linear discriminant analysis. In: Proceedings of the International Conference on Learning Representations (ICLR 2015), pp. 1–13 (2015)
3. Du, Y., Lu, X., Zeng, W., Hu, C.: A novel fuzzy linear discriminant analysis for face recognition. Intell. Data Anal. **22**(3), 675–696 (2018)
4. Juuti, M., Corona, F., Karhunen, J.: Stochastic discriminant analysis for linear supervised dimension reduction. Neurocomputing **291**, 136–150 (2018)
5. Liu, Q., Lu, H., Ma, S.: Improving kernel fisher discriminant analysis for face recognition. IEEE Trans. Circuits Syst. Video Technol. **14**(1), 42–49 (2004)
6. Stuhlsatz, A., Lippel, J., Zielke, T.: Feature extraction with deep neural networks by a generalized discriminant analysis. IEEE Trans. Neural Netw. Learn. Syst. **23**(4), 596–608 (2012)
7. Wang, M., Huang, J., Liu, M., Zhang, D.: Functional connectivity network analysis with discriminative hub detection for brain disease identification. Assoc. Adv. Artif. Intell. (AAAI) **33**, 1198–1205 (2019)
8. Wu, L., Shen, C., van den Hengel, A.: Deep linear discriminant analysis on fisher networks: a hybrid architecture for person re-identification. Pattern Recogn. **65**, 238–250 (2017)
9. Yang, J., Frangi, A.F., Yang, J., Zhang, D., Jin, Z.: KPCA plus LDA: a complete kernel fisher discriminant framework for feature extraction and recognition. IEEE Trans. Pattern Anal. Mach. Intell. **27**(2), 230–244 (2005)
10. Yu, M., Shao, L., Zhen, X., He, X.: Local feature discriminant projection. IEEE Trans. Pattern Anal. Mach. Intell. **38**(9), 1908–1914 (2016)
11. Zhou, Z.H., Feng, J.: Deep forest: towards an alternative to deep neural networks. In: Proceedings of the 27th International Joint Conference on Artificial Intelligence (IJCAI 2017), Melbourne, Australia, 19–25 August 2017, pp. 3553–3559 (2017)

HOSENet: Higher-Order Semantic Enhancement for Few-Shot Object Detection

Lingli Zhang[1,2] , Ke-Jia Chen[1,2(✉)] , and Xiaomeng Zhou[3]

[1] School of Computer Science,
Nanjing University of Posts and Telecommunications, Nanjing, China
{1218043304,chenkj}@njupt.edu.cn
[2] Jiangsu Key Laboratory of Big Data Security and Intelligent Processing,
Nanjing University of Posts and Telecommunications, Nanjing, China
[3] School of Communication and Information Engineering,
Nanjing University of Posts and Telecommunications, Nanjing, China
B17060324@njupt.edu.cn

Abstract. Few-shot object detection is to detect objects of novel categories from only a few annotated examples and has recently attracted attention. Existing methods focus on designing new training strategies on a widely used object detection network, while ignoring the importance of network representation ability for this problem. This paper proposes a novel network model named HOSENet, which introduces a higher-order semantic enhancement module (e.g. second-order pooling) in the forward process of the network. The entire network is first trained with base class data and then fine-tuned by support images to detect objects of novel classes. Experiments on PASCAL VOC and MS COCO show that the detection precision of our method in novel classes is far superior to other competitive methods without compromising the detection precision in base classes.

Keywords: Few-shot object detection · Semantic enhancement · Second-order pooling

1 Introduction

Object detection is one of the fundamental sub-areas in computer vision, which aims to locate and classify the objects of interest in images. It has been widely applied in many vision scenes, such as face detections [1], vehicle detections [2] and pedestrian detections [3]. The difficulty of this field lies mainly in the different appearances, shape and posture of various objects as well as the illumination and occlusion when imaging them.

Although the accuracy of the object detection algorithm [4–6] has been greatly improved with the introduction of deep learning, most of these algorithms rely heavily on massive labeled image data, which requires not only huge efforts to annotate, but also long training time. Obviously, this learning method is not as effective as human cognitive behaviors, which are able to identify a new object from a very small number of

Y. Peng et al. (Eds.): PRCV 2020, LNCS 12306, pp. 175–186, 2020.
https://doi.org/10.1007/978-3-030-60639-8_15

samples. The computer vision systems also need this ability to learn from a few samples, when some object categories are inherently scarce, such as endangered animals or rare medical data. However, the deep learning model will encounter overfitting and poor generalization problem if the convolution network is directly trained with only a few data.

Few-shot object detection [7–10] is recently developed to learn from few shots of data in a given category to detect the objects falling into that category (Fig. 1). Most existing few-shot object detection methods use the idea of meta-learning [7, 9] in the network. They first learn generalizable meta-features from base classes and then reweight the features for novel classes detection by producing the feature map from a few support samples.

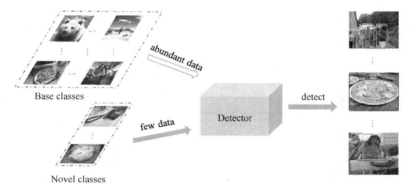

Base classes

Novel classes

Fig. 1. The illustration of few-shot object detection. The network is trained in a certain way by using sufficient base class data and few novel class data, and can detect both new classes and base classes.

Recently, a method based on fine-tuning is proposed [10]. After training the object detection network with the base classes data, the parameters for the last layer of the network are fine-tuned by using supporting images to detect the objects of unseen classes. Despite its simplicity, it achieves the state-of-art result.

The above methods mainly focus on the design of training methods but only use the basic object detection architecture, whose nonlinear modeling capability is far from enough for few-shot object detection. We know that humans can quickly identify new classes from a small number of samples due to their ability to extract image semantics. Similarly, for the problem of insufficient novel samples, it is particularly critical to enhance the network's semantic representation. Therefore, this paper introduces a second-order representation module in the forward process of convolution to learn the correlation between channels and then insert this module into the Faster R-CNN [4] framework to solve the problem of sample scarcity.

Our main contributions are as follows:

(1) We propose to introduce a semantic enhancement module into the convolution network, improving the network representation ability for few-shot object detection problems.

(2) We conduct two ablation experiments on the module position and configuration. The former analyzes the impact of the module on the detection results when it is inserted into the different layers of the convolution networks. The latter is to decide whether to use the first-order, second-order or hybrid-order pooling for semantic enhancement.

(3) Sufficient experiments verify that for few-shot object detection, enhancing the expressive capability of the network can greatly improve the detection precision especially when detecting novel classes, and our proposed method achieves the best results.

2 Related Work

General Object Detection. Currently, the mainstream object detection methods are based on deep learning, and are mainly divided into two categories: single-stage detectors and two-stage detectors. Single-stage detectors [5, 6, 15] do not generate region proposals. They generally treat all the locations on the image as potential regions, and then classify each region of interest as background or object and regress the bounding box of the objects. Two-stage detectors [4, 17] first generate a series of region proposals as samples through the RPN (region proposal network), then classify the samples and fine-tune the bounding boxes. The methods mentioned above require a large number of annotated images to train the network. If the samples of certain categories are naturally scared, or their annotations are difficult to obtain, direct training of the network with few samples will result in overfitting, thus deteriorating the model generalization and detection accuracy.

Few-shot Object Detection. Few-shot object detection is an emerging task that deals with the above situation. Given the supporting image S that contains a close-up of class C, the task is to detect the objects of class C in the query image Q and to mark them with bounding boxes. The pioneering methods in this field [7, 9] use the idea of meta-learning. Kang et al. [7] propose a model that first learns meta features from base classes, and then utilizes a few support examples to identify the meta features that are important and discriminative for detecting novel classes. The model can be adapted accordingly to transfer detection knowledge from the base classes to the novel ones. Yan et al. [9] present a method to achieve low-shot object detection and segmentation, which extends Faster/Mask R-CNN by proposing meta-learning over RoI (Region-of-Interest) features instead of a full image feature. Later, Fan et al. [18] propose an RPN attention module and a matching module. The RPN attention module improves the response to new classes to make the unseen objects become the region of interest. The matching module learns the relationship between the support images and the region of interest, effectively improving the detection accuracy. Recently, Wang et al. [10] propose a new network, which only fine-tunes the last layer of existing detectors on rare classes and achieves better detection. Following this work, our method also uses the Faster R-CNN object detection architecture, and the similar training method, but enhances the semantic representation of the feature extraction module.

Semantic Enhancement Modules. Researchers have discussed the modeling of first-order and higher-order representation in the convolutional network. In early methods [11, 12], the first-order information enhancement module is proposed for the image classification task. By modeling the first-order relationship (such as average pooling or max pooling) between the channels of the feature map, the weights of channels are calculated to reweight the original feature map. Subsequently, some researchers [13, 14] modeled the second-order relationship between channels to improve the nonlinear representation ability of the network, thereby further improving the classification accuracy. Inspired by the above work, this paper introduces the second-order and hybrid-order pooling module into the convolutional network. With an enhanced representation ability, the entire network can better detect the objects of new classes.

3   HOSENet

This section introduces the overall architecture of our model (as shown in Fig. 2), and the embedded HOSE module (Fig. 3) that enhances the semantic representation of the network. In the experiments, we try first-order pooling, second-order pooling and hybrid-order pooling which integrates both and finally choose the second-order pooling as the embedded module of the network.

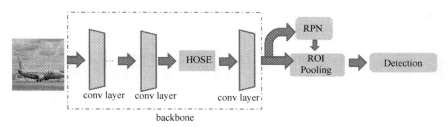

Fig. 2. Overall network architecture. A higher-order semantic enhancement (HOSE) module is inserted in the end of backbone to enhance the nonlinear representation of the network.

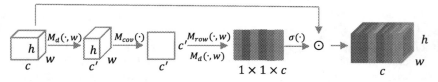

Fig. 3. The second-order semantic enhancement module, where σ denotes the sigmoid function, M_d is the operation of dimensionality reduction, and is the element-wise multiplication.

3.1   Network Overview

Our detection model is based on Faster R-CNN [4], which is a two-stage object detector, containing the backbone, the RPN (region proposal network), the RoI pooling and the detection module for classifying the object categories and predicting the bounding box coordinates. The backbone is responsible for extracting features from images, which are learned from base classes can also be transferred to feature extraction on new classes.

Due to the limited number of novel classes, we introduce higher-order semantic enhancement module into the backbone to improve nonlinear capability of convolution networks. The backbone is based on Resnet-101 as shown in Table 1.

Table 1. HOSENet with Resnet-101 architecture

	Output	Layer
conv1	112×112	$7 \times 7, 64$, Stride $= 2$
pool1	56×56	max pool, 3×3, Stride $= 2$
conv2_x		$\begin{bmatrix} 1 \times 1, 64 \\ 3 \times 3, 64 \\ 1 \times 1, 256 \end{bmatrix} \times 3$
conv3_x	28×28	$\begin{bmatrix} 1 \times 1, 128 \\ 3 \times 3, 128 \\ 1 \times 1, 512 \end{bmatrix} \times 4$
conv4_x	14×14	$\begin{bmatrix} 1 \times 1, 256 \\ 3 \times 3, 256 \\ 1 \times 1, 1024 \end{bmatrix} \times 22$ HOSE
conv5_x	7×7	$\begin{bmatrix} 1 \times 1, 512 \\ 3 \times 3, 512 \\ 1 \times 1, 2048 \end{bmatrix} \times 3$

3.2 Higher-Order Semantic Enhancement Module (HOSE)

In the image classification task, the use of higher-order correlation between the channels to reweight convolution feature maps helps to improve the representation power of the network, and the module can be easily inserted into any position between convolutions layers. This inspires us to apply this module into the few-shot object detection task.

Suppose the input feature map is $F \in R^{C \times H \times W}$. Firstly, the channel information of the feature map is aggregated by 1×1 convolution to generate a compressed feature map: $F' \in R^{C' \times H \times W}$, which can extract valid information and reduce the complexity of subsequent operations. Secondly, the correlation between each two channels is calculated to produce covariance matrix $M_{cov} \in R^{C' \times C'}$, where the element in the i-th row and the j-th column represents the correlation between the i-th channel and the j-th channel. Thirdly, the M_{cov} is transformed into a vector by row-wise convolution M_{row} and dimensionality reduction is performed. Afterwards, the sigmoid function is used as a nonlinear activation to obtain a vector $M_c \in R^{C \times 1 \times 1}$ where each element represents the importance of each channel. Finally, a dot product is performed between the vector and the original feature map. The above steps make full use of the connections between the channels of the feature map, reminding the network to pay more attention to certain channels when facing different images with different objects. The internal process of the module is shown in Fig. 3.

We also propose a hybrid-order semantic enhancement module that integrates first-order pooling with second-order pooling in parallel. The original feature map is reweighted by multiplying the weight coefficients M_h they learned. In our experiments, both the second-order and hybrid-order pooling modules are verified to enhance the nonlinearity representation of the convolutional network, so they are referred to as higher-order semantic enhancement modules in this paper.

In addition, the insertion position of the module is studied, that is, in the middle or the end of the backbone. Through the experiment, it is determined that the module is better inserted in the back end. We will discuss in detail in the experiment in Sect. 4.

3.3 Training Strategy for the Network

We follow the training strategy in [10], which outperforms the meta-learning methods by a large margin. This training strategy is divided into two steps.

The first step is to use base classes to train the network with the following loss function [4]:

$$L = L_{rpn} + L_{cls} + L_{loc} \tag{1}$$

where L_{rpn} is the loss for generating region proposals, L_{cls} is cross-entropy loss function for the box classification, and the L_{loc} is smooth L1 loss for bounding box regression.

The second step is to fine-tune the last layer of the network. We input a subset of base classes and novel classes to fine-tune the network, which is a collection containing a few shots of each class in the entire dataset. It allows the network to learn to detect new classes while not compromising the ability to detect base classes. When the network is fine-tuned, the classification loss function changes from the cross entropy loss function in the first step to the cosine similarity loss function [23], which makes the detection precision of the base classes decreases less, especially when the number of training samples is small.

4   Experiments

In order to evaluate the HOSE module, we conducted three experiments, including the impact of different insertion positions of the module on the result, the impact of different pooling configurations of the module (first-order, second-order and hybrid-order) on the result, and the comparison of our method with other state-of-art methods. All the comparative methods are implemented in the PyTorch framework.

4.1   Settings

We evaluate all few-shot object detection methods with two widely-used object detection benchmark datasets: PASCAL VOC [20, 24] and MS COCO [21]. The PASCAL VOC dataset contains 20 classes, from which 15 classes are randomly selected to build base classes and the remaining classes are novel classes. The COCO dataset contains 80 classes, and we divide them into 60 classes as base classes and 20 classes as novel classes.

Our implementation is based on Faster R-CNN [4]. The hyper-parameters follow the settings in [10]. A learning rate of 0.02 is used during base training and 0.001 during few-shot fine-tuning. The training data of the experiment is composed of all the annotations of base classes and a few annotations of novel classes. For the latter, we randomly select k annotated samples from each class, where k indicates how many shots you want.

For evaluation metrics, we adopt average precision over all IOU thresholds (AP), AP at IOU thresholds 0.5(AP50) and AP at IOU thresholds 0.75(AP75). To make the results more convincing, the experiment was repeated three times by random sampling, and the results are averaged for the final evaluation.

4.2   Ablation Studies

Impact of Different Insertion Positions. Consider that Resnet-101 has many residual blocks, we put the module in the middle and the end of the network respectively.

Figure 4 shows the detection results of the method inserting HOSE module in different positions, where the middle level represents the method with the HOSE module embedded after the second residual stage of Resnet-101, and the high level represents the method with the HOSE module embedded after the third residual stage of Resnet-101. The results show that, embedding HOSE module at the end of the Resnet-101 is better than embedding it in the middle. This indicates that when the representation obtained by the convolutional network is more sufficient, the calculation of higher-order correlation is more accurate and thus the semantic enhancement module is more effective. Moreover, if the module is embedded in the middle of the network, the subsequent convolutions may weaken the nonlinear representation ability of this module, thereby, reducing the detection precision. Therefore, it is placed at the end of Resnet-101 in our method.

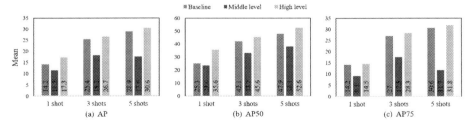

Fig. 4. Few-shot detection performance for novel classes on PASCAL VOC dataset. The method variants with the HOSE module in different positions of Resnet-101 are compared.

Impact of Different Module Configurations. We use average pooling as first-order pooling module follows [8], which is average pooling followed by convolution and non-linear activations to capture channel dependency, and then to rescale the channel for data recalibration. The hybrid-order pooling module is to multiply the weights learned in the first-order pooling and the weights learned in the second-order pooling module to obtain the weight values of each channel of the feature map, and then to reweight the original feature map.

Figure 5 presents the comparison results. When there is only one shot, the hybrid-order pooling module performs better than others. As the number of shots increases, the impact of the second-order pooling module is greater than that of the hybrid-order pooling module. Based on the overall results, our method uses second-order pooling module as the HOSE module for comparison with other related few-shot object detection methods.

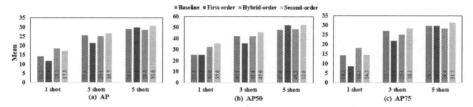

Fig. 5. Few-shot detection performance for novel classes on PASCAL VOC dataset. The method variants with the HOSE module in different configurations are compared.

In summary, the ablation experiments show that the use of the correlation between the channels at the end of the network can effectively enhance the semantic representation of the convolutional network, which can greatly improve the overall performance of few-shot object detection.

4.3   Comparison with Competing Networks

We compare our method with existing methods FSRW [7], MetaDet [19],Meta R-CNN [9], and FsDet [10]. The first three use the training strategy of meta-learning and the last one is based on fine-tuning as we do.

Results on PASCAL VOC. We report the average AP50 for the novel classes on PAS-CAL VOC in Table 2. Our method uses Resnet-101 with the semantic enhancement module as the backbone. Meta R-CNN and FsDet use Resnet-101 as their backbone. The first three models are based on meta-learning [7, 9, 19], and the rest are based on fine-tuning. Since the novel samples are randomly selected, the results of a single experiment are not stable. In order to make the experimental results more convincing, we repeat the experiment three times by random sampling, and used the average value for comparison. From the Table 2, the result shows that our model significantly outperforms the others, especially when the labeled images are extremely scarce. In the one-shot case, our method even has a 7.6% improvement on AP50 compare to FsDet, which once performed best in all methods. For more detailed comparisons, the average AP50 values for the base classes on PASCAL VOC of FSRW, FsDet and HOSENet are listed in Table 3. It shows that, like FsDet, our method also maintains a good performance in the detection of base classes.

Table 2. Few-shot detection performance (AP50) on the novel classes of the PASCAL VOC dataset.

Method/shots	1	3	5	10
FSRW [7]	14.8	26.7	33.9	47.2
MetaDet [19]	18.9	30.2	36.8	49.6
Meta R-CNN [9]	19.9	35.0	45.7	51.5
FsDet [10]	25.3	42.1	47.9	52.8
HOSENet (ours)	**32.9**	**47.4**	**52.6**	**54.9**

Table 3. Few-shot detection performance (AP50) on the base classes of the PASCAL VOC dataset.

Method/shots	1	3	5	10
FSRW [7]	66.4	64.8	63.4	63.6
FsDet [10]	77.6	77.3	77.4	77.5
HOSENet (ours)	**79.2**	**77.9**	**77.8**	**77.5**

Results on MS COCO. Similarly, we report the average AP, AP75 for the 20 novel classes with 10 shots and 30 shots on MS COCO in Table 4. The experiment is repeated three times by random sampling to output the average value. The comparison result is similar to that on the PASCAL VOC dataset, that is, after using the semantic enhancement module, the detection performance of novel classes is significantly improved and the performance of detecting bases classes is not compromised, especially when there are few samples. (see Table 5).

Table 4. Few-shot detection performance on AP, AP75 for novel classes on MS COCO dataset.

Method/shots	10		30	
	AP	AP75	AP	AP75
FSRW [7]	5.6	4.6	9.1	7.6
MetaDet [19]	7.1	6.1	11.3	8.1
Meta R-CNN [9]	8.7	6.6	11.1	10.8
FsDet [10]	9.1	8.8	12.1	12.0
HOSENet (ours)	**10.0**	**9.1**	**14.0**	**14.0**

Table 5. Few-shot detection performance on AP, AP75 for the base classes on MS COCO dataset.

Method/shots	1		3		10	
	AP	AP75	AP	AP75	AP	AP75
FsDet [10]	31.9	34.3	32.0	35.1	32.4	35.7
HOSENet (ours)	**31.9**	**34.4**	**32.2**	**34.9**	**32.4**	**35.3**

Finally, we compare some detection results of FsDet and HOSENet for 10 shots novel classes on MS COCO in Fig. 6. The first two lines are the detection results of FsDet, and the next two lines are of ours.

Fig. 6. Examples of 10-shots detection results of novel classes on MS COCO. The novel classes contain person, motorcycle, train, cat, sheep, chair, dining table, bicycle, airplane, boat, dog, cow, couch, tv, car, bus, bird, horse, bottle, and potted plant.

It can be seen that for the images in the first row, FsDet cannot detect the objects of new classes (i.e. motorcycle, airplane, train and birds, respectively), while HOSENet can detect them. For the images in the second row, both HOSENet and FsDet can detect the object, but the bounding box given by HOSENet is more accurate. However, there are some cases where objects of new classes are not detected in both methods. For example, for the images in the first and third columns of the second row, both methods fail to detect the new class of person.

5 Conclusion

In this work, we proposed an effective approach for few-shot object detection by exploiting the higher-order semantic enhancement module in the backbone of a Faster R-CNN network. Specifically, the second-order or hybrid-order pooling module is inserted into the convolutional network. The experiments show that with enhanced representation ability, the performance of detecting new class objects can be significantly improved in a fine-tuning-based training strategy. This implies that for the few-shot objection detection problem, the improvement of image understanding could be more important than the improvement of training methods, which will be studied in depth in future work.

References

1. Qin, H., Yan, J., Li, X., Hu, X.: Joint training of cascaded CNN for face detection. In: IEEE Conference on Computer Vision and Pattern Recognition, pp. 3456–3465 (2016)
2. Sun, Z., Bebis, G., Miller, R.H.: On-road vehicle detection: a review. IEEE Trans. Pattern Anal. Mach. Intell. **28**(5), 694–711 (2006)
3. Dollar, P., Wojek, C., Schiele, B., Perona, P.: Pedestrian detection: a benchmark. In: IEEE Conference on Computer Vision and Pattern Recognition, pp. 304–311 (2009)
4. Ren, S., He, K., Girshick, R., Sun, J.: Faster R-CNN: towards real-time object detection with region proposal networks. In: Advances in Neural Information Processing Systems, pp. 91–99 (2015)
5. Redmon, J., Farhadi, A.: Yolov3: An incremental improvement. In: arXiv preprint arXiv: 1804.02767 (2018)
6. Liu, W., et al.: SSD: single shot multibox detector. In: Leibe, B., Matas, J., Sebe, N., Welling, M. (eds.) ECCV 2016. LNCS, vol. 9905, pp. 21–37. Springer, Cham (2016). https://doi.org/10.1007/978-3-319-46448-0_2
7. Kang, B., Liu, Z., Wang, X., Yu, F., Feng, J., Darrell, T.: Few-shot object detection via feature reweighting. In: Proceedings of the IEEE International Conference on Computer Vision, pp. 8420–8429 (2019)
8. Schwartz, E., Karlinsky, L., Shtok, J., Harary, S., Marder, M., Pankanti, S., et al.: Repmet: Representative-based metric learning for classification and one-shot object detection. In: arXiv preprint arXiv:1806.04728 (2018)
9. Yan, X., Chen, Z., Xu, A., Wang, X., Liang, X., Lin, L.: Meta R-CNN: towards general solver for instance-level low-shot learning. In: Proceedings of the IEEE International Conference on Computer Vision, pp. 9576–9585 (2019)
10. Wang, Y., Thomas, E., Trevor, D., Joseph, E., Yu, F.: Frustratingly Simple Few-Shot Object Detection.In: arXiv preprint arXiv:2003.06957 (2020)
11. Hu, J., Shen, L., Sun, G.: Squeeze-and-excitation networks. In: Proceedings of the IEEE Conference on Computer Vision and Pattern Recognition, pp. 7132–7141 (2018)
12. Woo, S., Park, J., Lee, J.-Y., Kweon, I.S.: CBAM: convolutional block attention module. In: Ferrari, V., Hebert, M., Sminchisescu, C., Weiss, Y. (eds.) ECCV 2018. LNCS, vol. 11211, pp. 3–19. Springer, Cham (2018). https://doi.org/10.1007/978-3-030-01234-2_1

13. Wang, Q., Li, P., Zhang, L.: G2DeNet: global Gaussian distribution embedding network and its application to visual recognition. In: Proceedings of the IEEE Conference on Computer Vision and Pattern Recognition, pp. 2730–2739 (2017)

14. Gao, Z., Xie, J., Wang, Q., Li, P.: Global second-order pooling convolutional networks . In: Proceedings of the IEEE Conference on Computer Vision and Pattern Recognition, pp. 3024–3033 (2019)

15. Joseph, R., Ali, F.: Yolo9000: better, faster, stronger. In: Proceedings of the IEEE Conference on Computer Vision and Pattern Recognition, pp. 7263–7271 (2017)

16. He, K., Zhang, X., Ren, S., Sun, J.: Deep residual learning for image recognition. In: Proceedings of the IEEE Conference on Computer Vision and Pattern Recognition, pp. 770–778 (2016)

17. Girshick, R.: Fast R-CNN In: Proceedings of the IEEE International Conference on Computer Vision, pp. 1440–1448 (2015)

18. Fan, Q., Zhuo, W., Tang, C.K., Tai, Y.W.: Few-Shot Object Detection with Attention-RPN and Multi-Relation Detector. In: arXiv preprint arXiv:1908.01998 (2020)

19. Wang, Y.-X., Ramanan, D., Hebert, M.: Meta-learning to detect rare objects. In: Proceedings of the IEEE International Conference on Computer Vision, pp. 9925–9934 (2019)

20. Everingham, M., Van Gool, L., Williams, C.K., Winn, J., Zisserman, A.: The PASCAL Visual Object Classes Challenge 2007 (VOC2007) Results (2007)

21. Lin, T.-Y., et al.: Microsoft COCO: common objects in context. In: Fleet, D., Pajdla, T., Schiele, B., Tuytelaars, T. (eds.) ECCV 2014. LNCS, vol. 8693, pp. 740–755. Springer, Cham (2014). https://doi.org/10.1007/978-3-319-10602-1_48

22. Chen, W.-Y., Liu, Y.-C., Kira, Z., Wang, Y.-C.F., Huang, J.-B.: A closer look at few-shot classification. In: arXiv preprint arXiv:1904.04232 (2019)

23. Vinyals, O., Blundell, C., Lillicrap, T., Kavukcuoglu, K., Wierstra, D.: Matching networks for one shot learning. In: Neural Information Processing Systems, pp. 3630–3638 (2016)

24. Everingham, M., Ali Eslami, S.M., Van Gool, L., Williams, C.K.I., Winn, J., Zisserman, A.: The pascal visual object classes challenge: a retrospective. Int. J. Comput. Vis. **111**(1), 98–136 (2015)

Multi-model Network for Fine-Grained Cross-Media Retrieval

Jiemi Bai[1], Yazhou Yao[1], Qiong Wang[1(✉)], Yichao Zhou[1], Wankou Yang[2], and Fumin Shen[3]

[1] Nanjing University of Science and Technology, Nanjing 210094, China
wangq@njust.edu.cn
[2] Southeast University, Nanjing 210096, China
[3] University of Electronic Science and Technology of China, Chendu 611731, China

Abstract. With the development of Internet, the forms of web data are rapidly increasing. However, existing cross-media retrieval methods mainly focus on coarse-grained, which is far from being satisfied in practical application. In addition, the heterogeneity gap among different types of media tends to result in inconsistent data representation, so the measuring similarity is quite challenging. In this work, we propose a novel multi-modal network for fine-grained cross-media retrieval. Specifically, our model consists of two networks, including proprietary networks and the common network. The proprietary network is designed as a single feature extraction network for each media to extract unique features for obtaining precise media feature representation. The common network is designed to extract common features of four different types of media. Comprehensive experiments demonstrate the effectiveness of our proposed approach. The source code and models of this work have been made public available at: https://github.com/fgcmr/fgcmr.

Keywords: Fine-grained cross-media retrieval · Proprietary network

1 Introduction

With the development of the Internet, various types of media data have shown explosive growth on the web [1–12]. Images, texts, audios, and videos are becoming the main form of people to know the world and cross-media retrieval is becoming increasingly popular especially with the development of deep learning [13–19]. Different from the traditional retrieval works [20–24] which usually focus on the single media type, as shown in Fig. 1, the cross-media retrieval deals with various types of media. However, the existing cross-media retrieval task usually concentrates on coarse-grained, which is far from being satisfied in

This work was supported by the National Natural Science Foundation of China (No. 61976116, 61773117), Fundamental Research Funds for the Central Universities (No. 30920021135), and the Primary Research & Development Plan of Jiangsu Province - Industry Prospects and Common Key Technologies (No. BE2017157).

Y. Peng et al. (Eds.): PRCV 2020, LNCS 12306, pp. 187–199, 2020.
https://doi.org/10.1007/978-3-030-60639-8_16

188   J. Bai et al.

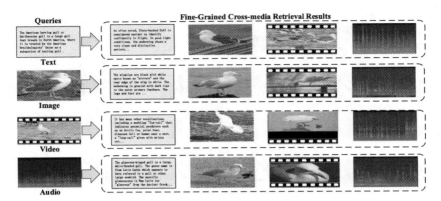

Fig. 1. Examples of fine-grained cross-media retrieval, where the audio data is visualized by spectrogram.

practical applications. Compared to coarse-grained cross-media retrieval, fine-grained cross-media retrieval has a wide range of research needs and application scenarios in both industry and academia.

Fine-grained cross-media retrieval is a challenging task due to the small inter-class and large intra-class appearance variance. In addition, the heterogeneity gap is another challenging problem which is far from being solved in cross-media retrieval [25–29]. To this end, a lot of efforts have been devoted to researching cross-media retrieval. For example, [47] proposed an important method to get a common semantic space by maximizing the correlation between two media data. Unfortunately, the local structure of data in each media and the structure matching between different media types are ignored. Another branch of works, deep neural networks (DNN) have established semantic association of different media types data through the advantage of feature extraction ability. [46,52] utilized a cross-media shared layer to map a common semantic space. Besides, [39] shared a uniform deep model FGCrossNet which learns four types (image, video, text, and audio) media without discriminative treatments. Sharing the layer and network reduced the heterogeneity gap to a certain extent, but these models ignore the local structure of data in different media types and don't concentrate on each media special feature.

In this paper, we proposed a simple yet effective approach to narrow the heterogeneity gap and extract the media feature precisely. Our work is motivated by the following observations: 1) Existing fine-grained cross-media retrieval methods can't fully consider the problems of large intra-class differences and small inter-class differences among fine-grained subcategories. 2) Current fine-grained cross-media method only adopts to share common network so that can't consider the special attributes of media. 3) For narrowing the heterogeneity gap, most methods make two different features closer from each dimension of the sample but don't consider the sample feature distribution of different media types comprehensively. Our proposed approach consists of two networks, including the proprietary network and the common network. Specifically, the proprietary network is designed for each media specific distinctive attributes to extract more representative features. The common network is designed for four different media types to obtain common feature representation to minimize the heterogeneity gap. Extensive experiments on public available data set demonstrate the effectiveness of our proposed approach.

2   Related Work

2.1   Cross-Media Retrieval

The task of this paper concerns the heterogeneity gap of the cross-media retrieval framework. At present, most research of the cross-media retrieval concerns two media types, the seldom research aims at more than two media types. The early works address the above problems by utilizing the graph regulation methods [31,51,55,58,64–67]. [58] proposed a joint graph regularized heterogeneous metric learning algorithm, which integrates firstly the structure of five media types into a joint graph regularization. With the development of computing ability, the complexity of neural network can be computed and some based on the DNN methods of cross-media retrieval can be solved. [31] proposed a method to regularize cross-media neural networks, which takes five input types.

2.2   Fine-Grained Recognition

Since our task aims at fine-grained aspect, which distinguish the categories with difficulty by ordinary method, fine-grained recognition can help us to extract the feature [50,63]. However, the current researches on the fine-grained aims merely at the visual recognition. For other media types, we used some methods of the visual recognition for reference. The early work is based on strongly supervised methods, which is utilizing bounding boxes or part annotations besides the image-level labels [31,35,56,57,59]. But these increased the cost of manual annotation. Researches on weakly supervised methods had solved the problem, which is only utilizing image-level labels [30,33,34,38,43,45,53,55,60,61]. These works obtained the remarkable results and it will be the trend of the fine-grained recognition.

2.3   Relations to Benchmark

Up to now, the benchmark [39] is the first and only work research on fine-grained cross-media retrieval. The author constructed a new benchmark for fine-grained cross-media retrieval, which consist of 200 fine-grained subcategories of the "Bird", and contains four media types, including image, text, video and audio. The benchmark deals four media data in the same format before they come in the common network simultaneously, but it doesn't consider the proprietary feature of media lead to the experimental effect of benchmark is not good enough. That motivated us to design a proprietary model for different types media, which

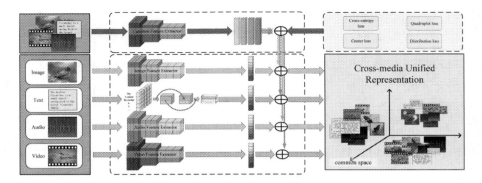

Fig. 2. An overview of our multi-model network for fine-grained cross-media retrieval.

extracts more precise proprietary features. In addition, our common network is a little similar to the benchmark network. Benchmark network introduced three loss functions to narrow heterogeneity gap and our common network introduces four loss functions to narrow heterogeneity gap.

3   The Proposed Approach

In this section, we present our method for fine-grained cross-media retrieval. As shown in Fig. 2, our framework consists of two groups of networks, which combine with each other to get the final feature representation for better performance.

3.1   Proprietary Network

Due to different types of media usually have inconsistent distribution and representation, we design different proprietary network for each media to extract special feature.

Image Proprietary Network: For image and video, we introduce bilinear network [43] which is widely used in fine-grained visual recognition as their proprietary network to extract features. For video data, we extract 25 frames from each video at regular intervals. The frames of video are similar to images so we choose the same proprietary network for them. The image proprietary network is composed of two CNNs. Two different streams represent different features obtained by CNN, and the features are operated through bilinear, then get the final feature representation through the pooling and full connection layers. The aim of bilinear is through the outer product we can acquire the relevances between two outputs and part-feature interactions to get the more accurate feature representation. We adopt a location in the image and the input image as l and i, respectively, and then the bilinear between them is $b(l, i)$. Its definition is as follows:

$$b(l, i) = E_a(l, i)^T E_b(l, i), \tag{1}$$

where E_a and E_b are the two functions that extract features (CNN a and CNN b). We calculate the value at each location of the image and then through the pooling function we converge them to form the image feature, which is defined as

$$P(i) = \sum_{l \in L} b(l, i). \tag{2}$$

L is the location area in the image. By the $P(i)$, we obtain the image feature and through the full connection layer to get the final feature representation.

Text Proprietary Network: In existing fine-grained cross-media retrieval, text data are usually converted to char embedding and processed with the convolutional neural network. Nevertheless, char embedding lacks rich semantic information and abandons the pre-training ecosystem of plug and play. In this work, we introduce Attention-Based Bidirectional Long Short-Term Memory Networks (Att-BLSTM) [62] as text proprietary network to capture the most important semantic information in a sentence. The text proprietary model is composed of four layers, the first layer receives the input sentences. The second layer transforms words into word vectors. The third layer is used to obtain deeper feature representation and extract text feature widely. The last layer multiplies generated the weight matrix and the deeper feature to get the sentence feature representation.

The input of the text proprietary network are the sentences. We denote each sentence as $T = [t_1, t_2, \cdots, t_n]$, where t_i is the word vector of the i-th word in the sentence and n is the length of the sentence. Through the pre-trained word embedding, each word t_i of the sentence is expressed as specific word vector e_i. So the input is transformed into $E = [e_1, e_2, \cdots, e_n]$. Then the sequence T enter the Long Short-Term Memory (LSTM) layer. In this text proprietary network, we choose a bidirectional LSTM to process the sequence which contains forward LSTM and backward LSTM. Then we combine with two outputs of LSTM on the basis of each element e_i to obtain the output vector. The output vector h_i can be defined as:

$$h_i = \overrightarrow{LSTM}(e_i) \oplus \overleftarrow{LSTM}(e_i). \tag{3}$$

We denote the output vectors set of LSTM layer as H and H can be represented as $H = [h_1, h_2, \ldots, h_n]$, in which n is the length of the sentence.

In virtue of some vectors have more important influence on sentences, so we introduce the attention layer to find these vectors and let them make a big difference. The main core of attention layer is attention neural network. It's widely used in natural language processing and image processing. And our aim is to obtain a weight matrix γ assigns large weight to important features. The weight of attention layer is obtained by:

$$\gamma = \text{softmax}(w^T \tanh(H)) \tag{4}$$

$$f = H\gamma^T, \tag{5}$$

where w is a trained weight matrix and the value of attention score after softmax is represented by γ. Finally H multiplies with weight to get the sentence vector representation f.

Audio Proprietary Network: For audio, we choose VGG-16 [48] as audio's proprietary network. The spectrogram of each audio compose audio data, the size of which is 448×448. The spectrogram contain the frequency of noise besides birds singing because the birds singing of environment always mix with the sound of the wind, water and other birds' singing.

3.2   Common Network

The common network is designed for four media types simultaneously to make four media types as similar as possible and then extract features. We choose FGCrossnet as our basic deep model and introduce a new loss function to reduce heterogeneity gap. Before entering the network, the preprocessing of media data expect text data is the same as above proprietary network. In order to let all media types share the same network, text data is transformed from one-dimensional vector into two-dimensional vector to keep the same input format as other media types. We consider four loss functions of the network from cross-entropy, center, quadruplet and distribution. Cross-entropy loss makes the network better distinguish the characteristics of different categories. Center loss makes the samples of same fine-grained subcategory have the same features. Quadruplet loss lets the distance of different categories as far as possible. Distribution loss lets the same subcategory of different media types have the same feature distribution.

3.3   Combined Network

Finally, we combine the output of proprietary network and the common network for each media. The combined network not only considers the special features of the media, but also considers the features when the four media types are as close as possible. The formula is as follows:

$$f = \alpha * f_{pro} + (1 - \alpha) * f_{com}. \tag{6}$$

Each media has to go through the same calculation to get the final unified representation f based on the integration of different features. In which f_{pro} is the feature of fully connected layer of the proprietary network and f_{com} is the feature of fully connected layer of the common network, α is the weight of proprietary network.

4   Experiments

4.1   Datasets and Evaluation Metrics

Datasets: We evaluate our approach on the benchmark fine-grained cross-media retrieval data set, PKU FG-XMedia [39]. It's the first and only data set on fine-grained cross-media retrieval direction. PKU FG-XMedia is composed of 200 fine-grained subcategories of birds and contains four media types (image, text, video and audio). The data set contains more than 50,000 instances including 11,788 image instances, 18,350 video instances, 8,000 text instances, and 12,000 audio instances. Furthermore, its data coming from many different websites result in different data quality, which increases the difficulty of retrieval.

Evaluation Metrics: We adopt mean average precision (MAP) as the evaluation metrics of retrieval. MAP considers both the ranking and accuracy of the results, which is widely used in cross-media retrieval. We first compute average precision (AP) of returned results by a query sample. Then calculating the average of AP values of all queries to obtain the final MAP.

4.2   Retrieval Tasks

Multi-modality Fine-grained Cross-media Retrieval: Any media type of four types as input data to retrieval all media data and return all instances of same semantics results. For example, if users submit the image of "Herring Gull", they will obtain samples about the video of "Herring Gull", the text of "Herring Gull" and the audio of "Herring Gull" in the test data, which is called image retrieval all media types (I→All). The remaining three media types as query samples can be called V→All (video), T→All (text), A→All (audio).

Bi-modality Fine-grained Cross-media Retrieval: Any media type of four types as input data to retrieval the other media data and return all instances of same semantics results. For example, if users submit the image of "Herring Gull", they will obtain samples about the video of "Herring Gull", which is called image retrieval video (I→V). Similarly, video retrieval image is represented as V→I. So bi-modality fine-grained cross-media retrieval contains: I→V, I→T, I→A, V→I, V→T, V→A, T→I, T→V, T→A, A→I, A→V, A→T.

4.3   Baselines

To verify the effectiveness of our approach, we compare our method with state-of-the-art methods on PKU FG-XMedia data set. The following cross-media retrieval methods as our baselines: FGCrossNet [39], MHTN [40], ACMR [49], JRL [58], GSPH [44], CMDN [46], SCAN [42], GXN [37]. We use these methods to carry out multi-modality fine-grained cross-media retrieval and bi-modality fine-grained cross-media retrieval.

4.4   Implementation Details

Input: For proprietary network, different media data are processed to input proprietary network separately to get better performance. For common network, four media data need to input the network simultaneously. In addition, for quadruplet loss function, four different media input should belong to three categories.

Table 1. Comparison of the bi-modality fine-grained cross-media retrieval performance on PKU FG-XMedia data set.

Methods	I→T	I→A	I→V	T→I	T→A	T→V	A→I	A→T	A→V	V→I	V→T	V→A	Average
Ours	**0.355**	**0.629**	**0.660**	**0.409**	**0.324**	**0.335**	**0.643**	**0.287**	**0.515**	**0.706**	**0.335**	**0.544**	**0.478**
FGCrossNet [39]	0.210	0.526	0.606	0.255	0.181	0.208	0.553	0.159	0.443	0.629	0.195	0.437	0.366
MHTN [40]	0.116	0.195	0.281	0.124	0.138	0.185	0.196	0.127	0.290	0.306	0.186	0.306	0.204
ACMR [49]	0.162	0.119	0.477	0.075	0.015	0.081	0.128	0.028	0.068	0.536	0.138	0.111	0.162
JRL [58]	0.160	0.085	0.435	0.190	0.028	0.095	0.115	0.035	0.065	0.517	0.126	0.068	0.160
GSPH [44]	0.140	0.098	0.413	0.179	0.024	0.109	0.129	0.024	0.073	0.512	0.126	0.086	0.159
CMDN [46]	0.099	0.009	0.377	0.123	0.007	0.078	0.017	0.008	0.010	0.446	0.081	0.009	0.105
SCAN [42]	0.050	–	–	0.050	–	–	–	–	–	–	–	–	0.050
GXN [37]	0.023	–	–	0.035	–	–	–	–	–	–	–	–	0.029

Table 2. Comparison of the multi-modality fine-grained cross-media retrieval performance on PKU FG-XMedia data set.

Methods	I→All	T→All	V→All	A→All	Average
Ours	**0.637**	**0.333**	**0.577**	**0.509**	**0.514**
FGCrossNet [39]	0.549	0.196	0.416	0.485	0.412
MHTN [40]	0.208	0.142	0.237	0.341	0.232
GSPH [44]	0.387	0.103	0.075	0.312	0.219
JRL [58]	0.344	0.080	0.069	0.275	0.192
CMDN [46]	0.321	0.071	0.016	0.229	0.159
ACMR [49]	0.245	0.039	0.041	0.279	0.151

Training Strategy: For image and video proprietary network training, we adopt a two-training strategy. First step, we only train the fully connected layer parameters in case of fixing other layers parameters. Second step, we train all layer parameters based on the first step model. We choose the Adam optimizer. The batch size is set to 128 in the first step and 64 in the second step. The learning rate is set to 0.001 in the first step and 0.0001 in the second step. For text proprietary network, we train the word embedding firstly and the dimension of word embedding is set to 100. In the experiment, we choose AdaDelta optimizer. The batch size is set to 32 and the learning rate is set to 1.0. At the second, third, forth layer, we select the drop rate 0.3, 0.3, 0.5 respectively. For audio proprietary network training, we use the pre-trained model VGG-16 to fine-tune the network. We choose the SGD optimizer. The batch size is set to 64 and the learning rate is 0.01. For common network, we adopt a three-training strategy. First step, we only use the image data to train network in case of using the pre-trained model ResNet50 to initialize the network. Second step, we use all media types data to train network with cross-entropy loss function based on the first step model. Final step, we train all media types data with center loss, quadruplet loss and distribution loss based on the second step model.

4.5   Experimental Results

Table 1 presents the scores of MAP result of various approaches about bi-modality fine-grained cross-media retrieval on PKU FG-XMedia data set. By observing Table 1, we can notice that on all bi-modality fine-grained cross-media retrieval, our method achieves best performance than other approaches, which has increased by about 11% on average. FGCrossNet achieves the second best performance than other compared methods, which uses a uniform depth model to learn the four media data simultaneously. It indicates the advantages of modeling four media types simultaneously. But the special characteristics of individual media type are not fully considered. MHTN is superior to other compared methods expect FGCrossNet and our approach, which uses transfer learning to jointly transfer knowledge from single-modal in source domain to all modalities in target domain to promote cross-media common representation learning [40]. The explanation is that transfer knowledge to other media types is useful to obtain better feature representation. ACMR has a better performance in image and text retrieval, which is based on adversarial learning. However, the performances of other bi-modality fine-grained cross-media retrievals are not very well may because of the generalization of its network is not strong. SCAN and GXN have the lower scores of MAP, mainly due to they need more additional image and text retrieval information so they are not suited to our benchmark. SCAN and GXN need the correspondence between image regions and text words, which is through sentence describe what happened in the image.

Table 2 presents the scores of MAP result of various approaches about multi-modality fine-grained cross-media retrieval on PKU FG-XMedia dataset. From the results, we can observe that our method obtains the highest scores of MAP. The method in this paper exceeds all the methods of comparative test. The average retrieval accuracy is promoted from 0.412 to 0.514. That indicates that our method is more effective in fine-grained cross-media retrieval. Only FGCross-Net and MHTN methods can deal with multiple media types data simultaneously, which have a better performance than other methods. By way of contrast, FGCrossNet is lack of consideration of media special attributes and MHTN is lack of model analysis of the small differences between subcategories.

From Table 1 and 2, we can observe that our method achieves the best performance not only in bi-modality fine-grained cross-media retrieval, but also in multi-modality fine-grained cross-media retrieval. Here are several reasons: (1) Our method considers each media special attributes through setting up a separate feature extractor for each media. (2) Our method can extract more similar features to reduce the heterogeneity gap through dealing with four media types data at the same time. (3) Our method considers intra-class, inter-class and intra-media variance comprehensively through introducing four different loss functions.

5   Conclusion

In this work, we studied the problem of fine-grained cross-media retrieval and proposed a new approach to solve feature representation for fine-grained cross-media retrieval. Our key idea is to design two network for fine-grained cross-media retrieval, one is proprietary network for each media to extract special feature, the other is common network for all media types to extract common feature features. Experiments demonstrated the effectiveness of our approach.

References

1. Yao, Y., et al.: Towards automatic construction of diverse, high-quality image dataset. IEEE Trans. Knowl. Data Eng. **32**(6), 1199–1211 (2020)
2. Lu, J., et al.: HSI Road: a hyper spectral image dataset for road segmentation. In: IEEE International Conference on Multimedia and Expo, pp. 1–6 (2020)
3. Hua, X., et al.: A new web-supervised method for image dataset constructions. Neurocomputing **236**, 23–31 (2017)
4. Yao, Y., et al.: Exploiting web images for dataset construction: a domain robust approach. IEEE Trans. Multimedia **19**(8), 1771–1784 (2017)
5. Zhang, J., et al.: Extracting visual knowledge from the internet: making sense of image data. In: International Conference on Multimedia Modeling, pp. 862–873 (2016)
6. Shen, F., et al.: Automatic image dataset construction with multiple textual metadata. In: IEEE International Conference on Multimedia and Expo, pp. 1–6 (2016)
7. Yao, Y., et al.: A domain robust approach for image dataset construction. In: ACM International conference on Multimedia, pp. 212–216 (2016)
8. Yao, Y., et al.: Exploiting web images for multi-output classification: from category to subcategories. IEEE Trans. Neural Netw. Learn. Syst. **31**(7), 2348–2360 (2020)
9. Shu, X., et al.: Personalized age progression with bi-level aging dictionary learning. IEEE Trans. Pattern Anal. Mach. Intell. **40**(4), 905–917 (2018)
10. Yao, Y., et al.: Bridging the web data and fine-grained visual recognition via alleviating label noise and domain mismatch. In: ACM International Conference on Multimedia (2020)
11. Sun, Z., et al.: CRSSC: salvage reusable samples from noisy data for robust learning. In: ACM International Conference on Multimedia (2020)
12. Zhang, C., et al.: Data-driven meta-set based fine-grained visual recognition. In: ACM International Conference on Multimedia (2020)
13. Liu, H., Yao, Y., Sun, Z., Li, X., Jia, K., Tang, Z.: Road segmentation with image-LiDAR data fusion in deep neural network. Multimed. Tools Appl. **1**, 1–16 (2019). https://doi.org/10.1007/s11042-019-07870-0
14. Liu, H., Han, X., Li, X., Yao, Y., Huang, P., Tang, Z.: Deep representation learning for road detection using Siamese network. Multimed. Tools. Appl. **78**(17), 24269–24283 (2018). https://doi.org/10.1007/s11042-018-6986-1
15. Xu, M., et al.: Deep learning for person reidentification using support vector machines. Adv. Multimed. (2017)
16. Chen, T., et al.: Classification constrained discriminator for domain adaptive semantic segmentation. In: IEEE International Conference on Multimedia and Expo, pp. 1–6 (2020)

17. Ding, L., et al.: Approximate kernel selection via matrix approximation. In: IEEE Transactions on Neural Networks and Learning Systems (2020)
18. Shu, X., et al.: Hierarchical long short-term concurrent memory for human interaction recognition. In: IEEE Transactions on Pattern Analysis and Machine Intelligence (2019)
19. G.-S. Xie, et al., SRSC: Selective, robust, and supervised constrained feature representation for image classification. IEEE Trans. Neural Netwk. Learn. Syst. (2019)
20. Sun, Z., et al.: Dynamically visual disambiguation of keyword-based image search. In: International Joint Conference on Artificial Intelligence, pp. 996–1002 (2019)
21. Yang, W., et al.: Discovering and distinguishing multiple visual senses for polysemous words. In: AAAI Conference on Artificial Intelligence, pp. 523–530 (2018)
22. Hu, B., et al.: PyRetri: A PyTorch-based Library for Unsupervised Image Retrieval by Deep Convolutional Neural Networks. arXiv preprint arXiv:2005.02154 (2020)
23. Gu, Y., et al.: Clustering-driven unsupervised deep hashing for image retrieval. Neurocomputing **368**, 114–123 (2019)
24. Wang, W., et al.: Set and rebase: determining the semantic graph connectivity for unsupervised cross modal hashing. In: International Joint Conference on Artificial Intelligence, pp. 853–859 (2020)
25. Huang, P., et al.: Collaborative representation based local discriminant projection for feature extraction. Digit. Signal Process. **76**, 84–93 (2018)
26. Zhang, J., et al.: Extracting privileged information from untagged corpora for classifier learning. In: International Joint Conference on Artificial Intelligence, pp. 1085–1091 (2018)
27. Yao, Y., et al.: Extracting multiple visual senses for web learning. IEEE Trans. Multimed. **21**(1), 184–196 (2019)
28. Yao, Y., et al.: Extracting privileged information for enhancing classifier learning. IEEE Trans. Image Process. **28**(1), 436–450 (2019)
29. Yang, W., et al.: Exploiting textual and visual features for image categorization. Pattern Recogn. Lett. **117**, 140–145 (2019)
30. Branson, S., et al.: Bird species categorization using pose normalized deep convolutional nets. arXiv preprint arXiv, 1406.2952 (2014)
31. Castrejon, L., et al.: Learning aligned cross-modal representations from weakly aligned data. In: IEEE Conference on Computer Vision and Pattern Recognition, pp. 2940–2949 (2016)
32. Chen, W., et al.: Beyond triplet loss: a deep quadruplet network for person re-identification. In: IEEE Conference on Computer Vision and Pattern Recognition, pp. 403–412 (2017)
33. Cui, Y., et al.: Large scale fine-grained categorization and domain-specific transfer learning. In: IEEE Conference on Computer Vision and Pattern Recognition, pp. 4109–4118 (2018)
34. Fu, J., et al.: Look closer to see better: Recurrent attention convolutional neural network for fine-grained image recognition. In: IEEE Conference on Computer Vision and Pattern Recognition, pp. 4438–4446 (2017)
35. Gavves, E., et al.: Local alignments for fine-grained categorization. Int. J. Comput. Vis. **111**(2), 191–212 (2015)
36. Gretton, A., et al.: A kernel two-sample test. J. Mach. Learn. Res. **13**(1), 723–773 (2012)
37. Gu, J., et al.: Look, imagine and match: Improving textual-visual cross-modal retrieval with generative models. In: IEEE Conference on Computer Vision and Pattern Recognition, pp. 7181–7189 (2018)

38. He, X., et al.: Fine-grained image classification via combining vision and language. In: IEEE Conference on Computer Vision and Pattern Recognition, pp. 5994–6002 (2017)
39. He, X., et al.: A new benchmark and approach for fine-grained cross-media retrieval. In: ACM International Conference on Multimedia, pp. 1740–1748 (2019)
40. Huang, X., et al.: Mhtn: Modal-adversarial hybrid transfer network for cross-modal retrieval. IEEE Trans. Cybernet. (2018)
41. Kim, J., et al.: Learning semantics with deep belief network for cross-language information retrieval. In: Proceedings of COLING 2012: Posters, pp. 579–588 (2012)
42. Lee, K.H., et al.: Stacked cross attention for image-text matching. In: European Conference on Computer Vision, pp. 201–216 (2018)
43. Lin, T.Y., et al.: Bilinear cnn models for fine-grained visual recognition. In: IEEE International Conference on Computer Vision, pp. 1449–1457 (2015)
44. Mandal, D., et al.: Generalized semantic preserving hashing for n-label cross-modal retrieval. In: IEEE Conference on Computer Vision and Pattern Recognition, pp. 4076–4084 (2017)
45. Peng, Y., et al.: Object-part attention model for fine-grained image classification. IEEE Trans. Image Process. **27**(3), 1487–1500 (2017)
46. Peng, Y., et al.: Cross-media shared representation by hierarchical learning with multiple deep networks. In: IJCAI, pp. 3846–3853 (2016)
47. Rasiwasia, N., et al.: A new approach to cross-modal multimedia retrieval. In: ACM International Conference on Multimedia, pp. 251–260 (2010)
48. Simonyan, K., et al.: Very deep convolutional networks for large-scale image recognition. arXiv preprint arXiv, 1409.1556 (2015)
49. Wang, B., et al.: Adversarial cross-modal retrieval. In: ACM International Conference on Multimedia, pp. 154–162 (2017)
50. Zhang, C., et al.: Web-supervised network with softly update-drop training for fine-grained visual classification. In: AAAI Conference on Artificial Intelligence, pp. 12781–12788 (2020)
51. Xie, G., et al.: Attentive region embedding network for zero-shot learning. In: IEEE Conference on Computer Vision and Pattern Recognition, pp. 9384–9393 (2019)
52. Wang, C., et al.: Deep semantic mapping for cross-modal retrieval. In: IEEE International Conference on Tools with Artificial Intelligence, pp. 234–241 (2015)
53. Wang, Y., et al.: Learning a discriminative filter bank within a cnn for fine-grained recognition. In: IEEE Conference on Computer Vision and Pattern Recognition, pp. 4148–4157 (2018)
54. Wei, Y., et al.: Cross-modal retrieval with cnn visual features: A new baseline. IEEE Trans. Cybernet. **47**(2), 449–460 (2016)
55. Xiao, T., et al.: The application of two-level attention models in deep convolutional neural network for fine-grained image classification. In: IEEE Conference on Computer Vision and Pattern Recognition, pp. 842–850 (2015)
56. Xie, S., et al.: Hyper-class augmented and regularized deep learning for fine-grained image classification. In: IEEE Conference on Computer Vision and Pattern Recognition, pp. 2645–2654 (2015)
57. Yang, S., et al.: Unsupervised template learning for fine-grained object recognition. In: Advances in Neural Information Processing Systems, pp. 3122–3130, (2012)
58. Zhai, X., et al.: Learning cross-media joint representation with sparse and semisupervised regularization. IEEE Trans. Circuits Syst. Video Technol. **24**(6), 965–978 (2014)

59. Zhang, N., et al.: Part-based r-cnns for fine-grained category detection. In: European Conference on Computer Vision, pp. 834–849 (2014)
60. Zhang, Y., et al.: Weakly supervised fine-grained categorization with part-based image representation. IEEE Trans. Image Process. **25**(4), 1713–1725 (2016)
61. Zheng, H., et al.: Learning multi-attention convolutional neural network for fine-grained image recognition. In: IEEE International Conference on Computer Vision, pp. 5209–5217 (2017)
62. Zhou, P., et al.: Attention-based bidirectional long short-term memory networks for relation classification. In: Annual Meeting of the Association for Computational Linguistics, pp. 207–212 (2016)
63. Zhang, C., et al.: Web-supervised network for fine-grained visual classification. In: IEEE International Conference on Multimedia and Expo, pp. 1–6 (2020)
64. Xie, G., et al.: Region graph embedding network for zero-shot learning. In: European Conference on Computer Vision (2020)
65. Zhou, T., et al.: Motion-attentive transition for zero-shot video object segmentation. In: AAAI Conference on Artificial Intelligence (2020)
66. Luo, H., et al.: SegEQA: video segmentation based visual attention for embodied question answering. In: IEEE Conference on Computer Vision, pp. 9667–9676 (2019)
67. Wang, W., et al.: Target-aware adaptive tracking for unsupervised video object segmentation. The DAVIS Challenge on Video Object Segmentation on CVPR workshop (2020)

Extraction of Spectral-Spatial 3-Dimensional Homogeneous Regions from Hyperspectral Images and Its Application to Fast Classification

Yanbin Cai, Zhuliang Geng, Yating Liang, and Peng Fu[✉]

School of Computer Science and Engineering, Nanjing University of Science and Technology,
Nanjing 210094, China
fupeng@njust.edu.cn

Abstract. Hyperspectral images have been widely applied to various fields due to the high spectral and spatial resolution. However, the vast amounts of spectral and spatial information also bring difficulties in hyperspectral image processing, where the efficiency is one of the biggest challenges. To address this challenge, we propose a method to extract the spectral-spatial 3-dimensional homogeneous regions (SS3DHRs) from hyperspectral images. First, highly correlated neighbor spectral bands are selected based on the correlation coefficients between adjacent bands; Based on the sub-band selection, a superpixel segmentation method is improved for hyperspectral images to gather the spatial information; Combining the spectral sub-bands and spatial superpixels, the SS3DHRs are collected from the 3-deminsion hyperspectral data cube. The SS3DHR can be processed as a unit for the subsequent applications, which may significantly reduce the redundant data and thus raise the efficiency. In experiment part, the extracted SS3DHRs are applied for hyperspectral image classification, where the experimental results demonstrate the effectiveness and efficiency of the proposed method.

Keywords: Hyperspectral image · Band selection · Superpixel segmentation · Spectral distance · Classification

1 Introduction

With the development of the hyperspectral imaging technologies, the spectral and spatial resolution of hyperspectral images has been improved significantly. Benefit from the high image resolution, hyperspectral images have been widely used in various applications, such as classification [1, 2], target detection [3, 4], change detection [5, 6], and so on.

Hyperspectral imaging is a spectral imaging acquisition where each pixel of the image is employed to acquire a set of images within certain spectral bands. Thus, a

This work was in part supported by the National Nature Science Foundation of China under Grant no. 61801222, and in part supported by the Fundamental Research Funds for the Central Universities under Grant no. 30919011230, and in part supported by the JiangSu Undergraduate Training Program for Innovation and Entrepreneurship under Item no. 20190288126Y.

© Springer Nature Switzerland AG 2020
Y. Peng et al. (Eds.): PRCV 2020, LNCS 12306, pp. 200–207, 2020.
https://doi.org/10.1007/978-3-030-60639-8_17

third spectral dimension is added to the traditional 2-dimension spatial images. The 3-dimension data cube can provide abundant spectral and spatial information for hyperspectral image applications. However, the efficiencies of applications are reduced when dealing with the huge amount of data. To solve this problem, some researchers attempted to select a part of valuable spectral bands from the third dimension. Luo et al. [7] proposed an information-assisted density peak index method for the band selection. The intra-band information entropy is introduced by using a density-based clustering approach, and meanwhile the band distance is integrated with channel proximity to control the compactness of local density. Wang et al. [8] designed an optimal clustering framework for the band selection, which can obtain the optimal clustering result for a particular form of objective function under a reasonable constraint. To further raise the processing speed of hyperspectral image applications, the reduction of spatial dimension has also been widely studied. Tarabalka et al. [9] improved a watershed segmentation method to define information about spatial structure, and then the segmented regions are processed for hyperspectral image classification. In recent years, an image processing technology—superpixel segmentation has been investigated. Superpixel segmentation algorithms can segment an image into small local regions adhering tightly to the boundaries. In [10–14], superpixels are adopted to boost various kinds of hyperspectral image applications. In spite of many researches have been studied for the dimension reduction, there are still three main challenges: (1) How to design an effective and fast approach to reduce the spectral dimension? (2) How to make the superpixel segmentation methods more suitable for hyperspectral images? (3) How to comprehensively reduce the spectral and spatial dimensions to further speed up subsequent applications? To address the abovementioned challenges, we propose a spectral-spatial 3-dimensional homogeneous region (SS3DHR) extraction method for hyperspectral images. First, a simple producer is established to select the highly correlated neighbor spectral bands; then, an improved superpixel model is proposed for the hyperspectral image segmentation; finally, the SS3DGRs are extracted by combing the spectral sub-band set and the spatial superpixel segmentation. Experimental results demonstrate the effectiveness and efficiency of the proposed SS3DHR for hyperspectral image classification.

2   Method

The proposed SS3DHR extraction method is composed of two main steps. First, highly correlated neighbor spectral bands are selected based on the correlation coefficient of adjacent bands; then, spatial homogeneous superpixels are generated by using an improved superpixel segmentation algorithm; finally, the SS3DHRs are extracted by combing the above-mentioned two main steps. Based on the SS3DHRs, subsequent hyperspectral image applications can be processed efficiently. The framework of the proposed method is displayed in Fig. 1.

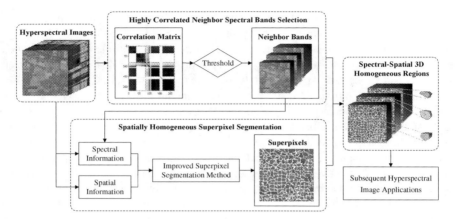

Fig. 1. Framework of the proposed method.

2.1 Highly Correlated Neighbor Spectral Bands Selection

For a given hyperspectral image $I(m, n, z)$ with spatial size m × n and spectral bands number z, the correlation coefficients between the adjacent bands are first calculated according to the following equation:

$$C(P_i, P_j) = \frac{m \times n \times \sum P_i \times P_j - \sum P_i \sum P_j}{\sqrt{m \times n \times \sum P_i^2 - \left(\sum P_i\right)^2} \sqrt{m \times n \times \sum P_j^2 - \left(\sum P_j\right)^2}} \quad (1)$$

where C represents the correlation coefficient between the two spectral bands, P_i and P_j denote the pixel values with the vector number $m \times n$ in the ith and jth bands, respectively. In practice, the spectral resolution of hyperspectral images is usually very high, thus the correlation coefficients between the neighbor spectral bands are very large. In this case, the highly similar bands can be gathered and processed together. However, the spectral curve might be changed obviously in some special wavelength, thus a threshold is required to group the spectral bands into several sub-band sets. In the proposed method, the highly correlated neighbor spectral bands are selected according the following procedure: Step1: input the first spectral band into Set 1; Step 2: calculate the correlation coefficient between the first band and the next adjacent band P_j; Step 3: if the correlation coefficient is larger than the predefined threshold t, the adjacent band P_j is grouped into the same sub-band set as the prior band; otherwise, put the band P_j into a new sub-band set; Step 4: repeat the steps 2 and 3 to assign each spectral band into a proper sub-band set. By using the proposed method, a hyperspectral image can be divided into several sub-band sets, where the spectral bands in each sub-band set are highly similar.

2.2 Spatially Homogeneous Superpixel Segmentation

In order to utilize the local spatial information, an improved superpixel segmentation method is designed for hyperspectral images. Superpixels are the spatially homogeneous local patches in an image, where the proposed superpixel segmentation method is improved from SLIC (simple linear iterative clustering) [15]. SLIC is a widely-used superpixel segmentation method focused on natural images. To make the superpixel segmentation method more effective on hyperspectral images, a novel spectral distance is designed in the proposed method. In our method, a spatial image (one band) of the hyperspectral image is first divided into k regular grids with equal edge size s, and then the k initial cluster centers are formulated with spatial coordinates and spectral signatures. After the cluster center initialization, each pixel is assigned to the closest cluster center based on their distances as follows:

$$D(x_i, c_k) = \sqrt{D_{Spetral}(x_i, c_k)^2 + \varpi^2 \left(D_{Spatial}(x_i, c_k) \big/ s\right)^2} \tag{2}$$

where $D(x_i, c_k)$ denotes the distance between the pixel x_i and the cluster center c_k, and it is composed of a spectral distance $D_{Spetral}(x_i, c_k)$ and a spatial distance $D_{Spatial}(x_i, c_k)$. The parameter ϖ is introduced to weigh the relative importance of the spectral and spatial distance. In the proposed superpixel segmentation method, the spectral distance is calculated as:

$$D_{Spetral}(x_i, c_k) = \sum_{l=1}^{L} \frac{\left|\bar{X}_i(l) - \bar{C}_k(l)\right|}{\bar{X}_i(l) + \bar{C}_k(l)} \tag{3}$$

where L is the number of the sub-band sets, $\bar{X}_i(l)$ and $\bar{C}_k(l)$ denote the mean value of the l th sub-band set for the pixel x_i and c_k, respectively. Regarding the spatial distance, the Euclidean distance is adopted as:

$$D_{Spatial}(x_i, c_k) = \sqrt{(m_i - m_k)^2 + (n_i - n_k)^2} \tag{4}$$

After the pixel assignment, the cluster centers are updated by calculating the mean vectors of all the pixels belonging to the cluster. Repeat the pixel assignment and cluster center update, and superpixels are generated if no further changes of the labels. Combining the highly correlated neighbor spectral bands selection and spatially homogeneous superpixel segmentation, the SS3DHRs are extracted.

3   Experimental Results

3.1   Hyperspectral Dataset Description

In our experiments, we exploit the classical hyperspectral image—Indian Pines (IPs) data set to evaluate the performance of the proposed method. The IPs hyperspectral image is obtained by the airborne visible/infrared imaging spectrometer (AVIRIS) over Indian Pines test site in North-western Indiana. AVIRIS is a widely-used hyperspectral imaging sensor, which produces hyperspectral images with 220 spectral bands ranging from 0.2 μm to 2.4 μm. After removing the spectral bands covering the region of water absorption 104-108, 150-163 and 220, the total band number of IPs data is 200. The spatial size of the utilized IPs data is 145 × 145 pixels with the resolution 20 m per pixel. The IPs data set contains 16 reference land cover classes, where the ground truth map is displayed in Fig. 2 (Table 1).

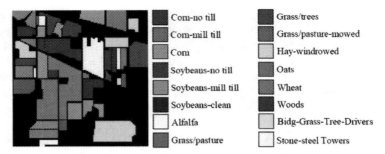

Fig. 2. Ground truth of the Indian Pines hyperspectral image.

Table 1. Number of samples of different classes in the Indian Pines data set.

Class	Name	Numbers
1	Alfalfa	46
2	Corn-notill	1428
3	Corn-mintill	830
4	Corn	237
5	Grass-pasture	483
6	Grass-trees	730
7	Grass-pasture-mowed	28
8	Hay-windrowed	478
9	Oats	20
10	Soybean-notill	972
11	Soybean-mintill	2455
12	Soybean-clean	593
13	Wheat	205
14	Woods	1265
15	Buildings-Grass-Trees-Drives	386
16	Stone-Steel-Towers	93
Total number		10249

3.2 Spatially Superpixel Segmentation Results

The superpixel segmentation is crucial for the accurate spatial information extraction. In the proposed superpixel segmentation method, there are two important parameters which may affect the segmentation results. The first one is the number of sub-band set L, and another one is the weight parameter ϖ. The number of the sub-band sets L is decided by the threshold t. When the threshold t is small, more neighbor bands may be grouped into a sub-band set, thus the number of the sub-band set will be small. In this situation, the image processing speed will be fast, but the superpixel segmentation accuracy might be reduced as less spectral information is exploited. When the threshold t is large, the converse is true. To balance the efficiency and accuracy, the threshold t is optimized as 0.9 in our experiments. Regarding the parameter ϖ, it may affect the boundary adherence and the regularity of the superpixel shape. If the parameter ϖ is too small, the spectral distance will play the dominant role. The segmented superpixels might be unregular in shape in this case, as shown in Fig. 3(a), where the parameter ϖ equals 0.0001. If the weight parameter is too large, the spatial distance will be dominate, thus the generated superpixels can not adhere tightly to the image local boundaries. As displayed in Fig. 3(c), the segmented superpixels are degraded as regular blocks with a large weight parameter $\varpi = 0.5$. Comprehensively consider the importance of the spectral and spatial distance, the parameter ϖ is set as 0.005 in our experiments, where the superpixel segmentation results is shown in Fig. 2(b). Compared with Fig. 3(a) and (c), the superpixel segmentation results in Fig. 2(b) are compactness and exhibiting more accurate boundary adherence.

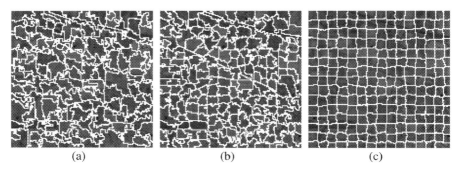

(a)   (b)   (c)

Fig. 3. Superpixel segmentation results with different weight parameter ϖ. (a) $\varpi = 0.0001$, (b) $\varpi = 0.005$, (c) $\varpi = 0.5$.

3.3 Hyperspectral Image Classification Based on SS3DHR

In this experiment part, we apply the SS3DHR for the hyperspectral image classification. Hyperspectral image classification is a classical and meaningful application in practice. In traditional classification methods, different types of classifiers are designed to group each pixel into one specific class. For a given hyperspectral image with spatial size m × n and spectral bands number z, we need to deal with m × n spatial pixels with z spectral features for each pixel. In this manner, the processing speed might be very slow. When we extract SS3DHRs from the hyperspectral image, a SS3DHR is regarded as a unit for the classification, where the number of spectral and spatial features is greatly reduced, and thus the efficiency of the hyperspectral image classification will be increased. Among the traditional hyperspectral image classification methods, the random forests (RF) method [16] usually produces good performance due to its decision-making mechanism. For a better comparison, we apply the RF method for hyperspectral image classification based on the single pixel and SS3DHR, respectively. The size of SS3DHR is important for the hyperspectral image classification. When the size of SS3DHR is large, i.e., the whole hyperspectral image is divided into only a few SS3DHRs, the classification process will be very quick. However, some spectral and spatial information might be lost in this case, thus the classification accuracy might be decreased. On the other hand, if the size of SS3DHR is small, the classification accuracy will be increased, and at the meantime, the processing time will be increased. Based on the experimental performance, we set the initial size of superpixel $s = 5$ and the sub-band selection threshold $t = 0.9$. In this experiment, we simply use the mean value of the spectral and spatial features in each SS3DHR for the classification. For the two compared methods, we randomly select 10% of all the samples for training and the rest for testing, where the classification maps are shown in Fig. 4. Figure 4(a) displays the classification map based on the traditional pixel-wise RF method, where the classification overall accuracy (OA) is 0.748; Fig. 4(b) shows the classification map with the SS3DHR-based RF method, where the classification OA is 0.716. Compared with the two methods, the classification accuracy of the SS3DHR-based RF method is slightly lower than the pixel-based RF method. However, regarding the processing speed, the pixel-based RF method costs 3.64 s and the SS3DHR-based method costs 1.18 s, where all the experiments are performed using MATLAB (R2016a) on a PC with 3.5 GHz CPU and 32 GB RAM. Obviously, the proposed SS3DHR-based method is much faster than the traditional pixel-based method. It is worth noting that in our experiment, we simply use the mean value of each SS3DHR for the classification. In real hyperspectral image applications, other statistical characteristics of SS3DHRs can be exploited for a more accurate classification. In reality, some applications are efficiency first and some others are accuracy first. When dealing with the real hyperspectral image applications, we can set different sizes of the SS3DHR to achieve various requirements.

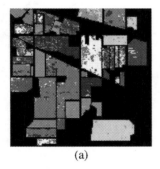
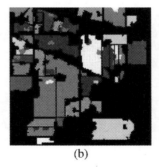

(a)   (b)

Fig. 4. Classification maps with the pixel-based and SS3DHR-based RF methods. (a) pixel-based RF method, (b) SS3DHR-based RF method.

4   Conclusion

In this paper we propose a new method to extract SS3DHRs from hyperspectral images. First, correlation coefficients between the two adjacent bands are calculated; based on the correlation coefficients, the highly correlated neighbor spectral bands are selected with a simple procedure. For the spatial local region extraction, an improved superpixel segmentation method is proposed with a new spectral distance. Combining the spectral sub-band sets and the spatial superpixels, the SS3DHRs are finally generated from hyperspectral images. In experiment part, the extracted SS3DHRs are applied for hyperspectral image classification, where the experimental results demonstrate the effectiveness and efficiency of the proposed method. Furthermore, the proposed SS3DHR extraction method can also help to speed up various subsequent hyperspectral image applications.

References

1. Alam, F.I., Zhou, J., Liew, A.W.C., Jia, X., Chanussot, J., Gao, Y.: Conditional random field and deep feature learning for hyperspectral image classification. IEEE Trans. Geosci. Remote Sens. **57**(3), 1612–1628 (2018)
2. Yang, X., Ye, Y., Li, X., Lau, R.Y.K., Huang, X.: Hyperspectral image classification with deep learning models. IEEE Trans. Geosci. Remote Sens. **56**(9), 5408–5423 (2018)
3. Yang, S., Shi, Z.: Hyperspectral image target detection improvement based on total variation[J]. IEEE Trans. Image Process. **25**(5), 2249–2258 (2016)
4. Wu, K., Xu, G., Zhang, Y., Bo, D.: Hyperspectral image target detection via integrated background suppression with adaptive weight selection. Neurocomputing **315**, 59–67 (2018)
5. Ertürk, A., Ertürk, S., Plaza, A.: Unmixing with SLIC superpixels for hyperspectral change detection. In: 2016 IEEE International Geoscience and Remote Sensing Symposium (IGARSS). pp. 3370–3373 (2016)
6. Wang, Q., Yuan, Z., Du, Q., Li, X.: GETNET, a general end-to-end 2-D CNN framework for hyperspectral image change detection. IEEE Trans. Geosci. Remote Sens. **57**(1), 3–13 (2019)
7. Luo, X., Xue, R., Yin, J.: Information-assisted density peak index for hyperspectral band selection[J]. IEEE Geosci. Remote Sens. Lett. **14**(10), 1870–1874 (2017)
8. Wang, Q., Zhang, F., Li, X.: Optimal clustering framework for hyperspectral band selection[J]. IEEE Trans. Geosci. Remote Sens. **56**(10), 5910–5922 (2018)
9. Tarabalka, Y., Chanussot, J., Benediktsson, J.A.: Segmentation and classification of hyperspectral images using watershed transformation. Pattern Recogn. **43**(7), 2367–2379 (2010)
10. Huang, Z., Li, S.: From difference to similarity: a manifold ranking-based hyperspectral anomaly detection framework. IEEE Trans. Geosci. Remote Sens. **57**(10), 8118–8130 (2019)
11. Xu, Q., Fu, P., Sun, Q., Wang, T.: A fast region growing based superpixel segmentation for hyperspectral image classification. In: Lin, Z., et al. (eds.) PRCV 2019. LNCS, vol. 11858, pp. 772–782. Springer, Cham (2019). https://doi.org/10.1007/978-3-030-31723-2_66
12. Fang, L., Zhuo, H., Li, S.: Super-resolution of hyperspectral image via superpixel-based sparse representation. Neurocomputing **273**, 171–177 (2018)
13. Karaca, A.C., Güllü, M.K.: Superpixel based recursive least-squares method for lossless compression of hyperspectral images. Multidimens. Syst. Signal Process. **30**(2), 903–919 (2018). https://doi.org/10.1007/s11045-018-0590-4
14. Fu, P., Sun, X., Sun, Q.: Hyperspectral image segmentation via frequency-based similarity for mixed noise estimation. Remote Sens. **9**(12), 1237 (2017)
15. Achanta, R., Shaji, A., Smith, K., Lucchi, A., Fua, P., Süsstrunk, S.: SLIC superpixels compared to state-of-the-art superpixel methods. IEEE Trans. Pattern Anal. Mach. Intell. **34**(11), 2274–2282 (2012)
16. Amini, S., Homayouni, S., Safari, A., Darvishsefat, A.A.: Object-based classification of hyperspectral data using Random Forest algorithm. Geo-spatial Inf. Sci. **21**(2), 127–138 (2018)

Feature-Less Stitching of Cylindrical Cable for Surface Inspection of Cable-Stayed Bridges

Kehui Zhang, Langming Zhou, and Changyan Xiao[✉]

College of Electrical and Information Engineering, Hunan University, Changsha 410000, Hunan, China
xcy19722@hotmail.com

Abstract. Cable surface inspection based on vision is of great significance for the scientific management and maintenance of cable-stayed bridges. In this paper, a set of automatic multi-view cable surface image acquisition system is designed in which images from four fixed cameras are wirelessly transmitted to a service computer, and then stitched together on the service computer. Clustering segmentation and histogram matching are used to enhance the collected images, and a new cylindrical objective image stitching algorithm is designed to obtain the distorted unwrapping cylindrical image from the 360° capture of the cable. The precise location and measurement of the damage can be achieved quickly by combining with the prior geometric dimensions of the cable. The tests on several groups of bridge data show that the proposed algorithm can solve the problem of feature-less cylindrical cable surface and the problem of complex background conditions, which is of great significance for quickly determining the position and distribution of cable surface damage.

Keywords: Cable inspection · Cylindrical back-projection · Image stitching

1 Introduction

Cable-stayed bridge has become the main type of long-span bridges, and the number of cable-stayed bridges rapidly increased [1]. Cables are the most important load-bearing components in cable-stayed bridges, are exposed to the air, wind, rain and sunshine for a long time. The polyethylene (PE) pipe is the protective layer for the cable. With age, the PE pipe will meet with fatigue, corrosion, and their coupled effects [2]. The long-accumulated damage causes internal steel wires to break, causing serious traffic accidents such as the collapse of the can tho bridge in 2007. The conventional inspection method of the cable is mainly to inspect the surface of the cable by using a lifting vehicle or a lifting trolley. This method has strong subjectivity, low detection efficiency, and is dangerous to the inspectors [3–5]. Therefore, it is crucial for developing an automatic image-based surface damage detection system to assess the conditions of the cables [6].

Nowadays, many tunnel inspection systems using computer vision have been developed to improve the efficiency and scientific management. [6] presented a damage detection algorithm which combines image enhancement techniques with principal component analysis (PCA) algorithm. They developed an image enhancement method together

Y. Peng et al. (Eds.): PRCV 2020, LNCS 12306, pp. 208–216, 2020.
https://doi.org/10.1007/978-3-030-60639-8_18

with a noise removal technique. Then the images are projected into PCA sub-space to identify and localize damage in cable surface of cable-stayed bridges. [7] developed the imaging and inspection of the Deep Tunnel Sewerage System (DTSS). They created cylindrical images captured by a novel 360° revolving camera system and developed a geometrical relationship to combine the camera trajectory with scene geometry to automatically create a panoramic view of the tunnel. [8] proposed an modified scale-invariant feature transform (SIFT) algorithm for stitching defect images from multiple perspectives.

According to these research, we design a set of automatic multi-view cable surface image capture and process device. The cables' surface images are completely captured by four cameras and then transmitted wirelessly to principal computer. Next, due to the collected data has the characteristics of less surface texture and complex background environments, clustering segmentation and histogram matching are done with the data to enhance image quality in the principal computer. Thereafter, we use cylindrical back-projection to unwrap cylindrical surface images and stitch them.

In summary, we make the following contributions:

- We develop an efficient segmentation algorithm combined clustering segmentation with histogram matching to separate the background from the stay cable.
- We infer a geometric relation between 2D captured cylindrical image and unwrapped image.
- We propose an image stitching algorithm aimed at some feature-less objects like stay cable, tunnel, bridge deck and so on.

2   Related Work

2.1   Image Segmentation

Image segmentation is a basic and key technology in the field of image processing. Its purpose is to separate the object from the background and provide a basis for subsequent processing such as object detection and accurate positioning. Some current methods are edge detection segmentation, region-based segmentation, clustering segmentation, etc. [9]. We use clustering segmentation and edge detection segmentation in this paper. Edge detection segmentation segments an image by detecting the edge of different areas. The edge are often detected by derivative operations, and derivatives are calculated using differential operator, like the Roberts gradient, Sobel operator, Laplacian, etc. [10]. Then is the clustering segmentation. Clustering is to divide the data set into several subsets according to the similarity among the elements. Clustering segmentation is an unsupervised statistical method, which does not require training samples and plays an important role in the application of image segmentation [11–15].

2.2   Cylindrical Back-Projection

For some cylindrical or spherical objects that are consisted of surface, distortion of surface texture will occur in the process of imaging. The solution to this problem is to project the image texture onto a two-dimensional plane that to convert the cylindrical projection into a plane projection, which is a process of cylindrical back-projection. [16] proposed a universal back-projection formula for all curved objects. Other recent methods [17, 18] were proposed by adding auxiliary rectangles, applying to the recognition of cylinder QR code.

2.3 Image Stitching

Image stitching is to use image processing technology to find overlapped parts in the images with overlapping areas, match the images, and then merge into an image. Feature-based registration method is the mainstream of image stitching algorithm. Image features can include corners, outlines, textures, or other special structures. Harris corner [19], FAST [20], SIFT [21], and SURF [22] features are often used in feature registration. There are two chief problems in stitching of cylindrical cables: registration of feature-less data under different four viewpoints.

3 Method

In this paper, the cable surface images of cable-stayed bridge captured by four CCD cameras are stored in the computer. For the stored images, we firstly separate the cable surface from the background, and then rectify the distorted cable surface. Lastly, we present an algorithm based on grid partitioning to stitch them. Figure 1 is shown the process of cable surface stitching.

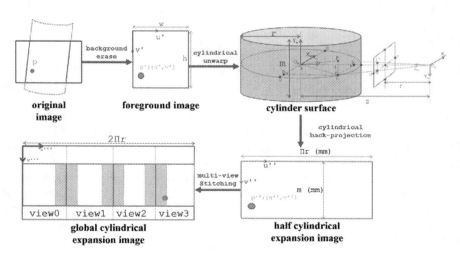

Fig. 1. The process of cable surface stitching.

3.1 Cable Surface Segmentation

In the acquisition images of bridge cables, all images are in the scene having parallax with the cable surface foreground and background. If only a single global transformation is used for modeling, it is easy to cause image distortion. Figure 2 shows the result of feature matching under parallax. Therefore, it is necessary to separate the cable surface from the background in our image processing.

Fig. 2. The result of feature matching under parallax shows different homography.

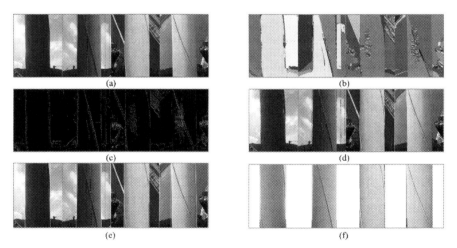

Fig. 2. The process of cable surface segmentation: (a) original image; (b) pre-segmentation; (c) edge detection; (d) line-segment detection; (e) border fitting; (f) background erase and color harmony.

In the segmentation between the cable surface and the background, it is prone to errors if the method of line-segment detection is simply used on account of the possible existence of rubber drainage tubes on the surface of the cable and complex background. If the images are segmented by the method of line-segment detection, some of which will segment the rubber drainage tubes and the background.

Therefore, the clustering segmentation and Flood Fill algorithm are selected to background segmentation, which can reserve the shape information of cable surface and remove its background. In image processing, starting from a starting node, the adjacent nodes are extracted or filled with different colors. After the segmentation, the brightness of four images from four cameras is different due to the parallax of the four cameras, which is adjusted by histogram matching. The specific algorithm steps are as follows:

(1) Pre-segment images by mean-shift and flood fill.
(2) Get edges from pre-segmented images and original images by Canny operator.
(3) Detect lines by using Line Segment Detector (LSD) algorithm.
(4) The outer rectangular bounds are employed to fit lines.
(5) Erase background.
(6) Adjust brightness by histogram matching algorithm.

3.2 Cylindrical Back-Projection

The image of cable surface acquired by CCD digital cameras is cylindrical surface image. In order to meet the requirements of subsequent image stitching, the cable image needs to be expanded into a plane image. We first build an ideal cylindrical back-projection model, assuming that the image of cable surface is an ideal cylindrical image, and is expanded along the direction of the generating line.

The imaging geometric relationship is shown in Fig. 3.:

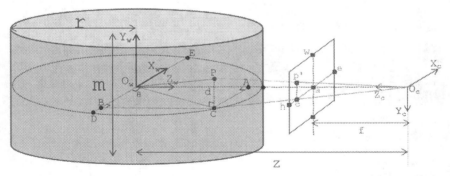

Fig. 3. Cylindrical back-projection: A 3D point $P(x, y, z)$ in a cylinder with known radius is projected onto a planar image as a point $p'(u, v)$.

A 3D point $P = [x, y, z]^T$ can be projected into a 2D pixel of a planar image, $p' = [u, v]^T$, using some known parameters: cylinder's radius, r, focal length, f, the distance from camera's optical center O_c to the center of the cylinder, z. This 3D point can be calculated as follows:

$$\begin{bmatrix} x \\ y \\ z \end{bmatrix} = \begin{bmatrix} r \cos \theta \\ d \\ r \sin \theta \end{bmatrix} \tag{1}$$

where angle, θ, and height of the point, d are two parameters in cylinder. Based on similar triangle theory, we can derive the following equations in the condition of the center of the cylinder locates at the center of the image:

$$\frac{m}{h} = \frac{2r}{w} = \frac{z}{f} \tag{2}$$

$$\frac{PC}{p'c} = \frac{AC}{ac} \tag{3}$$

The following formula is derived by Eq. (2) and Eq. (3):

$$\begin{cases} v = \left\{ \dfrac{h}{2} - \dfrac{zd}{2r(z - r|\sin\theta|)}w \right. \\[2ex] u = \begin{cases} \dfrac{w}{2}(1 - \dfrac{z|\cos\theta|}{z-r|\sin\theta|}), & \theta \geq -\dfrac{\pi}{2} \\[2ex] \dfrac{w}{2}(1 + \dfrac{z|\cos\theta|}{z-r|\sin\theta|}), & \theta < -\dfrac{\pi}{2} \end{cases} \end{cases} \tag{4}$$

An example of the result of cylindrical back-projection is shown in Fig. 4.:

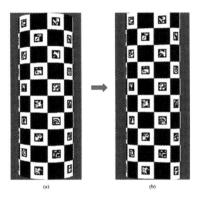

Fig. 4. (a) 2D cylindrical image. (b) unwrapped image after cylindrical back-projection.

3.3   Image Stitching

In order to obtain a complete cylinder image, we need to stitch the acquired image. However, the image features are relatively simple and there are few local similar regions so that we choose SIFT algorithm to accurately extract key points and realize image matching through feature point description. SIFT feature extraction is very time-consuming and difficult to achieve real-time computing speed, so the SiftGPU, namely to achieve real-time calculation is obtained by using the graphics acceleration rate, aimed at 400 × 300 size of image point extraction can be achieved with the same speed of 4 frames per second, basically guarantee the real-time positioning effect, can complete the processing of large amount of data in a short time.

In addition, the key to the success of stitching is to obtain uniform matching points. However, due to the lack of features and uneven distribution on the cable surface, the SIFT feature matching algorithm guided by global homography cannot obtain uniform matching points. We use the following methods to solve this problem: first, the cylindrical back-projection formula in 3.2 is used to expand the cylindrical image to obtain more features; Second, APAP (As Projective As Possible) algorithm [27] is used to divide the image into multiple local image blocks by mesh, and the uniformity of each local image block is obtained to make the matching points more uniform and more dense. Figure 5 is shown the the comparison between two different feature matching methods.

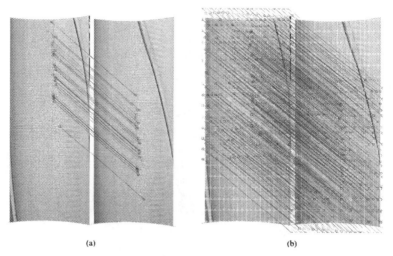

(a)   (b)

Fig. 5. (a) SIFT feature matching algorithm guided by homography. (b) feature matching algorithm guided by local homography with meshing.

4   Experimental Results

To verify the key parts of the above, we tested several groups of data collected from a cable-stayed bridge in Kaifu district, Changsha city named Hongshan Bridge as shown in Fig. 6. Hongshan Bridge is the world's largest span cable-stayed bridge without back cable pylon, and the only concrete pylon bridge with a height of more than 100 m [24].

The experimental result is shown in Fig. 7, demonstrating that the proposed algorithm are suitable for cable detection.

Fig. 6. The data collection in Hongshan Bridge.

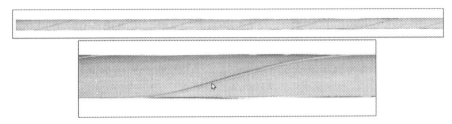

Fig. 7. The result of stitching.

5   Conclusion

In this work, we conduct digital image processing on the damage detection of cable surface. The core work is the splicing of the cable surface with few features to obtain a complete cable image. The preliminary work is to extract the cable area from the complex background accurately, then we use the derived cylindrical back-projection formula to rectify it. Finally through a series of feature extraction and matching, stitching, we obtain a complete cylinder image. The experimental results show that the system can be applied to actual cable detection.

References

1. Li, F., Hui, S.J., Ou, T.: The state of the art in structural health monitoring of cable-stayed bridges. J. Civ. Struct. Health Monit. **6**(1), 1–255 (2015)
2. Mehrabi, F.B., Armin, F.: In-service evaluation of cable-stayed bridges, overview of available methods and findings. J. Bridg. Eng. **11**(6), 716–724 (2006)
3. Tabataba, H.F.: Inspection and maintenance of bridge stay cable systems-NCHRP Synthesis 353. Washington, DC 20001, USA: Transportation Research Board (2005)
4. Eadon Consulting Homepage. http://www.eadonconsulting.co.uk/ProjectHumberBridgeMainCableInspection.aspx. Accessed 06 April 2020
5. StartTribune. http://www.startribune.com/local/minneapolis/141318253.html. Accessed 06 April 2020

6. HNF, Ho., Kim, K.D.S., Park, Y.S.T.: An efficient image-based damage detection for cable surface in cable-stayed bridges. NDT and E Int. **58**, 18–23 (2013)
7. Pahwa, F., Ramanpreet Singh, S.: Feature-less Stitching of Cylindrical Tunnel (2018)
8. Xinke, L.F., Chao, G.S., Yongcai, G.T.: Cable surface damage detection in cable-stayed bridges using optical techniques and image mosaicking. Opt. Laser Technol. **110**, 36–43 (2018)
9. Yuheng, F., Song Hao, S., Yan, T.: Image Segmentation Algorithms Overview(2017)
10. Kundu, M.K.F., Pal, S.K.S.: Thresholding for edge detection using human psychovisual phenomena. Pattern Recogn. Lett. **4**(6), 433–441 (1986)
11. Macqueen, J.F.: Some methods for classification and analysis of multivariate observations. In: Proceedings of Berkeley Symposium on Mathematical Statistics & Probability. (1965)
12. Johnson, S.F.: Hierarchical clustering schemes. Psychometrika **32**(3), 241–254 (1967)
13. Son, L.H.F., Cuong, B.C.S., Lanzi, P.L.T.: A novel intuitionistic fuzzy clustering method for geo-demographic analysis. Expert Syst. Appl. **39**(10), 9848–9859 (2012)
14. Luxburg, U.V.F.: A tutorial on spectral clustering. Stat. Comput. **17**(4), 395–416 (2007)
15. Ester, M., Kriegel, H.P., Sander, J., Xu, X.: A density-based algorithm for discovering clusters in large spatial databases with noise. In: KDD. pp. 226–231 (1996)
16. Minghua, F., Xu, L.V.S., Wang, T.: Universal back-projection algorithm for photoacoustic computed tomography. Phys. Rev. E Stat. Nonlin. Soft Matter Phys. **71**(12), 016706 (2005)
17. Li, X.F., Shi, Z.S., Guo, D.T.: Reconstruct argorithm of 2D barcode for reading the QR code on cylindrical surface. In: International Conference on Anti-counterfeiting, IEEE (2013)
18. Lay, K.T.F., Wang, L.J.S., Wang, C.H.T.: Rectification of QR-code images using the parametric cylindrical surface model. In: International Symposium on Next-generation Electronics, IEEE (2015)
19. Harris, F.C., Stephens, S.: A combined corner and edge detector. In: Proceedings 4th Alvey Vision Conference. pp. 147–151 (1988). https://doi.org/10.5244/c.2.23
20. Rosten, E., Drummond, T.: Machine learning for high-speed corner detection. In: Leonardis, A., Bischof, H., Pinz, A. (eds.) ECCV 2006. LNCS, vol. 3951, pp. 430–443. Springer, Heidelberg (2006). https://doi.org/10.1007/11744023_34
21. Lowe, D.G.F.: Distinctive image features from scale-invariant keypoints. Int. J. Comput. Vis. **60**(2), 91–110 (2004)
22. Bay, H., Tuytelaars, T., Van Gool, L.: SURF: speeded up robust features. In: Leonardis, A., Bischof, H., Pinz, A. (eds.) ECCV 2006. LNCS, vol. 3951, pp. 404–417. Springer, Heidelberg (2006). https://doi.org/10.1007/11744023_32
23. Julio, Z.F., Tat-Jun Chin, S., Michael, S., Brown, T.: As-projective-as-possible image stitching with moving DLT. In: Computer Vision & Pattern Recognition (2013)
24. Wikipedia "Hongshan Bridge(Changsha)" entry.https://zh.wikipedia.org/w/indexphp?title=%E6%B4%AA%E5%B1%B1%E5%A4%A7%E6%A1%A5_(%E9%95%BF%E6%B2%99)&oldid=57604554. Accessed 06 April 2020

Principal Semantic Feature Analysis
with Covariance Attention

Yuliang Chen[1], Yazhou Liu[1(✉)], Pongsak Lasang[2], and Quansen Sun[1]

[1] The School of Computer Science and Engineering,
Nanjing University of Science and Technology, Nanjing 210094, China
{cyl_1024,yazhouliu}@njust.edu.cn
[2] The Panasonic R&D Center Singapore, Singapore 469332, Singapore

Abstract. In this work, we present a new module for semantic segmentation. This new module is designed as a plug in module for the backbone networks to further boosting the segmentation performance using the principal semantic feature analysis with covariance attention. Specifically, the spatial and channel covariance attention module are designed respectively, which can filter noisy regions and help the CNN to adaptively extract the dominant semantic content. By using the proposed covariance attention modules, a covariance attention architecture is built over FCN. Experimental results demonstrate the substantial benefits brought by the proposed covariance attention scheme, and show that the covariance attention mechanism is feasible and effective for improving the accuracy of semantic segmentation.

Keywords: Image segmentation · Neural networks · Covariance matrices · Hand-engineered features

1 Introduction

Semantic segmentation, as a fundamental task in computer vision, has been extensively studied whose goal is to assign a semantic label for every pixel of image. The success of many artificial intelligence applications, such as autonomous driving [5, 6], medical diagnosis [9, 10], and image synthesis [12], are conditioned on the accuracy of the semantic segmentation. One of the breakthroughs is fully convolutional network [2] (FCN) and many state-of-the-art models are based on it [16, 17], and remarkable progress has been made.

Recent efforts have revealed the importance of the attention modules for deep convolutional neural networks [18, 19] based computer vision tasks. In general, for a visual recognition task, the neural network is often required to emphasize certain content while avoiding distracting information. However, features extracted by the attention module can produce more consistent semantic information [9].

In order to extract attention features, many attention modules based on dependencies between data distribution have been designed for complex visual analysis. Non-local neural networks [22] utilized a self-attention mechanism to captured long-range dependence

---

Y. Chen—Student as the first author.

Y. Peng et al. (Eds.): PRCV 2020, LNCS 12306, pp. 217–229, 2020.
https://doi.org/10.1007/978-3-030-60639-8_19

information between pixels in image for object detection. Yuan et al. [23] employed an object context pooling attention scheme to learn the object context map recording the similarities between all the pixels and the associated pixel for scene parsing. Specially, in SeNet [24], Squeeze-and-Excitation block has been proposed to improve the quality of dominant semantic representations by explicitly modeling the interdependencies between channel features.

While the above-mentioned attention modules are successful in improving recognition precision for vision tasks, the following aspects are ignored: (1) The original feature usually needs to be transformed into another feature with different shape, such as scalars, vectors, to adapt to the calculation process of dependence, which may interfere with the original spatial distribution of the feature to some degree. (2) Because of the high volume of the feature maps, the computation of attention matrices or tensor encoding dependencies normally needs high time and space complexity. (3) Currently, all the features are learnt by network and the hand-engineered feature, which might provide complementary information and human prior for learning, are abandoned.

In this work, we formulate the dependencies over local and global clues as a covariance matrix projection process to obtain more consistent feature representation. More specially, motivated by design of image covariance matrix presented on the 2DPCA [25], we proposed the feature covariance matrix (FCM), which can be directly constructed via using the original feature map without breaking the its spatial structure. Using the FCM to explicitly model the dependencies between the data distribution of feature maps, the feature representation ability can be further enhanced. Sine there is no feature shape conversion, the proposed approach, feature covariance attention mechanism, has lower space and time complexity. It enables CNN network to further enhance the dominating feature and filter irrelevant information.

Specifically, we incorporate the covariance matrix analysis, which is an important tool in classic pattern recognition, into the design of attention modules for deep semantic segmentation framework. Similar to [17], the spatial and channel covariance attention modules are built by adopting the proposed FCM. In addition, inspired by the covariance descriptor [26] which exploits the dependency between heterogeneous statistics to boost the accuracy of classic vision tasks before the deep learning era, we further integrate the FCM with the hand-engineered features and expect richer spatial feature representation can be obtained by combining the handcraft and learnt features.

The overview of the proposed framework is depicted in Fig. 1 and the contribution of this work lies in following aspects:

1) A novel approach that exploits covariance matrix to generate attention matrix encoding the dependencies over local and global clues is proposed, which has lower complexity.
2) Base on this method, the spatial and channel covariance attention modules are designed respectively, which can boost the accuracy of semantic segmentation.
3) An edge attention feature is presented to retain and enhance the presentation of details by combining the hand-engineered and learnt features.
4) We build a network named CANet for scene parsing by using above mentioned covariance attention modules and achieve very competitive performance on multiple challenging datasets.

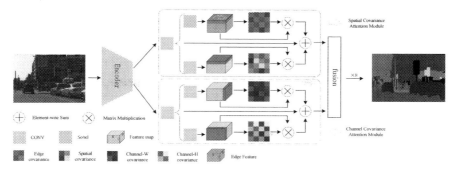

Fig. 1. Overview of the proposed covariance attention semantic segmentation network.

2 Related Work

Semantic Segmentation. Recent CNN based methods [11, 27, 28] for semantic segmentation have achieved significant success. Fully convolutional network (FCN) [2], firstly replace the last full connected layer with a deconvolution layer to get high level and highly refined features. In [10], the features in encoder are used to combined to the output features within decoder. The atrous convolution proposed by Chen et al. [13, 29, 30] can expand the receptive field without extra parameters. In DeepLab v2 [30], the researchers proposed a multi-scale robust semantic segmentation method, atrous space pyramid pool (ASPP). Later, DeepLab v3 [31] utilized image-level features over ASPP to encode global context. Similar to ASPP, Zhao et al. [14] used pyramid pooling module to integrate the information of local and global context in the PSPNet. Yang et al. [31] skillfully combined ASPP module with dense connection presented in Densenet [32] to model DenseASPP for wider field of perception.

Attention Mechanism. Attention model had been proved very successful in many vision tasks, including detection [22], classification [24] and the semantic segmentation [16, 17]. There are basically two strategies to build attention mechanism so far: (1) learn a weight vector or mask by a new branch in neural network to reconstruct raw features toward emphasizing the important local regions or channel-wise features and filtering irrelevant information, such as [24, 33] (2) explicitly model rich contextual dependencies over local and global features, in which, generally, a huge correlation matrix or tensor between each spatial point features is calculated to enhance the dominating semantic extraction relative to inessential information, such as [16, 17, 22].

Covariance Analysis. Covariance is widely used in the areas of pattern recognition, computer vision and signal processing, etc. [34, 35]. One of the most famous of these is principal component analysis (PCA), a feature extraction and data representation technique. In order to reduce the expenditure of memory and time in traditional PCA, Yang et al. [25] propose the 2DPCA that can obtain "image covariance matrix", directly from original images. Zhang et al. [36] explicitly indicated 2DPCA which is working in the column direction of images when evaluating a image covariance matrix. For efficient image representation and recognition, the researchers develop the 2DPCA, i.e. (2D)2PCA [36] by simultaneously considering the row and column directions. In this paper, we refer to 2DPCA, and exploits feature covariance matrices to model the dependencies over local and global clues.

3   Covariance Attention Mechanism

In the section, the design of spatial and channel covariance attention module is firstly provided. Then, the architecture of the proposed network (CANet) for semantic segmentation will be descried.

3.1   Spatial Covariance Attention Module

The spatial covariance attention module (SCA) is designed to focus the principal spatial distribution. As shown in Fig. 2, let $\mathcal{M} \in \mathbb{R}^{C \times W \times H}$ be the output features at the last layer of encoder. Two convolution layers comprised of 1×1 filter are applied on $\mathcal{M}$, individually, to generate two feature maps $\mathcal{F} \in \mathbb{R}^{C' \times W \times H}$ and $\mathcal{B} \in \mathbb{R}^{C \times W \times H}$. Split $\mathcal{F}$ according to the channel dimension, $\mathcal{F} = \left\{ P^1, P^2, P^3 \cdots P^{C'} \right\}$. Inspired by image covariance matrix [25], we define the following matrix:

$$Cov^s = \frac{1}{C'} \sum_{i=1}^{C'} \left(P^i - \bar{P} \right)^T \left(P^i - \bar{P} \right) \tag{1}$$

where $P^i \in \mathbb{R}^{W \times H}$ and $\bar{P} = \frac{1}{C'} \sum_i^{C'} P^i$. The matrix $Cov^s \in \mathbb{R}^{H \times H}$ is called the spatial feature covariance matrix. By applying softmax on the spatial feature covariance matrix Cov^s, corresponding spatial covariance attention matrix $\mathcal{H}^s \in \mathbb{R}^{H \times H}$ can be denoted by

$$\mathcal{H}^s_{i,j} = \frac{exp\left(Cov^s_{i,j} \right)}{\sum_{j=1}^{H} exp\left(Cov^s_{i,j} \right)} \tag{2}$$

where $Cov^s_{i,j}$ represents the (i, j) element in covariance matrix Cov^s, and $\mathcal{H}^s_{i,j}$ can be consider as the dependency between the ith column and the jth column of spatial

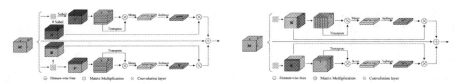

Fig. 2. The detail of spatial covariance attention module.   **Fig. 3.** The detail of channel covariance attention module.

features. Then, the original feature maps are reconstructed by performing matrix multi-plication between the spatial covariance matrix $\mathcal{H}^s$ and channel features $\mathcal{B}$ for extracting the dominant semantic information, and thus the primary spatial feature (PSF) $\mathcal{M}^*$ can be expressed by

$$\mathcal{M}^*_j = \lambda\mathcal{B}_j \cdot \mathcal{H}^s + \mathcal{M}_j \tag{3}$$

where the $\mathcal{M}^*_j$ is the attention feature of jth channel feature in $\mathcal{M}$ after covariance matrix projection. λ is a learnable coefficient of weight.

Additionally, although primary spatial information contained in feature map can be captured by the spatial covariance attention operation, some elaborate details of raw fea-ture, including boundary, texture and so on, are also crucial to the semantic segmentation. To enrich the representation of primary spatial feature extracted by attention modules, an edge attention feature that is generated by combining the hand-engineered edge features with a covariance attention operation is proposed. To be specific, two fixed convolution kernels, Sobel, are used on learnt feature provided by neural network to extract edge features, and then they are enhanced by covariance attention operation.

First of all, two tensors, $T \in \mathbb{R}^{C' \times W \times H}$ and $N \in \mathbb{R}^{C \times W \times H}$, are obtained by two convolution layers with 1 x 1 filter on the feature $\mathcal{M}$, respectively. Then, first-derivative $\left\{T'_x, T'_y\right\}$ and $\left\{N'_x, N'_y\right\}$ in X direction and Y direction of T and N are computed respectively by two Sobel filters and their magnitude T' and N' can be calculated by Eqs. (4) and (5). As shown in the top part of Fig. 2, similar to the feature $\mathcal{F}$, the edge feature magnitude T' will be feed to the spatial feature covariance matrices generator for an edge covariance attention matrix $\mathcal{H}^{edge} \in \mathbb{R}^{H \times H}$.

$$T' = \sqrt{T'^2_x + T'^2_y} \tag{4}$$

$$N' = \sqrt{N'^2_x + N'^2_y} \tag{5}$$

Finally, the edge attention feature (EAF) E after covariance matrix projection can be denoted by

$$E_j = N'_j \cdot \mathcal{H}^{edge} \tag{6}$$

where the E_j and N'_j indicate j^{th} channel feature in E and N', respectively. Since the edge attention feature E is essentially spatial feature, it is merged with the primary spatial features $\mathcal{M}^*$ by summation. Therefore, $\mathcal{M}'$, the spatial attention features with edge attention can be formulated as:

$$\mathcal{M}'_j = \lambda\mathcal{B}_j \cdot \mathcal{H}^s + \mu N'_j \cdot \mathcal{H}^{edge} + \mathcal{M}_j \tag{7}$$

where both of λ and μ are the learnable coefficient of weight. Using the covariance matrix to model the dependencies over local and global clues, the spatial attention module is capable of producing more consistent semantic information.

3.2 Channel Covariance Attention Module

In order to highlight the primary channel feature and reduce irrelevant channel information, a channel covariance attention module (CCA) utilizing channel covariance matrices is proposed. Different from the SE block [24], where all contextual semantic information within each channel panel is coarsely aggregated to a scalar value, these elements in each column or row in feature map are taken as the presentation of each channel feature and then the average of a series of covariance matrices is used as the final relation descriptor between each channel. The details are shown in the Fig. 3, follow the design of spatial covariance attention module, two convolution layers with 1×1 filter are utilized on $\mathcal{M}$ so that two original feature maps $\mathfrak{R}'$ and $\mathfrak{R}''$ can be obtained for evaluating a pair of channel covariance matrices with different direction. Then we fix channel direction and split $\mathfrak{R}', \mathfrak{R}''$ according to the width and height dimensions of feature, respectively. Let $\mathfrak{R}' = \{ \boldsymbol{Q}^1, \boldsymbol{Q}^2, \boldsymbol{Q}^3 \cdots \boldsymbol{Q}^W \}$ and $\mathfrak{R}'' = \{ \boldsymbol{R}^1, \boldsymbol{R}^2, \boldsymbol{R}^3 \cdots \boldsymbol{R}^H \}$, $\boldsymbol{Q}^i \in \mathbb{R}^{H \times C}$ and $\boldsymbol{R}^i \in \mathbb{R}^{W \times C}$. Two analogous channel feature covariance matrices can be separately presented by

$$Cov^{C_0} = \frac{1}{W} \sum_{i=1}^{W} \left(\boldsymbol{Q}^i - \bar{\boldsymbol{Q}} \right)^T \left(\boldsymbol{Q}^i - \bar{\boldsymbol{Q}} \right) \tag{8}$$

$$Cov^{C_1} = \frac{1}{H} \sum_{i=1}^{H} \left(\boldsymbol{R}^i - \bar{\boldsymbol{R}} \right)^T \left(\boldsymbol{R}^i - \bar{\boldsymbol{R}} \right) \tag{9}$$

where the Cov^{C_0} and $Cov^{C_1} \in \mathbb{R}^{C \times C}$. Then, softmax operation:

$$\mathcal{H}_{i,j}^{C_0} = \frac{\exp\left(Cov_{i,j}^{C_0} \right)}{\sum_{j=1}^{C} \exp\left(Cov_{i,j}^{C_0} \right)} \tag{10}$$

$$\mathcal{H}_{i,j}^{C_1} = \frac{\exp\left(Cov_{i,j}^{C_1} \right)}{\sum_{j=1}^{C} exp\left(Cov_{i,j}^{C_1} \right)} \tag{11}$$

in which $\mathcal{H}_{i,j}^{C_0}$ or $\mathcal{H}_{i,j}^{C_1}$ denotes a correlation coefficient of ith channel relative to j^{th} channel feature. Using the channel covariance attention matrices to map $\mathfrak{R}'$ and $\mathfrak{R}''$ respectively, the channel attention feature can be easily acquired. Then the two channel attention features are fused by simply adding. Thus, the channel attention feature (CAF) can be presented by following equation:

$$\mathcal{M}_j^\circ = \gamma \left(\mathfrak{R}_j' \cdot \mathcal{H}^{C_0} + \mathfrak{R}_j'' \cdot \mathcal{H}^{C_1} \right) + \mathcal{M}_j \tag{12}$$

where $\mathcal{M}_j^\circ$ expresses the jth channel attention feature in $\mathcal{M}$ after covariance matrix projection, and the γ is trainable scalar quantity. By using the covariance matrix to model the dependencies between channel features, the channel covariance attention module is able to obtain more consistent semantic information. Illustrative examples are represented in Figs. 4 and 5, which show the contributions of different modules.

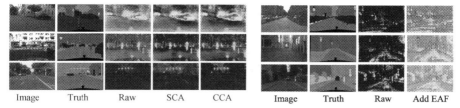

Image   Truth   Raw   SCA   CCA   Image   Truth   Raw   Add EAF

Fig. 4. Visualization of the effects of the attention module on Cityscapes validation set. The 1, 2 and 3 rows respond respectively to the 'car', 'person', 'traffic light' class.

Fig. 5. Visualization result of the effects of edge attention feature for a certain channel feature.

3.3 Network Architecture

As illustrated in Fig. 1, we adopt the popular convolutional network ResNet as our backbone and two dilated convolutions with stride 2 and 4 are severally employed in the last two ResNet block for high resolution feature maps. Then, the spatial and channel covariance attention module are directly joint in parallel. By leveraging the two proposed covariance attention modules, the dominant semantic representations contained in the features that produced by backbone will be adaptively captured, which is helpful to enhanced original feature. And the majority of fusion module consists of three convolution layers with 1×1 filter, in which two filters are employed to handle the output of spatial and channel covariance attention modules, respectively. Then, an element-wise sum operation is performed on the spatial attention features and channel attention features. The final fusion features will be used to generate pixel-level prediction.

4 Experiments

In this section, the CANet is evaluated on multiple datasets and the experimental results, including the qualitative and quantitative analysis of the proposed attention module, are presented.

4.1 Dataset

Cityscapes is a semantic segmentation dataset focusing on urban street scenes from 50 cities, which contains 5000 images with pixel-level labels and 20000 images with coarse annotations that involve 19 categories of both objects and stuffs. Each image has
    2048×1024 pixels. In this work, only the finely annotated images are used and they are divided into 2,975/500/1,525 images for training, validation, and testing, respectively.
    Pascal Context contains 4,998 images for training and 5105 images for testing, all of which have detailed semantic labels for wholes scenes. This dataset is re-annotated from Pascal VOC and there are 60 classes used for evaluation.
    ADE20K, as a scene parsing benchmark, provides more than 20k fully annotated images with 150 objects and stuff categories. It has up to 1,038 different image-level labels in total. The dataset is divided as follows: 20,210 for training, 2,000 for validation and 3,000 for testing.

4.2 Training and Implementation Details

Following [16], the prediction outputs of CANet are bilinearly upsampled by a factor of 8 to calculate the loss. The standard SGD optimizer is used to optimize the entropy loss, and the momentum coefficient is set to 0.9 and the weight decay coefficient is set to 0.0001. All the models are initialized using ImageNet pre-trained weights. In addition, we adopt a poly learning rate policy: the initial learning rate is multiplied by $1 - \left(\frac{iter}{\max\_iter}\right)^{power}$ with power = 0.9. The base learning rates for cityscapes and ADE20k are set to 0.003, and for Pascal-context, it is set to 0.001. We randomly crop a 375×375 region on Pascal-context and ADE20k. For cityscapes, training images are randomly cropped to 768×768 region. Owing to the limited GPU memory, the batch size is set to 6.

4.3 Ablation Studies for Attention Modules

In the ablation study, extensive experiments are conducted on the validation set of Cityscapes for verifying the validity of proposed covariance attention module. The individual component of the full attention module is added to the dilated network one by one. To further explore the effectiveness of the edge attention features, it was separated from spatial attention features for better observation. The results are summarized in Table 1. PSF, CAF, EAF respectively represent primary spatial feature, channel attention feature, edge attention feature. We can observe that the proposed covariance attention modules can remarkably improve the accuracy of scene parsing by leveraging feature covariance matrix to model the dependencies over local and global information to focus the primary semantic information. Compare with the bassline, denoted as dilated FCN, incorporating the primary spatial feature (PSF) improves the performance by 2.5–3.5% no matter how deep the backbone network is, which proves the effectiveness of the primary spatial feature. In additional, by adding the channel attention feature to baseline, the performance is also improved by 2–3%. When these two types of attention feature are integrated together into the dilated FCN, the result represents an improvement of 4.5% for ResNet50. Even for a deeper network (Res101), a remarkable improvement is also yielded.

In addition, it can be observed that a substantial benefit, about 0.7%, can be obtained by using of edge attention features, which make the proposed network achieving 80.02% and 81.18% for different backbone respectively. In general, the results of ablation study demonstrate the feature covariance matrix has ability to model the dependencies over the local and global clues to further improves the performance of semantic segmentation.

4.4 Complexity Analysis

In recent work [17], a spatial attention module (PAM) and a channel attention module (CAM) are proposed, which are able to aggregate contextual information to model dependence for every pixel. To demonstrate the benefits and limitations between it and the proposed attention modules, their performances are evaluated on the Pascal-Context test.

For a fair comparison, both of the model are trained with same training set and parameters. The accuracies are shown in Table 2, and time complexity and space complexity

Table 1. Ablation study on the validation set of cityscapes

Method	BaseNet	PSF	CAF	EAF	mIoU
Dilated FCN	Res50				74.92
CANet	Res50		✓		76.83
CANet	Res50	✓			78.51
CANet	Res50	✓	✓		79.43
CANet	Res50	✓	✓	✓	80.02
Dilated FCN	Res101				77.41
CANet	Res101		✓		80.32
CANet	Res101	✓			79.97
CANet	Res101	✓	✓		80.41
CANet	Res101	✓	✓	✓	81.18

are also given in Table 3. It can be observed that the both methods have comparable performance in terms of mIoU. More specifically, the proposed attention modules perform slightly better than PAM and CAM for the deeper backbone and worse for the shallower backbone. But PAM is about an order of magnitude higher in time complexity than the proposed SCA and about two orders of magnitude higher in space complexity. For channel attention module, the CCA has the same time complexity and spatial complexity compared with CAM. From the accuracy point of view, CCA slightly outperforms the CAM when the backbone is Res101.

Table 2. Performance comparison of attention modules on Pascal context test.

Method	mIoU (Res50)	mIoU (Res101)
PAM	48.28	50.33
SCA (proposed)	48.22	49.85
CAM	48.21	49.81
CCA (proposed)	47.74	50.13
PAM + CAM	48.50	49.95
SCA + CCA	48.17	50.53

Table 3. Complexity analysis of attention methods.

Method	T(n)	S(n)
PAM	$O\left((WH)^2 C\right)$	$O\left((WH)^2\right)$
SCA	$O\left(WH^2 C\right)$	$O\left(H^2\right)$
CAM	$O\left(C^2 WH\right)$	$O\left(C^2\right)$
CCA	$O\left(C^2 WH\right)$	$O\left(C^2\right)$

From Table 3, another interesting aspect can be observed is that the proposed attention modules are more friendly to the high-resolution semantic segmentation, since its space and time complexity growth is much smaller than the baseline when the input feature map size became bigger (W and H).

4.5   Comparison with the State-of the-Arts

The benchmarking results against the state-of-the-art segmentation networks are presented in Tables 4 and 5. In addition, since the proposed method is relevant to DANet [17] that recently achieves very impressive results on multiple benchmark datasets, we retrain the DANet model with same training parameters and environment as proposed CANet for a fair comparison.

Table 4. ADE20K performance.

Methods	BaseNet	mIoU
SegNet [1]		21.64
FCN [2]		29.39
DilatedNet [4]		32.31
CiSS-Net [8]	Res50	42.56
RefineNet [11]	Res152	40.70
PSPNet [14]	Res101	41.96
UperNet [15]	Res101	42.66
DSSPN-Softmax [21]	Res101	42.03
DANet [17] (Retrained)	Res101	42.74
CANet (Proposed)	Res101	43.01

Table 5. Pascal-context performance.

Methods	BaseNet	mIoU
FCN-8 s [2]		37.8
ParseNet [3]		40.4
VeryDeep [7]		44.5
CiSS-Net [8]		48.7
RefineNet [11]	Res152	47.3
DeepLab-v2 [13]	Res101	45.7
PSPNeta [14]	Res101	47.8
EncNet [20]	Res101	51.7
DANet (Published)	Res101	52.6
DANet (Retrained)	Res101	50.0
CANet (Proposed)	Res101	50.5

ADE20k Datasets. Table 4 shows the comparison results on the validation set of ADE20k. Among all the compared algorithms, the proposed CANet performs the best, even if the dataset has very complicated scenes and diverse objects. Note that we do not use additional data and online hard example mining (OHEM) strategy in this experiment, which can further boost the performance.

Pascal-Context Datasets. To further demonstrate the generality of the proposed attention method for semantic segmentation, we carry out experiments on Pascal-context, and the results are listed in Table 5. Compared with the published methods, we can obviously observe that the CANet built by using the covariance attention method obtains a competitive accuracy without bells and whistles. Moreover, CANet has better result rather than DANet under the same training set, which further indicates that the proposed covariance attention mechanism can effectively capture the dominant semantic information via employing feature covariance matrix as a projection to promote segmentation effect.

4.6 Qualitative Analysis

We further adopt the method introduced in [37] to visualize the changes in the features before and after the covariance attention module. As shown in Fig. 4, for each input image, we randomly select a class as a representative and show its corresponding raw features and attention features in columns 2, 3 and 4. Intuitively, by modeling the dependencies over local and global information, the covariance attention feature finally aggregate richer contextual information compared with the original features. For example, in the third row, the traffic light area reacts more sensitively after attention processing, and the sky, which is often the background of traffic lights, becomes brighter.

In addition, we specifically visualize the impact of EAF. As shown in Fig. 5, When the original features are fused with EAF features, the edges within features become clearer, which helps the CANet to improve the segmentation effect.

5 Conclusion

In this paper, a new attention module, covariance attention, has been presented in this work. This new method is attractive for the following reasons: 1) Covariance matrix is used as a new attention module to model the global and local dependency for the feature maps and the local and global dependency is formulated as simple matrix projection process; 2) Since covariance matrix can encode the joint distribution information for the heterogeneous yet complementary statistics, the hand-engineered features is combined with the learnt feature effectively using covariance matrix to boosting the segmentation performance; 3) A semantic segmentation framework based covariance attention mechanism is proposed and very competitive performance have been obtained.

References

1. Badrinarayanan, V., Kendall, A., Cipolla, R.: SegNet: a deep convolutional encoder-decoder architecture for image segmentation. IEEE Trans. Pattern Anal. Mach. Intell. **39**, 2481–2495 (2017)
2. Jonathan Long, E.S., Darrell, T.: Fully convolutional networks for semantic segmentation. In: IEEE Conference on Computer Vision and Pattern Recognition 2015, pp. 3431–3440 (2015)
3. Liu, W., Rabinovich, A., Berg, A.C.: ParseNet: looking wider to see better. In: IEEE Conference on Computer Vision and Pattern Recognition (2015)
4. Fisher Yu, V.K.: Multi-scale context aggregation by dilated convolutions. arXiv:1511.07122 (2016)
5. Wei, J., He, J., Zhou, Y., Chen, K., Tang, Z., Xiong, Z.: Enhanced object detection with deep convolutional neural networks for advanced driving assistance. IEEE Trans. Intell. Transp. Syst. 1–12 (2019)
6. Teichmann, M., Weber, M., Zöllner, M., Cipolla, R., Urtasun, R.: MultiNet: real-time joint semantic reasoning for autonomous driving, In:IEEE Intelligent Vehicles Symposium, pp. 1013–1020 (2018)
7. Zifeng Wu, C.S., van den Hengel, A.: Bridging category-level and instance-level semantic image segmentation. arXiv:1605.06885 (2016)
8. Zhou, Y., Sun, X., Zha, Z., Zeng, W.: Context-reinforced semantic segmentation. In: The IEEE Conference on Computer Vision and Pattern Recognition, pp. 4046–4055 (2019)

9. Sinha, A., Dolz, J.: Multi-scale guided attention for medical image segmentation. In: IEEE Conference on Computer Vision and Pattern Recognition (2019)
10. Ronneberger, O., Fischer, P., Brox, T.: U-Net: convolutional networks for biomedical image segmentation. In: Navab, N., Hornegger, J., Wells, William M., Frangi, Alejandro F. (eds.) MICCAI 2015. LNCS, vol. 9351, pp. 234–241. Springer, Cham (2015). https://doi.org/10.1007/978-3-319-24574-4_28
11. Lin, G., Milan, A., Shen, C., Reid, I.: RefineNet: multi-path refinement networks for high-resolution semantic segmentation. In: IEEE Conference on Computer Vision and Pattern Recognition, pp. 5168–5177 (2017)
12. Wang, T., Liu, M., Zhu, J., Tao, A., Kautz, J., Catanzaro, B.: High-resolution image synthesis and semantic manipulation with conditional GANs. In: IEEE/CVF Conference on Computer Vision and Pattern Recognition, pp. 8798–8807 (2018)
13. Chen, L., Papandreou, G., Kokkinos, I., Murphy, K., Yuille, A.L.: DeepLab: semantic image segmentation with deep convolutional nets, atrous convolution, and fully connected CRFs. IEEE Trans. Pattern Anal. Mach. Intell. **40**, 834–848 (2018)
14. Zhao, H., Shi, J., Qi, X., Wang, X., Jia, J.: Pyramid scene parsing network. In: IEEE Conference on Computer Vision and Pattern Recognition, pp. 6230–6239 (2017)
15. Tete Xiao, Y.L., Zhou, B., Jiang, Y., Sun, J.: Unified perceptual parsing for scene understanding. In: IEEE Conference on Computer Vision and Pattern Recognition (2018)
16. Huang, Z., et al.: CCNet: criss-cross attention for semantic segmentation. In: IEEE International Conference on Computer Vision (2019)
17. Fu, J., et al.: Dual attention network for scene segmentation. In: IEEE Conference on Computer Vision and Pattern Recognition (2019)
18. Lecun, Y., Bottou, L., Bengio, Y., Haffner, P.: Gradient-based learning applied to document recognition. Proc. IEEE **86**, 2278–2324 (1998)
19. Krizhevsky, A., Sutskever, I., Hinton, G.E.: ImageNet classification with deep convolutional neural networks. In: Conference and Workshop on Neural Information Processing Systems, pp. 1097–1105 (2012)
20. Zhang, H., et al.: Context encoding for semantic segmentation. In: IEEE/CVF Conference on Computer Vision and Pattern Recognition, pp. 7151–7160 (2018)
21. Liang, X., Xing, E., Zhou, H.: Dynamic-structured semantic propagation network. In: IEEE/CVF Conference on Computer Vision and Pattern Recognition, pp. 752–761 (2018)
22. Wang, X., Girshick, R., Gupta, A., He, K.: Non-local neural networks. In: IEEE/CVF Conference on Computer Vision and Pattern Recognition, pp. 7794–7803 (2018)
23. Yuan, Y., Wang, J.: OCNet: object context network for scene parsing. In: IEEE Conference on Computer Vision and Pattern Recognition (2018)
24. Hu, J., Shen, L., Sun, G.: Squeeze-and-excitation networks. In: IEEE/CVF Conference on Computer Vision and Pattern Recognition, pp. 7132–7141 (2018)
25. Jian, Y., Zhang, D., Frangi, A.F., Jing-yu, Y.: Two-dimensional PCA: a new approach to appearance-based face representation and recognition. IEEE Trans. Pattern Anal. Mach. Intell. **26**, 131–137 (2004)
26. Tuzel, O., Porikli, F., Meer, P.: Pedestrian detection via classification on Riemannian manifolds. IEEE Trans. Pattern Anal. Mach. Intell. **30**, 1713–1727 (2008)
27. Zhao, H., Qi, X., Shen, X., Shi, J., Jia, J.: ICNet for real-time semantic segmentation on high-resolution images. arXiv:1704.08545 (2018)
28. Wang, Y., et al.: LEDNet: a lightweight encoder-decoder network for real-time semantic segmentation. arXiv:1905.02423 (2019)
29. Chen, L.C., Papandreou, G., Schroff, F., Adam, H.: Rethinking atrous convolution for semantic image segmentation. arXiv:1706.05587 (2017)

30. Chen, L.C., Papandreou, G., Kokkinos, I., Murphy, K., Yuille, A.L: Semantic Image Segmentation with deep convolutional nets and fully connected CRFs. arXiv:1412.7062 (2014)
31. Yang, M., Yu, K., Zhang, C., Li, Z., Yang, K.: DenseASPP for semantic segmentation in street scenes. In: IEEE/CVF Conference on Computer Vision and Pattern Recognition, pp. 3684–3692 (2018)
32. Huang, G., Liu, Z., van der Maaten, L., Weinberger, K.Q.: Densely connected convolutional networks. In: IEEE Conference on Computer Vision and Pattern Recognition, pp. 2261–2269 (2017)
33. Zhao, H., et al.: PSANet: point-wise spatial attention network for scene parsing. In: Ferrari, V., Hebert, M., Sminchisescu, C., Weiss, Y. (eds.) ECCV 2018. LNCS, vol. 11213, pp. 270–286. Springer, Cham (2018). https://doi.org/10.1007/978-3-030-01240-3_17
34. Kirby, L.S.M.: Application of the KL procedure for the characterization of human faces. IEEE Trans. Pattern Anal. Mach. Intell. **12**, 103–108 (1990)
35. Jain, A.K., Duin, R.P.W., Jianchang, M.: Statistical pattern recognition: a review. IEEE Trans. Pattern Anal. Mach. Intell. **22**, 4–37 (2000)
36. Zhang, D., Zhou, Z.-H.: $(2D)^2$PCA: two-directional two-dimensional PCA for efficient face representation and recognition. Neurocomputing **69**, 224–231 (2005)
37. Selvaraju, R.R., Cogswell, M., Das, A., Vedantam, R., Parikh, D., Batra, D.: Grad-CAM: visual explanations from deep networks via gradient-based localization. Int. J. Comput. Vision **128**(2), 336–359 (2019). https://doi.org/10.1007/s11263-019-01228-7

Hierarchical Fusion for Gender Recognition Based on Hand Images

Manman Zhang, Yuchun Fang$^{(\boxtimes)}$ ⓘ, and Yifan Li

School of Computer Engineering and Science, Shanghai University, Shanghai, China
ycfang@shu.edu.cn

Abstract. Human hand features contain rich gender information. In this paper, we propose an approach to boost the performance of a given gender recognizer through the hierarchical fusion of human hand subspace features, texture features, and geometric features. We first use Eigenhand to extract subspace features. Hand texture features are obtained by applying local binary mode histograms. Then we get the geometric features by calculating the length ratio of fingers other than the thumb. We call this method as Geometric Gender Descriptor of Hand Images (GGD-H). Based on these, we use the serial strategy to fuse texture features and geometric features in the feature-level and then fuse its classification result with the subspace feature in the decision-level. The fused vectors are used as the final feature vectors to feed the support vector machine for our gender recognition task. Through this method, we can obtain gender difference features from multiple directions, thereby enhancing the perception ability of the same human hand in different scenes and the robustness of feature expression. The final experimental results show that the method proposed in this paper achieves a recognition rate of 0.988 on 11K Hands, exceeding the common hand gender recognition scheme.

Keywords: Gender recognition · Hand feature · Decision-level fusion

1 Introduction

Gender recognition is one of the foremost research topics of pattern recognition and computer vision. Computer systems with gender recognition capabilities have essential applications in many areas such as criminal investigations [6] and human handprints research in archeology [10]. Many studies show that the hand contains many unique features that can reveal gender information [1]. Compared with facial image models that vary with facial expressions and other factors, the variability of hand images is much lower. So the hand is suitable for feature extraction and model building [1]. As we all know, it is difficult for human eyes to

The work is supported by the National Natural Science Foundation of China under Grant No.: 61976132 and the National Natural Science Foundation of Shanghai under Grant No.: 19ZR1419200.

Y. Peng et al. (Eds.): PRCV 2020, LNCS 12306, pp. 230–241, 2020.
https://doi.org/10.1007/978-3-030-60639-8_20

judge the gender identity of an object based on the hand. However, from another perspective, its hard-to-recognize nature can precisely protect the privacy of users in the identity recognition system. Moreover, the storage requirements of hand images are low, and the amount of calculation is small. Therefore, it has significant advantages over facial recognition as well as a wide range of application needs and research prospects.

At present, there is no mainstream algorithm framework and widely used large data sets for gender recognition based on hand features. The lack of this point makes the relevant research lack a unified comparison standard. So the current research results are few. Researchers [3,11,12] often use hand textures or geometric features for gender recognition. The acquisition of texture features is relatively simple. But the surface of the hand skin is affected by many factors, which can easily interfere with gender recognition. Geometric features contain distinct information about gender differences, but they are not easy to obtain and may have geometric errors. Both methods have advantages and disadvantages. Currently, there is no research scheme to fuse the above two features.

In this paper, we propose an approach to boost the performance of a given gender recognizer through the multi-directional fusion of features. The illustration of our method is shown in Fig. 1. We first use the Eigenhand method to extract subspace features and use the Support Vector Machine (SVM) to do the first recognition. After that, the extracted textures and geometric features are fused in the feature-level for the second recognition. The feature-level fusion of texture features and geometric features can obtain more distinctive gender information than single features and reduce the redundant information caused by the correlation between the two features sets. For the extraction of texture features, we adopt the Local Binary Patterns Histograms (LBPH) method [2] for processing. We extract geometric features by the Geometric Gender Descriptor of Hand Images (GGD-H). Finally, the decision-level fusion is performed on the above two classification results to get the outcome. We further adopt decision-level fusion not only to sufficiently synthesize the performance results of the three kinds of features but also to avoid the problem of excessively high dimensions caused by the fusion of the three types of features at the feature layer. Therefore, we can enhance the robustness of the recognition system, improve the efficiency of the algorithm as well as facilitate the development of real-time recognition system.

The major contributions of our work are summarized as follows.

- We use Eigenhand, LBPH, and GGD-H for hand multi-directional feature extraction. The method improves the perception ability of the recognition scheme in different scenarios of the same hand and enhances the robustness of the system.
- We perform feature-level fusion and decision-level fusion for gender recognition to enhance the fault tolerance of the recognition method. The final accuracy is better than the recognition accuracy obtained by other existing studies.

The rest of this paper is organized as follows. In Sect. 2, we briefly review related work. Section 3 provides an overview of related feature extraction methods.

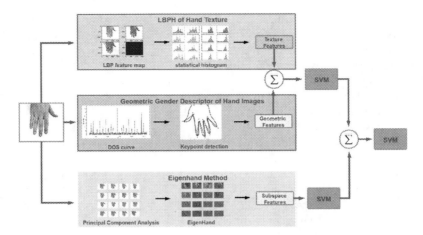

Fig. 1. The framework of hierarchical fusion algorithm for hand based gender recognition.

Section 4 proposes our fusion principles. In Sect. 5, we introduce the data set used and the experimental results of gender recognition. Finally, this paper is concluded in Sect. 6.

2   Related Work

Gender recognition is a binary classification problem, which requires to determine whether the given information belongs to a man or a woman according to certain feature information.

Brown *et al.* [5] proposed that the ratio of the index finger to ring finger contained gender information. The ratio of the index finger to the ring finger of the female is greater than that of males.

The geometric characteristics of the hand mainly include the palm width, length, and aspect ratio. And secondly, the relationship between the index finger and the ring finger. Most hand gender recognition base on geometric features. Amayeh *et al.* [3] used the region and boundary features based on Zernike moments and Fourier descriptors to describe the geometric features of the hand shape. Wu and Yuan [11] used palm width and palm aspect ratio for feature extraction. Xie *et al.* [12] proposed a gender recognition method based on the hand back skin texture (HBST), which describes the gender difference well.

In recent years, the use of Convolutional Neural Networks (CNN) has led to many state-of-the-art results in image classification problems. Mahmoud Afifi [1] used CNN to perform multi-level and complete feature extraction on hand images, which is an extremely practical feature extraction scheme.

3   Hand Image Representation

For high-dimensional original images, the correlation between each dimensional feature is high. The PCA method can not only reduce the data dimension but also remove the correlation among features. In this way, we can improve the efficiency of the algorithm and obtain an uncorrelated low-dimensional feature space. Besides, for the human hand, as a part of the body, it has texture features. When it is extracted as a geometric image, it also has geometric features. Therefore, obtaining multiple feature sets at the same time can enhance the robustness of feature expression. In this way, the recognition result will not cause errors due to the detection deviation of a specific feature.

3.1   Eigenhand

Principal Component Analysis (PCA) is an extremely effective data dimensionality reduction method.

The basic formula of PCA is a linear transform as shown in Eq. (1).

$$X_n \cdot W_k^T = X_k \tag{1}$$

For hand image, each row in the original data set X_n can represent a piece of hand information. X_k is the data of X_n reduced from n-dimension to k-dimension. $W_k^T \in \mathbb{R}^{k \times n}$. Each row of W_k^T can be regarded as a sample. That is, each row can also represent a type of hand information, which we name as "Eigenhand". Therefore, each row of W_k^T represents a principal component, and the importance of the samples decreases from the first row to the k-th row.

Eigenhand method can not only extract the subspace feature, but also reduce the dimension of the feature vector efficiently, which significantly improves the efficiency of the algorithm. The reshaped "Eigenhand" is shown in Fig. 2.

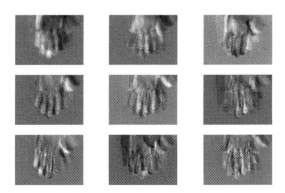

Fig. 2. The examples of Eigenhand

3.2   LBPH of Hand Texture

Local Binary Patterns Histograms (LBPH) method is proposed by Ahonen
et al. [2]. It is based on LBP operator [7]. For a pixel, the LBP value can generate
by comparing it with the pixels in its n neighbors as shown in Eq. (2).

$$lbp(x, y)_k = \begin{cases} 1 & (gray(x, y)_k \leq gray(x, y)) \\ 0 & (gray(x, y)_k > gray(x, y)) \end{cases} \tag{2}$$

where k represents the k-th point around the center pixel. Then the LBP value
of the center pixel can be obtained by Eq. (3).

$$lbp(x, y) = \sum_{k=0}^{neighbors-1} lbp(x, y)_k \cdot 2^n \tag{3}$$

We use 8-sample LBP operator, the feature value of each pixel is precisely
between 0 and 255, so an LBP map can be formed. Based on the LBP map,
the image can be divided into $x \times y$ sub-regions. For each region, a statistical
histogram h_i can be obtained. $H = [h_1, h_2...h_m]$, $j = 1, 2, ...m$, $m = x \times y$ is the
final LBPH vector of Hand Texture (LBPH-HT).

3.3   Hand Geometry Descriptor

Just like the human face, the hand also has geometric characteristics that contain
rich gender information. But these features cannot be included in the texture
feature. Therefore, it is considered to superimposing the texture feature and the
geometric feature of the hand to obtain a new combined feature vector.

Geometric Gender Descriptor of Hand Images. Dean R. Snow [9] and
Brown *et al.* [5] found that the ratio of the index finger to the ring finger of the
female is higher than that of male. The core idea of our method which we refer
to as Geometric Gender Descriptor of Hand Images (GGD-H) is to obtain the
gender difference by calculating the finger length ratios.

According to the keypoint detection method, the geometric position (x, y)
of each keypoint can be obtained. We record the geometric position of the i-th
point as $s_i = (x_i, y_i)$, $i = 1, ..., 10$. The lengths of the index finger (d_2), middle
finger (d_3), ring finger (d_4) and little finger (d_5) can be obtained by calculating
$|s_3 - s_4|$, $|s_5 - s_6|$, $|s_7 - s_8|$, $|s_9 - s_{10}|$. Finally we can get the ratio of index
finger to ring finger, middle finger and ring finger to little finger according to
Eq. (4).

$$\begin{cases} r_1 = \frac{d_2}{d_4} \\ \\ r_2 = \frac{d_2}{d_5} \end{cases} \tag{4}$$

Geometric Keypoints Detection of Hand Images. To form the GGD-H, we need to get the hand keypoints. The following are our two solutions.

DOS Curvature Method. The keypoint detection is mainly divided into the hand contour extraction, calculation of the curvature of each point on the outline, and obtain the keypoint whose curvature is the extreme value. We use the Difference of Slopes (DOS) [4] method to detect the curvature of the hand contours. For any point on the outline, we take two points before and after ω points from this point and form two vectors v_1, v_2 with this point. Then we calculate the angle θ between them as shown in Eq. (5) and obtain the contour curvature map of the hand.

$$\theta = \arccos\left(\frac{v_1 \cdot v_2}{|v_1|\,|v_2|}\right) \tag{5}$$

Theoretically, the extreme value protrusion on the contour curvature map corresponds to the part of the edge of the hand contour that changes suddenly, namely the fingertip or finger valley. So the keypoints of the hand can be found by extracting the extreme point.

OpenPose Method. The keypoint detection of the hand is mainly to find the geometric position of the fingertips, finger valleys, and the joint parts between them. The CMU Perceptual Computing Lab of Carnegie Mellon University released the OpenPose keypoint detection model [8]. We use this model for keypoints detection.

According to the Eq. (5), the curvature graph is shown in Fig. 3 when ω is 20.

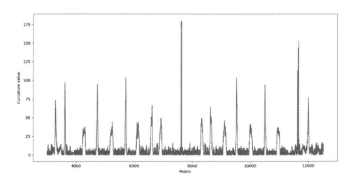

Fig. 3. The curvature of hand contour points

However, due to the dense point set of the hand contour, the feature points can not be extracted by selecting the top 10 points directly, as shown in Fig. 4(a). Therefore, we extract the extreme points of curvature at intervals of 40 to achieve the filtering effect. According to the above curvature diagram, the keypoint positions of each fingertip and finger valley can be obtained, as shown in Fig. 4(b). Figure 4(c) shows the detection results of OpenPose method. Then the length of each finger and its ratio r_1, r_2 can be calculated in order.

(a)   (b)   (c)

Fig. 4. Detection results of the two schemes: (a) Failed extraction example of scheme 2, (b) Successful extraction example of scheme 2, (c) Extraction example of scheme 2

4   Hierarchical Fusion Scheme

We get Eigenhand, LBPH-HT, and GGD-H in Sect. 3. Among them, the subspace feature covers the overall characteristics, while the texture feature and geometric feature separately cover certain aspects of the hand. Therefore, we fuse the texture and the geometric feature in the feature-level to get a more comprehensive feature. However, in order not to make the fused feature dimension too high, we do not blend the subspace feature in the feature-level. Instead, we use it and the above fused feature for the first recognition and then merge the recognition results of the two in the decision-level.

4.1   Feature-Level Fusion

We denote $M \in \mathbb{R}^{m \times n}$ as the LBPH-HT matrix, where m is the number of samples, n is the feature dimension of LBPH-HT for each sample, and $D \in \mathbb{R}^{m \times r}$ is the geometric feature matrix, where r is equal to 2, corresponding to the r_1 and r_2 values of each sample. Then we perform feature-level fusion on LBPH-HT and GGD-H by Eq. (6).

$$F = M \oplus D \tag{6}$$

The fused feature matrix is $F \in \mathbb{R}^{m \times f}$ where $f = n + r$.

4.2   Decision-Level Fusion

After feature-level fusion, the class prediction probability matrix of the classifier for each sample is recorded as X_1. The probability matrix obtained by the subspace feature is marked as X_2. To have more accurate decision-making, we perform second-level decision fusion after the first-level decision, according to Eq. (7).

$$X = (1-\lambda)X_1 + \lambda X_2 \tag{7}$$

Where λ needs to take the appropriate threshold according to the recognition effect of X_1 and X_2. Each dimension vector of X is x_i, then the input sample set of the decision level classifier SVM is $\{(x_1, y_1), ...(x_l, y_l)\}$, $i = 1, ..., l$, where $y_i = \{0, 1\}$, 0 means female, 1 means male.

4.3 Classifiers

In order to enhance the reliability of the experiment, we experimented with three different classifiers: Support Vector Machines (SVM), Linear Discriminant Analysis (LDA), and XGBoost.

In the case of SVM, we divided the boundary between male and female by finding a maximum margin hyperplane. SVM has excellent generalization ability. Its own optimization goal is to minimize structural risk, not empirical risk. Therefore, through the concept of margin, a structured description of the data distribution is obtained, thus reducing the requirements for data scale and data distribution. In the case of LDA, we need to find the best discriminant vector space to maximize the distance between the two types of data and minimize the distance within the class. XGBoost is a tree integration model. It will integrate the classification results of many tree models to get the final result. So it is usually better than the Decision Tree.

5 Experiments

5.1 Dataset

Our experimental data comes from the 11k Hands data set [1] published by York University. The data set covers 11076 hand images (1600×1200 pixels) from 190 experimental subjects between 18 and 75 years old, including the palm and back of the hand. The metadata records associated with each image include ID, gender, age, skin color, and a set of captured hand information such as right or left hand, hand side (dorsal or palm), and logical indicators. The logical indicators refer to whether the hand image contains accessories, nail polish, or irregularities. Our study focuses on gender recognition. Therefore, we selected metadata records other than age as the initial data recordset.

5.2 Feature Extraction and Result Analysis

For the original dimension of 11076×12288, we retained 95% of the principal component variance. The data size after dimensionality reduction is 11076×314. We set the first 7500 samples of the data set as the training set and the remaining 3576 samples as the test set. For the Eigenhand method, the accuracy of the test set is 0.984. We can see that the Eigenhand includes the main features of the hand image very well and improve the classification recognition efficiency of the classifier. In the meanwhile, effective dimensionality reduction has dramatically improved the effectiveness of the algorithm.

Then, we take the obtained LBPH-HT feature directly as the input of the SVM classifier. The input dimension of the data is 11076×3776. Then we divided the data set by the same ratio. The accuracy of the test set is 0.941. We can see in Table 1 that the LBPH-HT, as a general feature extraction method, has strong applicability and flexibility in each recognition scenario. It is a generally reliable scheme.

Table 1. The recognition effect of each feature

	Eigenhand	LBPH-HT	GGD-H (DOS)	GGD-H (OpenPose)
Accuracy	**0.984**	0.941	0.622	0.642
AUC	**0.996**	0.972	0.461	0.527

According to the Sect. 3.3, we can obtain the detection results d_2, d_3, d_4, d_5 and r_1, r_2 respectively. We can see the results of DOS method and OpenPose method in Table 1. Due to the development of hand keypoint detection technology, the extraction of individual geometric features still can be improved.

5.3 Hierarchical Fusion

Feature-Level Fusion. Next, we superimpose the acquired GGD-H and LBPH-HT dimensions as input to the classifier. The feature vector dimension expanded from 11076×3776 to 11076×3378. Under the same SVM parameter settings, the effect is as follows.

Table 2. Recognition effect of feature-level fusion

	LBPH-HT+GGD-H (DOS)	LBPH-HT+GGD-H (OpenPose)
Accuracy	0.873	**0.947**
AUC	0.948	**0.979**

As shown in Table 2, the performance of the LBPH-HT fused with GGD-H is significantly improved. Compared with the two geometric feature extraction schemes, the improvement effect of OpenPose is better with the accuracy of 0.947. The geometric features of the hand can help improve the accuracy of the model.

The effectiveness of feature-level fusion can be further verified by the pure OpenPose method result in Table 1.

Table 3. Comparison of the recognition results of each classifier

	SVM	LDA	XGBoost
Accuracy	**0.947**	0.900	0.877
Precision	**0.945**	0.8492	0.920
Recall	**0.910**	0.889	0.735
F1-score	**0.928**	0.866	0.817

We further use this fusion feature to verify its effectiveness on different classifiers, as shown in Table 3. It shows that the recognition results of the SVM classifier are best on the accuracy and other evaluation standards. The ROC curves of each classifier on the test set are shown in Fig. 5.

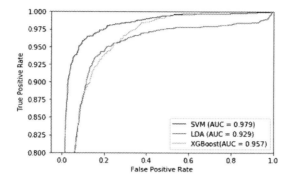

Fig. 5. ROC curves of all classifiers

Decision-Level Fusion. Although the fused features achieve good results on the SVM classifier, it is still lower than the 0.984 recognition rate of the Eigenhand method. To further improve the recognition rate, we choose SVM as the second-level decision classifier and perform decision-level fusion according to the theory in Sect. 4.2. Table 4 shows that the final recognition rate increase from 0.984 to 0.988. The performance of the two fusions is shown in the Fig. 6.

Table 4. The effect of two fusion operations

	Feature-level fusion	Decision-level fusion
Accuracy	0.947	**0.988**
Precision	0.945	**0.976**
Recall	0.910	**0.982**
F1-score	0.928	**0.979**
AUC	0.980	**0.998**

At present, there are relatively few research results on gender recognition of hand features. Many research results are based on different data sets, and most of them are small samples. So these results are not comparable in terms of recognition rate. Therefore, we can only make a qualitative comparison between our method and some methods in recent years, as shown in Table 5.

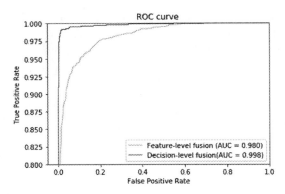

Fig. 6. ROC of two fusion operations and PCA method

Table 5. Method comparison

	Amayeh et al. [3]	Wu and Yuan [11]	Afifi [1]	Ours
Feature	Zernike moments and Fourier descriptors	Palm aspect ratio	CNN+LBP	Fusion of subspace, texture and geometric features
Classifier	LDA	PSSVM	SVM	SVM
Number of data set	40	30	190	190
Data set size	40	180	11076	11076
Accuracy	0.980	0.850	0.973	**0.988**

6 Conclusion

In this paper, we investigate the problem of gender recognition based on hand features. Based on the observation of the biological characteristics of the hand, we extract hand feature sets by three different feature expressions of Eigenhand, LBPH-HT, and GGD-H. After that, we perform two fusions. For the first time, feature-level fusion is achieved on LBPH-HT and GGD-H to enhance the fault tolerance of the feature expression. In the second time, we perform decision-level fusion on the recognition results to improve the robustness of the model. In our work, we use classic and efficient algorithms such as PCA and LBPH, and our intuitive and concise geometric feature calculation methods to ensure the efficiency of our algorithm. The two fusion operations enable us to achieve a high accuracy comparable to the CNN method.

Gender recognition research based on hand features is an emerging research direction and field. With the application needs of biometrics and identity authentication in recent years, as well as the better availability and privacy protection of hands, The field has made great progress. The work of this paper is based on this background. In future work, we will further explore more geometric features and examine the possibility of application in this field.

References

1. Afifi, M.: 11k hands: gender recognition and biometric identification using a large dataset of hand images. Multimedia Tools Appl. **78**(15), 20835–20854 (2019)
2. Ahonen, T., Matas, J., He, C., Pietikäinen, M.: Rotation invariant image description with local binary pattern histogram fourier features. In: Salberg, A.-B., Hardeberg, J.Y., Jenssen, R. (eds.) SCIA 2009. LNCS, vol. 5575, pp. 61–70. Springer, Heidelberg (2009). https://doi.org/10.1007/978-3-642-02230-2_7
3. Amayeh, G., Bebis, G., Nicolescu, M.: Gender classification from hand shape, pp. 1246–1252. IEEE (2008)
4. Boreki, G., Zimmer, A.: Hand geometry: a new approach for feature extraction, pp. 149–154 (2005)
5. Brown, W.M., Hines, M., Fane, B.A., Breedlove, S.M.: Masculinized finger length patterns in human males and females with congenital adrenal hyperplasia. Horm. Behav. **42**(4), 380–386 (2002)
6. Kanchan, T., Krishan, K.: Anthropometry of hand in sex determination of dismembered remains - a review of literature. J. Forensic Leg. Med. **18**(1), 14–17 (2011)
7. Ojala, T., Pietikainen, M., Maenpaa, T.: Multiresolution gray-scale and rotation invariant texture classification with local binary patterns. IEEE Trans. Pattern Anal. Mach. Intell. **24**(7), 971–987 (2002)
8. Simon, T., Joo, H., Matthews, I., Sheikh, Y.: Hand keypoint detection in single images using multiview bootstrapping. In: 30th IEEE Conference on Computer Vision and Pattern Recognition, pp. 4645–4653. IEEE (2017)
9. Snow, D.R.: Sexual dimorphism in upper palaeolithic hand stencils. Antiquity **80**(308), 390–404 (2006)
10. Wang Ze, W.G.: Using digital technology to determine the gender identity in prehistoric handprint rock paintings. J. Art Coll. Inner Mongolia Univ. **3**, 49–55 (2014)
11. Wu, M., Yuan, Y.: Gender classification based on geometry features of palm image. Sci. World J. (2014)
12. Xie, J., Zhang, L., You, J., Zhang, D., Qu, X.: A study of hand back skin texture patterns for personal identification and gender classification. Sensors **12**(7), 8691–8709 (2012)

Person Search via Anchor-Free Detection and Part-Based Group Feature Similarity Estimation

Qing Liu🆔, Keyang Cheng$^{(\boxtimes)}$🆔, and Bin Wu🆔

School of Computer Science and Communication Engineering,
Jiangsu University, Zhenjiang 212013, China
kycheng@ujs.edu.cn

Abstract. In order to solve the problems of insufficient accuracy of pedestrian bounding boxes in person search and large-scale person matching. A novel person search framework is proposed, which includes: (1) A multi-layer cascade heatmap mechanism (MCHM) is proposed, which aggregates pedestrian features by multi-layer heatmaps cascaded and improves the accuracy of the pedestrian bounding box by optimizating the offset between the center of the bounding box and the center point. (2) A learnable part-based pedestrian feature weight calculation module is proposed, which can learn the weight of the part according to the importance of the part-based feature instead of manually set hyperparameters. (3) A group feature correlation graph convolution network (GFCGCN) is proposed, which can calculate the similarity between group pedestrian features and provide a more accuracy end to end person search work. Some ablation studies and comparative experiments on datasets CUHK-SYSU, PRW show that our model can effectively achieve more accuracy pearch search with accuracy of 88.7% rank-1 and 78.2% mAP.

Keywords: Pedestrian re-identification · Convolutional neural network · Graph convolutional network · Person detection · Person search

1 Introduction

Pedestrian re-identification is a research, which aims at matching a target person with a gallery of images. Although numerous methods have been proposed, most methods [1,2] mainly rely on external object detectors to detect pedestrians in the video, and then perform pedestrian matching on candidate sets. These methods treat detection and re-identification as two separate tasks. Existing pedestrian detectors inevitably produce false detections, missing detections, and misalignments, which will harm the final searching performance significantly. Person search has recently emerged as the task of finding a person, provided as a cropped exemplar, in a gallery of non-cropped images [3,4].

© Springer Nature Switzerland AG 2020
Y. Peng et al. (Eds.): PRCV 2020, LNCS 12306, pp. 242–254, 2020.
https://doi.org/10.1007/978-3-030-60639-8_21

Although the existing works have tried to address these bottlenecks, these works generally have the following deficiencies: 1) They ignored the impact of inaccurate pedestrian bounding boxes on pedestrian re-identification. The general object detection model usually detects multiple categories of objects. Obviously the wrong category will have a negative impact on pedestrian re-identification. 2) The simple pedestrian feature similarity measurement method is not robust enough. Both Euclidean distance and Consine distance require the feature to be highly robust. The information such as background texture has a great influence on distance measurement. 3) They do not consider the influence of the crowd features around the target pedestrian on the re-identification results. If the pedestrians around the target all appear in another camera, the confidence that the target appears in that camera should be higher.

To address this deficiency, an novel person search framework is proposed in this paper. This framework uses MCHM to aggregate information on features to reduce the impact of background texture in pedestrian features and improve the accuracy of pedestrian bounding boxes. And the GFCGCN module is proposed to calculate group pedestrian features so as to achieve more accurate pedestrian re-identification. The contributions of our model are as follows:

(1) A multi-layer cascade heatmap mechanism (MCHM) is proposed, which aggregates pedestrian features by multi-layer heatmaps and improves the accuracy of the pedestrian bounding box by optimizating the offset between the center of the bounding box and the center point.
(2) A learnable part-based pedestrian feature weight calculation module is proposed, which can learn the weight of the part according to the importance of the part-based feature instead of manually set hyperparameters.
(3) A group feature correlation graph convolution network (GFCGCN) is proposed, which can calculate the similarity between group pedestrian features and provide a more accuracy end to end person search work.

2   Related Work

In this section we first introduce prior art on the two separate tasks of person detection and person re-identification, and then introduce the person search.

2.1   Pedestrian Detection and Re-identification

In recent years, convolutional neural networks (CNNs) at pedestrian detection joint learning the classification model and the features in an end-to-end fashion [5]. Commonly used pedestrian detection models can be divided into single-stage models and two-stage models. While single-stage object detectors [6,7] are preferable for runtime performance, the two-stage strategy of Faster R-CNN remains the more robust general solution [8], versatile to tailor region proposals to custom scene geometries [9] and to add multi-task branches [10,11].

Person re-identification aims to associate pedestrians over non-overlapping cameras. Most previous methods try to address this task on two directions, i.e., feature representation and distance metric learning. Some methods design different kinds of hand-crafted features to achieve certain success on small datasets. But these methods are limited for large-scale searching. While there are two main trends in the modern CNN model learning: (1) by Siamese networks and contrastive losses; (2) by ID classification with crossentropy losses. In the first, pairs [12,13], triplets [14] or quadruplets [15] are used to learn a corresponding number of Siamese networks, by pushing or pulling the same or the different person ids, respectively. In the second, [16] define as many classes as people IDs, train classifiers with a cross-entropy loss, and take the network features as the embedding metric during inference.

2.2   Person Search

Person search is a recently introduced problem of matching a probe person bounding box against a set of gallery whole scene images [17]. Some methods [11,26] design online learning object functions to learn large number of identities in the training set and achieve great performance on recent person search datasets. However, these methods only employ individual appearance for verification, which ignores the underlying relationship between individuals in the scene. This is challenging due to the uncontrolled false alarms, misdetections, and misalignment emerging in the auto-detection process. The multi-scale matching problem turns out a more severe challenge in person search.

3   Method

In this section, the proposed person search framework will be introduced in detail. Firstly, a backbone network is used to encode features and a multi-layer cascaded heatmap mechanism (MCHM) is proposed to makes the bounding box more accurate, where the center point and bounding box are continuously optimized by training. Secondly, a group feature correlation graph convolution network (GFCGCN) is applied to output the similarity estimation. The overall structure of the framework is shown in Fig. 1.

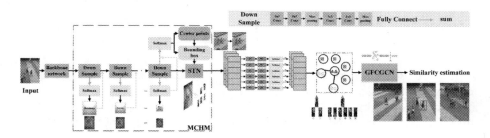

Fig. 1. An overview of the person search framework.

3.1 Multi-layer Cascade Heatmap and Anchor-Free Based Detection

The ResNet is adopted as backbone to extract deep features and then a multi-layer cascaded heatmap mechanism (MCHM) is proposed for further information aggregation. In a non-cropped image, it usually contains rich background information, which brings difficulty to detection and re-identification. To reduce the interference caused by background information, the MCHM outputs heatmaps from different levels of features to make the model pay more attention to pedestrians. Finally, multi-level heatmaps use up-sampling layers to aggregate information on features to achieve more accurate pedestrian detection.

The common object detection model obtains the bounding box coordinates by regression, which depends on the quality of the regression. However, this method usually causes information loss or contains too much noise information such as contains too much background information. In this case, the anchor-free based detect head of the MCHM not only outputs the bounding box coordinates, but also outputs the center point of the pedestrian. During the training steps, the bounding box is continuously corrected by minimizing the offset between the center point and the center of bounding box as shown in Fig. 2. Hence the center point is responsible for localizing the objects more precisely. Note that the benefits for pedestrian detection performance may be marginal. But it is critical for pedestrian re-identification because the pedestrian features extracted according to bounding box.

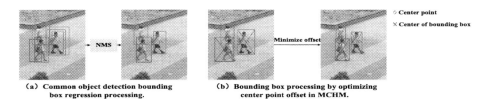

(a) Common object detection bounding box regression processing. (b) Bounding box processing by optimizing center point offset in MCHM.

Fig. 2. An example diagram of pedestrian bounding box optimization process.

For each box $b^i = (x_1^i, y_1^i, x_2^i, y_2^i)$ in the image, assign the object center (c_x^i, c_y^i) as $c_x^i = \frac{x_1^i + x_2^i}{2}$ and $c_y^i = \frac{y_1^i + y_2^i}{2}$, respectively. Then its location is obtained by dividing the stride $(\tilde{c_x^i}, \tilde{c_y^i}) = (\lfloor \frac{c_x^i}{4} \rfloor, \lfloor \frac{c_y^i}{4} \rfloor)$. Then the heatmap response at the location (x, y) is computed as $M_{xy} = \sum_{i=1}^{N} exp^{-\frac{(x - \tilde{c_x^i})^2 + (y - \tilde{c_y^i})^2}{2\sigma_c^2}}$. Where N denotes the number of objects in the image and σ_c denotes the standard deviation. The loss function is defined as pixel-wise logistic regression with focal loss:

$$L_{heatmap} = -\frac{1}{N} \sum_{xy} \begin{cases} (1 - \hat{M}_{xy})^\alpha log(\hat{M}_{xy}), & if \quad M_{xy} = 1; \\ (1 - \hat{M}_{xy})^\beta (\hat{M}_{xy})^\alpha log(1 - \hat{M}_{xy}) & otherwise \end{cases} \quad (1)$$

where $\hat{M}$ is the estimated heatmap, and α, β are the parameters.

Assume the outputs of the bounding box size and the offset as $\hat{S} \in R^{W*H*2}$ and $\hat{O} = R^{W*H*2}$, respectively. For each GT box $b^i = (x_1^i, y_1^i, x_2^i, y_2^i)$ in the image, we can compute its size as $s^i = (x_2^i - x_1^i, y_2^i - y_1^i)$. Similarly, the GT offset can be computed as $o^i = (\frac{c_x^i}{4}, \frac{c_y^i}{4}) - (\lfloor \frac{c_x^i}{4} \rfloor, \lfloor \frac{c_y^i}{4} \rfloor)$. Denote the estimated size and offset at the corresponding location as $\hat{S}^i$ and $\hat{O}^i$, respectively. Then we enforce l_1 loss for the two outputs:

$$L_{box} = \sum_{i=1}^{N} ||o^i - \hat{o}^i||_1 + ||s^i - \hat{s}^i||_1 \qquad (2)$$

3.2   Pedestrian Feature Extraction and Group Feature Proposal

Once the bounding box obtained, the pedestrians can be extracted by a STN module. Because there is a gap between feature coordinates and image coordinates. The STN module can correct this gap by affine transformation. As individual features are not sufficient for real world retrieval task, a group features is employed to help calculate the weights of the part-based features. Suppose the set of persons which appear on both probe and gallery scenes as positive feature pairs. The way of judging whether two features belong to the same person is to compute the similarity between the feature pairs. x_i^r, x_j^r is denoted as the $r-th$ part from feature i and j. As shown in Fig. 3, consider different feature parts, the final similarity $s(i, j)$ can be represented as the summation of different parts:

$$s(i, j) = \sum_{r=1}^{R} *w_r dist(x_i^r, x_j^r) \qquad (3)$$

where dist denotes the Euclidean distance between x_i^r, x_j^r, R is the number of part (R = 6 in our framework). w_r is the contribution weight of the $r-th$ feature part.

Fig. 3. A diagram of pair of features similarity weights calculation.

Because the weights of different parts are significantly different across samples, due to possible occlusions, different viewpoints and lighting conditions. The weight w_r will have a impact on the final similarity. In this case, the model uses two fully connected layers and a Softmax layer to output a learnable weights w_r. It takes in R pairs of feature vectors, and the Softmax layer output R normalized weights. Given an object pair (i, j), the corresponding label $y = 1$ if these two samples belong to the same person, otherwise $y = 1$, the loss function is as follows:

$$L = \begin{cases} 1 - s(i,j) & if \quad y = 1 \\ max(0, s(i,j) + \alpha) & if \quad y = -1 \end{cases} \qquad (4)$$

This loss term builds a margin α between positive and negative pairs, and thus safeguards the discrminativeness of the embedded features.

3.3   Group Feature Correlation Graph Learning

For a given image pair A, B. The motivation is to determine whether the target in image A also appears in image B. Therefore, assuming that a target is captured in image A and B, respectively. All need to be done is to determine whether these two targets are the same pedestrian. For probability events, if most of the pedestrians around the target in A appear in B, then there is a higher confidence that the target also appears in image B. Based on this, we fully consider the impact of crowd on pedestrian re-identification.

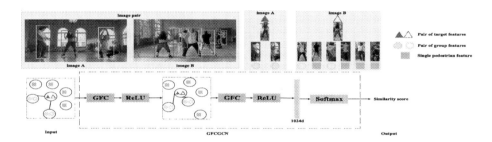

Fig. 4. An overview of the model structure of the GFCGCN network.

Given K pedestrian group feature pairs $(A_i, B_i), i \in 1, .., K$. A graph is dessigned to jointly take the target pairs and the K group feature pairs as well as single features (only appear in one image) into consideration as shown in Fig. 4. In this graph, the target pedestrian node is the center of the graph, which is connected to all the group feature nodes for information aggregation and node weight updation.

Assume the graph mentioned above is denoted by G. Where $G = (V, E)$, V represents the N-dimensional feature vector, and E represents the edge set of the graph. Each node is assigned with a pair of features $(X_{A_j}, X_{B_j}), j \in 0, ..., K$. Suppose the images have K group feature pairs, then $N = K + 1$. We define $X \in R^{N*2d}$ and $A \in R^{N*N}$, where d is the pedestrian feature dimension. A denote the adjacent matrix associated with graph G and it can be expressed by the following formula:

$$A_{i,j} = \begin{cases} 1 & if \quad i = 1 \quad or \quad j = 1 \quad or \quad i = j, \\ 0 & otherwise \end{cases} \qquad (5)$$

where $i, j \in 1, ..., N$. To better implement the model, the adjacency matrix A is normalized for the ease of learning. The adjacency matrix A can be seen as a stack of $\{A_1, ..., A_T\}$, each A_t is normalized symmetrically by $A_t = \Lambda_t^{-\frac{1}{2}} * \hat{A}_t * \Lambda_t^{-\frac{1}{2}}$. Where $\hat{A}_t = A_t + I$ and Λ_t is the diagonal node degree matrix of $\hat{A}_t$. $\hat{A}$ and Λ are used to denote the stack of $\hat{A}_t$ and Λ_t, respectively. Finally, a group feature correlation graph convolution network (GFCGCN) is proposed to update the weights and output the similarity. The network structure can be shown in Fig. 4 and the layer-wise GFCGCN propagates as follows:

$$GFC(V^l, A)^{l+1} = \sigma(\Lambda^{-\frac{1}{2}} * \hat{A} * \Lambda^{-\frac{1}{2}} * V^{(l)} * W^{(l)})) \tag{6}$$

where $V^{(l)}$ is the outputs of the $l - th$ layer, and $V^{(0)} = X$ as input. $W^{(l)}$ is the learnable parameters and σ is the ReLU activation function. Finally, a fully connected layer is applied to merge all the vertices into a 1024-dimensional feature vector. And a binary Softmax layer is employed supervise network training.

4   Experiments and Analysis

4.1   Datasets

CUHK-SYSU. The CUHK-SYSU dataset [17] consists of 18184 images, labeled with 8,432 identities and 96,143 pedestrian bounding boxes (23,430 boxes are ID labeled). The images, captured in urban areas by hand-held cameras or from movie snapshots, vary largely in viewpoint, lightning, occlusion and background conditions.

PRW. The PRW dataset [4], acquired in a university campus from six cameras, consists of 11,816 images with 43,110 bounding boxes (34,304 boxes are ID labeled) and 932 identities. Compared to CUHK-SYSU, PRW is with features less images and IDs but many more bounding boxes per ID (36.8, against 2.8 in CUHK-SYSU), which makes it more challenging.

4.2   Implementation Details

An ImageNet pretrained ResNet-50 model is applied as a backbone. The model is trained 60 epochs with the Adam optimizer and a starting learning rate of 0.001. The learning rate is reduced by 10% every 10 epochs. All the training images are resized to 512 ∗ 128. Besides, a standard data augmentation including rotation, scaling and color jittering is applied to enhance data. The model is implemented on Pytorch, trained and tested on two Tesla P100 GPUs.

4.3   Multi-layer Cascade Heatmap Mechanism

The MCHM is used to detect pedestrians and extract features from non-cropped images. In order to verify the effectiveness of the MCHM, some relevant comparative experiments and ablation experiments are conducted.

Table 1. A comparison of accuracy between the proposed MCHM and common object detection methods.

Methods	AP	AP_{50}	AP_{60}
Faster RCNN [19]	26.8	46.7	36.7
RGB-D Faster RCNN [18]	36.7	59.5	38.9
MFI-SSD [20]	45.0	63.5	46.7
CornerNet [21]	47.6	63.7	53.1
CenterNet [22]	52.4	64.8	56.3
MCHM (ours)	**56.8**	**70.1**	**57.1**

Comparative Experiments. Some common object detection models propose bounding boxes by anchor, which can be summarized as anchor-based methods. Besides, there are some key point-based methods called anchor-free methods. In this section, the proposed MCHM is compared with two sorts of common object detection methods on the datasets CUHK-SYSU. Among them, RGB-D Faster RCNN [18], Faster RCNN [19] and MFI-SSD [20] are employed as the anchor-based method. And CornerNet [21] and CenterNet [22] are employed as the anchor-free method. The results are shown in Table 1. It can be seen from the experiment that the proposed MCHM is significantly better than the common object detection model under the optimization of the pedestrian center point.

Ablation Study. Different level of cascaded heatmap layers (3, 4, 5 layers) are applied to the ablation experiment to explore the performance of the MCHM. The experimental results are shown Fig. 5.

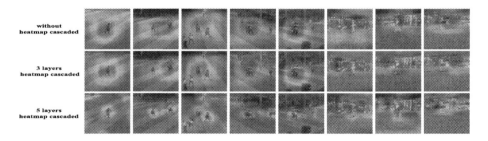

Fig. 5. The effect of MCHM on pedestrian feature extraction. The greater the number of cascaded heatmaps, the more clustered pedestrian features are and the less background information it contains.

Without using MCHM, the model can only learn some roughly information around pedestrian features. Usually that features contain too much background information, which is harmful to pedestrian re-identification. The MCHM can

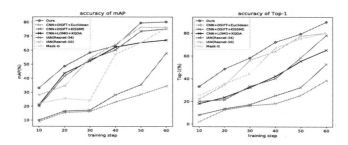

Fig. 6. The accuracy of the comparative experiments on performance.

effectively reduce the background information. In addition, the different level of cascaded heatmaps also has a greater impact on the information aggregation. Compared with the 3-layer heatmaps cascaded, The 5-layer structure can more accurately pay attention to the pedestrians. Overall, the MCHM can effectively aggregate pedestrian information.

4.4 Group Feature Correlation Graph Learning

Conventionally Euclidean distance is used to calculate the similarity between features. However, this approach often requires the features to be very robust. The quality of the features affects the similarity result directly. This model combines target pedestrians and contextual pedestrians to form group features to calculate the similarity between features. The proposed GFCGCN model measures the correlation between group features and finally makes a comprehensive similarity estimation. Similarly, some research is carried out to explore the effectiveness of the model. Firstly, some comparative experiments are performed between the GFCGCN and some existing metric learning methods. Secondly, some ablation experiments are conducted to study the influence of this part on the final results. Finally, some person search results of this framework is demonstrated (Fig. 6).

Comparative Experiments. The model is compared with some previous metric learning re-identification models such as IAN [11], Dis-GCN [25], as well as some other hand-crafted features such as DSIFT [23] and LOMO [24]. The experimental quantification results are shown in Table 2.

Ablation Study. In this subsection, the MCHM and GFCGCN are combined to do ablation experiments to explore the person search result. The ablation experiments use different backbone networks, GFCGCN and conventional distance formula. The quantitative results are described in Table 3. In addition, the influence of the number of group features K can be qualitatived in Fig. 7 and Fig. 8. It can be seen from the curve in the figure that for five pedestrians appearing at the same time, this model can significantly improve the effect of pedestrian search. Therefore, this framework can be more suitable for person search in crowded scenes.

Table 2. Quantitative results of some comparative experiments.

Methods	Datasets	mAP (%)	Top-1 (%)	Top-5 (%)	Top-10 (%)
CNN+DSIFT+Euclidean [23]	CUHK-SYSU	34.5	38.6	45.8	56.8
CNN+DSIFT+KISSME [23]	CUHK-SYSU	47.8	53.5	60.8	78.9
CNN+LOMO+XQDA [24]	CUHK-SYSU	68.9	74.1	80.2	89.9
IAN(Resnet-34)IAN [11]	CUHK-SYSU	73.1	78.0	84.2	92.4
IAN(Resnet-50)IAN [11]	CUHK-SYSU	75.0	80.5	86.0	96.3
Dis-GCN [25]	CUHK-SYSU	75.8	80.1	90.2	93.2
Ours	CUHK-SYSU	**78.2**	**88.7**	**94.8**	**96.5**
CNN+DSIFT+Euclidean [23]	PRW	17.4	23.8	32.3	40.7
CNN+DSIFT+KISSME [23]	PRW	18.4	25.8	30.1	40.2
CNN+LOMO+XQDA [24]	PRW	20.4	23.1	34.8	43.8
IAN(Resnet-34)IAN [11]	PRW	23.0	50.8	60.8	74.7
IAN(Resnet-50)IAN [11]	PRW	35.8	56.7	65.3	75.8
Dis-GCN [25]	PRW	40.5	56.8	62.4	70.0
Ours	PRW	**57.8**	**72.3**	**80.5**	**86.4**

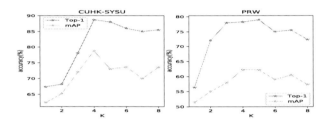

Fig. 7. The impact of group feature size K on performance.

Table 3. An ablation study of the proposed MCHM mechanism on dataset CUHK-SYSU.

Model structure	Similarity estimation	mAP (%)	Top-1 (%)
Resnet-34+MCHM	**GFCGCN**	**73**	**78.3**
Resnet-34+MCHM	Euclidean distance	58.1	63
Resnet-34+MCHM	cosine distance	56.3	59.8
Resnet-50+MCHM	**GFCGCN**	**78.2**	**88.7**
Resnet-50+MCHM	Euclidean distance	68.4	71.6
Resnet-50+MCHM	Cosine Distance	65.3	70.9

Fig. 8. The performance of different crowd sizes in the experiment. The red bounding box represents the target pedestrian, the green bounding box represents the surrounding crowd, and the blue bounding box represents the pedestrian first appeared in the scene. (Color figure online)

5   Conclusion

In this work, a novel person search framework with a MCHM module and GFCGCN module is proposed. The framework combines pedestrian detection and re-identification as one task and significantly improves the person search result. Instead of identifying target independently, the framework combines the surrounding crowds to form group features for re-identification. The framework has been verified on public datasets and achieved better re-identification results. It can be used to implement an end-to-end person search work in the surveillance system.

Acknowledgments. This research is supported by National Natural Science Foundation of China (61972183, 61672268) and National Engineering Laboratory Director Foundation of Big Data Application for Social Security Risk Perception and Prevention.

References

1. Wu, D., Zhang, K., Zheng, S.J., et al.: Random occlusion recovery for person re-identification. J. Imaging Sci. Technol. **63**(3), 30405-1–30405-9 (2019)
2. Wu, Q., Dai, P., Chen, P., et al.: Deep adversarial data augmentation with attribute guided for person re-identification. Signal Image Video Process. 1–8 (2019). https://doi.org/10.1007/s11760-019-01523-3
3. Liu, H., Feng, J., Jie, Z., et al.: Neural person search machines. In: Proceedings of the IEEE International Conference on Computer Vision, pp. 493–501 (2017)
4. Zheng, L., Zhang, H., Sun, S., et al.: Person re-identification in the wild. In: Proceedings of the IEEE Conference on Computer Vision and Pattern Recognition, pp. 1367–1376 (2017)
5. Guo, S., Bai, Q., Zhou, X.: Foreign object detection of transmission lines based on faster R-CNN. In: Kim, K.J., Kim, H.-Y. (eds.) Information Science and Applications. LNEE, vol. 621, pp. 269–275. Springer, Singapore (2020). https://doi.org/10.1007/978-981-15-1465-4_28

6. Lin, T.Y., Goyal, P., Girshick, R., et al.: Focal loss for dense object detection. In: Proceedings of the IEEE International Conference on Computer Vision, pp. 2980–2988 (2017)
7. Redmon, J., Farhadi, A.: YOLO9000: better, faster, stronger. In: Proceedings of the IEEE Conference on Computer Vision and Pattern Recognition, pp. 7263–7271 (2017)
8. Durkee, M.S., Sibley, A., Ai, J., et al.: Improved instance segmentation of immune cells in human lupus nephritis biopsies with Mask R-CNN. In: Medical Imaging 2020: Digital Pathology, vol. 11320, p. 1132019. International Society for Optics and Photonics (2020)
9. Jiang, H., Li, S., Liu, W., et al.: Geometry-aware cell detection with deep learning. MSystems 5(1) (2020)
10. Hasan, I., Tsesmelis, T., Galasso, F., et al.: Tiny head pose classification by bodily cues. In: 2017 IEEE International Conference on Image Processing (ICIP), pp. 2662–2666. IEEE (2017)
11. Xiao, J., Xie, Y., Tillo, T., et al.: IAN: the individual aggregation network for person search. Pattern Recogn. **87**, 332–340 (2019)
12. Jiang, M., Li, C., Kong, J., et al.: Cross-level reinforced attention network for person re-identification. J. Vis. Commun. Image Represent. 102775 (2020)
13. Şerbetçi, A., Akgül, Y.S.: End-to-end training of CNN ensembles for person re-identification. Pattern Recognit. 107319 (2020)
14. Zhao, C., Lv, X., Zhang, Z., et al.: Deep fusion feature representation learning with hard mining center-triplet loss for person re-identification. IEEE Trans. Multimedia (2020)
15. Zhang, C., Yue, J., Qin, Q.: Deep quadruplet network for hyperspectral image classification with a small number of samples. Remote Sens. **12**(4), 647 (2020)
16. Ye, M., Shen, J., Lin, G., et al.: Deep Learning for Person Re-identification: A Survey and Outlook. arXiv preprint arXiv:2001.04193 (2020)
17. Xiao, T., Li, S., Wang, B., et al.: Joint detection and identification feature learning for person search. In: Proceedings of the IEEE Conference on Computer Vision and Pattern Recognition, pp. 3415–3424 (2017)
18. Zhu, X., Chen, C., Zheng, B., et al.: Automatic recognition of lactating sow postures by refined two-stream RGB-D faster R-CNN. Biosyst. Eng. **189**, 116–132 (2020)
19. Mai, X., Zhang, H., Jia, X., et al.: Faster R-CNN with classifier fusion for automatic detection of small fruits. IEEE Trans. Autom. Sci. Eng. (2020)
20. Zhou, J., Chen, B., Zhang, J., et al.: Multi-scales feature integration single shot multi-box detector on small object detection. In: MIPPR 2019: Pattern Recognition and Computer Vision, vol. 11430, p. 114300E. International Society for Optics and Photonics (2020)
21. Law, H., Deng, J.: Cornernet: detecting objects as paired keypoints. In: Proceedings of the European Conference on Computer Vision (ECCV), pp. 734–750 (2018)
22. Duan, K., Bai, S., Xie, L., et al.: Centernet: keypoint triplets for object detection. In: Proceedings of the IEEE International Conference on Computer Vision, pp. 6569–6578 (2019)
23. Zhao, R., Ouyang, W., Wang, X.: Unsupervised salience learning for person re-identification. In: CVPR, pp. 3586–3593 (2013)
24. Kostinger, M., Hirzer, M., Wohlhart, P., Roth, P.M., Bischof, H.: Large scale metric learning from equivalence constraints. In: CVPR, pp. 2288–2295 (2012)

25. Ktena, S.I., et al.: Distance metric learning using graph convolutional networks: application to functional brain networks. In: Descoteaux, M., Maier-Hein, L., Franz, A., Jannin, P., Collins, D.L., Duchesne, S. (eds.) MICCAI 2017. LNCS, vol. 10433, pp. 469–477. Springer, Cham (2017). https://doi.org/10.1007/978-3-319-66182-7_54
26. He, Z., Zhang, L.: End-to-end detection and re-identification integrated net for person search. In: Jawahar, C.V., Li, H., Mori, G., Schindler, K. (eds.) ACCV 2018. LNCS, vol. 11362, pp. 349–364. Springer, Cham (2019). https://doi.org/10.1007/978-3-030-20890-5_23

Adaptive Model Updating Correlation Filter Tracker with Feature Fusion

Jingjing Shao[1], Lei Xiao[1], and Zhongyi Hu[2(✉)]

[1] College of Computer Science and Artificial Intelligence, Wenzhou University,
Wenzhou, China
194511981406@stu.wzu.edu.cn, xiaolei@wzu.edu.cn
[2] Intelligent Information Systems Institute, Wenzhou University, Wenzhou, China
hujunyi@163.com

Abstract. Aiming at the poor accuracy of a single feature in the challenging scenarios, as well as the failure of tracking caused by partial or complete occlusion and background clutter, a correlation filter tracking algorithm based on feature fusion and model adaptive updating is proposed. On the basis of the background-aware correlation filter, the proposed algorithm firstly introduces the CN feature and integrates with the HOG feature to improve the accuracy of tracking. Then, the Average Peak-to-Correlation Energy (APCE) is introduced, and the results of object tracking are fed back to the tracker through the ratio changes. The tracker is adaptively updated, which improves the robustness of the algorithm to occlusion and background clutter. Finally, the proposed algorithm is experimented on the self-build ship dataset. The experimental results show that the algorithm can adapt well to complex scenes, such as object occlusion and background clutter. Compared to the state-of-the-art trackers, the average precision of the proposed tracker is improved by 2.3%, the average success rate is improved by 2.9%, and the average speed is about 18 frames per second.

Keywords: Object tracking · Correlation filter · Feature fusion · Model updating · APCE

1 Introduction

Visual tracking is one of the key technologies in the computer vision field, and it has wide application prospects in video surveillance, human-computer interaction, medical diagnosis and so on. With the continuous deepening of research [1–4], visual tracking has made some progress in stages, but it is difficult to accurately locate the tracked object due to the interference factors such as partial

---

J. Shao—Student.

This study was financially supported by the Natural Science Foundation of Zhejiang Province Major Project (LZ20F020004), the Science and Technology Plan Major Science and Technology Projects of Wenzhou (ZY2019020), the Natural Science Foundation of Zhejiang Province (LY16F020022), and Wenzhou Science and Technology Planning Project (S20180017) of China.

© Springer Nature Switzerland AG 2020
Y. Peng et al. (Eds.): PRCV 2020, LNCS 12306, pp. 255–265, 2020.
https://doi.org/10.1007/978-3-030-60639-8_22

or complete occlusion, rotation, motion blur and so on. Therefore, there are still great challenges in building a robust tracker.

In recent years, mainstream tracking algorithms are divided into two categories, namely correlation filter and deep learning [5–8]. Among them, correlation filters (CFs) [9–12] algorithm is a classical algorithm for object tracking, which is favored by researchers because of its fast speed [13]. Bolme et al. [14] proposed MOSSE filter, which was the first time to introduce CF into object tracking with extremely fast speed. Based on that, Henriques et al. [15] introduced a Gaussian kernel function for acceleration, and extended the single-channel grayscale feature to the multi-channel Histogram of Oriented Gradient (HOG) to improve the tracking accuracy. Li et al. [16] proposed a scale adaptive multi-feature fusion tracker, adding the HOG feature and CN feature [17] on the basis of gray feature to improve the overall performance of the tracker. Besides, Danelljan et al. [18] proposed three-dimensional filter, one-dimensional scale filter and two-dimensional translation filter. This precise scale estimation method can be combined with any other tracking algorithm without scale estimation, and won the first place in the VOT2014 [19] competition. Since the methods based on CF are affected by the boundary effect, in order to overcome this problem, Danelljan et al. [20] added spatial regularization to suppress it, so that the search area can be expanded, and Gauss-Seidel was used to solve the filter to simplify the calculation. The models of the above algorithms are not effective for tracking targets with deformation and motion blur. Bertinetto et al. [21] complemented the HOG feature and color histogram feature, which was robust to motion blur, illumination and deformation, and added scale to the HOG to improve the accuracy of the tracker. Galoogahi et al. [22] used the negative samples generated by real shifts to include a larger search area and real background, and proposed an ADMM-based optimization method to reduce the computation. In recent years, deep learning-based methods have become more and more popular. Wang et al. [23] proposed a lightweight end-to-end training network, DCFNet, which simultaneously learns deep features and performs filtering processes. Wu et al. [24] learned the multi-level same-resolution compressed (MSC) features, which effectively incorporate both deep and shallow features for efficient online tracking, in an end-to-end offline manner.

The above methods have achieved good tracking effects in terms of accuracy and robustness. However, in the case of complex scenes, such as partial or complete occlusion, background clutter, etc., the problem of tracking loss will still occur. For this reason, in the framework of background-aware correlation filters (BACF), the following improvements have been made: (1) The use of excellent features is the basis for accurate tracking. A single feature have defects in accuracy. Considering the method of feature fusion to improve the accuracy of tracking, and adding the CN feature on the basis of HOG feature. (2) In order to better solve the problem of tracking failure caused by occlusion, the APCE method is introduced in the online update stage to adaptively update model to improve the robustness of the tracker.

2   The Tracker

The proposed algorithm is based on the background-aware correlation filter, which combines the features of HOG and CN, and introduces APCE [25] for adaptive model update, thereby improving the tracking algorithm's robustness to occlusion and background clutter. The framework of the proposed algorithm is shown in Fig. 1.

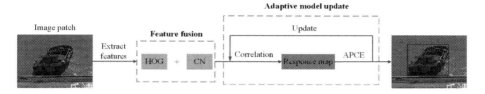

Fig. 1. Framework of the Proposed Algorithm. For each input image patch, first extract the HOG and CN features from the prediction area, fuse the two, and then obtain the corresponding response map through correlation filtering. APCE is introduced to adaptively update the model to determine whether to update the model at the current frame. Finally, update the model at the appropriate frame.

2.1   Background-Aware Correlation Filters

The background-aware correlation filters significantly increases the number of samples based on the traditional CF method and improves the sample quality through cropping operator, and has good real-time tracking. Therefore, We make improvements on the basis of background perception related filters in order to improve the accuracy of the algorithm. The basic objective function [26] of CF is:

$$E(\mathrm{h}) = \frac{1}{2}\|\mathrm{y} - \sum_{k=1}^{K}\mathrm{h}_k \star \mathrm{x}_k\|_2^2 + \frac{\lambda}{2}\sum_{k=1}^{K}\|\mathrm{h}_k\|_2^2 \qquad (1)$$

where y is the desired output response, x_k and h_k represents the kth channel of the vectorized image and filter respectively. λ is a regularization constant, and $\star$ is the spatial correlation operator. Equation 1 is the form of a single sample. When we use D cyclic samples, it becomes the following form:

$$E(\mathrm{h}) = \frac{1}{2}\sum_{j=1}^{T}\|\mathrm{y}(j) - \sum_{k=1}^{K}\mathrm{h}_k{}^{\top}\mathbf{P}\mathrm{x}_k[\triangle\tau_j]\|_2^2 + \frac{\lambda}{2}\sum_{k=1}^{K}\|\mathrm{h}_k\|_2^2 \qquad (2)$$

the size of the sample x changes from D to T, which is much larger. Use the larger sample to generate a cyclic sample. $[\triangle\tau_j]$ is the circular shift operator. Then we need to extract the middle part of the size D. This step is replaced by

P, which is a $D \times T$ binary matrix. **P** can be calculated in advance, and it is a constant matrix.

Taking advantage of the fast solution of the cyclic samples in the frequency domain, the expression is transformed into the frequency domain. The formula is as follows:

$$E(\mathrm{h}, \hat{\mathrm{g}}) = \frac{1}{2}\|\hat{\mathbf{y}} - \hat{\mathbf{X}}\hat{\mathrm{g}}\|_2^2 + \frac{\lambda}{2}\|\mathrm{h}\|_2^2$$
$$s.t. \quad \hat{\mathrm{g}} = \sqrt{T}(\mathbf{FP}^\top \otimes \mathbf{I}_K)\mathrm{h} \tag{3}$$

where $\hat{\mathbf{X}} = [\mathrm{diag}(\hat{\mathbf{x}}_1)^\top, ..., \mathrm{diag}(\hat{\mathbf{x}}_K)^\top]$ and $\hat{}$ refers to the Discrete Fourier Transform (DFT) of a signal. $\hat{\mathrm{g}}$ is a $KT \times 1$ auxiliary variable and $\hat{\mathrm{g}} = [\hat{\mathrm{g}}_1^\top, ..., \hat{\mathrm{g}}_K^\top]$. h is defined as $\mathrm{h} = [\mathrm{h}_1^\top, ..., \mathrm{h}_K^\top]$ of size $KD \times 1$. The DFT of one-dimensional signal α is expressed as $\hat{\alpha} = \sqrt{T}\mathbf{F}\alpha$, $\mathbf{F}$ is an $T \times T$ orthogonal Fourier transform matrix. $\mathbf{I}_K$ is a $K \times K$ identity matrix ($\mathbf{PP}^\top = \mathbf{I}$), $\otimes$ refers to the Kronecker product.

Finally, the optimization solution of Eq. 3 is mainly used to put the constraint term into the optimization function by using the Augmented Lagrangian Method (ALM) [27].

$$L(\hat{\mathrm{g}}, \mathrm{h}, \hat{\zeta}) = \frac{1}{2}\|\hat{\mathbf{y}} - \hat{\mathbf{X}}\hat{\mathrm{g}}\|_2^2 + \frac{\lambda}{2}\|\mathrm{h}\|_2^2$$
$$+ \hat{\zeta}^\top(\hat{\mathrm{g}} - \sqrt{T}(\mathbf{FP}^\top \otimes \mathbf{I}_K)\mathrm{h}) \tag{4}$$
$$+ \frac{\mu}{2}\|\hat{\mathrm{g}} - \sqrt{T}(\mathbf{FP}^\top \otimes \mathbf{I}_K)\mathrm{h}\|_2^2$$

where $\hat{\zeta} = [\hat{\zeta}_1^\top, ..., \hat{\zeta}_K^\top]$ and μ is a penalty factor. Equation 4 can be solved iteratively using Alternating Direction of Method of Multipliers (ADMM) [27] technology, and $\hat{\mathrm{g}}$ and h are optimized and solved separately.

2.2   Feature Fusion

The single feature has defects in accuracy. Considering the method of feature fusion to improve tracking accuracy, CN feature is added to the basis of HOG feature.

Color-Naming (CN). CN is an 11-dimensional color space feature that maps the 3-dimensional color features of the RGB space to black, blue, brown, gray, green, orange, pink, purple, red, white, and yellow. CN can separate objects of different colors, and it can distinguish objects and backgrounds with significant color difference and similar texture shapes.

The CN adopts the adaptive color attribute algorithm to map the RGB space to the 11-dimensional color space with obvious discrimination to obtain the 11-dimensional color feature vector, which is then mapped into the 10-dimensional subspace, reducing the dimension from 11 to 10 dimensions. Therefore, HOG and CN are serially combined into **M**, assuming that the vectors of HOG and CN are $\mathbf{H}_i(i = 1, 2, ..., 31)$ and $\mathbf{C}_j(j = 1, 2, ..., 10)$, respectively. $\mathbf{H}_i$ and $\mathbf{C}_j$ represent the i-th channel HOG and the j-th channel CN of the image respectively, then

$M = [H_1\ H_2\ ...\ H_{31}\ C_1\ C_2\ ...\ C_{10}]$, the 31-channel HOG and the 10-channel CN extracted from the training image patch are serially fused to obtain the 41-channel M.

HOG emphasizes the edge information of the image, while CN focuses on color information. The two features are complementary and improve the performance of the filter. Although the idea is simple, the performance improvement is very promising.

2.3 Adaptive Model Update

In the process of model tracking, the appearance and scale of the object will change. Figure 2 shows the object occlusion during tracking. If the tracker is updated at Fig. 2(b), the model may drift or even lose the object. In order to adapt to the changes of the tracking model, the maximum response value and the APCE are introduced to determine when the model will be updated. The formula is as follows:

$$\text{APCE}(t) = \frac{|F_{\max}(t) - F_{\min}(t)|^2}{\text{mean}(\sum_{w,h} (F_{w,h}(t) - F_{\min}(t))^2)} \tag{5}$$

where $F_{\max}$, $F_{\min}$ and $F_{w,h}$ represent the maximum response, minimum response and current frame response value, respectively. When the target is occluded or lost, APCE will suddenly decrease. In this case, the model is not updated to avoid model drift. Only when APCE and $F_{\max}$ are greater than the historical mean in a certain proportion, the model is updated, greatly reducing the model drift.

The online updating strategy of the model is still the same linear interpolation method as the traditional CF:

$$\hat{x}_{model}^{(f)} = (1 - \eta)\,\hat{x}_{model}^{(f-1)} + \eta\,\hat{x}^{(f)} \tag{6}$$

where η if a learning rate, $\hat{x}_{model}^{(f)}$ indicates the model at frame f.

(a) No occlusion   (b) Occlusion

Fig. 2. Two frames of a ship sequence on the self-build ship dataset. (a) is the 343th frame image of a ship video sequence, with the object in the red bounding box and not occluded by other ships or objects; (b) is the 520th frame image of a ship video sequence, with the object in the red bounding box and occluded by other ships. (Color figure online)

3   Experiments

In order to verify the reliability of the proposed tracker AMUMF (Adaptive Model Updating Correlation Filter Tracker with Feature Fusion), the self-build ship dataset was used for evaluation, and compared with 6 excellent correlation filter trackers, such as KCF, SAMF, STAPLE, STAPLE_CA, SRDCF, BACF.

3.1   Experimental Setup and Methodology

The experimental environment of the algorithm is MATLAB R2016a on Windows system. All experiments are completed on a desktop computer equipped with an Intel Core i5-9400 CPU at 2.90 GHz.

Experimental Dataset. The experimental data used in this research is a self-build ship dataset, which contains 60 ship video sequences. In order to better evaluate and analyze the advantages and disadvantages of the tracking method, 11 attributes such as illumination variation (IV), scale variation (SV), occlusion (OCC), deformation (DEF), motion blur (MB), fast motion (FM), in-plane rotation (IPR), out-of-plane rotation (OPR), out-of-view (OV), background clutters (BC) and low resolution (LW) are used to annotate the sequence, so as to classify these sequences. Figure 3 shows the 60 ship video tracking sequences.

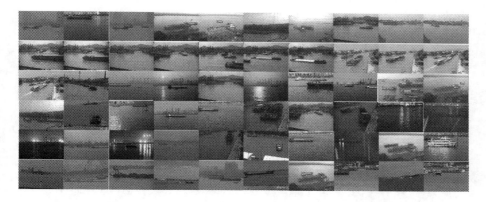

Fig. 3. Ship video tracking sequences. The blue box in the figure represents the tracked target. (Color figure online)

Parameter Settings. The specific parameters of the algorithm are set as: the thresholds of the maximum response and APCE in the adaptive model updating are 0.5 and 0.85, respectively. Other parameter settings are the same as the BACF.

3.2   Analysis

According to the evaluation method of the OTB [28], the one-pass evaluation (OPE) method is adopted. And there are two evaluation criteria selected, i.e. precision plot and success plot.

Table 1. Comparison of overall performance of 7 trackers on self-build ship dataset (/%). The best results are shown in bold, and the second-ranked is underlined.

	AMUMF	BACF	STAPLE_CA	SRDCF	STAPLE	SAMF	KCF
Success rate	**78.2**	<u>75.3</u>	55.6	69.0	57.6	63.1	59.2
Precision	**68.9**	<u>66.6</u>	37.4	57.4	39.4	48.2	39.4

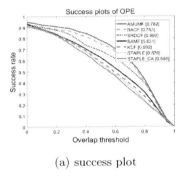
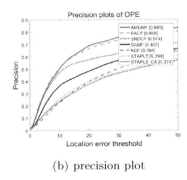

(a) success plot   (b) precision plot

Fig. 4. Comparison of success plot and precision plot of 7 trackers on self-build ship dataset.

Quantitative Analysis. We tested AMUMF on the self-build ship dataset and compared with other 6 trackers. Table 1 shows the success rate and precision of AMUMF and other 6 trackers. It can be seen that the success rate and precision of AMUMF are 78.2% and 68.9%, respectively, and the best results are obtained. This is 2.9% and 2.3% higher than BACF without feature fusion and adaptive model update. Figure 4 shows the corresponding precision curve and success rate curve of the 7 tracker. Figure 5 shows a comparison of success plot based on video attributes. It can be seen that AMUMF performs well at low resolution (LR), background clutter (BC), scale variation (SV), in-plane rotation (IPR), occlusion (OCC), out-of-plane rotation (OPR). Especially under OCC, the success rate of AMUMF is 5.3% higher than BACF.

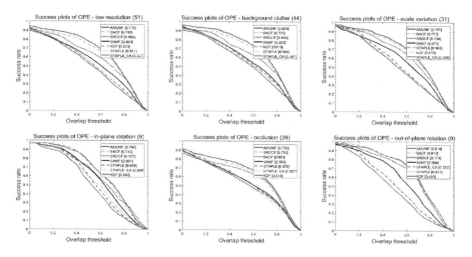

Fig. 5. Attribute-based evaluation. Comparison of success plot of 7 trackers on self-build ship dataset. AMUMF outperforms other trackers in these 6 video attributes.

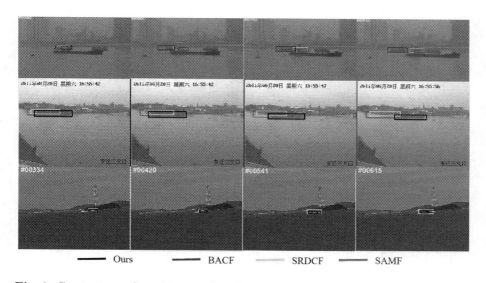

Fig. 6. Comparison of tracking results of 4 types of trackers. Each row represents a sequence.

Qualitative Analysis. We selected 3 representative video sequences for qualitative analysis, and compared AMUMF with BACF, SRDCF and SAMF as shown in Fig. 6. All three sequences under LR. In addition, the 1st sequence under OCC and BC. Only AMUMF continues to track accurately, other trackers are lost. The 2nd sequence under IPR, AMUMF and BACF can continue to track accurately, other trackers produce drift. In the 3rd sequence, two ships with the same appearance intersect. In addition to BACF, other trackers can track the target, but only AMUMF can accurately locate.

4 Conclusion

In the framework of background-aware correlation filter, a correlation filter tracker based on feature fusion and model adaptive updating is proposed. Two kinds of features are extracted, HOG and CN, and they are serially fused to obtain the final response map, so that the object is accurately located. We also introduce a high-confidence model updating to adaptively update tracking model, which effectively improves the robustness of the tracker to occlusion and background clutter. Experiments on the self-build ship dataset prove that the proposed AMUMF is superior to other trackers in terms of precision and success rate. In the future, we will further research the optimization of algorithms and how to improve the real-time tracking.

References

1. Xiao, L., Xu, M., Hu, Z.: Real-time inland CCTV ship tracking. Math. Probl. Eng. **2018**(2018), 1–10 (2018)
2. Xiao, L., Wang, H., Hu, Z.: Visual tracking via adaptive random projection based on sub-regions. IEEE Access **6**, 41955–41965 (2018)
3. Hu, Z., Xiao, L., Teng, F.: An overview of compressive trackers. Int. J. Signal Process. Image Process. Pattern Recogn. **9**(9), 113–122 (2016)
4. Hu, Z., Zou, M., Chen, C., Wu, Q.: Tracking via context-aware regression correlation filter with a spatial-temporal regularization. J. Electron. Imaging **29**(2), 023029 (2020)
5. Yun, S., Choi, J., Yoo, Y., Yun, K., Young Choi, J.: Action-decision networks for visual tracking with deep reinforcement learning. In: Proceedings of the IEEE Conference on Computer Vision and Pattern Recognition, pp. 2711–2720 (2017)
6. Valmadre, J., Bertinetto, L., Henriques, J., Vedaldi, A., Torr, P.H.: End-to-end representation learning for correlation filter based tracking. In: Proceedings of the IEEE Conference on Computer Vision and Pattern Recognition, pp. 2805–2813 (2017)
7. Ma, C., Huang, J.B., Yang, X., Yang, M.H.: Hierarchical convolutional features for visual tracking. In: Proceedings of the IEEE International Conference on Computer Vision, pp. 3074–3082 (2015)
8. Bertinetto, L., Valmadre, J., Henriques, J.F., Vedaldi, A., Torr, P.H.S.: Fully-convolutional Siamese networks for object tracking. In: Hua, G., Jégou, H. (eds.) ECCV 2016. LNCS, vol. 9914, pp. 850–865. Springer, Cham (2016). https://doi.org/10.1007/978-3-319-48881-3_56
9. Mueller, M., Smith, N., Ghanem, B.: Context-aware correlation filter tracking. In: Proceedings of the IEEE Conference on Computer Vision and Pattern Recognition, pp. 1396–1404 (2017)
10. Li, F., Tian, C., Zuo, W., Zhang, L., Yang, M.H.: Learning spatial-temporal regularized correlation filters for visual tracking. In: Proceedings of the IEEE Conference on Computer Vision and Pattern Recognition, pp. 4904–4913 (2018)

11. Lukezic, A., Vojir, T., Cehovin Zajc, L., Matas, J., Kristan, M.: Discriminative correlation filter with channel and spatial reliability. In: Proceedings of the IEEE Conference on Computer Vision and Pattern Recognition, pp. 6309–6318 (2017)

12. Tang, M., Yu, B., Zhang, F., Wang, J.: High-speed tracking with multi-kernel correlation filters. In: Proceedings of the IEEE Conference on Computer Vision and Pattern Recognition, pp. 4874–4883 (2018)

13. Henriques, J.F., Caseiro, R., Martins, P., Batista, J.: Exploiting the circulant structure of tracking-by-detection with kernels. In: Fitzgibbon, A., Lazebnik, S., Perona, P., Sato, Y., Schmid, C. (eds.) ECCV 2012. LNCS, vol. 7575, pp. 702–715. Springer, Heidelberg (2012). https://doi.org/10.1007/978-3-642-33765-9_50

14. Bolme, D.S., Beveridge, J.R., Draper, B.A., Lui, Y.M.: Visual object tracking using adaptive correlation filters. In: 2010 IEEE Computer Society Conference on Computer Vision and Pattern Recognition, pp. 2544–2550 (2010)

15. Henriques, J.F., Caseiro, R., Martins, P., Batista, J.: High-speed tracking with kernelized correlation filters. IEEE Trans. Pattern Anal. Mach. Intell. **37**(3), 583–596 (2014)

16. Li, Y., Zhu, J.: A scale adaptive kernel correlation filter tracker with feature integration. In: Agapito, L., Bronstein, M.M., Rother, C. (eds.) ECCV 2014. LNCS, vol. 8926, pp. 254–265. Springer, Cham (2015). https://doi.org/10.1007/978-3-319-16181-5_18

17. Danelljan, M., Shahbaz Khan, F., Felsberg, M., Van de Weijer, J.: Adaptive color attributes for real-time visual tracking. In: Proceedings of the IEEE Conference on Computer Vision and Pattern Recognition, pp. 1090–1097 (2014)

18. Danelljan, M., Häger, G., Khan, F., Felsberg, M.: Accurate scale estimation for robust visual tracking. In: British Machine Vision Conference, pp. 1–11. BMVA Press (2014)

19. Kristan, M., et al.: The visual object tracking VOT2014 challenge results. In: Agapito, L., Bronstein, M.M., Rother, C. (eds.) ECCV 2014. LNCS, vol. 8926, pp. 191–217. Springer, Cham (2015). https://doi.org/10.1007/978-3-319-16181-5_14

20. Danelljan, M., Hager, G., Shahbaz Khan, F., Felsberg, M.: Learning spatially regularized correlation filters for visual tracking. In: Proceedings of the IEEE International Conference on Computer Vision, pp. 4310–4318 (2015)

21. Bertinetto, L., Valmadre, J., Golodetz, S., Miksik, O., Torr, P.H.: Staple: complementary learners for real-time tracking. In: Proceedings of the IEEE Conference on Computer Vision and Pattern Recognition, pp. 1401–1409 (2016)

22. Kiani Galoogahi, H., Fagg, A., Lucey, S.: Learning background-aware correlation filters for visual tracking. In: Proceedings of the IEEE International Conference on Computer Vision, pp. 1135–1143 (2017)

23. Wang, Q., Gao, J., Xing, J., Zhang, M., Hu, W.: DCFNet: discriminant correlation filters network for visual tracking. arXiv preprint arXiv:1704.04057 (2017)

24. Wu, Q., Yan, Y., Liang, Y., Liu, Y., Wang, H.: DSNet: deep and shallow feature learning for efficient visual tracking. In: Jawahar, C.V., Li, H., Mori, G., Schindler, K. (eds.) ACCV 2018. LNCS, vol. 11365, pp. 119–134. Springer, Cham (2019). https://doi.org/10.1007/978-3-030-20873-8_8

25. Wang, M., Liu, Y., Huang, Z.: Large margin object tracking with circulant feature maps. In: Proceedings of the IEEE Conference on Computer Vision and Pattern Recognition, pp. 4021–4029 (2017)

26. Kiani Galoogahi, H., Sim, T., Lucey, S.: Multi-channel correlation filters. In: Proceedings of the IEEE International Conference on Computer Vision, pp. 3072–3079 (2013)

27. Boyd, S., Parikh, N., Chu, E., Peleato, B., Eckstein, J.: Distributed optimization and statistical learning via the alternating direction method of multipliers. Found. Trends Mach. Learn. **3**(1), 1–122 (2011)
28. Wu, Y., Lim, J., Yang, M.H.: Object tracking benchmark. IEEE Trans. Pattern Anal. Mach. Intell. **37**(9), 1834–1848 (2015)

A Deep Tracking and Segmentation Approach for Soccer Videos Visual Effects

Shenhui Peng[1], Li Song[1,2(✉)], Jun Ling[1], Rong Xie[1], Song Xu[3], and Lin Li[3]

[1] Institute of Image Communication and Network Engineering,
Shanghai Jiao Tong University, Shanghai, China
{shenhuipeng,song_li,lingjun,xierong}@sjtu.edu.cn
[2] MoE Key Lab of Artificial Intelligence, AI Institute,
Shanghai Jiao Tong University, Shanghai, China
[3] MIGU Co., Ltd., Shanghai, China
{xusong,lilin}@migu.cn

Abstract. The applications of deep learning algorithm in sports contain enormous potential. Specifically, in soccer, tracking algorithm could record the tracks of players, which could play as an assistant to assess team performance and evaluate strategies. Moreover, through segmentation model, we could extract semantic attributes of players. This auxiliary information may contribute to the special visual effects processing in broadcasting or entertainment area. Unlike general tracking tasks, soccer videos contain much more cases of deformation, blur, and occlusion. In this paper, we propose a novel model which could combine tracking and segmentation together. A novel deformable cross-similarity correlation (DF_CORR) is adopted to estimate the deformation of players. A new soccer tracking dataset is established to evaluate the performance of top-ranked trackers in soccer videos. In soccer tracking dataset, our model outperforms the state-of-the-art trackers whose accuracy is decreased significantly compared with the general tracking tasks. Moreover, our extensive experiments show comparable segmentation performance against SiamMask, while running in a real-time speed of 36.2FPS.

Keywords: Soccer · Visual tracking · Semantic segmentation · Deep learning

1 Introduction

Visual tracking is a fundamental and critical topic within computer vision area. Initialized by the location of target in the first frame of a video, the tracker is designed to evaluate the correspondences of target and estimate its position in

This work was supported by MoE-China Mobile Research Fund Project (MCM20180702), National Key R&D Project of China (2019YFB1802701), the 111 Project (B07022 and Sheitc No. 150633) and the Shanghai Key Laboratory of Digital Media Processing and Transmissions.

© Springer Nature Switzerland AG 2020
Y. Peng et al. (Eds.): PRCV 2020, LNCS 12306, pp. 266–277, 2020.
https://doi.org/10.1007/978-3-030-60639-8_23

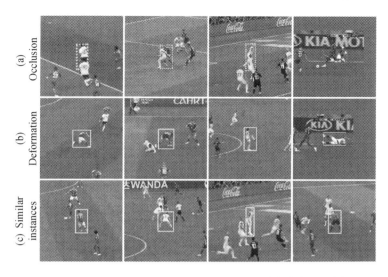

Fig. 1. Visualization of the soccer tracking dataset. Line (a), (b) and (c) show some situations about occlusion, deformation and similar instances respectively. The bounding box with dot lines means that target is occluded by other objects.

the following frames. In soccer, tracking algorithm could record every movement of players. Coaches and teams could use these attributes to assess team performance and evaluate strategies. In recent years, with the popularization of smart devices, deep learning applications have evolved rapidly in sports game broadcasting and entertainment area. With the auxiliary target semantic attributes, the impression of video could be significantly enhanced by adding special visual effects. Thus, it is practical and feasible to narrow the gap between visual object tracking and video object segmentation.

Currently, most of the state-of-the-art visual tracking approaches tend to involve template-based strategy [1,2] which consists of a template branch and a detection branch. Such algorithm is efficient to estimate the paralleled shift and scale change of target. As for high-dimensional transformations such as in-plane rotation and deformation, this algorithm may not appropriate.

Unlike the visual tracking tasks, the video object segmentation (VOS) algorithm is designed to obtain binary per-pixel segmentation mask which expresses whether or not a pixel belongs to the target [3]. This property could contribute to handling high-dimensional transformations of target. Unfortunately, some studies [4] show that top-ranked object segmentation approaches perform poorly in short-term tracking tasks. The background clutter could significantly decrease the performance of models and these segmentation errors will lead to an irrecoverable tracking failure.

Different from general tracking tasks, there exist some unique features in soccer videos. First, the videos contain a panoramic view of the entire field. The scale of player is around 100 pixels. Such low resolution will cause

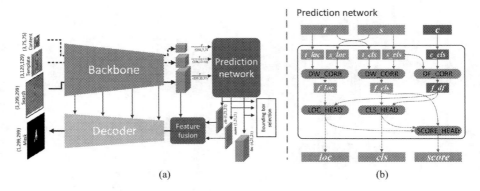

Fig. 2. (a) Overall structure of encoder-decoder pipeline. (b) The details of prediction network. The network will utilize the feature from deformable cross-similarity correlation operation to localize and classify targets. DW_CORR and DF_CORR refer to depth-wise correlation and deformable correlation respectively.

segmentation failure. Second, there are several similar instances around the target, since players wear only two kinds uniforms. It is a tough task for model's discriminative ability. Third, there exist much more high-dimensional transformations than normal task such as deformation, rotation, occlusion and blur as shown in Fig. 1. These features will degrade the performance of pre-trained model. Due to the lack of semantic mask labels in existing soccer videos, we could not easily implement transfer learning based on top-ranked strategies.

To address these problems, a novel deformable cross-similarity correlation (DF_CORR) is proposed to estimate the deformation of the targets as shown in Fig. 2(b). DF_CORR could predict the target deformation by destroying the spatial structure of the template patches intentionally, which could help to predicting the high-dimensional transformation. An encoder-decoder structure combined with feature fusion network is adopted to achieve tracking and segmentation in a one-shot processing. The main contributions of this paper are summarized as follows:

- We propose a novel deformable cross-similarity correlation which could predict the target deformation by destroying the spatial structure of the template patches pixel-by-pixel.
- We propose an encoder-decoder structure combined with a feature fusion network, which could achieve tracking and segmentation in one shot processing with a real-time speed.
- We propose a soccer tracking dataset which is more challenging than general tasks and can evaluate or expand the scope of the existing trackers.

2   Related Work

2.1   Visual Tracking

Discriminative correlation filters tracker (DCF) [5] prove robust target localization performance depends on the model's discriminative ability between the target and the background. Kernelized correlation filters tracker (KCF) [6] proposes a good way to speed up processing by converting the convolution operation to an element-wise product via the Fourier domain. Furthermore, there exist some approaches which attempt to use convolution operation instead of correlation operation to achieve end-to-end training [7]. Meanwhile a mini-batch of scaled images is generated to predict target scale. The critical bottleneck of these algorithms is bounding box estimation. To address this problem, SiamRPN [1] and its succeeding works [2,8,9] propose a siamese network followed by a Region Proposal Network, which is inspired by the Faster R-CNN [10] in object detection filed. To improve the target-background discriminative ability during tracking procedure, [11] develops an end-to-end tracking architecture, capable of fully exploiting both target and background appearance information for target model prediction derived from a discriminative learning loss.

2.2   Video Object Segmentation

Traditionally, tracking tasks tend to achieving a real-time processing speed and are designed to mainly focus on the target position rather than to represent the shape and contour. Conversely, video object segmentation algorithms have been more concerned with an accurate representation of the object of interest [3].

One branch is to process video frames one by one, independently. [12] only adopts the ground-truth mask provided in the first frame with a pretrained fully-convolutional network for classification without any temporal information. Some strategies [13,14] attempt to compute optical flow between frames to exploit temporal information. By performing spatio-temporal MRF model, [14] proposes an algorithm that alternates between a temporal fusion operation and a mask refinement feed-forward CNN, progressively inferring the results of video object segmentation.

Currently, some strategies [9,15] attempt to combine tracking and segmentation together. In particular, SiamMask adopts SiamRPN to do bounding box regression. Segmentation mask is computed from a subarea of whole feature maps guided by top-ranked bounding box. Benefited by this structure, SiamMask could achieve real-time processing.

3   Methodology

3.1   Siamese-Based Feature Extraction

Visual tracking tasks could be described as an similarity comparison problem via Siamese network [1]. Feature maps from search image and template image are extracted through Siamese network respectively. In detail, in *template branch*, template features $\phi(t)$ is extracted from template image t which is a subarea of first frame guided with initial target bounding box. Meanwhile, the *search branch* will extract search features $\phi(s)$ from search image s which is a subimage from current frame. The backbone network $\phi(\cdot)$ shares parameters between two branches as described in Siamese network. The similarity between template and search features could be evaluated as:

$$f_i(s, t) = \varphi_i^{feature}(\phi(s)) \star \varphi_i^{kernel}(\phi(t)), \, s, t \in \mathbb{R}^2 \tag{1}$$

where $\star$ denotes the cross-correlation operation. i denotes cls or loc. $\varphi_{cls}^{feature}$, $\varphi_{loc}^{feature}$, φ_{cls}^{kernel} and φ_{loc}^{kernel} are four adjustment network which is aimed to covert common features from backbone into specific task space.

3.2 Deformable Cross-Similarity Correlation

The cross correlation module is proposed by [5,6] which utilizes hand-crafted features. Subsequently, CREST [7] proves that deep network could replace correlation operation by convolution operation to achieve end-to-end training. In SiamRPN, cross-correlation is extended to embed much higher level information such as anchors, by adding a huge convolutional layer to scale the channels [2].

Traditional cross correlation layer has the ability to assess the location of target. Because of the invariance of spatial structure of template batches, cross correlation could not adapt to predict high-dimensional transformation of deformable targets. To tackle this, deformable cross-similarity correlation is proposed. This operation could predict the target deformation by destroying the spatial structure of the template patches intentionally, which can be described as performing cross correlation operation and evaluate the similarity pixel by pixel through spatial dimension. As shown in Fig. 2(b), the prediction head will use the feature from depth-wise correlation and deformable correlation to implement foreground-background classification and target localization.

The normal Siamese-based tracker tends to utilize a template patch which is two times larger than the target itself to ensure a balance of foreground-background discrimination. Based on this, we propose a content pipeline which is only concentrated on target feature as shown in Fig. 2. The feature maps from content c will be split through H and W followed by a flatten operation and then convolved with s as shown in Fig. 3(c). The experiment results prove that this intentional operation could significantly improve the capacity of model to predict target deformation.

Different from SiamRPN++ [2], we do not utilize dense anchor based method to achieve bound box regression. The output of the network will directly lead to the location. Since the target will span several pixels, these pixels at the center of the target will be more discriminating than those at the boundary with a posteriori inference. The *score* map is more like a 2D gaussian distribution localized in the center of the target. The *cls* and *score* will be utilized not

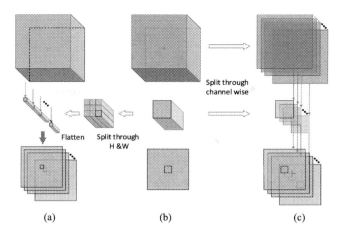

Fig. 3. Sketch of different cross correlation operations. (a) Deformable cross-similarity correlation could predict the target deformation by destroying the spatial structure of the template patches intentionally, which can be described as performed cross correlation operation pixel by pixel through spatial dimension. (b) Cross correlation predicts a single channel similarity map between target template and search patches in SiamFC [16]. (c) Depth-wise cross correlation (Group convolution) predicts multichannel correlation features between a template and search patches [2].

only to select the best predicted bounding box, but also to filter the background cluster of the feature map in feature fusion network.

3.3   Feature Fusion Network

The details of feature fusion network are shown in Fig. 4. The raw feature maps s from backbone without correlation operation is much larger than cls and $score$. A huge depth-wise convolution, whose size is same with c, followed by $1 \times 1conv$ is adopted to increase its receptive field without huge computation cost. Sine not all information in feature map contributes to semantic segmentation, and similar instances or background interference could cause segmentation failure. Tracker output could instruct the model to concentrate on target feature itself as a selective mask.

We state a training objective as follows:

$$\mathcal{L}_{seg} = \frac{1}{N} \sum_{i=1}^{N} \mathcal{L}_{seg}(p_{seg}^i, t_{seg}^i), \tag{2}$$

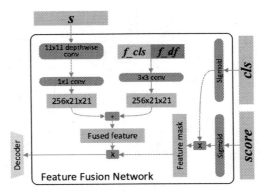

Fig. 4. Flow chart of feature fusion network. Results from tracker output could instruct the decoder to suppress the background interference.

$$\mathcal{L}_{track} = \frac{1}{N}\sum_{i=1}^{N}\mathcal{L}_{cls}(p_{cls}^{i}, t_{cls}^{i}) + \frac{\lambda_1}{N}\sum_{i=1}^{N}t_{cls}^{i}\mathcal{L}_{loc}(p_{loc}^{i}, t_{loc}^{i})$$

$$+ \frac{\lambda_2}{N}\sum_{i=1}^{N}t_{cls}^{i}\mathcal{L}_{score}(p_{score}^{i}, t_{score}^{i}) \qquad (3)$$

where N is the total position number in cls. $\mathcal{L}_{seg}$ denote the binary cross entropy loss. $\mathcal{L}_{cls}$ and $\mathcal{L}_{score}$ denote the focal loss. $\mathcal{L}_{loc}$ denote the IoU loss.

4   Experiment

4.1   Soccer Tracking Datatset

Sine we want to achieve a unification in tracking and segmentation for soccer videos, a soccer tracking dataset is established for evaluation. The sources of soccer match videos are downloaded from Internet. In each soccer match video, we cut around 10 short clips, with each containing around 6 tracking tracks. We implement a semi-automatic strategy to label the dataset. For each track, we first adopt a state-of-the-art tracking algorithm to do pre-labeling. When the overlap rate is unsatisfactory or the bounding box can not fit target well, we manually correct the bounding box frame by frame. For the requirement of some training and testing cases or certain application scenarios, we also add the occlusion label if 30% or more area of the target is occluded by other objects.

Specifically, this soccer tracking dataset contains total 1268 instance tracks. 374048 frames are labeled with bounding boxes associated with occlusion label. Some potential traits of the dataset are extracted by observing all the tracks:

- The resolution of the target is generally lower than the existing tracking and semantic segmentation datasets.

– Occlusion occurs promiscuously as shown in Fig. 1(a). The model should be capable of deducing the overall target feature from unblocked area.
– Compared with general datasets, soccer players are more prone to be deformed when they run and hit the ball, as shown in Fig. 1(b).
– The biggest difference is that there exist many similar instances around the target as shown in Fig. 1(c). Tracking soccer players requires models to be more discriminative than general tasks.

Obviously, these above features tend to occur in a kind of combination, which is a tough challenge for the application of pre-trained models. This dataset will be released in the future.

4.2   Implementation Details

The hardware environment for training is a workstation with 2 TITANxp GPUs. The hardware for all of the tests below is a workstation with i7-9700K CPU, 16 GB RAM and GTX 1080 Ti GPU. We choose ResNet50 [17] and InceptionV3 [18] as feature extract backbone which is pre-trained on ImageNet [19] for object classification. We adopt COCO [20], YoutubeVIS [21] and GOT-10k [22] as tracking datasets. Sine COCO and YoutubeVIS contain the mask label of target, we could use them to train the decoder. We utilize alternate training strategy for tracking and segmentation. The optimizer is ADAM with fine-tuned last two layers of the backbone.

4.3   Tracking Results

Our soccer tracking dataset's assessment criteria are very similar to OTB benchmark [23]. Higher Area Under the Curve (AUC) means higher overlap rate between prediction and labels. Note that we devide the whole soccer tracking data set into training, validation and test sets. There are total 1016 training tracks, 126 validation tracks and 126 testing tracks in soccer tracking dataset which contain all of the tough situation mentioned above. We select several state-of-the-art trackers SiamRPN [1], SiamRPN++ [2], SiamMask [9], ATOM [24], UpdateNet [25] and DAT [26]. SiamRPN++ and SiamRPN are fine-tuned in soccer tracking training set. Table 1 presents the results soccer tracking dataset. Our tracker ranks the first in AUC and precision. The other trackers' AUC drops rapidly since they could not fit the target deformation well. Besides, our tracker is leading the rank of AUC even without fine-tuned in training set as shown in Table 2. These results illustrate a higher adaptability of deformable cross-similarity correlation to handle deformation tasks.

4.4   Segmentation Results

To evaluate the segmentation performance in soccer tracking dataset, we select SiamMask which could achieve tracking and segmentation in one-shot processing

Table 1. Tracking results in soccer tracking dataset. AUC is short for area under the curve of overlap rate. Precision@20 means the predict center position deviation is no more than 20 pixels. (F) refers to fine-tuned in soccer tracking training set.

Method	AUC	Precision@20	FPS
Ours_InceptionV3(F)	**0.726**	0.921	36.2
Ours_ResNet50(F)	<u>0.713</u>	**0.932**	40.8
SiamRPN(F)	0.700	0.891	**77.7**
SiamRPN++(F)	0.690	<u>0.930</u>	30.2
UpdateNet	0.656	0.824	<u>42.0</u>
SiamMask	0.650	0.794	36.6
DAT	0.634	0.924	0.317
ATOM	0.606	0.783	28.6

Fig. 5. Segmentation results of our method, SiamMask and Deeplab in soccer tracking dataset. The inpus for Deeplab is shown in the upper left corner of the forth row.

like our strategy and Deeplab [27]. Because the semantic information is not annotated in soccer tracking dataset, we train Deeplab model on a basketball player segmentation dataset for transfer learning. Figure 5 illustrates input search image and segmentation results. Compared with the other two approaches, we limit the range of input area in Deeplab to reduce background interference. Deeplab predicts the players' edge more precisely, but it can not distinguish similar instances even with strict constraints. Our work shows comparable performance against SiamMask and has a better discriminative ability between similar instances.

4.5 Ablation Studies

In order to evaluate the contribution of each component, we test different backbone structure and construct a tracker without deformable cross-similarity

Table 2. Ablation experiment results in soccer tracking dataset. The tracking performance will degrade significantly without the contribution of DF_CORR.

Method	AUC	Precision@20
InceptionV3(F)	**0.726**	0.921
InceptionV3	0.717	0.911
ResNet50(F)	0.713	**0.932**
InceptionV3 w/o DF_CORR	0.664	0.838

correlation as the baseline. The results in Table 2 prove that ResNet50 is apt to fit precision and InceptionV3 tends to improve AUC. We have to be aware that there is a dramatic performance degradation both in precision and AUC without DF_CORR, which means that proposed DF_CORR operation could contribute to predicting the target deformation and promotes tracking accuracy.

4.6   Visual Effects in Soccer Videos

In recent years, with the popularization of smart devices, deep learning applications have evolved rapidly in sports game broadcasting and entertainment area. With the auxiliary target semantic attributes, the impression of video could be significantly enhanced by adding special visual effects. Figure 6 shows some demos of effects. By obtaining the location and semantic information of the target, we could automatically layer the image, just like some functions in Photoshop. Such utils could make the special effects more vivid, and reduce labor burden.

Fig. 6. Visualization of some visual effects. Line (a) shows the halo circle under the player's feet. Line (b) shows the effect of adding a light column. These effects utils are still in early versions

5 Conclusion

In this paper, we present a novel tracking-segmentation network which could predict target location and segmentation mask in one-shot processing. A deformable cross-similarity correlation is adopted to fit the deformation of the target. Our encoder-decoder structure combined with feature fusion network shows comparable segmentation performance against SiamMask. A soccer tracking dataset is established to evaluate and expand the scope of applications. Extensive experiments show our strategy outperforms state-of-the-art trackers in AUC and could achieve real-time speed. Such strategy could be also adopted in visual effects adding for soccer videos.

References

1. Li, B., Yan, J., Wu, W., Zhu, Z., Hu, X.: High performance visual tracking with Siamese region proposal network. In: Computer Vision and Pattern Recognition, pp. 8971–8980. IEEE (2018)
2. Li, B., Wu, W., Wang, Q., Zhang, F., Xing, J., Yan, J.: Siamrpn++: evolution of Siamese visual tracking with very deep networks. In: Computer Vision and Pattern Recognition, pp. 4282–4291. IEEE (2019)
3. Perazzi, F., Pont-Tuset, J., McWilliams, B., Van Gool, L., Gross, M., Sorkine-Hornung, A.: A benchmark dataset and evaluation methodology for video object segmentation. In: Computer Vision and Pattern Recognition, pp. 724–732. IEEE (2016)
4. Kristan, M., et al.: The sixth visual object tracking VOT2018 challenge results. In: Leal-Taixé, L., Roth, S. (eds.) ECCV 2018. LNCS, vol. 11129, pp. 3–53. Springer, Cham (2019). https://doi.org/10.1007/978-3-030-11009-3_1
5. Bolme, D.S., Beveridge, J.R., Draper, B.A., Lui, Y.M.: Visual object tracking using adaptive correlation filters. In: Computer Vision and Pattern Recognition, pp. 2544–2550. IEEE (2010)
6. Henriques, J.F., Caseiro, R., Martins, P., Batista, J.: High-speed tracking with kernelized correlation filters. IEEE Trans. Pattern Anal. Mach. Intell. **37**(3), 583–596 (2014)
7. Song, Y., Ma, C., Gong, L., Zhang, J., Lau, R., Yang, M.H.: Crest: convolutional residual learning for visual tracking. In: International Conference on Computer Vision, pp. 2555–2564. IEEE (2017)
8. Zhu, Z., Wang, Q., Li, B., Wu, W., Yan, J., Hu, W.: Distractor-aware Siamese networks for visual object tracking. In: Ferrari, V., Hebert, M., Sminchisescu, C., Weiss, Y. (eds.) ECCV 2018. LNCS, vol. 11213, pp. 103–119. Springer, Cham (2018). https://doi.org/10.1007/978-3-030-01240-3_7
9. Wang, Q., Zhang, L., Bertinetto, L., Hu, W., Torr, P.H.: Fast online object tracking and segmentation: a unifying approach. In: Computer Vision and Pattern Recognition, pp. 1328–1338. IEEE (2019)
10. Ren, S., He, K., Girshick, R., Sun, J.: Faster R-CNN: towards real-time object detection with region proposal networks. In: Neural Information Processing Systems, pp. 91–99 (2015)
11. Bhat, G., Danelljan, M., Gool, L.V., Timofte, R.: Learning discriminative model prediction for tracking. In: International Conference on Computer Vision, pp. 6182–6191. IEEE (2019)

12. Maninis, K.K., et al.: Video object segmentation without temporal information. IEEE Trans. Pattern Anal. Mach. Intell. **41**(6), 1515–1530 (2018)
13. Perazzi, F., Khoreva, A., Benenson, R., Schiele, B., Sorkine-Hornung, A.: Learning video object segmentation from static images. In: Computer Vision and Pattern Recognition, pp. 2663–2672. IEEE (2017)
14. Bao, L., Wu, B., Liu, W.: CNN in MRF: video object segmentation via inference in a cnn-based higher-order spatio-temporal MRF. In: Computer Vision and Pattern Recognition, pp. 5977–5986. IEEE (2018)
15. Ci, H., Wang, C., Wang, Y.: Video object segmentation by learning location-sensitive embeddings. In: Ferrari, V., Hebert, M., Sminchisescu, C., Weiss, Y. (eds.) ECCV 2018. LNCS, vol. 11215, pp. 524–539. Springer, Cham (2018). https://doi.org/10.1007/978-3-030-01252-6_31
16. Bertinetto, L., Valmadre, J., Henriques, J.F., Vedaldi, A., Torr, P.H.S.: Fully-convolutional Siamese networks for object tracking. In: Hua, G., Jégou, H. (eds.) ECCV 2016. LNCS, vol. 9914, pp. 850–865. Springer, Cham (2016). https://doi.org/10.1007/978-3-319-48881-3_56
17. He, K., Zhang, X., Ren, S., Sun, J.: Deep residual learning for image recognition. In: Computer Vision and Pattern Recognition, pp. 770–778. IEEE (2016)
18. Szegedy, C., Vanhoucke, V., Ioffe, S., Shlens, J., Wojna, Z.: Rethinking the inception architecture for computer vision. In: Computer Vision and Pattern Recognition, pp. 2818–2826. IEEE (2016)
19. Deng, J., Dong, W., Socher, R., Li, L.J., Li, K., Fei-Fei, L.: Imagenet: a large-scale hierarchical image database. In: Computer Vision and Pattern Recognition, pp. 248–255. IEEE (2009)
20. Lin, T.-Y., et al.: Microsoft COCO: common objects in context. In: Fleet, D., Pajdla, T., Schiele, B., Tuytelaars, T. (eds.) ECCV 2014. LNCS, vol. 8693, pp. 740–755. Springer, Cham (2014). https://doi.org/10.1007/978-3-319-10602-1_48
21. Yang, L., Fan, Y., Xu, N.: Video instance segmentation. In: International Conference on Computer Vision, pp. 5188–5197. IEEE (2019)
22. Huang, L., Zhao, X., Huang, K.: Got-10k: a large high-diversity benchmark for generic object tracking in the wild. IEEE Trans. Pattern Anal. Mach. Intell. 1 (2019)
23. Wu, Y., Lim, J., Yang, M.H.: Object tracking benchmark. IEEE Trans. Pattern Anal. Mach. Intell. **37**(9), 1834–1848 (2015)
24. Danelljan, M., Bhat, G., Khan, F.S., Felsberg, M.: Atom: accurate tracking by overlap maximization. In: Computer Vision and Pattern Recognition, pp. 4660–4669. IEEE (2019)
25. Zhang, L., Gonzalez-Garcia, A., Weijer, J.V.D., Danelljan, M., Khan, F.S.: Learning the model update for Siamese trackers. In: International Conference on Computer Vision, pp. 4010–4019. IEEE (2019)
26. Pu, S., Song, Y., Ma, C., Zhang, H., Yang, M.H.: Deep attentive tracking via reciprocative learning. In: Neural Information Processing Systems, pp. 1931–1941 (2018)
27. Chen, L.C., Papandreou, G., Kokkinos, I., Murphy, K., Yuille, A.L.: DeepLab: semantic image segmentation with deep convolutional nets, atrous convolution, and fully connected CRFs. IEEE Trans. Pattern Anal. Mach. Intell. **40**(4), 834–848 (2017)

Towards More Robust Detection for Small and Densely Arranged Ships in SAR Image

Jingpu Wang$^{(\boxtimes)}$, Youquan Lin, Long Zhuang, and Jie Guo

Nanjing Research Institute of Electronics Technology, Jiangsu, China
13813085308@163.com

Abstract. Independent of sunlight and weather conditions, synthetic aperture radar (SAR) imagery is widely applied to detect ships in marine surveillance. This paper proposes a ship target detection algorithm, aiming at the missing and false detection in the situation of dense ship targets in SAR images. Firstly, we used the spatial pyramid pooling (SPP) to enhance the feature extraction capability in different scales. Then, we modified the regression loss function with three factors of center distance, overlap area and length-width ratio to reduce the error of location. Finally, we proposed double threshold soft non maximum suppression (DTSOFT-NMS) to reduce the missing detections for dense ships. The experimental results reveal that our model exhibits excellent performance on the open SAR-ship-dataset and improves average precision (AP) by 6.5% compared with the baseline YOLOv3 model.

Keywords: Synthetic Aperture Radar (SAR) · Spatial pyramid pooling · Double threshold soft non maximum suppression (DTSOFT-NMS)

1 Introduction

Synthetic Aperture Radar (SAR) is an active microwave sensor that can emit electromagnetic waves and receive echo signals for active imaging without being restricted by external conditions such as weather and light. With all-weather characteristics, SAR images have been widely applied in civil and military fields, and ship target detection based on SAR images has become one of the hot research issues.

Traditional ship detection methods were mainly based on the different backscattering characteristics between ship targets and the ocean. Common methods include constant false alarm rate (CFAR) [1], polarization decomposition, wavelet decomposition and template method. With the development of SAR, the resolution of SAR images has further improved, and traditional detection methods are difficult to meet the current actual needs in terms of accuracy and detection efficiency.

Due to the strong feature extraction capabilities of convolutional neural network (CNN), deep learning has achieved great success in object detection. Object detection algorithms based on CNN can be divided into two categories: one is a two-stage object detection algorithm, also known as a region-based algorithm, includes Fast R-CNN [2], Faster R-CNN [3], R-FCN [3], and Mask R-CNN [5]. The other one is a one-stage

© Springer Nature Switzerland AG 2020
Y. Peng et al. (Eds.): PRCV 2020, LNCS 12306, pp. 278–289, 2020.
https://doi.org/10.1007/978-3-030-60639-8_24

object detection algorithm, also called regression-based object detection algorithm. This type of algorithm directly generates the category and position information of the target. Representative algorithms include SSD [6], RetinaNet [7] and YOLO series [8–10].

Zhang [11] proposed a ship target detection algorithm based on visual attention via CNN coupling cascade (3c2n-guided), which can improve detection performance and greatly reduce missed detection and false positives. CHEN [12] proposed a target detection network combined with attention mechanism to accurately locate the target in complex scenarios. JIAO [13] proposed a densely connected multiscale neural network based on Fasters R-CNN framework to solve the multiscale multi-scene SAR ship detection problem. Miao [14] proposed a detection method that the traditional CFAR was combined with Faster R-CNN network to improve the detection performance. Yang [15] proposed a new detection method based on SSD, which combines context information fusion, migration model learning and SSD, and conducts training and testing on the public SSDD dataset. In the past years, deep learning has made great progress in ship target detection of SAR images, but there are still many problems.

Because the SAR system has multiple imaging modes and imaging resolutions, ships have different sizes in SAR images. Current detection methods have poor detection effect on multi-size ships, especially small-size ships. Besides, in the dense areas with small adjacent spaces between ships, the detection strategy of CNN will lead to missing and false detection.

In this paper, we propose a method for multiscale and densely arranged ships detection in SAR Image. The main contributions of this paper are as follows.

1. We chose YOLOv3 as the baseline and use the SPP module to improve the feature extraction capability of the network;
2. We redesigned the regression loss function based on CIOU [17], which can reduce the scale sensitivity and improve the location accuracy;
3. We proposed DTSOFT-NMS into the final processing to improve the detection effect for dense ships;
4. Based on the public SAR ship dataset [18] proposed by the Chinese Academy of Sciences, we use k-means clustering to redesign the anchor boxes, train and test the proposed network.

The organization of this paper is as follows. Section 2 relates to the related work of the baseline model. Section 3 illustrates our proposed method and network structure. Section 4 introduces the dataset used in our experiments and describes the experimental details and results. Section 5 presents the conclusion.

2 Related Work

In 2016, You Only Look Once (YOLO) has been introduced which unlocks the potential of real-time performance. YOLO uses regression method to predict the coordinates of the bounding boxes and achieve the classification of the targets. YOLO applies a grid approach for bounding box detection and adds a SoftMax layer to directly predict the object class. The detection speed reaches 30 FPS, but YOLO has serious location error, causing poor detection accuracy. YOLOv2 modified the feature extraction network with Darknet-19, uses Anchor Box to predict targets and applies Batch Normalization (BN) for training. Compared with YOLO, the running speed and detection accuracy of YOLOv2 have been significantly improved. YOLOv3 uses the deep residual network Darknet-53 to extract features and achieve multi-scale prediction, which make the accuracy and speed meet the actual engineering needs.

The network structure of YOLOv3 is composed of feature extraction network darknet-53 and feature pyramid network (FPN).

Inspired by residual network, darknet-53 has 53 convolution layers, which can be divided into 5 residual convolution blocks. Each convolution block is composed of multiple residual units. The residual operation is carried out by the input and two digital cumulative modeling (DBL) units to construct the residual unit. Every DBL cell contains convolution layer, BN, and leaky ReLU activation functions. By introducing residual units, the depth of the network can be increased to reduce the gradient vanishing problem.

The FPN of YOLOv3 is shown in Fig. 1. In the process of detection, the input image will be down-sampled for 5 times, and the last three feature maps are fused with larger feature maps by up-sampling. When the input image size is 416×416, the detection is performed on the scales of $13 \times 13, 26 \times 26$ and 52×52.

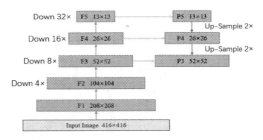

Fig. 1. The feature pyramid network of YOLOv3

Feature maps of different scales contain different feature information. The small maps can usually provide deep semantic information, while large maps contain more targets position information, especially small targets information. Thus, YOLOv3 fuses feature maps of different scales to improve multiscale detection ability. It can not only detect large size targets, but also optimize the detection of small size targets.

3   Methods

3.1   Spatial Pyramid Pooling

In this paper, we use SPP module to improve the extraction ability of local and global features between multiscale ships in SAR images.

The SPP module is based on the SPP-net [16], which is used to solve the problem of different image input to fixed size output to the full connection layer. The application of SPP in our method can alleviate the over-sensitivity of the convolutional layer for position information. It can realize the fusion of local feature and global feature at the feature maps level.

The structure of SPP is shown in Fig. 2, which is all pooling layers. SPP does not change the size of feature maps before and after processing. There are three sizes of pooling kernel, with $3 \times 3, 5 \times 5, 7 \times 7$, and then superimpose the initial feature map to obtain the new feature map. We add three SPP modules before the detectors of the three scale feature maps in the baseline model. The new structure of our models is shown as Fig. 3.

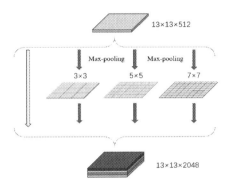

Fig. 2. The structure of SPP

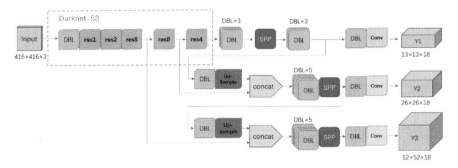

Fig. 3. The structure we proposed

In SAR images, the shapes of ships are different, so ships in the same image may have multiple sizes. Besides, the same ship in various resolution images has different sizes. Therefore, multiscale ship is a difficult problem for SAR ship detection. YOLOv3 performed object detection on three scale feature maps and each of which is responsible for different ships.

Usually, the feature maps of 52×52 detect small ships and $26 \times 26, 13 \times 13$ feature maps detect big ships. Three SPP modules before three detectors can improve the detection effect of multiscale ship.

3.2 Loss Function

The loss function of the original YOLOv3 network is composed of four parts, including center coordinate error, width-height error, confidence error and classification error. Among them, center coordinate error and width and height coordinate error are calculated by mean square error (MSE). However, the various target sizes will lead to the large fluctuation of the mean square error loss, which will eventually affect the positioning accuracy of the target and reduce the detection accuracy.

Inspired by the CIOU [17], we modify the loss function with overlapping area, distance penalty and length-width penalty to reduce scale sensitivity enhance positioning accuracy. The new loss function is defined as follows:

$$L_{reg} = 1 - IoU + \rho^2(b, \ b^{gt}) + \alpha v \tag{1}$$

$$IoU = \frac{|B \bigcap B^{gt}|}{|B \bigcup B^{gt}|} \tag{2}$$

where B is the target prediction box and B^{gt} is the ground true box; b and b^{gt} represent the center points of the target prediction box and the true box; ρ is the Euclidean distance between the two points.

In the length-wide penalty, α is the coefficient used to balance the proportion and v is used to measure the proportion consistency between the anchor frame and the target frame. Their formulas are as follows:

$$v = \frac{4}{\pi^2}\left(arctan\frac{w^{gt}}{h^{gt}} - arctan\frac{w}{h}\right)^2 \tag{3}$$

$$\alpha = \frac{1}{(1 - IoU) + 1} \tag{4}$$

where w, h are the width and height of prediction box and w^{gt}, h^{gt} are the width and height of the ground true box.

It can be seen that the length-wide penalty is invariant with respect to the scale. When calculating the loss between the prediction box and the real box, our loss function not only considers the distance between the overlapping area and the center point, but also adds the influencing factor of the aspect ratio of length and width. Therefore, the new loss function can achieve better regression for different distances, directions, areas and achieve faster convergence speed and convergence effect.

3.3 DTSOFT-NMS

Most detection algorithms use Non maximum suppression (NMS) to suppress overlapping boxes. The idea of NMS is to regard the box with the highest score as the best detection result and then remove all other prediction boxes with high overlap degree. The mathematical expression of this algorithm is as follows:

$$S_f = \begin{cases} S_i, & U_{IoU}(Max, \ B_i) < N_t \\ 0, & U_{IoU}(Max, \ B_i) \geq N_t \end{cases} \tag{5}$$

where S_i is the original score of the prediction box; S_f is the final score of the prediction box; Max is the prediction box with the highest score, B_i is the other boxes to be checked except Max; U_{IOU} (Max, Bi) is the IOU of B_i and Max and N_t is the threshold of IOU.

When the overlap of the two boxes is small, IOU is less than the threshold value. According to the NMS algorithm, the two prediction boxes belong to different target detection results, so two boxes are both reserved. When the overlap of two boxes is very high and IOU is greater than the threshold, as shown in Fig. 4(a), the NMS algorithm will determine that these two boxes are the detection results of the same target, so it will clear the confidence score of the box with a low score, so as to complete the filtering of the repeated prediction box.

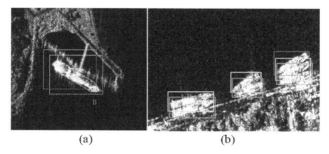

(a)   (b)

Fig. 4. An illustration of dense ships

However, when the ships are very closed, the prediction box of two different targets have large IOU and the prediction box with lower score will be cleared, resulting in a missed detection. As shown in Fig. 4 (b), the distance between ships is very close in port area. The prediction boxes of different colors represent the detection results of different ship targets. However, according to the NMS algorithm, only the prediction box of the target with the highest score will be kept, and multiple ships will be regarded as the same target, resulting in missed detection.

Therefore, we propose DTSOFT-NMS algorithm to solve the detection problem of dense targets. The core idea of DTSOFT-NMS algorithm is to reduce the score of the prediction box with large overlap by using the penalty strategy, and to reserve the prediction box belonging to different targets to the greatest extent, so as to reduce the missed detection.

In the DTSOFT-NMS algorithm, the penalty function of the confidence score is calculated as (6). The confidence score of the prediction box with small overlap will not decay. However, the prediction box with large overlap attenuates greatly, but does not clear directly. When the IOU of the prediction box is really large,the score will be remove. Such a strategy can keep the prediction box of densely distributed ship targets and reduce the missed detection.

$$S_f = \begin{cases} 0 & , N_{t1} \leq U_{IoU}(Max, B_i) \\ S_i e^{-\frac{IoU(Max, B_i)^2}{N_{t2}}} & , N_{t2} \leq U_{IoU}(Max, B_i) < N_{t1} \\ S_i & , \quad U_{IoU}(Max, B_i) < N_{t2} \end{cases} \tag{6}$$

4 Experiments

4.1 The Experiment Platform

All experiments were implemented on a workstation with an Intel(R) Core (TM) i5-9400F @2.90 GHz, an NVIDIA RTX 2070 GPU and the Pytorch framework. The initial learning rate of the network was set to 0.001 for the first 400 iterations and 0.0001 for the last 400 iterations. The batch size is 8. The optimization algorithm used Adam, with beta1 of 0.9, beta2 of 0.999 and epsilon of 1e-8.

4.2 Dataset

In this paper, we use the SAR-ship-Dataset 18 published by Chinese academy of sciences to train and test the deep neural network. A total of 102-scene gaofen-3 and 108-scene sentinel-1 SAR images are used to build a SAR ship target deep learning sample library. The dataset contains 43819 ship slices and their label information, while the target to be detected is only one class. Thus, it is sufficient for training our network. In our experiment, 80% of the samples were used as the training set and 20% of the sample data were testing set.

4.3 Evaluation Metrics

The evaluation metrics used in this paper are average accuracy (AP) to quantitatively evaluate the detection effect of the model. The average accuracy is defined as follows:

$$AP = \int_0^1 P(R)dR \tag{7}$$

$$R = \frac{X_{TP}}{X_{TP} + X_{FN}} \tag{8}$$

$$P = \frac{X_{TP}}{X_{TP} + X_{FP}} \tag{9}$$

where P is the detection accuracy; R is the target recall rate; X_{TP} represents the number of targets correctly detected; X_{FN} represents the number of targets not detected; X_{FP} represents the number of targets that were incorrectly checked out.

4.4 Ablution Study

To illustrate the effectiveness of the proposed approaches, we compared the effects of different versions of our method on the AP through step-by-step experiments based on the SAR-ship-Dataset. The various approaches incorporated into our model, as mentioned above are shown in Table 1.

Table 1. Ablution Study

Method	SPP	CIOU	DTSOFT-NMS	AP (%)
Proposed	√			86.22%
		√		85.43%
			√	83.56%
		√	√	86.96%
	√		√	87.86%
	√	√		89.75%
	√	√	√	89.92%

We add three SPP modules before the detectors of the three scale feature maps in the YOLOv3 network to enhances the feature extraction capability of network at different scales and the AP of the model is improved by 2.9%.

By using the CIOU loss into the loss function, the sensitivity of the network to different ship target scales is reduced, and the AP of the model is improved by 2.06%.

By contrast, using DTSOFT-NMS improves the AP of the model by only 0.17%. The reason for this lesser improvement maybe that the dataset contains relatively few densely arranged ship samples. Based on the approaches mentioned above, the final AP of the proposed model is 89.92% on the SAR-Ship-Dataset.

4.5 Results Analysis

In Fig. 5, we can see that as the number of training iterations increases, the loss value gradually decreases and converges to a low value. Finally, the loss function converges to 0.69.

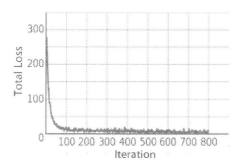

Fig. 5. Loss function curve during the training process

The application of SPP in YOLOv3 network can realize the fusion of local feature and global feature at the feature graph level, so as to enrich the expression ability. As shown in Fig. 6, the blue line represents the precision of our detection algorithm and the yellow line represents the original YOLOv3. By comparing, we can find that adding SPP structure can improve AP by 2.96%.

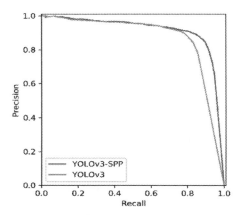

Fig. 6. The PR curves of YOLOv3 and YOLOv3-SPP

The ship detection results obtained under the different environmental conditions represented in the SAR-Ship-Dataset were analyzed, as shown in Fig. 7. We show several detection results for SAR ships.

The first row shows results for ship detection against complex backgrounds. We find that the proposed method can effectively distinguish targets from their backgrounds. The second row shows results for the detection of ships of different sizes. It is clear that the proposed algorithm achieves an improved detection effect for small ships and big ships with a lower missed detection rate. The third row shows densely arranged ships detection. It can be seen that the algorithm proposed in this paper can effectively distinguish closely spaced ships.

To better illustrate the effectiveness of the proposed DTSOFT-NMS, we select a sub-dataset from sentinel-1 SAR images with densely arranged ships. It contains 32 images and 108 ship targets. The result in Table 2 shows that DTSOFT-NMS can reduce some missed detection compare with original NMS.

4.6 Comparison with Other Methods

In this section, our proposed method is quantitatively compared with several mainstream object detection models based on deep learning in terms of the AP and detection speed. The results are shown in Table 3.

It is apparent that the proposed method achieves the best AP of 89.92% on the SAR-Ship-Dataset compared with other one-stage object detection methods including SSD and original YOLOv3. The use of SPP and DTSOFT-NMS increase the computation of

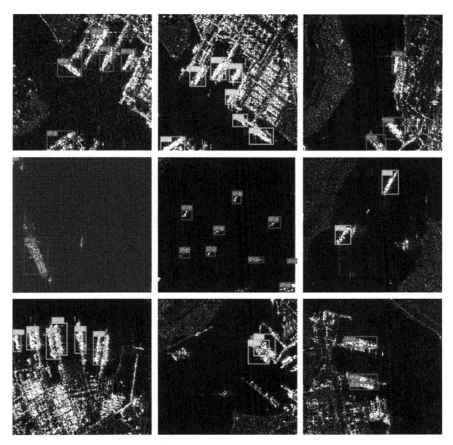

Fig. 7. Experimental results

Table 2. Densely arranged ships detection

Methods	Detected ships	True ships	False alarms	Missed ships	Precision	Recall
NMS	114	94	20	14	82.45%	87.03%
DTSOFT-NMS	122	100	22	8	81.96%	92.59%

the network. Thus, our algorithm is not as fast as original YOLOv3. But its detection time of one image is 30.14 ms, which is sufficient for real-time detection. Compared with the two-stage object detection algorithm Faster R-CNN, the time cost of the proposed algorithm is only 30% of that of Faster R-CNN, and the AP of our method is 1% higher than Faster R-CNN.

Table 3. Compare with other methods

Method	Backbone	AP	Time
Faster R-CNN	Resnet-101	88.74%	93.29 ms
SSD	VGG16	78.62%	31.52 ms
YOLOv3	Darknet-53	83.36%	26.74 ms
Proposed	Darknet-53 with SPP	89.92%	30.14 ms

5 Conclusion

In this paper, we proposed a detection algorithm based on YOLOv3 for ship detection in SAR image and verified our method on the public SAR-Ship-Dataset. Based on the YOLOv3, this paper used the spatial pyramid pooling structure to improve the feature extraction capability of the network. Then, the regression loss function based on CIOU is proposed to make the network converge faster and obtain higher positioning accuracy during training. Finally, this paper proposed DTSOFT-NMS to reduce the missed and false detection in the process of intensive target detection. The detection time of a single image in the SAR-Ship-Dataset is 30.14 ms and the AP is 89.92%, which is significantly improved for densely arranged ships and multiscale ships.

References

1. Xing, X.W., Chen, Z.L., Zou, H.X., et al.: A fast algorithm based on two-stage CFAR for detecting ships in SAR images. In: The 2nd Asian-Pacific Conference on Synthetic Aperture Radar, pp. 506–509 (2010)
2. Girshick, R.: Fast R-CNN, IEEE International Conference on Computer Vision, pp. 1440–1448 (2015)
3. Ren, S., He, K., Girshick, R., Sun, J.: Faster R-CNN: Towards real-time object detection with region proposal networks. IEEE Trans. Pattern Anal. and Mach. Intell. **39**(6), 1137–1149 (2017)
4. Dai, J., Li, Y., He, K., Sun, J.: R-FCN: object detection via region-based fully convolutional networks. In: Advances in Neural Information Processing Systems, pp. 379–387 (2016)
5. Kaiming, H., Georgia, G., Piotr, D., et al.: Mask R-CNN. In: IEEE Transactions on Pattern Analysis and Machine Intelligence, pp. 2961–2969 (2017)
6. Liu, W., Anguelov, D., Erhan, D., Szegedy, C., Reed, S., Fu, C.-Y., Berg, Alexander C.: SSD: single shot multibox detector. In: Leibe, B., Matas, J., Sebe, N., Welling, M. (eds.) ECCV 2016. LNCS, vol. 9905, pp. 21–37. Springer, Cham (2016). https://doi.org/10.1007/978-3-319-46448-0_2
7. Lin, T.-Y., Goyal, P., Girshick, R., He, K., Dollar, P.: Focal loss for dense object detection. In: IEEE Transactions on Pattern Analysis and Machine Intelligence, no. 99, pp. 2980–2988 (2017)
8. Redmon, J., Divvala, S., Girshick, R., Farhadi, A.: You only look once: unified, real-time object detection. In: IEEE Conference on Computer Vision and Pattern Recognition, pp. 779–788 (2016)
9. Redmon, J., Farhadi, A.: Yolo9000: better, faster, stronger. In: IEEE Conference on Computer Vision and Pattern Recognition, pp. 6517–6525 (2017)

10. Redmon, J., Farhadi, A.: Yolov3: an incremental improvement. arXiv:1804.02767 (2018)
11. Zhao, J., Zhang, Z., Yu, W., et al.: A cascade coupled convolutional neural network guided visual attention method for ship detection from SAR images. IEEE Access **6**, 50693–50708 (2018)
12. Chen, C., He, C., Hu, C., Pei, H., Jiao, L.: A deep neural network based on an attention mechanism for SAR ship detection in multiscale and complex scenarios. IEEE Access **7**, 104848–104863 (2019)
13. Jiao, J., et al.: A densely connected end-to-end neural network for multiscale and multi scene SAR ship detection. IEEE Access **6**, 20881–20892 (2018)
14. Kang, M., Leng, X., Lin, Z., Ji, K.: A modified faster R-CNN based on CFAR algorithm for SAR ship detection. In: 2017 International Workshop on Remote Sensing with Intelligent Processing, pp. 1–4 (2017)
15. Long, Y.A.N.G., Juan, S.U., Xiang, L.I.: Ship detection in SAR images based on deep convolutional neural network. Syst. Eng. Electr. **41**(9), 1990–1997 (2019)
16. He, K., Zhang, X., Ren, S., Sun, J.: Spatial pyramid pooling in deep convolutional networks for visual recognition. In: Fleet, D., Pajdla, T., Schiele, B., Tuytelaars, T. (eds.) ECCV 2014. LNCS, vol. 8691, pp. 346–361. Springer, Cham (2014). https://doi.org/10.1007/978-3-319-10578-9_23
17. Zheng, Z., Wang, P., Liu, W.: Distance-IoU loss: faster and better learning for bounding box regression. arXiv:1911.08287 (2019)
18. Wang, Y., Wang, C., Zhang, H., Dong, Y., Wei, S.: A SAR dataset of ship detection for deep learning under complex backgrounds. Remote Sens. **11**(7), 765–769 (2019)

SLPNet: Towards End-to-End Car License Plate Detection and Recognition Using Lightweight CNN

Wei Zhang[1], Yaobin Mao[1(✉)], and Yi Han[2]

[1] Nanjing University of Science and Technology, Nanjing 210094, China
edward.w.zhang@outlook.com, maoyaobin@njust.edu.cn
[2] Zhejiang Huayun Information Technology Co. LTD, Hangzhou 310008, China
hanyiuse@163.com

Abstract. As a core component, Automatic License Plate Recognition (ALPR) plays an important role in modern Intelligent Transportation System (ITS). Due to the complexity in real world, many existing license plate detection and recognition approaches are not robust and efficient enough for practical applications, therefore ALPR still a challenging task both for engineers and researchers. In this paper, a Convolutional Neural Network (CNN) based lightweight segmentation-free ALPR framework, namely SLPNet is established, which succinctly takes license plate detection and recognition as two associated parts and is trained end-to-end. The framework not only accelerates the processing speed, but also achieves a better match between the two tasks. Other contributions includes an anchor-free LP localization network based on corners using a novel MG loss is proposed and a multi-resolution input image strategy is adopted for different tasks to balance the operation speed and accuracy. Experimental results on CCPD data set show the effectiveness and efficiency of our proposed approach. The resulting best model can achieve a recognition accuracy of 98.6% with only 3.4M parameters, while the inference speed is about 25 FPS on a NVIDIA Titan V GPU. Code is available at https://github.com/JackEasson/SLPNet_pytorch.

Keywords: License plate detection · License plate recognition · End to end training · Segmentation-free · Lightweight CNN

1 Introduction

Automatic License Plate Recognition (ALPR) plays an important role in Intelligent Transportation System (ITS) which is widely used in traffic management, intelligent surveillance and parking management in large cities [20], hence, attracts considerable research attentions in recent two decades.

Even significant progress has been made especially with the help of deep learning in recent years, ALPR in nature environment is still a challenging task,

---

W. Zhang—Student author.

Y. Peng et al. (Eds.): PRCV 2020, LNCS 12306, pp. 290–302, 2020.
https://doi.org/10.1007/978-3-030-60639-8_25

due to the complexity and uncertainty of the environment and uncontrollable photographing conditions. Generally, an ALPR task can be divided into three subtasks according to its operating process, namely license plate detection, character segmentation and character recognition. With the popular of deep learning, a new trend to combine some of the above steps into one is appeared. For instance, character segmentation is merged with recognition in certain methods that directly predict the whole license plate numbers in form of sequences, which is known as segmentation-free license plate recognition [12,22,28]. Furthermore, some recent work has utilized a single network to simultaneously localize license plates and recognize the characters in a single forward inference. Such network not only can be trained end-to-end, but can avoid intermediate error accumulation [20] as well. However, it suffers from a dilemma of image resolution that the network operates at, since large images can get high recognition accuracy while at the expense of slow operation speed.

In this paper, a lightweight deep network named as SLPNet is proposed for ALPR task. The network architecture is based on lightweight fully convolutional network and designed specifically to reduce the intermediate error as well as reduce the impact from the unbefitting resolution of input images at the same time. The whole ALPR framework is segmentation-free and can be trained end-to-end with the advantages of high accuracy and low computational cost. The main contributions of the paper are summarized as follows:

(1) An anchor-free method for license plate detection based on corners instead of regions is introduced. As the detected corners can provide more geometrical distortion information that will be further used for perspective correction, license plate recognition can benefit from this design.
(2) A Multiple Constraints Gaussian Distance loss (MG loss for short) is put forward to improve corner localization precision, which is demonstrated to make license plate detector's training more stable and efficient.
(3) To further improve the recognition rate, we integrate a multi-resolution strategy into the end-to-end network architecture. For different subtask networks, an image is decomposed into different resolutions that are utilized as respect inputs.

The rest of the paper is organized as follows. Section 2 provides a brief review on traditional and current methods for ALPR. The proposed approach is presented in Sect. 3 by detailing the detection and recognition subtasks, network architectures as well as the loss function design. The experimental results are reported and discussed in Sect. 4. Finally, in Sect. 5 a conclusion is drawn.

2   Related Work

There are two kinds of approaches to perform ALPR in terms of feature extraction, i.e., manual feature extraction and automatic feature extraction. Generally, manual features are extracted by image analysis while automatic feature extractions are depended upon machine learning.

2.1 Methods Based on Manual Features

Traditional ALPR systems are usually based on manual features and can be divided into three separate subtasks, namely license plate detection, character segmentation and character recognition.

Existing license plate detection methods can be roughly classified into four categories based on the features they used, namely edge based [26], color based [1], texture based [25] and character feature based [17] method. Edge and color are obvious features that are easily affected by the variation of illumination, while texture and character features can provide more fine distributed information thus are more robust.

Traditional segmentation methods based on manual features often uses pixel connectivity and some prior knowledge of characters [2] to perform segmentation. They are simple but lack of robustness. Meanwhile, some more sophisticated approaches like character-contour-based methods [4] are complex and slow.

For license plate recognition task, each segmented character is subject to classification by Optical Character Recognition (OCR) technique. Template matching [5] is a simple and straightforward way, but lack of reliability. Thus, many new methods extract more efficient features with some advanced filters like Gabor filter [10] to further improve recognition accuracy.

2.2 Methods Based on Machine Learning

Shallow Learning. Early ALPR systems use shallow machine learning to substitute manual feature selection and classification in detection, segmentation and recognition subtasks. Classic machine learning algorithms such as SVM, AdaBoost are combined with different composite features conveniently to achieve better performance. In [8], an initial set of possible character regions are obtained by AdaBoost classifiers and then passed to a support vector machine (SVM) where noncharacter regions are rejected. For license plate recognition, in addition to SVM, many classifiers can be employed to recognize characters with effective feature extraction such as ANN [9].

Deep Learning. With the remarkable development of deep learning in recent years, detection and recognition tasks can reach better precision and robustness with the help of deep neural networks, freeing people from manual feature selection. Rayson Laroca et al. [11] used a one-stage detector to efficiently localize license plate regions, while Z. Selmi et al. [19] use simple convolutional neural network (CNN) to complete the task of single character classification and the method achieves high recognition accuracy.

Moreover, some current state-of-the-art ALPR frameworks adopt segmentation free methods and the whole network can be trained end to end, which leads to an efficient learning process and achieves excellent performance. Typical work such as RPNet [22] adopts a simple CNN as backbone for license plate detection and employs fully connected layers to classify characters in each detected image.

Besides, in [12], Bidirectional Recurrent Neural Networks (BRNNs) with Connectionist Temporal Classification (CTC) [6] are adopted to label the sequential data without character separation, leading to a high recognition accuracy. To further accelerate the process speed, S. Zherzdew et al. [28] use lightweight CNNs to extract features and train the model with CTC loss [6].

3   SLPNet

Different from aforementioned methods described in Sect. 2, an ALPR framework called Skip-shuffle License Plate Network (SLPNet) based on lightweight fully convolutional networks (FCNs) is proposed here. As illustrated in Fig. 1, our approach divides the whole framework into two associated parts: detection part and recognition part. Our method completes the detection task by localizing four corners of each license plate (LP). Then, the detected and cropped LP region will be processed by perspective correction to effectively reduce the recognition difficulty. For LP recognition, we treat it as a sequence labeling problem similar to LPRNet. Although the networks in different part are designed separately, the network as a whole can be trained end to end and be optimized with a joint detection and recognition loss function. In the following subsections, we will give a detailed description about each components.

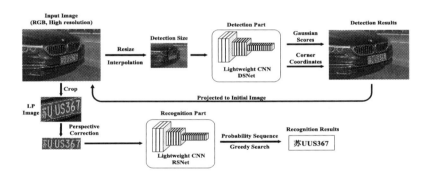

Fig. 1. The overall structure of our ALPR framework.

3.1   Detection Subnetwork

Corner Localization. To perform anchor-free object detection, an effective box is proposed in FSAF [29] where area inside is regarded as effective region (or positive region). Only cells from effective region are subject to object coordinates regression. Our approach also utilizes effective region to localize LP corners as shown in Fig. 2. We represent a LP region with an ordinary quadrangle according to the positions of its four corners, rather than the shape of a straight rectangle.

Similar to FSAF, two shrunk factors, δ_1 and δ_2 ($\delta_1 < \delta_2$) are selected to obtain effective boxes and ignore boxes as illustrated in Fig. 2(b). In our method,

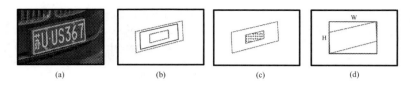

(a)   (b)   (c)   (d)

Fig. 2. An example of generating effective regions from ground truth. (a) A license plate with ground-truth bounding box; (b) Different types of boxes: bounding box (green), ignore box (black) and effective box (red); (c) Cells in positive area; (d) Rectangular bounding (blue) determined by a bounding box (green). (Color figure online)

three pairs of shrunk factors, namely $(0.8, 1.2)$, $(1.0, 1.5)$ and $(0.6, 0.9)$ are used respectively for different size of LP images. Large shrunk factor pairs are used for small size LPs while smaller ones for large size LPs. That means more attention is paid to LPs with small size in training process. We divide the LPs into three classes namely small, middle and large, according to its size using k-means clustering based on the size of the rectangular bounding box determined by ground-truth corners as shown in Fig. 2(d).

Nonlinear Transformation. Each ground-truth bounding box consists of 4 corners and is represented by a vector $g = (x_1, y_1, x_2, y_2, x_3, y_3, x_4, y_4)$. Our goal is to learn a nonlinear transformation that maps the network's output, $o = (t_{x_1}, t_{y_1}, t_{x_2}, t_{y_2}, t_{x_3}, t_{y_3}, t_{x_4}, t_{y_4})$ to the ground-truth g. Each (t_{x_i}, t_{y_i}) $i \in \{1, 2, 3, 4\}$ is the offset of the i-th corner to the cell center (x, y).

$$t_{x_i} = \sqrt[3]{\frac{x_i - 2^l(x + 0.5)}{z}}, t_{y_i} = \sqrt[3]{\frac{y_i - 2^l(y + 0.5)}{z}}. \tag{1}$$

where l is the pyramid level and z is an integer factor that shrinks the output ranges. Equation (1) first maps the coordinate (x, y) to the input image, then compute the offsets between the projected coordinates and g and regularizes the results with a cube root function.

Confidence of Detection Output. The normalized 2D Gaussian function is used to work out a score between the ground-truth corners and those predicted from the detection network. The score, also known as confidence, is denoted as Gaussian Scores that is described in formula (2).

$$G(x,y) = Ae^{-\left(\frac{(x-x_0)^2}{2\sigma_x^2} + \frac{(y-y_0)^2}{2\sigma_y^2}\right)},$$
$$\sigma_x = \alpha W, \sigma_y = \alpha H, A = 1. \tag{2}$$

where (x, y) is a pair of the predicted corner coordinates and (x_0, y_0) is the ground-truth. As shown in Fig. 2(d), W and H are the width and the height of a rectangular bounding box determined by the 4 corners. Moreover, a scale factor $\alpha \in (0, 1)$ is used to control the Gaussian variance.

Detection Network. A lightweight fully convolutional network called DSNet is proposed for LP detection, where ShuffleNetv2 [16] units are adopted as basic blocks and skip connection [7] between different basic blocks is added as illustrated in Fig. 3. To further improve the performance, several Global context (GC) blocks [3] are employed to enhance the ability of feature representation. Specially, a stem block [21] for spatial downsampling is utilized in DSNet and complex feature maps are generated by three pyramid feature maps in different stages. In our implementation, the input image size is set to 512×512, thus, the output map size is 32×32. Therefore, the pyramid level l is equivalent to 4 and z can be set to 128 in formula (1).

The Loss Function for Detection Network. The targets of the detection network are to work out the nonlinear transformation of LP corners by regression and get high Gaussian Scores in cells from all regions. To achieve the goal, we proposed a Multiple Constraints Gaussian Distance loss (MG loss) inspired by CIoU loss [27]. The MG loss for each cell in positive regions is defined as

$$L_{MG} = (1 - Conf) + \frac{\rho^2(b, b^{gt})}{c^2} + \alpha v + \beta d,$$

$$\alpha = \frac{v}{(1 - Conf) + v},$$

$$v = \frac{2}{\pi^2}[(\arctan \frac{w_1}{h_1} - \arctan \frac{w_1^{gt}}{h_1^{gt}}) + (\arctan \frac{w_2}{h_2} - \arctan \frac{w_2^{gt}}{h_2^{gt}})], \qquad (3)$$

$$\beta = \frac{d}{(1 - Conf) + d}, d = \sqrt[2]{\frac{1}{4} \sum_{i=1}^{4} (Gs_i - Gs_i^{gt})^2}.$$

where $Conf$ represents LP confidence that is averaged from four corners' real Gaussian Scores in a cell. $\rho(b, b^{gt})$ and c are distance and scale factor that are similar to CIoU loss. We enumerate four corners on each LP clockwise, therefore, in above formula (w_1, h_1) are worked out from Corner 1 and Corner 3. $(w_1^{gt}, h_1^{gt}, w_2^{gt}, h_2^{gt})$ are widths and heights generated by corners from groundtruth. Gs_i represents the predicted Gaussian Score of the i-th Corner, while Gs_i^{gt} represents a real Gaussian Score.

From formula (3), one can see MG loss consists of four terms: localization loss, distance loss, bounding shape loss and corners dispersion loss. The first item is the main loss and the others are constraints to make the learning process more stable and efficient. Considering the imbalance between positive and negative samples, we calculate Gaussian Score loss in each cell with Focal loss [14].

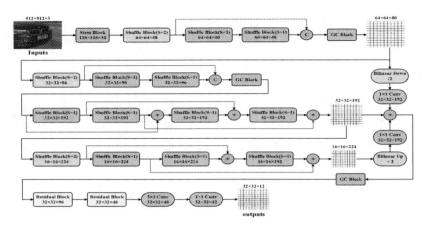

Fig. 3. The network structure of DSNet. '*C*' represents concatenation operation while '+' represents addition. '*S*' is the stride of convolutions. Bilinear interpolation is used to resize the pyramid features.

3.2 Recognition Subnetwork

Sequence Prediction. Since the number of characters on Chinese LP is unfixed depending on the type of vehicle, we treat the LP recognition as a sequence labeling problem by use of CTC employed in LPRNet [28], a kind of segmentation-free license plate recognition method.

Recognition Network. The recognition network, named as RSNet is also based on ShuffleNetv2 units, as illustrated in Fig. 4, where four kinds of blocks, $PDB(a) \sim PDB(d)$ for different spatial downsampling are designed. These blocks benefit from the parallel structure and are good at extracting features through downsampling. Like DSNet mentioned above, the intermediate feature maps are augmented with the global context embedding blocks (GCE blocks) [15]. To improve recognition performance, we also aggregate feature maps from different pyramid levels to get more complex maps.

The Loss Function for Recognition Network. Since the RSNet is based on CTC, the CTC loss is adopted in training recognition network.

3.3 Network Cascade

For end-to-end training, we need to link up the two separated networks and process the detection and recognition tasks sequentially. The predicted corners from DSNet are used to crop LP regions from raw images, which will be fed into RSNet then. Thus RSNet can be compatible with DSNet better, leading to less intermediate error propagation and achieving higher recognition accuracy. Since the DSNet uses small size images to perform detection, the detected and cropped LP regions are not suitable for recognition. To solve the problem, we re-map the

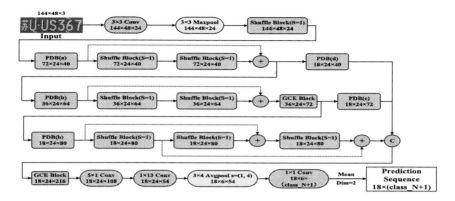

Fig. 4. The network structure of RSNet. '$PDB(a) \sim PDB(d)$' are four kinds of parallel downsampling blocks with different pooling layers or different strides. '$class\_N$' is the total number of character classes. '$Mean$' is an operator that averages the output maps in the 2nd dimension.

detected region to raw image and perform perspective correction according to the LP's coordinates of the DSNet's output. This multi-resolutions strategy can improve recognition accuracy effectively.

We utilize NMS [18] to get final detection results according to the predicted confidence. When decoding the recognition outputs, greedy search is applied on the output sequence of the RSNet to get the class with max probability.

4 Experiments

4.1 Chinese Licence Plate Dataset

In this section, the experimental results are reported. The performance of our method is compared with other state-of-the-art models in terms of a Chinese City Parking Dataset (CCPD)[22]. The CCPD dataset is the largest publicly available labeled license plate dataset in China by far. We randomly select 60,000 images from CCPD for experiment. As usual, all the images are subsequently split into 3 subsets in the proportion of 8:1:1 respectively for training, validation and testing.

4.2 Training Details

All training experiments are performed by pytorch on a NVIDIA Titan V GPU with 12 GB memory. We set Gaussian scale factor to 0.2 in training and use Adam optimizer with batch size of 22. The initial learning rate is set to 0.005. We drop the learning rate every epoch with exponential decay and the decay factor is 0.98. To make the training more stable, the whole process is consists of two stages. In the first phase, the detection network is trained until the average

Gaussian Scores of predicted corners in validation set is greater than 0.6, then the two subnetworks are trained jointly. In this stage, a loss function compound with both detection and recognition losses as described in formula (4) are used, where λ is a weighted factor, and $\lambda \in (0,1)$:

$$L_{lp} = L_{det} + \lambda L_{reg} \tag{4}$$

Here, λ is set to 0.5. The model is trained jointly for 70 epochs in total. All hyperparameters are adjusted through experiments and can be optimized further.

4.3   Experimental Results

During the training process, no data augmentation is performed and only the 48,000 raw images in the training set are utilized. We evaluate the performance of the proposed SLPNet with other publicly reported models on Chinese LP recognition. The rule for the calculation of the recognition accuracy is described as follows: only when a LP is detected successfully ($IoU \geq 0.5$ or $GaussianScore \geq 0.6$) and all the characters of the LP on an image are correctly recognized, the result is considered to be correct. The recognition accuracy for different methods is illustrated in Table 1.

Table 1. Experiment results of different methods on CCPD.

Model	End-to-end	Accuracy	Frame	Parameters
HyperLPR [24]	No	78.8%	–	11M
MTCNN+LPRNet [23, 28]	No	91.8%	12FPS	3.6M
RPNet [22]	Yes	93.4%	58FPS	210M
SLPNet(ours)	Yes	98.6%	25FPS	3.4M

Other three publicly available models that we compared with are Hyper-LPR [24], MTCNN+LPRNet [23] and RPNet [22]. HyperLPR [24] is an open source Chinese license plate detection and recognition framework with high speed. The framework use a mixture of deep neural networks and classic image processing algorithms to perform detection, segmentation and recognition. MTCNN+LPRNet [23] is another open source lightweight ALPR framework based on LPRNet. It uses MTCNN to detect license plate and uses LPRNet, a segmentation-free method to perform recognition. RPNet [22] is an excellent end-to-end LP recognition model that first issued the CCPD dataset. As one can see, compared with other models, the SLPNet achieves the highest recognition accuracy at 98.6% with the least parameters and the inference speed arrives at 25 FPS. Some detection and recognition results are shown in Fig. 5, from which one finds that even under clutter scenes and uneven illumination condition our approach still can get stable results.

Fig. 5. Some typical detection and recognition results on CCPD with our SLPNet model.

To demonstrate the feasibility of the proposed MG loss, two other loss functions, namely Gaussian loss which is the first term of MG loss, and traditional smooth L1 loss are also adopted for comparison in term of LP detection. The detection accuracy with different loss functions is illustrated in Table 2.

Table 2. Detection accuracy with different loss functions.

Loss function	MG loss	Gaussian loss	Smooth L1 loss [11]
Detection accuracy	98.2%	97.7%	96.3%

LP detection can be viewed as a special nature scene text detection problem and DB [13] is a popular real-time scene text detector recently. As the last experiment, DB is utilized to detect LPs, compared with our SLPNet. The detection results are shown in Table 3. It's obvious that our SLPNet is much more efficient in LP detection task. And for better LP detection and recognition performance, an appropriative LP detector is considered to be necessary.

Table 3. Detection results on CCPD with different detectors.

Model	Precision	Recall	F1-Score
DB [13]	44.55%	81.00%	57.49%
SLPNet(ours)	99.87%	99.25%	99.56%

4.4 Performance Analysis

It should be noted that a segmentation-free ALPR framework with end-to-end training not only leads to easier learning process but also achieves a better balance between different subnetworks. The integration of the recognition subnetwork with the detection subnetwork can make the framework more consistent and matched for each other in end-to-end training. The networks in our method are built based on lightweight convolutional blocks and enhancement modules with low computational cost so that we can improve the detection and recognition performance with less parameters. Compared with RPNet which is also trained on CCPD, our method needs only a subset of the dataset while achieves higher recognition accuracy, owing to the multi-resolution training strategy and corner based detector with MG loss.

5 Concluding Remarks

In this paper, we introduce SLPNet, a segmentation-free end-to-end framework for efficient license palate detection and recognition, which can achieve up to 98.6% recognition accuracy. The model is based on lightweight convolutional networks therefore it can run fast and the total parameters are only 3.4M. To raise the detection rate, the proposed detection subnetwork uses corners instead of regions to locate license plates and a new MG loss function is introduced. The perspective transformation is utilized to correct LP images so that character recognition rate is improved. To gain better performance, a multi-resolutions strategy is adopted without adding any computational cost nearly. Compared with existing ALPR methods, our approach exhibits a noteworthy performance and great potentiality for LP detection and recognition.

References

1. Ashtari, A.H., Nordin, M.J., Fathy, M.: An iranian license plate recognition system based on color features. IEEE Trans. Intell. Transp. Syst. **15**(4), 1690–1705 (2014)
2. Busch, C., Domer, R., Freytag, C., Ziegler, H.: Feature based recognition of traffic video streams for online route tracing. IEEE Veh. Technol. Conf. **3**, 1790–1794 (1998)
3. Cao, Y., Xu, J., Lin, S., Fangyun, W., Hu, H.: GCNET: non-local networks meet squeeze-excitation networks and beyond (2019). https://arxiv.org/pdf/1904.11492.pdf
4. Capar, A., Gokmen, M.: Concurrent segmentation and recognition with shape-driven fast marching methods. In: 18th International Conference on Pattern Recognition, vol. 1, pp. 155–158 (2006)
5. Comelli, P., Ferragina, P., Granieri, M.N., Stabile, F.: Optical recognition of motor vehicle license plates. IEEE Trans. Veh. Tech. **44**(4), 790–799 (2002)
6. Graves, A., Fernández, S., Gomez, F.: Connectionist temporal classification: labelling unsegmented sequence data with recurrent neural networks. In: International Conference on Machine Learning, pp. 369–376 (2006)
7. He, K., Zhang, X., Ren, S., Sun, J.: Deep residual learning for image recognition. In: IEEE Conference on Computer Vision & Pattern Recognition, pp. 770–778 (2016)
8. Ho, W.T., Lim, H.W., Tay, Y.H.: Two-stage license plate detection using gentle adaboost and sift-svm. In: Asian Conference on Intelligent Information & Database Systems, pp. 109–114 (2009)
9. Jiao, J., Ye, Q., Huang, Q.: A configurable method for multi-style license plate recognition. Pattern Recogn. **42**(3), 358–369 (2009)

10. Kimand, M.K., Kwon, Y.B.: Recognition of gray character using gabor filters. Proc. Int. Conf. Inform. Fusion. **1**, 419–424 (2002)
11. Laroca, R., Severo, E., Zanlorensi, L.A.: A robust real-time automatic license plate recognition based on the yolo detector. In: 2018 International Joint Conference on Neural Networks, pp. 6005–6014 (2018)
12. Li, H., Wang, P., Shen, C.: Toward end-to-end car license plate detection and recognition with deep neural networks. IEEE Trans. Intell. Transp. Syst. **20**(3), 1126–1136 (2019)
13. Liao, M., Wan, Z., Yao, C., Chen, K., Bai, X.: Real-time scene text detection with differentiable binarization. In: AAAI Conference on Artificial Intelligence, pp. 11474–11481 (2020)
14. Lin, T.Y., Goyal, P., Girshick, R., He, K.: Focal loss for dense object detection (2017). https://arxiv.org/pdf/1708.02002.pdf
15. Liu, W., Rabinovich, A., Berg, A.C.: Parsenet: looking wider to see better (2015). https://arxiv.org/pdf/1506.04579.pdf
16. Ma, N., Zhang, X., Zheng, H.T., Sun, J.: Shufflenetv2: practical guidelines for efficient CNN architecture design. In: 2018 European Conference on Computer Vision, pp. 116–131 (2018)
17. Matas, J., Zimmermann, K.: Unconstrained licence plate and text localization and recognition. In: IEEE Intelligent Transportation Systems, pp. 572–577 (2005)
18. Neubeck, A., Gool, L.J.V.: Efficient non-maximum suppression. In: International Conference on Pattern Recognition, pp. 20–24 (2006)
19. Selmi, Z., Halima, M.B., Alimi, A.M.: Deep learning system for automatic license plate detection and recognition. In: 14th IAPR International Conference on Pattern Recognition, pp. 1132–1138 (2017)
20. Tian, Y., Lu, X., Li, W.: License plate detection and localization in complex scenes based on deep learning. In: The 30th Chinese Control And Decision Conference, pp. 6569–6574 (2018)
21. Wang, R.J., Li, X., Ling, C.X.: PELEE: a real-time object detection system on mobile devices. In: 32nd Conference on Neural Information Processing Systems, pp. 7466–7476 (2018)
22. Xu, Z., Yang, W., Meng, A., Lu, N., Huang, L.: Towards end-to-end license plate detection and recognition: a large dataset and baseline. In: 2018 European Conference on Computer Vision, pp. 261–277 (2018)
23. Xue, X.: A two stage lightweight and high performance license plate recognition in MTCNN and LPRNet (2019). https://github.com/xuexingyu24/License_Plate_Detection_Pytorch
24. Zeussees: High performance chinese license plate recognition framework (2020). https://github.com/zeusees/HyperLPR
25. Zhang, H., Jia, W., He, X., Qiang, W.: Learning-based license plate detection using global and local features. In: International Conference on Pattern Recognition. vol. 2, pp. 1102–1105 (2006)
26. Zheng, D., Zhao, Y., Wang, J.: An efficient method of license plate location. Pattern Recogn. Lett. **26**(15), 2431–2438 (2005)
27. Zheng, Z., Wang, P., Liu, W., Li, J., Ren, D.: Distance-IOU loss: faster and better learning for bounding box regression. In: AAAI Conference on Artificial Intelligence, vol. 34, pp. 12993–13000 (2020)

28. Zherzdev, S., Gruzdev, A.: LPRNet license plate recognition via deep neural networks (2018). https://arxiv.org/pdf/1806.10447.pdf
29. Zhu, C., He, Y., Savvides, M.: Feature selective anchor-free module for single-shot object detection. In: IEEE Conference on Computer Vision and Pattern Recognition (CVPR), pp. 840–849 (2019)

Hierarchical Representations with Discriminative Meta-filters in Dual Path Network for Tracking

Fei Xie[1], Ning Wang[2], Yuncong Yao[1], Wankou Yang[1(✉)], Kaihua Zhang[2], and Bo Liu[3]

[1] School of Automation, Southeast University, Nanjing 210096, China
{220191672,wkyang}@seu.edu.cn
[2] School of Automation, Nanjing University of Information Science and Technology, Nanjing 210044, China
[3] JD.com, Beijing, China

Abstract. In visual tracking task, accuracy and robustness are critical issues for achieveing remarkable performance. In this paper, we propose a novel dual path network with discriminative meta-filters and hierachical representations to solve these issues. We first design geometrically sensitivity pathway (GESP) and geographical sensitivity pathway (GASP) as two subtasks for target classification and scale estimation. GASP mainly includes powerful discriminative meta-filters to find coarse location of target and GESP can refine region of interests online while adapt the appearance model to the target swiftly. Then, a dual path network is developed in a online and offline framework. Specifically, meta-filters are trained offline in order to gain meta-knowledge of similar tracking scenes. Finally, we present three suggestions on deigning modern tracker. Extensive experiments on VOT2018 datasets verify the superior performance of proposed method compared with other state-of-the-arts, achieving expected average overlap (EAO) of 0.467.

Keywords: Object tracking · Meta-filter · Dual path network · Online learning · Hierarchical deep features

1 Introduction

Generic visual tracking is a crucial task in computer vision aiming at locating the specific continuous object in video. Limited information, usually the first

---

This work is supported in part by National Major Project of China for New Generation of AI (No. 2018AAA0100400), in part by the Natural Science Foundation of China under Grant nos. 61773117, 61876088, the Primary Research & Development Plan of Jiangsu Province - Industry Prospects and Common Key Technologies under Grant No. BE2017157.
F. Xie—He is currently working toward the Master degree in the School of Automation, Southeast University.

© Springer Nature Switzerland AG 2020
Y. Peng et al. (Eds.): PRCV 2020, LNCS 12306, pp. 303–315, 2020.
https://doi.org/10.1007/978-3-030-60639-8_26

annotation is provided during visual object tracking. One unique characteristic of generic object tracking is that no prior knowledge (e.g., the object class) about the object, as well as its surrounding environment, is allowed [1]. The quality of localization and scale estimation of target are the most influential factor of the performance.

Recently, localization and scale estimation tend to be two subtasks of the tracking problem [2]. Before the deep learning methods, trackers based on Discriminative Correlation Filter (DCF) [3,4] framework took dominant positions in tracking method. Traditional correlation trackers suffer from inefficiency and low accuracy due to its inherent flaw. It is natural that the deep learning ways are applied to other computer vision tasks, such as object detection, semantic segmentation and visual tracking. SiamFC [5], introduces siamese learning paradigm into visual object tracking, though it employs brutal multi-scale test which is inaccurate and inefficiency [2]. Then, SiamRPN tracker family [6–8] perform an accurate and efficient target scale estimation by introducing the Region Proposal Network (RPN) [9]. However, the pre-defined anchor settings not only introduce ambiguous similarity, but also demand huge prior-knowledge about target. SiamFC++ [10] adopts the anchor-free regression and classification style based on Siamese learning paradigm, it still heavily rely on the sufficient prior-knowledge about target. Motivated by the aforementioned analysis, we propose three suggestions on designing modern visual object trackers:

- **Balance between online learning and offline training:** The breakthroughs on object detection provide a better way to replace multi-scale estimation in object tracking. For example, RPN [9] structure achieves astonishing accuracy in SiamRPN [6]. Because siamese formulation does not provide a powerful discriminative model, we highly recommend that online learning needs to be well-designed. The Correlation based trackers [3,4,11,12] are able to tackle with online model updation. However, the problems of model drift and insufficient training of online model result in low accuracy.
- **Fully utilization of Multi-level deep convolutional features:** Deep model should be trained for robustness, while the shallow model should emphasize accurate target localization [13]. We highly recommend that deep and shallow models should be emphasized equally in order to have better robustness performance. Even though the high quality training data is crucial for the success of end-to-end representation [7], we argue that models designed for both deep and shallow features can reduce the burden of offline training.
- **Online searching strategies are highly recommended in scale estimation branch:** Both the RPN structure from Faster-RCNN [9] or one-stage anchor-free detection from FCOS [14] output the coordinates of target directly without online searching strategy. We strongly consist that it cannot tackle with severe appearance deformation and complex scenes. In our work, we choose the IoU-Net [15] prediction proposed by Atom [2] as our scale estimation branch. It can perform online searching strategy when the coarse location of target is determined.

2   Related Work

Generic object tracking can be divided into two frameworks: Tracking framework and detection framework. Generally, tracking framework trackers are mainly based on correlation filters. MOSSE [4] proposes a CF tracker by learning a minimum output sum of squared error for target appearance and calculate in Fourier domain. KCF [3] adopts ridge regression and circulant matrix to facilitate the speed of calculation in Fourier domain. C-COT [16] converts feature maps of different resolutions into a continuous spatial domain to achieve better accuracy. The subsequent ECO [17] has better efficiency by removing the redundant correlation filters.

ATOM [2] tracker adopts IoU-Net [15] and online learning to classify the target and estimate the scale. Online learning and offline training are combined together. ATOM achieves better robustness performance than Siamese-based trackers. However, it still lack of multi-level deep convolutional features fusion and its online learning is totally independent of offline training which can be further improved. DiMP [18] combines online training and offline training together.

SiamRPN and its succeeding works [3,4,11,12] modifies a Region Proposal Network after a siamese network. They have direct bounding box regression ability thanks to extensive offline training. However, the robustness still suffers from the weak discriminative ability of siamses-based detection networks. The pre-defined anchors of Region Proposal Network (RPN) [9] also need to be well-designed. Even though the SiamFC++ [10] adopts an anchor-free style for bounding box regression, its performance still heavily rely on extensive offline training and robustness cannot be improved as much as accuracy.

3   Proposed Method

Two meta-filters in Geographical Sensitivity Pathway (GASP) are trained to have more discriminative power between foreground and background. The geometrically sensitivity pathway (GESP) focus more on the appearance model of the object in order to estimate the scale accurately.

3.1   Dual Path Network

The whole pipeline of our tracker consists of two meta-filters and a Box Fast Adaption Module. Hierarchical feature representations are used for two meta-filters in order to achieve better performance on localizations. Similar to the object segmentation in [19], the Box Fast Adaption Module can have accurate object outline estimation after the localization process (Fig. 1).

3.2   Multi-hierarchical Independent Discriminati Filters in Online Learning

Inspired by discriminative correlation filter (DCF) approaches, we formulate our learning objective based on L^2 classification error. Each sample x_k contains D

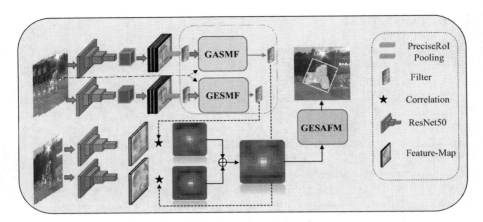

Fig. 1. Pipeline of Dual Pathway Network. GASMF stands for meta-filtes in Geographical Sensitivity Pathway. GESMF stands for meta-filtes in Geometrically Sensitivity Pathway. GESAFM is Appearance Fast Adaption Module in Geometrically Sensitivity Pathway

feature channels $x_j^1 \ x_j^2, \ldots, x_j^D$, extracted from the same image patch, where k is the index of the samples. Assume that $f = \{f_d\}_{d=1:D}$ is a set of D channel features. The correlation filters algorithm can be formulated as:

$$\arg\min_{f} \sum_{k=1}^{K} \|\phi(x_k, f) - y_k\|_{L^2}^2 + \lambda \sum_{d=1}^{D} \|f\|_{L^2}^2 \tag{1}$$

where x_k is the cyclic shift sample of the x_k and y_k is the Gaussian response label. The optimization problem in Eq. (1) can be solved efficiently in the Fourier domain.

$$\phi(x_k, f) = \sum_{d=1}^{D} f_d * x_k^d \tag{2}$$

In our work, we try to combine the online optimization with offline training, thus we approximate the loss with a quadratic function and optimize it by backward propagation instead of Fast Fourier Transform (FFT).

In this section, the discriminative learning loss is described in details. The input to our model predictor D consists of a training set $S_{\text{train}} = \{(x_j)\}_{j=1}^{n}$ of deep feature maps $x_j \in \mathcal{X}$ generated by the backbone network F. During online tracking, correlation filter is optimized to generate a target model $f = D(S_{\text{train}})$. The model f is defined as the filter weights of a convolutional layer. The maximum value of the model output should localize the center of target.

$$L(f) = \frac{1}{|S_{\text{train}}|} \sum_{(x) \in S_{\text{wain}}} \|r(x * f, c)\|^2 + \|\lambda f\|^2 \tag{3}$$

Here, $*$ denotes convolution and λ is a regularization factor. The function $r(s,c)$ computes the residual at every spatial location based on the target confidence scores $s = x * f$ and the ground-truth target center coordinate c. In Eq. (1), $r(s,c) = s - y_c$, traditional correlation filter trackers optimize the residuals between response and the Gaussian target scores. Thus, the difference of target and distractor response usually represents the discriminative ability of the correlation filters. However, during online tracking, background noise and distractors are far more abundant than our target resulting in imbalance of the positive and negative samples.

In order to learn a more discriminative filter, it is common to have a weight matrix in the learning loss. In our work, We employ a hinge-like loss in r, clipping the scores at zero as $\max(0,s)$ in the background region. Thus, the filter is more focus on the hard negative distractors instead of easy negative samples. We believe that it could contribute to a more discriminative filter and efficiency online optimization.

$$r(s,c) = v_c \cdot (m_c s + (1 - m_c) \max(0, s) - y_c) \tag{4}$$

The mask m_c modifies the spatial weight of scores, having values in the interval $m_c(t) \in [0,1]$ at each spatial location $t \in \mathbb{R}^2$.

In our work, we use convolutional layers D to generate the filter $f = D(S_{\text{train}})$ by implicitly minimizing the error (3).

$$f^{(i+1)} = f^{(i)} - \alpha \nabla L\left(f^{(i)}\right) \tag{5}$$

Instead of minimizing the error (3) in Fourier domain, we approximate the error with a quadratic function and directly employ gradient descent optimization using a step length α.

$$
\begin{aligned}
L(f) \approx \tilde{L}(f) = \frac{1}{2}\left(f - f^{(i)}\right)^{\mathrm{T}} Q^{(i)} \left(f - f^{(i)}\right) \\
+ \left(f - f^{(i)}\right)^{\mathrm{T}} \nabla L\left(f^{(i)}\right) + L\left(f^{(i)}\right)
\end{aligned}
\tag{6}
$$

Here, the filter variables f and $f^{(i)}$ are seen as vectors and $Q^{(i)}$ is positive definite square matrix. The steepest descent is adopted in order to achieve a fast convergence performance. By solving $\frac{\mathrm{d}}{\mathrm{d}\alpha}\tilde{L}\left(f^{(i)} - \alpha \nabla L\left(f^{(i)}\right)\right) = 0$, we could find the step length α.

$$\alpha = \frac{\nabla L\left(f^{(i)}\right)^{\mathrm{T}} \nabla L\left(f^{(i)}\right)}{\nabla L\left(f^{(i)}\right)^{\mathrm{T}} Q^{(i)} \nabla L\left(f^{(i)}\right)} \tag{7}$$

In this work, We set $Q^{(i)} = \left(J^{(i)}\right)^{\mathrm{T}} J^{(i)}$, where $J^{(i)}$ is the Jacobian of the residuals at $f^{(i)}$. This design of positive definite square matrix $Q^{(i)}$ involves with second-order gradient descent of residuals at $f^{(i)}$ which can contribute to a fast and efficient convergence.

Compared to the traditional correlation filter (CF) algorithms, We treat the hierarchical features differently. Because the shallow and deep features are both critical to the localization and classification, we train a set of independent filters for each feature. The decomposition of the function of two filters are beneficial to the overall performance. Conventional CF algorithms with one single filter is usually difficult to tackle with both classification and localization tasks during online tracking leading to model drift and insufficient online learning.

3.3 Filter Generations in Meta-learning Style

The motivation of our learning algorithm is that discriminative filters for similar visual objects in arbitrary background have amounts of sharing weights. Filters for objects with the same high-level semantic information should be robust towards changes, motion blur, scale variations, etc. To extract useful sharing filter weights in similar tracking scenes, we separate scene-independent information through offline training (Fig. 2).

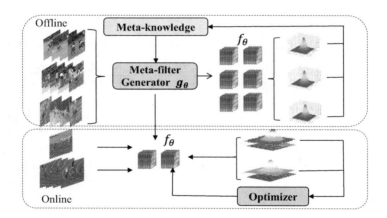

Fig. 2. Multi-hierarchical independent discriminative filters combined with online learning and offline training framework.

Algorithm 1. Meta-filters $f^{(i)}$ in offline training
___
Training samples $S_{\text{train}} = \{(x_j, c_j)\}_{j=1}^n$,test samples $S_{\text{test}} = \{(x_j, c_j)\}_{j=1}^m$

repeat

    $f^{(i)} \leftarrow \text{FilterGen}(S_{\text{train}})$ # Generate filters

    $\nabla L\left(f^{(i)}\right) \leftarrow \text{FilterGrad}\left(f^{(i)}, S_{\text{test}}\right)$ #Apply filters

    $\text{FilterGen} \leftarrow \text{BackProp}(\nabla L\left(f^{(i)}\right))$ # Update FilterGen

until N_{iter}
___

With these sharing weights stored in convolutional networks to generate meta-filters, our online discriminative model for classification can be adapted to the specific objects fastly. We introduce a network module called filter generation network g_θ. It consists of two convolutional layer and a precise ROI pooling. During offline training, the $S_{\text{train}} = \{(x_j, c_j)\}_{j=1}^{n}$, composed of several tracklets, are used to generate meta-filters through averaging the pooled feature maps. And then, the test samples $S_{\text{test}} = \{(x_j, c_j)\}_{j=1}^{m}$ are applied with generated filters to optimize the filter generation network.

Details of our meta-filters in Geographical Sensitivity Pathway (GASP) and Geometrically Sensitivity Pathway (GESP) are show in Fig. 3. ResNet-50 Block3 features in different stage are passed to a convolutional block (Cls). Regions defined by the input bounding boxes are then pooled to a fixed size using Precise Pooling layers. After a convolutional block, the weights of filter are generated to perform as convolutional block for features of searching image. Online optimizers optimize weights of filters during online tracking while offline optimizers try to learn meta-knowledges of filter-generation.

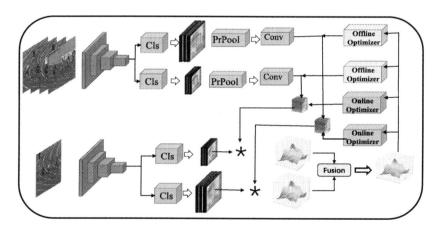

Fig. 3. Full architecture of our discriminative meta-filters. Pseudo-siamese network is not shown here for simplicity.

3.4 Appearance Fast Adaption Module

After the coarse spatial location of target is figured out, we need a subnetwork to acquire the accurate localization of target. In this work, we adopt an independent IoU-Nets [15] with template feature modulation. We train our independet IoU-Net [15] with template feature modulation for measuring the differences between proposals and ground truth. Full architecture can be viewed in Fig. 4.

The template features x_0 and searching area features x are extracted by modulation branch and test branch. The bounding box annotion A_0 is as extral

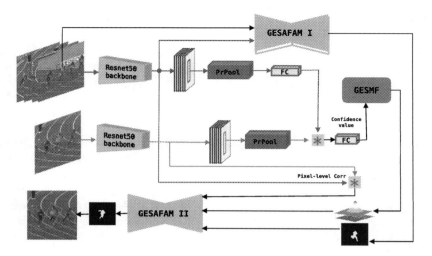

Fig. 4. Full architecture of our Appearance Fast Adaption Module (AFAM I) and Appearance Fast Adaption Module II (AFAM II).

Algorithm 2. Online searching strategy in Box Fast Adaption Module

---

Random Proposals Generation $P_i = \text{GaussionRandom}(A_{i-1})$, $\sum_1^k i = 8$
$N_{\text{iter}} = 0$, step length $= \alpha$
repeat
    $f^{(i)} \leftarrow \text{testBranFexactor}(P_i)\#$ Feature extraction for proposal
    $M_{vec} \leftarrow \text{ModuBran}(x_0, A_t)\#$ Modulation convolutional kernels
    $BoxConVal^{(i)} = g\left(c\left(x_0, A_0\right) \cdot z(x, A)\right)\#$Confidence value of proposal
    $p \leftarrow \text{BackProp}(\nabla(BoxConVal^{(i)})$ # Optimization through Backward Propagation
    $P_i \leftarrow P_i + \alpha p\#$ Proposal updation
    $N_{\text{iter}} = N_{\text{iter}} + 1$
until $N_{\text{iter}} = 10$

---

modulation information for generating box confidence value. The modulation information $c(x_0, A_0)$ is added to the test branch as convolution kernel. The feature representation of search area $z(x, A)$ has strong spatial correspondence with the searching frame. Thus it could reflect the spatial coordinate difference between template and test frame.

$$\text{BoxConVal}(A) = g\left(c\left(x_0, A_0\right) \cdot z(x, A)\right) \tag{8}$$

During online tracking, we apply another online searching strategy to maximun the confidence value with bounding box optimization. We use Gaussian distribution and previous position of target to generate initial proposals. For each proposel, we obtain the confidence value through the Box Fast Adaption Module. By backward propagation to obtain the gradient of confidence value, we optimize the length and center position of current proposal directly. Details are shown in Algorithm 2.

Appearance Fast Adaption Module II (AFAM II) provides pixel-level target information. We use the features extracted from ResNet-50. For the first frame and ground truth, we obtain the pseudo-mask for the target from the AFAM II. Then the extra information from Appearance Fast Adaption Module I (AFAM I) and pseudo-mask are concatenated together. The refinement network will output the final appearance estimation. Although the AFAM II are pretrained on training segmentation sequences from Youtube-VOS, yet it is not design for the segmentation task. During training, we use bounding box labels as inputs to predict target mask. So it should be considered as target appearance estimation, not instance segmentation.

4 Experiments

Meta-filters in Geographical Sensitivity Pathway (GASMF) and Box Fast Adaption Module in Geometrically Sensitivity Pathway (GESBFAM) are firstly trained jointly with ImageNet pretrained weights. Because ImageNet pretrained models are for classification task which may not suitable for tracking, we firstly train the GASMP and GESBFAM 40 epochs in the training splits of TrackingNet [20], LaSOT [21], GOT10k [1] and COCO [22] datasets to adapt backbone to tracking task. Then, we add the meta-filters in Geometrically Sensitivity Pathway (GESMF) to train another 30 epochs for a more discrminative power model. We train our model by sampling 26,000 frame-pairs per epoch. We use ADAM [23] with learning rate decay of 0.2 every 10th epoch. We use features extracted from the third block from Resnet. We set the kernel size of the meta-filters to 64 * 4 * 4. Appearance Fast Adaption Module (AFAM) in Geometrically Sensitivity Pathway are pre-trained on 3471 training segmentation sequences from Youtube-VOS [24].

4.1 Ablation Studies

We compared the performance of different combinations in Resnet50. ResNet-50 Block3 features in different stage. If we select adjacent layers, more redundancy and interference will be introduced into our tracking framework, thus causing the performance degradation. From the Table 2 the best performance achieved is from the layer3a and layer3e. When using two meta-filters, the EAO comes to 0.455, which demonstrates the effectiveness of two filters. The Box Fast Adaption Module improves accuracy a lot which is 0.652 comparing to 0.597 (Table 1).

4.2 Results on Several Benchmarks

VOT2018 [25] datasets consist of 60 test sequences. With no training dataset provided, VOT is the most challenging benchmark for tracking which has topics including fast motion, occlusion, etc. We tested our tracker on this benchmark

Table 1. VOT2018-comparison with different settings.

Tracker	Settings0	Settings1	Settings2	Settings3	Settings4	Settings5
Filters	Single	Single	Two	Two	Two	Two
Features	Block2e	Block3e	Block2e+3e	Block3e+3f	Block3a+3e	Block3a+3e
Box adaption	–	–	–	–	–	Yes
A ↑	0.595	0.596	0.596	0.597	0.596	0.652
R ↓	0.172	0.168	0.157	0.165	0.155	0.155
EAO ↑	0.421	0.435	0.455	0.448	0.457	0.467

and present the results in Table 2 and Table 3. To the best of our knowledge, we achieves an EAO of 0.467 on VOT2018 (Kristanetal, 2018) and EAO of 0.334 on VOT2019 benchmark which is the new state-of-the-art performance. Our tracker also can run at 30 FPS in Nvidia GeForce 1080ti which is still very competitive (Tables 4 and 5).

Table 2. VOT2018-comparison with state-of-the-art trackers. The top three results are in red, **blue** and green fonts. Best viewed in color display.

Tracker	Ours	SiamFC++ [10]	ATOM [2]	SiamRPN++ [8]	DaSiamRPN [7]	ECO [17]
Where	–	AAAI20	CVPR19	CVPR19	ECCV18	CVPR17
A ↑	0.652	0.587	0.590	**0.600**	0.586	0.484
R ↓	0.155	**0.183**	0.204	0.234	0.276	0.276
EAO ↑	0.467	**0.426**	0.401	0.411	0.383	0.280

Table 3. VOT2019 realtime-comparison with state-of-the-art trackers. The top three results are in red, **blue** and green fonts. Best viewed in color display.

Tracker	Ours	SiamBAN [26]	DiMP [18]	SiamRPN++ [8]	SiamMask [27]	SiamCRF [25]
Where	–	CVPR20	CVPR19	CVPR19	ECCV18	CVPR17
A ↑	0.636	**0.602**	0.582	0.580	0.594	0.549
R ↓	0.276	0.396	**0.371**	0.482	0.461	0.346
EAO ↑	0.334	**0.327**	0.321	0.283	0.287	0.262

Table 4. GOT-10K-comparison with state-of-the-art trackers. The top three results are in red, **blue** and green fonts. Best viewed in color display.

Tracker	Ours	SiamFC++	ATOM [2]	SiamRPN++ [8]	SiamFCv2 [5]	ECO [17]
Where	–	AAAI20	CVPR19	CVPR19	ICCV19	CVPR17
AO ↑	60.0	**59.5**	55.6	51.8	37.4	31.6
SR(0.5) ↑	71.6	**69.5**	63.4	61.8	40.4	30.9
SR(0.75) ↑	**46.0**	47.9	40.2	32.5	14.4	11.1

Table 5. OTB-15-The top three results are in red, **blue** and green fonts. Best viewed in color display.

Tracker	Ours	ATOM [2]	SiamRPN++ [8]	DaSiamRPN [7]	ECO [17]
Where	–	CVPR19	CVPR19	ECCV18	CVPR17
Success ↑	67.7	66.9	69.6	65.8	**69.1**

5 Conclusions

In this paper, we propose three suggestions on designing modern visual object trackers. We combine offline training and online learning of discriminative filters together. The meta-learning ways are stressed and successfully applied in object tracking. The meta-knowledge of the filter generations on similar tracking scenes are learned through convolutional network. Gradient descent optimization is carefully designed to adapt our filters to unseen objects efficiently. Moreover, a pseudo-siamese network structure enpowers the discriminative ability of our meta-filters. Our tracker can perform online searching strategies to find the best object bounding box. The balance of online searching and offine training helps us to achieve better results with less training resource.

References

1. Huang, L., Zhao, X., Huang, K.: Got-10k: a large high-diversity benchmark for generic object tracking in the wild. In: IEEE Transactions on Pattern Analysis and Machine Intelligence (2019)
2. Danelljan, M., Bhat, G., Khan, F.S., Felsberg, M.: Atom: accurate tracking by overlap maximization. In: Proceedings of the IEEE Conference on Computer Vision and Pattern Recognition, pp. 4660–4669 (2019)
3. Henriques, J.F., Caseiro, R., Martins, P., Batista, J.: High-speed tracking with kernelized correlation filters. IEEE Trans. Pattern Anal. Mach. Intell. **37**(3), 583–596 (2014)
4. Bolme, D.S., Beveridge, J.R., Draper, B.A., Lui, Y.M.: Visual object tracking using adaptive correlation filters. In: 2010 IEEE Computer Society Conference on Computer Vision and Pattern Recognition, pp. 2544–2550. IEEE (2010)
5. Bertinetto, L., Valmadre, J., Henriques, J.F., Vedaldi, A., Torr, P.H.S.: Fully-convolutional siamese networks for object tracking. In: Hua, G., Jégou, H. (eds.) ECCV 2016. LNCS, vol. 9914, pp. 850–865. Springer, Cham (2016). https://doi.org/10.1007/978-3-319-48881-3_56
6. Li, B., Yan, J., Wu, W., Zhu, Z., Hu, X.: High performance visual tracking with siamese region proposal network. In: Proceedings of the IEEE Conference on Computer Vision and Pattern Recognition, pp. 8971–8980 (2018)
7. Zhu, Z., Wang, Q., Li, B., Wu, W., Yan, J., Hu, W.: Distractor-aware siamese networks for visual object tracking. In: Proceedings of the European Conference on Computer Vision (ECCV), pp. 101–117 (2018)
8. Li, B., Wu, W., Wang, Q., Zhang, F., Xing, J., Yan, J.: Siamrpn++: evolution of siamese visual tracking with very deep networks. In: Proceedings of the IEEE Conference on Computer Vision and Pattern Recognition, pp. 4282–4291 (2019)

9. Ren, S., He, K., Girshick, R., Sun, J.: Faster R-CNN: towards real-time object detection with region proposal networks. In: Advances in Neural Information Processing Systems, pp. 91–99 (2015)

10. Xu, Y., Wang, Z., Li, Z., Ye, Y., Yu, G.: Siamfc++: towards robust and accurate visual tracking with target estimation guidelines. arXiv preprint arXiv:1911.06188 (2019)

11. Sun, C., Wang, D., Lu, H., Yang, M.-H.: Correlation tracking via joint discrimination and reliability learning. In: Proceedings of the IEEE Conference on Computer Vision and Pattern Recognition, pp. 489–497 (2018)

12. Valmadre, J., Bertinetto, L., Henriques, J., Vedaldi, A., Torr, P.H.S.: End-to-end representation learning for correlation filter based tracking. In: Proceedings of the IEEE Conference on Computer Vision and Pattern Recognition, pp. 2805–2813 (2017)

13. Bhat, G., Johnander, J., Danelljan, M., Khan, F.S., Felsberg, M.: Unveiling the power of deep tracking. In: Proceedings of the European Conference on Computer Vision (ECCV), pp. 483–498 (2018)

14. Tian, Z., Shen, C., Chen, H., He, T.: FCOS: fully convolutional one-stage object detection. In: Proceedings of the IEEE International Conference on Computer Vision, pp. 9627–9636 (2019)

15. Jiang, B., Luo, R., Mao, J., Xiao, T., Jiang, Y.: Acquisition of localization confidence for accurate object detection. In: Proceedings of the European Conference on Computer Vision (ECCV), pp. 784–799 (2018)

16. Danelljan, M., Robinson, A., Shahbaz Khan, F., Felsberg, M.: Beyond correlation filters: Learning continuous convolution operators for visual tracking. In: Leibe, B., Matas, J., Sebe, N., Welling, M. (eds.) European Conference on Computer Vision, vol. 9909, pp. 472–488. Springer, Cham (2016). https://doi.org/10.1007/978-3-319-46454-1_29

17. Danelljan, M., Robinson, A., Shahbaz Khan, F., Felsberg M.: Eco: efficient convolution operators for tracking. In: Proceedings of the IEEE Conference on Computer Vision and Pattern Recognition, pp. 6638–6646 (2017)

18. Bhat, G., Danelljan, M., Van Gool, L., Timofte, R.: Learning discriminative model prediction for tracking. In: Proceedings of the IEEE International Conference on Computer Vision, pp. 6182–6191 (2019)

19. Lukezic, A., Matas, J., Kristan, M.: D3s-a discriminative single shot segmentation tracker. In: Proceedings of the IEEE/CVF Conference on Computer Vision and Pattern Recognition, pp. 7133–7142 (2020)

20. Muller, M., Bibi, A., Giancola, S., Alsubaihi, S., Ghanem, B.: Trackingnet: a large-scale dataset and benchmark for object tracking in the wild. In: Proceedings of the European Conference on Computer Vision (ECCV), pp. 300–317 (2018)

21. Fan, H., et al.: Lasot: a high-quality benchmark for large-scale single object tracking. In: Proceedings of the IEEE Conference on Computer Vision and Pattern Recognition, pp. 5374–5383 (2019)

22. Lin, T.-Y., et al.: Microsoft coco: common objects in context. In: Fleet, D., Pajdla, T., Schiele, B., Tuytelaars, T. (eds.) European Conference on Computer Vision, vol. 8693, pp. 740–755. Springer, Cham (2014). https://doi.org/10.1007/978-3-319-10602-1_48

23. Kingma, D.P., Ba, J.: Adam: a method for stochastic optimization. arXiv preprint arXiv:1412.6980 (2014)

24. Xu, N., et al.: Youtube-vos: sequence-to-sequence video object segmentation. In: Proceedings of the European Conference on Computer Vision (ECCV), pp. 585–601 (2018)

25. Matej Kristan, et al.: The seventh visual object tracking vot2019 challenge results. In: Proceedings of the IEEE International Conference on Computer Vision Workshops (2019)
26. Chen, Z., Zhong, B., Li, G., Zhang, S., Ji, R.: Siamese box adaptive network for visual tracking. arXiv preprint arXiv:2003.06761 (2020)
27. Wang, Q., Zhang, L., Bertinetto, L., Hu, W., Torr, P.H.S.: Fast online object tracking and segmentation: a unifying approach. In: Proceedings of the IEEE Conference on Computer Vision and Pattern Recognition, pp. 1328–1338 (2019)

Weakly Supervised Pedestrian Attribute Recognition with Attention in Latent Space

Mingjun Sun[1,2], Hua Yang[1,2(✉)], and Guangtao Zhai[1,2]

[1] Institution of Image Communication and Network Engineering, Shanghai Jiao Tong University, Shanghai, China
{sun-mj,hyang,zhaiguangtao}@sjtu.edu.cn
[2] Shanghai Key Lab of Digital Media Processing and Transmission, Shanghai, China

Abstract. Pedestrian attribute recognition is a key problem in intelligent surveillance. Relations between attributes and human body structures or relations among attributes are beneficial to attribute recognition, while the annotations are just image-level binary labels. In this work, we propose a novel pedestrian attribute recognition network that takes advantage of latent attribute localizations and local attribute relations to improve the performance of pedestrian attribute recognition. Our method generates latent attribute localization maps by weakly-supervised learning in latent attribute localization (LAL) module. These latent attribute localization maps are fed into the local attribute attention (LAA) module to extract local attributes, and local attributes are interacted with each other with the attention mechanism. Extensive experiments made on the publicly pedestrian attribute datasets of PETA and RAP show that our model outperforms previous methods.

Keywords: Pedestrian attribute recognition · Latent attribute localization · Local attribute attention

1 Introduction

Pedestrian attribute recognition aims to extract semantic descriptions from target person image, including low-level descriptions (e.g., wearing, hairstyle) and high-level ones (e.g., gender, age). Pedestrian attribute recognition is one of the active research areas in computer vision because of its wide applications in intelligent video surveillance systems. Accurate attributes recognition also benefits other applications such as person re-identification and person retrieval.

This work was supported in part by National Natural Science Foundation of China (NSFC, Grant No. 61771303), Science and Technology Commission of Shanghai Municipality (STCSM, Grant Nos. 19DZ1209303, 18DZ1200102, 18DZ2270700, 20DZ1200203), and SJTUYitu/Thinkforce Joint laboratory for visual computing and application.
Mingjun Sun is a student.

In recent years, researchers pay more attention to solve the pedestrian attribute recognition problem. At first, pedestrian attribute recognition mainly relies on hand-crafted features such as color and texture histograms [1,2]. Recently, methods based on deep learning achieve great success, which formulates it as a multi-label classification problem [3]. These methods can roughly be divided into spatial attention methods and label relation methods. Spatial attention methods are proposed to pay more attention to discriminative local features of pedestrian [4–6], which are proved useful due to the existing relations between pedestrian attribute and attribute location on the human body. Label relation methods are proposed to exploit semantic relations to assist attribute recognition [7,8], which improve the performance of pedestrian attribute recognition by considering the dependency and conflicts of labels.

The main problems of existing methods are at least one of the following: (1) Ideally, one attribute corresponds to one specific region according to spatial attention methods. However, the relation between attributes and the human body is quite complicated, and these regions could be disconnected. (2) Some pedestrian attributes are predicted from shared feature vectors, which is beneficial to train the feature extractor. However, classifier needs to handle more redundant features when predicting a specific attribute. For example, The attribute *female* is inferred from many low level features (long hair, dress style), and it doesn't correspond to a specific human body region.

To address these problems, we propose to learn the latent attributes localization and handle local attributes with attention. The target attributes can be seen as combinations of latent attributes, which could be related to more precise regions and easier to represent. In our method, the latent attribute features are first extracted with spatial constraints. And then target attributes are predicted with attention mechanism to model relations among latent attributes.

Different from previous methods, we propose the latent attribute localization module (LAL) to localize the latent attributes, and generated latent attribute localization maps are used to extract local attributes. Then local attribute attention (LAA) methods are adopted to model relations among local attributes. The final predictions are obtained through a voting scheme to output the maximum predictions among different feature levels. The proposed framework is end-to-end trainable and requires only image-level annotations.

The main contributions of this work are as follows:

1. We propose a framework to handle latent attribute localizations and local attribute relations simultaneously in a weakly supervised manner. The latent attribute localization (LAL) module is proposed to localize discriminative latent attributes with image-level labels.
2. The local attribute attention (LAA) method is proposed to process latent attributes simultaneously and model relations among local attributes.
3. We conduct extensive experiments on publicly available pedestrian attribute datasets PETA and RAP, and our method outperforms previous methods.

2   Related Work

Many works have been proposed in the field of pedestrian attribute recognition. At first, pedestrian attribute recognition mainly relies on hand-crafted features such as color and texture histograms [1,2]. Recent years CNN-based approaches make great success in pedestrian attribute recognition and outperform most of traditional methods. The problem is formulated as a multi-label classification problem [3]. These methods can roughly be divided into spatial attention methods and label relation methods.

Spatial attention methods are proposed to pay more attention to discriminative local features of pedestrian [4–6,9], which are proved useful due to the existing relations between pedestrian attribute and attribute location on the human body. Liu et al. [4] propose a multi-directional attention model to learn multi-scale attentive features for pedestrian analysis, which extracts attention maps with convlution methods. Fabbri et al. [5] propose a generative adversial models,which uses features extracted from different human body parts. Li et al. [9] combine pose estimation and spatial transform network to extract local features. Tang et al. [6] extract features from different regions with spatial tranform network. Li et al. [10] model the spatial relations by simply dividing the image into rigid grids. However, these methods try to learn spatial constraints for all attributes, which is unnecessary and hard to learn.

Label relation methods are proposed to exploit semantic relations to assist attribute recognition [7,8], which improve the performance of pedestrian attribute recognition by considering the dependency and conflicts of labels. Wang et al. [7] propose a CNN-RNN network to exploit the relations among attributes. Zhao et al. [8] divide the attributes into several groups and attempt to explore the intra-group and inter-group relationships. However, these methods are mainly reply on pre-defined rules and don't take advantage of relations between attributes and human body regions.

3   Methods

The overview of our proposed framework is illustrated in Fig. 1. The proposed framework consists of a backbone network, Latent Attribute Localization (LAL) modules and Local Attribute Attention (LAA) modules applied to different feature levels. The key idea of this work is to take advantage of latent attribute localizations and local attributes relations to improve the effect of pedestrian attribute recognition.

3.1   Network Architecture

Formally, given an input pedestrian image along with its corresponding labels $y = [y^1, y^2, ..., y^C]^T$ where C is the total number of attributes, and y^c is a binary label that indicates the presence of c-th attribute if $y^c = 1$. We adopt the Inception-V3 [11] as the backbone network in our framework.

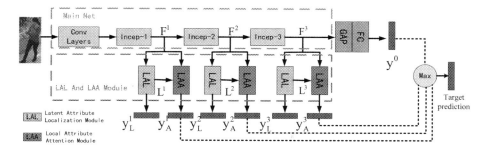

Fig. 1. Overview of the proposed framework. The latent attribute localization (LAL) modules and local attribute attention (LAA) modules are appended after different feature levels to handle attributes of different levels. Outputs from different branches are trained with intermediate supervision ways. Inference process is shown in dashed line, predictions from different levels vote for final prediction in a maximum way. The prediction output from LAL modules are not involved in inference process.

Pedestrian attribute recognition deals with attributes of various levels, so we extract features after different inceptions to take both low-level details feature and high-level semantics feature into account. For each pedestrian image, we can obtain feature representation $F^i, i = 1, 2, 3$ from 3 different levels in backbone network. We then conduct the LAL module and LAA module with different level features F^i. LAL module extracts latent attribute localization maps, and these maps are fed into LAA module. In LAA module, the local attributes are first extracted with features F^i and latent attribute localization maps. The relations among local attributes are modeled in LAA modules by applying corresponding weights on local attributes, and then full connect layers are used to make target attribute predictions. Attributes predictions are generated from different feature levels. During the inference process, final predictions are predicted by voting the maximum prediction among predictions made in different levels. The whole pipeline is shown in Fig. 1.

3.2 Latent Attribute Localization Module

As mentioned in Sect. 1, previous spatial attention methods extract regions related to specific attributes to improve performance. There are two things that should be considered. First, regions related to target attributes could be disconnected and hard to learn. Second, there are correlations among the attributes. Thus it is not suitable to learn each attribute location independently. To address such problems, we propose to learn latent attributes locations.

The details of the latent attribute localization module are shown in Fig. 2. The latent attribute localization method is motivated by weakly supervised detection and localization method [12–14]. Given input $F^i, i \in 1, 2, 3$, stacked convolution layers with kernel size equals to 1 are used to extract latent attribute localization maps z^i. In our experiment, the convolution layer number is set as 3. The kernel number of last convolution layer referred to as N_i equals the number of the latent attributes. The pixel value of c-th channel at position (h, w) of latent attribute localization maps are referred as $z_{c,h,w}^i$. The extracted latent attribute localization maps are then spatially normalized to put more attention on discriminative regions, and we get normalized latent attribute localization maps $a^i \in \mathbb{R}^{H_L^i \times W_L^i \times N^i}$. The nomalization process of $a_{c,h,w}^i$ is shown as Eq. 1:

$$a_{c,h,w}^i = \frac{exp(z_{c,h,w}^i)}{\sum_{h,w} exp(z_{c,h,w}^i)} \tag{1}$$

Following [14], the feature maps F^i are concurrently passed to convolution layers followed by the sigmoid function. This branch is used to decrease the influence when the latent attribute is absent. The parameter in this branch is set the same as previous branch. And we can get the latent attribute confidence maps s^i from this branch. The output from two branches are element-wise multiply to get the final latent attribute localization map L^i as Eq. 2:

$$L^i = a^i \odot s^i \tag{2}$$

And we get the latent attribute localization maps L^i that the LAL module learns to represent in the weakly supervised method. Then we convert latent attribute into target attribute predictions for training. The localization map L^i is fed through pooling layers and full connect layers to make target attribute prediction, and predictions from LAL Module on i-th level are referred as y_L^i. These predictions are not involved in reference process.

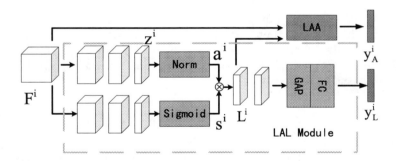

Fig. 2. Details of LAL module

3.3 Local Attribute Attention Module

The LAA module is proposed to handle relations among local attributes and make target attribute recognition. The target attribute can be seen as combination of attributes, and attention mechanisms are adopted in LAA module to handle the relations. The details are shown in Fig. 3.

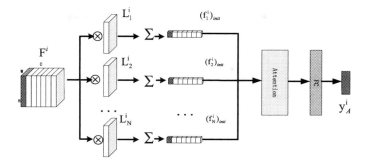

Fig. 3. Details of LAA module

With feature maps F^i from i-th level and extracted latent attribute localization maps L^i, we can extract local attributes vectors, shown as Eq. 3:

$$(f_n^i)_{init} = \sum_{h,w} L_n^i \odot F^i, n \in 1, 2, ..., N \tag{3}$$

To make local attributes distinguishable from each other, we adopt hard position encoding methods [15]. For the n-th local attribute, it computes cosine and sine functions of different wavelengths and adds to the local attribute vector [15]. The position encodered n-th local attribute is referred as f_n^i. With input set of N local attributes $f_1^i, f_2^i, ...f_N^i$, the ralation feature r_n^i of the whole local attribute set with respect to the n-th local attribute f_n^i is computed as follows:

$$r_n^i = \sum_{m=1}^{N} w_{mn}^i \cdot \phi_W^i(f_m^i) \tag{4}$$

where ϕ_W^i is learnable linearly transform, w_{mn}^i is the relation weight that indicates the ralation of m-th local attribute and n-th local attribute. And w_{mn}^i in computed as Eq. 5:

$$(w_{mn}^i)_{init} = \frac{\phi_K^i(f_n^i) \cdot \phi_Q^i(f_m^i)}{\sqrt{d_k}}$$
$$w_{mn}^i = \frac{(w_{mn}^i)_{init}}{\sum_k (w_{kn}^i)_{init}} \tag{5}$$

where ϕ_K^i, ϕ_Q^i are learnable linearly transform. d_k is the dimention of local attribute vector. It calculates the similarity of f_n^i in key space and f_m^i in query space as the relation between n-th local attribute and m-th local attribute. Then the relation features are passed through full connect layers to get the target attribute predictions y_A^i.

3.4   Loss Function

We adopt the weighted binary cross-entropy loss function [3] as the loss function, formulated as follows:

$$Loss(\hat{y}, y) = -\frac{1}{K} \sum_{c=1}^{K} e^{-p^c} (y^c log(\hat{y}^c) + (1 - y^c) log(1 - \hat{y}^c)) \qquad (6)$$

where y^c is the ground truth label which represents whether target person has c-th attribute, $\hat{y}$ denotes the output label probability of the c-th label, p^c denotes the positive ratio of c-th attribute in the training set, K denotes the number of target attributes. As stated in [3], the usage of weights of positive samples in the loss function can alleviate the problem of imbalanced label distribution. All predictions(y^0, y_L^i, y_A^i) are trained with the above weighted entropy loss. During inference process, predictions(y^0, y_A^i) from different levels vote for final prediction in a maximum way. The predictions(y_L^i) from LAL module are just for auxiliary training.

4   Experiments

4.1   Implementation

Datasets. For evaluations, we used two publicly available pedestrian attribute datasets: (1) **PETA Dataset** [16]. The dataset consists of 19000 person images. Following [3], we divide the whole dataset into three nonoverlapping partitions: 9500 for model training, 1900 for verification, and 7600 for model evaluation. And we choose 35 attributes that positive ratio is higher than 5% in our experiment. (2) **RAP Dataset** [17]. The dataset has 41585 images drawn from 26 indoor surveillance cameras. Following the official protocol [17], we split the whole dataset into 33,268 training images and 8,317 test images. We choose 51 attributes that positive ratio higher than 1% in our experiments.

Evaluation Metrics. We adopt two types of metrics for evaluation [17]: (1) Label-based metric: we calculate the **mA** as the mean of positive accuracy and negative accuracy for each attribute. (2) Instance-based metric: we adopt four well-known criteria: **accuracy, precision, recall** and **F1 score**.

Implementation Details. Our model is implemented with Pytorch. Inception-V3 model pretrained from the ImageNet image classification task is adopted as the backbone network. We train the network with batch size equals to 32. The initial learning rate of training is 1e–4 and changes to 1e–5 after 10 epochs. The optimization algorithm used in training is Adam optimization algorithm [18].

Table 1. Performance comparisons against previous methods on RAP dataset. Best results are in **bold**, and second best results are underlined.

Method	Metric				
	mA	Acc	Prec	Recall	F1
ACN [19]	69.66	62.61	80.12	72.26	75.98
DeepMar [3]	73.79	62.02	74.92	76.21	75.56
JRL [7]	77.81	–	78.11	78.98	78.58
GRL [8]	81.20	–	77.70	**80.90**	79.29
RA [20]	81.16	–	79.45	79.23	79.34
HP-Net [4]	76.12	65.39	77.33	78.79	78.05
PGDM [9]	74.31	64.57	78.86	75.9	77.35
Ours	**83.22**	**69.91**	**80.56**	80.04	**80.29**

Table 2. Performance comparisons against previous methods on PETA dataset. Best results are in **bold**, and second best results are underlined.

Method	Metric				
	mA	Acc	Prec	Recall	F1
ACN [19]	81.15	73.66	84.06	81.26	82.64
DeepMar [3]	82.89	75.07	83.68	83.14	83.41
JRL [7]	85.67	–	86.03	85.34	85.42
GRL [8]	86.70	–	84.34	**88.82**	86.51
RA [20]	86.11	–	84.69	88.51	86.56
HP-Net [4]	81.77	76.13	84.92	83.24	84.07
PGDM [9]	82.97	78.08	86.86	84.68	85.76
Ours	**86.91**	**78.62**	**87.10**	87.04	**87.06**

Table 3. Performance comparisons on RAP dataset when gradually adding proposed component to the baseline model.

Dataset	PETA		RAP	
Component	Metric			
	mA	F1	mA	F1
Baseline	80.63	82.58	75.80	77.78
LAL(single level)	82.48	83.07	77.59	79.24
LAL+LAA(single level)	84.11	83.48	80.37	79.40
LAL+LAA(multi-level)	**86.29**	**85.02**	**83.22**	**80.29**

4.2   Evaluation

In this section, we compare the performance of our proposed method against several previous methods. We choose the representative methods of three catagories mentioned in Sect. 1 and compare proposed method with them: (1) Holistic methods including ACN [19] and DeepMAR [3]. (2) Label relation methods including JRL [7], GRL [8] and RA [20]. (3) Spatial attention methods including HP-Net [4] and PGDM [9].

Tabel 1 and Tabel 2 shows the comparison results of the proposed methods against previous methods on PETA dataset and RAP dataset. The result suggests that our proposed method achieves superior performance compared with existing works. On RAP dataset, our proposed method achieves significant performance. The proposed method achieves the best performance on mA, accuracy, precision, and F1. The mA and F1 can be selected as main metrics to evaluate the methods performance for classification problems. The results suggest that our proposed method achieves superior performances compared with existing methods. The precision and recall metrics are mutually exclusive. Moreover, our method aims to extract more precise local features with complex network structures, which improves the credibility of results. So the precision of our method is quite high. Besides, our method has a faster speed compared with relation-based network [7,8] for its parallel structure. The improvement of performance on PETA dataset is not so obvious. In fact, there are more attributes at high levels depend on human and object interaction in the RAP dataset (e.g., action, attachment), which are better improved by the proposed methods. The comparison details are shown in Sect. 4.4.

4.3   Ablation Study

To validate our contributions, we further perform ablation studies on the PETA dataset and RAP dataset. We choose mA and F1 score as the representation of label-based and instance-based metrics to evaluate the effect. The result with the Inception-V3 net is chosen as the baseline method for comparison. In the baseline method, the input image is passed through convolution layers, pooling layers and full connect layers to make attribute recognition.

As shown in Table 3, based on Inception-v3 as the baseline network, we gradually add modules on it to analyze the effect. (1) **Latent Attribute Localization Module** We first evaluate the contribution of the LAL module by appending LAL Module after backbone convolution layers. The predictions from LAL modules are selected as target attribute predictions. The mA and F1 both increase, which demonstrates the spatial regularization of latent attributes is effective. And the improvement is quite obvious. (2) **Local Attribute Attention Module** We evaluate the impact of the LAA module by appending the LAA module with the LAL module at the same time. The improvement of mA is quite significant, which demonstrates the attention mechanism on local attributes is useful. The effect is further improved by handling the relationship among attributes with LAA module. (3) **Modules on Different Levels** Then we evaluate the effect of applying the LAL module and the LAA Module on multiple feature levels. We apply the proposed Modules after Incep-1, Incep-2, and Incep-3 and get final predictions with the voting max scheme. The multi-level framework is proved useful because it takes advantage of features extracted from different levels. Considering the model complexity, LAL and LAA modules are applied on 3 levels in our method.

4.4   Improvements on Different Attributes

Table 4. 5 attributes with the greatest improvement in RAP dataset

Attribute	Improvement
Attach paper bag	12.4%
Upbody suitup	12.1%
Shoes cloth	11.5%
Attach handbag	11.3%
Bold head	9.6%

Compared with Baseline methods, the proposed methods increase the accuracy better in the following attributes on RAP dataset: attach paper bag (increase 12.4%), upbody suitup (increase 12.1%), shoes cloth (increase 11.5%), attach handbag (increase 11.3%), bold head (increase 9.6%), as shown in Table 4. We can find that these attributes are more focused on understanding the structure of the human body. These attributes are local attributes or combinations of local attributes. On th contrary, the improvements on abstract attributes like age and body shape are not so obvious. Our proposed method improves the ability to understand human body structure by learning latent attributes. Thus our method performs better on recognition of the attributes related with human body.

5   Conclusion

In this work, we propose a novel end-to-end model for pedestrian attribute recognition by learning latent attributes. Latent Attribute Localization (LAL) module learns the relation between latent attributes and human bodies, and the localization maps are used to extract local attributes. Relations among local attributes are modeled with Local Attribute Attention (LAA) methods. Our proposed model outperforms a wide range of existing pedestrian attribute recognition methods. Extensive experiments demonstrate the effects of proposed modules LAL and LAA.

References

1. Layne, R., Hospedales, T.M., Gong, S., Mary, Q.: Person re-identification by attributes. In: BMVC, vol. 2, p. 8 (2012)
2. Liu, C., Gong, S., Loy, C.C., Lin, X.: Person re-identification: what features are important? In: Fusiello, A., Murino, V., Cucchiara, R. (eds.) ECCV 2012. LNCS, vol. 7583, pp. 391–401. Springer, Heidelberg (2012). https://doi.org/10.1007/978-3-642-33863-2_39
3. Li, D., Chen, X., Huang, K.: Multi-attribute learning for pedestrian attribute recognition in surveillance scenarios. In: 2015 3rd IAPR Asian Conference on Pattern Recognition (ACPR), pp. 111–115. IEEE (2015)
4. Liu, X., et al.: Hydraplus-net: attentive deep features for pedestrian analysis. In: Proceedings of the IEEE international conference on computer vision, pp. 350–359 (2017)
5. Fabbri, M., Calderara, S., Cucchiara, R.: Generative adversarial models for people attribute recognition in surveillance. In: 2017 14th IEEE International Conference on Advanced Video and Signal Based Surveillance (AVSS), pp. 1–6. IEEE (2017)
6. Tang, C., Sheng, L., Zhang, Z., Hu, X.: Improving pedestrian attribute recognition with weakly-supervised multi-scale attribute-specific localization. In: Proceedings of the IEEE International Conference on Computer Vision, pp. 4997–5006 (2019)
7. Wang, J., Zhu, X., Gong, S., Li, W.: Attribute recognition by joint recurrent learning of context and correlation. In: Proceedings of the IEEE International Conference on Computer Vision, pp. 531–540 (2017)
8. Zhao, X., Sang, L., Ding, G., Guo, Y., Jin, X.: Grouping attribute recognition for pedestrian with joint recurrent learning. In: IJCAI, pp. 3177–3183 (2018)
9. Li, D., Chen, X., Zhang, Z., Huang, K.: Pose guided deep model for pedestrian attribute recognition in surveillance scenarios. In: 2018 IEEE International Conference on Multimedia and Expo (ICME), pp. 1–6. IEEE (2018)
10. Li, Q., Zhao, X., He, R., Huang, K.: Visual-semantic graph reasoning for pedestrian attribute recognition. In: Proceedings of the AAAI Conference on Artificial Intelligence, vol. 33, pp. 8634–8641 (2019)
11. Ioffe, S., Szegedy, C.: Batch normalization: accelerating deep network training by reducing internal covariate shift. arXiv preprint arXiv:1502.03167 (2015)
12. Bilen, H., Vedaldi, A.: Weakly supervised deep detection networks. In: Proceedings of the IEEE Conference on Computer Vision and Pattern Recognition, pp. 2846–2854 (2016)
13. Sarafianos, N., Xu, X., Kakadiaris, I.A.: Deep imbalanced attribute classification using visual attention aggregation. In: Proceedings of the European Conference on Computer Vision (ECCV), pp. 680–697 (2018)
14. Zhu, F., Li, H., Ouyang, W., Yu, N., Wang, X.: Learning spatial regularization with image-level supervisions for multi-label image classification. In: Proceedings of the IEEE Conference on Computer Vision and Pattern Recognition, pp. 5513–5522 (2017)
15. Vaswani, A., et al.: Attention is all you need. In: Advances in Neural Information Processing Systems, pp. 5998–6008 (2017)
16. Deng, Y., Luo, P., Loy, C.C., Tang, X.: Learning to recognize pedestrian attribute. arXiv preprint arXiv:1501.00901 (2015)
17. Li, D., Zhang, Z., Chen, X., Ling, H., Huang, K.: A richly annotated dataset for pedestrian attribute recognition. arXiv preprint arXiv:1603.07054 (2016)

18. Kingma, D.P., Ba, J.: Adam: a method for stochastic optimization. arXiv preprint arXiv:1412.6980 (2014)
19. Sudowe, P., Spitzer, H., Leibe, B.: Person attribute recognition with a jointly-trained holistic CNN model. In: Proceedings of the IEEE International Conference on Computer Vision Workshops, pp. 87–95 (2015)
20. Zhao, X., Sang, L., Ding, G., Han, J., Di, N., Yan, C.: Recurrent attention model for pedestrian attribute recognition. In: Proceedings of the AAAI Conference on Artificial Intelligence, vol. 33, pp. 9275–9282 (2019)

Intra-Camera Supervised Person Re-ID by Tracklet Level Classifier

Yan Bai, Weiquan Huang, and Yin Wang[✉]

Tongji University, Shanghai, China
{yan.bai,weiquanh,yinw}@tongji.edu.cn

Abstract. In this work, we propose a novel method to perform intra-camera supervised person re-ID by Tracklet Level Classifier (TLC). The key idea of our method is to train classifiers for every intra-camera ID, which is tracklet level, compared with camera level of previous works. By training tracklet level classifiers, we make the backbone learned to extract intra-camera invariant representations. With the fine-trained classifiers, we mine and exploit latent inter-camera ID matching pairs easily. Previous works needs two stages and relies on complicated rules to match inter-camera pairs while we simplify the training strategy to only one stage and do not need a complex design to match tracklet over cameras. Extensive experiments and ablation studies on three large re-ID datasets show that our simple and effective TLC method achieve state-of-the-art among all the intra-camera supervised person re-ID methods.

Keywords: Person Re-ID · Weakly supervised · Tracklet

1 Introduction

Person re-Identification (re-ID) is a hot research topic nowadays for its application in surveillance camera systems. It is a sub-topic in image retrieval in computer vision. Person re-ID aims to match the same pedestrian across non-overlapping camera views [1–5]. With the develop of deep learning, especially deep convolutional neural network (CNN) [6,7], the performance of person re-ID has been improved significantly [8–11]. Conventional supervised learning methods depend on the well annotated labels to extract camera-view invariant features for person images due to CNN is a data driven method. However, the labels are annotated manually, which is very time- and money-consuming. As the number of cameras grows, the workload of matching ID across the cameras increases by a quadratic scale, making it harder to annotate person re-ID labels than other computer vision task, e.g. classification and detection. The high cost of labeling limits the fully supervised person re-id methods, so other methods are needed that do not need such labels.

There have been some unsupervised or weakly supervised person re-ID methods to address this problem. We roughly classify them to four categories: (1) Domain adaption methods [12–14] (2) Unsupervised clustering methods [15,16], (3) Semi-supervised methods [17], (4) Intra-camera supervised or tracklet supervised methods [18–21]. Domain adaption methods regard different datasets as different domains and aim to

© Springer Nature Switzerland AG 2020
Y. Peng et al. (Eds.): PRCV 2020, LNCS 12306, pp. 328–342, 2020.
https://doi.org/10.1007/978-3-030-60639-8_28

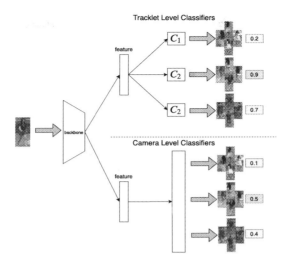

Fig. 1. Illustration of *Tracklet Level Classifier* (TLC) method and *Camera Level Classifier* (CLC) method for intra-camera supervised person re-ID learning. Given a feature extracted by the backbone network, TLC builds independent classifiers for every tracklets, a.k.a in-camera IDs. A tracklet classifier predicts whether a feature belongs to a tracklet. Compared with common used CLC methods build classifiers for every camera and each classifier predict which tracklet the feature belongs to. The probability is determined by all logits in a camera in CLC while by only the tracklet's logit itself in TLC, which makes a more precise tracklet classifiers and thus giving a easy way to matching latent cross-camera tracklet pairs for performance improvements.

transfer the source domain trained model to the target domain in a domain adaption manner. Unsupervised clustering methods generate labels automatically for target domain data by means of clustering algorithms and then fine tune a model from source domain with the generated labels for target domain. Semi-supervised methods use a slight amount of labels for training and then using techniques to generate labels for unlabeled data, then train on generated labels. Intra-camera supervised or tracklet supervised methods are recently proposed promising methods that using intra-camera annotated labels or tracklet label generated by existing tracking methods which perform the best among all these 4 categories.

Domain adaption and unsupervised clustering methods rely on a fine trained model from the source domain, which is not scalable and also performs mediocrely. Semi-supervised methods still rely on high costing cross camera labels and perform better with more labels, which can be regarded as special case of fully supervised learning. Intra-camera and tracklet supervised methods are similar because they only rely on intra-camera labels. The difference is that tracklet supervised methods use label generated by tracking methods [22–25] and need original videos to perform the tracking, while intra-camera use human annotated intra-camera label, which is easier and cheaper to get than cross-camera labels. They are two versions of trade-off between label quality and cost. Intra-camera methods achieve promising performance recently. However, UTAL [19] and TAUDL [18] need a large mini-batch (384) at run time for matching the latent cross-camera tracklet pairs, which may occupy at least ve 1080-Ti GPUs in training, UGA [21] builds a complicated graph algorithm to match cross-camera tracklet pairs. Besides, these methods are made of two stage training, which may be cumbersome.

We also observe that all above intra-camera methods use multi-task on camera level classifiers, as illustrated in lower part of Fig. 1, in which circumstances the predicted probability for a feature to every tracklet are determined by not only its logit but also logits of the same camera. That makes it hard to tell if a feature belongs to a tracklet. In multi-label learning methods [26,27], classifiers for every class are trained to fit the demand that an image may have multi labels. In intra-camera supervised person re-ID, a feature may belong to multi tracklet in different camera, makes it possible to adopt some concept from multi-label classification.

To address above problems and inspired by multi-label classification, we propose a *Tracklet Level Classifier* (TLC) architecture, compared with existing works' *Camera Level Classifier* (CLC). The difference between TLC and CLC is illustrated in Fig. 1. Our TLC method trains classifier for every tracklet and for leveraging the intra-camera label and inter-camera underlying tracklet pairs we propose two components of training objective loss function: **Intra-Camera Tracklet Classification** and **Inter-Camera Tracklet Classification**. The two losses take effect all at the same one stage and our model runs at a normal mini-batch size, e.g. 60.

Intra-Camera Tracklet Classification helps the model to exploit the intra-camera labels, which makes a feature extractor that extracts intra-camera invariant features for pedestrian images and a serial of classifiers for every tracklet.

Inter-Camera Tracklet Classification is designed to mine and exploit the latent matching pairs of tracklet across different cameras. With the help of precise classifier trained by Intra-Camera Tracklet Classification, we match the pairs only by a hyperparameter threshold, which makes our model simple and effective. To sum up, our contributions of this paper is:

- We propose an intra-camera supervised person re-ID architecture named *Tracklet Level Classifier* (TLC), which is simple yet effective. Without huge hardware burden or complex matching strategy, our TLC method achieves competitive performances.
- Inspired by multi-task learning, we adopt building independent classifiers for every tracklet instead of every camera, which fits the intra-camera supervised person re-ID scenario better and is proved by analysis and experiments.
- Extensive experiments and ablation studies on three popular person re-ID datasets demonstrate the effectiveness of proposed TLC.

2   Related Word

2.1   Supervised Person Re-ID

Most existing person re-ID models are trained by supervised learning methods on inter-camera pairwise ID labelled training data [8–11,28,29]. Training on large labelled data boosts their performance by hard association of inter-camera images and makes them perform state-of-the-art. However, supervised learning based re-id methods suffer from significant performance decrease when lacking a large scale of inter-camera identity labelled training data, which is labour intensive and very expensive. This limits their usability and scalability for realistic applications [20].

2.2 Semi-supervised Person Re-ID

Semi-supervised learning methods [17, 30] are to self-learn knowledge from large unlabelled training data based on the knowledge learned from labelled training data. Therefore, they decrease the need of labelled training data and achieve a trade-off between model scalability and re-id accuracy. However, these methods still need some expensive cross-view pairwise labelling. To get rid of exhaustively collecting a large set of labelled training data for every application domain, lots of unsupervised learning based person re-id models are proposed [14, 15, 18, 19, 21, 31].

2.3 Unsupervised Person Re-ID

Unsupervised domain adaptation based methods and unsupervised clustering based methods are two typical methods on unsupervised learning person re-id models. Unsupervised domain adaptation [14] based methods handle unlabeled target domain by leveraging information from the labelled data in source domain. Unsupervised clustering methods [15, 31] use the unsupervised clustering algorithms to obtain the pseudo labels of target domain data and fine tune the source domain model with pseudo labels on target domain [21]. Both of the adaptation methods and cluster methods rely on the similarity between the different domain.

2.4 Intra-Camera Supervised Person Re-ID

Intra-camera or tracklet supervised methods [18–21] are recently proposed promising methods for person re-ID, which only need intra-camera labels and achieve pretty good performance. Without the need of inter-camera pair matching label, the annotation is much easier and cheaper to get and even can be generated by trackers. Intra-camera supervised methods usually use multi-task learning for every camera, and design complex rules to match the latent inter-camera matching pairs to boost the performance. Some methods [18, 19] match the tracklet pairs in mini-batches, and thus demand for several GPUs during training time.

2.5 Multi-label Classification

Multi-label classification is a type of classification that every instance may have several labels instead of only one label in conventional classification, e.g. classification of all the objects in an images is a typical multi-label classification. Multi-label classification methods usually transform multi-label into multiple binary label for classification. [26, 27]. The problem of intra-camera supervised person re-ID is similar to multi-label classification. If we regard every intra-camera label, a.k.a tracklet, as a class, then every instance may have multiple labels because a pedestrian may appear in multiple camera. How to assign the label would be the key to adopt multi-label classification in intra-camera classification method.

3 Problem Formulation

In *intra-camera supervised* (ICS) person re-identification scenario, suppose we have person images from K cameras. For each camera $k \in \{1, 2, ..., K\}$, we annotate each person image $I_i^k \in \{I_1^k, I_2^k, ..., I_{N_k}^k\}$ with an intra-camera identity label $y_j^k \in \{y_1^k, y_2^k, ..., y_{T_k}^k\}$, where N_k is the number of person images and T_k is the number of intra-camera identities in camera k. Generally speaking, when annotating person images with intra-camera identities, the most common method is using a detector to generate the bounding boxes and using a tracker to make a tracklet ID for each bounding box. We adopt the Sparse Space-Time Tracklet (SSTT) [18] sampling and assume that each intra-camera ID belongs to different person ID, so we mark every intra-camera ID as a *tracklet*. For clarity, the tracklet ID is annotated respectively, which means for any given cameras p and q and any tracklet ID y_i^p and y_j^q, we do not know whether they belong to the same person or not.

Compare with conventional fully supervised person re-ID problem, the most different and challenging point is that in ICS scenario we do not have the inter-camera tracklet matching information, which decreases the annotation costs yet increases the difficulty of modeling. A good method to problem should effectively take use of the tracklet ID and have a proper way to mining the latent inter-camera tracklet matching pairs.

4 Proposed Method

We propose a novel method that can better exploit the tracklet ID and find latent inter-camera tracklet matching, by designing Tracklet Level Classifiers (TLC). We model the problem as a tracklet classification problem and propose two objective functions to exploit the intra-camera tracklet label and inter-camera tracklet matching: (1) *Intra-Camera Tracklet Classification* that takes use of intra-camera label (Sect. 4.1), (2) *Inter-Camera Tracklet Classification* that mines the latent tracklet matching pairs across cameras (Sect. 4.2).

An overview of our proposed TLC method is shown in Fig. 2. The model consists of a backbone network for feature extraction and several *Tracklet Classifier*. Each tracklet classifier is designed to classify whether a feature belongs to this tracklet. With the intra-camera and inter-camera tracklet classification method, our model is capable of better exploiting the intra-camera label and automatically find latent inter-camera tracklet matching pairs.

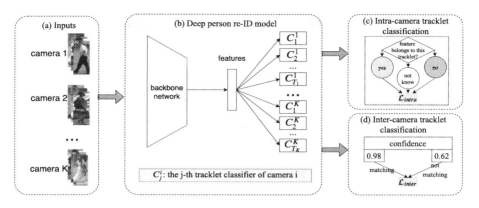

Fig. 2. Overview of proposed Tracklet Level Classifier(TLC) method. **(a)** Given images from different cameras, the labels are annotated independently over these cameras. **(b)** TLC method is designed to train classifiers for every within-camera ID, a.k.a tracklet, we train the TLC model by two components of loss function: **(c)** *Intra-camera tracklet classification* is designed to exploit the given supervision of intra-camera label. We set the tracklet classifiers' label by the intra-camera label and then calculate the loss to train the model. **(d)** *Intra-camera tracklet classification* mines the latent matching pairs across cameras. Due to the preciseness of tracklet classifiers, we judge the matching with only as simple rule of threshold. With the novel design of tracklet classifier and two dedicated objective loss functions, our TLC model learns the intra-camera supervision information effectively.

4.1 Intra-Camera Tracklet Classification

Our model is based on a multi-task learning strategy [19] in tracklet level, compared with previous works [18–21] in camera level. Using multi-task learning aims to better exploit the great generalization ability of convolutional neural networks due to the common knowledge is shared by all these tasks.

The tracklet classifier is a classifier that predict whether a feature belongs to specific tracklet, so it naturally is a binary classifier. For C_j^i, the j-th tracklet classifier from the i-th camera, it contains a weight vector $\boldsymbol{W}_j^i$ and a sigmoid function that maps any real number to a probability, defined as $\sigma(x) = \frac{1}{1+e^{-x}}$. Most common used objective function of binary classifier is Binary Cross-Entropy loss, which is formulated as:

$$\mathcal{L}_b = -(y \log(\hat{p}) + (1 - y) \log(1 - \hat{p})) \tag{1}$$

where $y \in \{1, 0\}$ is the label that indicates true or false of a classification and $\hat{p}$ is the predicted probability of C_j^i. According to the practice, we call the real number that feeds to $\sigma(\cdot)$ as a *logit*. Due to the vanishing gradient problem of $\sigma(\cdot)$, the logit should be near 0. So we adopt an $L2$-norm strategy to get the logit and formulate the $\hat{p}$ as:

$$\hat{p}_j^i = \sigma\left(\frac{\boldsymbol{F} \cdot \boldsymbol{W}_j^i}{||\boldsymbol{F}|| \cdot ||\boldsymbol{W}_j^i||} \cdot s\right) \tag{2}$$

where $\boldsymbol{F}$ is the feature extracted by the backbone network and $\boldsymbol{W}_j^i$ is the weight vector for C_j^i and s is a hyper-parameter used to adjust the scale of the logit.

In *intra-camera tracklet classification* scenario, the key factor is to assign a proper label for the classifiers. Given a feature F_k^p extracted from an image from the k-th tracklet of camera p, due to the prior that a tracklet is unique within a camera, we get $y = 1$ for C_k^p and $y = 0$ for $C_{q \neq k}^p$. But for any tracklet classifier from other cameras, we do not know whether the image is from that tracklet or not in this stage. The way to assign label for unknown tracklet classifier is illustrated in Sect. 4.2, and for now we just leave that loss to 0 and then our intra-camera tracklet classification loss function for C_j^i is:

$$\mathcal{L}_{intra}^{i,j} = \mathcal{L}_b^{i,j} \cdot \mathbb{1}(\text{feature from camera i}) \tag{3}$$

where $\mathbb{1}(\cdot)$ is an indicator function that returns $\{1, 0\}$ when given $\{\text{True, False}\}$. Due to the unequal distributed label that $y = 1$ appears far less that $y = 0$, we average the two conditions respectively and give the complete loss function as:

$$\mathcal{L}_{intra} = \frac{1}{\sum^{y=1} 1} \sum^{y=1} \mathcal{L}_{intra}^{i,j} + \frac{1}{\sum^{y=0} 1} \sum^{y=0} \mathcal{L}_{intra}^{i,j} \tag{4}$$

with the *intra-camera tracklet classification* learning, we can train a model that exploit all the intra-camera supervise information, and get a good feature extractor for person re-ID and get a predictor for each tracklet to tell whether a feature belongs to it.

4.2 Inter-Camera Tracklet Classification

Person re-ID relies highly on the feature consistency across different camera views. While we do not have the inter-camera labels, it is key for ICS person re-ID to mining the latent inter-camera tracklet matching pairs and exploit that information for the model training. Previous works [18–20] have designed lots of complicated rules to match the pairs, while thanks to the good design of our model, it is very easy for us to find good pairs for *inter-camera tracklet classification* training. By performing the *intra-camera tracklet classification*, we get a serial of tracklet classifiers that could predict if a feature belongs to it. Due to the shared information between camera are similar and the good generalization ability of deep network, it makes it possible for the classifier to predict even a feature from another camera.

Specifically, given a feature F_j^i and a tracklet classifier C_n^m, where $m \neq i$. We can easily calculate the predicted probability by Eq. 2, but here we use a different way:

$$\hat{p}^{i,j,m,n} = \sigma\left(\frac{\boldsymbol{W}_j^i \cdot \boldsymbol{W}_n^m}{||\boldsymbol{W}_j^i|| \cdot ||\boldsymbol{W}_n^m||} \cdot s\right) \tag{5}$$

where $\boldsymbol{W}_b^a$ is weight vector for C_b^a. Review Eq. 2, we can easily find $\frac{F}{||F||}$ and $\frac{W_j^i}{||W_j^i||}$ are two unit vectors and their product is the cosine between two vectors. The cosine is max when two vectors are same, which makes the predicted probability max. A fine-trained classifier should have a largest average probability to all the features that belong to it, so we assume the classifier's weight vector $\frac{W_j^i}{W_j^i}$ as a representative vector for all the features that belong to it. To have a more precise matching and minimize the random error of a feature, we use $\boldsymbol{W}_j^i$ as a representative vector for all features that belong to it.

Here we use a threshold hyper-parameter t and when $\hat{p}^{i,j,m,n} > t$, we regard it as a matched pair between C_j^i and C_n^m. Our *inter-camera tracklet classification* is performed when any F_j^i matches C_n^m, which means tracklet C_j^i are found to be in camera m. Then we set label $y^{i,j,m,k} = \mathbb{1}(\hat{p}^{i,k,m,k} > t)$ for all the tracklet classifiers C_k^m of camera m. We do **not** set the $y^{i,j,m,k}$ if F_j^i do not match any tracklet in camera m, because when our tracklet classifiers are not fine-trained, the latent matching pair's predicted probability might be lower than t, and it harms the model to use those false negative labels. Our *inter-camera tracklet classification* uses the objective loss:

$$\mathcal{L}_{inter}^{i,j,m,n} = \mathcal{L}_b^{i,j,m,n} \cdot \mathbb{1}(y^{i,j,m,n} \text{ is set}) \tag{6}$$

like Eq. 4, we average the losses respectively between positive and negative conditions, and the complete version of inter-camera tracklet classification loss is:

$$\mathcal{L}_{inter} = \frac{1}{\sum^{y=1} 1} \sum^{y=1} \mathcal{L}_{inter}^{i,j,m,n} + \frac{1}{\sum^{y=0} 1} \sum^{y=0} \mathcal{L}_{inter}^{i,j,m,n} \tag{7}$$

With *inter-camera tracklet classification* learning, our model can continuously find and exploit the inter-camera matching pairs to improve the person re-ID performance.

4.3 Model Objective Loss Function

We use together the intra and inter camera objective loss functions and get the model objective loss function:

$$\mathcal{L} = \mathcal{L}_{intra} + \mathcal{L}_{inter} \tag{8}$$

4.4 Theoretical Analysis

Compared with previous used common Camera Level Classifier methods [20], the largest difference of our TLC uses a different way to predict the probability of an image to a tracklet. Previous Camera Level Classifier methods use a soft-max function:

$$\hat{p}_{softmax}^i = \frac{e^{L_i}}{\sum^j e^{L_j}} \tag{9}$$

where L_i is the logit of tracklet i. While we use sigmoid function $\sigma(\cdot)$ to generate the prediction probability:

$$\hat{p}_{sigmoid}^i = \frac{1}{1 + e^{-L_i}} \tag{10}$$

The reason that we use sigmoid other than soft-max is that by using sigmoid we build an *absolute classifier* compared with soft-max a *relative classifier*. We call it *absolute classifier* because the probability in Eq. 10 is determined only by logit L_i, compared with *relative classifier* in Eq. 9, the probability is determined by all the logits of that camera.

Supposing we have two groups of of logits $[4, 2, 2]$ and $[-2, -4, -4]$, the probability calculated by sigmoid is $[0.98, 0.88, 0.88]$ and $[0.12, 0.02, 0.02]$. But their probability by soft-max are the same, which is $[0.78, 0.11, 0.11]$. Due to soft-max only focus on relative values among all the logits, it cannot distinguish if the logits are all low or all high.

By adopting sigmoid function and using Tracklet Level Classifier, we can directly maximize the positive logits and minimize the negative logits, which we believe can improve the model training performance instead of optimizing the relative values of logits. At the same time, with *absolute classifiers* of tracklets, we can simply use the predicted probability to match inter-camera pairs, instead of designing complicated rules. The effects are discussed in Sect. 5.3.

5   Experiment

5.1   Experimental Setup

Datasets. To evaluate we proposed TLC method, all experiments are evaluated on three existing large-scale re-id datasets, i.e., Market-1501 [32], DukeMTMC-reID [33,34], and MSMT17 [35].The Market-1501 dataset is collected by a total of six cameras, including 5 high-resolution cameras and one low-resolution camera. DukeMTMC-reID is a subset of the DukeMTMC dataset. 1,812 identities were captured by 8 cameras in it. MSMT17 is the largest re-ID dataset. A total of 15 cameras are used, including 12 outdoor cameras and 3 indoor cameras.

The details of the these datasets including the number of ID and images are shown in Table. 1.

Table 1. The details of datasets used in our experiments.

Dataset	ID	Cam	Tain	Test	Images
Market-1501	1,501	6	751	750	32,668
DukeMTMC-reID	1,812	8	702	1,110	36,411
MSMT17-V2	1,467	15	1,041	3,060	126,441

Evaluation Protocol. Cumulative Matching Characteristic (CMC) [32] curve and the mean average precision (mAP) were used in all our experiments to evaluate the performance of our method. All the test data in the test set to evaluate the model performance, and all results reported in this paper are under the single-query setting without post-processing.

Implementation Details. In practice, we used an ImageNet pre-trained ResNet-50 as the backbone network of our TLC model. In order to balance the model training across camera views, we used a CPK sampling method, which randomly selects K number of images for P number of identities from C number of camera views for each mini-batch.

Table 2. Comparing TLC with the state-of-the-art methods on the image person re-ID dataset.

Category	Method	Market-1501				DukeMTMC-reID				MSMT17			
		Rank-1	Rank-5	Rank-10	mAP	Rank-1	Rank-5	Rank-10	mAP	Rank-1	Rank-5	Rank-10	mAP
Clustering	BUC [15]	61.9	73.5	78.2	29.6	40.4	52.5	58.2	22.1	–	–	–	–
	TSSL [16]	71.2	–	–	43.3	62.2	–	–	38.5	–	–	–	–
Semi-supervised	MVC [17]	72.2	–	–	49.6	52.9	–	–	33.6	–	–	–	–
Domain adaption	MAR [12]	67.7	–	–	40.0	67.1	–	–	48.0	–	–	–	–
	ECN [13]	75.1	–	91.6	43.0	63.3	–	80.4	40.4	30.2	–	46.8	10.2
	SGG++ [14]	86.2	94.6	96.5	68.7	76.0	85.8	89.3	60.3	–	–	–	–
Intra-Camera	TAUDL [18]	63.7	–	–	43.2	61.7	–	–	43.5	28.4	–	–	12.5
	UTAL [19]	69.2	–	–	46.2	62.3	–	–	44.6	31.4	–	–	13.1
	MTML [20]	85.3	–	96.2	65.2	71.1	–	86.9	50.7	44.1	–	63.9	18.6
	UGA [21]	87.2	–	–	70.3	75.0	–	–	53.3	49.5	–	–	21.7
	TLC (Our Method)	88.1	95.3	97.2	71.1	79.2	88.7	91.9	60.6	46.1	59.5	65.1	19.1
Supervised	DG-Net [36]	94.8	–	–	86.0	86.6	–	–	74.8	77.2	87.4	90.5	52.3
	st-ReID [9]	97.2	99.3	99.5	86.7	94.0	97.0	97.8	82.8	–	–	–	–

The C, P, K are set to be 3, 5, 4, making up and mini-batch of 60. The training images are resized to 256×128 in pixel for all datasets. Adam optimizer is adopted for training our TLC model, with a learning rate of 3.5e−4. The hyper-parameter scale s is set to be 4 and hyper-parameter threshold t is set to be 0.95.

5.2   Comparisons to the State-of-the-Art

In this section, we evaluate our method compared with the proposed method with some state-of-the-art person re-ID methods, specifically compared with four similar intra-camera learning based methods on Market-1501, DukeMTMC-reID and MSMT17. The performances of these methods are shown in Table. 2.

We compared the proposed TLC model with 12 existing state-of-the-art methods, which can be grouped into 5 categories: (1) Clustering methods including BUC [15] and TSSL [16]; (2) Semi-supervised learning methods, including MVC [17]; (3) Domain adaption based methods including MAR [12], ECN [13] and SGG++ [14]; (4) Intra-camera supervised methods including TAUDL [18], UTAL [19], MTML [20] and UGA [21]; (5) State-of-the-art Supervised learning methods including DG-Net [36] and st-ReID [9].

The results are shown in Table. 2. Clustering based unsupervised methods, i.e. BUC and TSSL, and semi-supervised method, i.e. MVC remains a long way to reach a comparable performance in the table. The Domain adaption based methods are comparable with intra-camera supervised ones, while our proposed TLC method achieves the best performance among all non-fully-supervised methods in Market-1501 and DukeMTMC-reID datasets. In MSMT17, TLC has a comparable performance. The supervised methods remain unreachable for all other categories of methods, with the supervision of expensive and time consuming cross-camera annotated label. However, our TLC method decreases the gap between supervised and non-fully-supervised methods. The scalability and low cost of getting labels makes our TLC a promising method in practice.

5.3 Ablation Study

Component Analysis. Our TLC method consists of two components, $\mathcal{L}_{intra}$ and $\mathcal{L}_{inter}$, that help the model to learn intra-camera and inter-camera knowledge of person re-ID. They are all necessary to our TLC method, but technically we can use $\mathcal{L}_{intra}$ only to train a model that learns only the intra-camera information. In order to find out how they contribute to the model performance, we experiment on it and results are shown in Table. 3. We find that: (1) Models trained with only $\mathcal{L}_{intra}$ achieve competitive performance, indicating that our $\mathcal{L}_{intra}$ leverages the shared knowledge behind all these tracklet classifier to train the model; (2) With both $\mathcal{L}_{intra}$ and $\mathcal{L}_{inter}$, our model runs even better, indicating that $\mathcal{L}_{intra}$'s inter-camera matching pairs help mine and exploit the information of cross-camera person re-ID, even though there may be some false matching pairs.

Table 3. Effective analysis on two components of TLC method: **TLC-intra** is trained with only $\mathcal{L}_{intra}$ and **TLC** is trained with both $\mathcal{L}_{intra}$ and $\mathcal{L}_{inter}$.

Method	Rank-1	Rank-5	Rank-10	mAP
Market1501				
TLC-intra	80.4	90.9	93.7	58.2
TLC	**88.1**	**95.3**	**97.2**	**71.1**
DukeMTMC-reID				
TLC-intra	74.9	85.5	88.6	55.8
TLC	**79.2**	**88.7**	**91.9**	**60.6**

Tracklet Level Classifier vs. Camera Level Classifier. We have discussed the difference between *Tracklet Level Classifier* (TLC) and *Camera Level Classifier* (CLC) in Sect. 4.4. In order to compare their effects, we set two groups of experiments, one uses sigmoid as TLC and the other uses soft-max as CLC, all experiments are supervised only with intra-camera information and do not match cross-camera pairs for a fair comparison. As shown in Table. 4, our TLC methods perform better in ICS scenario and we believe the improvement is from the *absolute classifier* which makes preciser classifier for every tracklet.

Discussion of Hyper-Parameter. We use two hyper-parameter for model training, s the scale in Eq. 5 and t the threshold in Sect. 4.2.

Scale. Scale is set to adjust the activation area of sigmoid function. A good choice of scale should cover all the domain of input and output and make sure their gradient do not disappear. The domain of input is $[-1, 1]$ and the domain if output is $[0, 1]$. We draw the sigmoid function with different scale in Fig. 3 (a). We can find that when scale = 1 or 2, it cannot cover the output domain of $[0, 1]$, while when scale = 8, inputs that larger than 0.75 or lower than -0.75 will suffer from zero gradient and scale = 4 is a proper choice.

Table 4. Comparison between *Tracklet Level Classifier* (TLC-intra) and *Camera Level Classifier* (CLC-intra) on intra-camera supervised person re-ID.

Method	Rank-1	Rank-5	Rank-10	mAP
Market1501				
CLC-intra	75.9	89.3	92.8	50.4
TLC-intra	**80.4**	**90.9**	**93.7**	**58.2**
DukeMTMC-reID				
CLC-intra	71.5	83.6	87.2	51.0
TLC-intra	**74.9**	**85.5**	**88.6**	**55.8**

Threshold. A proper threshold is significant for *Inter-camera Tracklet Classification*, for a threshold too large decrease the opportunity of finding matching pairs while one too low increase the false positive rate. In order to quest on how the threshold affects the matching pairs, we calculate a serial of precision and recall of matching pairs on different thresholds, as shown in Fig. 3 (b). It can be see that when threshold is 0.95, the precision and recall reach a balance with both at high values.

Pair Matching Dynamics. To further quest on the effect of Pair matching in **Inter-Camera Tracklet Classification**, we tracked the evolving dynamics of matching pairs during the model training process. As shown in Fig. 3(c), the precision of pair matching goes down slightly while the recall increase significantly with the training step grows. With the intra-camera loss and the generalization ability of backbone network, the model learns to extract some inter-camera invariant features, in which situation the inter-camera loss find and exploit the inter-camera matching pair to further optimize the model. With the supervision of the two losses, the model gets maturer and maturer during training time.

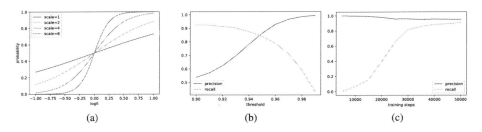

(a)   (b)   (c)

Fig. 3. (a) Sigmoid function with different scales. The optimal scale is 4 where the input and output domain are all covered with a good gradient. **(b)** Precision and recall of matching pairs on different threshold. The choosing of threshold is a trade-off between precision and recall of matching pairs. **(c)** Dynamics of self-discovered inter-camera identity matching pairs during model training on the Market-1501 dataset.

6 Conclusion

In this work, we propose a novel method to perform intra-camera supervised called *Tracklet Level Classifier*. The key idea of TLC is training classifiers for every intra-camera ID. Our TLC method is simple yet effective, which simplify previous works' two stage training strategy to one stage and do not rely on complex matching rules compared with them. Theoretical analysis and ablation studies prove our superiority to existing Camera Level Classifier methods. Extensive experiments show that our method achieves state-of-the-art among other intra-camera supervised methods and comparable to fully supervised methods.

References

1. Gray, D., Tao, H.: Viewpoint invariant pedestrian recognition with an ensemble of localized features. In: Forsyth, D., Torr, P., Zisserman, A. (eds.) ECCV 2008. LNCS, vol. 5302, pp. 262–275. Springer, Heidelberg (2008). https://doi.org/10.1007/978-3-540-88682-2_21
2. Prosser, B.J., Zheng, W.S., Gong, S., Xiang, T., Mary, Q.: Person re-identification by support vector ranking. In: BMVC, **2**, p. 6 (2010)
3. Zheng, W.S., Gong, S., Xiang, T.: Reidentification by relative distance comparison. IEEE Trans. Pattern Anal. Mach. Intell. **35**(3), 653–668 (2012)
4. Liao, S., Hu, Y., Zhu, X., Li, S.Z.: Person re-identification by local maximal occurrence representation and metric learning. In: Proceedings of the IEEE Conference on Computer Vision and Pattern Recognition, pp. 2197–2206 (2015)
5. Zhu, X., Bhuiyan, A., Mekhalfi, M.L., Murino, V.: Exploiting Gaussian mixture importance for person re-identification. In: 2017 14th IEEE International Conference on Advanced Video and Signal Based Surveillance (AVSS), pp. 1–6. IEEE (2017)
6. He, K., Zhang, X., Ren, S., Sun, J.: Deep residual learning for image recognition. In: Proceedings of the IEEE Conference on Computer Vision and Pattern Recognition, pp. 770–778 (2016)
7. Szegedy, C., Ioffe, S., Vanhoucke, V., Alemi, A.A.: Inception-v4, inception-resnet and the impact of residual connections on learning. In: Proceedings of the Thirty-First AAAI Conference on Artificial Intelligence, pp. 4278–4284 (2017)
8. Zheng, F., et al.: Pyramidal person re-identification via multi-loss dynamic training. In: Proceedings of the IEEE Conference on Computer Vision and Pattern Recognition, pp. 8514–8522 (2019)
9. Wang, G., Lai, J., Huang, P., Xie, X.: Spatial-temporal person re-identification. Proceedings of the AAAI Conference on Artificial Intelligence. **33**, pp. 8933–8940 (2019)
10. Zhang, Z., Lan, C., Zeng, W., Chen, Z.: Densely semantically aligned person re-identification. In: Proceedings of the IEEE Conference on Computer Vision and Pattern Recognition, pp. 667–676 (2019)
11. Zheng, M., Karanam, S., Wu, Z., Radke, R.J.: Re-identification with consistent attentive siamese networks. In: Proceedings of the IEEE Conference on Computer Vision and Pattern Recognition, pp. 5735–5744 (2019)
12. Yu, H.X., Zheng, W.S., Wu, A., Guo, X., Gong, S., Lai, J.H.: Unsupervised person re-identification by soft multilabel learning. In: Proceedings of the IEEE Conference on Computer Vision and Pattern Recognition, pp. 2148–2157 (2019)
13. Zhong, Z., Zheng, L., Luo, Z., Li, S., Yang, Y.: Invariance matters: exemplar memory for domain adaptive person re-identification. In: Proceedings of the IEEE Conference on Computer Vision and Pattern Recognition, pp. 598–607 (2019)

14. Fu, Y., Wei, Y., Wang, G., Zhou, Y., Shi, H., Huang, T.S.: Self-similarity grouping: a simple unsupervised cross domain adaptation approach for person re-identification. In: Proceedings of the IEEE International Conference on Computer Vision, pp. 6112–6121 (2019)
15. Lin, Y., Dong, X., Zheng, L., Yan, Y., Yang, Y.: A bottom-up clustering approach to unsupervised person re-identification. Proceedings of the AAAI Conference on Artificial Intelligence. 33, pp. 8738–8745 (2019)
16. Wu, G., Zhu, X., Gong, S.: Tracklet self-supervised learning for unsupervised person re-identification. In: AAAI Conference on Artificial Intelligence, pp. 12362–12369 (2020)
17. Xin, X., Wang, J., Xie, R., Zhou, S., Huang, W., Zheng, N.: Semi-supervised person re-identification using multi-view clustering. Pattern Recogn. 88, 285–297 (2019)
18. Li, M., Zhu, X., Gong, S.: Unsupervised person re-identification by deep learning tracklet association. In: Proceedings of the European conference on computer vision (ECCV), pp. 737–753 (2018)
19. Li, M., Zhu, X., Gong, S.: Unsupervised tracklet person re-identification. IEEE Trans. Pattern Anal. Mach. Intell. 42(7), 1770–1782 (2020)
20. Zhu, X., Zhu, X., Li, M., Murino, V., Gong, S.: Intra-camera supervised person re-identification: a new benchmark. In: Proceedings of the IEEE International Conference on Computer Vision Workshops, pp. 1079–1087 (2019)
21. Wu, J., Yang, Y., Liu, H., Liao, S., Lei, Z., Li, S.Z.: Unsupervised graph association for person re-identification. In: Proceedings of the IEEE International Conference on Computer Vision, pp. 8321–8330 (2019)
22. Bertinetto, L., Valmadre, J., Henriques, J.F., Vedaldi, A., Torr, P.H.S.: Fully-convolutional siamese networks for object tracking. In: Hua, G., Jégou, H. (eds.) ECCV 2016. LNCS, vol. 9914, pp. 850–865. Springer, Cham (2016). https://doi.org/10.1007/978-3-319-48881-3_56
23. Guo, Q., Feng, W., Zhou, C., Huang, R., Wan, L., Wang, S.: Learning dynamic siamese network for visual object tracking. In: Proceedings of the IEEE International Conference on Computer Vision, pp. 1763–1771 (2017)
24. Nam, H., Han, B.: Learning multi-domain convolutional neural networks for visual tracking. In: Proceedings of the IEEE conference on computer vision and pattern recognition, pp. 4293–4302 (2016)
25. Wang, L., Ouyang, W., Wang, X., Lu, H.: STCT: Sequentially training convolutional networks for visual tracking. In: Proceedings of the IEEE Conference on Computer Vision and Pattern Recognition, pp. 1373–1381 (2016)
26. Nam, J., Kim, J., Loza Mencía, E., Gurevych, I., Fürnkranz, J.: Large-scale multi-label text classification—revisiting neural networks. In: Calders, T., Esposito, F., Hüllermeier, E., Meo, R. (eds.) ECML PKDD 2014. LNCS (LNAI), vol. 8725, pp. 437–452. Springer, Heidelberg (2014). https://doi.org/10.1007/978-3-662-44851-9_28
27. Babbar, R., Schölkopf, B.: Dismec: distributed sparse machines for extreme multi-label classification. In: Proceedings of the Tenth ACM International Conference on Web Search and Data Mining, pp. 721–729 (2017)
28. Hou, R., Ma, B., Chang, H., Gu, X., Shan, S., Chen, X.: Interaction-and-aggregation network for person re-identification. In: Proceedings of the IEEE Conference on Computer Vision and Pattern Recognition, pp. 9317–9326 (2019)
29. Chen, T., et al.: ABD-Net: attentive but diverse person re-identification. In: Proceedings of the IEEE International Conference on Computer Vision, pp. 8351–8361 (2019)
30. Wang, H., Zhu, X., Xiang, T., Gong, S.: Towards unsupervised open-set person re-identification. In: 2016 IEEE International Conference on Image Processing (ICIP), pp. 769–773. IEEE (2016)
31. Wu, J., et al.: Clustering and dynamic sampling based unsupervised domain adaptation for person re-identification. In: 2019 IEEE International Conference on Multimedia and Expo (ICME), pp. 886–891. IEEE (2019)

32. Zheng, L., Shen, L., Tian, L., Wang, S., Wang, J., Tian, Q.: Scalable person re-identification: a benchmark. In: Proceedings of the IEEE International Conference on Computer Vision, pp. 1116–1124 (2015)
33. Zheng, Z., Zheng, L., Yang, Y.: Unlabeled samples generated by GAN improve the person re-identification baseline in vitro. In: Proceedings of the IEEE International Conference on Computer Vision, pp. 3754–3762 (2017)
34. Ristani, E., Solera, F., Zou, R., Cucchiara, R., Tomasi, C.: Performance measures and a data set for multi-target, multi-camera tracking. In: Hua, G., Jégou, H. (eds.) ECCV 2016. LNCS, vol. 9914, pp. 17–35. Springer, Cham (2016). https://doi.org/10.1007/978-3-319-48881-3_2
35. Wei, L., Zhang, S., Gao, W., Tian, Q.: Person transfer GAN to bridge domain gap for person re-identification. In: Proceedings of the IEEE Conference on Computer Vision and Pattern Recognition, pp. 79–88 (2018)
36. Zheng, Z., Yang, X., Yu, Z., Zheng, L., Yang, Y., Kautz, J.: Joint discriminative and generative learning for person re-identification. In: Proceedings of the IEEE Conference on Computer Vision and Pattern Recognition, pp. 2138–2147 (2019)

Multi-level Prediction with Graphical Model for Human Pose Estimation

Yilei Chen, Xuemei Xie[✉], Lihua Ma, Jiang Du, and Guangming Shi

School of Artificial Intelligence, Xidian University, Xi'an 710071, China
xmxie@mail.xidian.edu.cn

Abstract. More and more complex Deep Neural Networks (DNNs) are designed for the improvement of human pose estimation task. However, it is still hard to handle the inherent ambiguities due to diversity of postures and occlusions. And it is difficult to meet the requirements for the high accuracy of human pose estimation in practical applications. In this paper, reasoning-based multi-level predictions with graphical model for single person human pose estimation is proposed to obtain the accurate location of body joints. Specifically, a multi-level prediction using cascaded network is designed with recursive prediction according to three different levels from easy to hard joints. At each stage, multi-scale fusion and channel-wise feature enhancement are employed for stronger contextual information to improve capacity of feature extraction. Heatmaps with rich spatial and semantic information are refined by explicitly constructing graphical model to learn the structure information for inference, which can implement the interactions between joints. The proposed method is evaluated on LSP dataset. The experiments show that it can achieve highly accurate results and outperform state-of-the-art methods.

Keywords: Pose estimation · Multi-scale fusion · Multi-level prediction · Graphical model · Information propagation

1 Introduction

In recent years, deep learning techniques have made significant progress and successfully tackled classic computer vision problems. As one of the most challenging fundamental tasks in computer vision, human pose estimation can be applied to many important tasks such as action recognition, human tracking and Human-Computer Interaction (HCI). Due to the diverse variations in body postures, clothing appearance and occlusion, results of many human pose estimation methods are still far from state-of-the-art performance.

Existing approaches can be classified into two categories: regression based and detection based. Regression based methods [1–3] generally try to map the input image to the output joints and directly produce joint coordinates. Detection based methods [4–12] intend to generate a likelihood heatmap for each joint,

This work was supported in part by the National Natural Science Foundation of China (No. 61836008, 61632019).
The first author is a student.

Y. Peng et al. (Eds.): PRCV 2020, LNCS 12306, pp. 343–355, 2020.
https://doi.org/10.1007/978-3-030-60639-8_29

and locate the joint as the keypoint by computing the argmax of pixel values in the heatmap. Conventionally, human pose estimation methods based on DNNs are mainly performed by multi-stage architectures [10–12]. These networks produce predictions in stages to output a refined pose without deeply mining the difficulty levels for joints. For example, occluded or unclear joints can distort predictions of the more clear joints. In order to make the more certain joints positively influence the less certain ones, we propose a cascade Multi-level Prediction Network (MPN), whose predictions are refined by a grphical model. In MPN, with the initial prediction generated in the first stage, the subsequent stages output refined predictions sequentially. We start with the most reliable joints, go through subsequent stages to predict the less reliable joints. Our proposed multi-stage network which is trained in an end-to-end manner, defines difficulty levels of different joints explicitly, and provides reasoning priors for less reliable joints implicitly.

Moreover, a central problem is that the per-joint is evaluated independently and the internal structures of the pose should not be ignored. In other words, the joint connections are not well exploited. To make use underlying spatial structure of human pose, we utilize Graph Neural Network (GNN) built on a graphical structure to learn the joint dependences. On the predefined graph, each node is associated with its neighboring joints. Spatial relations is captured through edge construction. Message aggregation and updation among nodes and edges contributes to the final precise prediction. Our method considers not only the joint location but also the spatial correlation of joints, and it is also end-to-end trainable. We show that our approach produces competitive results on the standard pose estimation benchmarks: Leeds Sports Pose (LSP) dataset [24].

2   Related Work

2.1   Single-Person Pose Estimation

The problem of single-person pose estimation aims at recognizing and localizing the joints in the images. Benefiting from the development of the DNNs based methods, significant progress has been made in the field. The general approach tackling this problem use DNNs to learn feature representations for obtaining score maps or the locations of joints. Some methods [1–3] directly employ deep features to regress joint positions. Only using joints coordinate lacks robustness, in order to provide more supervision information, Tompson et al. [4] first employ heatmap to indicate the ground truth. Each heatmap channel which occupied with a 2D Gaussian distribution is correspond to a joint, and is centered at the actual joint location. Moreover, Papandreous et al. [5] propose an improved representation of the joint location, which is a combination of binary activation heatmap and corresponding offset. Heatmap has since come overwhelmingly dominant amongst pose estimation models, and most of the recent research is based on heatmap representation.

It is essential to facilitate an excellent network architecture and to make better use of the input information. Rafi et al. [6] design a network with multi-scale inputs based on GoogleNet [7]. Xiao et al. [8] add deconvolutional layers to ResNet [9]. In terms of multi-stage style for human pose refinement, some work [10–12] get predictions from coarse to fine through a cascade architecture via intermediate supervision. Especially, Hourglass uses a residual module as the component unit, processing repetitive down-sampling and up-sampling to exploit multi-scale features. Sun et al. [13] propose a novel High-Resolution Network (HRNet) with repeated multi-scale feature fusions to keep high-resolution representation of features across the whole network, reduce the quantization error caused by down-sampling and get higher accuracy.

Different from above which attempt to fit detected body joints into models, Gkioxari et al. [16] propose a convolutional Recurrent Network in a fixed order to output joint location one by one following a chain model. The output of each step depends on both the input image and the previously output. Lifshitz et al. [17] also define a sequential prediction order. They both allow for the easy cases to be processed first while the harder cases are processed last, and as a result use the contextual information from the joints predicted before them.

Intermediate supervision and recursive prediction have proven their significant advantages among recent methods. Meantime, Hourglass Module [11] which process multi-scale feature fusions and residual learning with repetitive down-sampling and up-sampling, has achieved better performance. In this work, we leverage advantages of these methods, and define a difficulty levels for sequential prediction on account of the degree of certainty. We propose a Multi-level Prediction Network with ResNet-50 and Hourglass Module as sub-networks to capture features across different scales and learn the spatial configuration implicitly.

2.2 Graphical Models and Inference

Excellent network architecture shall produce robust feature representation, but the constrains of human pose estimation are not involved as relation priors. In most previous work, to model joint relations, the pictorial structures are used to define the deformable configurations by spring-like connections between pairs of joints. Tompson et al. [4] create a fully connected graph on body parts and perform an approximate Markov Random Field (MRF) over the spatial relations. However, for datasets cover a large range of the possible poses, the distribution of spatial locations might be less effective. Yang et al. [15] explicitly incorporate human pose priors including body part mixture types and standard quadratic deformation constrains into models with a tree structure or a loopy structure. Lifshitz et al. [17] discretize the image into log-polar bins centered around the augmented 30 joints and use the VGG-based network to get a probability distribution over the log-polar bins, indicating the relative joint location. Ning et al. [18] inject the external knowledge representation into a fractal network by a learned projection matrix to guide network training process. Chu et al. [19] propose the a bi-directional tree structured model through adopting geometrical transform kernels to capture the spatial relationships of joints from feature maps. Zhang et al. [22] propose Pose Graph Neural Network (PGNN) to learn a structured representation of human pose as a graph. Direct message passing between different joints is enabled and spatial relation is captured.

To better estimate the human pose, network architecture design with large parameters are expected on account of the articulated structure of the human body. In our work, we use a joint detector combined with a graphical model to infer the precise location. By compensating for the large variation of the poses, we empirically learning the spatial configuration.

3   Proposed Method

The overall architecture is illustrated in Fig. 1. It includes two steps: Multi-level Pose Network (MPN) for heatmap detection and Pose Inference using Graphical model.

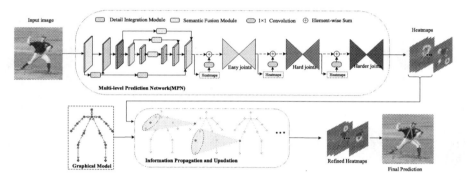

Fig. 1. The proposed network architecture, consisting of two parts: Multi-level Prediction Network (MPN) for Heatmap Detection and Pose Inference using Graphical Model.

3.1   Cascade Multi-level Pose Network

MPN is constructed by a multi-stage style where the previous stage serves as a prior for the output of heatmaps on the next stage. Considering the reliability of different joints, MPN predicts all the joints from the most reliable joints to the less certain ones, and thus defines three levels from easy joints to hard joints. Each stage focus on a subset of joints.

Multi-scale Feature Fusion Module. Previous methods [11,12] have proved that the use of multi-scale information is crucial for feature learning. As we all know, features with larger scales are benefit for localization, and features with lower scales are discriminative for classification. We aim to extract more representative features through down-sampling process. In order to localize, some work use a U-shape architecture, larger scale features from shallower layers are down-sampled. Therefore, the lost information can hardly be recovered in the up-sampling or deconvolution procedure. Skip connections are used for element-wise summation or concatenation for feature fusion.

In this work, we learn from the convolutional attention mechanism, and design Semantic Fusion Module (SFM) and Detail Integration Module (DIM) for enhancement of discrimination and localization respectively. As shown in Fig. 2, we adopt ResNet-50 as the basic module for feature extraction. 1×1 convolutional operation and Rectified Linear Unit (ReLU) are applied to make high-resolution features and low-resolution features have consistent channels. The feature maps in lower scales are up-sampled by bilinear interpolation in SFM to obtain the higher semantics of lower-resolution features. The feature maps in higher scales are down-sampled by max-pooling to complement details of higher-resolution features. SFM and DIM both utilize element-wise sum operation for the final feature fusions.

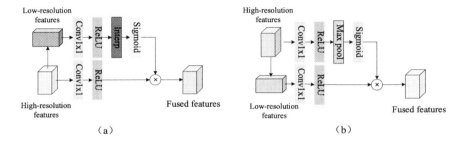

Fig. 2. Two modules for multi-scale feature fusions: (a) Semantic Fusion Module. (b) Detail Integration Module.

Multi-level Pose Prediction. Simply increasing the feature representation by feature fusions is not novel. Multi-stage methods aim to iteratively refine the predictions and produce increasingly effective results. An issue arises that the accuracies are unbalanced. For example, the head is always with high accuracy, the ankle is always with low accuracy. To avoid the negative influences, let some joints that are easier to locate predict in the former stage, some joints that are more difficult to locate predict in the latter stage.

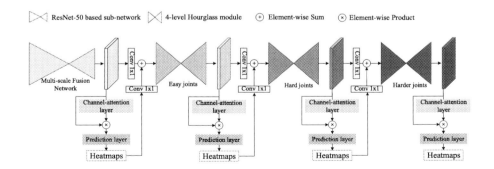

Fig. 3. Multi-level pose prediction network.

Based on the accuracies of all joints, this work defines difficulty levels of different joints. At each stage, use the previous predicted joints as a prior to implicitly predict the other joints. Divide the joints into three levels: easy joints (head, neck), hard joints (shoulders, hips), harder joints (elbows, wrists, knees, ankles). The multi-scale fusion network has a so-called effective feature representation and output heatmaps of all joints. Then a cascade three-stage module get sequentially predictions starting from easy joints to hard joints secondly, and to the harder joints finally (see Fig. 3). Each module consists of a hourglass module and 1×1 convolutional operation for feature fusions and intermediate supervision. Based on channel attention mechanism [23], we employ global average pooling operation to perform channel-wise calculation of features which denotes the channel-wise importance. The importance measurement is followed by normalization, and it is the weights for all channels which should be multiplied by original features. The channel-wise attention layers without parameters can help enhance discriminability and alleviate redundant interference.

3.2 Inference Mechanism Based on Graphical Model

We perform the inference using the predefined graphical model through GNN to learn structured information. Each joint serves as a node, and each edge represents the local contextual relations between joints. With the guidance of graphical model and information aggregating from the associated joints, the contextual structure information is propagated to refine each joint at steps.

Construction of Graphical Model. Graphical model can be constructed to model relations on different joints of human skeleton. Thus, the human pose can be expressed as a graphical model with joints as nodes and joint connections as edges. Our work utilizes graphical model to explicitly represent the spatial relationships among joints. GNN is a general neural network architecture defined according to a graph structure $G = (V, E)$ where V is the set of K nodes and E are edges. As for the pose estimation task, each node represents a body joint and each edge denotes the relationships of neighboring joints. Besides, the nodes can be constructed with arbitrary topologies to make the internal structure be well exploited.

In general, the fully connected graphical model that connect each joint to every other joint seems to be the ideal choice. However, due to the flexibility of an articulated human body, not all body joints are related to each other. For example, the head may provide little information to locate the right ankle. While the joint inter-connectivity can provide reliable guidance of locating an ambiguous joints, the connection of those unrelated or weakly related joints shall no longer transfer effective features, or even transfer negative influence. To overcome the issue and simplify the graphical model, we utilize the tree graphical model. As shown in Fig. 1, tree structure is a natural skeleton structure with physical connections in human body. Thus, we argue that simple tree graphical model is powerful enough for inference, we also adopt the loopy graphical model to denote body structure. Based on the tree structure, the loopy graphical model adds edges with shoulder and wrist, hip and ankle, which may give more mutual information.

It should be noted that the graphical model is undirected that allows bi-directional influence for the inference mechanism. GNNs map the tree graphical model to outputs via information aggregation and propagation. Each node is represented by the heatmap learned from the joint detector. Each edge represents priors implied in the graphical model. For node k $(k \in k)$, the heatmap feature as the initial hidden state h_k^0:

$$h_k^0 = F_k(\Theta, I),\tag{1}$$

where F is the network form heatmap detection, Θ is the learnable parameter matrix of network, and I represents the input image.

Information Propagation in Neural Network. The predefined graphical model contains high-level semantics and provides important and explicit information on spatial locations. A graph based convolutional propagation can be applied to node k $(k \in k)$ via two steps at the propagation time t. First, node representations are transformed by a learnable parameter matrix which represent as the message passing function $M_t(\cdot)$. Second, these transformed node representations are gathered to node k from its neighboring nodes j $(j \in N(k))$ followed by a non-linear function, which is called node updation function $U_t(\cdot)$. The inference mechanism enforce the information propagation for the final feature representation on the graph.

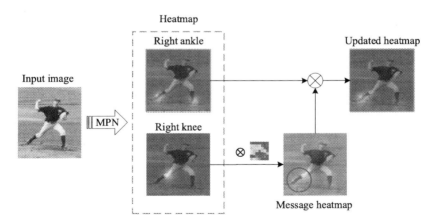

Fig. 4. Example of information propagation and updation mechanism from right knee to right ankle.

Each node collects information from its unary terms. As shown in Fig. 1, target node k with hidden state h_k^t, and one of its neighboring nodes j with hidden state h_j^t, the information propagated from its neighbors can be calculated as:

$$x_k^t = M_t(h_k^t, h_j^t | j \in N(k)). \tag{2}$$

After information collection, each node update its hidden state with node update function $U_t(\cdot)$. For target node k, $U_t(\cdot)$ can be expressed as:

$$h_k^t = U_t(h_k^{t-1}, x_k^t). \tag{3}$$

To reduce the negative influence by relevant nodes, this paper retains the hidden state of target node. Different convolution kernels have lead to different directions on the same feature map. Message can be passed between feature maps through the geometrical transform kernels. Hence, using convolutional operation to learn different kernels and sum up for different nodes. The message passing function $M_t(\cdot)$ is expressed as:

$$x_k^t = \sum_{j \in N(k)} W_{p,k} h_j^t + b_{p,k}, \tag{4}$$

where $W_{p,k}$ is the weights and $b_{p,k}$ is the bias term of node k for different convolution layers p.

The size of convolution kernels should also consider the joint displacement. To effectively integrate information and cover large joint displacement, we employ successive convolutions with 3×3 kernels. One is that the consistent 3×3 convolutions can approximate and enlarge receptive field. On the other hand, each convolution is followed by a non-linear layer which can propagate more effective information and reduce the number of parameters. After summing up different neighboring information, the transform kernels provide heatmap representations which is approximate to the exact distribution of target node. We only use element-wise product for node updation to refine the original state. The transformed kernels and the effect of element-wise product can be seen in Fig. 4.

4   Experiments

4.1   Dataset and Evaluation Metrics

Our approach is trained on LSP [24] training set and validated on the LSP dataset. The LSP dataset consists of 11k training images and 1k testing images from sports activities.

For evaluation, Percentage of Correct Keypoints (PCK) metric is used. An estimated joint is considered correct if its distance from ground truth is less than a fraction of the torso length.

4.2   Training Details

Data Augmentation. We train our models by adding the MPII [25] training dataset to the LSP dataset. During training, we crop the images with the target human centered at the images with roughly the same scale, and warp the image patch to the size 256×256. Then we randomly rotate ($\pm 30°$) and flip the images. We also perform random rescaling (0.75 to 1.25) to make the model more robust to scale change. During testing, we simply use the image size as the rough scale, and the image center as the rough position of the target human to crop the image patches for the LSP dataset.

Configurations. We train our model with Pytorch using the initial learning rate of 2.5×10^{-4}. The parameters are optimized by RMSprop algorithm. We train the model on the LSP dataset for 90 epochs and use mini-batches of size 16. The learning rate is decreased by 10 times of initial value after 60 epochs. We adopt the Mean-Squared Error (MSE) based loss function which calculate each joint predicted score map with its ground truth map. The heatmap outputs are all supervised via a MSE loss.

4.3   Results and Analyses

Table 1 gives a summary of results on the LSP dataset. We investigate our network with different network architectures that considering priors on spatial correlations of body joints. Our best performance is achieved by joint training the propose MPN and tree graphical model. As reported in Table 1, the proposed multi-stage network with difficulty levels get a 95.2% PCK, which is a 1.2% improvement compared to [22]. Next, the inference with graphical model, which provides explicit information and directions, further improves the perfomance by 0.4% PCK.

Table 1. Comparisons of PCK@0.2 score on the LSP dataset.

Method	Head	Sho.	Elbo.	Wri.	Hip	Knee	Ank.	Mean
Chu et al. [19]	-	-	-	-	-	-	-	80.8
Wei et al. [15]	-	-	-	-	-	-	-	81.1
Chu et al. [20]	-	-	-	-	-	-	-	83.1
Lifshitz et al. [17]	96.8	89.0	82.7	79.1	90.9	86.0	82.5	86.7
Chu et al. [21]	98.1	93.7	89.3	86.9	93.4	94.0	92.5	92.6
Ning et al. [18]	98.2	94.4	91.8	89.3	94.7	95.0	93.5	93.9
Zhang et al. [22]	98.4	94.8	92.0	89.4	94.4	94.8	93.8	94.0
Proposed MPN	98.5	95.3	94.0	92.3	**94.9**	96.5	94.6	95.2
Proposed MPN-GNN	**98.6**	**96.6**	**94.7**	**92.8**	94.8	**96.9**	**94.8**	**95.6**

Table 2. Comparisons of proposed MPN of PCK@0.2 score on the LSP dataset.

Method	Head	Sho.	Elbo.	Wri.	Hip	Knee	Ank.	Mean
Belagiannis et al. [26]	95.2	89.0	81.5	77.0	83.7	87.0	82.8	85.2
Pishchulin et al. [27]	97.0	91.0	83.8	78.1	91.0	86.7	82.0	87.1
Insanfutdinov et al. [28]	97.4	92.7	87.5	84.8	91.5	89.9	87.2	90.1
Wei et al. [10]	97.8	92.5	87.0	83.9	91.5	90.8	89.9	90.5
Bulat et al. [29]	97.2	92.1	88.1	85.2	92.2	91.4	88.7	90.7
Yang et al. [12]	98.3	94.5	92.2	88.9	94.4	95.0	93.7	93.9
Proposed MPN(w/o squential prediction)	98.3	94.7	92.8	89.1	94.8	95.3	93.5	94.1
Proposed MPN(w/o channel-wise attention)	**98.5**	94.9	93.5	91.9	**95.0**	96.1	94.4	94.9
Proposed MPN	98.5	**95.3**	**94.0**	**92.3**	94.9	**96.5**	**94.6**	**95.2**

Multi-level Prediction Network. We first evaluate the proposed MPN for single-person pose estimation on LSP dataset. Table 2 reports the ablation study on multi-level prediction, multi-scale feature fusions and channel-wise attention. The multi-scale fusions and channel-wise attention mechanism help to retain and enhance the spatial contextual information. Compared to the state-of-the-art performance, our network without sequential prediction has achieved 0.2% improvement. Even with the multi-level guidance and an channel-wise attention layer for cross stage aggregation, certain priors are effectively helpful for positive predictions which gives a improvement by 0.8% and 1.1% respectively.

Type of Graphical Model. Table 3 lists our investigation on the graphical model types. Both loopy model and tree model show more or less improvements on accuracy. The two types have the similar performance which indicates the availability of explicit inference mechanism.

Number of the Message Passing Layers. Table 3 also lists comparations on number of the message passing layers as it controls spread of information propagation. The accuracy increases by a small amount from $T = 1$ to $T = 3$ on both two types of graphical model.

Qualitative Results. Figure 5 shows some pose estimation results on LSP datasets. As seen, the inference mechnism is able to effectively correct false predictions on MPN. It indicates that the graphical model can encode relationships among joints and further propagate useful information. The bottom row of Fig. 5 also shows some failure cases which is marked as a red box. Perhaps the message passing layers give a negative transfer, and the joint is shifted to wrong location.

Table 3. Comparisons of different types of graphical models and message passing layers of PCK@0.2 score on the LSP dataset.

Experiments	Configurations				
Tree	√		√	√	
Loopy		√			
$T = 1$			√		
$T = 2$	√	√			
$T = 3$				√	
Mean	95.51	95.53	95.38	**95.60**	95.20

Fig. 5. Qualitative results on the LSP dataset. 1st row: results from proposed MPN. 2nd row: results from both proposed MPN-GNN.

5 Conclusion

This paper has incorporated the DNNs with explicit priors of difficulty level and relations among joints. Our network leverages the advantages of recursive architectures and the contextual information. Through multi-scale and cross stage feature fusions, the enhanced features are well exploited for sequential prediction from easy joints to hard joints. Then a graphical model is defined for explicit information propagation and inference to refine the results using GNNs. The combination of the two parts obtains state-of-the-art performance on LSP dataset. However, there are also many possible directions for future work. For example, how to incorporate information with more effective human body representation, especially for more diverse datasets like MPII.

References

1. Toshev, A., Szegedy, C.: Deeppose: human pose estimation via deep neural networks. In: IEEE Conference on Computer Vision and Pattern Recognition on Proceedings, Columbus, OH, USA, pp. 1653–1660. IEEE (2014)

2. Krizhevsky, A., Sutskever, I., Hinton, G.E.: Imagenet classification with deep convolutional neural networks. In: Advances in Neural Information Processing Systems 25 on Proceedings, pp. 1097–1105. NIPS, Lake Tahoe, Nevada, USA (2012)

3. Pfister, T., Simonyan, K., Charles, J., Zisserman, A.: Deep convolutional neural networks for efficient pose estimation in gesture videos. In: Cremers, D., Reid, I., Saito, H., Yang, Ming-Hsuan (eds.) ACCV 2014. LNCS, vol. 9003, pp. 538–552. Springer, Cham (2015). https://doi.org/10.1007/978-3-319-16865-4_35

4. Tompson, J.J., Jain, A., LeCun, Y., Bregler, C.: Joint training of a convolutional network and a graphical model for human pose estimation. In: Advances in Neural Information Processing Systems 27 on Proceedings, Montreal, Quebec, Canada, pp. 1799–1807. NIPS (2014)

5. Papandreou, G., et al.: Towards accurate multi-person pose estimation in the wild. In: IEEE Conference on Computer Vision and Pattern Recognition on Proceedings, Honolulu, Hawaii, pp. 4903–4911. IEEE (2017)

6. Rafi, U., Leibe, B., Gall, J., Kostrikov, I.: An efficient convolutional network for human pose estimation. In: British Machine Vision Conference on Proceedings, New York, pp. 109.1–109.11. BMVA Press (2016)

7. Szegedy, C., et al.: Going deeper with convolutions. In: IEEE Conference on Computer Vision and Pattern Recognition on Proceedings, Columbus, OH, USA, pp. 1–9. IEEE (2015)

8. Xiao, B., Wu, H., Wei, Y.: Simple baselines for human pose estimation and tracking. In: Ferrari, V., Hebert, M., Sminchisescu, C., Weiss, Y. (eds.) ECCV 2018. LNCS, vol. 11210, pp. 472–487. Springer, Cham (2018). https://doi.org/10.1007/978-3-030-01231-1_29

9. He, K., Zhang, X., Ren, S., Sun, J.: Deep residual learning for image recognition. In: IEEE Conference on Computer Vision and Pattern Recognition on Proceedings, Las Vegas, NV, USA, pp. 770–778. IEEE (2016)

10. Wei, S.E., Ramakrishna, V., Kanade, T., Sheikh, Y.: Convolutional pose machines. In: IEEE Conference on Computer Vision and Pattern Recognition on Proceedings, Las Vegas, NV, USA, pp. 4724–4732. IEEE (2016)

11. Newell, A., Yang, K., Deng, J.: Stacked hourglass networks for human pose estimation. In: Leibe, B., Matas, J., Sebe, N., Welling, M. (eds.) ECCV 2016. LNCS, vol. 9912, pp. 483–499. Springer, Cham (2016). https://doi.org/10.1007/978-3-319-46484-8_29

12. Yang, W., Li, S., Ouyang, W., Li, H., Wang, X.: Learning feature pyramids for human pose estimation. In: IEEE International Conference on Computer Vision on Proceedings, Venice, Italy, pp. 1281–1290. IEEE (2017)

13. Sun, K., Xiao, B., Liu, D., Wang, J.: Deep high-resolution representation learning for human pose estimation. In: IEEE Conference on Computer Vision and Pattern Recognition on Proceedings, Long Beach, CA, USA, pp. 5686–5696. IEEE (2019)

14. Sapp, B., Taskar, B.: Modec: Multimodal decomposable models for human pose estimation. In: IEEE Conference on Computer Vision and Pattern Recognition on Proceedings, Portland, OR, USA, pp. 3674–3681. IEEE (2013)

15. Yang, W., Ouyang, W., Li, H., Wang, X.: End-to-end learning of deformable mixture of parts and deep convolutional neural networks for human pose estimation. In: IEEE Conference on Computer Vision and Pattern Recognition on Proceedings, Las Vegas, NV, USA, pp. 3073–3082. IEEE (2016)

16. Gkioxari, G., Toshev, A., Jaitly, N.: Chained predictions using convolutional neural networks. In: Leibe, B., Matas, J., Sebe, N., Welling, M. (eds.) ECCV 2016. LNCS, vol. 9908, pp. 728–743. Springer, Cham (2016). https://doi.org/10.1007/978-3-319-46493-0_44

17. Lifshitz, I., Fetaya, E., Ullman, S.: Human pose estimation using deep consensus voting. In: Leibe, B., Matas, J., Sebe, N., Welling, M. (eds.) ECCV 2016. LNCS, vol. 9906, pp. 246–260. Springer, Cham (2016). https://doi.org/10.1007/978-3-319-46475-6_16

18. Ning, G., Zhang, Z., He, Z.: Knowledge-guided deep fractal neural networks for human pose estimation. IEEE Trans. Multimedia **20**(5), 1246–1259 (2017)

19. Chu, X., Ouyang, W., Li, H., Wang, X.: Structured feature learning for pose estimation. In: IEEE Conference on Computer Vision and Pattern Recognition on Proceedings, Las Vegas, NV, USA, pp. 4715–4723. IEEE (2016)

20. Chu, X., Ouyang, W., Li, H., Wang, X.: CRF-CNN: modeling structured information in human pose estimation. In: Advances in Neural Information Processing Systems 29 on Proceedings, Barcelona Spain, pp. 316–324. NIPS, Centre Convencions Internacional Barcelona (2016)

21. Chu, X., Yang, W., Ouyang, W., Ma, C., Yuille, A.L., Wang, X.: Multi-context attention for human pose estimation. In: IEEE Conference on Computer Vision and Pattern Recognition on Proceedings, Honolulu, HI, USA, pp. 5669–5678. IEEE (2017)

22. Zhang, H., et al.: Human pose estimation with spatial contextual information (2019). eprint arXiv:1901.01760

23. Hu, J., Shen, L., Albanie, S., Sun, G., Wu, E.: Squeeze-and-excitation networks. IEEE Trans. Pattern Anal. Mach. Intell. **42**(8), 2011–2023 (2020)

24. Johnson, S., Everingham, M.: Clustered Pose and Nonlinear Appearance Models for Human Pose Estimation. In: British Machine Vision Conference on Proceedings, pp. 1–11. DBLP, Aberystwyth, UK (2010). https://doi.org/10.5244/C.24.12

25. Andriluka, M., Pishchulin, L., Gehler, P., Schiele, B.: 2D human pose estimation: new benchmark and state of the art analysis. In: IEEE Conference on Computer Vision and Pattern Recognition on Proceedings, Columbus, OH, USA, pp. 3686–3693. IEEE (2014)

26. Belagiannis, V., Zisserman, A.: Recurrent human pose estimation. In: 12th IEEE International Conference on Automatic Face and Gesture Recognition on Proceedings, Washington, DC, USA, pp. 468–475. IEEE (2017)

27. Pishchulin, L., et al.: DeepCut: joint subset partition and labeling for multi person pose estimation. In: IEEE Conference on Computer Vision and Pattern Recognition on Proceedings, Las Vegas, NV, USA, pp. 4929–4937. IEEE (2016)

28. Insafutdinov, E., Pishchulin, L., Andres, B., Andriluka, M., Schiele, B.: DeeperCut: a deeper, stronger, and faster multi-person pose estimation model. In: Leibe, B., Matas, J., Sebe, N., Welling, M. (eds.) ECCV 2016. LNCS, vol. 9910, pp. 34–50. Springer, Cham (2016). https://doi.org/10.1007/978-3-319-46466-4_3

29. Bulat, A., Tzimiropoulos, G.: Human pose estimation via convolutional part heatmap regression. In: Leibe, B., Matas, J., Sebe, N., Welling, M. (eds.) ECCV 2016. LNCS, vol. 9911, pp. 717–732. Springer, Cham (2016). https://doi.org/10.1007/978-3-319-46478-7_44

Automatic Classification of Sleep Stages Based on Raw Single-Channel EEG

Kailin Xu[1,2] [ID], Siyu Xia[1(✉)], and Guang Li[2(✉)]

[1] Southeast University, Nanjing 210096, People's Republic of China
xia081@gmail.com
[2] Zhejiang University, Hangzhou 310027, People's Republic of China
guangli@zju.edu.cn

Abstract. Electroencephalogram (EEG) is a common signal for monitoring people's sleep quality. Manual sleep stage classification on EEG is a time-consuming task. In this paper, we design a model for automatic sleep stage classification based on raw single-channel EEG. This model can preserve the information, broaden the network and enlarge the receptive field as much as possible to extract appropriate time invariant features and classify sleep stage well. For the class-imbalanced problem in sleep stage classification, most of the exsisting methods rely on cross entropy loss and adjust model hyperparameters by experience, leading to poor performance. We implement a two-step training algorithm. The first is pre-training the model with the hyperparameters obtained by Bayesian Optimization after rebalancing datasets by over-sampling. The second is using feedback loss in model fine-tuning to reduce the impact of class-imbalanced problem. The loss weights dynamically change with the per-class F1-score which is used as feedback information. We evaluate our method on Fpz-Cz channel from the Sleep-EDF dataset. The overall accuracy, macro F1-score, Cohen's Kappa coefficient are 85.53%, 81.18%, 0.80 respectively, showing our method has better classification performance than the state-of-the-art methods and is an efficient tool for automatic sleep stage classification.

Keywords: Automatic sleep stage classification · Raw single-channel EEG · Deep learning · Feedback loss

1 Introduction

Sleep disorders are common in people and can lead to serious health problems that affect the quality of life [1]. Monitoring people's sleep quality has important implications for medical research and practice [2].

A polysomnography (PSG) records the physiological signals of a subject during sleep at night, which is composed by multiple signals such as electroencephalogram (EEG), electrocardiogram (ECG), electrooculogram (EOG),

---

The first author is a student.

© Springer Nature Switzerland AG 2020
Y. Peng et al. (Eds.): PRCV 2020, LNCS 12306, pp. 356–368, 2020.
https://doi.org/10.1007/978-3-030-60639-8_30

and electromyography (EMG) [3]. According to American Academy of Sleep Medicine (AASM) [4], sleep stages can be divided into wake (W), three non-rapid eye movement (NREM) stages (N1-N3), and rapid eye movement (REM). And stage N3 (also called Slow Wave Sleep) is divided into two distinct stages, N3 and N4 in Rechtschaffen and Kales (R&K) [5]. Most PSG recordings last at least eight hours. For sleep experts, manual sleep stage classification in such a long signal is a tedious task and highly dependent on the appropriate inter-rater agreement. Therefore, it is important to classify sleep stage automatically, which can avoid the human subjective bias in classification.

Automated sleep stage classification algorithms can be divided into two categories: the hand-engineered feature-based methods and the automated feature extraction-based methods. For the first category, methods extract features such as time, frequency and time-frequency domain features [6–10] for training. Because these methods only extracts features, they may lose most of the original information. As a result, these methods do not generalize well, especially given the nature of PSG recordings, where variability effects are caused by a number of factors, including patient and hardware differences, etc. For the second category, methods learn directly from the raw data, which may solve the limitation in handcrafted feature extraction. Recently, because some neural networks can be trained and optimized end-to-end, they are used both as feature extractors and classifiers. For example, [11] build model with stacked sparse autoencoders, [12] build model with convolutional neural networks. [13] build model with convolutional neural network and bidirectional recurrent neural network.

Considering the number of channels for neural network's input, we use single EEG channel which is cheap and ensures the subjects' sleep does not be affected by the instruments. In the methods investigated, the accuracy of sleep stage classification is not high for the method based on raw single-channel signal, especially for the Sleep-EDF dataset. Besides, the macro F1-score (MF1) and Cohen's Kappa coefficient (κ) do not exceed 0.8 in most studys [11–14], due to the serious imbalance of the dataset and the authors might do not pay more attention to the process of model optimization and use the loss which is not particularly suitable for imbalanced datasets.

At present, there is little methods using Bayesian Optimization [15] in sleep stage classification. However, Bayesian Optimization can find the optimal hyper-parameter according to the previous hyperparameter adjustment results, which has been successfully applied in the machine learning methods [16–19]. In addition, using the idea of back propagation in network training as reference, we propose feedback loss by employing the per-class F1-score of the model as feedback information to adjust the penalty weight of the loss function constantly.

Based on the two points above, we design a deep learning model with a two-step training method, including pre-training network with hyperparameters found by Bayesian Optimization and fine-tuning with feedback loss. This model can preserve the information, increase the number of network channels and enlarge the receptive field to the utmost extent while extracting time invariant features and being trained to classify the sleep stage. Through experiments, the two-step training method can effectively train our model end-to-end through back propagation, and feedback loss can make MF1 and κ exceed 0.8, decreasing bias towards the majority class caused by imbalanced datasets. Moreover, the model can be automatically trained without any hand-engineered features. It is important to note that feedback loss can be used in multiple fields where imbalanced dataset exits, besides sleep stage classification.

2 Methods

2.1 Model Architecture

The architecture of our model (see Fig. 1) consists of ten convolutional layers and two fully-connected layers. Small kernels of size 3×1 are used in every convolutional layer. In view of the rapid fluctuation characteristic of EEG signal, small convolution kernels can extract the subtle information, and realize the suppression of noise through the convolutional layer combination, because the bigger kernels such as 5×1 and 7×1 can be replaced by the combination of small kernels [20]. Moreover, the use of small convolution kernels can reduce the parameters and increase the nonlinearity of model while ensuring the receptive

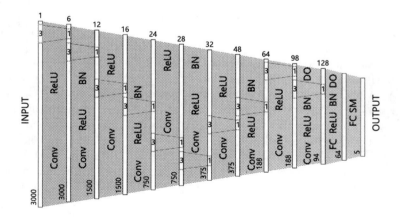

Fig. 1. Architecture of our model. Conv, Relu, BN, DO, FC and SM mean convolutional layer, rectified linear unit, batch normalization, dropout, fully-connected layer and softmax severally. The use order of the units in each layer is from bottom to top. The number at the bottom represents feature size, and the number at the top represents the number of channels.

field. Every two adjacent convolutional layers are a block, and perform five operations sequentially: convolution, applying the rectified linear unit (ReLU) [21] activation, convolution, applying ReLU activation, batch normalization [22]. The first convolution applies stride 1 and padding to ensure the input feature has the same size compared with output feature when the network width increases. The second convolution applies stride 2 and padding to take place of pooling [23]. In a block, using convolutional layers with stride 1 and stride 2 together can effectively increase the receptive field and broaden the network without losing much information. The adoption of batch normalization and ReLU activation can accelerate model convergence and prevent vanishing gradient. A regularization technique named dropout [24] is used to alleviate overfitting problems and will be removed from the model during testing to provide certain outputs.

2.2   Pre-training

In pre-training, the first step is replicating the minority class samples in the original training set until all sleep stages have the same number of samples to reduce the harm of skewed distribution. Then we train model on balanced dataset using a gradient-based optimizer called Adam [25] with hyperparameters found by Bayesian Optimization. Bayesian Optimization is a method for hyperparameter tuning, which searches for the optimal hyperparameter based on the previous results, leading to better performance than human expert-level for many algorithms [15]. In our algorithm, this method is used to adjust the hyperparameters of two dropout layers which means the rate at randomly dropping units along with their connections from the neural network during training. This hyperparameter is very important. If the hyperparameter is too large, only a small part of the model is left in the training process. If the hyperparameter is too small, the regularization effect is not obvious, and the model is still serious overfitting.

2.3   Fine-Tuning with Feedback Loss

In sleep stage classification, the class-imbalanced problem is serious but most methods do not choose to improve the loss function for solving it, resulting in low values of MF1 and κ. For methods test on Sleep-EDF dataset, the values mostly lower than 0.8 [11–14]. As a result, we specifically design a new algorithm to calculate loss named feedback loss, based on the thought of back propagation. The following is a detailed description (see Algorithm 1).

Algorithm 1. Fine-Tuning with Feedback Loss

Input: *model, data, target, multiplier*
Output: *model*
  for $i = 1$ to n **do**
    $output = model(data)$
    $fscore, mean, standard\_deviation = compute(output, target)$
    $fscore = (fscore - mean)/standard\_deviation$
    $weight = \max(-fscore + 1, multiplier \times fscore + 1)$
    $FeedbackLoss = CrossEntropyLoss(weight)$
    for $j = 1$ to n_below **do**
      $output = model(data)$
      $loss = FeedbackLoss(output, target)$
      $model = get\_model(model, loss)$
    end for
  end for
  return *model*

For the convenience, we introduce the process for one class that is identical to the processes for other classes. (a) Put *data* into *model* to obtain *output*, then compare the *output* and *target* by *compute* to obtain F1-score *fscore* for the class, and the *mean* and *standard deviation* for all per-class F1-scores; (b) Centralize *fscore* with zero-mean by the equation as follows:

$$fs = \frac{fs - m_f}{s_f} \tag{1}$$

where *fs*, m_f and s_f are the *fscore*, *mean* and *standard deviation* respectively; (c) Obtain *weight* by the equation as follows:

$$w = \max\left(-fs + 1, k_f fs + 1\right) \tag{2}$$

where w is *weight* of the class. k_f is *multiplier* of the F1-score, and the value lies between -0.25 and 0; (d) Reload *weight* as penalty weight on *CrossEntropyLoss* to obtain *FeedbackLoss*; (e) Use *FeedbackLoss* to obtain *loss* with *output* and *target* and optimize model parameters by *get_model* with *loss* for *n_below* epochs; (f) Return *model* after running step a,b,c,d,e for n times.

Feedback loss has the following characteristics. The first is selecting F1-score as feedback information. In imbalanced dataset, the accuracies of minority classes is easy to be affected by majority classes, resulting in there is not much difference between the accuracies of minority classes and majority classes. For recall and precision, both of them are important and an increase in one leads to a decrease in the other. As a result, F1-score is selected, because it is a combination of recall and precision. The second is using centralizing F1-score with zero-mean (see Eq. 1) and the computing method of weight (see Eq. 2) together. For the F1-scores, the sum of the positive values is equal to the negative value after centralizing. The method of getting weight by F1-score comes from Leaky ReLU (see Fig. 2). For a class with good recognition results, its F1-score is positive after centralizing. The slow weight adjustment with $\frac{dw}{dfs} = k_f$ leads to the suppression on effects of the class. In contrast, for a class with poor recognition results,

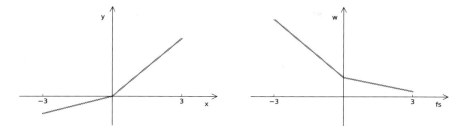

Fig. 2. Comparison between Leaky ReLU (left) and the calculation of weight in feedback loss (see Eq. 2) (right). For the important portion, like the positive part of x and the negative part of fs, the slope of curve stays the same, the absolute value is 1. For the unimportant portion, like the negative part of x and the positive part of fs, the slope of curve is small.

its F1-score is negative after centralizing. The rapid weight adjustment with $\frac{dw}{dfs} = -1$ and the gradient of weight will not decrease with the iteration, leading to the constant attention on effects of the class. This method can ignore the effects of the classes identified well appropriately and focus on classes identified badly, resulting in the improvement of MF1 and κ. The third is that the penalty weight is adjusted by the feedback repeatedly in the process of training and the latest weight is used to calculate feedback loss for model optimization, ensuring the model has a good performance. By the above statement, we find feedback loss can apply to most class-imbalanced problems.

3   Results

3.1   Data

In this study, two versions of the Sleep-EDF dataset [26,27] are used. The first version (Sleep-EDF-13) is contributed in 2013 with 61 polysomnograms, and the second version (Sleep-EDF-18) is contributed in 2018 with 197 polysomnograms. Both of them have two studies about the age effects on sleep in healthy individuals (SC = Sleep Cassette) and the temazepam effects on sleep in individuals with mild difficulty falling asleep (ST = Sleep Telemetry) separately. To get closer to normal, we choose the data in SC studies. Files in SC studies are obtained in healthy Caucasians aged 25–101, without any sleep-related medication. Each PSG lasts nearly 20 h, and contains two EEG signals from Fpz-Cz and Pz-Oz electrode locations, one EOG signal, one EMG signal and one event marker. Every two recordings are collected during two subsequent day-night periods on a subject. The EOG and EEG signals are sampled at 100 Hz. According to the R&K standard, sleep experts manually classify the recordings into one of the six classes, W, N1, N2, N3, N4, REM. The N3 and N4 classes are combined in one class named N3 on the basis of AASM. Because each SC file lasts nearly 20 h and we mainly focus on the sleep period, only the recordings between half hour before and after the sleep periods are retained. Table 1 presents the number of 30-s epochs and proportion for each sleep stage in two different versions.

Table 1. Number of 30-s epoches for sleep stages in two versions of Sleep-EDF.

Dataset	Type	W	N1	N2	N3	R
Sleep-EDF-13	Number	8,285	2,804	17,799	5,703	7,717
	Proportion (%)	19.58	6.63	42.07	13.48	18.24
Sleep-EDF-18	Number	65,951	21,522	69,132	13,039	25,835
	Proportion (%)	33.74	11.01	35.37	6.67	13.22

For Sleep-EDF-13 dataset, we test our method using k-fold cross-validation, where k was set to 14. In each fold, $n_f \mid k$ recordings are selected for testing model, and the remaining recordings are taken to train model, where n_f is the number of recordings in the dataset. This process is repeated k times. Then we combine the predicted sleep stages from all folds and compute the evaluation metrics. Sleep-EDF-18 has 153 recordings, from which 142 recordings are selected for training network, 8 recordings are validation set, and the remaining 3 recordings are test set. It is important to note that the training set, validation set and test set are divided into subjects, ensuring test subjects' epochs do not appear in training set, so we can verify our method on the unknown subjects. If all recordings are mixed before testing, the data of the same subject will appear in the training set and the test set. Although the model's performance improves [15], its practicability reduces. Our division ensures that the data of the same subject will not appear in training set and test set at the same time.

3.2 Experimental Design

In the pre-training, the Adam optimizer's learning rate is 0.001, and 30-s epoch is taken as a sample. After shuffling the oversampled dataset, 128 samples are loaded into the model as a mini-batch. The cost function is multiclass crossentropy. For hyperparameter tuning, we randomly search for five hyperparameter combinations at first, then use Bayesian Optimization, setting (0.3, 0.8) as the search space for hyperparameter of every dropout layer and choosing Gaussian Process [28] as internal regressor which is trained with training set comprised by the previous results. After that, search for multiple hyperparameter combinations by two parts: randomly initializing multiple combinations and searching for more combinations using L-BFGS-B [29]. Next, the mean m_p and standard deviation s_p of each combination are obtained by using regressor. Obtain the fitted value v at each combination by the equation as follows:

$$v = m_p + k_p s_p \qquad (3)$$

k_p is enlargement factor. Then return the combination corresponding to the maximum value which is the new hyperparameter combination. We repeat this search operation 15 times. In the fine-tuning, the Adam optimizer's parameter learning rate, mini-batch size are set 0.0002, 128 respectively. In addition, *multiplier*, *n_below*, *n* are −0.24, 8 and 6 separately.

3.3   Evaluation Metrics

Different metrics are used to evaluate the performance of our approach including per-class recall, overall accuracy (ACC), MF1 and κ.

$$ACC = \frac{\sum_{c=1}^{N} TP_c}{TF} \tag{4}$$

$$MF1 = \frac{\sum_{c=1}^{N} FS_c}{N} \tag{5}$$

TP_c is the true positive epoches of class c, TF is the number of epoches in the dataset, FS_c is the F1-score of class c, N is the number of classes.

3.4   Sleep Stage Classification Performance and Comparison

Table 2 shows confusion matrices obtained from Sleep-EDF-13 and Sleep-EDF-18 datasets respectively. It can be seen that true positive values in the main diagonals are higher than other values in the same rows and columns, meaning our method can accurately identify each classes in most case. Table 3 shows the comparison of our method with other state-of-the-art methods across overall accuracy, MF1 and κ. In terms of overall accuracy, our study performs better than the state-of-the-art algorithms compared. Moreover, for MF1 and κ, our method reaches the highest level, indicating that feedback loss is highly applicable to the current imbalanced dataset. In Sleep-EDF-13, κ reaches 0.80 (between 0.8 and 1), indicating almost complete agreement between the sleep experts and our method, and κ reaches 0.78 in Sleep-EDF-18 (between 0.61 and 0.80), indicating the agreement between the sleep experts and our method are substantial [30].

Table 2. Confusion matrix achieved by the proposed method.

True	Predicted (Sleep-EDF-13)					Predicted (Sleep-EDF-18)				
	W	N1	N2	N3	R	W	N1	N2	N3	R
W	**4482**	437	36	12	112	**1455**	128	3	1	4
N1	193	**1293**	278	3	345	39	**161**	64	0	25
N2	50	542	**11056**	371	608	2	102	**1021**	129	28
N3	2	5	225	**3216**	1	0	0	16	**369**	0
R	98	558	254	0	**4367**	20	97	4	0	**428**

Table 3. Classification performance comparison of our method with other methods by evaluation metrics.

Dataset	Method	ACC (%)	MF1 (%)	κ
Sleep-EDF-13	SleepEEGNet [14]	84.26	79.66	0.79
	DeepSleepNet [13]	82.0	76.9	0.76
	Tsinalis et al. [11]	78.9	73.7	
	Tsinalis et al. [12]	74.8	69.8	
	This study	**85.53**	**81.18**	**0.80**
Sleep-EDF-18	SleepEEGNet [14]	80.03	73.55	0.73
	This study	**83.84**	**77.36**	**0.78**

3.5 Method Analysis

In order to see the difference between the predictions of our method and the labels, we selects one file of test dataset named SC4001E0, and draws the predicted hypnogram and target hypnogram. We can find our method's judgments are the same as the labels in most epochs (see Fig. 3). For understand the confusion matrixs in Table 2 better, we visualize them (see Fig. 4). The value in the cell is the recall, the ratio of the number in corresponding cell of the confusion matrix to the number of 30-s epochs for corresponding sleep stage. For the recall values in the cells, there is a no identification error in many pairs including W-N3, N1-N3 and N3-R. In addition, the confusion matrix is almost symmetric across the diagonal proving class-imbalanced problem is eliminated to some extent.

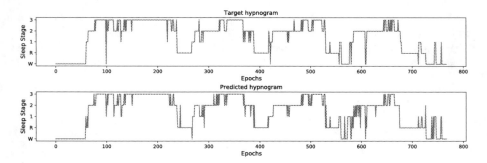

Fig. 3. Comparison of the target hypnogram (top) with the predicted hypnogram (bottom) in SC4001E0. The overall accuracy, MF1, κ, and the recall of N1 stage reach 87.63%, 83.32%, 0.84, and 60.38% respectively.

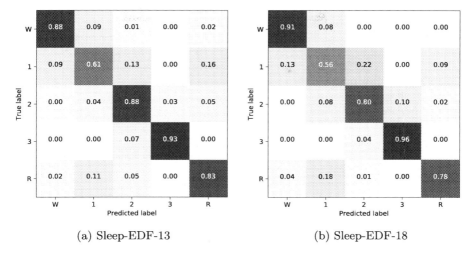

(a) Sleep-EDF-13   (b) Sleep-EDF-18

Fig. 4. Confusion matrix visualization based on recall. For each square, the larger the number, the darker the color. In both datasets, it can be seen the worst performance is noted for N1 stage, and the N3 stage reaches the best performance.

4   Discussion

As for the recognition difficulty of N1 stage, after analyzing the confusion matrix and consulting relevant information, we find that the reason for this result lies in the data itself. The first reason is the number of N1 stage is small compared with the other sleep stages. For example, the proportion of N1 stage in Sleep-EDF-13 and Sleep-EDF-18 are 6.63% and 11.01% respectively. The second reason is N1 stage is easy to confuse with REM. Because θ wave (frequency ranges from 4 hz to 8 hz) occurs only during REM stage and N1 stage [31], making the characteristics of N1 stage and REM stage are similar, leading to the failure of classifier in distinguishing the two stages.

In general, the larger the dataset, the better the performance of the classifier. However, we test our method on Sleep-EDF-13 and Sleep-EDF-18, finding this is not the case. Same problem appears in other study [32]. Therefore, the labels of Sleep-EDF-18 dataset may not be accurate enough due to the heavy workload of expert manual annotation. For κ, [33] find it between 0.48 and 0.89 by studying the agreement between two experts. Similarly, [34] find it between 0.72 and 0.85. These studies prove that manual classification is defective. As the era progresses, the manual classification may lag behind the automatic classification.

The class-imbalance data is particularly common in many fields, and the same situation appears in this study. We use two methods to solve this problem. The first is to simply copy the samples of minority classes to make them reach the same number as the majority classes. The second is mainly through the design of feedback loss. Except for these methods, we have tried other methods, such as random under-sampling and SMOTE over-sampling, but the performance of our method does not improve. There are many studies on the improvement of loss algorithm, including focal loss [35], dice loss [36], etc. We can combine their characteristics with feedback loss. For the current feedback loss, we directly use the F1-score as feedback information to adjust the weight. However, PID control is widely used in the feedback regulation. If we regard the error between the current F1-score and the expected F1-score as feedback information, put it into the PID controller [37] and use the output as penalty weight for feedback loss, our method may improve unexpectedly.

5   Conclusion and Future Work

We propose a deep learning model which can extract time invariant features and classify sleep stage under retaining information, widening the network and increasing the receptive field. We also implement a two-step training algorithm: pre-training model on oversampled datasets with model hyperparameters adjusted by Bayesian Optimization, and proposing feedback loss in fine-tuning to alleviate class-imbalanced problem. Experimental results show that our method outperforms the state-of-the-art methods on the sleep stage classification task. Since our model automatically learns from the original EEG, we believe that our method is a better way to implement sleep stage classification than the hand-engineering methods. When developing an automated system, imbalanced dataset often occurs, such as the arrhythmia detection by ECG and the epilepsy detection by EEG, and feedback loss can make a contribution in these areas.

References

1. Panossian, L., Avidan, A.: Review of sleep disorders. Med. Clin. N. Am. **93**(2), 407–425 (2009)
2. Wulff, K., Gatti, S., Wettstein, J., Foster, R.: Sleep and circadian rhythm disruption in psychiatric and neurodegenerative disease perspectives. Nat. Rev. Neurosci. **11**(8), 589–99 (2010)
3. Alickovic, E., Subasi, A.: Ensemble SVM method for automatic sleep stage classification. IEEE Trans. Instrum. Measure. **67**(6), 1258–1265 (2018)
4. Berry, R., et al.: AASM scoring manual updates for 2017 (version 2.4). Journal of clinical sleep medicine : JCSM : official publication of the American Academy of Sleep Medicine 13 (04 2017)
5. Hobson, J.A.: A manual of standardized terminology, techniques and scoring system for sleep stages of human subjects. Electroencephalogr. Clin. Neurophysiol. **26**(6), 644 (1969)
6. Chen, J., Valehi, A., Razi, A.: Predictive modeling of biomedical signals using controlled spatial transformation. arXiv: Signal Processing (2018)
7. Liu, Z., Sun, J., Zhang, Y., Rolfe, P.: Sleep staging from the eeg signal using multi-domain feature extraction. Biomed. Sig. Process. Control **30**, 86–97 (2016)
8. Maity, A., et al.: Multifractal detrended fluctuation analysis of alpha and theta EEG rhythms with musical stimuli. Chaos Solitons Fractals **81**, 52–67 (2015)

9. Shayegh, F., Sadri, S., Amirfattahi, R., Ansari-Asl, K.: A model-based method for computation of correlation dimension, lyapunov exponents and synchronization from depth-EEG signals. Comput. Meth. Programs Biomed. **113**, 323–337 (2013)
10. Zaeri-Amirani, M., Afghah, F., Mousavi, S.: A feature selection method based on shapley value to false alarm reduction in ICUS a genetic-algorithm approach. In: 2018 40th Annual International Conference of the IEEE Engineering in Medicine and Biology Society (EMBC), pp. 319–323 (2018)
11. Tsinalis, O., Matthews, P.M., Guo, Y.: Automatic sleep stage scoring using time-frequency analysis and stacked sparse autoencoders. Ann. Biomed. Eng. **44**(5), 1587–1597 (2016)
12. Tsinalis, O., Matthews, P.M., Guo, Y., Zafeiriou, S.: Automatic sleep stage scoring with single-channel EEG using convolutional neural networks. ArXiv (2016)
13. Supratak, A., Dong, H., Wu, C., Guo, Y.: Deepsleepnet: a model for automatic sleep stage scoring based on raw single-channel EEG. IEEE Trans. Neural Syst. Rehabil. Eng. **25**(11), 1998–2008 (2017)
14. Mousavi, S., Afghah, F., Acharya, U.R.: SleepEEGNet: automated sleep stage scoring with sequence to sequence deep learning approach. PLOS ONE **14**, e0216456 (2019)
15. Snoek, J., Larochelle, H., Adams, R.P.: Practical Bayesian optimization of machine learning algorithms. Adv. Neural Inf. Process. Syst. **4**, 2951–2959 (2012)
16. Al-Zuhairi, M., Pradhan, B., Lee, S.: Application of convolutional neural networks featuring Bayesian optimization for landslide susceptibility assessment. CATENA **86**, 104249 (2019)
17. Frean, M., Boyle, P.: Using Gaussian processes to optimize expensive functions. In: Wobcke, W., Zhang, M. (eds.) AI 2008. LNCS (LNAI), vol. 5360, pp. 258–267. Springer, Heidelberg (2008). https://doi.org/10.1007/978-3-540-89378-3_25
18. Liang, X.: Image-based post-disaster inspection of reinforced concrete bridge systems using deep learning with Bayesian optimization. Comput. Aid. Civ. Infrastruct. Eng. **34**(5), 415–430 (2019)
19. Martinez-Cantin, R., Freitas, N., Brochu, E., Castellanos, J., Doucet, A.: A Bayesian exploration-exploitation approach for optimal online sensing and planning with a visually guided mobile robot. Auton. Robots **27**, 93–103 (2009)
20. Simonyan, K., Zisserman, A.: Very deep convolutional networks for large-scale image recognition. arXiv 1409.1556 (2014)
21. Nair, V., Hinton, G.E.: Rectified linear units improve restricted Boltzmann machines. In: Proceedings of the 27th International Conference on Machine Learning, pp. 807–814 (2010)
22. Ioffe, S., Szegedy, C.: Batch normalization: accelerating deep network training by reducing internal covariate shift (2015). arXiv: Learning
23. Springenberg, J.T., Dosovitskiy, A., Brox, T., Riedmiller, M.: Striving for simplicity: the all convolutional net (2014). arXiv: Learning
24. Srivastava, N., Hinton, G., Krizhevsky, A., Sutskever, I., Salakhutdinov, R.: Dropout: a simple way to prevent neural networks from overfitting. J. Mach. Learn. Res. **15**, 1929–1958 (2014)
25. Kingma, D.P., Ba, J.: Adam: A method for stochastic optimization (2014). arXiv: Learning
26. Goldberger, A.L., et al.: Physiobank, physiotoolkit, and physionet components of a new research resource for complex physiologic signals. Circulation **101**(23), 215–220 (2000)

27. Kemp, B., Zwinderman, A.H., Tuk, B., Kamphuisen, H.A.C., Oberye, J.J.: Analysis of a sleep-dependent neuronal feedback loop: the slow-wave microcontinuity of the EEG. IEEE Trans. Biomed. Eng. **47**(9), 1185–1194 (2000)

28. Ounpraseuth, S.: Gaussian processes for machine learning. J. Am. Stat. Assoc. **103**, 429–429 (2008)

29. Morales, J., Nocedal, J.: Remark on "algorithm 778: L-BFGS-B: Fortran subroutines for large-scale bound constrained optimization". ACM Trans. Math. Softw. **38**(7), 1–4 (2011)

30. Hassan, A.R., Subasi, A.: A decision support system for automated identification of sleep stages from single-channel EEG signals. Knowl. Based Syst. **128**, 115–124 (2017)

31. Cheng, J.: Research of Sleep Staging Based on EEG Signals. Ph.D. thesis, Beiiing Institute of Technology (2015)

32. Sors, A., Bonnet, S., Mirek, S., Vercueil, L., Payen, J.: A convolutional neural network for sleep stage scoring from raw single-channel EEG. Biomed. Sig. Process. Control **42**, 107–114 (2018)

33. Stepnowsky, C., Levendowski, D., Popovic, D., Ayappa, I., Rapoport, D.: Scoring accuracy of automated sleep staging from a bipolar electroocular recording compared to manual scoring by multiple raters. Sleep Med. **14**(11), 1199–1207 (2013)

34. Wang, Y., Loparo, K., Kelly, M., Kalpan, R.: Evaluation of an automated single-channel sleep staging algorithm. Nat. Sci. Sleep **7**, 101–111 (2015)

35. Lin, T.Y., Goyal, P., Girshick, R., He, K., Dollar, P.: Focal loss for dense object detection. In: The IEEE International Conference on Computer Vision (ICCV), pp. 2999–3007, October 2017

36. Li, X., Sun, X., Meng, Y., Liang, J., Wu, F., Li, J.: Dice loss for data-imbalanced NLP tasks. ArXiv (2019)

37. Qin, Y., Sun, L., Hua, Q., Liu, P.: A fuzzy adaptive PID controller design for fuel cell power plant. Sustainability **10**(7), 2438 (2018)

Multi-Cue and Temporal Attention for Person Recognition in Videos

Wenzhe Wang, Bin Wu$^{(\boxtimes)}$, Fangtao Li, and Zihe Liu

Beijing Key Laboratory of Intelligent Telecommunication Software and Multimedia,
Beijing University of Posts and Telecommunications, Beijing 100876, China
{wangwenzhe,wubin,lift,ziheliu}@bupt.edu.cn

Abstract. Recognizing persons under unconstrained settings is challenging due to variation in pose and viewpoint, partial occlusion, and motion blur. Inference only by face-based recognition techniques would fail in these cases. Previous studies mainly focus on this problem on still images while they cannot handle the temporal variations in videos. In this work, we aim to tackle these challenges and propose a Multi-Cue and Temporal Attention (MCTA) framework to recognize persons in videos. For the spatial domain, we extract features from multiple visual cue regions and utilize a Multi-Cue Attention Module to integrate them. For the temporal domain, we adopt a Temporal Attention Module to combine the video frames, which is learned to assess the quality of different frames adaptively. By this means, MCTA can comprehensively explore the complementary information in spatial-temporal dimensions for person recognition in videos. Moreover, we introduce Character Recognition in Videos (CRV), a new video dataset for character recognition under challenging settings. Extensive experiments on CRV demonstrate the effectiveness of our proposed framework. Dataset with annotations and all codes used in this paper are publicly available at https://github.com/zhezheey/MCTA.

Keywords: Person recognition · Multiple cues · Spatial-temporal attention

1 Introduction

Recognizing persons in images or videos is frequently needed in practical scenarios. As a key task of the understanding of multimedia content and computer vision, person recognition has been widely studied and achieved great success in multiple settings, including face recognition [1,2], person re-identification [3,4], and speaker recognition [5]. Nonetheless, person recognition under unconstrained scenarios remains a challenging task and far from being well solved. Issues like great variation in pose and viewpoint, partial occlusion, motion blur, and noisy sounds in practice bring substantial difficulties for these methods.

Supported by the National Key R&D Program of China (2018YFC0831500), the National Natural Science Foundation of China (No. 61972047), and the NSFC-General Technology Basic Research Joint Funds (No. U1936220).

© Springer Nature Switzerland AG 2020
Y. Peng et al. (Eds.): PRCV 2020, LNCS 12306, pp. 369–380, 2020.
https://doi.org/10.1007/978-3-030-60639-8_31

Fig. 1. How do we recognize a person in videos when the face is invisible? The hairstyle, clothing, scene, and information of adjacent frames are significant.

To tackle the problem, a natural idea is to combine the information of multiple cues. Existing studies mainly focus on image-based condition, in which additional visual cues, such as hairstyles, clothing, or scenes, are utilized to recognize persons in photos when the faces are blurred or even cropped. They rely on concatenation [6,7], heuristics [8,9], or simple attention methods [10] to combine contextual cues from multiple regions and achieve better results on benchmarks.

However, compared to image-based person recognition, the video-based scenario attracts far less attention. Recognizing persons in unconstrained videos is a more challenging task with many practical applications, which needs both the spatial and temporal context to help recognition (see Fig. 1). Recently, iQIYI and ACM Multimedia held a challenge [11] towards multi-model person identification based on a large-scale video dataset and attracted hundreds of researchers [12–15] to come up with novel ideas. However, these studies are limited in two aspects: (1) The large dataset with high-quality faces (99.65% of videos contain clear faces) provides much richer information than practical scenarios that the model can get over 90% mAP only by face features [12], making other visual and multi-model cues almost useless. (2) These methods rely on simple concatenation [14,15] to integrate multi-cue information, and averaging [12] or heuristic rules [13,15] to model the temporal information separately, which are obviously over-simplified.

In this work, we propose a Multi-Cue and Temporal Attention (MCTA) framework for person recognition in videos. In the spatial domain, different regions of a person, including face and upper body, are detected and then input along with the whole image into specific Convolutional Neural Networks (CNNs) to extract the features. Moreover, we adopt a Multi-Cue Attention Module (MCAM) to adaptively combine multiple visual cues. In the temporal domain, we introduce a Temporal Attention Model (TAM), which applies an attention weighted average on the sequence of image features, to model the importance of different frames. Finally, our MCTA framework achieves person recognition by end-to-end spatial-temporal information learning in videos.

To evaluate our method under more diverse settings and facilitate the research, we construct a dataset named Character Recognition in Videos (CRV) by annotating 79 characters in 4,405 video clips from 34 movies or TV series. In particular, all of the actors or actresses play multiple roles in one movie or TV series, and there are certain differences between the characters. To simplify the problem, each video clip is limited to contain only one or one main person similar to [11]. Compared with the existing dataset [11], CRV is more challenging, which requires additional visual cues like hairstyles or clothing to complement facial information and recognize the characters. Experiments demonstrate that our proposed framework can significantly raise the performance on CRV.

In summary, the main contributions of this work include:

- We propose a novel MCTA framework to recognize persons in videos under unconstrained settings, which incorporates both the spatial and temporal information for person recognition.
- We introduce an MCAM to integrate features of visual cues from multiple regions, and a TAM to access the quality of different frames and combine them. These two modules together comprehensively explore the information of persons in videos.
- We construct CRV, a novel and challenging dataset for character recognition, to promote the research on person recognition in videos.
- Our framework achieves state-of-the-art performance on CRV, which demonstrates the effectiveness of our method.

2   Related Work

2.1   Face Recognition and Person Re-identification

As the most widely studied and applied direction of person recognition, face recognition algorithms [1,2] have achieved impressive results on verification and recognition tasks. The state-of-the-art method, ArcFace [2] achieved a face verification accuracy of 99.83% on LFW [16], which is even better than human-level performance. Another popular task is person re-identification [3,4], which aims at recognizing pedestrians across cameras within a relatively short period, where visual cues are likely to remain consistent. However, these methods are highly sensitive to environmental conditions and inadequate to handle the variations in social media photos or movies, where the faces and bodies are always invisible or blurred and the clothing may be changed.

2.2   Person Recognition in Photo Albums

Person recognition under unconstrained settings is the problem of interest in this work, which mainly focuses on the persons in photo albums. Zhang *et al.* [8] introduced the People In Photo Albums (PIPA) dataset and combined three visual recognizers on face, body, and poselet-level cues to recognize the persons. To further improve the performance on PIPA, some studies [6,7] paid attention to exploiting more visual cues, such as head [6,7], upper body [6,7], scene [6], pose [7], or other human attributes [6] respectively. Other studies [9,10] focused on combining visual cues and social context to exploit domain-specific information. Li *et al.* [9] exploited contextual cues at person, photo, and group levels and combined them with a heuristic rule to identify persons. Huang *et al.* [10] proposed a framework to couple social context learning with people recognition by a unified formulation, which integrated multiple visual cues adaptively and achieved state-of-the-art performance. However, simply applying image-based methods to recognize persons in videos would lose temporal information and require high computing cost.

2.3   Person Recognition in Unconstrained Videos

Person recognition in unconstrained videos attracts far less attention due to challenges as follows: (1) Lacking annotated video datasets. (2) The temporal and multi-model information of videos put forward higher requirements for algorithms. In particular, the Celebrity Video Identification Challenge [11] held in 2019 presented iQIYI-VID-2019, a large-scale video dataset for multi-modal person recognition. However, almost all (99.65%) videos in this dataset contain clear faces, which is much different from the real-world scenarios. The winning team [12] relied only on face features and achieved better results than others that combining multiple cues [14,15]. Moreover, most teams chose averaging [12] or heuristic rules [13,15] to aggregate the image features of a video in the competition, which are obviously over-simplified. Another noteworthy task is person search in videos. Huang *et al.* [17] proposed a framework to incorporate both the visual similarity and the identity invariance along a tracklet, and developed a new schedule to improve the reliability of propagation, which outperformed mainstream person re-id methods on this problem. However, this task is essentially different, where a clear portrait for each person is required.

3   The Proposed Framework

3.1   Overview

Given a video clip, the task of person recognition in videos is to recognize the identity in the clip, which is defined as a standard supervised classification task [8,11] that we train and test on the same set of identities.

In this work, we devise a Multi-Cue and Temporal Attention framework for this task. As shown in Fig. 2, the overall architecture of MCTA framework mainly contains four parts:

1) Multi-cue region detector and feature extractor. The framework takes as input one video clip which is sampled into F frames. To obtain regions of multiple visual cues, different body parts, including face, head, upper body, and

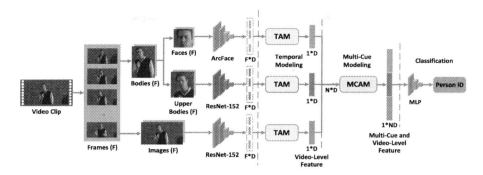

Fig. 2. The overall architecture of multi-cue and temporal attention framework.

the whole body are cropped from frames with region-specific detectors. Next, specific CNNs are adopted to extract the spatial features of these body regions and the whole images.

2) Temporal feature modeling. For each visual cue region, this stage computes a quality score vector for the feature of each sampled frame and aggregates them into a video-level representation. For this, a Temporal Attention Module (TAM) is adopted to adaptively learn the importance of different frames and integrate them.

3) Spatial feature modeling. This stage fuses the video features of different visual cues and forms the final video-level representation for each video clip. For this, a Multi-Cue Attention Module (MCAM) is learned to combine the visual cues from different regions with adaptive weights.

4) Classifier with video-level features. In this stage, the identity of person is predicted by the video-level feature. A three-layer Multilayer Perceptron (MLP) model is adopted for the final classification.

3.2 Temporal Attention Module

Unlike the task of action recognition in videos [18], the movement between frames has little effect on person recognition. For temporal information, we mainly consider the quality of video frames, which is represented by the quality of visual features here. To this end, we devise a Temporal Attention Module (TAM) as the quality predictor, which is inspired by [3], to compute the scores of different frames and aggregate them better.

For each visual cue region, the input of TAM is a feature matrix $\mathbf{X} \in \mathbb{R}^{F \times D}$, where F is the sampled frame number for a video clip and D is the length of feature vector. Then $\mathbf{X}$ is fed into a fully connected layer and a softmax layer, to get the quality score matrix $\mathbf{Z}$:

$$\mathbf{Y} = \mathbf{W}_F \mathbf{X} + \mathbf{b}, \tag{1}$$

$$\mathbf{z}_i = \frac{\exp(\mathbf{y}_i)}{\sum_{i=1}^{F} \exp(\mathbf{y}_i)}, \tag{2}$$

where $\mathbf{W}_F \in \mathbb{R}^{F \times F}$, $\mathbf{b} \in \mathbb{R}^F$ are parameters to be learned, and $\mathbf{y}_i$ denotes the i-th row of $\mathbf{Y}$. The i-th row of $\mathbf{Z}$, denoted by $\mathbf{z}_i$, is the quality score vector for the i-th frame. Finally, the output feature vector $\mathbf{o}$ is fused through:

$$\mathbf{o} = \sum_{i=1}^{F} \mathbf{z}_i \odot \mathbf{x}_i \,, \tag{3}$$

where $\odot$ is the element-wise multiplication operator, and $\mathbf{x}_i$ denotes the feature vector of the i-th frame. As a result, TAM calculates an attention weighted average on the sequence of image features, which aggregates complementary information from all frames in a video, and the influence of image regions with poor quality is compensated by other frames.

3.3   Multi-Cue Attention Module

Previous studies [6–10] have proved the utility of incorporating multiple visual cues for person recognition. The facial features play a decisive role when the front face is clear. However, in unconstrained videos, the face may be invisible due to the limited scope of camera or occlusion, where we have to resort to other visual cues like hairstyle, clothing, or scene to recognize its identity. Moreover, different visual cue regions always vary in contributions across instances. Inspired by [10], we utilize a Multi-Cue Attention Module (MCAM) to combine the features of multiple visual cues adaptively.

The MCAM is a neural network that takes the stacked features $\mathbf{X} \in \mathbb{R}^{N \times D}$ from all N visual cue regions as input, where N is the number of visual cues and D is the length of feature vector. N positive coefficients are yielded as the weights of different regions through a fully connected layer, a mean operation, and a softmax layer in MCAM:

$$\mathbf{Y} = \mathbf{W}_N \mathbf{X} + \mathbf{b} \,, \tag{4}$$

$$\bar{y}_i = \frac{1}{D} \sum_{j=1}^{D} y_{i,j} \,, \tag{5}$$

$$z_i = \frac{\exp(\bar{y}_i)}{\sum_{i=1}^{N} \exp(\bar{y}_i)} \,, \tag{6}$$

where $\mathbf{W}_N \in \mathbb{R}^{N \times N}$, $\mathbf{b} \in \mathbb{R}^N$ are learnable parameters, $\bar{y}_i$ denotes the average value of the i-th row $(y_{i,1}, y_{i,2}, ..., y_{i,D})$ of $\mathbf{Y}$, and z_i is the weight for the i-th visual cue region. Then the final feature vector of the i-th visual cue region is given by:

$$\mathbf{o}_i = z_i \mathbf{x}_i \,, \tag{7}$$

where the i-th row of $\mathbf{X}$, denoted by $\mathbf{x}_i$, is the raw feature vector of the i-th visual cue region. Finally, we get the output feature vector $\mathbf{o}$ by the concatenation of $\mathbf{o}_i$. As a result, MCAM can calculate a weighted score for each visual cue adaptively and combine them by weighted concatenation.

4 Experiments

4.1 New Dataset: Character Recognition in Videos

Existing datasets for person recognition under unconstrained settings are mainly based on still images [8, 10], while the video-based dataset is rare. To our best knowledge, the latest video dataset for person identification is iQIYI-VID [11], which contains about 200,000 videos of 10,034 identities collected from real online videos. However, as discussed in Sect. 1, almost all video clips in the dataset have clear appearances of faces, which makes it different from the practical scenarios and limited in use.

To facilitate related research and assess our method under more practical settings, we build a high-quality Character Recognition in Videos dataset from movies and TV series, dubbed as CRV. This dataset contains 4,405 video clips of 79 characters in total, which are played by 37 actors or actresses in 34 movies and TV series. The duration of video clips is in the range of 1 to 30 s, with 5.14 s on average. Particularly, all actors and actresses in CRV play multiple (from 2 to 4) roles with differences in one movie or TV series. We need to combine facial features with different visual cues, such as hairstyles, clothing, actions, or surrounding scenes, to identify the specific character in a video clip.

The construction process contains four steps as follows: (1) We first select more than 50 movies and TV series in which one person (or twins) plays multiple roles. Videos of characters acted by one person that are nearly indistinguishable by annotators are excluded. (2) We then ask ten annotators to segment video clips of the selected characters from the movies and TV series. The length of each clip is limited to 1 to 30 s to avoid redundant information while keeping the temporal information. Particularly, each video clip must contain only one or one main person. The scene in one clip should be fixed. Characters with less than 5 clips are discarded. (3) Each candidate video clip is labeled according to its character by another annotator and then checked twice to guarantee the accuracy of both the segmentation and the label. (4) At last, the dataset is randomly split into training, validation, and testing sets by the ratio of 4: 3: 3.

Figure 3 shows several examples of video frames in our dataset. It can be seen that recognizing characters, especially those played by the same actor or actress, is very challenging. Characters in CRV are diverse in views, poses, clothing, and scenes.

4.2 Implementation Details

Data Preprocessing and Feature Extraction. In the experiments, each video clip is uniformly sampled into 16 frames by time. We use three visual cue regions of each frame: face, upper body, and the whole image. For the sampled frames, the bodies of persons are first detected and cropped with Mask R-CNN [19] pre-trained on MS-COCO [20]. Then we separately adopt MTCNN [21] for face detection and alignment, and SSD [22, 23] pre-trained on Hollywood-Heads [24] to detect the heads inside the cropped body regions. The regions

Norbit / Rasputia Latimore
(Norbit)

Shen Wei / Black Robe Envoy
(Guardian)

Hiroko Watanabe / Itsuki Fujii
(Love Letter)

Hallie Parker / Annie James
(The Parent Trap)

Fig. 3. Examples of CRV dataset.

of upper bodies are obtained by simple geometric rules based on the region of heads and bodies. For feature extraction, we adopt ArcFace [2] pre-trained on MS1M-ArcFace [25] to extract face features, and ResNet-152 [26] pre-trained on ImageNet [27] as the feature extractor for other visual cue regions. The face features are duplicated four times and concatenated to have the same dimension (2,048-D) as the other ResNet [26] features. Each feature is then scaled by its maximum absolute value. In particular, we use zero vectors as the features for visual cue regions that are invisible or not detected.

Network Training. In our framework, the features of frames for multiple visual cues are extracted firstly, and the other parts of MCTA are trained end-to-end. A three-layer MLP with a hidden layer of 2,048 nodes is chosen for the final classification based on the validation set. We train MCTA with the cross-entropy loss at a learning rate of 0.0003. The MCTA is implemented based on the Keras framework.

4.3   Comparison with the State-of-the-Art Methods

As discussed in Sect. 2.3, person recognition in videos especially under unconstrained scenarios is a relatively new task, which attracts less attention and is far

Table 1. Comparison to the existing state-of-the-art and baseline methods on CRV.

Method	Module		Result(%)		
	Multi-cue modeling	Temporal modeling	mAP@100	mAP	Accuracy
Face + MLP [12]	–	Average	81.82	83.26	79.98
Multi + MLP	Concatenation	Average	85.71	87.14	87.85
MCTA-t	Concatenation	TAM	86.49	87.90	90.31
MCTA-mc	MCAM	Average pooling	86.63	88.04	90.44
MCTA	MCAM	TAM	**87.01**	**88.42**	**90.56**

from well being solved now. To validate the effectiveness of the proposed MCTA framework, we compare it with existing state-of-the-art and baseline methods on the CRV dataset. The mean Average Precision (mAP) and accuracy are calculated as the evaluation metrics. The details of these methods are as follows:

1) Face + MLP [12]. This method uses only face features [2] of frames as the input and adopts an MLP for classification. The averaged probability vector of frames in a video is used as its prediction result. The main idea is similar to [12], the winning team in [11], while we remove tricks like data augmentation and model ensemble. Therefore we consider this model as the state-of-the-art method for person recognition in videos.

2) Multi-Cue + MLP. This method uses the same framework with Face + MLP except that the concatenation of multi-cue features is taken as the input.

3) MCTA-t. This method is a simplified version of MCTA, which replaces the MCAM with simple concatenation to integrate the multi-cue features.

4) MCTA-mc. This method is another simplified version of MCTA, which replaces the TAM with average pooling to combine the features of different frames.

5) MCTA. This is the complete MCTA framework as described in Sect. 3, which adopts the MCAM to integrate features of multiple visual cues and the TAM to model the importance of different frames.

Analysis and Discussion. The results of these methods are listed in Table 1. We can observe that recognizing characters in CRV dataset is challenging. Compared with Face + MLP [12], Multi-Cue + MLP raises the performance by a considerable margin, which validates the significance of multiple visual cues. Moreover, by comparison of Multi-Cue + MLP, MCTA-t, MCTA-mc, and MCTA, we can find that the multi-cue and temporal attention module are both effective in this task. MCAM can generate adaptive weights for different visual cues, and as a result, the mAP is significantly improved. Although the temporal information between frames is poor in a short video clip, TAM gives slightly better performance than average pooling by exploiting the image quality of different frames. Furthermore, MCTA obtains the best performance on CRV, which demonstrates the complementary effect of multiple visual cues and temporal information in videos.

Table 2. Results (%) of different visual cues.

Cue	mAP	accuracy
Face (F)	83.18	86.68
Upper body (U)	72.37	80.09
Image (I)	63.81	74.79
F + U	86.26	89.85
F + I	87.30	89.38
U + I	75.59	82.16
F + U + I	**88.42**	90.56
F + Head + U + Body + I	88.35	**91.40**

4.4 Ablation Study

Significance of Multiple Visual Cues. Here we show the comparison of different visual cues. The results are listed in Table 2. For single visual cue, face feature achieves the highest mAP, which can be used to recognize the person in videos to a pair or group of characters and further recognize it by extra facial information such as expression and decoration. The combination of multiple visual cues can significantly improve the performance, which demonstrates the complementary effect of multi-cue information. Moreover, when we add the features of head and the whole body to MCTA, the performance changes insignificantly. The head feature can provide extra information when the face is invisible, while it's already covered by the upper body feature. Compared with upper body, the whole body always changes in views for different videos, which makes the information unclear.

Significance of Different Modules. Here we explore the effect of multi-cue modeling module, temporal modeling module, and MLP module in MCTA. Table 3 lists the results of the cross combination of different multi-cue information modeling methods, *i.e.*, simple concatenation, heuristic weighted concatenation (the weight ratio is set to F: U: I = 0.4: 0.3: 0.3 after experimental comparison) and MCAM, and different temporal information modeling methods, *i.e.*, average pooling, max pooling, LSTM, and TAM. From the results, we can find that MCAM and TAM perform better than all of the other methods, which proves that our MCTA framework can better explore the spatial-temporal characteristics of persons in unconstrained videos. Moreover, the mAP reduces from 88.42% to 73.92% when we remove the MLP module in MCTA, which reflects the necessity of the MLP part.

Analysis on Hyper Parameters. For the sampled frame number F, we compare the results of MCTA for $F = 1, 4, 8, 16$, and 24. The results are listed in Table 4. It shows that the mAP increases with the growth of input frames (from 71.72% to 88.42%) and reaches a smooth state when F is larger than 16 (88.42%

Table 3. Results (mAP, %) of the cross combination of different multi-cue and temporal information modeling methods.

	Concatenation	Heuristics	MCAM
Average pooling	87.52	87.81	88.04
Max pooling	86.09	86.16	86.18
LSTM	87.00	87.29	87.42
TAM	87.90	88.14	**88.42**

Table 4. Results (%) of different sampled frame number (F).

F	1	4	8	16	24
mAP	71.72	87.60	88.14	88.42	**88.50**

vs. 88.50%). Therefore, we finally choose $F = 16$ for the balance of accuracy and model complexity.

5   Conclusion

In this paper, we propose a new framework named MCTA for person recognition in videos, which utilizes an MCAM to adaptively combine the features of multiple visual cues and a TAM to aggregate the frame-level features by assessing the importance of different frames. We construct a novel dataset CRV from movies and TV series for the recognition of characters under challenging settings. Extensive comparing experiments and ablation studies on CRV show that our approach can learn a better spatial-temporal representation for person recognition in unconstrained videos.

References

1. Schroff, F., Kalenichenko, D., Philbin, J.: Facenet: A unified embedding for face recognition and clustering. In: CVPR, pp. 815–823 (2015)
2. Deng, J., Guo, J., Xue, N., Zafeiriou, S.: Arcface: Additive angular margin loss for deep face recognition. In: CVPR, pp. 4690–4699 (2019)
3. Song, G., Leng, B., Liu, Y., Hetang, C., Cai, S.: Region-based quality estimation network for large-scale person re-identification. In: AAAI, pp. 7347–7354 (2018)
4. Zheng, Z., Yang, X., Yu, Z., Zheng, L., Yang, Y., Kautz, J.: Joint discriminative and generative learning for person re-identification. In: CVPR, pp. 2138–2147 (2019)
5. Dehak, N., Kenny, P.J., Dehak, R., Dumouchel, P., Ouellet, P.: Front-end factor analysis for speaker verification. IEEE Trans. Audio Speech Lang. Process. **19**(4), 788–798 (2011)
6. Oh, S.J., Benenson, R., Fritz, M., Schiele, B.: Person recognition in personal photo collections. In: ICCV, pp. 3862–3870 (2015)
7. Kumar, V., Namboodiri, A., Paluri, M., Jawahar, C.V.: Pose-aware person recognition. In: CVPR, pp. 6223–6232 (2017)

8. Zhang, N., Paluri, M., Taigman, Y., Fergus, R., Bourdev, L.: Beyond frontal faces: Improving person recognition using multiple cues. In: CVPR, pp. 4804–4813 (2015)
9. Li, H., Brandt, J., Lin, Z., Shen, X., Hua, G.: A multi-level contextual model for person recognition in photo albums. In: CVPR, pp. 1297–1305 (2016)
10. Huang, Q., Xiong, Y., Lin, D.: Unifying identification and context learning for person recognition. In: CVPR, pp. 2217–2225 (2018)
11. Liu, Y., et al.: iQIYI celebrity video identification challenge. In: ACM MM, pp. 2516–2520 (2019)
12. Huang, Z., Chang, Y., Chen, W., Shen, Q., Liao, J.: Residualdensenetwork: a simple approach for video person identification. In: ACM MM, pp. 2521–2525 (2019)
13. Fang, X., Zou, Y.: Make the best of face clues in iQIYI celebrity video identification challenge 2019. In: ACM MM, pp. 2526–2530 (2019)
14. Dong, C., Gu, Z., Huang, Z., Ji, W., Huo, J., Gao, Y.: Deepmef: a deep model ensemble framework for video based multi-modal person identification. In: ACM MM, pp. 2531–2534 (2019)
15. Chen, J., Yang, L., Xu, Y., Huo, J., Shi, Y., Gao, Y.: A novel deep multi-modal feature fusion method for celebrity video identification. In: ACM MM, pp. 2535–2538 (2019)
16. Huang, G.B., Mattar, M., Berg, T., Learned-Miller, E.: Labeled faces in the wild: a database for studying face recognition in unconstrained environments. In: Workshop on Faces in 'Real-Life' Images: Detection, Alignment, and Recognition, pp. 1–14 (2008)
17. Huang, Q., Liu, W., Lin, D.: Person search in videos with one portrait through visual and temporal links. In: ECCV, pp. 425–441 (2018)
18. Wang, L., Xiong, Y., Wang, Z., Qiao, Y., Lin, D., Tang, X., Gool, L.V.: Temporal segment networks: towards good practices for deep action recognition. In: ECCV, pp. 20–36 (2016)
19. He, K., Gkioxari, G., Dollár, P., Girshick, R.: Mask r-cnn. In: ICCV, pp. 2961–2969 (2017)
20. Lin, T.Y., et al.: Microsoft COCO: Common Objects in Context. In: Fleet, D., Pajdla, T., Schiele, B., Tuytelaars, T. (eds.) ECCV 2014. LNCS, vol. 8693, pp. 740–755. Springer, Cham (2014). https://doi.org/10.1007/978-3-319-10602-1_48
21. Zhang, K., Zhang, Z., Li, Z., Qiao, Y.: Joint face detection and alignment using multitask cascaded convolutional networks. IEEE Sig. Process. Lett. **23**(10), 1499–1503 (2016)
22. Liu, W., et al.: SSD: single shot multibox detector. In: Leibe, B., Matas, J., Sebe, N., Welling, M. (eds.) ECCV 2016. LNCS, vol. 9905, pp. 21–37. Springer, Cham (2016). https://doi.org/10.1007/978-3-319-46448-0_2
23. Marin-Jimenez, M.J., Kalogeiton, V., Medina-Suarez, P., Zisserman, A.: LAEO-Net: revisiting people looking at each other in videos. In: CVPR, pp. 3477–3485 (2019)
24. Vu, T.H., Osokin, A., Laptev, I.: Context-aware cnns for person head detection. In: ICCV, pp. 2893–2901 (2015)
25. Guo, Y., Zhang, L., Hu, Y., He, X., Gao, J.: MS-Celeb-1M: a dataset and benchmark for large-scale face recognition. In: Leibe, B., Matas, J., Sebe, N., Welling, M. (eds.) ECCV 2016. LNCS, vol. 9907, pp. 87–102. Springer, Cham (2016). https://doi.org/10.1007/978-3-319-46487-9_6
26. He, K., Zhang, X., Ren, S., Sun, J.: Deep residual learning for image recognition. In: CVPR, pp. 770–778 (2016)
27. Deng, J., Dong, W., Socher, R., Li, L.J., Li, K., Fei-Fei, L.: Imagenet: A large-scale hierarchical image database. In: CVPR, pp. 248–255 (2009)

Zero-Shot Learning Based on Salient Region and Enhanced Semantics

Zongrong Pan and Anna Zhu$^{(\boxtimes)}$

Wuhan University of Technology, Wuhan, China
ZongrongPan@outlook.com, annakkk@live.com

Abstract. Zero-shot learning (ZSL) refer to recognizing the new class without training samples. Traditionally, the projection function learned from visual features to semantic features is used for object recognition. However, few works will focus on accurate feature representation of recognition objects. The human designed semantics are not discriminative and sufficient to recognize different and new classes. In this paper, we propose to use the image reconstruction to extract enhanced semantics (ES) on salient region of image. The salient region of image is encoded corresponding to predefined attributes and ES features. And then decoded to original image of salient region. The Lifted structure feature embedding (LSFE) is applied to make the extended features more discriminative. Softmax is used for classification thus makes ES features more accurate. Experiments on two benchmark datasets AwA2 and CUB, demonstrate the effectiveness of the proposed approach.

Keywords: Zero-shot learning · Object saliency detection · Lifted structure feature embedding · Enhanced semantic

1 Introduction

In recent years, deep learning models have made a great breakthrough in object recognition task [4]. However, the limitation of these models are also obvious. The supervised learning models are required to get enough training samples. Moreover, the training data needs to be labelled which is so expensive and training a deep convolutional neural networks (CNNs) from scratch is complex. It's easy to collect the daily object samples such as book, but it's hard to get enough pictures of rare animals or a newly identified specie. The achievements of these models are based on massive training data. Few models take few or no training samples for a given class in consider.

In our daily life, give children a single "horse" picture and tell them that the zebra is a horse but black and white stripes on it. When they see a picture of "zebra", they can combine the "horse" picture and the "black and white" semantic information, finally recognized the "zebra" easily. The conventional deep learning model is fail to work well on few or no training samples. On the contrary, humans are expert in recognizing objects without seeing any samples. Inspired by human's great ability of recognition, Lampert et al. [5] propose Zero-shot learning (ZSL) and use attributes as a bridge to transfer knowledge. In recent years, ZSL attracts a lot of interest of researches and industrials.

© Springer Nature Switzerland AG 2020
Y. Peng et al. (Eds.): PRCV 2020, LNCS 12306, pp. 381–393, 2020.
https://doi.org/10.1007/978-3-030-60639-8_32

Traditional object recognition seeks for a mapping from visual space to label space. Hence, the categories of training set are separate from those from test set. In other words, the training data and test data do not share the label space. ZSL aims to recognize the new class from unseen classes, based on the assumption that both seen classes and unseen classes share a common semantic space which we call the bridge. There are several types of bridges in ZSL. The first type is attributes. Attributes were introduced in [5] and enjoy a great popularity [7, 9, 11]. For example, attributes, such as black, has wings and furry are shared among different kinds of animals, which were used for building up the relationship between seen classes and unseen classes. The second type is word vector or text-description [14–16]. This type of bridge automatically mine the relationship of diverse animals via text corpus. The third type is hierarchies or taxonomies [7, 8]. The semantic concepts of different kinds would be extracted form taxonomies.

The semantic attributes as bridge are used for connecting seen classes and unseen classes. However, there are still some problems to be solved. Firstly, few existing models extract image feature manually or from a pre-trained on ImageNet [19] CNN model, which cannot do well in this task. Secondly, the semantic attributes is defined by human generally and may be not enough discriminative for the recognition task. Moreover, there is a huge different of attributes between different animals. For examples, tiger and pig both have the same semantic attributes "tail", but huge variations in visual, which is semantic gap. Thirdly, some existing models [27, 28] propose to seek for a mid-level semantic representation and introduce enhanced semantics. However, the enhanced semantics cannot describe object itself accurately.

In this paper, we propose a deep learning model to handle these issues in ZSL. We summarize our contributions on three folds: Firstly, in our model, we propose a framework to combine an autoencoder-decoder and a salient object detection model in order to find the mapping between the salient regions of image and enhanced semantics. Secondly, we make the extended semantics more discriminative via deep metric learning. Finally, we take the advantages of a CNN to find the abstract features and utilize image reconstruction to learn enhanced semantics which can describe object more accurately and reduce semantic gap.

2   Related Work

Zero-Shot Learning. Most of existing classic models can be divided into the following five types roughly. The first group is positive projection [7, 9, 15, 20]. These approaches project visual features of an instance onto semantic embedding to obtain a semantic representation. Then compare this semantic representation with all unseen semantic prototype representation and find the closest one as forecast result. The second type is class similarity [9, 22]. These methods usually train a classifier from seen classes to the label. Then utilize the relationship between the seen classes and unseen classes to transfer the knowledge and find the most similar one. The third one is mid-level semantic representation [25–28]. These work seek for a mid-level representation which will catch more important information of the data in place of semantic prototype representation. The forth type is back projection [29], which project the semantic representation onto visual space. Then compare visual features of the instance with visual prototype representation and find the closest one. The last one is based on data synthesis [11], whose models can utilize semantic prototype representation to synthesize visual features for unseen classes without real images and convert ZSL into conventional supervised problem. According to whether the unlabeled samples are used during the training period, ZSL can be roughly divided into two groups: inductive ZSL [9, 25, 29] and transductive ZSL [2, 3]. For the inductive ZSL, only samples of source domain classes are available during the training period. For the transductive ZSL, the labeled source samples and the unlabeled target samples are both utilized to accomplish the ZSL task. Depending on whether the source data is used for testing during the test phase, ZSL can also be categorized into two types: conventional ZSL [20, 22] and generalized ZSL [2, 23]. For the conventional ZSL, the most existing models assume that the test images come solely from the target classes. Different from this, generalized ZSL takes the training images from the source classes in consider during the test phase.

Salient Object Detection. Most existing models did not pay much attention on visual feature extraction. Image features from a pre-trained on ImageNet are learned, which is not optimal for a particular ZSL task. Taking it in consider, Li et al. [21] propose a zoom-net based on attention mechanism to optimize the image features. Different from [21], J. Fu et al. [10] consider that focusing on the region of object itself will contribute to object recognition. Salient object detection has gone through three waves. The first wave starts with Itti et al. [6]. After that, plenty of models are influence by them more or less. After three waves, some models [13, 24] based on deep convolutional neural network achieve great success. Inspired by this, our model will take the advantage of object salient detection to find the salient region which benefits to classification.

Lifted Structure Feature Embedding. The user-defined attributes are semantic prototype representation, but not exhaustive. They are not enough discriminative for ZSL task. Under such consideration, [21, 27, 28] propose to find a more discriminative mid-level semantic representation in place of user-defined. Generally, hinge loss or triplet loss is used to increase inter-class distance and reduce intra-class distance. However, both of them can't take full advantage of batches of deep learning. Different from them, our model introduces lifted structure feature embedding (LSFE) [12] based on deep metric learning, which optimize positive and negative samples distance via sample-pairs distance matrix.

Autoencoder. In previous works, Kodirov et al. [26] propose a semantic autoencoder (SAE) for ZSL. This model first projects visual features onto semantic attributes, and then adds constraints to make the semantic features reconstruct the visual features. Meanwhile, this model implemented through sparse coding and dictionary learning proves to alleviate domain shift problem effectively. In contrast to this, our model adopts convolutional autoencoder to catch non-linear features which can represent objects better. Besides, our model replaces semantic prototype features with enhanced semantic features and adds constrains. In general, autoencoder can be roughly divided into undercomplete autoencoders and overcomplete autoencoders. In our model, we utilize undercomplete autoencoder to learn the underlying structure of data.

3   The Proposed Method

3.1   Problem Definition

There are two sets in the training and test set respectively. Given the seen class set $\emptyset_s = \{X_s, Y_s, Z_s\}$ in the training set with c_s seen classes and n_s labeled samples. In another data set $\emptyset_u = \{X_u, Y_u, Z_u\}$ with c_u unseen classes and n_u unlabeled samples. Each sample x_i is d-dimensional visual representation vector. So it's easy to draw $X_s = [x_1, x_2, \ldots, x_{n_s}] \in R^{d \times n_s}$ and $X_u = [x_1, x_2, \ldots, x_{n_u}] \in R^{d \times n_u}$, which respectively represent seen and unseen class samples set. Z_s and Z_u represent the label of the seen and unseen class samples respectively, $Z_s \cap Z_u = \emptyset$. $Y_s \in R^{m \times n_s}$ and $Y_u \in R^{d \times n_u}$ represent m- dimensional semantic representation vector of seen and unseen class. Then semantic information of seen class Y_s is given and semantic information of unseen class Y_u is unknown (unlabeled). The target is predict Z_u by given the semantic information $P \in R^{m \times (c_s + c_u)}$.

3.2   Our Model

The framework of the proposed model consists of three main components: the salient detection network to detect salient object for image representation. Feature encoder to build the relationship between visual and enhanced semantic attributes and feature decoder to reconstruct the image from the semantic features. The whole framework is illustrated in Fig. 1.

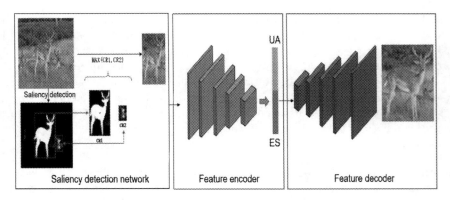

Fig. 1. The framework of the proposed model

Salient Detection Network. Our model takes the advantages of [13] to obtain the salient regions. Salient regions from the original images can be obtain by a series of short connections from the high-level features to low-level features. However, multiple salient regions will be obtained in the image. We select the maximum connected region as the final salient detection region X for it which contains more information of image and has higher probability on salient region.

Fig. 2. Reconstruction of training data in (a) and test data in (b)

Generally, the input image X' has multiple salient regions $x_1, x_2, \ldots\ldots, x_m$. We select the maximum connected region X as shown in Fig. 2. Therefore, we can obtain X as follow:

$$X = R_{max}(X') = \max(x_1, x_2, \ldots\ldots, x_m) \tag{1}$$

Feature Encoder and Visual-Semantic Mapping. Our model uses CNN to extract visual features from salient region X. We adopt a resnet_50 as feature subnet and use the final fully connected output as the visual embedding features. The salient region X is converted to a 2048 dimensional visual feature vector in this way. These processes can be represented as $\emptyset(X)$. It is vital to build the bridge from the visual embedding to the semantic embedding. In our model, we adopt fully connected layers to map the 2048 dimensional vector to D-dimensional (D depends on the dimension of user-defined attributes (UA) and enhanced semantic attributes in different datasets). The non-linear mapping W, which represents the projection parameters in fully connected layers, is used in our work. In addition, the semantic features y will be normalized as follow:

$$\varphi(y) = \frac{y}{||y||^2} \tag{2}$$

Furthermore, we extend the semantic attributes y' to make the different class more discriminative. Similarly, the semantic attributes contains UA and ES can be represented as $\varphi(y, y')$. The higher the compatibility score, the more the image matching the semantics. Finally, we can obtain the compatibility score of the hybrid model as follow:

$$F(X, y, y'; W) = \emptyset(X)^T W \varphi(y, y') \tag{3}$$

Feature Decoder. Through the above steps, we can obtain a $1 \times 1 \times D$ mid-level semantic representation. Firstly, the mid-level semantic representation converts to $1 \times 1 \times 256$ feature map by convolution. Next, we upsample by bilinear interpolation. Then we upsample by deconvolution and convolution in each stage. After that, our model can obtain a size $224 \times 224 \times 3$ image, which reconstructs to the original image.

3.3 Training

Given an image X, we can obtain the predicted semantic value by the learned non-linear mapping W. For the normalized enhanced semantic attributes $\varphi(y, y')$, we can obtain the compatibility score by inner product:

$$S = <W^T \emptyset(X), \varphi(y, y')> \tag{4}$$

For the user-defined semantic attributes, we attempt to seek for a mapping to align the predicted semantic attributes and the ground truth. This part is to train the attributes classifier. The softmax loss is used for training this part:

$$loss_1 = -\frac{1}{n} \sum_i^n \frac{\exp(<W^T X, \varphi(y)>)}{\sum_s \exp(<W^T X, \varphi(y)^s>)}, s \epsilon Y^S \tag{5}$$

For the extended semantic attributes, we aim to find the learned features to be more discriminative among different categories. Our model takes the advantages of LSFE, which can make full use of batches of deep learning. First, sample-pairs distance matrix D^2 can be obtained by Eq. (7) (It represents the distance between i-th sample and j-th sample.):

$$D_{i,j} = \left\| \varphi(y_i') - \varphi(y_j') \right\|_2^2 \tag{6}$$

Unlike the random selection of anchor points for triplet loss, LSFE selects positive pairs from a batch randomly. After that, our model utilizes nonrandom sampling to find the hardest negative edge from D^2, which contains the information needed for sub-gradient descent. LSFE loss can be written as follow:

$$loss_2 = \frac{1}{2|P|} \sum_{(i,j) \in P} max(0, J_{i,j})^2 \tag{7}$$

$$J_{i,j} = \log\left(\sum_{(i,k) \in N} exp\{m - D_{i,k}\} - \sum_{(j,l) \in N} exp\{m - D_{j,l}\}\right) + D_{i,j} \tag{8}$$

For the part of decoder, we expect the enhanced semantic attributes can be learned by image reconstruction. Moreover, our model can learn the principal components of the nonlinear combination of hidden features during the reconstruction process. Meanwhile, the learned enhanced semantic attributes can be more accurate for object feature representation. Hence, we use the MSE loss function to calculate the error between the reconstructed image and the real image:

$$loss_3 = min_{W,W'} \|X - W'WX\|_F^2 \tag{9}$$

After building the mapping from visual embedding features to enhanced semantic attributes, ZSL can be done with k nearest neighbors search in the enhanced semantic space. Actually, the loss function can be divided into three parts: user-defined attributes align loss, extended semantic LSFE loss and image reconstruction loss. Therefore, we can obtain the whole loss function as follow:

$$L = loss_1 + loss_2 + loss_3 \tag{10}$$

3.4 ZSL Prediction

It's easy to know the user-defined attributes, but the extended semantic attributes is unknown and no ground-truth. Actually, we consider that ZSL is a semi-supervised learning problem. The similar classes are more likely to have similar semantic information. So we assume that the extended semantic attributes between the seen class and unseen class also have the same similar relationship sim_s^u as the user-defined attributes between seen class semantic embedding Y_s and unseen class semantic embedding Y_u. Therefore, the similar relationship can be obtained as follow:

$$sim_s^u = argmin \left\| y_u - \sum sim_s^u y_s \right\|_2^2 + \alpha \|sim_s^u\|_2^2, s \in y_s \tag{11}$$

Our model utilizes the similar relationship sim_s^u to transfer knowledge from the user-defined attributes of the seen class y_s' to extended semantic attributes of the unseen class y_u':

$$y_u' = \sum sim_s^u y_s' \tag{12}$$

For new class classification, we can take the maximum of the compatibility score among the unseen class. Optimal solution for compatibility scores S^* among the unseen class can be obtained as follow:

$$S^* = \mathop{argmax}_{u \in y_u} \left(<W^T \emptyset(X)^T, \varphi(y, y')> \right) \tag{13}$$

4   Experiments

4.1   Datasets and Evaluation

We evaluate our proposed model on two representative ZSL datasets: Animals with Attributes2 (AwA2) [17] and Caltech-UCSD-Birds200-2111 (CUB) [18]. For AwA2 dataset, it contains 37322 images from 50 class of common animals. Each class in this dataset is labeled with 85-dim vector both continuous and binary to denote the attributes. We utilize the continuous 85-dimension class-level attributes as semantic space and adopt standard 40/10 split in our experiment. For CUB dataset, it dataset consists of 11788 images of 200 bird species. It has a fixed split for evaluation with 150 training classes and 50 test classes. Continuous 312-dimension class-level attributes provided in [18] are used.

To calculate the classification accuracy rate, zero-sample learning is the same as other single-label image classification, using the Top-1 accuracy rate. But if the accuracy rate is calculated as the average of all pictures, then the performance of the model depends largely on the densest classification. Finally, we select average per-class top-1 accuracy in GBU [17] setting as following:

$$acc = \frac{1}{||Z_u||} \sum_{c=1}^{||Z_u||} \frac{\#correct\ predictions\ in\ c}{\#samples\ in\ c} \tag{14}$$

4.2   Network Architecture Implement

The network architecture is composed of two parts: encoder and decoder. First, a size $224 \times 224 \times 3$ image is input into the network. After 5 stages, the 7×7 2048-dimensional features are obtained. Then through full connected layers, the features are mapping to $1 \times 1 \times D$ enhanced semantic feature representation, and this part is the encoder part. The decoder starts from the middle semantic feature representation. After 5 stages, it is reconstructed into a size $224 \times 224 \times 3$ image. The whole architecture is shown as Table 1.

4.3   Comparative Experiments

We set up four groups of experiments. The group 1 was performed on original image feature extraction and the enhanced semantics based on triplet loss (TL). Different from group 1, salient region is adopted in group 2. The group 3 and 4 are based on LSFE enhanced semantics, but different feature extraction. The results are shown in Table 2.

As shown in Table 2, group 2 is 2.3% and 3.4% higher than group 1 in AwA2 and CUB respectively. As for group 3 and group 4, salient region images outperform original images by 2.7% in AwA2 and 4.1% in CUB respectively. Therefore, for the same extended semantics method, we conclude that features extracted from salient region improve average per-class top-1 accuracy significantly.

For original image extraction, group3 is 3.1% and 2.2% higher than group1 in AwA2 and CUB respectively. By comparing group 2 and group 4, we can find that LSFE is

Table 1. The proposed model architecture

	Layers	Output	Convolution/pooling
Encoder	Input	224×224	
	en_conv1	112×112	$7 \times 7, 64$, stride 2
	en_conv2_x	56×56	3×3, max pooling, stride 2
			$\begin{bmatrix} 1 \times 1, 64 \\ 3 \times 3, 64 \\ 1 \times 1, 256 \end{bmatrix} \times 3$
	en_conv3_x	28×28	$\begin{bmatrix} 1 \times 1, 128 \\ 3 \times 3, 128 \\ 1 \times 1, 512 \end{bmatrix} \times 4$
	en_conv4_x	14×14	$\begin{bmatrix} 1 \times 1, 256 \\ 3 \times 3, 256 \\ 1 \times 1, 1024 \end{bmatrix} \times 6$
	en_conv5_x	7×7	$\begin{bmatrix} 1 \times 1, 512 \\ 3 \times 3, 512 \\ 1 \times 1, 2048 \end{bmatrix} \times 3$
	encoder	1×1	average pooling, D-dim, fc
decoder	de_conv1	7×7	1×1, 256 Convolution Bilinear upsampling
	de_conv2	14×14	3×3, 256 Deconvolution 3×3, 256 Convolution
	de_conv3	28×28	3×3, 256 Deconvolution 3×3, 256 Convolution
	de_conv4	56×56	3×3, 256 Deconvolution 3×3, 128 Convolution
	de_conv5	112×112	3×3, 128 Deconvolution 3×3, 64 Convolution
	decoder	224×224	3×3, 64 Deconvolution 3×3, 3 Convolution

3.5% and 2.9% higher than triplet loss in AwA2 and CUB respectively. The above experimental results demonstrate LSFE can improve the recognition accuracy in a great deal.

The training (see in Fig. 3(a)) and test data (see in Fig. 3(b)) belong to different categories, but the reconstruction process is actually an unsupervised learning process.

Table 2. The accuracy of the model on AwA2 and CUB under different extended semantics

Visual-semantic	Group	AwA2	CUB
Original image-TL	1	51.3	34.3
Salient region-TL	2	53.6	37.7
Original - LSFE	3	54.4	36.5
Salient region - LSFE	4	57.1	40.6

Table 3. Comparative experiments of proposed model based on reconstruction

Model	Feature extraction	AwA2	CUB
Non-reconstruction	Original image	54.4	36.5
Reconstruction	Original image	56.2	41.2
Non-reconstruction	Salient region	57.1	40.6
Reconstruction	Salient region	**59.4**	**47.3**

The input image is encoded and then reproduced by the decoder. After adding two constraints, the image cannot be well reconstructed, but the model can capture the main feature components of the object.

For the feature extraction from original image, the reconstruction model is 1.8% and 4.7% higher than non-reconstruction model in AwA2 and CUB respectively. For the feature extraction from salient region, by comparing reconstruction model with non-reconstruction model, we can find that the reconstruction model improve 2.3% and 6.7% in AwA2 and CUB respectively. Through the above two comparative analysis, we believe that the reconstruction hybrid model can improve the recognition accuracy effectively.

Following the evaluation setting of GBU setting [17], we compare our model with 7 other classic ZSL model in Table 4. We can see that on AwA2 and CUB, our result outperforms than DAP, CONSE, SYNC, SAE on AwA2, but lower than DEVISE, DEM, RN. For CUB dataset, our model outperforms than DAP, CONSE and SAE, but is inferior to DEVISE, SYNC, DEM and RN. However, our model takes the advantages of salient regions and LSFE. Moreover, we implement by convolutional autoencoder and decoder which make the model learn non-linear principal components of hidden features to describe object itself more accurately. Finally, through comparative experiments, we prove that our model can effectively improve predicted accuracy.

Table 4. Comparison with the results of the classic zero-sample learning model

Model	AwA2	CUB
DAP [9]	46.1	40.0
CONSE [22]	44.5	34.4
DEVISE [15]	59.7	52.0
SYNC [11]	46.6	55.6
SAE [26]	54.1	33.3
DEM [29]	67.1	51.7
RN [1]	64.2	55.6
OURS	59.4	47.3

5 Conclusion

In this paper, we propose an end-to-end deep learning model to accomplish ZSL task. Our model contains 3 parts: salient detection network, convolutional autoencoder and decoder. The salient detection network is to find the salient region of original image, which optimizes visual feature extraction. Convolutional autoencoder seeks for a mapping from salient region image to the enhanced attributes and make extended semantic attributes more discriminative. Finally, our model reconstructs salient region image by decoder to describe the object more accurately. The average per-class top-1 accuracy of our model is 59.4% and 47.3% in AwA2 and CUB respectively and these experimental results show the effectiveness of our model.

References

1. Sung, F., Yang, Y., Zhang, L., et al.: Learning to compare: relation network for few-shot learning. In: Proceedings of the IEEE Conference on Computer Vision and Pattern Recognition, pp. 1199–1208 (2018)
2. Song, J, Shen, C, Yang, Y, et al.: Transductive unbiased embedding for zero-shot learning. In: Proceedings of the IEEE Conference on Computer Vision and Pattern Recognition, pp. 1024–1033 (2018)
3. Yu, Y., Ji, Z., Li, X., et al.: Transductive zero-shot learning with a self-training dictionary approach. IEEE Trans. Cybern. **48**(10), 2908–2919 (2018)
4. He, K., Zhang, X., Ren, S., et al.: Deep residual learning for image recognition. In: IEEE Conference on Computer Vision and Pattern Recognition, pp. 770–778 (2016)
5. Lampert, C.H., Nickisch, H., Harmeling, S.: Learning to detect unseen object classes by between-class attribute transfer. In: IEEE Conference on Computer Vision and Pattern Recognition, pp. 951–958 (2009)
6. Itti, L., Koch, C., Niebur, E.: A model of saliencybased visual attention for rapid scene analysis. IEEE Trans. Pattern Anal. Mach. Intell. **20**(11), 1254–1259 (1998)
7. Akata, Z., Perronnin, F., Harchaoui, Z., et al.: Label-embedding for attribute-based classification. In: Proceedings of the IEEE Conference on Computer Vision and Pattern Recognition, pp. 819–826 (2013)

8. Rohrbach, M., Stark, M., Schiele, B.: Evaluating knowledge transfer and zero-shot learning in a large-scale setting. In: Proceedings of the IEEE Conference on Computer Vision and Pattern Recognition, pp. 1641–1648 (2011)
9. Lampert, C.H., Nickisch, H., Harmeling, S.: Attribute-based classification for zero-shot visual object categorization. IEEE Trans. Pattern Anal. Mach. Intell. **36**(3), 453–465 (2013)
10. Fu, J., Zheng, H., Mei, T.: Look closer to see better: recurrent attention convolutional neural network for fine-grained image recognition. In: Proceedings of the IEEE Conference on Computer Vision and Pattern Recognition, pp. 4438–4446 (2017)
11. Changpinyo, S., Chao, W.L., Gong, B., et al.: Synthesized classifiers for zero-shot learning. In: Proceedings of the IEEE Conference on Computer Vision and Pattern Recognition, pp. 5327–5336 (2016)
12. Oh Song, H., Xiang, Y., Jegelka, S., et al.: Deep metric learning via lifted structured feature embedding. In: Proceedings of the IEEE Conference on Computer Vision and Pattern Recognition, pp. 4004–4012 (2016)
13. Hou, Q., Cheng, M.M., Hu, X., et al.: Deeply supervised salient object detection with short connections. In: Proceedings of the IEEE Conference on Computer Vision and Pattern Recognition, pp. 3203–3212 (2017)
14. Elhoseiny, M., Saleh, B., Elgammal, A.: Write a classifier: Zero-shot learning using purely textual descriptions. In: Proceedings of the IEEE International Conference on Computer Vision, pp. 2584–2591 (2013)
15. Frome, A., et al.: Devise: Adeepvisual-semantic embedding model. In: Proceedings of Advances in Neural Information Processing Systems, pp. 2121–2129 (2013)
16. Lei Ba, J., Swersky, K., Fidler, S.: Predicting deep zero-shot convolutional neural networks using textual descriptions. In: Proceedings of the IEEE International Conference on Computer Vision, pp. 4247–4255 (2015)
17. Xian, Y., Lampert, C.H., Schiele, B., et al.: Zero-shot learning—a comprehensive evaluation of the good, the bad and the ugly. IEEE Trans. Pattern Anal. Mach. Intell. **41**(9), 2251–2265 (2018)
18. Wah, C., Branson, S.., Perona, P., et al.: Multiclass recognition and part localization with humans in the loop. In: International Conference on Computer Vision, pp. 2524–2531. IEEE (2011)
19. Russakovsky, O., Deng, J., Su, H., et al.: Imagenet large scale visual recognition challenge. Int. J. Comput. Vision **115**(3), 211–252 (2015)
20. Romera-Paredes, B., Torr, P.: An embarrassingly simple approach to zero-shot learning. In: International Conference on Machine Learning, pp. 2152–2161 (2015)
21. Li, Y., Zhang, J., Zhang, J., et al.: Discriminative learning of latent features for zero-shot recognition. In: Proceedings of the IEEE Conference on Computer Vision and Pattern Recognition, pp. 7463–7471 (2018)
22. Norouzi, M., et al.: Zero-shot learning by convex combination of semantic embeddings. arXiv preprint arXiv:1312.5650 (2013)
23. Chao, W.-L., Changpinyo, S., Gong, B., Sha, F.: An empirical study and analysis of generalized zero-shot learning for object recognition in the wild. In: Leibe, B., Matas, J., Sebe, N., Welling, M. (eds.) ECCV 2016. LNCS, vol. 9906, pp. 52–68. Springer, Cham (2016). https://doi.org/10.1007/978-3-319-46475-6_4
24. Long, J., Shelhamer, E., Darrell, T.: Fully convolutional networks for semantic segmentation. In: Proceedings of the IEEE Conference on Computer Vision and Pattern Recognition, pp. 3431–3440 (2015)
25. Kodirov, E., Xiang, T., Fu, Z., et al.: Unsupervised domain adaptation for zero-shot learning. In: Proceedings of the IEEE Conference on Computer Vision, pp. 2452–2460 (2015)
26. Kodirov, E., Xiang, T., Gong, S.: Semantic autoencoder for zero-shot learning. In: IEEE Conference on Computer Vision and Pattern Recognition, pp. 3174–3183 (2017)

27. Ye, M., Guo, Y.: Zero-shot classification with discriminative semantic representation learning. In: Proceedings of the IEEE Conference on Computer Vision and Pattern Recognition, pp. 7140–7148 (2017)
28. Jiang, H., Wang, R., Shan, S., et al.: Learning discriminative latent attributes for zero-shot classification. In: Proceedings of the IEEE International Conference on Computer Vision, pp. 4223–4232 (2017)
29. Zhang, L., Xiang, T., Gong, S.: Learning a deep embedding model for zero-shot learning. In: Proceedings of the IEEE Conference on Computer Vision and Pattern Recognition, pp. 2021–2030 (2017)

Improving Backbones Performance by Complex Architectures

Jinxin Shao[1], Yutao Hu[3], Zhen Liu[1], Teli Ma[1], and Baochang Zhang[1,2](✉)

[1] School of Automation Science and Electrical Engineering, Beihang University,
Beijing 100191, People's Republic of China
{shaojinxin,liuzhenbuaa,mt19868,bczhang}@buaa.edu.cn
[2] Shenzhen Academy of Aerospace Technology, Shenzhen 518057, Guangdong,
People's Republic of China
[3] School of Electronic and Information Engineering, Beihang University, Beijing 100191,
People's Republic of China
huyutao@buaa.edu.cn

Abstract. Recently, Convolution Neural Networks (CNNs) have achieved great success in computer vision. To further boost the performance, the depth of the backbone network is continuously increased, which improves the capacity of feature learning but also brings the heavy burden in computation. To address the issues, this paper introduces a complex convolution method to systematically improve the performance of the backbone network. Our contributions are three-fold: 1) the complex architecture backbone network can improve the classification performance without increasing or even reducing the number of parameters; 2) for the detection task, the complex architecture backbone network can improve the ability of feature map extraction, at the same time our joint bounding box generation method using both real and imaginary parts of complex features can obviously improve the object detection ability. 3) the proposed method has a strong generalization ability for both detection and classification tasks. We have achieved significant performance improvements in both classification and detection tasks, which validate the effectiveness of our methods.

Keywords: Complex architectures · Backbones performance · Complex feature map

1 Introduction

Backbone design is significant in the field of computer vision, especially for classification and detection tasks. In recent years, with the development of machine learning technology, the main methods of classification and object detection has changed from a feature-based method to a convolution neural networks (CNNs) based method [1, 2]. For the detection and classification tasks, both of them need to use a suitable and efficient backbone network to extract feature maps from the input image. Classification tasks often use fully connected layers to deal with the feature map and then calculate loss. For detection tasks, it is necessary to use information of feature maps and labels directly to calculate loss. For existing backbone networks, the direction of improvement could be concluded in two aspects: improving accuracy and saving parameters. To improve the network accuracy, the depth of the network is continuously increased, but for deep networks with more than 20 layers, there will be obvious degradation [3]. To address this issue, He et al. proposed residual neural network [4]. Inspired by the idea of residual learning, the use of identity mapping not only alleviates the problems of gradient explosion and gradient disappearance caused by the increase in network depth, but also avoids the degradation of the network and enables the network depth to reach thousands layers. The representative networks proposed under this idea are ResNet [5], ResNeXt [6] and Res2Net [7]; To meet the needs of the booming edge computing technology [8], small backbone networks with fewer layers have also been proposed. They can save lot parameters by reducing the number of convolution layers. The lightweight networks under this idea include MobileNet [9], ShuffleNet [10] and SqueezeNet [11]. Meanwhile, the pre-trained model of the backbone network of the classification task can be used for the detection task to improve the performance of the detection task. Therefore, it is of great importance to design a structure to balance number of parameters and prediction accuracy. In a summary, how to better improve the performance of the backbone network, that is, based on fewer backbone network parameters to obtain better-performing classification and detection results has very important research significance.

In this paper, we utilize complex structure to improve backbone network performance. Based on the existing complex convolution, complex batch normalization, complex ReLU and, complex weight initialization strategy [12], we follow the line of these algorithms and propose complex down sampling, complex dropout, etc. Using these complex architectures, we transform several backbone networks into complex networks, and proposed several methods of combining complex feature maps to evaluate the classification accuracy of the complex backbone networks. For the detection task, we use the YOLOv3 model to test the efficacy of the complex backbone network based on the VOC dataset. The backbone network we tested is not limited to darknet53: to test whether this change is effective for a wide range of backbone networks, we deleted 15 layers of residual blocks of darknet-53, that is, 30 convolution layers, showing that the algorithm has a lifting effect on existing backbone networks.

The main contributions of this work are as follows:

1) We show the employment of complex convolution backbone networks can improve the classification performance without increasing the amount of parameters, based on our effective combination of real and imaginary feature maps.
2) Extensive experiments demonstrate that the use of real and imaginary feature maps in the same framework can improve the detection accuracy.

2   Complex Convolution Neural Networks

Since complex numerical operations are mostly used in the field of signal analysis, most complex neural networks are applied to the speech signals for enhancing the phase information or predict spectrum. Trabelsi [12], which originally integrated a complex neural network, utilized a complex neural network to test the music transcription of the MusicNet dataset and the speech spectrum prediction and achieved good results. Choi [13] proposed the Deep Complex U-Net model for evaluation on a mixture of Voice Bank corpus and DEMAND database, which has been widely used by many deep learning models for speech enhancement. Pfeifenberger [14] estimates the complex weights by using the full potential of complex-valued LSTM, MLP, and directly obtains beamforming weights from complex-valued microphone array data. A complex-valued deep neural network for speech enhancement and source separation is proposed. It can be seen that most of the improvement work of complex neural networks is applied to speech signal processing, and this work attempts to use it in visual tasks. The composition principle of the complex neural network is almost the same, as shown below.

In this network, after the initialization of complex values, real and imaginary parts of the complex numbers are treated as logically different real-valued entities. By this way we can use real-valued algorithms to simulate complex number operations internally.

Note that the real part of the complex convolution kernel matrix is W_{real}, the imaginary part is W_{imag}, the real part of the input image vector is written as x_{real}, and the imaginary part is written as x_{imag}. In particular, the imaginary part here is represented by real numbers. In the convolution operation, the formula is written as follows:

$$(W * x)_{real} = W_{real} * x_{real} - W_{imag} * x_{imag}$$
$$(W * x)_{imag} = W_{imag} * x_{real} + W_{real} * x_{imag} \tag{1}$$

The '*' represents a two-dimensional real convolution operation. Expressed in matrix form as:

$$\begin{bmatrix} (W * x)_{real} \\ (W * x)_{imag} \end{bmatrix} = \begin{bmatrix} W_{real} & -W_{imag} \\ W_{imag} & W_{real} \end{bmatrix} * \begin{bmatrix} x_{real} \\ x_{imag} \end{bmatrix} \tag{2}$$

2.1 Complex Batch Normalization

For batch normalization of real data, only one-dimensional data needs to be converted into a normal distribution [15]. For complex data, real and imaginary part may have different variances, which will bring bias into the data. Therefore, we treat it as the two-dimensional data, and use the covariance matrix V to normalize the eccentricity of it. As shown in the Eq. (3), $x - E[x]$ refers to the deviation of the two-dimensional data from the center.

$$\tilde{x} = (V)^{-\frac{1}{2}}(x - E[x]) \tag{3}$$

Where the covariance matrix V ψ is denoted as:

$$V = \begin{pmatrix} V_{rr} & V_{ri} \\ V_{ir} & V_{ii} \end{pmatrix} = \begin{pmatrix} \text{Cov}(x_{real}, x_{real}) & \text{Cov}(x_{real}, x_{img}) \\ \text{Cov}(x_{img}, x_{real}) & \text{Cov}(x_{img}, x_{img}) \end{pmatrix} \tag{4}$$

Where V needs to satisfy the condition of positive semi-definite matrix to make the inverse matrix of V in the above formula be solvable. After mathematical derivation, the conditions to be met are $V_{rr} + V_{ii} = 1$, $V_{ri} = V_{ir} = 0$. Similarly, imitating the batch normalization formula of the real-value network, the input complex values are scaled and translated as follow:

$$BN(\tilde{x}) = \gamma \tilde{x} + \beta, \ \gamma = \begin{pmatrix} \gamma_{rr} & \gamma_{ri} \\ \gamma_{ir} & \gamma_{ii} \end{pmatrix} \tag{5}$$

where γ_{rr} and γ_{ii} are initialized to $\frac{1}{\sqrt{2}}$, γ_{ir}, γ_{ri} and β are initialized to 0.

2.2 Complex ReLU and Other Functions

The ReLU involved in our proposed module is a complex ReLU, which is also called the CReLU. It is a separate ReLU activation applied to both the real and imaginary parts of the neuron, defined as:

$$CReLU(x) = ReLU(x_{real}) + iRELU(x_{imag}) \tag{6}$$

When the real and the imaginary part are the same sign, that is, when the input complex number is in the first or third quadrant, the formula satisfies the Cauchy-Riemann equation obviously. A series of other complex methods also adopt this idea, first divide the real and imaginary part, then treat them as independent real data, such as complex pooling, complex sigmoid.

3 Methodology for Using Complex Structure in Object Classification and Object Detection

For classification and detection tasks, the final prediction depends on the feature map of backbone network. In the classification tasks, the final feature map is a one-dimensional vector. While in the detection tasks, a high-dimensional tensor is often used. Therefore, it can be said that the network for classification task is composed of the backbone network and classifier. While the network for detection task is composed of the backbone network and the object detection business part [16]. Therefore, for classification tasks, it is only necessary to design some simple rules to combine complex-valued low-dimensional feature maps, while for detection tasks, it is necessary to flexibly design the application of complex feature maps according to the characteristics of object detection business part.

Therefore, to produce one-dimensional feature map, we designed four functions to combine the obtained one-dimensional complex feature map. They are called magnitude, signed-magnitude, summation and absoluted-summation respectively. We also try to directly convert complex feature maps into real feature maps through convolution, and the experimental results show all of them improve the performance of backbone network. In the detection task, we use the complex feature map's real part, the combined feature map of the real and imaginary parts, and the fully complex feature map of both real and imaginary parts combined with non-maximum suppression method to improve the detection accuracy. These methods gain improved detection results by making better use of the information of complex feature maps.

3.1 Complex Feature Map Combination Method

Image classification is a basic problem in image understanding. There are lots of data sets for evaluating image classification effects, such as CIFAR-10/100 [17], Caltech-101/256 [18] and ImageNet [19]. With the great success of the AlexNet [20] deep convolutional neural networks based methods have begun to replace traditional hand-crafted algorithms, and a series of effective backbone networks have been proposed. Based on AlexNet, some improved backbone, such as DenseNet [21], GoogleNet [22], ResNet [4], VGG [23], SENet [24] and ShuffleNet [10] have been proposed and achieved great success. Some classification networks that combine CNN with traditional image processing methods, such as GCN [25], have also achieved good results.

In this work, we use several widely used backbones to verify the improved characteristics of complex convolution for classification in the CIFAR dataset. The basic VGG network structure is shown in Fig. 1; the other two backbone network improvement methods are similar to it. For the CIFAR task, a linearization layer with the width of 4096 is not required. Too many linearization layers is also the reason for the excessive network parameters. We found that removing these fully connected layers does not influence the performance of the network, but on the other hand, can reduce the number of parameters. Therefore, some experiments used this structure which removed a fully connected layer, as shown in the Fig. 1, the 1st chart.

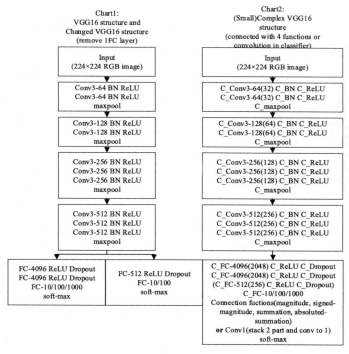

Fig. 1. The scheme of complex architectures for VGG16 network

Based on this motivation, we use complex architectures on the backbone network, which will double its parameter amount (see related work for the method). For the classifier part, if the fully-connected layer is also complex, some methods need to be designed to combine two one-dimensional vectors. And to verify whether this structure can improve performance when the parameter amount is constant or even reduced, we deleted half the number of convolution kernels to perform the same experiment, as shown in Fig. 1, the 2nd chart.

We design five different methods to combine the complex vectors after the complex classifier, four are mathematical calculation methods, the other one uses 1×1 convolution. Magnitude method just treats two parts as mathematical complex numbers, like $v_{real} + v_{imaginary}i$, which $v_{real}, v_{imaginary}$ are 2 vectors on behalf of the input. The combined result is v_{output}. So this method just calculates the magnitude of it, which is written as:

$$\text{Magnitude:} v_{output} = (v_{real}^2 + v_{imaginary}^2)^{\frac{1}{2}} \tag{7}$$

Obviously it is always positive, no negative number will occur. It does not matter for the sigmoid function. But in more fuzzy work, this kind of feature map may cause problem because the feature map should be signed. Therefore, an improved method called signed-magnitude is proposed. The major difference is that the symbolic function is used to make the results keep the same sign with the real part, so it can be written as:

$$\text{Signed - Magnitude:} v_{output} = (v_{real}^2 + v_{imaginary}^2)^{\frac{1}{2}} \times \text{sgn}(v_{real}) \tag{8}$$

So inspired by this idea, we designed 2 more combine function. In these method both the real and imaginary part are treated as 1-dimension data. So if we want a signed output vector, which just calculate the summation, written as:

$$\text{Summation:} v_{output} = v_{real} + v_{imaginary} \tag{9}$$

If we want positive results, just calculate the sum of absolute value, which we called absoluted-summation, written as:

$$\text{Absoluted - Summation:} v_{output} = |v_{real}| + |v_{imaginary}| \tag{10}$$

Also 1-by-1 Convolution Layer is used to combine the 2 part into one, it just like another full connection layer. Since the input tensor is a one-dimensional vector, just the kernel size equals 1 may suit for this work. The formula is shown in Eq. (11).

Conv1:use 1×1 convolution kernel to connect the real and imaginary part   (11)

3.2 Joint Bounding Box Generation Method of Complex Feature Map

Detection task is a middle-level problem in the field of computer vision, that is, it needs to understand the foreground and background of the image. So far, object detection methods based on deep learning can be divided into 2 categories: two-stage detection methods and one-stage detection methods. The two-stage detection methods delineate the detection area first, and then determine whether there are targets in the selected area. R-CNN [26], fast R-CNN [27], faster R-CNN [28] and SPP-Net [29] are representative two-stage detection methods.. One-step detection method uses intensive sampling directly from the feature map to obtain the prior frame, and then do classification and regression on the prior frame. Obviously, one-stage method has a faster detection speed. Representatives of such method are YOLO [30], SSD [31], OD-GCN [32], and RON [33]. With the introduction of YOLOv4 [34], this type of method has achieved a balance between detection accuracy and efficiency, becoming the main trend of future research.

Therefore, we take YOLOv3 as an example to study the improvement effect of complex architectures on the backbone network in our work. To judge whether this method will improve performance under the condition of no pre-training model or any backbone network, we deleted the 15-layer residual connection layer of darknet-53, that is, the 30-layer convolution layer. So it can also be called "dark-23" as a comparison in the experiments. In the same way, we also conduct the complex architectures on original darknet-53 backbone network. Because of the lack of pre-trained model about the complex darknet-53 backbone network, we repeatedly assign the original real darknet-53 pre-trained model to the real and imaginary parts of the complex model. This pre-trained model may not be ideal. If there is a better pre-trained model, a better detection effect may be achieved. And the effectiveness of improvement can be judged by the comparison of different feature map using methods.

The output part of the backbone network is shown in Fig. 2. The specific structure of the darknet-53 and "darknet-23" backbone network can also be clearly seen from this figure. In this part, we use complex convolution, complex batch normalization, complex leaky ReLU and others to replace the real-valued function, which is also clear in the diagram. So we just make the input image in which real part equal to its real and imaginary part equal to zero. After this module, we will get 3 complex 3-dimensional tensors. If it is regarded as a real tensor, it can be regarded as 4-dimensional tensor. How to use this tensor for training and testing is shown in Fig. 3.

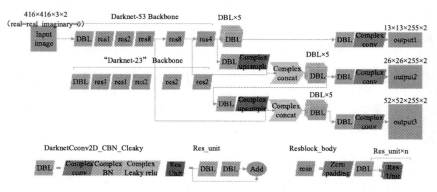

Fig. 2. The complex architectures for darknet-53 and "darknet-23"

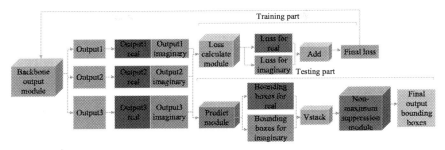

Fig. 3. The principle of training and testing using the output complex-value feature map

Here we conduct 3 experiments. First, we use the real part of the output complex feature map. Second, we use the signed-magnitude of the output complex feature map (see Eq. (8)). Last, we use both the real and imaginary parts of the output feature map for training and testing. For using real part of complex feature map and using signed-magnitude of complex feature map, these two methods actually convert the 4-dimensional output tensor into 3-dimensional tensor. Therefore, for subsequent processing parts, such as loss function calculation, or prediction box generation, there is no need to transform these parts. However, if we want to use both real and imaginary parts of complex tensor at the same time, it is equivalent to use two 3-dimensional feature maps. Here we treat the real and imaginary parts as two independent feature maps to calculate the loss function and generate the detection frame respectively. Then, the losses are combined together through summation. Therefore, the totally loss can be written as:

$$L_{final} = L_{real} + L_{imaginary} \qquad (12)$$

According to the principle of back propagation, it can be derived and the training parameters can feedback on the backbone network automatically. During the test process, the real and imaginary parts will simultaneously generate bounding boxes from the prediction model. At this time, we put the prediction frames generated by the two parts in the same stack, and use the non-maximum suppression method to remove the repeated prediction frames to obtain the most comprehensive result. Therefore, the final output of bounding boxes can be written as:

$$bboxes_{output} = bboxes_{real} \cup bboxes_{imaginary} \qquad (13)$$

Employing the designed method, the information of all complex feature maps can be used as much as possible by reducing the missed detection rate of objects.

4   Result Analysis

4.1   Result Analysis for Object Classification

In the experiments, we involve 3 different backbones, VGG16 [23], ResNet18 [4] and SENet18 [24] for comparison. It is clearly that the connection methods designed for complex architectures are obviously robust to classification problems. We designed 5 methods to connect the complex feature map and the final output result. All the classification results have been significantly improved, which shows that the complex backbone network has fundamentally improved the performance of the backbone network. When the number of parameters of the multiple backbone networks is doubled, from the experimental results of the three networks, a performance improvement of 2–4% points can be achieved. Among them, due to the parameter quantity's redundancy of the fully connected layer, the VGG16 network's parameter quantity can be basically unchanged. Similarly, when the number of parameters is reduced to the half by reducing the number of convolution channels, according to the statistical results of the three network experiments, a classification improvement of 1–2.5% points can be achieved (VGG16 network can be reduced to a quarter of the original parameter). The above experiments are sufficient to demonstrate that the improvement of the complex architectures backbone network is owing to the backbone itself (Table 1).

4.2   Result Analysis for Object Detection

In this part, we use the VOC data set to evaluate the improvement of the detection effect by complex architectures network. We employ VOC2007trainval dataset and VOC2012trainval dataset as the training data, while utilize VOC2007test dataset during the test. For the part without pre-trained model, "darknet-23" is used as backbone network. We tested non-complex convolution model and three variants of the complex convolution model (using only real part of feature map, using signed-magnitude of feature map, and using both real and imaginary parts of feature map) on it. Obviously, for any networks, the involvement of complex convolution on the backbone network will improve the feature map extraction ability. For the part with pre-trained model, since it is hard to train the complex darknet-53 pre-trained model on ImageNet dataset due to the hardware limitations, the real-number pre-trained model is loaded twice for both real and imaginary parts. Therefore, in this part, we don't compare the results with original darknet53 detection model. On the other hand, we compare the 3 improved models to evaluate the effectiveness of the improvement. We can see that under the condition of using the same complex convolution backbone network, employing full-feature map information (signed-magnitude) will improve the detection effect by nearly 5% points compared with only using real-part feature map. More than that, using the information of the real and imaginary feature map together by NMS will improve the detection effect by nearly 8% points compared with only using the real part of feature map. This shows that our proposed method, using non-maximum suppression methods to jointly apply real and imaginary feature maps for bounding box prediction, is effective in object detection task. Additionally, it can be applied to various detection models and tasks in the future (Table 2).

Table 1. CIFAR10 result by complex method.

cifar10 (%)	Baseline	Summation	Magnitude	Absoluted-summation	Signed-magnitude	Convolution1 × 1 (after classifier)
VGG16	84.44 (0)	88.25 (3.81)	88.75 (4.31)	88.76 (4.32)	88.65 (4.21)	88.55 (4.11)
Parameter numbers:	33.647 m	67.273 m	67.273 m	29.977 m (remove 1 FC layer)	29.977 m (remove 1 FC layer)	29.977 m (remove 1 FC layer)
ResNet18	88.04 (0)	90.23 (2.19)	90.06 (2.02)	90.21 (2.17)	90.68 (2.64)	90.38 (2.34)
Parameter numbers:	11.174 m	22.353 m	22.353 m	22.353 m	22.353 m	22.353 m
SENet18	87.4 (0)	90.24 (2.84)	90.38 (2.98)	90.11 (2.71)	90.34 (2.94)	90.47 (3.07)
Parameter numbers:	11.390 m	22.771 m	22.771 m	22.771 m	22.771 m	22.771 m

cifar10 (%)	Baseline	Small summation	Small magnitude	Small absoluted-summation	Small signed-magnitude	Small convolution1 × 1 (after classifier)
VGG16	84.44 (0)	87.24 (2.8)	86.79 (2.35)	86.85 (2.41)	87.08 (2.64)	87.76 (3.32)
Parameter numbers:	33.648 m	16.845 m	16.845 m	7.503 m (remove 1 FC layer)	7.503 m (remove 1 FC layer)	7.503 m (remove 1 FC layer)
ResNet18	88.04 (0)	89.14 (1.1)	89.05 (1.01)	89.04 (1)	89.85 (1.81)	89.16 (1.12)
Parameter numbers:	11.174 m	5.598 m	5.598 m	5.598 m	5.598 m	5.598 m
SENet18	87.4 (0)	88.98 (1.58)	89.14 (1.74)	89.47 (2.07)	89.76 (2.36)	89.08 (1.68)
Parameter numbers:	11.390 m	5.641 m	5.641 m	5.641 m	5.641 m	5.641 m

Table 2. VOC result by complex method.

VOC(mAP)	Baseline	Real	Signed-magnitude	Real and imaginary with NMS
"Darknet-23" (without pretrain model)	0.608413 (0)	0.617585 (0.009172)	0.622672 (0.014259)	0.631942 (0.023529)
Darknet-53 (load pretrain model both real and imaginary)	–	0.685472 (0)	0.731432 (0.04596)	0.759334 (0.073862)

5 Conclusion and Future Works

In this work, we use CIFAR and VOC datasets to verify the effectiveness of the complex architecture on the backbone network. Through theoretical analysis and experimental verification, the following conclusions can be obtained. The complex convolution architectures can improve the performance of feature extraction of backbone network and improve performance in either classification or object detection. Moreover, for classification tasks, classification performance can be improved without increasing or even reducing the number of parameters. In the object detection task, when the prediction frames jointly listed by the real and imaginary feature map are combined by using the non-maximum suppression, it will significantly improve the detection performance. Moreover, this method could be generalized for any detection task.

We plan to employ ImageNet and COCO data sets for more scientifically verifying the ability of complex architectures on improving the performance of the backbone network. Meanwhile, we expect to explore whether the joint detection method of real and imaginary parts based on non-maximum suppression has certain improvement capabilities on some special tasks, such as small target detection tasks or unclear target detection tasks, by reducing the missed detection rate for difficult-to-identify targets.

Acknowledgement. Baochang Zhang is also with Shenzhen Academy of Aerospace Technology, Shenzhen, China, and he is the corresponding author. He is in part Supported by Shenzhen Science and Technology Program (No. KQTD2016112515134654).

References

1. Jiao, L., Zhang, F., Liu, F., et al.: A survey of deep learning-based object detection. IEEE Access **7**, 128837–128868 (2019)
2. Hu, Y., Yang, Y., Zhang, J., et al.: Attentional kernel encoding networks for fine-grained visual categorization. IEEE Trans. Circ. Syst. Video Technol. (2020)
3. He, K., Zhang, X., Ren, S., Sun, J.: Identity mappings in deep residual networks. In: Leibe, B., Matas, J., Sebe, N., Welling, M. (eds.) ECCV 2016. LNCS, vol. 9908, pp. 630–645. Springer, Cham (2016). https://doi.org/10.1007/978-3-319-46493-0_38

4. He, K., Zhang, X., Ren, S., et al.: Deep residual learning for image recognition. In: Proceedings of the IEEE Conference on Computer Vision and Pattern Recognition, pp. 770–778 (2016)
5. He, K., Gkioxari, G., Dollár, P., et al.: Mask R-CNN. In: Proceedings of the IEEE International Conference on Computer Vision, pp. 2961–2969 (2017)
6. Xie, S., Girshick, R., Dollár, P., et al.: Aggregated residual transformations for deep neural networks. In: Proceedings of the IEEE Conference on Computer Vision and Pattern Recognition, pp. 1492–1500 (2017)
7. Gao, S., Cheng, M.M., Zhao, K., et al.: Res2net: a new multi-scale backbone architecture. IEEE Trans. Pattern Anal. Mach. Intell. (2019)
8. Chen, J., Ran, X.: Deep learning with edge computing: a review. Proc. IEEE **107**(8), 1655–1674 (2019)
9. Howard, A.G., Zhu, M., Chen, B., et al.: Mobilenets: efficient convolutional neural networks for mobile vision applications. arXiv preprint arXiv:1704.04861 (2017)
10. Zhang, X., Zhou, X., Lin, M., et al.: Shufflenet: an extremely efficient convolutional neural network for mobile devices. In: Proceedings of the IEEE Conference on Computer Vision and Pattern Recognition, pp. 6848–6856 (2018)
11. Iandola, F.N., Han, S., Moskewicz, M.W., et al.: SqueezeNet: AlexNet-level accuracy with 50x fewer parameters and < 0.5 MB model size. arXiv preprint arXiv:1602.07360 (2016)
12. Trabelsi, C., Bilaniuk, O., Zhang, Y., et al.: Deep complex networks. arXiv preprint arXiv:1705.09792 (2017)
13. Choi, H.S., Kim, J.H., Huh, J., et al.: Phase-aware speech enhancement with deep complex U-Net. arXiv preprint arXiv:1903.03107 (2019)
14. Pfeifenberger, L., Zöhrer, M., Pernkopf, F.: Deep complex-valued neural beamformers. In: ICASSP 2019-2019 IEEE International Conference on Acoustics, Speech and Signal Processing (ICASSP), pp. 2902–2906. IEEE (2019)
15. Ioffe, S., Szegedy, C.: Batch normalization: accelerating deep network training by reducing internal covariate shift. arXiv preprint arXiv:1502.03167 (2015)
16. Li, Z., Peng, C., Yu, G., et al.: Detnet: a backbone network for object detection. arXiv preprint arXiv:1804.06215 (2018)
17. Krizhevsky, A., Hinton, G.: Learning multiple layers of features from tiny images. In: Handbook of Systemic Autoimmune Diseases, vol. 1, no. 4 (2009)
18. Fei-Fei, L., Fergus, R., Perona, P.: Learning generative visual models from few training examples: an incremental bayesian approach tested on 101 object categories. In: 2004 Conference on Computer Vision and Pattern Recognition Workshop, p. 178. IEEE (2004)
19. Deng, J., Dong, W., Socher, R., et al.: Imagenet: a large-scale hierarchical image database. In: 2009 IEEE Conference on Computer Vision and Pattern Recognition, pp. 248–255. IEEE (2009)
20. Krizhevsky, A., Sutskever, I., Hinton, G.E.: Imagenet classification with deep convolutional neural networks. In: Advances in Neural Information Processing Systems, pp. 1097–1105 (2012)
21. Huang, G., Liu, Z., Van Der Maaten, L., et al.: Densely connected convolutional networks. In: Proceedings of the IEEE Conference on Computer Vision and Pattern Recognition, pp. 4700–4708 (2017)
22. Szegedy, C., Liu, W., Jia, Y., et al.: Going deeper with convolutions. In: Proceedings of the IEEE Conference on Computer Vision and Pattern Recognition, pp. 1–9 (2015)
23. Simonyan, K., Zisserman, A.: Very deep convolutional networks for large-scale image recognition. arXiv preprint arXiv:1409.1556 (2014)
24. Hu, J., Shen, L., Sun, G.: Squeeze-and-excitation networks. In: Proceedings of the IEEE Conference on Computer Vision and Pattern Recognition, pp. 7132–7141 (2018)
25. Luan, S., Chen, C., Zhang, B., et al.: Gabor convolutional networks. IEEE Trans. Image Process. **27**(9), 4357–4366 (2018)

26. Girshick, R., Donahue, J., Darrell, T., et al.: Rich feature hierarchies for accurate object detection and semantic segmentation. In: Proceedings of the IEEE Conference on Computer Vision and Pattern Recognition, pp. 580–587 (2014)
27. Girshick, R.: Fast R-CNN. In: Proceedings of the IEEE International Conference on Computer Vision, pp. 1440–1448 (2015)
28. Ren, S., He, K., Girshick, R., et al.: Faster R-CNN: towards real-time object detection with region proposal networks. In: Advances in Neural Information Processing Systems, pp. 91–99 (2015)
29. He, K., Zhang, X., Ren, S., et al.: Spatial pyramid pooling in deep convolutional networks for visual recognition. IEEE Trans. Pattern Anal. Mach. Intell. **37**(9), 1904–1916 (2015)
30. Redmon, J., Divvala, S., Girshick, R., et al.: You only look once: unified, real-time object detection. In: Proceedings of the IEEE Conference on Computer Vision and Pattern Recognition, pp. 779–788 (2016)
31. Liu, W., et al.: SSD: Single Shot MultiBox Detector. In: Leibe, B., Matas, J., Sebe, N., Welling, M. (eds.) ECCV 2016. LNCS, vol. 9905, pp. 21–37. Springer, Cham (2016). https://doi.org/10.1007/978-3-319-46448-0_2
32. Liu, Z., Jiang, Z., Wei, F.: OD-GCN object detection by knowledge graph with GCN. arXiv preprint arXiv:1908.04385 (2019)
33. Kong, T., Sun, F., Yao, A., et al.: Ron: reverse connection with objectness prior networks for object detection. In: Proceedings of the IEEE Conference on Computer Vision and Pattern Recognition, pp. 5936–5944 (2017)
34. Bochkovskiy, A., Wang, C.Y., Liao, H.Y.M.: YOLOv4: Optimal Speed and Accuracy of Object Detection. arXiv preprint arXiv:2004.10934 (2020)

Residual Attention SiameseRPN for Visual Tracking

Xu Cheng[✉], Enlu Li, and Zhangjie Fu

School of Computer and Software, Nanjing University of Information Science and Technology,
Nanjing 210044, China
xcheng@nuist.edu.cn

Abstract. Visual tracking demands to perform the accurate object location given the object state of the first frame. The existing methods have proposed various ways to handle the challenging problems, yet few of them take the relationship between shallow features and deep semantic features into account. Based on an extensive analysis, we first propose a residual attention SiameseRPN visual tracking method for accurate object state estimation, which introduces the correlation filter in a Siamese network framework. A novel loss function is presented to enhance the discriminative capability. Our approach is derived from three different loss terms that is capable of training a model in a few iterations. Second, we present channel attention mechanism to improve the tracking performance, which is offline trained to capture the general features in the tracking. Third, the proposed tracking model is trained in end-to-end manner and takes full advantage of both low-level representation for correlation filter and high-level semantic features for deep object representation by using multi-task learning strategy which can mine the relationship from both levels. Our approach benefits from two complementary effects. Finally, extensive evaluation and ablation studies demonstrate the effectiveness of the proposed tracking approach. Our tracker achieves state-of-the-art performance on five challenging benchmarks, which proves great potentials in balancing accuracy and speed.

Keywords: Surveillance · Deep learning · Correlation filter · Siamese network · Attention

1 Introduction

Visual tracking is one of the fundamental tasks in computer vision, and has many practical applications, such as human-computer interaction, action recognition, scene understanding, visual navigation, automatic driving and so on. Although much progress has been done in the past decade, it still remains challenging for a tracker to work at a high speed and is robust to complex scenarios including occlusion, illumination variations, low resolution, background clutter, and motion blur.

Recent deep learning based trackers and correlation filter based trackers have shown great potential for robust and fast tracking. Although basic CF has a high running speed due to their element-wise multiplications using Fast Fourier Transform. For complex

© Springer Nature Switzerland AG 2020
Y. Peng et al. (Eds.): PRCV 2020, LNCS 12306, pp. 407–419, 2020.
https://doi.org/10.1007/978-3-030-60639-8_34

scenarios, however, the accuracy of basic CF trackers often drops considerably. Deep network model has been widely developed to improve tracking performance due to their strong feature representation. Most existing approaches rely on the amount of the training data. The deep network model is extensively trained on large benchmarks offline and aggressively learned the object sequences online. These approaches have achieved very good results on some recent challenges.

Despite all these significant progress, most trackers suffer from several weaknesses and still can't attain consummate results. First, the training datasets are far smaller than other visual datasets such as ImageNet. The insufficient training data may cause the deep network model ineffective when facing all kinds of tracking challenges. Second, deep features learned offline can't adapt to specific object or unseen categories well during the tracking. Third, model updating schemes from these methods inevitably affect the network model adaptability, which degrades the tracking accuracy and increases the computationally expensive. These limitations lead to inferior accuracy.

To tackle the above limitations, our contributions can be summarized as follows.

(1) We propose a residual attention SiameseRPN method for visual tracking, which is an end-to-end deep network architecture. A novel loss function from three different aspects is presented to enhance the discriminative capability. Correlation filter layer and semantic feature layer are used to mine the relationship both low-level and high-level features in multi-task learning framework.
(2) An effective attention mechanism is utilized within the Siamese network architecture, which offline learns feature representations to adapt online object tracking.
(3) Numerous experimental results on five challenging benchmarks show that the proposed tracking method achieves state-of-the-art performance.

The rest of the paper is organized as follows. In Sect. 2, we review related work of existing object tracking algorithms. Section 3 briefly introduces the generative adversarial network. In Sect. 3.3, we introduce our approach for visual tracking. In Sect. 4, we present experimental results in two tracking benchmarks. Finally, Sect. 5 concludes this paper.

2   Related Work

There are extensive surveys of visual tracking in literature [1, 2]. We mainly discuss the representative trackers based on deep learning and correlation filters.

Deep Learning Tracking. Deep learning has been widely used to improve tracking performance. Some tracking methods combine deep learning models with correlation filters such as HCF [3], DeepSRDCF [4], ECO [5]. Another method formulates tracking task as a classification or regression problem, including CNN-SVM [6], DeepTrack [7], FCNT [8], TSN [9]. The advantage of these trackers is that they utilize the superior representation power of deep features. However, tracking speed is reduced due to online updating of the deep network model.

Recently some deep model based approaches are trained on videos offline and used to track the object online through an end-to-end deep network learning such as MDNet

[10], CFNet [11], RTT [12], ACFN [13]. The aforementioned problems have been most successfully addressed by Siamese network architecture [14–20]. SINT [14] formulates tracking task as a verification problem and trains a Siamese network model for object matching during the tracking. Similar methods include SiamFC [15], SiamRPN++[16], SiamRPN [17], SiamMN [18], Deeper and Wider Network [19] and SINT++ [20], etc. The VITAL tracker [21] generates hard samples by using adversarial learning and leverages the class imbalance with an effective loss. These methods advance the development of end-to-end deep network model and achieve the promising results on some challenging benchmarks. However, deep network model may suffer from over-fitting due to deficiency of training data.

Correlation Filter Tracking. Recent advances of correlation filter (CF) have achieved great success in terms of speed and accuracy [22–34]. We arrange these algorithms in a hierarchy and classify them into two categories: Basic correlation filter based trackers and regularized correlation filter based trackers. Some basic CF trackers have been developed to boost performance in tracking by using scale estimation [23], spatial constraints [24], reducing boundary effects [25], and long-term tracking [26]. However, basic CF trackers are limited in their detection range since they require the filter size and patch size to be equal. To address this issue, several regularized CF based trackers are proposed, including SRDCF [24], STRCF [27], ACFN [13], DeepSRDCF [4], ECO [5], DMSRDCF [28], C-COT [29], DCFNet [31], CSR-DCF [30], SAMF [32], MCPF [33], ATOM [34], etc. Among others, some trackers combine CF with deep features, which have shown significant improvement.

In this work, we focus on residual attention SiameseRPN for visual tracking. Different from the goals of the above mentioned approaches, our multi-stream network architecture is proposed to address the problem of object drift by using attention mechanism in the sequences.

3   The Proposed Tracking Method

In this section, we first introduce the network architecture of the proposed approach, and then give a detailed training process and loss function. Finally, we apply our model to visual tracking task.

3.1   ResNet Based Siamese Tracking Method

The Siamese network based object tracking methods [14] formulate visual tracking as a matching problem between the object template and the search area. The similarity measure is learned from Siamese deep network structure. The object state is usually given in the first frame of the sequence and can be used as object template z. The goal is to find the most similar candidates from the following frame x.

$$f(z, x) = \phi(z) * \phi(x) + b \qquad (1)$$

where $\phi()$ is a semantic embedding space; $f()$ denotes a similarity function; b is bias.

Furthermore, SiameseRPN [17] is trained with a ResNet-50 backbone by the spatial aware sampling scheme, which can overcome the translation invariance, asymmetrical features for classification and regression.

Network Architecture: We utilize ResNet-50 as base network architecture and feature extractor. Different from the original ResNet which has a large stride of 32 pixels, we reduce the strides at the last block from 16 pixels and 32 pixels to 8 pixels by modifying the conv5 block to have unit spatial stride. We crop the center 7×7 regions as the object template features to reduce a heavy computational burden on the correlation module. Furthermore, we fine-tune ResNet to improve the tracking accuracy. The parameters of deep network model are jointly trained in an end-to-end manner. The flowchart of the proposed tracker is shown in Fig. 1.

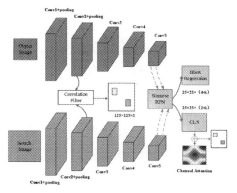

Fig. 1. The flowchart of the proposed tracking method. The intermediate layers in the common feature extractor have been omitted for clarity.

For the ResNet-50 network model, we utilize multi-level features extracted from the first residual block and the last residual block for tracking, respectively. The outputs of two-level features (Conv1, Conv5) are denoted as $F_1(\cdot)$, and $F_5(\cdot)$, respectively. On the one hand, object localization is obtained using the correlation filter working on the low-level fine-grained representations. Correlation filter is carried out as a differentiable layer. On the other hand, high-level semantic features are extracted from Conv5 and fed into the SiameseRPN module to achieve classification and regression tasks.

Offline Training: ResNet-50 deep network architecture is pre-trained on the training datasets of ImageNet, COCO, ImageNet DET and ImageNet VID to learn a general object feature for object representation. We employ single scale images with 127 pixels for template patches and 255 pixels for searching regions, respectively. To enhance the capacity to distinguish distracters of deep network model, we randomly obtain shifting and scaling following a uniform distribution on the search image as data augmentation techniques.

Our network model is trained with stochastic gradient descent (SGD). We use a warm-up learning rate of 0.001 for first 10 iterations to optimize the RPN branches.

For the last 15 iterations, the whole network is end-to-end trained with learning rate exponentially decayed from 0.005 to 0.0005. Weight decay of 0.0005 and momentum of 0.9 are used. We first train our model with for 5 warm-up epochs with learning rate linearly increased from 10^{-7} to 2×10^{-3}, then use a cosine annealing learning rate schedule for the rest of 45 epochs.

Discriminative Loss Function: The training loss is optimized from three different aspects. First, the classification loss and regression loss are written as follows.

$$L_{SiamRPN} = \frac{1}{N_{pos}} \sum_{x,y} L_{cls}(p_{x,y}, c^*_{x,y}) + \frac{\lambda}{N_{pos}} \sum_{x,y} 1\{c^*_{x,y} > 0\} \cdot L_{reg}(t_{x,y}, t^*_{x,y}) \qquad (2)$$

where L_{cls} denotes the focal loss for the object classification; L_{reg} is the IoU loss for the object location; we assign 1 to $c^*_{x,y}$ if it is a positive sample; N_{pos} is the number of positive samples; $1_{\{.\}}$ denotes the indicator function that takes 1 if the condition holds and takes 0 if not. $t_{x,y}$ and $t^*_{x,y}$ stands for the object position and ground-truth, respectively.

Second, different from CF that uses hand-crafted features for visual tracking, we develop to learn low-level feature representation fitting a CF. The features are obtained by a low-level convolutional layer of CNN model. The loss function is designed by

$$L_{low} = \|g(x) - y\|_2^2 = \|\mathbf{X}\mathbf{w} - y\|_2^2 \qquad (3)$$

where x is a search image; $\mathbf{X}$ is the circulate matrix of x for the search image patch; $\mathbf{w}$ is the learned CF.

Third, the high-level semantic is used to measure the similarities between the object template and the search image. The problem can be further written as the minimization of the following logistic loss.

$$L_{high} = \sum_{x,z} \log(1 + \exp(-y(x, z)f(x, z))) \qquad (4)$$

The whole network is trained from end-to-end based on a multi-task learning strategy. The final loss can be overall formulated as follow.

$$L = L_{SiamRPN} + L_{low} + L_{high} \qquad (5)$$

3.2 Channel Attention

A convolutional feature channel corresponds to certain visual information. In some certain circumstances, some feature channels are more important than others. The channel attention scheme is to keep the adaptation ability of deep network model to adapt the object appearance changes. To share a common attention, we propose a channel attention scheme to assist the object location.

The architecture of the channel attention is shown in Fig. 2, which is composed by a dimension reduction layer, a ReLU and a dimension increasing layer with sigmoid activation. Given a set of M channel features $F = [f_1, f_2, \ldots, f_M]$, the output of attention net is obtained by computing channel-wise re-scaling on the input in Eq. (7) where β is the parameter of the channel attention.

$$\bar{q}_i = \beta_i \cdot q_i \qquad i = 1, 2, \ldots, d \qquad (6)$$

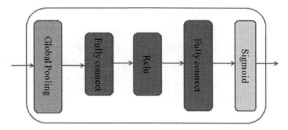

Fig. 2. The architecture of channel attention mechanism.

3.3 Online Tracking

Given the first frame with annotation, we utilize data augmentation strategies to construct an initial training set containing 20 positive samples. The object template is then obtained using the ResNet network architecture which is fine-tuned with the initial training set.

In the tracking phase, when a new input video frame arrives, we crop some large search patches centered at the previous target position with multiple scales. These search patches are fed into the ResNet-50 to get their feature representations. The fine-grained object representation is fed into the correlation filter layer. The semantic representation is evaluated based on the channel attention mechanism. The object candidate states $x = \{x_1, x_2, \ldots, x_N\}$ are randomly drawn based on the object position in the last frame. The candidates are estimated by finding the maximum of the fused correlation response in Eq. (7).

$$x_t^* = \underset{i=1,\ldots,M}{\arg\max} S(x_i) \tag{7}$$

The candidate with the maximum object confidence score is considered as the tracking result. The parameters of deep network model are updated every 20 frames using positive and negative samples collected in previous tracking frames.

4 Experiments

4.1 Implementation Details

In this work, the proposed method is carried out in Python using Tensorflow and Keras deep learning libraries. We test our tracker on a PC machine with an Intel i7 CPU (32G RAM) and an NVIDIA GTX 1080Ti GPU (11G memory), which runs in real-time with 24.8 frames per second (fps). The quantitative analysis and ablation studies are evaluated in this section.

In the initial training phase, the convergence loss threshold is set to 0.02 and the maximum iteration number is 50. For the Siamese network framework, we use the initial object of the first frame as the object template and crop the search region with 3 times of the object size from the current frame. For the scale evaluation, we generate a proposal pyramid with three scales, i.e., 45/47, 1, and 45/43 times of the previous object size.

4.2   Overall Performance

We evaluate our approach with other competing trackers on five challenging tracking benchmarks, including OTB-100 [40], UAV123 [41], VOT2018 [35], LaSOT [37] and TrackingNet [45]. The proposed approach is compared with the state-of-the-art trackers, including the correlation filter based trackers, such as SRDCF [24], MCPF [33], C-COT [29], ECO [5], and STRCF [27]; the non-real-time deep tracking algorithms such as MDNet [10], CREST [38], LSART [43], VITAL [21], and DAT [44]; and the real-time deep learning tracking methods such as ACT [42], SiamFC [15], ATOM [34], CFNet [11], SiamRPN++ [16], LTMU [36], DaSiamRPN [39], and UPDT [46]. In the following, we will report the quantitative analysis on these benchmarks.

OTB-100: Table 1 shows the success overlap rate in the dataset. Among the compared trackers, our tracker obtains an AUC score of 68.1%, competitive with UPDT tracking method.

UAV123: The benchmark includes 123 low altitude aerial videos captured from a UAV. The AUC score on this benchmark is reported in Table 1. SiamRPN++ achieves an AUC score of 61.3%. Our tracker significantly outperforms SiamRPN++ and obtains AUC score of 63.4%.

Table 1.   State-of-the-art trackers on OTB-100 and UAV123 benchmarks in terms of AUC score.

	ECO	CCOT	DaSiam RPN	ATOM	UPDT	MDNet	SiamRPN++	Our
OTB-100	64.3	68.2	65.8	66.9	69.2	67.8	68.9	68.1
UAV123	50.6	51.3	58.6	64.4	54.5	52.5	61.3	63.4

VOT2018: We evaluate our tracker on this challenging dataset which consists of 60 video sequences. Accuracy and robustness are used as measures to evaluate the tracking performance. EAO (Expected Average Overlap) is obtained to rank trackers. Results are given in Table 2. We can see that SiamRPN++ achieves the best performance in terms of accuracy. However, it obtains inferior robustness compared with ACT and ATOM. Our tracker has a 15.1% lower failure rate, while achieving compatible accuracy.

Table 2.   Comparison of state-of-the-art trackers on VOT2018 benchmark.

	DAT	ACT	DaSiam-RPN	ATOM	UPDT	SiamRPN	SiamRPN++	Our
Accuracy	0.505	0.519	0.586	0.590	0.536	0.586	0.600	0.587
Robustness	0.140	0.201	0.276	0.204	0.184	0.276	0.234	0.151
EAO	0.385	0.356	0.383	0.401	0.378	0.383	0.414	0.440

LaSOT: We evaluate the proposed tracker on this dataset consisting of 280 sequences which have longer sequences with an average of 2500 frames per sequence. Therefore, it is important to adapt the object appearance variations. Figure 3 shows the success rate plot. ATOM tracker employs the pre-trained ResNet-18 to discriminate the object from the background. Our approach uses end-to-end trained method and further improves the performance with an AUC score of 51.6%. The experiment evaluations demonstrate that model adaption capabilities of the proposed tracking method on video sequences.

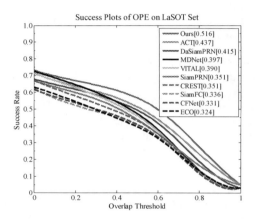

Fig. 3. Success plot on the LaSOT benchmark.

TrackingNet: We carry out our approach on the large-scale TrackingNet dataset. Table 3 shows the tracking evaluation results. SiamRPN++ reports a satisfied AUC score of 73.3%. Our method achieves AUC score of 74.1% with the same ResNet-50 as in SiamRPN++.

Table 3. Comparison of state-of-the-art trackers on TrackingNet benchmark.

	ECO	CFNet	MDNet	CSRDCF	UPDT	SiamFC	SiamRPN++	Our
AUC	55.4	57.8	60.6	53.4	61.1	57.1	73.3	74.1
P	49.2	53.3	56.5	48.0	55.7	53.3	69.4	69.6
P_{norm}	61.8	65.4	70.5	62.2	70.2	66.3	80.0	80.4

4.3 Ablation Studies

We conduct ablation evaluation to verify the contributions of different components and different layer features using OTB-100 and VOT2018 benchmarks. Table 4 shows the AUC scores of each variation.

Table 4. Ablation study of our method on OTB-100 and VOT2018 benchmarks. L_4 and L_5 denote conv4 and conv5, respectively. Finetune is whether the backbone is trained offline.

BackBone	L_4	L_5	Finetune	OTB-100	VOT2018
AlexNet				0.666	0.355
ResNet-50	√		√	0.679	0.347
		√	√	0.675	0.337
ResNet-50	√	√		0.676	0.392
	√	√	√	**0.700**	**0.408**

Feature Selection. The choice of features from different layers plays a significant role in visual tracking. The number of parameters and type of network layers directly affect the speed and accuracy of the tracking algorithms. First, different deep network architectures are evaluated on two popular benchmarks. AlexNet and ResNet-50 are used as backbones to verify the tracking performance. The AUC score is shown in Table 4. Our tracker and SiamRPN++ can benefit from the deeper layers network architecture. In addition, the tracking performance can obtain a great improvement by finetuning the backbone. Furthermore, the experiment results show that conv4 alone obtains a satisfying performance with 0.347 in EAO. Deeper layer and shallow layer perform with 5% drops. We combine conv4 and conv5 to obtain the improvements.

Effectiveness of Different Components. The proposed tracking approach consists of SiamRPN (S), correlation filter layer (CF), and channel attention module (A). To evaluate the importance of different components, we carry out the following variants: (1) ours (S) is our tracker merely using SiamRPN to track the object in every frame; (2) ours (CF) stands for our method by combining low-level correlation filter and high-level semantic representation to obtain the object location in every frame; (3) our tracker (A) denotes the proposed tracker with the channel attention module; and (4) our (S+CF+A) is the final tracker. The effectiveness of different components is evaluated in Table 5.

Table 5. Effectiveness of different components for our tracking algorithm.

Tracker	F-score	Pr	Re	fps
Ours (S)	0.553	0.551	0.541	34.7
Ours (CF)	0.583	0.584	0.557	30.6
Ours (A)	0.597	0.607	0.565	21.4
Ours (S+CF+A)	0.603	0.613	0.596	24.8

Table 5 shows the experimental results of the variants and illustrates that all components can boost the tracking performance. Removal of the channel attention module from the proposed model causes a 6.4% performance drop, while removal of the correlation filter layer reduces the performance by 7.8%. The accuracy of both variants is comparable to the original SiamRPN, while the number of failures increases. Therefore, the attention module is crucial for a robust object selection strategy during the tracking process. The proposed tracking approach leads to a 9.5% EAO and a 8.7% accuracy reduction. Therefore, it benefits from the rotated bounding box estimation.

5   Conclusions

In this paper, we propose an end-to-end deep network architecture for visual object tracking. Channel attention mechanism is introduced to the Siamese network framework. Low-level features are used to learn correlation filter and high-level semantic features are used to deep object representation. Then both two-level features are jointly represented in multi-task learning framework. The loss function is designed to optimize the deep network parameters. Experiments show that the proposed tracking approach significantly improves tracking performance in terms of accuracy and speed. In future work, we plan to incorporate spatial-temporal attention module representation in our model framework to further improve its effectiveness.

Acknowledgments. This work is supported in part by the National Natural Science Foundation of China (Grant No. 61802058, 61911530397); in part by the Equipment Advance Research Foundation Project of China (Grant No. 61403120106); in part by the Project funded by the China Postdoctoral Science Foundation (Grant No. 2019M651650); in part by the Startup Foundation for Introducing Talent of Nanjing University of Information Science and Technology (Grant No. 2018r057).

References

1. Li, P., Wang, D., Wang, L., et al.: Deep visual tracking: review and experimental comparison. Pattern Recogn. **76**, 323–338 (2018)
2. Zhu, P., Wen, L., Du, D., et al.: Vision meets drones: past, present and future. arXiv preprint arXiv:2001.06303 (2020)
3. Ma, C., Huang, J., Yang, X., et al.: Hierarchical convolutional features for visual tracking. In: Proceedings of the IEEE International Conference on Computer Vision, Piscataway, NJ, pp. 3074–3082. IEEE (2015)
4. Danelljan, M., Hager, G., Khan, F., et al.: Convolutional features for correlation filter based visual tracking. In: Proceedings of the IEEE International Conference on Computer Vision Workshops, Piscataway, NJ, pp. 58–66. IEEE (2015)
5. Danelljan, M., Bhat, G., Khan, F., et al.: ECO: efficient convolution operators for tracking. In: Proceedings of the IEEE Conference on Computer Vision and Pattern Recognition, Piscataway, NJ, pp. 6638–6646. IEEE (2017)
6. Hong, S., You, T., Kwak, S., et al.: Online tracking by learning discriminative saliency map with convolutional neural network. In: International Conference on Machine Learning, New York, NY, pp. 597–606. ACM (2015)
7. Li, H., Li, Y., Porikli, F.: DeepTrack: learning discriminative feature representations online for robust visual tracking. IEEE Trans. Image Process. **25**(4), 1834–1848 (2015)

8. Wang, L., Ouyang, W., Wang, X., et al.: Visual tracking with fully convolutional networks. In: Proceedings of the IEEE International Conference on Computer Vision, Piscataway, NJ, pp. 3119–3127. IEEE (2015)
9. Teng, Z., Xing, J., Wang, Q., et al.: Robust object tracking based on temporal and spatial deep networks. In: Proceedings of the IEEE International Conference on Computer Vision, Piscataway, NJ, pp. 1144–1153. IEEE (2017)
10. Nam, H., Han, B.: Learning multi-domain convolutional neural networks for visual tracking. In: Proceedings of the IEEE Conference on Computer Vision and Pattern Recognition, Piscataway, NJ, pp. 4293–4302. IEEE (2016)
11. Valmadre, J., Bertinetto, L., Henriques, J., et al.: End-to-end representation learning for correlation filter based tracking. In: Proceedings of the IEEE Conference on Computer Vision and Pattern Recognition, Piscataway, NJ, pp. 2805–2813. IEEE (2017)
12. Cui, Z., Xiao, S., Feng, J., et al.: Recurrently target-attending tracking. In: Proceedings of the IEEE Conference on Computer Vision and Pattern Recognition, Piscataway, NJ, pp. 1449–1458. IEEE (2016)
13. Choi, J., Chang, H., Yun, S., et al.: Attentional correlation filter network for adaptive visual tracking. In: Proceedings of the IEEE Conference on Computer Vision and Pattern Recognition, Piscataway, NJ, pp. 4807–4816. IEEE (2017)
14. Tao, R., Gavves, E., Smeulders, A.: Siamese instance search for tracking. In: Proceedings of the IEEE Conference on Computer Vision and Pattern Recognition, Piscataway, NJ, pp. 1420–1429. IEEE (2016)
15. Bertinetto, L., Valmadre, J., Henriques, João F., Vedaldi, A., Torr, P.H.S.: Fully-convolutional siamese networks for object tracking. In: Hua, G., Jégou, H. (eds.) ECCV 2016. LNCS, vol. 9914, pp. 850–865. Springer, Cham (2016). https://doi.org/10.1007/978-3-319-48881-3_56
16. Li, B., Wu, W., Wang, Q., et al.: SiamRPN++: evolution of Siamese visual tracking with very deep networks. In: Proceedings of the IEEE Conference on Computer Vision and Pattern Recognition, Piscataway, NJ, pp. 4282–4291. IEEE (2019)
17. Li, B., Yan, J., Wu, W., et al.: High performance visual tracking with Siamese region proposal network. In: Proceedings of the IEEE Conference on Computer Vision and Pattern Recognition, Piscataway, NJ, pp. 8971–8980. IEEE (2018)
18. Zhou, W., Wen, L., Zhang, L., et al.: SiamMan: Siamese motion-aware network for visual tracking. arXiv preprint arXiv:1912.05515 (2019)
19. Zhang, Z., Peng, H.: Deeper and wider Siamese networks for real-time visual tracking. In: Proceedings of the IEEE Conference on Computer Vision and Pattern Recognition, Piscataway, NJ, pp. 4591–4600. IEEE (2019)
20. Wang, X., Li, C., Luo, B., et al.: SINT++: robust visual tracking via adversarial positive instance generation. In: Proceedings of the IEEE Conference on Computer Vision and Pattern Recognition, Piscataway, NJ, pp. 4864–4873. IEEE (2018)
21. Song, Y., Ma, C., Wu, X., et al.: VITAL: visual tracking via adversarial learning. In: Proceedings of the IEEE Conference on Computer Vision and Pattern Recognition, Piscataway, NJ, pp. 8990–8999. IEEE (2018)
22. Henriques, J., Caseiro, R., Martins, P., et al.: High-speed tracking with kernelized correlation filters. IEEE Trans. Pattern Anal. Mach. Intell. 37(3), 583–596 (2014)
23. Danelljan, M., Häger, G., Khan, F., et al.: Discriminative scale space tracking. IEEE Trans. Pattern Anal. Mach. Intell. 39(8), 1561–1575 (2016)
24. Danelljan, M., Hager, G., Khan, F., et al.: Learning spatially regularized correlation filters for visual tracking. In: Proceedings of the IEEE International Conference on Computer Vision, Piscataway, NJ, pp. 4310–4318 IEEE (2015)
25. Kiani Galoogahi, H., Sim, T., Lucey, S.: Correlation filters with limited boundaries. In: Proceedings of the IEEE Conference on Computer Vision and Pattern Recognition, Piscataway, NJ, pp. 4630–4638. IEEE (2015)

26. Ma, C., Yang, X., Zhang, C., et al.: Long-term correlation tracking. In: Proceedings of the IEEE Conference on Computer Vision and Pattern Recognition, Piscataway, NJ, pp. 5388–5396. IEEE (2015)
27. Li, F., Tian, C., Zuo, W., et al.: Learning spatial-temporal regularized correlation filters for visual tracking. In: Proceedings of the IEEE Conference on Computer Vision and Pattern Recognition, Piscataway, NJ, pp: 4904–4913. IEEE (2018)
28. Gladh, S., Danelljan, M., Khan, F., et al.: Deep motion features for visual tracking. In: 23rd International Conference on Pattern Recognition, Piscataway, NJ, pp. 1243–1248. IEEE (2016)
29. Danelljan, M., Robinson, A., Shahbaz Khan, F., Felsberg, M.: Beyond correlation filters: learning continuous convolution operators for visual tracking. In: Leibe, B., Matas, J., Sebe, N., Welling, M. (eds.) ECCV 2016. LNCS, vol. 9909, pp. 472–488. Springer, Cham (2016). https://doi.org/10.1007/978-3-319-46454-1_29
30. Lukezic, A., Vojir, T., Zajc, L., et al.: Discriminative correlation filter with channel and spatial reliability. In: Proceedings of the IEEE Conference on Computer Vision and Pattern Recognition, Piscataway, NJ, pp. 6309–6318. IEEE (2017)
31. Wang, Q., Gao, J., Xing, J., et al.: DCFNet: discriminant correlation filters network for visual tracking. arXiv preprint arXiv:1704.04057 (2017)
32. Li, Y., Zhu, J.: A scale adaptive kernel correlation filter tracker with feature integration. In: Agapito, L., Bronstein, M.M., Rother, C. (eds.) ECCV 2014. LNCS, vol. 8926, pp. 254–265. Springer, Cham (2015). https://doi.org/10.1007/978-3-319-16181-5_18
33. Zhang, T., Xu, C., Yang, M.: Multi-task correlation particle filter for robust object tracking. In: Proceedings of the IEEE Conference on Computer Vision and Pattern Recognition, Piscataway, NJ, pp. 4335–4343. IEEE (2017)
34. Danelljan, M., Bhat, G., Khan, F., et al. Atom: accurate tracking by overlap maximization. In: Proceedings of the IEEE Conference on Computer Vision and Pattern Recognition, Piscataway, NJ, pp. 4660–4669. IEEE (2019)
35. Kristan, M., et al.: The sixth visual object tracking VOT2018 challenge results. In: Leal-Taixé, L., Roth, S. (eds.) ECCV 2018. LNCS, vol. 11129, pp. 3–53. Springer, Cham (2019). https://doi.org/10.1007/978-3-030-11009-3_1
36. Dai, K., Zhang, Y., Wang, D., et al.: High-performance long-term tracking with meta-updater. In: Proceedings of the IEEE Conference on Computer Vision and Pattern Recognition. IEEE, Piscataway, NJ (2020)
37. Fan, H., Lin, L., Yang, F., et al.: LaSOT: a high-quality benchmark for large-scale single object tracking. In: Proceedings of the IEEE Conference on Computer Vision and Pattern Recognition, Piscataway, NJ, pp. 5374–5383. IEEE (2019)
38. Song, Y., Ma, C., Gong, L., et al.: CREST: convolutional residual learning for visual tracking. In: Proceedings of the IEEE International Conference on Computer Vision, Piscataway, NJ, pp. 2555–2564. IEEE (2017)
39. Zhu, Z., Wang, Q., Li, B., Wu, W., Yan, J., Hu, W.: Distractor-aware Siamese networks for visual object tracking. In: Ferrari, V., Hebert, M., Sminchisescu, C., Weiss, Y. (eds.) ECCV 2018. LNCS, vol. 11213, pp. 103–119. Springer, Cham (2018). https://doi.org/10.1007/978-3-030-01240-3_7
40. Wu, Y., Lim, J., Yang, M.: Object tracking benchmark. IEEE Trans. Pattern Anal. Mach. Intell. 37(9), 1834–1848 (2015)
41. Mueller, M., Smith, N., Ghanem, B.: A benchmark and simulator for UAV tracking. In: Leibe, B., Matas, J., Sebe, N., Welling, M. (eds.) ECCV 2016. LNCS, vol. 9905, pp. 445–461. Springer, Cham (2016). https://doi.org/10.1007/978-3-319-46448-0_27
42. Chen, B., Wang, D., Li, P., Wang, S., Lu, H.: Real-time 'actor-critic' tracking. In: Ferrari, V., Hebert, M., Sminchisescu, C., Weiss, Y. (eds.) ECCV 2018. LNCS, vol. 11211, pp. 328–345. Springer, Cham (2018). https://doi.org/10.1007/978-3-030-01234-2_20

43. Sun, C., Wang, D., Lu, H., et al.: Learning spatial-aware regressions for visual tracking. In: Proceedings of the IEEE Conference on Computer Vision and Pattern Recognition, Piscataway, NJ, pp. 8962–8970. IEEE (2018)

44. Pu, S., Song, Y., Ma, C., et al.: Deep attentive tracking via reciprocative learning. In: Neural Information Processing Systems, pp. 1931–1941. MIT Press, Cambridge (2018)

45. Müller, M., Bibi, A., Giancola, S., Alsubaihi, S., Ghanem, B.: TrackingNet: a large-scale dataset and benchmark for object tracking in the wild. In: Ferrari, V., Hebert, M., Sminchisescu, C., Weiss, Y. (eds.) ECCV 2018. LNCS, vol. 11205, pp. 310–327. Springer, Cham (2018). https://doi.org/10.1007/978-3-030-01246-5_19

46. Bhat, G., Johnander, J., Danelljan, M., Khan, F.S., Felsberg, M.: Unveiling the power of deep tracking. In: Ferrari, V., Hebert, M., Sminchisescu, C., Weiss, Y. (eds.) ECCV 2018. LNCS, vol. 11206, pp. 493–509. Springer, Cham (2018). https://doi.org/10.1007/978-3-030-01216-8_30

The Devil is in the Detail: Deep Feature Based Disguised Face Recognition Method

Shumin Zhu[1], Jianjun Qian[1(✉)], Yangwei Dong[1], and Waikeung Wong[2]

[1] PCA Lab, Key Lab of Intelligent Perception and Systems for High-Dimensional Information of Ministry of Education, and Jiangsu Key Lab of Image and Video Understanding for Social Security, School of Computer Science and Engineering, Nanjing University of Science and Technology, Nanjing, China
{zhushumin,csjqian,dongyangwei}@njust.edu.cn
[2] Institute of Textiles and Clothing, The Hong Kong Polytechnic University, Hong Kong, China
calvin.wong@polyu.edu.hk

Abstract. Face recognition have been developed rapidly, launched by the breakthrough of Deep learning based face representation method. However, disguised face verification in the wild is still a challenge problem. To address this issue, we propose a novel deep feature based disguised face recognition scheme (DDFR). DDFR introduces the multi-scale residual network with AM-softmax loss for learning face representation. In training stage, we put the different occlusions (mask, sunglasses and scarf et al.) on clean face images to enhance the diversity of training set. Meanwhile, both aligned face image and un-aligned face image are combined to improve the discriminative power of feature representation for disguised face verification. Experimental results demonstrate that the proposed method achieves the better results than state-of-the-art methods on the DFW (disguised face in the wild) set.

Keywords: Disguised face verification · Face alignment · Add face occlusion · Multi-scale feature extraction

1 Introduction

Face recognition has been a hot topic in the field of computer vision and has been widely used in public security and finance. Recently, Deep learning based representation methods achieve the landmark breakthrough in face recognition [3,10,12,18]. This achievement is mainly due to the applications of a better convolutional neural network architecture [6,13,16,17] and a more restrictive loss function. However, disguised face recognition is still a challenging problem

This work was supported by the National Science Fund of China under Grant Nos. 61876083, 61876084, U1713208, and Program for Changjiang Scholars. Hong Kong Scholar program, The Hong Kong Polytechnic University (YZ2K).

© Springer Nature Switzerland AG 2020
Y. Peng et al. (Eds.): PRCV 2020, LNCS 12306, pp. 420–431, 2020.
https://doi.org/10.1007/978-3-030-60639-8_35

since it contains personal unconscious or conscious cover of the face to hide personal identity, as well as one person imitates another to deceive the face recognition system. In this task, large intra-class distance and small inter-class distance make unconstrained disguised face verification very difficult. To solve this problem, there are many works have been developed, ranging from the sparse representation based method to deep convolutional neural network, in the past decades.

In the literature, Wright et al. [19] proposed a sparse representation based method to handle face recognition with real-disguise. To further improve the robustness of the sparse model, robust sparse representation models are developed for robust face recognition by using M-estimator to characterize the error term [7,21]. Subsequently, Yang et al. [20] employed nuclear norm to describe the error term and presented a novel nuclear norm based matrix regression model to solve facial images contain disguise or occlusion. Qian et al. [11] introduced the low rank regularized term to ridge regression for solving disguise face recognition. However, these methods overlooked open sets of subjects, which is limited for real world applications. To facilitate the research of unrestricted disguised faces recognition, Kushwaha et al. [9] proposed a novel Disguised Faces in the Wild (DFW) dataset. DFW is mainly used to evaluate the performance of various methods in dealing with disguised face recognition. The authors of DFW also organize a competition [14] in conjunction with the International Conference on Computer Vision and Pattern Recognition (CVPR) 2018. Based on DFW, many deep learning based methods are developed to solve disguised face verification. Zhang et al. [23] proposed a two-stage training approach for this task. At the first stage, they employed generic aligned face images and unaligned face images to train two 64-layer DCNNs [10] in conjunction with AM-Softmax [18]. At the second stage, PCA is used to obtain the low dimensional compact feature representation. Smirnov et al. [15] proposed a new deep embedding learning method for disguised face recognition. They used general face images to train AEFANet with Auxiliary Embedding. Bansal et al.[1] combined ResNet-101 [6] and Inception-ResNet-v2 [16] with L2-constrained Softmax for handling disguised face verification (Fig. 1).

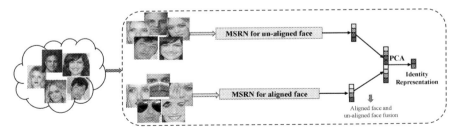

Fig. 1. The proposed Deep feature based disguised face recognition scheme (DDFR).

However, above mentioned methods use nearly ten million face images to learn the face representation model. To further improve the disguised face verification performance, this paper presents a novel method for capturing intrinsic feature of face image with little training data as possible. The main contributions of our work are as follows:

- We propose a multi-scale residual (MSR) block to capture more detailed facial features for improving the performance to distinguish the person and its impersonator.
- We put the different face occlusions (mask, sunglasses and scarf et al.) on clean face images to enhance the diversity of training set and increases the intra-class differences of training set.
- We combine the feature of aligned face image and unaligned face image to improve the discriminative power of feature representation for disguised face verification. Experimental results on the DFW dataset demonstrate that our method achieves better performance than state-of-the-art methods.

2   Deep Disguised Face Verification Framework

In this section, we introduce the multi-scale feature representation method for obtaining more detailed facial features. The facial occlusion synthesis scheme is proposed to enrich the diversity of training set. Finally, we fuse the features of aligned facial image and unaligned facial image into one disguised face verification framework.

2.1   Multi-scale Feature Extraction to Capture More Facial Details

As well known, ResNet is a good tool to capture the image feature with deeper network [6]. Based on this, W. Liu et al. developed ResNet-like block by combining 3×3 convolution kernels with a residual unit for face representation and achieved remarkable results [10].

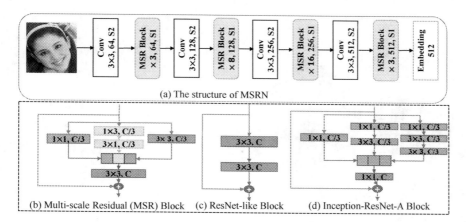

Fig. 2. (a) Our MSRN structure. (b) Our Multi-scale Residual Block. (c) ResNet-like Block. (d) Inception-ResNet-A Block.

ResNet-like block motivates us to design the multi-scale residual block for matching faces with intentional and unintentional disguises. Compared with ResNet-like block, our multi-scale residual block includes convolution kernels of different sizes and employs the different nonlinear transform to combine the features. The schema of Multi-scale Residual block has to be shown in Fig. 2 (b). Suppose that there are 64 feature maps and the size of each feature map is 80×80 as the input of this block. Based on this, we convolved the feature map with three different scale convolution kernels. The number of convolution kernels for each scale is 21. Then, we connect the convolved feature maps together and use 3×3 convolution kernels to further represent the connected feature maps. Here, the number of convolution kernels is 64. Finally, all the convolved feature and the input feature maps are added together as a whole.

The main difference between Multi-scale Residual block and Inception-ResNet block is that the Inception-ResNet block draws the idea of Network-in-Network to reduce dimension, it should use a 1×1 size convolution kernel before and after using a multi-scale convolution kernel. The structure of Inception-Resnet-A is shown in Fig. 2 (d). In addition, Multi-scale Residual block can capture rich facial feature than Inception-ResNet block. The experiments in Sect. 3 also support our view.

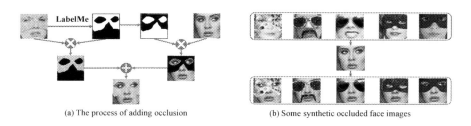

(a) The process of adding occlusion   (b) Some synthetic occluded face images

Fig. 3. The pipeline of facial occlusion synthesis and some synthesized results.

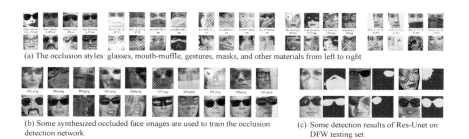

(a) The occlusion styles: glasses, mouth-muffle, gestures, masks, and other materials from left to right.

(b) Some synthesized occluded face images are used to train the occlusion detection network.

(c) Some detection results of Res-Unet on DFW testing set.

Fig. 4. The various of occlusion styles. Our synthesize training set of Res-Unet and some detection results of Res-Unet.

2.2 Facial Occlusion Synthesis

In the task of disguised face recognition, the person can wear glasses and masks to weaken its distinguishability, and the pretender can also increase the similarity between itself and the imitated person in this way. In DFW, it is easy to lead the overfitting problem since the diguised face images in the training set are limited. In general, most methods use a large number of face images without disguise in training stage. To fit the complex face data in the disguised face test set, the previous work only expand the number and category of face images in the training set.

Here, we just put occlusions on partial of clean face images to enrich the diversity of the training set for improving the disguised face verification performance. To ensure the authenticity of facial occlusion synthesis, we employs the LabelMe to obtain the occlusion part and then synthesized the facial occlusion images as shown in Fig. 3 (a). Specifically, we collect face images with different occlusions from the training set of DFW, MAFA [5] and NUST-RF dataset. All face images are aligned by using the same face alignment method. There are 285 occlusion styles, including glasses (55), mouth-muffle (53), masks (53), gestures (55), and other materials (69). Some face images shows in Fig. 4(a). Then, the LabelMe is employed to mark the occlusion's position in the face image. Subsequently, we can obtain the synthesed occlusion face by combing the occlusion and clean face images.

To add an appropriate proportion of face images with occlusion to the training set, we propose that the rate of occluded face images in the training set should be consistent with the test set. We introduce an occlusion detection network (the backbone is Res18-Unet [4]) to detect whether the images in the test set are occluded and further count the proportion of face images containing occlusion in the test set. And then we randomly select the same proportion face images of each person in the training set to put occlusion on them. For occlusion detection network, the training set is composed of various occlusion styles and 100,000 face images without occlusion from CASIA-WebFace [22] dataset. Some occlusion detection results of the network on DFW are shown in Fig. 4 (c).

2.3 Aligned and Unaligned Face Feature Fusion

It is known that aligned face images have the advantage of eliminating posture changes compared to unaligned face images, which makes the model pay more attention to details such as facial texture. However, the unaligned face images possibly contain some irregular discriminative information. We think that the irregular discriminative information of unaligned face image and the discriminative information of aligned face image can be combined together to further improve the discriminative power of face representation. Finally, PCA is then used to achieve the low-dimensional feature vector.

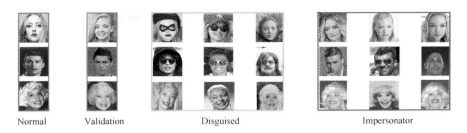

| Normal | Validation | Disguised | Impersonator |

Fig. 5. Some example images of Disguised Faces in the Wild. It consists of four kinds of images: normal, validation, disguised and impersonator.

3   Experiments

3.1   Disguised Faces in the Wild Dataset

The Disguised Faces in the Wild (DFW) data set includes 11157 images from 1000 different identities. Among them, the training set contains 3386 face images of 400 people. And the test set includes 7,771 face images of 600 people. Most of the identities have four kinds of face images: Normal, Validation, Disguised and Impersonator. These four types images are shown in Fig. 5.

There are three pre-defined protocols. Protocol 1 aims to evaluate the performance of face verification methods under impersonation only. There are 25,046 pairs of face images for this protocol. In Protocol 2, the given face verification methods are evaluated for disguises via obfuscation only. The number of image pairs is 9,041,283 for this protocol. And Protocol 3 is used for evaluating the given methods on the whole dataset. The total number of image pairs for this protocol is 9,066,329.

3.2   Experimental Settings

Mini Training Set. The mini training set is designed to facilitate the ablation study. This training set is composed of CASIA-WebFace [22], PubFig [8] and the training set of DFW [9]. We removed the identities overlap between training set and testing set strictly according to provided identity names. More details for removing the overlap identities can be found in AM-Softmax paper [18]. The final generic mini training dataset includes 10,397 identities and 444,895 (0.44M) face images.

Big Training Set. This set is an extension of the mini training set. The expanded images are all from the VGGFace2 data set [2]. Compared with Mini training set, there are another 8047 persons and each person have about 200 face images. The final expanded big training set includes 18444 people and about 2,039,485 (2.04M) face images.

Training Setting. In our experiments, all face images are resized to 160×160. AM-Softmax is used in our model. The parameters m (cosine margin constrain) is 0.35 and s (norm-scale of features) is 30. The batch size is 24 for the mini training set and the large training set. For the mini training set, the learning rate starts from 0.1 and is divided by 10 at 140K, 180K iterations and the maximum iterations is 200K. For the large training set, to make full use of this data set, the learning rate starts from 0.1 and is divided by 10 at 700K, 900K iterations, and the maximum iterations are 1M. In addition, we use random horizontal flipping for data augmentation. For PCA, we select the first 250 eigenvectors to form the projection matrix.

3.3   Experimental Results

We compare our verification results on DFW with state-of-the-art methods published in DFW competition. The compared results are shown in the Table 1. And the ROC curves on the mini training set and the big training set are shown in the Fig. 6.

We can see from the Table 1 that when the training set is 0.44M, our method outperforms Occlusionface on three protocols. Specifically, our method is 13% higher than the Occlusionface at 0.1% FAR on Protocol-1. On Protocol-2 and Protocol-3, our method is 6% and 7% higher than the Occlusionface at 1%FAR and 0.1%FAR, respectively. It is worth mentioning that even with 0.44M training set, our method is nearly 6% higher than UMDNet whose training set is 5.6M at 0.1% FAR on Protocol-1. And the results of our method on other protocols are also similar with UMDNets.

When the number of face images in the training set increases to 2.04M, the results of our method on three protocols are obviously better than UMD-Nets. Compared with the AEFRL whose training set is 8.3M, even through our methods is 0.5% lower than AEFRL at 0.1% FAR on Protocol-1, our method is 2%–4% higher than AEFRL at FARs on Protocol-2 and Protocol-3. Compared with MiRA-Face whose training set is 7.6M, our method is nearly 1% higher than MiRA-Face at 1% FAR and is 9% higher at 0.1% FAR on Protocol-1. On Protocol-2 and Protocol-3, the results of our method are only a tiny difference with MiRA-Face. Overall, our method outperforms other state-of-the-art methods with less training set.

3.4   Ablation Study

Effect of Face Alignment Scale. We use two different face alignment scales (one contains hair and face contours, and the other remove these two parts) to train our multi-scale residual network model, and compare their verification results on DFW test set.

From Table 2, we can see that the alignment scale of the face without hair shows a clear advantage on three protocols. On Protocol-1, the alignment scale of without hair part is nearly 5% higher than the face images with hair part at 0.1% FAR. On Protocol-2 and Protocol-3, the alignment scale without hair is

Table 1. Verification accuracy (%) of our method and other published methods on three protocols

Method	Protocol-1 GAR		Protocol-2 GAR		Protocol-3 GAR	
	@1%FAR	@0.1%FAR	@1%FAR	@0.1%FAR	@1%FAR	@0.1%FAR
AEFRL (8.3M)	**96.80**	57.64	87.82	77.06	87.90	75.54
MiRA-Face (7.6M)	95.46	51.09	**90.65**	80.56	**90.62**	79.26
UMDNets (5.6M)	94.28	53.27	86.62	74.69	86.75	72.90
Occlusionface (0.52M)	93.44	46.21	80.45	66.05	80.80	65.34
DDFR (0.44M)	93.86	59.23	86.14	73.43	86.35	72.47
DDFR (2.04M)	96.30	**60.84**	90.19	**80.61**	90.30	**79.57**

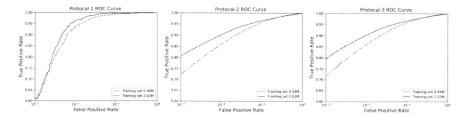

Fig. 6. ROC Curves for three protocols on the mini training set and the big training set

2%–5% higher than that with hair at FARs. In general, the face images without the hair part can achieve better performance than that with hair cause it can remove the interference of hairstyle and beard.

Table 2. Verification accuracy (%) of our model trained with different face align scale

Method	Protocol-1 GAR		Protocol-2 GAR		Protocol-3 GAR	
	@1%FAR	@0.1%FAR	@1%FAR	@0.1%FAR	@1%FAR	@0.1%FAR
With hair scale	91.30	51.55	80.47	64.52	80.86	63.88
Without hair scale	**92.87**	**55.48**	**83.09**	**69.07**	**83.38**	**68.50**

Effect of Different Feature Extraction Block. To compare the feature extraction capabilities of MSR block, Resnet-like block and Inception-Resnet-A block, we prepare three different models. The first one is MSRN, then we replace the MSR block in the MSRN with Resnet-like block and Inception-Resnet-A block to construct Resnet-like model and Inception-Resnet-A model, respectively. The structures of these three blocks are shown in Fig. 2. And the verification results of the three models on the DFW test set are shown in Table 3 (Fig. 7).

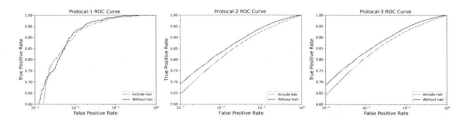

Fig. 7. ROC curves for three protocols with different face alignment scale

We can find that whether on Protocol-1 (impersonation) or Protocol-2 (obfuscation), MSR Block has stronger feature representation capabilities than the other two network Blocks. Especially on Protocol-1, the model with the MSR Block is 4.5% higher than the ResNet-like Block model and 7% higher than the Inception-Resnet-A Block model at 0.1% FAR (Fig. 8).

Table 3. Verification accuracy (%) of ResNet-Like Block (R-L Block), Inception-Resnet-A Block (I-R-A Block), and our MSR Block on three protocols.

Method	Protocol-1 GAR		Protocol-2 GAR		Protocol-3 GAR	
	@1%FAR	@0.1%FAR	@1%FAR	@0.1%FAR	@1%FAR	@0.1%FAR
R-L Block	90.88	50.98	81.48	67.14	81.89	66.72
I-R-A Block	91.93	48.44	81.77	67.51	82.11	67.04
MSR Block	**92.71**	**55.48**	**83.09**	**69.07**	**83.38**	**68.50**

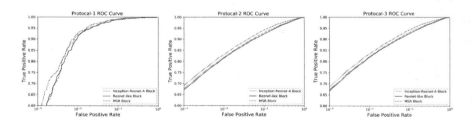

Fig. 8. ROC curves of different blocks on three protocols

Effect of Increasing the Rate of Occluded Face Images. To understand the impact of different proportions of synthetic occluded face images in the training set on the model's disguised face verification results, we compare three different rates 5%, 20% and 100%. Among them, 5% is also the rate of occlusion images in the DFW test set.

We can see that when we keep the rate of the occluded face images in training set consistent with the proportion of the occluded face images in the DFW test set, the effectiveness of our model on disguised face verification tasks is improved. However, when we increase the proportion of synthetic occluded face images synthesized in the training set to 20% and 100%, the verification performance of the model on the DFW test set decreases. On Protocol-1, our model trained with 5% synthetic occluded face images is 2% higher than the model trained with ordinary face images at 0.1% FAR. On Protocol-2 and Protocol-3, our model trained with 5% synthetic occluded face images is 1% higher than the model trained with ordinary face images at 0.1% FAR. Therefore, it is effective to add synthetic occlusion face images to the training set. However, the ratio of the added synthetic occlusion images should be consistent with the occlusion image ratio in the test set (Table 4).

Table 4. Verification accuracy (%) of our model trained with or with out occlusion faces

Method	Protocol-1 GAR		Protocol-2 GAR		Protocol-3 GAR	
	@1%FAR	@0.1%FAR	@1%FAR	@0.1%FAR	@1%FAR	@0.1%FAR
Ordinary face	93.64	57.20	85.95	72.47	86.09	71.65
Occluded face (20%)	93.03	55.66	86.11	72.46	86.26	71.58
Occluded face (100%)	91.54	52.85	82.52	65.77	82.74	65.06
Occluded face (5%)	**93.86**	**59.24**	**86.14**	**73.43**	**86.35**	**72.48**

Effect of Facial Feature Fusion. We train aligned MSRN model and unaligned MSRN model by using aligned and unaligned face images, respectively. Then, we merge the aligned and unaligned face features in the embedding layer and use PCA to achieve the compressed feature. The verification results of these three methods on the DFW test set are shown in the Table 5. And the ROC curves of these three methods on the three protocols are shown in Fig. 9. On Protocol-1, the feature fusion method is 4% higher than aligned MSRN model and 2% higher than unaligned MSRN model at 0.1% FAR. On Protocol-2, the feature fusion method is 3% higher than aligned MSRN model and 5% higher than unaligned MSRN model at 1% FAR. And at 0.1% FAR, the GAR of feature fusion method is 4% higher than aligned MSRN model and 8% higher than unaligned MSRN model. In general, the feature fusion method has more advantages than the single feature method.

Table 5. Verification accuracy (%) of our model trained with aligned face images, unaligned face images and feature fusion of both.

Method	Protocol-1 GAR		Protocol-2 GAR		Protocol-3 GAR	
	@1%FAR	@0.1%FAR	@1%FAR	@0.1%FAR	@1%FAR	@0.1%FAR
Unalign	92.70	57.20	80.75	64.91	81.03	64.21
Align	92.87	55.48	83.09	69.06	83.38	68.50
Fusion+PCA	**93.86**	**59.24**	**86.14**	**73.43**	**86.35**	**72.47**

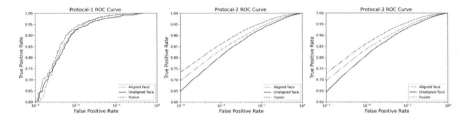

Fig. 9. ROC curves of aligned face feature, unaligned face feature and feature fusion of both for three protocols

4 Conclusion

In this article, we propose a multi-scale residual network and a method of adding real occlusion to the training set for disguised face verification. Compared with the deep metric learning method, our proposed method enriches the facial feature by using multi-scale residual blocks and increases the diversity of training set samples by adding real occlusion on a clean face. Another advantage is that our method uses fewer training samples and achieves better results than the state-of-the-art methods. In future work, we will further investigate how to design efficient deep face representation model with little training set as possible. It is always interesting in developing attention neural network to handle disguised face representation.

References

1. Bansal, A., Ranjan, R., Castillo, C.D., Chellappa, R.: Deep features for recognizing disguised faces in the wild. In: CVPR Workshops, pp. 10–16 (2018)
2. Cao, Q., Shen, L., Xie, W., Parkhi, O.M., Zisserman, A.: VGGFace2: a dataset for recognising faces across pose and age. In: 2018 13th IEEE International Conference on Automatic Face & Gesture Recognition (FG 2018), pp. 67–74. IEEE (2018)
3. Deng, J., Guo, J., Xue, N., Zafeiriou, S.: ArcFace: additive angular margin loss for deep face recognition. In: CVPR, pp. 4690–4699 (2019)
4. Diakogiannis, F.I., Waldner, F., Caccetta, P., Wu, C.: ResUNet-a: a deep learning framework for semantic segmentation of remotely sensed data. ISPRS J. Photogramm. Remote Sens. **162**, 94–114 (2020)

5. Ge, S., Li, J., Ye, Q., Luo, Z.: Detecting masked faces in the wild with LLE-CNNs. In: CVPR, pp. 2682–2690 (2017)
6. He, K., Zhang, X., Ren, S., Sun, J.: Deep residual learning for image recognition. In: CVPR, pp. 770–778 (2016)
7. He, R., Zheng, W.S., Hu, B.G.: Maximum correntropy criterion for robust face recognition. TPAMI **33**(8), 1561–1576 (2010)
8. Kumar, N., Berg, A.C., Belhumeur, P.N., Nayar, S.K.: Attribute and simile classifiers for face verification. In: ICCV, pp. 365–372. IEEE (2009)
9. Kushwaha, V., Singh, M., Singh, R., Vatsa, M., Ratha, N., Chellappa, R.: Disguised faces in the wild. In: CVPR Workshops, pp. 1–9 (2018)
10. Liu, W., Wen, Y., Yu, Z., Li, M., Raj, B., Song, L.: SphereFace: deep hypersphere embedding for face recognition. In: CVPR, pp. 212–220 (2017)
11. Qian, J., Yang, J., Zhang, F., Lin, Z.: Robust low-rank regularized regression for face recognition with occlusion. In: CVPR Workshops, pp. 21–26 (2014)
12. Ranjan, R., Castillo, C.D., Chellappa, R.: L2-constrained softmax loss for discriminative face verification. arXiv preprint arXiv:1703.09507 (2017)
13. Simonyan, K., Zisserman, A.: Very deep convolutional networks for large-scale image recognition. arXiv preprint arXiv:1409.1556 (2014)
14. Singh, M., Singh, R., Vatsa, M., Ratha, N.K., Chellappa, R.: Recognizing disguised faces in the wild. IEEE Trans. Biometrics Behav. Identity Sci. **1**(2), 97–108 (2019)
15. Smirnov, E., Melnikov, A., Oleinik, A., Ivanova, E., Kalinovskiy, I., Luckyanets, E.: Hard example mining with auxiliary embeddings. In: CVPR Workshops, pp. 37–46 (2018)
16. Szegedy, C., Ioffe, S., Vanhoucke, V., Alemi, A.A.: Inception-v4, inception-ResNet and the impact of residual connections on learning. In: AAAI (2017)
17. Szegedy, C., et al.: Going deeper with convolutions. In: CVPR, pp. 1–9 (2015)
18. Wang, F., Cheng, J., Liu, W., Liu, H.: Additive margin softmax for face verification. IEEE Signal Process. Lett. **25**(7), 926–930 (2018)
19. Wright, J., Yang, A.Y., Ganesh, A., Sastry, S.S., Ma, Y.: Robust face recognition via sparse representation. TPAMI **31**(2), 210–227 (2008)
20. Yang, J., Luo, L., Qian, J., Tai, Y., Zhang, F., Xu, Y.: Nuclear norm based matrix regression with applications to face recognition with occlusion and illumination changes. TPAMI **39**(1), 156–171 (2016)
21. Yang, M., Zhang, L., Yang, J., Zhang, D.: Regularized robust coding for face recognition. IEEE Trans. Image Process. **22**(5), 1753–1766 (2013). https://doi.org/10.1109/TIP.2012.2235849
22. Yi, D., Lei, Z., Liao, S., Li, S.Z.: Learning face representation from scratch. arXiv preprint arXiv:1411.7923 (2014)
23. Zhang, K., Chang, Y.L., Hsu, W.: Deep disguised faces recognition. In: CVPR Workshops, pp. 32–36 (2018)
24. Zhang, K., Zhang, Z., Li, Z., Qiao, Y.: Joint face detection and alignment using multitask cascaded convolutional networks. IEEE Signal Process. Lett. **23**(10), 1499–1503 (2016)

Piecewise Hashing: A Deep Hashing Method for Large-Scale Fine-Grained Search

Yimu Wang[1] , Xiu-Shen Wei[2(✉)] , Bo Xue[1] , and Lijun Zhang[1]

[1] State Key Laboratory for Novel Software Technology,
Nanjing University, Nanjing, China
{wangym,xueb,zhanglj}@lamda.nju.edu.cn
[2] PCA Lab, Key Lab of Intelligent Perception and Systems for High-Dimensional
Information of Ministry of Education, and Jiangsu Key Lab of Image and Video
Understanding for Social Security, School of Computer Science and Engineering,
Nanjing University of Science and Technology, Nanjing, China
weixs@njust.edu.cn

Abstract. Fine-grained search task such as retrieving subordinate categories of birds, dogs or cars, has been an important but challenging problem in computer vision. Although many effective fine-grained search methods were developed, with the amount of data increasing, previous methods fail to handle the explosive fine-grained data with low storage cost and fast query speed. On the other side, since hashing sheds its light in large-scale image search for dramatically reducing the storage cost and achieving a constant or sub-linear time complexity, we leverage the power of hashing techniques to tackle this valuable yet challenging vision task, termed as *fine-grained hashing* in this paper. Specifically, our proposed method consists of two crucial modules, *i.e.*, the bilinear feature learning and the binary hash code learning. While the former encodes both local and global discriminative information of a fine-grained image, the latter drives the whole network to learn the final binary hash code to present that fine-grained image. Furthermore, we also introduce a novel multi-task hash training strategy, which can learn hash codes of different lengths simultaneously. It not only accelerates training procedures, but also significantly improves the fine-grained search accuracy. By conducting comprehensive experiments on diverse fine-grained datasets, we validate that the proposed method achieves superior performance over the competing baselines.

Keywords: Fine-grained image retrieval · Deep hashing · Multi-task learning · Large-scale methods

1 Introduction

As a fundamental and challenging problem in computer vision, fine-grained image analysis (FGIA) [26] has been an active research area for several decades.

Y. Wang—Is a student.

© Springer Nature Switzerland AG 2020
Y. Peng et al. (Eds.): PRCV 2020, LNCS 12306, pp. 432–444, 2020.
https://doi.org/10.1007/978-3-030-60639-8_36

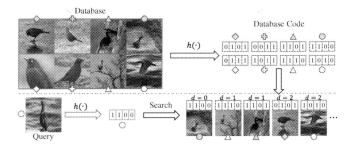

Fig. 1. An example of the *fine-grained hashing* problem. Its goal is to retrieve images belonging to multiple subordinate categories of a super-category (*i.e.*, birds). Each fine-grained image will be mapped into a binary code by a hashing function $h(\cdot)$. We return the images which are in the same variety as the query image since they have smaller hamming distance d. Here, different shapes of markers represent different subordinate categories.

One of the central tasks in FGIA is fine-grained image search [25,28], whose goal is to retrieve images belonging to multiple subordinate categories of a super-category (*e.g.*, different species of birds, different models of cars, or different kinds of clothes) and then return the images which are in the same variety as the query image, cf. Fig. 1. Due to small inter-class variations and large intra-class variations caused by the fine-grained nature, it is desirable to capture the discriminative parts of these fine-grained objects to form a powerful image representation.

During the booming of deep learning, recent years have witnessed effective progress of fine-grained search using deep learning techniques. Some trials [25] employed the pre-trained CNN models to unsupervisedly locate fine-grained objects and then obtain the deep features for image search. Later, to break through the limitation of unsupervised fine-grained search by pre-trained models, some works [29,30] tended to discovery novel loss functions under the supervised metric learning paradigm. Although these previous methods obtain good image search accuracy, they still cannot handle the *large-scale* fine-grained data with low storage cost and fast query speed, especially for the significant increment of data amount in the deep learning era. Moreover, with the rapidly explosive growth of vision data, more and more large-scale fine-grained datasets [3,8,9,24] are proposed recently. In consequence, the demand of handling large-scale data for fine-grained search methods has increased dramatically.

To deal with the large-scale data amount challenge, in machine learning, approximate nearest neighbor (ANN) search [1,2] has attracted much attention in recent years. Among ANN search methods, hashing [6,16,27] has been an active and representative subarea, which is able to map the data points to binary codes with hash functions by preserving the similarity in the original space of the data points. Thanks to the binary hash code representation, the storage cost for the large-scale data can be drastically reduced, and also the time complexity can be constant or sub-linear.

Therefore, in this paper, to alleviate the large-scale fine-grained search problem, we investigate a novel problem, *i.e.*, *fine-grained hashing*, as a step towards efficient and effective large-scale fine-grained image search. To realize fine-grained hashing, we propose an end-to-end trainable network which is inspired by a state-of-the-art fine-grained recognition backbone model, *i.e.*, bilinear pooling [14] and is also tailored for the fine-grained nature. Specifically, as shown in Fig. 2, our proposed method consists of a bilinear feature learning module and a hash code learning module. The former can encode the discriminative information of a fine-grained image by utilizing both global and local streams into the intermediate feature vector, while the latter plays the role to drive the whole network training and obtain the final binary hash codes.

The key novelty of our method is the "piecewise" and "synergistic" proposals. First, based on the outer product operation in bilinear pooling [14], the obtained feature can be viewed as a set of sub-vectors, and thus, each of which implicitly attends to one part of the image. We perform the part-level local stream mechanism upon these part-level sub-vectors "piece by piece" ("piecewise" proposal) to capture the discriminative cues for effectively representing fine-grained parts (*e.g.*, "tufted heads", "red-yellow stripe"), which favors the fine-grained nature. Additionally, a parallel global stream is also designed to obtain the global-level image feature. By aggregating both local (part-level) and global (object-level) information, the final image representation can be fed into the hash code learning module. Second, in order to accelerate the training process, we simultaneously train hash functions of different lengths in a novel multi-task hash training framework. Hash functions of different lengths share the convolutional layers with each other to learn feature representations (the "synergistic" proposal), while saving 70% training and inference time (with four functions) both theoretically and practically.

Empirical results on five fine-grained datasets, *i.e.*, *CUB* [21], *Aircraft* [17], *NABirds* [8], *VegFru* [9], and *Food101* [3] show that our piecewise hashing method significantly outperforms competing baselines, including the deep or non-deep hashing state-of-the-arts. Meanwhile, we perform our proposed method on a popular generic dataset, *i.e.*, *CIFAR-10* [11], to demonstrate that our method is able to achieve the best search accuracy when facing the generic images.

The main contributions of this paper are summarized as:

- We study the practical and challenging fine-grained hashing problem and propose an end-to-end trainable network with a novel multi-task hash training strategy to deal with the large-scale fine-grained search problem.
- We devise a novel piecewise hashing method consisting a bilinear feature learning module and a hash code learning module. The former resorts to the special structure of the bilinear CNN features to learn discriminative image features by leveraging both the global and local streams, while the latter drives the whole network training and returns the final fine-grained hash codes. Besides, the proposed multi-task strategy improves the efficiency and accuracy by synergistically learning the common layers across hash functions of different lengths.

– We conduct experiments on five fine-grained datasets and one generic dataset. Empirical results show that our proposed network outperforms previous hashing methods for both fine-grained images and generic images.

2   Related Work

2.1   Fine-Grained Image Search

Fine-grained image search has attracted increasing research attention in recent years, where the instances are within a subordinate category and different in slight patterns. In the literature, [28] is the first attempt to use handcraft features for fine-grained image search. Inspired by the power of deep learning, some unsupervised and supervised deep fine-grained image search methods have been proposed. Selective convolutional descriptor aggregation (SCDA) [25] is the first work on deep fine-grained image search, which directly discovers the discriminative parts in the images unsupervisedly. [29] defines the fine-grained search as a deep metric learning problem and tries to learn discriminative representations by designing specific loss functions. At the same time, with the rapid growth of fine-grained visual data, more and more large-scale fine-grained datasets have been proposed containing abundant labeled images, to name a few, *RPC* [24], *Cars* [5], *DA-Retail* [23] and *NABirds* [8], facilitating further research. Nevertheless, these previous fine-grained search work cannot handle large-scale fine-grained data, as they represented the images with *high-dimensional real-valued* vectors.

2.2   Hashing

Hashing is a widely used method for ANN search in large scale image retrieval with encouraging efficiency in both speed and storage. Existing hashing methods can be roughly categorized into unsupervised and supervised hashing. Unsupervised hashing methods [18] learn hash functions from unlabeled data, *e.g.*, LSH [6], SH [27] and ITQ [7]. Supervised hashing methods [22] attempt to leverage supervised information (*e.g.*, similarity matrix or label information) to improve the quality of hash codes, *e.g.*, KSH [16] and SDH [19]. Inspired by powerful feature representations learning with deep neural networks [20], the deep supervised hashing [10] adopting deep learning to generate high-level semantic features has been proposed. Deep Pairwise-Supervised Hashing (DPSH) [13] preserves relative similarity between image triplets straightly by integrating feature learning and hash functions in an end-to-end manner. Further, HashNet [4] tackles the data imbalance problem between similar and dissimilar pairs and alleviates this drawback by adjusting the weights of similar pairs. To additionally accelerate the training procedure, several asymmetric deep hashing methods are proposed, *i.e.*, Asymmetric Deep Supervised Hashing (ADSH) [10], which only learns the hash function for query points to alleviate time-consuming. However, previous hash methods were designed for generic images, which were not capable for *fine-grained* images search.

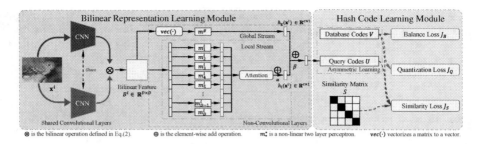

Fig. 2. Overview structure of our proposed piecewise hashing, which consists of two main modules, *i.e.*, bilinear representation learning and hash code learning. For bilinear representation learning, we develop both global (object-level) and local (part-level) stream to discover discriminative information for fine-grained objects by leveraging the bilinear features tailored for fine-grained images. Specifically, for better capturing different but subtle part features, we perform the part-level attention on the local stream. For hash code learning, three loss functions corresponding to three hash learning principles are employed, and the network can be trained in an end-to-end fashion with only image-level supervisions.

3   Proposed Method

In this section, we elaborately introduce our piecewise hashing method. Besides, for improving the efficiency, we also propose a simple yet effective multi-task training strategy to jointly learn hash functions of different code-lengths while sharing the convolutional layers of the bilinear representation learning module.

Vectors and scalars are bold lower case italic letters and lower case italic letters, such as $\boldsymbol{\alpha}$ and c. The i-th element of a vector $\boldsymbol{\alpha}$ and the i,j-th element of a matrix S are represented as $\boldsymbol{\alpha}_i$ and $S_{i,j}$, while the i-th column and j-th row of a matrix S are presented as $S_{*,i}$ and $S_{j,*}$. Assume that we have m training data points and n database points denoted as $X = \left\{\mathbf{x}^i\right\}_{i=1}^{m}$ and $Y = \left\{\mathbf{y}^j\right\}_{j=1}^{n}$. The pairwise supervised information is denoted as $S \in \{-1, +1\}^{m \times n}$ in training. If point $\mathbf{x}^i$ and point $\mathbf{y}^j$ are similar, $S_{ij} = +1$, otherwise $S_{ij} = -1$. Under this condition, the goal of supervised hashing is to learn binary hash codes $\boldsymbol{code} \in \{-1, +1\}^{c}$ for each point, and its corresponding hashing function $h(\cdot; \Theta)$, where c is the target length of binary code. Additionally, we use $U = \left\{\mathbf{u}^i\right\}_{i=1}^{m} \in \{-1, +1\}^{m \times c}$ and $V = \left\{\mathbf{v}^j\right\}_{j=1}^{n} \in \{-1, +1\}^{n \times c}$ to denote the learned binary hash codes for training points and database points. The hash codes have to preserve the similarity S between each point, which means the Hamming distance between $\mathbf{u}^i$ and $\mathbf{v}^j$ should be as small as possible if $S_{ij} = +1$.

3.1   Our Piecewise Hashing

Our model is shown in Fig. 2, which contains two modules: the bilinear representation learning module and the hash-code learning module. We will elaborate them as follows.

Bilinear Representation Learning Module. In this part, we introduce our network architecture in detail. For clarity, all the equations take one data point $\mathbf{x}^i$ as input.

Bilinear pooling (BCNN) [14] is an effective architecture tailored for the fine-grained task. BCNN represents an image as an outer product of features derived from two CNNs and discovers localized feature interactions. Inspired by BCNN, we acquire bilinear representations and further generate hash codes based on them. Specifically, we assume the outputs of two CNNs in BCNN are re-organized into $f_A(\mathbf{x}^i) \in \mathbb{R}^{D \times L}$ and $f_B(\mathbf{x}^i) \in \mathbb{R}^{D \times L}$, where D denotes the dimension of the outputs and L denotes the spatial locations. Then, we define the bilinear feature at location l as:

$$\text{bilinear}(l, \mathbf{x}^i, f_A, f_B) = f_A(l, \mathbf{x}^i) f_B(l, \mathbf{x}^i)^\top . \tag{1}$$

The sum pooling aggregates the bilinear combination of features across all locations in the image to obtain the global bilinear representation $B^i \in \mathbb{R}^{D \times D}$ as follows:

$$B^i = \sum_{l=1}^{L} \text{bilinear}(l, \mathbf{x}^i, f_A, f_B) . \tag{2}$$

After obtaining the bilinear representation B^i, the global stream and the local stream are paralleled to derive outputs. In the global stream, the D^2-dimension bilinear representation B^i is mapped to a c-dimension binary-like output. A straightforward solution to realize this mapping is employing a multi-layer perception (MLP) $m(\cdot)$. Therefore, we derive the global (object-level) binary-like output via an MLP $m^g(\cdot) : \mathbb{R}^{D^2} \to \mathbb{R}^c$ as:

$$h_g(\mathbf{x}^i) = m^g(\text{vec}(B^i)) , \tag{3}$$

where $\text{vec}(\cdot)$ vectorizes a $D \times D$ matrix to a $D^2 \times 1$ vector.

The key novelty of our method is "piecewise" in the local stream. Based on the outer product operation above, the bilinear representations can be viewed as a set of sub-vectors (pieces), and thus, each of which implicitly attends to a highly localized image feature (*e.g.*, "tufted heads" and "red-yellow stripe"). Hence, in the local stream, we apply the part-level attention ("piecewise") to emphasize the discriminative pieces and understate the pointless pieces over the bilinear representation. Specifically, we locally re-organize the bilinear representation B^i by column and map each column to a local binary-like output via an MLP. Then we derive the local (part-level) bilinear binary-like output from the convex combination of local binary-like outputs for each column as follows:

$$\begin{aligned} h_l(\mathbf{x}^i) = \text{Attention}(\boldsymbol{\alpha}, B^i) = \sum_{d=1}^{D} \boldsymbol{\alpha}_d m_d^l(B_{*,d}^i) \\ \text{s.t. } \sum_{d=1}^{D} \boldsymbol{\alpha}_d = 1 , \end{aligned} \tag{4}$$

where $\boldsymbol{\alpha} \in \mathbb{R}^{D \times 1}$ is a learnable parameter, and $m_d^l(\cdot) : \mathbb{R}^D \to \mathbb{R}^c$ represents the MLP for the d-th column.

The overall binary-like output $h_{soft}(\mathbf{x}^i; \Theta)$ is also the convex combination of the global bilinear binary-like output $h_g(\mathbf{x}^i)$ and the local bilinear binary-like output $h_l(\mathbf{x}^i)$. Discrete hash codes $h(\mathbf{x}_i; \Theta)$ are derived by employing sign$(\cdot)$ on the overall binary-like output as follows:

$$h(\mathbf{x}_i; \Theta) = \text{sign}(h_{soft}(\mathbf{x}^i; \Theta)) = \text{sign}(\beta h_g(\mathbf{x}^i) + (1 - \beta) h_l(\mathbf{x}^i))$$
$$\text{s.t. } 0 \leq \beta \leq 1, \tag{5}$$

where β is a learnable parameter, and Θ is the parameters of the whole network including $\boldsymbol{\alpha}$ and β.

Hash-Code Learning Module. In this section, we design our objective function by three following principles including: preserving the similarity of data points in the original space, distributing the codes to uniformly fulfill the code space, and generating compact binary codes. In total, these principles correspond to three loss functions to drive the training procedure of the whole network.

To preserve the similarity S during the training procedure, the previous work [10,15] achieves it by minimizing the ℓ_2 loss between similarity S_{ij} and inner product of query-database binary code pairs $\mathbf{u}^i \mathbf{v}^{j\top}$. As all the hash codes are discrete, it is hard to complete the training of hash functions. Previous works always employ discrete function sign$(\cdot)$ to calculate the hash codes $\mathbf{u}_i$ and utilize sigmoid$(\cdot)$ or tanh$(\cdot)$ functions to approximate sign$(\cdot)$. The common similarity loss they minimize is as:

$$\begin{aligned} J_{S0} &= \sum\nolimits_{i=1}^{m} \sum\nolimits_{j=1}^{n} \|\mathbf{u}^i \mathbf{v}^{j\top} - c S_{ij}\|^2 \\ &= \sum\nolimits_{i=1}^{m} \sum\nolimits_{j=1}^{n} \|\text{sign}(h_{soft}(\mathbf{x}^i; \Theta)) \mathbf{v}^{j\top} - c S_{ij}\|^2 \\ &\approx \sum\nolimits_{i=1}^{m} \sum\nolimits_{j=1}^{n} \|\tanh(h_{soft}(\mathbf{x}^i; \Theta)) \mathbf{v}^{j\top} - c S_{ij}\|^2 . \end{aligned} \tag{6}$$

Nevertheless, employing such non-linear functions would inevitably slow down or even affect the convergence of the network [12]. To ease such a problem, we directly optimize the real-valued network outputs instead of the approximation of $\mathbf{u}_i$, and impose a regularizer, *i.e.*, the quantization loss J_Q, on the real-valued network outputs to approach the desired discrete values. Particularly, we present the J_S and the regularizer J_Q as:

$$J_S = \sum\nolimits_{i=1}^{m} \sum\nolimits_{j=1}^{n} \|h_{soft}(\mathbf{x}^i; \Theta) \mathbf{v}^{j\top} - c S_{ij}\|^2 , \tag{7}$$

$$J_Q = \sum\nolimits_{i=1}^{m} \|\mathbf{u}^i - h_{soft}(\mathbf{x}^i; \Theta)\|^2 . \tag{8}$$

One of the advantages of hash is storage efficiency. This brings another goal to accomplish, that is, hash-codes points should uniformly fulfill the 2^c code space. Hence, for fully utilizing each bit of hash codes, we set a term J_B to make each bit of the hash codes be balanced on all the points. Ideally, if we sum up all the hash-codes, the results should be $\mathbf{0}$. This term can be presented as:

$$J_B = \sum\nolimits_c \left[\left| \sum\nolimits_{i=1}^{m} \mathbf{u}_c^i \right| + \left| \sum\nolimits_{j=1}^{n} \mathbf{v}_c^j \right| \right] . \tag{9}$$

The final objective function we minimize is as follows:

$$\min_{V,\Theta} J = J_S + \lambda J_Q + \mu J_B$$
$$\text{s.t. } V \in \{-1, +1\}^{n \times c},$$

(10)

where λ and μ are hyper-parameters, which are set to 200 and 0.1 in all the experiments.

3.2 Multi-task Hash Training Strategy

The another key novelty of our method is the "synergistic" proposal. Multi-task hash training is a simple but effective training strategy. In the previous deep hashing work [4,10], though the convolutional layers can be shared, they still separately train models of different code-lengths, while in shallow hash methods [6,16], they usually utilize the same features to train several hash functions of different lengths.

To reduce the training time and the redundancy among different hash functions, we propose a multi-task hash training strategy enabling us to learn hash functions of different code-lengths simultaneously. We split the network into the convolutional layers, and the non-convolutional layers, $i.e.$, the global and local stream, which are directly related to the code-length shown in Fig. 2. Specifically, hash functions of different code-lengths contain the non-convolutional layers, while sharing the convolutional layers with each other ("synergistic"). When we learn four hash functions of 12 bits, 24 bits, 32 bits, and 48 bits concurrently, the corresponding objective function becomes as:

$$\min_{V_{12}, V_{24}, V_{32}, V_{48}, \Theta} J_{mul} = J_{12} + J_{24} + J_{32} + J_{48}$$
$$\text{s.t. } V_{12} \in \{-1, +1\}^{n \times 12}, V_{24} \in \{-1, +1\}^{n \times 24},$$
$$V_{32} \in \{-1, +1\}^{n \times 32}, V_{48} \in \{-1, +1\}^{n \times 48},$$

(11)

where J_{12}, J_{24}, J_{32} and J_{48} are the objective functions where we set $c = 12, 24, 32, 48$ in Eq. (10). Note that, there is no hyper-parameter between these four terms, which reveals our multi-task learning strategy is not tricky.

Multi-task hash training strategy enables us to learn intermediate representations and hash functions of different code-lengths simultaneously, saving computation time and memory space. As the code length grows, the model would contain more parameters and then get prone to overfitting. This strategy can help overcome such problems by sharing parts of parameters on the whole network thus benefiting the training procedure. Additionally, the theoretical analyses in Sect. 3.4 prove that our strategy is efficient.

3.3 Learning Algorithm

In this section, we present an alternating learning algorithm to learn V and Θ of Eq. (10). The pseudo codes of our algorithm can be found in Appendix.

The parameters are learned alternatively, which means we update one parameter with another parameter fixed. The details are presented below.

Normally, we are only given the database points $Y = \{\mathbf{y}^j\}_{j=1}^{n}$ and the pairwise supervised information S between them, we can learn hash-codes and hash functions by sampling a subset or the whole set of Y as the query set X for training, i.e., $X \subseteq Y$. To accelerate the training procedure, we construct a subset X of database Y instead of using the whole database. Hence, we only consider the item related to V in J_B of Eq. (10).

Learn Θ with V fixed. When V is fixed, we learn and update the parameter Θ of our neural network by back-propagation (BP) algorithm.

Specifically, for each query point $\mathbf{x}^i$ in the query points, the gradient can be calculated as:

$$\frac{\partial J}{\partial \mathbf{z}^i} = \frac{\partial J_S}{\partial \mathbf{z}^i} + \lambda \frac{\partial J_Q}{\partial \mathbf{z}^i} + \mu \frac{\partial J_B}{\partial \mathbf{z}^i} = \sum_{i=1}^{m} \left[\left(\mathbf{z}^i \mathbf{v}^{j\top} - cS_{ij} \right) \mathbf{v}^j + \lambda(\mathbf{u}^i - \mathbf{z}^i) \right] , \quad (12)$$

where $\mathbf{z}^i = h_{soft}(\mathbf{x}^i, \Theta)$. Once we have the $\frac{\partial J}{\partial \mathbf{z}^i}$, we can compute $\frac{\partial J}{\partial \Theta}$ based on $\frac{\partial J}{\partial \mathbf{z}^i}$ using the chain rule and the back propagation algorithm to update Θ.

Learn V with Θ fixed. When Θ is fixed, we can easily reformulate the Eq. (10) as follows:

$$\min_{V} J(V) = \mathrm{tr}\left(V \left[Z^\top V Z^\top - 2cZ^\top S - 2\lambda \bar{V}^\top \right] \right) + \mu \sum_{k=1}^{c} |\mathbf{1} \cdot V_{*,k}| + \epsilon$$
$$\text{s.t. } V \in \{-1, +1\}^{n \times c}, \quad (13)$$

where $Z = \left[\mathbf{z}^1, \mathbf{z}^2, \ldots, \mathbf{z}^m \right] \in [-1, +1]^{m \times c}$, $\mathbf{1} = [1, 1, \ldots, 1] \in \{1\}^{1 \times n}$, $\bar{V} = \{\bar{V}_j = \mathbf{I}(\mathbf{y}^j \in X)\mathbf{u}^j\}_{j=1}^{n \times 1} \in \mathbb{R}^{n \times c}$, $\mathbf{I}(\cdot)$ is the indicator function, and "ϵ" is a constant independent of V. For convenience, we let $Q = (Z^\top V Z^\top - 2cZ^\top S - 2\lambda \bar{V}^\top)^\top$. Then we can rewrite this problem as:

$$\min_{V} J(V) = \mathrm{tr}\left(VQ^\top \right) + \mu \sum_{k=1}^{c} |\mathbf{1} \cdot V_{*,k}| + \epsilon$$
$$\text{s.t. } V \in \{-1, +1\}^{n \times c}. \quad (14)$$

To ease this problem, we update one bit a time. We alternatively update one column of V with the other columns fixed. Hence, the optimal solution of this bit by bit problem is:

$$J(V_{*k}) = \begin{cases} = -\operatorname{sign}(Q_{*,k} + \mu \mathbf{1}^\top), & \mathbf{1} \cdot V_{*,k} \geq 0 \\ = -\operatorname{sign}(Q_{*,k} - \mu \mathbf{1}^\top), & \mathbf{1} \cdot V_{*,k} < 0 \end{cases} . \quad (15)$$

3.4   Out-of-Sample Extension and Model Analyses

After completing the learning procedure, hash-codes for all the database points can be easily generated. As for the point $\mathbf{x}_q$ in query points, we can use $\mathbf{u}^q = \operatorname{sign}(h_{soft}(\mathbf{x}^q, \Theta))$ to generate binary hash-codes. The total computational complexity for training our piecewise hashing is $O(n)$. For the training complexity, while the complexity of separate training without multi-task is $O_4(n) \approx 4O(n)$, multi-task strategy accelerates model training by saving 70% time cost with the complexity of $O_{4multi-task}(n) \approx 1.2O(n)$. Thus, our proposal of simultaneously training hash functions of different lengths is theoretically efficient.

4   Experiments

In this section, we evaluate the performance of our proposed method on both fine-grained and generic datasets, and then compare with state-of-the-art approaches.

We evaluate the performance of our proposed method on six datasets, *i.e.*, CUB [21], Aircraft [17], NABirds [8], VegFru [9], Food101 [3] and CIFAR-10 [11]. For the above datasets, we follow the standard split proposed with these datasets, while two images will be treated as a ground-truth similar pair if they share the same label. We evaluate the retrieval performance by adopting two evaluation metrics: mean Average Precision (mAP) and Precision-recall Curves (PR Curves) based on lookup. Details are available in Appendix. All the data are reported with average values running five times. We compare our deep piecewise hashing method with several state-of-the-art hashing methods, including shallow methods, *i.e.*, LSH [6], ITQ [7], SH [27], SDH [19] and KSH [16], and deep supervised methods, *i.e.*, DSH [15], DPSH [13], HashNet [4], and ADSH [10]. For all deep hashing methods, we use raw images resized to 224×224 as inputs. For traditional shallow methods, we extract 4096-dimensional deep features by the VGG-16 model pre-trained with ImageNet to conduct fair comparisons. Besides, for all the state-of-the-art hashing methods, we prefer to employ the hyper-parameters introduced in their papers.

The mAP results on six datasets are presented in Table 1. Additional results of PR Curves and Top-5K mAP on all the datasets are available in Appendix.

Search Accuracy on the Fine-Grained Datasets: Our proposed method with the multi-task learning strategy outperforms other hashing methods across different code-lengths on fine-grained datasets. Specifically, the mAP of our piecewise hashing obtains relative improvements over the next-best state-of-the-art methods of 14.36%, 37.52%, 32.82%, and 24.30% on CUB. We notice that similar improvements are achieved on other fine-grained datasets.

Search Accuracy on the Generic Dataset: Moreover, as shown in Table 1, our piecewise hashing still outperforms other hashing methods on the generic dataset, *i.e.*, *CIFAR-10*, obtaining significant increment of 1.90%, 4.53%, 2.85%, and 2.57% for different lengths of hash codes, respectively.

Table 1. Comparisons of mAP w.r.t. different number of bits on six datasets, *CIFAR-10*, *CUB*, *Aircraft*, *NABirds*, *VegFru* and *Food101*. Best in bold.

Method	Backbone	CIFAR-10				CUB				Aircraft			
		12 bits	24 bits	32 bits	48 bits	12 bits	24 bits	32 bits	48 bits	12 bits	24 bits	32 bits	48 bits
LSH	–	0.1162	0.1215	0.1224	0.1244	0.0152	0.0235	0.0288	0.0415	0.0169	0.0219	0.0238	0.0282
SH	–	0.1316	0.1289	0.1287	0.1274	0.0666	0.0809	0.0893	0.1048	0.0328	0.0385	0.0404	0.0428
ITQ	–	0.1544	0.1607	0.1630	0.1656	0.0855	0.1196	0.1376	0.1549	0.0438	0.0528	0.0582	0.0605
KSH	–	0.2353	0.2563	0.2669	0.2763	0.1125	0.1502	0.1722	0.1954	0.0557	0.0738	0.0814	0.0892
SDH	–	0.1746	0.2140	0.2115	0.2362	0.0964	0.1442	0.1491	0.1827	0.0489	0.0636	0.0690	0.0765
DPSH	ResNet50	0.6872	0.7024	0.7281	0.7437	0.0685	0.0885	0.1008	0.1148	0.0874	0.1087	0.1354	0.1394
DSH	ResNet50	0.7230	0.7644	0.7746	0.7920	0.1360	0.1899	0.2237	0.2744	0.0814	0.1066	0.1221	0.1445
HashNet	ResNet50	0.7261	0.7614	0.7858	0.7950	0.1203	0.1777	0.1993	0.2213	0.1491	0.1775	0.1942	0.2032
ADSH	ResNet50	0.6599	0.7413	0.7590	0.7672	0.0209	0.1002	0.2997	0.4535	0.0924	0.2314	0.3204	0.4278
Ours	VGG-16	**0.7451**	**0.8097**	**0.8143**	**0.8207**	**0.2796**	**0.5651**	**0.6279**	**0.6956**	**0.4392**	**0.5662**	**0.5997**	**0.6296**

Method	Backbone	NABirds				VegFru				Food101			
		12 bits	24 bits	32 bits	48 bits	12 bits	24 bits	32 bits	48 bits	12 bits	24 bits	32 bits	48 bits
LSH	–	0.0064	0.0096	0.0132	0.0201	0.0077	0.0117	0.0147	0.0207	0.0158	0.0199	0.0221	0.0279
SH	–	0.0258	0.0437	0.0470	0.0660	0.0258	0.0437	0.0549	0.0660	0.0410	0.0480	0.0494	0.0532
ITQ	–	0.0351	0.0591	0.0668	0.0782	0.0306	0.0569	0.0711	0.0866	0.0559	0.0748	0.0831	0.0939
KSH	–	0.0396	0.0645	0.0768	0.0915	0.0353	0.0006	0.0923	0.1096	0.0804	0.0954	0.1040	0.1099
SDH	–	0.0327	0.0637	0.0814	0.0945	0.0384	0.0659	0.0868	0.1085	0.0719	0.1048	0.1167	0.1295
DPSH	ResNet50	0.0159	0.0225	0.0241	0.0380	0.0375	0.0541	0.0731	0.0931	0.0795	0.1059	0.1370	0.2025
DSH	ResNet50	0.0139	0.0225	0.0304	0.0392	0.0537	0.0786	0.0970	0.1119	0.1225	0.2392	0.2643	0.2961
HashNet	ResNet50	0.0157	0.0242	0.0276	0.0351	0.0726	0.1157	0.1284	0.1568	0.2186	0.3222	0.3515	0.4109
ADSH	ResNet50	0.0124	0.0998	0.1782	0.3041	0.0838	0.2460	0.3679	0.5285	0.0296	0.0499	0.1605	0.4835
Ours	VGG-16	**0.0940**	**0.2619**	**0.3419**	**0.4093**	**0.2974**	**0.5525**	**0.6058**	**0.6674**	**0.4177**	**0.5896**	**0.6284**	**0.6666**

5   Conclusion

In this paper, we presented a piecewise hashing method for the novel fine-grained hashing task. One of the key contributions was the local stream with piecewise part-level attention on bilinear representations to capture the discriminative cues for effectively representing fine-grained parts. Besides, our proposed multi-task training strategy can decrease the training and inference time while concurrently learned several hash functions and improving the search accuracy. Experimental results on diverse fine-grained datasets and the generic dataset showed the superiority of our method. In the future, we would like to explore novel fine-grained hashing methods under the unsupervised setting.

References

1. Andoni, A., Indyk, P.: Near-optimal hashing algorithms for approximate nearest neighbor in high dimensions. Commun. ACM **51**(1), 117 (2008)
2. Andoni, A., Razenshteyn, I.: Optimal data-dependent hashing for approximate near neighbors. In: STOC, pp. 793–801 (2015)
3. Bossard, L., Guillaumin, M., Van Gool, L.: Food-101 – mining discriminative components with random forests. In: Fleet, D., Pajdla, T., Schiele, B., Tuytelaars, T. (eds.) ECCV 2014. LNCS, vol. 8694, pp. 446–461. Springer, Cham (2014). https://doi.org/10.1007/978-3-319-10599-4_29

4. Cao, Z., Long, M., Wang, J., Yu, P.S.: HashNet: deep learning to hash by continuation. In: ICCV, pp. 5608–5617 (2017)
5. Gebru, T., Krause, J., Wang, Y., Chen, D., Deng, J., Fei-Fei, L.: Fine-grained car detection for visual census estimation. In: AAAI, pp. 4502–4508 (2017)
6. Gionis, A., Indyk, P., Motwani, R.: Similarity search in high dimensions via hashing. In: VLDB, pp. 518–529 (1999)
7. Gong, Y., Lazebnik, S., Gordo, A., Perronnin, F.: Iterative quantization: a procrustean approach to learning binary codes for large-scale image retrieval. TPAMI **35**(12), 2916–2929 (2012)
8. Horn, G.V., Branson, S., Farrell, R., Haber, S.: Building a bird recognition app and large scale dataset with citizen scientists: The fine print in fine-grained dataset collection. In: CVPR, pp. 595–604 (2015)
9. Hou, S., Feng, Y., Wang, Z.: VegFru: a domain-specific dataset for fine-grained visual categorization. In: ICCV, pp. 541–549 (2017)
10. Jiang, Q.Y., Li, W.J.: Asymmetric deep supervised hashing. In: AAAI, pp. 3342–3349 (2018)
11. Krizhevsky, A., Hinton, G.: Learning multiple layers of features from tiny images. Technical report, Citeseer (2009)
12. Krizhevsky, A., Sutskever, I., Hinton, G.E.: ImageNet classification with deep convolutional neural networks. In: NeurIPS, pp. 1097–1105 (2012)
13. Li, W.J., Wang, S., Kang, W.C.: Feature learning based deep supervised hashing with pairwise labels. arXiv preprint arXiv:1511.03855 (2015)
14. Lin, T.Y., RoyChowdhury, A., Maji, S.: Bilinear convolutional neural networks for fine-grained visual recognition. TPAMI **40**(6), 1309–1322 (2018)
15. Liu, H., Wang, R., Shan, S., Chen, X.: Deep supervised hashing for fast image retrieval. In: CVPR, pp. 2064–2072 (2016)
16. Liu, W., Wang, J., Ji, R., Jiang, Y.G., Chang, S.F.: Supervised hashing with kernels. In: CVPR, pp. 2074–2081 (2012)
17. Maji, S., Kannala, J., Rahtu, E., Blaschko, M., Vedaldi, A.: Fine-grained visual classification of aircraft. Technical report (2013)
18. Mu, Y., Wright, J., Chang, S.-F.: Accelerated large scale optimization by concomitant hashing. In: Fitzgibbon, A., Lazebnik, S., Perona, P., Sato, Y., Schmid, C. (eds.) ECCV 2012. LNCS, vol. 7572, pp. 414–427. Springer, Heidelberg (2012). https://doi.org/10.1007/978-3-642-33718-5_30
19. Shen, F., Shen, C., Liu, W., Tao Shen, H.: Supervised discrete hashing. In: CVPR, pp. 37–45 (2015)
20. Simonyan, K., Zisserman, A.: Very deep convolutional networks for large-scale image recognition. arXiv preprint arXiv:1409.1556 (2014)
21. Wah, C., Branson, S., Welinder, P., Pietro Perona, S.B.: The Caltech-UCSD Birds-200-2011 Dataset. Technical report CNS-TR-2011-001, California Institute of Technology (2011)
22. Wang, J., Kumar, S., Chang, S.F.: Semi-supervised hashing for scalable image retrieval. In: CVPR, pp. 3424–3431 (2010)
23. Wang, Y., Song, R., Wei, X.S., Zhang, L.: An adversarial domain adaptation network for cross-domain fine-grained recognition. In: WACV, pp. 1228–1236 (2020)
24. Wei, X.S., Cui, Q., Yang, L., Wang, P., Liu, L.: RPC: a large-scale retail product checkout dataset. arXiv preprint arXiv:1901.07249 (2019)
25. Wei, X.S., Luo, J.H., Wu, J., Zhou, Z.H.: Selective convolutional descriptor aggregation for fine-grained image retrieval. TIP **26**(6), 2868–2881 (2017)
26. Wei, X.S., Wu, J., Cui, Q.: Deep learning for fine-grained image analysis: a survey. arXiv preprint arXiv:1907.03069 (2019)

27. Weiss, Y., Torralba, A., Fergus, R.: Spectral hashing. In: NeurIPS, pp. 1753–1760 (2009)
28. Xie, L., Wang, J., Zhang, B., Tian, Q.: Fine-grained image search. TMM **17**(5), 636–647 (2015)
29. Zheng, X., Ji, R., Sun, X., Wu, Y., Huang, F., Yang, Y.: Centralized ranking loss with weakly supervised localization for fine-grained object retrieval. In: IJCAI, pp. 1226–1233 (2018)
30. Zheng, X., Ji, R., Sun, X., Zhang, B., Wu, Y., Huang, F.: Towards optimal fine grained retrieval via decorrelated centralized loss with normalize-scale layer. In: AAAI, pp. 9291–9298 (2019)

MHASiam: Mixed High-Order Attention Siamese Network for Real-Time Visual Tracking

Lei Pu[1], Xinxi Feng[2], Zhiqiang Hou[3]([✉]), Wangsheng Yu[2], Yufei Zha[4], and Zhiqiang Jiao[1]

[1] Graduate College, Air Force Engineering University, Xi'an 710077, China
[2] Information and Navigation College, Air Force Engineering University, Xi'an 710077, China
[3] School of Computer Science and Technology, Xi'an University of Posts and Telecommunications, Xi'an 710121, China
hzq@xupt.edu.com
[4] School of Computer Science, Northwestern Polytechnical University, Xi'an 710072, China

Abstract. Fully Convolutional Siamese network (SiamFC) has demonstrated high performance in the visual tracking field, but the learned CNN features are redundant and not discriminative to separate the object from the background. To address the above problem, this paper presents a Mixed High-order Attention Siamese network (MHASiam) for real-time object tracking. We first proposes a High-order Attention (HA) module that is integrated into the Siamese network to select the features in the channel domain. Especially, a high-order attention module is followed by the last layer of the network, and this benefit to obtain the higher-order representation for improving the discriminate ability of the model. Then, a First-order Attention (FA) module is introduced to further enhance the richness of attention knowledge, which is combined with the HA module in a parallel manner. Finally, the GOT10k data set is employed to train our Mixed High-order Attention Siamese network (MHASiam) to improve the target representation ability. Experimental results show that the proposed algorithm improves the accuracy by 9.4% and the success rate by 4.8% compared with the SiamFC tracker.

Keywords: Visual tracking · Siamese network · High-order attention · Channel attention

1 Introduction

Visual tracking is one of the difficult problems in the field of computer vision, and it is the basis for achieving more advanced visual understanding and scene

The first author is a student. This research has been supported by National Natural Science Foundation of China (No. 61571458, No. 61703423).

Y. Peng et al. (Eds.): PRCV 2020, LNCS 12306, pp. 445–456, 2020.
https://doi.org/10.1007/978-3-030-60639-8_37

analysis [24]. Visual tracking technology is widely used in video surveillance, human-computer interaction, robotics, video editing and auto-driving. Under the condition that only the initial position and target scale are given, the visual tracking task is supposed to achieve continuous and stable tracking of moving targets in subsequent frames. Despite much progress have been made in recent year, long-term stable visual tracking is still a challenging problem because of the complex factors such as scale variation, target rotation, deformation, rapid motion, illumination changes and similar object interference.

Recently, the visual tracking algorithms based on the Siamese network have achieved a good balance between tracking accuracy and speed, which has attracted extensive attention and research [2,11,17,30]. The tracking algorithm based on the Siamese network regards the tracking as a similarity matching task. By learning the general features of the target offline on a very large data set, the Siamese-based algorithm only use the initial frame as a template for online tracking. Since there is no online model update, the tracking methods based on Siamese network have obvious speed advantage compared to other deep learning based tracking algorithms.

However, almost all of the Siamese-based trackers only exploit the first-order image representation, which limits the nonlinear modeling capability of Siamese networks. Recently, modeling of the higher statistics has attracted great attention in a variety of computer visual tasks, such as object recognition [9], person re-identification [5], semantic segmentation [4], and fine-grained image categorization [18]. In order to further improve the feature representation and discrimination ability of the Siamese network, this paper explores the high-order image representation and further proposes a tracking algorithm based on the Mixed High-order Attention Siamese network (MHASiam).

MHASiam is a twofold Siamese network comprised of a first-order attention branch and a high-order attention branch. Each branch is a Siamese network computing the similarity scores between the target image and a detection image. For the first-order branch, we adopt the Squeeze-and-Excitation module [14] to compute channel-wise weights based on global average pooling. For the high-order branch, we propose a high-order attention module that is motivated by SENet [14] to exploit higher-order statistics for more discriminative image representations based on covariance pooling. In order to maintain the end-to-end fashion and the number of parameters, the two branches are joint trained and the similarity score maps are combined. Evaluations show that our tracker outperforms the baseline algorithm by a large margin on OTB2015 benchmark and achieves comparable performance with many state-of-the-art trackers.

2   Related Works

In this section, we made a review about the recent proposed methods that are most relevant to our tracker. A comprehensive review of the tracking methods is beyond the scope of the paper, and surveys of this field can be found in [6,19,24].

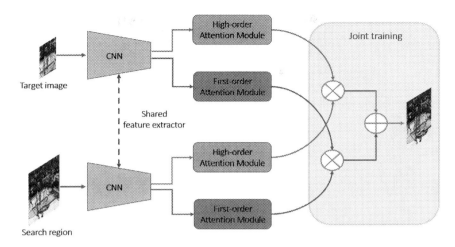

Fig. 1. The architecture of the proposed MHASiam network. The MHASiam is composed of a shared CNN feature extractor (indicated by green box in the figure) and two attention branches (first-order attention and high-order attention). The two branches are joint trained and combined in testing time. When a pair of target image and search image flow into the network, two types of attentions are extracted, the output of each branch is a response map. We obtain the target location by searching the final response map after sum operation. (Color figure online)

2.1 Deep Feature Based Trackers

Since the deep feature has superior representation and generalization ability, it is a worthwhile research to migrate it to the tracking task. At present, the research on the combination of deep features and tracking tasks can be divided into two directions. First, the offline pre-training network is used for feature extraction [8,12,20,23]. For example, Ma *et al.* [20] extracted features from the VGG19 network. The combination of multi-layer deep features and correlation filter algorithms yields significant performance improvements. Qi *et al.* [23] also proposed a similar method, except that the Hedge algorithm is used to adaptively fuse multi-layer features. Moreover, UPDT [3] and ECO [7] constructed trackers based on the continuous convolution filters. The other direction is offline training with online fine-tuning. The representative work is the MDNet algorithm [22], which treat tracking as a classification problem, and learnt an offline deep feature extractor and then online updated the classifier by adding some learnable fully-connected layers to perform online tracking.

2.2 Siamese Network Based Trackers

Siamese architectures have recently attracted considerable attention in visual tracking due to computational efficiency and robustness. Using Siamese Network for object tracking starts with SINT [25], which formulated visual tracking as a verification learning problem and trained a Siamese architecture offline to learn a similarity metric. The most well-known work is the SiamFC [2], which brought cross correlation to enable a much larger search area. Notable improvements have been made to this popular tracker to address several limiting issues [10,11,17,26, 30,31]. CFNet [26] attempts to introduce correlation filter into Siamese network to speed up tracking. SA-Siam [11] aims to improve SiamFC with heterogeneous features. SiamDW [31] incorporates the deeper and wider network and achieves better performance. To cope with the scale changes, several RPN-based trackers are further investigated [17,32].

3   Mixed High-Order Attention Siamese Network

We propose a mixed high-order attention Siamese network for real-time visual tracking. Figure 1 shows the network architecture of the proposed MHASiam tracker. As shown in Fig. 1, the proposed architecture consists of a Siamese network for feature extraction and two attention branches. In this section, we describe the proposed method in detail. Firstly, a brief introduction to the fully convolutional Siamese network is presented, Then we demonstrate the proposed MHA module. Finally, the two attention branches are joint trained and complement each other. By this way, a large performance improvement is achieved.

3.1   Fully Convolutional Siamese Network

The essence of the Siamese-based visual tracking methods is similarity learning in an embedding space. Given the first frame as an exemplar, the goal is to find a most similar instance from subsequent frames. The key problem is how to learn a powerful matching function. Assuming that the template image is set as z, and the candidate image is set as x, and f represent the similar function. First, the template feature $\varphi(x)$ and the candidate feature $\varphi(z)$ are extracted through the network, and the similarity score map can be recorded as $f(\varphi(x), \varphi(z))$. When the two images are of the same size, the similarity becomes a value, and when search image is much larger, the similarity measure becomes a response map. The Siamese network tracking algorithm calculates the similarity by adding a cross-correlation layer, as shown below:

$$f(z,x) = \varphi(z) * \varphi(x) + b \tag{1}$$

where b is an offset term that represents the same real value at each location, $*$ indicating a cross-correlation operation. The maximum value of the response map is the position corresponding to the target. A large number of positive and negative sample pairs and logistic regression loss functions are used to train the network. The training details can be found in the paper [2].

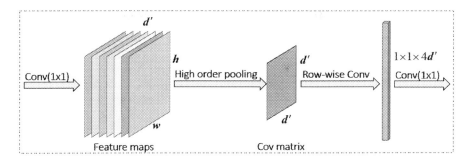

Fig. 2. High-order Attention (HA) module. The feature maps are shown as feature dimensions, e.g. h, w, d' denotes a feature map with channel number d', height h and width w. The high order pooling operation is performed on all the feature maps.

3.2 High-Order Attention Module

The attention module has achieved great success in the field of natural language processing, especially the self-attention mechanism, which greatly promoted the development of natural language processing. In recent years, a large amount of research work has applied the attention mechanism to the field of computer vision from the remarkable image recognition to the panoramic segmentation. In order to improve the ability of the Siamese network to discriminate specific targets, this paper proposes a MHA module that combines FA attention and HA attention.

The multi-channel feature of deep neural networks can achieve more robust representation of the target, but almost all network structures treat the significance of each channel equally. However, different channels play different roles in tracking different targets. Therefore, in certain circumstance some feature channels are more significant than the others. For the tracking task of a specific target, only the responses of some channels may be useful, and the responses of other channels can be viewed as interference.

Figure 2 illustrated the diagram of the proposed high-order attention module. Assume that the output feature tensor of the last convolutional layer is represented as $X \in R^{h \times w \times d}$, where d is the number of feature channels, h, w is the height and width of feature map. In order to reduce the computational burden, we firstly use 1×1 convolution to decrease the channel number from d to d'. Then we adopt high order pooling to obtain a $d' \times d'$ covariance matrix that denotes the channel correlations. The covariance matrix is described as :

$$
\begin{bmatrix}
\text{cov}\,(X_1, X_1) & \text{cov}\,(X_1, X_2) & \cdots & \text{cov}\,(X_1, X_{d'}) \\
\text{cov}\,(X_2, X_1) & \text{cov}\,(X_2, X_2) & \cdots & \text{cov}\,(X_2, X_{d'}) \\
\vdots & \vdots & \ddots & \vdots \\
\text{cov}\,(X_{d'}, X_1) & \text{cov}\,(X_{d'}, X_2) & \cdots & \text{cov}\,(X_{d'}, X_{d'})
\end{bmatrix}
\tag{2}
$$

where X_i denotes the feature map of i-th channel, $i \in [1, d']$. The i-th row of the resulting covariance matrix can be regarded as statistical correlation of

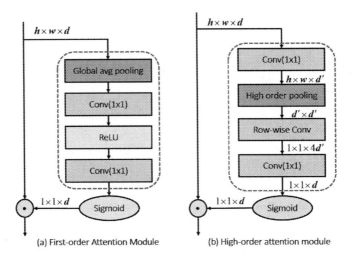

Fig. 3. The comparison of the first-order attention module and the high-order attention module. The feature maps are shown as the shape of their tensors, e.g. $h \times w \times d$ denotes a feature map with channel number d, height h and width w. "$\odot$" denotes element-wise multiplication. The global average pooling operation is performed on each feature map while the high order pooling performed on all the feature maps. The green boxes denote convolution operation. (Color figure online)

channel i with all channels. For the $d' \times d'$ matrix, we use row-wise convolution to transform the matrix to a $4d'$-dimensional vector. Then we perform another 1×1 convolution to decrease the vector dimension from $4d'$ to d, and this time we use the sigmoid function as a nonlinear activation. Finally, the obtained d-dimension vector is performed dot product with the output feature tensor $X \in R^{h \times w \times d}$ in the channel dimension.

For the first-order attention module, we borrow the channel attention module from SENet [14] (as shown in Fig. 3(a)) to model the relationship between the high-level feature channels. By this way, the weight vector β about the importance of the channel feature can be obtained. Given a set of d channel features $\mathbf{Z} = [\mathbf{z}_1, \mathbf{z}_2, \ldots, \mathbf{z}_d]$, we execute channel-wise re-scaling on the input as bellow:

$$\tilde{\mathbf{z}}_i = \beta_i \cdot \mathbf{z}_i \quad i = 1, 2, \ldots, d \tag{3}$$

3.3   Joint Training and Testing

In order to maintain the end-to-end fashion and reduce the computation complexity, the two branches are joint trained and tested. The final response map is computed by weighted averaging of the two branches:

$$f_{\mathrm{Final}}(z, x) = \lambda f_{\mathrm{FA}}(z, x) + (1 - \lambda) f_{\mathrm{HA}}(z, x) \tag{4}$$

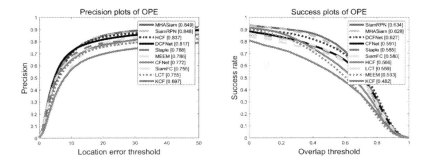

Fig. 4. Distance precision and overlap success plots on the OTB2015 [28] dataset.

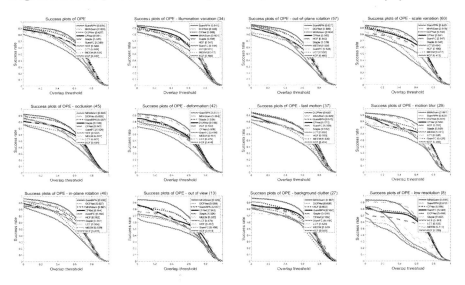

Fig. 5. Attribute-based distance precision metric on the OTB2015 dataset [28].

where f_{Final} denotes the final response map. f_{FA}, f_{HA} are the response map from the first-order attention branch and high-order attention branch. λ is the weighting parameter to balance the importance of the two branches, which is set to 0.5 in this paper. After obtaining the position of the tracked target, we use multi-scale inputs to deal with scale changes as same as SiamFC.

4   Experiments

In this section, we first present the implementation details, and then evaluate the proposed tracker on a large benchmark dataset [28] containing 100 videos with comparison to some related and state-of-the-art methods.

4.1   Implementation Details

The proposed tracker is implemented on PyTorch 1.1.0 framework and all the experiments are executed on a PC with a Intel Core i5-8400 2.8GHz CPU with a GeForce GTX Titan Xp GPU. The average testing speed of MHASiam is 65 fps. Similar to SiamFC, we use AlexNet [16] as base network and trained offline on the GOT10k dataset [15]. To adapt to the scale variation, we search on three scales during evaluation and testing.

4.2   Datasets and Compared Trackers

In the OTB2015 dataset, this paper compares the proposed tracker with 10 other related and state-of-the-art tracking algorithms, including:

(1) methods based on the combination of pre-training deep feature and correlation filter: HCF [20]
(2) correlation filter methods based on traditional hand-crafted features: staple [1], LCT [21], KCF [13]
(3) tracking methods based on end-to-end learning: CFNet [26], DCFNet [27], SiamFC [2], SiamRPN [17]
(4) other methods: MEEM [29]

4.3   Overall Performance

Figure 4 shows the comparison curve of tracking accuracy and success rate of the algorithm and comparison algorithm in this paper. Compared with the baseline algorithm SiamFC, the proposed method has significant performance improvement both in accuracy and success rate, the accuracy is increased by 9.4%, and the success rate is increased by 4.8%. Compared with DCFNet and CFNet which are also based on SiamFC, the algorithm in this paper still has better performance, mainly due to the application of attention mechanism and training on large-scale datasets. Compared to the recently published SiamRPN tracker, our method demonstrates a little advantage on precision plot.

4.4   Attribute-Based Evaluation

In order to analyze the performance of the algorithm in various tracking scenarios in detail, 11 annotation attributes in OTB2015 data set are used to analyze the algorithm. Figure 5 lists the tracking success rate of each algorithm under

MHASiam ····· SiamFC ····· CFNet ——— SiamRPN

Fig. 6. Sample tracking results on challenging image sequences (from top to down are *MotorRolling, Skiing, Matrix, CarScale, Human9*). We show some tracking results of SiamFC [2], CFNet [26], SiamRPN [17] as well as the proposed algorithm.

each attribute respectively, with red representing the optimal result and green representing the suboptimal result. It can be seen from Fig. 5 that the algorithm in this paper achieves the optimal or suboptimal tracking results on almost all attributes.

4.5   Quantitative Evaluation

Figure 6 shows the qualitative comparisons with the performing tracking methods: SiamFC [2], CFNet [26], SiamRPN [17] and the proposed method on five challenging image sequences including *Human3, MotorRolling, Human9, Skiing, Freeman4*.

Qualitative analysis mainly compares the algorithm in this paper and three related algorithms. As shown in Fig. 6, the algorithm in this paper can well deal with these complex scenes. Especially in *Human3* and *reeman4* video sequences, the performance is even better than the latest SiamRPN algorithm, which proves the effectiveness of the algorithm.

(1) **Background clutter:** the interference of similar objects has always been one of the difficult problems of Siam-based algorithms, especially in the *Free-man4* video sequence, on the one hand, the tracking target is very small, on the other hand, there are a large number of similar objects in the background, making most of the algorithms difficult to track the target successfully. SiamRPN and CFNet algorithm both lose the target in #76 frames, but the algorithm in this paper benefits from the introduction of attention mechanism, which can achieve continuous and stable tracking of the target.

(2) **Low resolution:** when the resolution of tracking target is low, the available information is extremely effective, making the visual tracking task more difficult. In *human3* and *skiing*, the resolution of the target is low, which requires the Siamese network to have better feature expression ability. In *skiing*, only our method and SiamRPN can achieve continuous tracking, both SiamFC and CFNet lose the target at about #18 frames. Compared with SiamRPN, this algorithm has better tracking accuracy. Through the use of high-order feature information, the trained Siamese network has a better representation ability in dealing with small targets, and has a better performance in tracking accuracy and success rate.

(3) **Illumination change:** the illumination change will cause the target pixel information to change greatly. In *motorrolling* and *human9*, the target has obvious illumination change. There is also a large degree of target rotation variation in *motorrolling*. SiamFC and CFNet both lose the target in two video sequences, and the algorithm in this paper can track the target stably in these complex scenes through attention mechanism, and has better tracking accuracy than other algorithms.

5   Conclusions

This paper presents a Siamese network tracking algorithm based on mixed high-order attention mechanism. By adding first order attention module and high order attention module to the basic network structure of SiamFC, the discrimination ability of the model is improved, and the model is trained on the large-scale data set GOT10k. The experimental results also prove the effectiveness of the algorithm, and the performance of all 11 attributes is improved obviously. In the future, we will consider the introduction of higher-order attention in earlier layers and more complex CNN models.

References

1. Bertinetto, L., Valmadre, J., Golodetz, S., Miksik, O., Torr, P.H.S.: Staple: complementary learners for real-time tracking. In: 2016 IEEE Conference on Computer Vision and Pattern Recognition, CVPR 2016, Las Vegas, NV, USA, 27–30 June 2016, pp. 1401–1409 (2016). https://doi.org/10.1109/CVPR.2016.156
2. Bertinetto, L., Valmadre, J., Henriques, J.F., Vedaldi, A., Torr, P.H.S.: Fully-convolutional siamese networks for object tracking. In: Hua, G., Jégou, H. (eds.) ECCV 2016. LNCS, vol. 9914, pp. 850–865. Springer, Cham (2016). https://doi.org/10.1007/978-3-319-48881-3_56

3. Bhat, G., Johnander, J., Danelljan, M., Shahbaz Khan, F., Felsberg, M.: Unveiling the power of deep tracking. In: The European Conference on Computer Vision (ECCV), September 2018
4. Carreira, J., Caseiro, R., Batista, J., Sminchisescu, C.: Semantic segmentation with second-order pooling. In: Fitzgibbon, A., Lazebnik, S., Perona, P., Sato, Y., Schmid, C. (eds.) ECCV 2012. LNCS, vol. 7578, pp. 430–443. Springer, Heidelberg (2012). https://doi.org/10.1007/978-3-642-33786-4_32
5. Chen, B., Deng, W., Hu, J.: Mixed high-order attention network for person re-identification. In: The IEEE International Conference on Computer Vision (ICCV) (2019)
6. Chen, Z., Hong, Z., Tao, D.: An experimental survey on correlation filter-based tracking. arXiv preprint arXiv:1509.05520 (2015)
7. Danelljan, M., Bhat, G., Khan, F.S., Felsberg, M.: ECO: efficient convolution operators for tracking. In: Conference on Computer Vision and Pattern Recognition, CVPR, pp. 6931–6939 (2017)
8. Danelljan, M., Hager, G., Shahbaz Khan, F., Felsberg, M.: Convolutional features for correlation filter based visual tracking. In: The IEEE International Conference on Computer Vision (ICCV) Workshops, December 2015
9. Gao, Z., Xie, J., Wang, Q., Li, P.: Global second-order pooling convolutional networks. In: Proceedings of the IEEE Conference on Computer Vision and Pattern Recognition, pp. 3024–3033 (2019)
10. Guo, Q., Feng, W., Zhou, C., Huang, R., Wan, L., Wang, S.: Learning dynamic siamese network for visual object tracking. In: The IEEE International Conference on Computer Vision (ICCV), October 2017
11. He, A., Luo, C., Tian, X., Zeng, W.: A twofold siamese network for real-time object tracking. In: Proceedings of the IEEE Conference on Computer Vision and Pattern Recognition, pp. 4834–4843 (2018)
12. He, Z., Fan, Y., Zhuang, J., Dong, Y., Bai, H.: Correlation filters with weighted convolution responses. In: Proceedings of the IEEE Conference on Computer Vision and Pattern Recognition, pp. 1992–2000 (2017)
13. Henriques, J.F., Caseiro, R., Martins, P., Batista, J.: High-speed tracking with kernelized correlation filters. IEEE Trans. Pattern Anal. Mach. Intell. **37**(3), 583–596 (2015)
14. Hu, J., Shen, L., Sun, G.: Squeeze-and-excitation networks. In: Proceedings of the IEEE conference on computer vision and pattern recognition, pp. 7132–7141 (2018)
15. Huang, L., Zhao, X., Huang, K.: Got-10k: a large high-diversity benchmark for generic object tracking in the wild. IEEE Tran. Pattern Anal. Mach. Intell. 1 (2018)
16. Krizhevsky, A., Sutskever, I., Hinton, G.E.: Imagenet classification with deep convolutional neural networks. In: NIPS2012, pp. 1106–1114 (2012). http://papers.nips.cc/paper/4824-imagenet-classification-with-deep-convolutional-neural-networks
17. Li, B., Yan, J., Wu, W., Zhu, Z., Hu, X.: High performance visual tracking with siamese region proposal network. In: Proceedings of the IEEE Conference on Computer Vision and Pattern Recognition, pp. 8971–8980 (2018)
18. Li, P., Xie, J., Wang, Q., Zuo, W.: Is second-order information helpful for large-scale visual recognition? In: Proceedings of the IEEE International Conference on Computer Vision, pp. 2070–2078 (2017)
19. Li, P., Wang, D., Wang, L., Lu, H.: Deep visual tracking. Pattern Recogn. **76**(C), 323–338 (2018). https://doi.org/10.1016/j.patcog.2017.11.007

20. Ma, C., Huang, J., Yang, X., Yang, M.: Hierarchical convolutional features for visual tracking. In: International Conference on Computer Vision, ICCV, pp. 3074–3082 (2015). https://doi.org/10.1109/ICCV.2015.352
21. Ma, C., Yang, X., Zhang, C., Yang, M.H.: Long-term correlation tracking. In: Proceedings of the IEEE Conference on Computer Vision and Pattern Recognition, pp. 5388–5396 (2015)
22. Nam, H., Han, B.: Learning multi-domain convolutional neural networks for visual tracking. In: Conference on Computer Vision and Pattern Recognition, CVPR, pp. 4293–4302 (2016)
23. Qi, Y., et al.: Hedged deep tracking. In: IEEE Conference on Computer Vision and Pattern Recognition, CVPR, pp. 4303–4311 (2016)
24. Smeulders, A.W.M., Chu, D.M., Cucchiara, R., Calderara, S., Dehghan, A., Shah, M.: Visual tracking: an experimental survey. IEEE Trans. Pattern Anal. Mach. Intell. 36(7), 1442–1468 (2014). https://doi.org/10.1109/TPAMI.2013.230
25. Tao, R., Gavves, E., Smeulders, A.W.M.: Siamese instance search for tracking. In: The IEEE Conference on Computer Vision and Pattern Recognition (CVPR) (2016)
26. Valmadre, J., Bertinetto, L., Henriques, J.F., Vedaldi, A., Torr, P.H.S.: End-to-end representation learning for correlation filter based tracking. In: IEEE Conference on Computer Vision and Pattern Recognition, CVPR, pp. 5000–5008 (2017). https://doi.org/10.1109/CVPR.2017.531
27. Wang, Q., Gao, J., Xing, J., Zhang, M., Hu, W.: DCFNET: discriminant correlation filters network for visual tracking. CoRR abs/1704.04057 (2017), http://arxiv.org/abs/1704.04057
28. Wu, Y., Lim, J., Yang, M.: Object tracking benchmark. IEEE Trans. Pattern Anal. Mach. Intell. 37(9), 1834–1848 (2015)
29. Zhang, J., Ma, S., Sclaroff, S.: MEEM: robust tracking via multiple experts using entropy minimization. In: Fleet, D., Pajdla, T., Schiele, B., Tuytelaars, T. (eds.) ECCV 2014. LNCS, vol. 8694, pp. 188–203. Springer, Cham (2014). https://doi.org/10.1007/978-3-319-10599-4_13
30. Zhang, Y., Wang, L., Qi, J., Wang, D., Feng, M., Lu, H.: Structured siamese network for real-time visual tracking. In: The European Conference on Computer Vision (ECCV), September 2018
31. Zhang, Z., Peng, H.: Deeper and wider siamese networks for real-time visual tracking. In: Proceedings of the IEEE Conference on Computer Vision and Pattern Recognition, pp. 4591–4600 (2019)
32. Zhu, Z., Wang, Q., Li, B., Wu, W., Yan, J., Hu, W.: Distractor-aware siamese networks for visual object tracking. In: The European Conference on Computer Vision (ECCV), September 2018

Deep Face Recognition Based on Penalty Cosface

Shuoyan Lin[1], Jianxiong Tang[1], Zhanxiang Feng[1], and Jianhuang Lai[1,2(✉)]

[1] School of Data and Computer Science, Sun Yat-Sen University, Guangzhou, China
{linshuoy,tangjx6}@mail2.sysu.edu.cn,
{fengzhx7,stsljh}@mail.sysu.edu.cn
[2] School of Information Science, Xinhua College of Sun Yat-Sen University,
Guangzhou, People's Republic of China

Abstract. Face recognition has achieved great progress because of the advancement of deep convolutional neural networks (CNNs) techniques. The Softmax loss is one of the most popular loss function for deep learning models. In many situations, face images are captured in the unconstrained environments with changing poses and illuminations, making face recognition very challenging because of the dramatic appearance variations. The models trained with the Softmax loss may fail to extract discriminative information for the face images with extreme illumination or pose conditions. Recently, Cosface has been proven effective for improving the generalization ability of the Softmax loss. Derived from Cosface, we propose a novel method named Penalty Cosface to address the unconstrained face recognition challenges and learn discriminative features. Specifically, we design a variant of Cosface that remove radial variations by penalizing ℓ_2-normalized constraints of the features and weights. Therefore, the discriminative ability of the Penalty Cosface is guaranteed by the large margin of the Cosface, and the penalty term is beneficial to simplifying the gradient calculations. Experimental results show that the Penalty Cosface improves the discriminative power of deep networks and outperforms the other variants of Softmax loss.

Keywords: Deep learning · Face recognition · Cosface · Penalty methods

1 Introduction

Face recognition (FR) is an important biometric technology for identity authentication [10], and is widely used in various areas, such as military, financial, security and so on. In the past decades, researchers have made great efforts to improve the performance of FR techniques. Early studies focus on designing hand-crafted features [15,25] whose performance degrades significantly when dealing with unconstrained face images in practical applications. Recently, deep learning methods [4,7,16] have achieved great breakthroughs in a wide range of

ⓒ Springer Nature Switzerland AG 2020
Y. Peng et al. (Eds.): PRCV 2020, LNCS 12306, pp. 457–469, 2020.
https://doi.org/10.1007/978-3-030-60639-8_38

research topics. Specially, with the emergence of large-scale face datasets, convolutional neural networks (CNNs) [7] has significantly improved the performance of FR. The Softmax loss is widely applied in many classification tasks including FR. Note that many face images are captured in the wild with changing illuminations, imaging qualities and poses. Therefore, face recognition is very challenging and the robustness of FR approach is essential for applications. However, the Softmax loss is insufficient to learn the discriminative features [8, 9] when dealing with face images of extreme illuminations and poses conditions.

Recently, many researchers have tried to develop new methods to improve the discriminative power of deep models. The basic idea of the recent proposals is to improve the discriminative ability of deep networks by maximizing the inter-class variance while minimizing the intra-class variance. Many works design loss functions to improve the discriminative ability. Wen et al. propose Center loss [22] to narrow the gaps between the samples and the corresponding class centers, leading to more distinguishing features in Euclidean space. In Cosface [21], Wang et al. propose the Large Margin Cosine Loss that introduces an additive angular margin to concentrate the samples on a hypersphere borderline. Similar to Cosface, Arcface [3] and A-Softmax [8] also introduce angular margins to the decision boundaries of the origin Softmax loss and improve the discriminative ability.

In this paper, we design a novel method named Penalty Cosface to learn the discriminative features for deep face recognition. Our idea is to design a loss function that remove the radial variations by penalizing the features and weights to satisfy the ℓ_2-normalization constraints. We first transform the Cosface into the equivalent constraint problem by establishing the equality relationships between the objectives and the corresponding ℓ_2-normalized vectors. Then, we give a relaxation term of the constraint problem. After that, we decompose the relaxation term using the penalty method [12]. The proposed approach leads to a suitable expression for optimization, and the gradient calculations are simplified. Experimental results show that the proposed method performs well on popular face datasets, indicating the discriminative ability of the loss function.

The main contributions of this paper can be summarized below:

- We propose a novel approach named Penalty Cosface (p-Cosface) to learn discriminative features for deep face recognition.
- The penalty terms of the p-Cosface enforce the weights and features to satisfy the ℓ_2-normalization constraints, so that Cosface can be approximated without any ℓ_2-normalized operators. During model training, each class of samples converges to an adaptive center near the hypersphere borderline.
- Experimental results demonstrate that the p-Cosface performs well on the popular face datasets and improves the discriminative ability of deep networks in both Euclidean and angular space.

2   Related Work

Deep Face Recognition. With the development of deep learning techniques [4,7,16] and the emergence of large-scale face datasets, face recognition (FR) has achieved a great progress in recent years. Taigman et al. are the first to conduct FR and propose a novel model named DeepFace [19] which train a 9 layers CNN to learn the face representation and achieve 97.345% accuracy on Label Faces in the Wild (LFW) dataset [5]. Schroff et al. propose the FaceNet [13] that introduces the triplet loss to learn deep features in Euclidean space. Notably, the FaceNet is trained on an extremely large-scale face dataset with 100M-200M training samples from about 8M different identities and the performance of FaceNet is superior over other method. Despite DeepFace and FaceNet, many other approaches are also proposed to improve the performance of the deep CNNs on FR [17,18].

Softmax Margin Loss Functions. Softmax loss is one of the most popular loss function for learning deep models. Given an input feature $\mathbf{x}_i \in \mathbf{R}_{d\times 1}$ with its corresponding label $y_i \in \{1, 2, \ldots, C\}$, the definition of Softmax loss is

$$L = -\sum_{i=1}^{N} \log \frac{e^{w_{y_i}^\top x_i}}{\sum_{j=1}^{C} e^{w_j^\top x_i}}, \tag{1}$$

where $\boldsymbol{w}_j \in \boldsymbol{R}_{d\times 1}$ denotes the weight vector. Although the Softmax loss is proven effective in many deep networks, it is insufficient to extract discriminative features in many challenging situations [8,9,21].

To improve the discriminative ability of Softmax loss, Wang et al. propose the Normalized Softmax loss (NSL) [21]. In Eq. 1, the inner product of $\boldsymbol{w}_j$ and $\boldsymbol{x}p_i$ is equivalent to

$$\boldsymbol{w}_j^\top \boldsymbol{x}_i = \|\boldsymbol{w}_j\|_2 \|\boldsymbol{x}_i\|_2 \cos\theta_{ji}, \tag{2}$$

where θ_{ji} is the angle between $\boldsymbol{w}_j$ and $\boldsymbol{x}_i$. Then, the formulas of the ℓ_2-normalization of $\boldsymbol{w}_j$ and $\boldsymbol{x}_i$ can be written as:

$$\widetilde{\boldsymbol{w}}_j = \frac{\boldsymbol{w}_j}{\|\boldsymbol{w}_j\|_2}, j = 1, 2, \ldots, C, \quad \tilde{\boldsymbol{x}}_i = \frac{\boldsymbol{x}_i}{\|\boldsymbol{x}_i\|_2}, i = 1, 2, \ldots, N. \tag{3}$$

By substituting Eq. 3 into Eq. 1, the NSL can be written as:

$$L = -\sum_{i=1}^{N} \log \frac{e^{s\widetilde{w}_{y_i}^\top x_i}}{\sum_{j=1}^{C} e^{s\widetilde{w}_j^\top x_i}} = -\sum_{i=1}^{N} \log \frac{e^{s\cos\theta_{y_i i}}}{\sum_{j=1}^{C} e^{s\cos\theta_{ji}}}, \tag{4}$$

where $s > 0$ is the scaling factor that guarantees the features to be separable in the angular space.

But Eq. 4 only considers the classification results and ignores the margin between features from different classes. The discriminative power of the NSL is not enough because features around the classification boundary may be affected by factors such as illuminations and poses, resulting in wrong classification decisions. Wang et al. proposes the following Cosface loss [21] to tackle with this problem, which can be formulated as:

$$L = -\sum_{i=1}^{N} \log \frac{e^{s(\cos \theta_{y_j i} - m)}}{e^{s(\cos \theta_{y_j i} - m)} + \sum_{j \neq y_i}^{C} e^{s \cos \theta_{ji}}}. \tag{5}$$

In Eq. 5, $m > 0$ denotes the margin. We can obtain a more discriminant classifier by enforcing a margin to separate features between different identities.

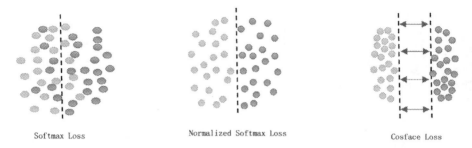

Softmax Loss   Normalized Softmax Loss   Cosface Loss

Fig. 1. The binary classification illustrations of different loss functions in cosine space. Dashed line denotes the decision boundary, and empty spaces in Cosface refer to the margins.

Figure 1 illustrates the binary classification examples in cosine space. It is obvious that Cosface loss separates the feature distributions by enforcing the margin in the cosine space.

In analogy with Cosface, Arcface [3] introduces an additive angle margin for closer intra-class distance, and A-Softmax [8] scales the angle between weights and features by a multiplicative margin. Many other Softmax margin loss functions are also proposed to enhance the discriminative ability of deep networks [9,20].

3   Penalty Cosface

3.1   Motivation

Cosface is a variant of Softmax loss in cosine space which improves the discriminative ability of deep models by compressing the inter-class variance using the margin factor. However, the features learned by Cosface are only suitable for cosine/angular space. Center loss [22] is designed to narrow the gaps between the samples and the corresponding class centers, and generates more distinguishing features in Euclidean space because features of different classes are separated by different centers. But the centers are proportional to the number of classes, resulting in numerous training parameters and slow convergence.

In this paper, we propose the Penalty Cosface (p-Cosface) approach which integrates the penalty method into the Cosface loss to learn discriminant models without introducing additional parameters and training consumptions. We first transform the Cosface loss into the equivalent constraint problem by establishing the equality constraints between the features/weights and the corresponding ℓ_2-normalized vectors. Then, we give a relaxation form of the constraint problem. After that, we decompose the relaxation form by using the penalty method. The penalty terms of the p-Cosface approach enforce the ℓ_2-norms of weights and features to be closed to 1, so that we can approximate the Cosface loss without using ℓ_2-normalized operators. In p-Cosface, samples from different classes are trained to converge to an adaptive center that is near the hypersphere borderline. Consequently, the p-Cosface combines the advantage of both the Cosface and the penalty terms to enhance the discriminative power of deep networks.

3.2   Penalty Cosface

The optimization problem of Cosface can be written as:

$$\min_{\theta} - \sum_{i=1}^{N} \log \frac{e^{s(\cos\theta_{y_i i} - m)}}{e^{s(\cos\theta_{y_i i} - m)} + \sum_{j \neq y_i}^{C} e^{s\cos\theta_{ji}}}. \tag{6}$$

Obviously, Eq. 6 is equivalent to the following constrained problem:

$$\min_{\boldsymbol{w},\boldsymbol{x}} \ - \log \frac{e^{s\left(\boldsymbol{w}_{y_i}^{\top} \boldsymbol{x}_i - m\right)}}{e^{s\left(\boldsymbol{w}_{y_i}^{\top} \boldsymbol{x}_i - m\right)} + \sum_{j \neq y_i}^{C} e^{s\boldsymbol{w}_j^{\top} \boldsymbol{x}_i}}$$
$$\text{s.t. } \boldsymbol{w}_j = \frac{\boldsymbol{w}_j}{\|\boldsymbol{w}_j\|_2}, j = 1, 2, \ldots, C,$$
$$\boldsymbol{x}_i = \frac{\boldsymbol{x}_i}{\|\boldsymbol{x}_i\|_2}, i = 1, 2, \ldots, N. \tag{7}$$

Under the constraints of Eq. 7, the optimal solutions $\boldsymbol{w}_{y_i}, \boldsymbol{x}_i$ are ℓ_2-normalized vectors. When the training process converges, we can learn the Cosface without space projection. But Eq. 7 involves the fractional optimization, which will result in complex gradient calculations. As a result, we design the following optimization problem:

$$\min_{\boldsymbol{w},\boldsymbol{x}} \ - \log \frac{e^{s\left(\boldsymbol{w}_{y_i}^{\top} \boldsymbol{x}_i - m\right)}}{e^{s\left(\boldsymbol{w}_{y_i}^{\top} \boldsymbol{x}_i - m\right)} + \sum_{j \neq y_i}^{C} e^{s\boldsymbol{w}_j^{\top} \boldsymbol{x}_i}}$$
$$\text{s.t. } \boldsymbol{w}_j = \frac{\boldsymbol{w}_j^k}{\|\boldsymbol{w}_j^k\|_2}, j = 1, 2, \ldots, C,$$
$$\boldsymbol{x}_i = \frac{\boldsymbol{x}_i^k}{\|\boldsymbol{x}_i^k\|_2}, i = 1, 2, \ldots, N, \tag{8}$$
$$k = 0, 1, \ldots,$$

where k denotes the last iteration and $\boldsymbol{w}_j^k, \boldsymbol{x}_i^k$ are the corresponding solutions. Equation 8 have a better structure than Eq. 7 from the view of optimization.

Clearly, $\boldsymbol{w}_j^k$ and $\boldsymbol{x}_i^k$ are constant terms in current iteration. By applying penalty method, we obtain the optimization problem of the Penalty Cosface (p-Cosface) approach:

$$
\begin{aligned}
\min_{\boldsymbol{w},\boldsymbol{x}} \quad & -\sum_{i=1}^{N} \log \frac{e^{s\left(\boldsymbol{w}_{y_i}^{\top}\boldsymbol{x}_i - m\right)}}{e^{s\left(\boldsymbol{w}_{y_i}^{\top}\boldsymbol{x}_i - m\right)} + \sum_{j \neq y_i}^{C} e^{s\boldsymbol{w}_j^{\top}\boldsymbol{x}_i}} + \frac{\beta_1}{2C}\sum_{j=1}^{C}\left\|\boldsymbol{w}_j - \frac{\boldsymbol{w}_j^k}{\|\boldsymbol{w}_j^k\|_2}\right\|_2^2 \\
& + \frac{\beta_2}{2N}\sum_{i=1}^{N}\left\|\boldsymbol{x}_i - \frac{\boldsymbol{x}_i^k}{\|\boldsymbol{x}_i^k\|_2}\right\|_2^2, \\
& k = 0, 1, \ldots,
\end{aligned}
\tag{9}
$$

where $\beta_1 > 0, \beta_2 > 0$ are the penalty parameters. Let $\mathcal{L}$ denotes the objective function in Eq. 9, $\mathcal{L}_s = -\sum_{i=1}^{N} \log \frac{e^{s\left(\boldsymbol{w}_{y_i}^{\top}\boldsymbol{x}_i - m\right)}}{e^{s\left(\boldsymbol{w}_{y_i}^{\top}\boldsymbol{x}_i - m\right)} + \sum_{j \neq y_i}^{C} e^{s\boldsymbol{w}_j^{\top}\boldsymbol{x}_i}}$, $f(\boldsymbol{z}) = \frac{\boldsymbol{z}}{\|\boldsymbol{z}\|_2}$. Then the brief expression of the objective function of p-Cosface can be written as follows:

$$
\mathcal{L} = \mathcal{L}_s + \frac{\beta_1}{2C}\sum_{j=1}^{C}\|\boldsymbol{w}_j - f(\boldsymbol{w}_j^k)\|_2^2 + \frac{\beta_2}{2N}\sum_{i=1}^{N}\|\boldsymbol{x}_i - f(\boldsymbol{x}_i^k)\|_2^2.
\tag{10}
$$

The penalty terms minimize the error between $\boldsymbol{w}_j, \boldsymbol{x}_i$ and the last normalized solutions. Therefore, when the training algorithm of Eq. 9 converges, Eq. 8 can obtain the approximate solutions as the solutions from Eq. 9 [12].

By solving Eq. 9, the p-Cosface can obtain the solutions similar to Cosface in Euclidean space. Under the influence of penalty terms, the features converge to the adaptive centers which are near the hypersphere borderline. Therefore, the features guided by Eq. 9 may benefit from the advantages of both the Cosface and the penalty terms.

Compared with Cosface, the proposed approach does not need any ℓ_2-normalized operators, which is helpful for simplifying the gradient calculations. Furthermore, the gradient of the penalty terms is easy to calculate. We can train the p-Cosface model in an end-to-end manner by combining the back-propagation and penalty method, and the gradient expressions of p-Cosface loss 10 on $\boldsymbol{w}_{y_i}$, $\boldsymbol{x}_i$ are given bellow:

$$
\begin{aligned}
\frac{\partial \mathcal{L}}{\partial \boldsymbol{w}_{y_i}} = {} & s(e^{-\mathcal{L}_{s,i}} - 1)\boldsymbol{x}_i + \beta_1(\boldsymbol{w}_{y_i} - f(\boldsymbol{w}_{y_i}^k)) \\
& + \sum_{l \neq i}^{N} \frac{se^{s\boldsymbol{w}_{y_i}^{\top}\boldsymbol{x}_l}}{e^{s(\boldsymbol{w}_{y_l}^{\top}\boldsymbol{x}_l - m)} + \sum_{j \neq y_l}^{C} e^{s\boldsymbol{w}_j^{\top}\boldsymbol{x}_l}}\boldsymbol{x}_l,
\end{aligned}
\tag{11}
$$

$$
\frac{\partial \mathcal{L}}{\partial \boldsymbol{x}_i} = s(e^{-\mathcal{L}_{s,i}} - 1)\boldsymbol{w}_{y_i} + \beta_2(\boldsymbol{x}_i - f(\boldsymbol{x}_i^k)).
\tag{12}
$$

4 Experiments

4.1 Feature Visualization

Parameter Setting. In this part, we select 8 different identities from CASIA-Webface [23] consisting of $3,552$ faces to train the model for feature visualization. We adopt VGG [16] as the backbone for this experiment. Notably, the dimension of the features is set to 2 to visualize the feature distribution. During training, we adopt Adam [6] as the optimizer, set the initial learning rate as 0.1, and decrease the learning rate by 10% for each epoch. The involved facial images are detected and aligned by MTCNN [24]. Finally, we compare the p-Cosface with the Softmax loss and Cosface to prove the effects of the proposed approach. For Cosface and p-Cosface, we set $m = 0.3$ and $s = 64$.

In the following experiments, the maximum of penalty parameter is set to 1000. We check the penalty terms for weight $\frac{1}{C}\sum_{j=1}^{C}\|w_j - f(w_j^k)\|_2^2$ and feature $\frac{1}{N}\sum_{i=1}^{N}\|x_i - f(x_i^k)\|_2^2$ every 100 iterations, if a penalty term is larger than 2, the corresponding penalty parameter is updated by $\beta = \beta \times 2$. The initial values of β_1, β_2 are set to 1.

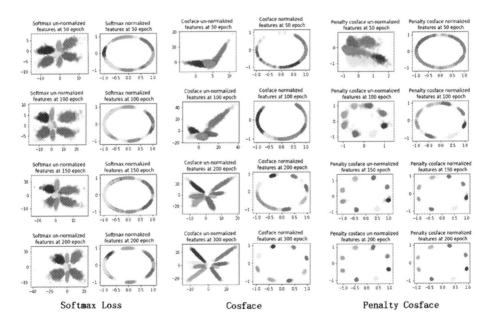

Fig. 2. Feature distributions of Softmax loss, Cosface loss and Penalty Cosface loss in Euclidean/Angular space.

Experimental Results. Figure 2 illustrates the visualization of different loss functions. For each loss function in Fig. 2, the left column displays the feature distributions in Euclidean space while the right column is the situation in angular space. Both p-Cosface and Cosface show the discriminating capability in angular space, while Softmax loss is insufficient to learn the discriminative features. Figure 2 also shows that the proposed p-Cosface learns the discriminative features both in Euclidean and angular space after 150 epochs training. It demonstrates that the proposed p-Cosface has the ability to learn the discriminative features. Compared with Cosface, the proposed p-Cosface separates the features in a quick convergence. Meanwhile, each sample is clustered to the corresponding center adaptively and the distributions are similar to the situation in angular spaces.

4.2   Face Recognition

In this section, we evaluate the proposed p-Cosface on $1:1$ face recognition experiments.

Datasets. We train the model on CASIA WebFace [23] with $10,575$ identities, $494,414$ images. LFW [5] is a standard face verification testing dataset for unconstraint face recognition. It includes $13,233$ face images from $5,749$ identities collected from website. We report the results of the comparison loss functions on $6,000$ pairs testing images on LFW. Besides the LFW dataset, we also report the performance of the proposed p-Cosface on VGG2_FP [1], CPLFW [3], CALFW [3], AGEDB_30 [11], CFP_FF [14] and CFP_FP [14]. All of the datasets for training and testing are summarized in Table. 1.

Table 1. Face datasets for training and testing.

Datasets	#Identites	#Image	#Pairs
CASIA-Webface	10,575	494414	–
LFW	5,749	13,233	6k
CFP-FP	500	7,000	7k
CFP-FF	500	7,000	7k
AGEDB-30	568	16,488	6k
CPLFW	5,749	11,652	6k
CALFW	5,749	12,174	6k
VGG2_FP	–	–	5k

Parameter Setting. The experiments are implemented in python 3.6 with Tensorflow 2.0. The involved facial images are detected and aligned using MTCNN. We adopt MobileFaceNet [2] and ResNet-50 [4] as the backbones and Adam as the optimizer. By setting the initial learning rate to 0.01, we decrease the learning rate by 10% for each epoch. For Cosface and p-Cosface, we set $m = \{0, 0.3\}$ and $s = 64$. During testing, we adopt the backbone features of 512 dimensions from the embedding network as the output for each face. All of the models are trained in 20 epochs, a total of $72,000$ steps.

Experimental Results. Table 2 and Table 3 illustrate the experimental results for the comparisons between p-CosFace, Cosface and Softmax using Mobile-FaceNet and ResNet-50 as backbones on several popular face recognition benchmarks. For all of the loss functions, the first row gives the accuracies evaluated by ℓ_2 distance and cosine distance respectively. For p-Cosface, the third row gives the accuracies evaluated by ℓ_2-distance in Euclidean space. Except the third row results of p-Cosface, all of the results are evaluated after ℓ_2-normalization.

Table 2 and Table 3 show that the proposed p-Cosface ($m = 0.3$) achieves the best performance for most of the cases and learns much discriminant features compared with the Softmax loss.

Table 2. The performance of comparison methods on MobileFaceNet.

	LFW	AGEDB_30	CALFW	CPLFW	CFP_FF	CFP_FP	VGG2_FP
Softmax loss	94.93%	79.03%	82.77%	72.13%	93.66%	81.57%	82.38%
	94.93%	80.40%	83.05%	74.52%	94.07%	82.07%	82.42%
Cosface	94.93%	79.38%	83.50%	74.98%	93.70%	**83.47%**	83.44%
m = 0	95.07%	80.80%	83.65%	76.48%	94.34%	**83.47%**	83.42%
Cosface	97.13%	84.85%	87.45%	77.33%	96.67%	79.66%	82.28%
m = 0.3	97.30%	85.33%	88.43%	78.43%	96.86%	81.14%	82.52%
p-Cosface	95.98%	76.60%	83.98%	73.57%	92.75%	81.34%	83.26%
m = 0	95.98%	79.85%	84.07%	76.32%	93.74%	81.66%	82.74%
	95.40%	77.62%	83.40%	73.00%	93.45%	80.20%	81.06%
p-Cosface	96.68%	**85.82%**	88.13%	**80.08%**	96.06%	80.40%	83.76%
m = 0.3	**97.65%**	85.73%	**88.93%**	80.02%	**97.04%**	**83.47%**	**85.08%**
	97.15%	83.52%	87.22%	78.17%	95.80%	79.51%	82.40%

Table 3. The performance of comparison methods on ResNet-50.

	LFW	AGEDB_30	CALFW	CPLFW	CFP_FF	CFP_FP	VGG2_FP
Softmax loss	94.77%	80.47%	83.13%	75.15%	94.67%	81.00%	82.62%
	95.87%	81.20%	83.82%	75.27%	95.07%	82.56%	82.98%
Cosface	96.70%	80.73%	85.73%	77.62%	95.47%	84.27%	83.24%
m = 0	96.72%	82.12%	85.80%	78.27%	95.56%	84.29%	84.06%
Cosface	97.13%	**86.70%**	87.40%	81.10%	95.97%	81.84%	82.48%
m = 0.3	97.83%	86.42%	88.97%	81.10%	**97.03%**	82.86%	83.90%
p-Cosface	97.72%	84.98%	88.55%	79.95%	96.96%	84.34%	84.34%
m = 0	97.82%	85.67%	88.68%	80.83%	96.96%	**84.39%**	84.44%
	97.33%	85.38%	87.45%	79.92%	96.43%	83.24%	84.18%
p-Cosface	97.82%	85.68%	89.23%	79.77%	96.76%	83.80%	**85.24%**
m = 0.3	**97.87%**	86.43%	**89.28%**	81.65%	96.86%	83.74%	85.10%
	97.77%	85.80%	88.66%	80.80%	96.66%	83.09%	84.50%

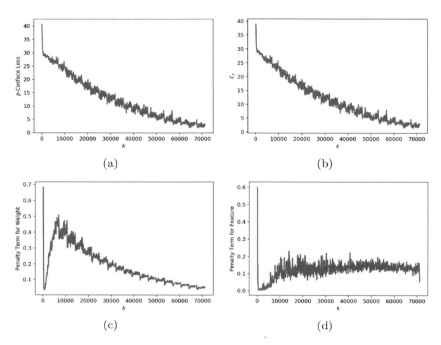

Fig. 3. The plots of (a) the value of p-Cosface loss versus the number of iterations, (b) the value of $\mathcal{L}_s$ of Eq. 10 versus the number of iterations, (c) penalty term $\frac{1}{C} \sum_{j=1}^{C} \|w_j - f(w_j^k)\|_2^2$ versus the number of iterations, (d) penalty term $\frac{1}{N} \sum_{i=1}^{N} \|x_i - f(x_i^k)\|_2^2$ versus the number of iterations.

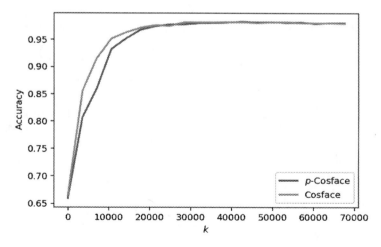

Fig. 4. The plots of the accuracy versus the number of iterations on LFW dataset, with $m = 0.35$, ResNet-50.

Figure 3 shows the declines of the function values of the proposed p-Cosface on CASIA-Webface. For each figure in Fig. 3, the abscissa denotes the number of iterations and ordinate is the function value. From Fig. 3(a) and Fig. 3(b), we find that the value of $\mathcal{L}_s$ of Eq. 10 is close to value of the p-Cosface loss $\mathcal{L}$ when the training process is terminated. As shown in Fig. 3(c) and Fig. 3(d), both of the penalty terms $\frac{1}{C}\sum_{j=1}^{C}\|w_j - f(w_j^k)\|_2^2$ and $\frac{1}{N}\sum_{i=1}^{N}\|x_i - f(x_i^k)\|_2^2$ decrease and converge as the number of iterations growing. It indicates the convergence the proposed p-Cosface.

Figure 4 plots the changes of accuracy of the Cosface and p-Cosface with $m = 0.35$. It shows that both the Cosface and p-Cosface have the similar convergence tendencies.

5 Conclusions

In this paper, we propose the p-Cosface to address the unconstrained face recognition challenges and learn discriminative features. By penalizing the ℓ_2-normalization constraints of the features and weights, the proposed p-Cosface can improve the discriminative power of the deep models. The experimental results show that the proposed Penalty Cosface performs well and indicates a discriminative capability in both Euclidean and angular space.

Acknowledgment. This project was supported by the Key Areas Research and Development Program of Guangdong Province (2019B010155003) and NSFC (61876104, 61902444).

References

1. Cao, Q., Shen, L., Xie, W., Parkhi, O.M., Zisserman, A.: Vggface2: a dataset for recognising faces across pose and age. In: Proceedings of the IEEE International Conference on Automatic Face & Gesture Recognition, pp. 67–74 (2018)
2. Chen, S., Liu, Y., Gao, X., Han, Z.: MobileFaceNets: efficient CNNs for accurate real-time face verification on mobile devices. In: Zhou, J., et al. (eds.) CCBR 2018. LNCS, vol. 10996, pp. 428–438. Springer, Cham (2018). https://doi.org/10.1007/978-3-319-97909-0_46
3. Deng, J., Guo, J., Xue, N., Zafeiriou, S.: Arcface: additive angular margin loss for deep face recognition. In: Proceedings of the IEEE Conference on Computer Vision and Pattern Recognition, pp. 4690–4699 (2019)
4. He, K., Zhang, X., Ren, S., Sun, J.: Deep residual learning for image recognition. In: Proceedings of the IEEE Conference on Computer Vision and Pattern Recognition, pp. 770–778 (2016)

5. Huang, G.B., Ramesh, M., Berg, T., Learned-Miller, E.: Labeled faces in the wild: a database for studying face recognition in unconstrained environments. Technical Report 07–49, University of Massachusetts, Amherst (2007)
6. Kingma, D.P., Ba, J.: Adam: a method for stochastic optimization. In: International Conference on Learning Representations. vol. abs/1412.6980 (2015)
7. Krizhevsky, A., Sutskever, I., Hinton, G.E.: Imagenet classification with deep convolutional neural networks. In: Advances in Neural Information Processing Systems, pp. 1097–1105 (2012)
8. Liu, W., Wen, Y., Yu, Z., Li, M., Raj, B., Song, L.: Sphereface: deep hypersphere embedding for face recognition. In: Proceedings of the IEEE Conference on Computer Vision and Pattern Recognition, pp. 212–220 (2017)
9. Liu, W., Wen, Y., Yu, Z., Yang, M.: Large-margin softmax loss for convolutional neural networks. In: Proceedings of the International Conference on Machine Learning, pp. 507–516 (2016)
10. Masi, I., Wu, Y., Hassner, T., Natarajan, P.: Deep face recognition: a survey. In: Proceedings of the SIBGRAPI Conference on Graphics, Patterns and Images, pp. 471–478 (2018)
11. Moschoglou, S., Papaioannou, A., Sagonas, C., Deng, J., Kotsia, I., Zafeiriou, S.: Agedb: the first manually collected, in-the-wild age database. In: Proceedings of the IEEE Conference on Computer Vision and Pattern Recognition Workshops, pp. 51–59 (2017)
12. Nocedal, J., Wright, S.: Numerical Optimization. Springer, New York (2006). https://doi.org/10.1007/978-0-387-40065-5
13. Schroff, F., Kalenichenko, D., Philbin, J.: Facenet: a unified embedding for face recognition and clustering. In: Proceedings of the IEEE Conference on Computer Vision and Pattern Recognition, pp. 815–823 (2015)
14. Sengupta, S., Chen, J.C., Castillo, C., Patel, V.M., Chellappa, R., Jacobs, D.W.: Frontal to profile face verification in the wild. In: Proceedings of the IEEE Winter Conference on Applications of Computer Vision, pp. 1–9 (2016)
15. Shan, C., Gong, S., McOwan, P.W.: Facial expression recognition based on local binary patterns: a comprehensive study. Image Vis. Comput. $27(6)$, 803–816 (2009)
16. Simonyan, K., Zisserman, A.: Very deep convolutional networks for large-scale image recognition. CoRR abs/1409.1556 (2015)
17. Sun, Y., Chen, Y., Wang, X., Tang, X.: Deep learning face representation by joint identification-verification. In: Advances in Neural Information Processing Systems, pp. 1988–1996 (2014)
18. Sun, Y., Liang, D., Wang, X., Tang, X.: Deepid3: face recognition with very deep neural networks. CoRR abs/1502.00873 (2015)
19. Taigman, Y., Yang, M., Ranzato, M., Wolf, L.: Deepface: closing the gap to human-level performance in face verification. In: Proceedings of the IEEE Conference on Computer Vision and Pattern Recognition, pp. 1701–1708 (2014)
20. Wang, F., Cheng, J., Liu, W., Liu, H.: Additive margin softmax for face verification. IEEE Signal Process. Lett. $25(7)$, 926–930 (2018)
21. Wang, H., et al.: Cosface: large margin cosine loss for deep face recognition. In: Proceedings of the IEEE Conference on Computer Vision and Pattern Recognition, pp. 5265–5274 (2018)
22. Wen, Y., Zhang, K., Li, Z., Qiao, Yu.: A discriminative feature learning approach for deep face recognition. In: Leibe, B., Matas, J., Sebe, N., Welling, M. (eds.) ECCV 2016. LNCS, vol. 9911, pp. 499–515. Springer, Cham (2016). https://doi.org/10.1007/978-3-319-46478-7_31

23. Yi, D., Lei, Z., Liao, S., Li, S.Z.: Learning face representation from scratch. CoRR abs/1411.7923 (2014)
24. Zhang, K., Zhang, Z., Li, Z., Qiao, Y.: Joint face detection and alignment using multitask cascaded convolutional networks. IEEE Signal Process. Lett. **23**(10), 1499–1503 (2016)
25. Zhao, G., Pietikainen, M.: Dynamic texture recognition using local binary patterns with an application to facial expressions. IEEE Trans. Pattern Anal. Mach. Intell. **29**(6), 915–928 (2007)

Hierarchical Representation Learning of Dynamic Brain Networks for Schizophrenia Diagnosis

Jiashuang Huang, Xu Li, Mingliang Wang, and Daoqiang Zhang[(✉)]

College of Computer Science and Technology, MIIT Key Laboratory of Pattern
Analysis and Machine Intelligence, Nanjing University of Aeronautics
and Astronautics, Nanjing 210029, China
dqzhang@nuaa.edu.cn

Abstract. The dynamic brain network (DBN) consists of a set of time-varying states (i.e., functional connectivity matrixes), and has revolutionized the field of brain network analysis through captured dynamic evolution patterns. However, current representation methods of DBN, which use hand-crafted features, are still not data-driven, and cannot effectively learn the hierarchically organized temporal nature of DBN. To address this issue, we propose a novel hierarchical representation learning (HARL) method for dynamic brain networks in the framework of graph convolutional networks. Specifically, we first define a graph model, whose nodes represent time-varying states, node features are determined by states' functional connectivity matrix, and edges learn according to node features. Then, based on this model, we build a HARL module to learn the representation of DBN under different levels. Each level consists of a convolution layer and a pooling layer. Through a convolution layer, node features can be updated according to their neighbor nodes, which can better learn the representation of each state. In the pooling layer, according to both node features and graph topology, we select some important nodes (i.e., states) from the whole graph to form a coarsened graph, which would be further input to the next level. In each level, the representation of DBN is generated by aggregating these node features. These representations will be input to the fully connected layer for disease prediction. Experiments on a real schizophrenia dataset demonstrate the effectiveness and advantages of our proposed method.

Keywords: Dynamic brain network · Hierarchical representation learning · Schizophrenia

---

This study was supported by the National Key Research and Development Program of China (Nos. 2018YFC2001600, 2018YFC2001602, 2018ZX10201002) and the National Natural Science Foundation of China (Nos. 61861130366, 61732006, 61876082). Jiashuang Huang and Xu Li contribute equally to this article and should be considered co-first authors.

Y. Peng et al. (Eds.): PRCV 2020, LNCS 12306, pp. 470–479, 2020.
https://doi.org/10.1007/978-3-030-60639-8_39

1   Introduction

In recent years there has been a surge of interest in analyzing dynamic brain networks (DBN) that can reveal dynamic functional connections of the human brain system [1]. Previous studies [2–4] about brain network usually used a single functional connectivity matrix. Recent evidence suggests that functional connections may not be stationary over time [5]. To explored the dynamics of brain connectivity from resting-state fMRI, DBN has been proposed to model temporal variability of those connectivities. DBN usually consists of a set of functional connectivity matrices, where each matrix is calculated based on overlapping subseries of fMRI time series by using a sliding window correlation. Thus, the DBN has been used in the diagnosis of brain diseases, such as Schizophrenia [6], and Alzheimers disease [7].

Traditional representation methods of DBN use hand-crafted features. These methods can be simply divided into two categories: 1) node-level based measurement and 2) state-based measurement. In the first category, the author defines a node-level measurement to describe the time-varying property of node of DBN. For example, Zhang et al.[8] suggest using the Pearson correlation of functional connectivities between each time-varying state to calculate this measurement. Braun et al. [9] define this measurement based on the average number of the DBN's community changes over time. The advantage of these methods is that they can directly transform the DBN into feature vectors that can be analyzed by using machine learning technologies. In the second category, the author uses K-means clustering to detect representative connectivity patterns from all time-varying states of the DBN. For example, Damaraju et al. [10] use these representative connectivity patterns to reveal transient states of dysconnectivity in schizophrenia. Nomi et al. [11] use the most frequently occurring representative connectivity patterns of DBN to clarify the functional role of the insular cortex and its subdivisions. However, both node-level based measurements and state-based measurements are still not data-driven. These hand-crafted features have limited generalization capability, which will affect the performance of the model in the identification of brain diseases.

Recently, some deep-learning based models have been proposed to learn the representation of DBN. These methods benefit from powerful nonlinear expression ability of neural networks, which can obtain more effective features than traditional methods. For example, Jie et al. [12] use three layers of convolutional operation to characterize temporal properties of DBN, and improves the performance of Alzheimers disease identification compared with traditional methods. Kam et al. [13] apply 3D convolutional neural networks (CNN) to extract features of DBN for early MCI detection. Besides CNN, recurrent neural networks (RNN) also have been used in the DBN analysis. In these methods, the author considers DBN as a sequence of data (consisted of a set of time-varying states), and uses the RNN to capture dynamic patters between these states. However, existing methods are designed to learn the representation of the whole sequence or individual states, thus neglecting hierarchically organized temporal nature of

DBN. Characterizing hierarchically properties of DBN can help reveal cognitive performance and the mechanism of brain diseases [14].

To address these issues, we propose a novel hierarchical representation learning (HARL) method for dynamic brain networks. The main idea of this paper is to build a deep neural network framework that can learn hierarchical representation of DBN. As shown in Fig. 1, firstly, we use the sliding window correlation to construct DBN and then model this DBN with a graph, where nodes represent each state of DBN, and edges represent correlations between states. Based on this graph, we design the HARL module to learn the representation of DBN at different levels. Each level consists of a graph convolutional layer, a graph pooling layer and a readout layer. These layers are used to update node features, select nodes, and aggregate node features respectively. Through this module, we can learn representations of DBN under three levels. Finally, we combine these representations to identify brain diseases by using a fully-connected layer. Experiments on a real schizophrenia dataset demonstrate the effectiveness and advantages of our proposed method.

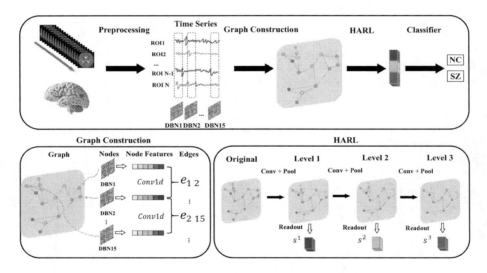

Fig. 1. Architecture of the proposed hierarchical representation learning (HARL) method for dynamic brain networks. There are three modules in our framework (i.e., a graph construction module, a HARL module, and a classier module). The HARL module consists of a convolution layer, a pooling layer, and a readout layer.

2   Materials and Preprocessing

We use a total of 120 subjects, including 67 normal controls (21 female (F)/46 male (M), aged 34.82 ± 11.28), and 53 Schizophrenia (11 female (F)/42 male (M), aged 36.75 ± 13.68), with rs-fMRI data from COBRE database. The data are acquired by Siemens Trio 3T scanner. The scan parameters of rs-fMRI are as follows: TR = 2000 ms, TE = 30 ms, flip angle = 90°, voxel size = 3.75 × 3.75 × 3.75 mm³. All rs-fMRI images are pre-processed by using the SPM 8 in the DPARSF toolbox version 2.0. The resulting volumes have 145 time points and are parcellated into 90 regions of interest (ROIs) using the AAL atlas. The DBN are constructed by using the sliding-window method which divides the time-series into a regular number of windows. In this paper, we set the length of window as 20 time points, and the step to move window as 9 time points. Then, there are 15 time-varying states in each DBN.

3   Proposed Method

Our proposed framework consists of two parts: Graph construction module (described in Sect. 3.1), and HARL module (described in Sect. 3.2). Let $A \in \mathbb{R}^{|\Omega| \times |\Omega|}$ represent the functional connectivity matrix, where Ω is the set of nodes corresponding to brain regions. Then, the DBN is defined as a set of functional connectivity matrixes $\mathbf{A} = \{A(1), A(2), .., A(T)\}$, where T is the number of the slice window. Mathematically, our task can be describe as obtaining a mapping $F : \mathbf{A} \rightarrow y$, where y is the label of subject (i.e., patient or normal control).

3.1   Graph Construction

In our work, we utilize a graph $G = \{V, E\}$ to model the DBN. In this graph, the node set $V = \{v_i, i = 1, 2, ..., T\}$ represents all states of DBN. The feature vector x_i on a node i is defined according the corresponding functional connectivity matrix $A(i)$. Due to this matrix is symmetric, we only use its upper triangular elements to construct the feature vector. The edge set $E = \{e_{ij}, i = 1, 2, ..., T, j = 1, 2, ..., T\}$ represents the connection between each node. In our model, this connection means the relationship between each state of DBN. Although these states are obtained in sequence, it is quite complicated to define the exact relationship between them due to the complicated biological mechanism of brain system. To simplify this graph model, we assume that each state may have potential connections with other states. Then, we set this model as a fully connected graph, and the weight of edges can be learned automatically by node features:

$$e_{ij} = conv1d(W x_i || W x_j), \tag{1}$$

where $conv1d$ represents 1D convolution operator, W is the learnable weight matrix that improves the feature expression ability, and $||$ is the concatenation operator. The value of e_{ij} indicates whether the connection is strong or weak. To make it easy to compare the connection weight of different nodes, we normalize these weights by utilizing the following function:

$$\alpha_{ij} = \frac{\exp(conv1d(W x_i || W x_j))}{\sum_{k \in N_i} \exp(conv1d(W x_i || W x_k))}, \tag{2}$$

where N_i means the neighbour nodes of node i. This normalized weight can further guide the graph convolutional for leaning node representation in the HARL module.

3.2   HARL

Our proposed HARL consists of three levels, and each level contains a graph convolutional layer, a graph pooling layer and a readout layer. The graph convolution layer is used to update node features for obtaining more effective representation. The graph pooling layer is used to select important nodes for generating a coarsened graph, which would be further input into the next level. A readout layer is used to generate a global representation of DBN from the individual representation of states. Three levels correspond to three hierarchical levels that extract features of the DBN at different scales. The detail of this module is described as follows:

Graph Convolution Layer: Convolution operation on graphs can be defined in either the spectral or non-spectral domain. In this work, we utilize the widely used graph convolution based on neighborhood aggregation [15]. It can be formulated as:

$$x_i^{l+1} = \sigma(\alpha_{ii}^l W_c {x_i}^l + \sum\nolimits_{j \in N_i^l} \alpha_{ij}^l W_c x_j^l), \tag{3}$$

where x_i^l is the node feature of l-th layer and $W_c \in \mathbb{R}^{F \times F'}$ is the convolution weight with input feature dimension F and output dimension F'. σ represents an nonlinear activation function.

Graph Pooling Layer: The pooling layer on the graph can be considered as a downsampling operation. Similar to the image data, the graph would be smaller through the pooling layer. In this work, we want to select important nodes (i.e., states) for generating a global representation of DBN. Then we use self-attention scores [16] to measure the importance of nodes. These scores are obtained by considering both graph features and topology in the framework of graph convolution formula. The self-attention score of node i at l-th level is calculated as follows.

$$z_i^l = \sigma(\alpha_{ii}^l W_p {x_i}^l + \sum\nolimits_{j \in N_i^l} \alpha_{ij}^l W_p x_j^l), \tag{4}$$

where $W_p \in \mathbb{R}^{1 \times F}$ is a learnable weight matrix. z_i^l is a scalar.

After obtaining self-attention scores, we adopt a simple node selection method proposed by [17]. The pooling ratio $k \in (0, 1]$ is a hyperparameter that determines the number of nodes to keep. Then, the top $[kT]$ nodes are selected based on the value of $Z = \{z_1, z_2, ..., z_T\}$.

$$idx = top - rank[Z, kT], \quad Z_{mask} = Z_{idx} \tag{5}$$

where $top - rank$ is the function that returns the indices of the top $[kT]$ values, idx is an indexing operation and Z_{mask} is the feature attention mask.

We get the coarsened node feature matrix X' by applying the following function: $X' = X \odot Z_{mask}$, where X is the original node feature matrix, and $\odot$ is the broadcasted elementwise product.

Readout Layer: This layer is a kind of global pooling layer that aggregates node features to generate the graph representation. In our work, the readout layer is used to learn the global representation of DBN from the individual representation of states. The equation of this aggregation operation is as follows:

$$s^l = \frac{1}{N} \sum_{i=1}^{N} x_i^l || \max_{i=1}^{N^l} x_i^l, \qquad (6)$$

where N^l is the number of nodes of graph at leave l, x_i^l is the feature vector of i-th node at leave l, and $||$ denotes concatenation. This operation can learn a global representation of DBN at each level, and the dimensions of these representations are consistent.

3.3   Implementation

The model is trained to minimize the cross-entropy cost between the ground-truth label Y and the predicted label $\hat{Y}$. The predicted label is obtained through a dense layer that connects S to $\hat{Y}$ by applying the softmax function: $\hat{Y} = soft \max(f(W_f \odot S))$, where $\odot$ operator represents element-wise multiplication, W_f is a learnable weight matrix, $S = s^1, s^2, s^3$. We describe the details of the learning process below: we use the training subjects DBN (i.e., $\mathbf{A}$) as input, and their corresponding labels as output. Then, the DBN are model as a graph G, and fed into the HARL module. The pooling ratio of the HARL set as 0.8 empirically. Our model is implemented in Pytorch using an NVIDIA Titan X GPU with 12 GB memory.

4   Experiments

4.1   Experimental Settings

We evaluate the performance of the proposed model for identifying brain disease in 120 subjects. The performance is evaluated by measuring the classification accuracy, sensitivity (i.e., the proportion of patients that are correctly predicted), and specificity (i.e., the proportion of normal controls that are correctly predicted) via 10-cross validation.

We compare it with state-of-the-art methods, including 1) traditional methods: SVM-based method (denoted as SVM), temporal variability method [8] (denoted as TV) and sparse temporally dynamic method [18] (denoted as STD). 2) deep learning based methods: weighted correlation kernels convolutional neural networks [12] (denoted as wck-CNN) and Spatial-Temporal Convolutional-Recurrent Network [19] (denoted as ST-CRN). We briefly introduce these methods. In the SVM-based method, we construct a functional connection matrix, and extract its upper triangle elements as features. These features would be input into the SVM classifier for predicting the label. In the TV method, we first extract node-level measurements to generate feature vectors, and then input these features into the SVM classifier. In the STD method, we first construct DBN based a spares learning method, and calculate clustering coefficients as feature vectors, and then input these features into the SVM classifier. In the wck-CNN method, we extract the representation of DBN by using three convolutional layers, and input this representation into a fully-connected layer. In the ST-CRN method, the representation of states are obtained by using the convolutional layer, and these representations are further input into the recurrent neural network for generating a global representation, which can be used to predict the label of subjects by using a fully-connected layer.

4.2   Results

The experimental results of all methods are summarized in Table 1. As can be seen, our method generally outperforms the competing methods. More specifically, our method achieves much higher accuracy (i.e., 91.6%), while the accuracy of the competing deep learning based methods are 84.1% and 88.3%. The possible reasons could be listed as follows 1) Both wck-CNN and ST-CRN use a single feature vector to represent the DBN, thus neglect the hierarchically organized temporal nature of DBN. In our method, we design the HARL module to learn the representation of DBN at different levels, which could improve the performance of the classification task. 2) Both wck-CNN and ST-CRN input the representation of DBN directly to the full connection layer instead of using a preprocessing layer (i.e., readout layer). The readout layer can significantly reduce the dimension of representations of DBN. In this work, we only have 120 subjects, so the overfitting problem should be considered in training the model.

Table 1. Performance Comparison of the Proposed and Competing Methods.

Methods	ACC (%)	SPE (%)	SEN(%)
SVM	67.5	62.8	74.0
TV [8]	65.8	64.2	68.0
STD [18]	70.8	65.7	78.0
wck-CNN [12]	84.1	82.8	88.0
ST-CRN [19]	88.3	87.1	90.0
Our method	**91.6**	**90.0**	**94.0**

4.3   Effectiveness of Hierarchical Representations

We further evaluate the effectiveness of hierarchical representations by a statistical test and a classification task. In the statistical test, we assess the statistical

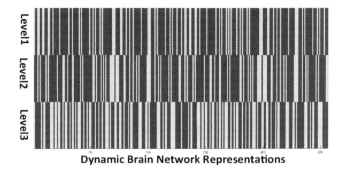

Fig. 2. The group difference of representations of DBN under different levels between normal controls and Schizophrenia. Here, p-values more than 0.05 are set to 1 (denoted as yellow lines). (Color figure online)

significance of these hierarchical representations (i.e., s^1,s^2, and s^3) between normal controls and Schizophrenia. The results are reported in Fig. 2. From this figure, we can see that there are significant differences on each level, and the third level has the most significant differences. This result suggests that the representation of DBN under different levels may provide discrepant information to identify Schizophrenia. In the classification task, we test the performance of our proposed method under different hierarchical levels. The results are reported in Fig. 3. We observe that the value of specificity increases with the number of levels, which further suggests the effectiveness of hierarchical representations.

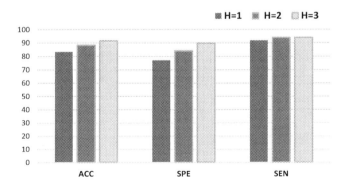

Fig. 3. Performance of the proposed method using different levels.

5   Conclusion

In summary, we define a graph model for characterizing the dynamic brain network, and propose a novel hierarchical representation learning method in the framework of graph convolutional networks for extracting DBN features under different hierarchical levels. This is completely different from the conventional methods, which only represent the DBN by using a single level. Results on 120 subjects from COBRE database demonstrate that our proposed method improves the classification performance compared with state-of-the-art methods.

References

1. Kucyi, A., Hove, M.J., Esterman, M., et al.: Dynamic brain network correlates of spontaneous fluctuations in attention. Cereb. Cortex **27**(3), 1831–1840 (2017)
2. Huang, J., Wang, M., Xu, X., et al.: A novel node-level structure embedding and alignment representation of structural networks for brain disease analysis. Med. Image Anal. **65**, 101755 (2020)
3. Wang, M., Zhang, D., Huang, J., et al.: Identifying autism spectrum disorder with multi-site fMRI via low-rank domain adaptation. IEEE Trans. Med. Imaging **39**(3), 644–655 (2019)
4. Wang, M., Huang, J., Liu, M., et al.: Functional connectivity network analysis with discriminative hub detection for brain disease identification. In: Proceedings of the AAAI Conference on Artificial Intelligence, vol. 33, pp. 1198–1205 (2019)
5. Chang, C., Metzger, C.D., Glover, G.H., et al.: Association between heart rate variability and fluctuations in resting-state functional connectivity. Neuroimage **68**, 93–104 (2013)
6. Sun, Y., Collinson, S.L., Suckling, J., et al.: Dynamic reorganization of functional connectivity reveals abnormal temporal efficiency in schizophrenia. Schizophr. Bull. **45**(3), 659–669 (2019)
7. Crdova-Palomera, A., Kaufmann, T., Persson, K., et al.: Disrupted global metastability and static and dynamic brain connectivity across individuals in the Alzheimers disease continuum. Sci. Rep. **7**, 40268 (2017)
8. Zhang, J., Cheng, W., Liu, Z., et al.: Neural, electrophysiological and anatomical basis of brain-network variability and its characteristic changes in mental disorders. Brain **139**(8), 2307–2321 (2016)
9. Braun, U., Schfer, A., Bassett, D.S., et al.: Dynamic brain network reconfiguration as a potential schizophrenia genetic risk mechanism modulated by NMDA receptor function. Proc. Natl. Acad. Sci. **113**(44), 12568–12573 (2016)
10. Damaraju, E., Allen, E.A., Belger, A., et al.: Dynamic functional connectivity analysis reveals transient states of dysconnectivity in schizophrenia. NeuroImage Clin. **5**, 298–308 (2014)
11. Nomi, J.S., Farrant, K., Damaraju, E., et al.: Dynamic functional network connectivity reveals unique and overlapping profiles of insula subdivisions. Hum. Brain Mapp. **37**(5), 1770–1787 (2016)
12. Jie, B., Liu, M., Lian, C., Shi, F., Shen, D.: Developing novel weighted correlation kernels for convolutional neural networks to extract hierarchical functional connectivities from fMRI for disease diagnosis. In: Shi, Y., Suk, H.I., Liu, M. (eds.) Machine Learning in Medical Imaging. MLMI 2018. LNCS, vol. 11046. Springer, Cham (2018). https://doi.org/10.1007/978-3-030-00919-9_1
13. Kam, T.E., Zhang, H., Jiao, Z., et al.: Deep learning of static and dynamic brain functional networks for early MCI detection. IEEE Trans. Med. Imaging **39**, 478–487 (2019)

14. Vidaurre, D., Smith, S.M., Woolrich, M.W.: Brain network dynamics are hierarchically organized in time. Proc. Natl. Acad. Sci. **114**(48), 12827–12832 (2017)
15. Kipf, T.N., Welling, M.: Semi-supervised classification with graph convolutional networks. arXiv preprint arXiv:1609.02907 (2016)
16. Lee, J., Lee, I., Kang, J.: Self-attention graph pooling. arXiv preprint arXiv:1904.08082 (2019)
17. Gao, H., Ji, S.: Graph u-nets. arXiv preprint arXiv:1905.05178 (2019)
18. Wee, C.Y., Yang, S., Yap, P.T., et al.: Sparse temporally dynamic resting-state functional connectivity networks for early MCI identification. Brain Imaging Behav. **10**(2), 342–356 (2016). https://doi.org/10.1007/s11682-015-9408-2
19. Wang, M., Lian, C., Yao, D., et al.: Spatial-temporal dependency modeling and network hub detection for functional MRI analysis via convolutional-recurrent network. IEEE Trans. Biomed. Eng. **67**, 2241–2252 (2019)

Graph-Temporal LSTM Networks
for Skeleton-Based Action Recognition

Hongsheng Li, Guangming Zhu$^{(\boxtimes)}$, Liang Zhang, Juan Song, and Peiyi Shen

School of Computer Science and Technology, Xidian University, Xi'an, People's
Republic of China
gmzhu@xidian.edu.cn

Abstract. Human action recognition is one of the challenging and
active research fields. Recently, spatio-temporal graph convolutions for
skeleton-based action recognition have attracted much attention. Several strategies, such as temporal downsampling, convolution striding,
and temporal pooling, are used to handle long action sequences. Recurrent neural networks are typically used for the processing of sequential data. In this paper, we propose a deep architecture that combines
spatio-temporal graph convolution and graph-temporal long short-term
memory (GT-LSTM) for skeleton-based human action recognition. Initially, topology-learnable spatio-temporal graph convolutions are applied
to learn the local spatio-temporal features of graph nodes and adaptively evolve graph topologies. Then, GT-LSTM successively performs
the spatio-temporal feature fusion with the node sequence and the temporal dimension, for the final recognition. Experimental results on the
NTU RGB+D and Kinetics-Skeleton datasets demonstrate that the proposed architecture can effectively perform graph node information aggregation, graph topology evolution, and spatio-temporal graph feature
fusion. liu2017skeleton.

Keywords: Human action recognition · Graph convolution · LSTM

1   Introduction

Human action recognition plays an important role in various computer vision
fields, such as video surveillance and human-computer interaction, and has
become an active research field in the recent years [9,20,21]. The 3D skeleton data represent human body structures by using a set of 3D coordinates of
human joints. Human skeleton can be naturally represented by a graph, where
each node represents one human joint. Therefore, human actions can be viewed
as spatio-temporal graphs.

Supported by the National Natural Science Foundation of China under Grant No.
61702390, the Fundamental Research Funds for the Central Universities under Grant
JB181001, the Key Research and Development Program of Shaanxi Province under
Grant No.2018ZDXMGY-036.

Y. Peng et al. (Eds.): PRCV 2020, LNCS 12306, pp. 480–491, 2020.
https://doi.org/10.1007/978-3-030-60639-8_40

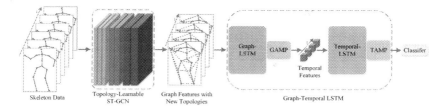

Fig. 1. Overview of the proposed architecture. Topology-learnable ST-GCN learns the local spatio-temporal features from the skeleton graph data. Graph-LSTM fuses the graph features with the node sequence, and GAMP aggregates all graph nodes. Temporal-LSTM fuses the features with the temporal dimension. GAMP denotes "graph average-maximum pooling", and TAMP denotes "temporal average-maximum pooling".

In the recent years, convolutional neural networks (CNNs) have been generalized into graph domains, and graph convolutional networks (GCNs) have emerged for the applications based on irregular graph-like structures. Spatial temporal GCNs (ST-GCNs) have been proposed for skeleton-based action recognition [22]. Different strategies are designed for graph partitioning, and multiple spatio-temporal graph convolution layers are applied to generate higher-level feature maps gradually on the graph. On the basis of ST-GCN, a two-stream adaptive ST-GCN (2s-AGCN) is proposed [13] by learning new connections beyond the handcrafted graph. Two extra learnable components are added to the handcrafted adjacency matrix to evolve the graph adaptively. These works disassemble the spatio-temporal graph convolution into spatial graph and temporal convolutions in a similar way to the spatiotemporal convolutional block "R(2+1)D" [19]. However, they still suffer from the inherent drawbacks of 3DCNNs (e.g.., using temporal downsampling, convolution striding, or temporal pooling) when dealing with long sequential data.

Recurrent neural networks (RNNs), which are typically used for the processing of sequential data, can be applied to graph sequences. The straightforward way is to fuse the graph node features before feeding them into the RNN/LSTM, such that the RNN/LSTM only focuses on the feature fusion with the temporal dimension. An attention-enhanced graph convolutional LSTM (AGC-LSTM) is constructed by replacing the spatial convolutions in ConvLSTM with graph convolutions for skeleton-based action recognition [14]. These methods perform LSTM operations on each graph node in succession to the graph convolution or node feature fusion operations. The graph topologies are maintained in the recurrent process of LSTM.

In the present study, we propose a novel approach that combines spatio-temporal graph convolution and LSTM for long action sequence recognition (Fig. 1). **First**, a topology-learnable ST-GCN is constructed to learn the local spatio-temporal features and adaptively evolve the graph topologies. This topology-learnable ST-GCN component neither should be deep nor should have a large temporal receptive field, because it is designed to learn the local spatio-temporal features. **Then**, a graph-temporal LSTM (GT-LSTM) component is constructed to fuse the features successively with the node sequence and the temporal dimension. The GT-LSTM component focuses on the graph-temporal global feature learning. The experimental results demonstrate that the topology-learnable ST-GCN can effectively aggregate the node information and evolve the graph topologies, and the GT-LSTM can efficiently fuse the learned local spatio-temporal features.

The main contributions of this work are visible in three aspects: **1**) a deep architecture that combines spatio-temporal graph convolution and LSTM to effectively learn the local and global spatio-temporal features. **2**) a GT-LSTM that provides a new way to successively fuse spatio-temporal graph features with the graph node sequence and the temporal dimension. **3**) the achievement of state-of-the-art performances on two large-scale datasets for skeleton-based action recognition.

2   Related Work

The recent approaches to construct graph convolutional networks can generally be categorized into **spectral** and **spatial** perspectives. The spatial perspective directly defines convolutions on the graph within the k-step neighbors. Yan et al. [22] constructed ST-GCN for skeleton-based action recognition. Spatial graph and temporal convolutions were successively implemented on the graphs and temporal edges. Instead of performing graph convolutions on the entire skeleton graphs, part-based GCNs were constructed for skeleton-based action recognition [10, 18]. Fixing the handcrafted graph topologies over all layers are not optimal. Thus, Shi et al. [13] proposed adaptive graph convolutional networks to learn the data-dependent graph topologies. In addition to the human joint information, the shapes and locations of body bones are important for skeleton-based action recognition. Zhang et al. [24] constructed graph edge convolutional networks to investigate bone information from skeleton data for action recognition. AGC-LSTM replaces the spatial convolutions in ConvLSTM with graph convolutions for skeleton-based action recognition [14]. Moreover, action recognition methods which focus on the incomplete skeletons [16], action-specific latent dependencies [6], and directed graphs [12], have been also proposed in the recent years. Besides, RNNs, which are typically used for the processing of sequential data, can also be applied to graph sequences. Many LSTM-based methods [5, 7, 8, 11, 15, 23] for skeleton-based action recognition have been proposed.

In this study, we propose a novel approach for the fusion of graph features with the node sequence and the temporal dimension. The graph features are learned by a topology-learnable ST-GCN. The experimental results demonstrate that the learned graph features have fully used the node information and the graph topologies, such that the graph topologies do not need to be maintained in the recurrent process of LSTM.

3   GT-LSTM Networks

3.1   Pipeline Overview

The proposed architecture consists of two components (as illustrated in Fig. 1): **topology-learnable ST-GCN** and **GT-LSTM**. The former is designed to learn the local spatio-temporal graph features and adaptively evolve the graph topologies. It fully uses the graph node information and the graph topologies. As such, the latter does not need to maintain the graph topologies in the recurrent process. The latter successively fuses the learned local spatio-temporal features with the node sequence and the temporal dimension to extract the global spatio-temporal features.

3.2   Topology-Learnable ST-GCN

A modified architecture based on ST-GCN [22] and 2s-AGCN [13] is constructed, namely denoted as the topology-learnable ST-GCN. The topology-learnable graph convolution is defined as follows:

$$\mathbf{f}_{out} = \sum_{k}^{K_v} \mathbf{W}_k \mathbf{f}_{in}(\mathbf{A}_k + \mathbf{B}_k), \tag{1}$$

where $\mathbf{f}_{in}$ and $\mathbf{f}_{out}$ denote the feature maps, $\mathbf{A}_k$ is the normalized physical adjaceny matrix, $\mathbf{B}_k$ is the learnable topology matrix which is initialized using $\mathbf{A}_k$ (instead of initialized to 0 as in 2s-AGCN [13]), and $\mathbf{W}_k$ is the weight matrix to evolve the features. The spatial configuration partitioning [22] is used; thus, $K_v = 3$.

The topology-learnable ST-GCN component is designed to learn the local spatio-temporal features. Thus, this component neither should be deep nor should have a large temporal receptive field. A seven-layer network is constructed. The temporal kernel size of the temporal convolutions was originally set to 9 in [13,22]. This study demonstrates the non-necessity to pursue a large temporal field in the topology-learnable ST-GCN because the GT-LSTM component will further fuse the features with the temporal dimension.

3.3   GT-LSTM

The topology-learnable ST-GCN component outputs graph sequences with new node features and topologies, whereas the GT-LSTM component fuses the features with the node sequence and the temporal dimension (Fig. 2).

Graph-LSTM. Initially, Graph-LSTM performs feature fusion with the node sequence. Let $\mathbf{X} = \{X_{t,i} | i = 1, \cdots, N, t = 1, \cdots, T\}$ denote the new node features, where N is the node count and T is the temporal length. The fusion operations are performed on the graph nodes at each time step, as shown in Eqs. (2)–(7). The fusion process is conducted with the node sequence.

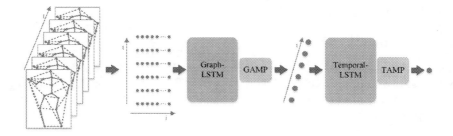

Fig. 2. Overview of the GT-LSTM architecture. Only the output features of the topology-learnable ST-GCN component are fed into GT-LSTM for the subsequent feature fusion.

$$i_{t,i} = \sigma(\tilde{W}_{xi}X_{t,i} + \tilde{W}_{hi}H_{t,i-1} + \tilde{b}_i), \tag{2}$$

$$f_{t,i} = \sigma(\tilde{W}_{xf}X_{t,i} + \tilde{W}_{hf}H_{t,i-1} + \tilde{b}_f), \tag{3}$$

$$o_{t,i} = \sigma(\tilde{W}_{xo}X_{t,i} + \tilde{W}_{ho}H_{t,i-1} + \tilde{b}_o), \tag{4}$$

$$G_{t,i} = \tanh(\tilde{W}_{xc}X_{t,i} + \tilde{W}_{hc}H_{t,i-1} + \tilde{b}_c), \tag{5}$$

$$C_{t,i} = f_{t,i} \circ C_{t,i-1} + i_{t,i} \circ G_{t,i}, \tag{6}$$

$$H_{t,i} = o_{t,i} \circ \tanh(C_{t,i}), \tag{7}$$

where $i_{t,i}$, $f_{t,i}$, $o_{t,i}$ are the gates of LSTM with the node sequence i at each time step t. $H_{t,i}$ and $C_{t,i}$ are the hidden and the cell states. $G_{t,i}$ is the candidate memory. $\tilde{W}_{x*}$, $\tilde{W}_{h*}$ and $\tilde{b}_*$ are the parameters of the linear operations in LSTM. $\sigma(\cdot)$ and $\tanh(\cdot)$ are the activation functions. Graph-LSTM can be bidirectional and have more than one layer. The corresponding details are ignored for simplicity.

GAMP. Graph-LSTM generates outputs at each recurrent step. A natural way of fusing the outputs is *global average pooling*. In this work, the readout layer is used, as shown in Eq. (8). *Global maximum pooling* is also performed for augmentation. The augmentation operation, however, can double the dimension of the fused features, which increases the parameters of the Temporal-LSTM. Therefore, a linear operation is performed to maintain the size of the fused features.

$$\overline{H}_t = \frac{1}{N}\sum_{i=1}^{N} H_{t,i} \parallel \max_{i=1}^{N} H_{t,i}. \tag{8}$$

$$X_t = W_G\overline{H}_t + b_G, \tag{9}$$

where $\overline{H}_t$ is the augmented feature, and W_G and b_G are the parameters to keep the channel size of X_t the same as $H_{t,i}$.

Temporal-LSTM. The Graph-LSTM and GAMP layers aggregate the graph into one single node, and the spatio-temporal features are fused into temporal feature sequences. Therefore, the Temporal-LSTM only needs to perform regular LSTM operations.

$$i_t = \sigma(\hat{W}_{xi}X_t + \hat{W}_{hi}H_{t-1} + \hat{b}_i), \tag{10}$$

$$f_t = \sigma(\hat{W}_{xf}X_t + \hat{W}_{hf}H_{t-1} + \hat{b}_f), \tag{11}$$

$$o_t = \sigma(\hat{W}_{xo}X_t + \hat{W}_{ho}H_{t-1} + \hat{b}_o), \tag{12}$$

$$G_t = \tanh(\hat{W}_{xc}X_t + \hat{W}_{hc}H_{t-1} + \hat{b}_c), \tag{13}$$

$$C_t = f_t \circ C_{t-1} + i_t \circ G_t, \tag{14}$$

$$H_t = o_t \circ \tanh(C_t), \tag{15}$$

where i_t, f_t, o_t are the gates of LSTM with the time step t. H_t and C_t are the hidden and the cell states. G_t is the candidate memory. $\hat{W}_{x*}$, $\hat{W}_{h*}$ and $\hat{b}_*$ are the parameters of the linear operations in LSTM.

TAMP. A similar pooling strategy to GAMP is performed on the outputs of Temporal-LSTM, as shown in Eq. (16). The GAMP and TAMP layers fix the size of the fused features regardless if the Graph-LSTM and Temporal-LSTM are unidirectional or bidirectional.

$$Y = W_T(\frac{1}{T}\sum_{t=1}^{T} H_t \parallel \max_{t=1}^{T} H_t) + b_T. \tag{16}$$

where W_T and b_T are the parameters of the linear operation to keep the channel size of Y the same as H_t.

3.4   GT-LSTM Networks

The numbers of output channels for each layer of the topology-learnable ST-GCN are set to 64, 64, 64, 128, 128, 256, and 256, and the default temporal kernel size is 5. The channel numbers of the inputs and the hidden states in GT-LSTM are 256, and two bidirectional LSTM layers are stacked by default in the Graph-LSTM and Temporal-LSTM.

Two-stream networks are constructed on the basis of the joint and the bone information [13]. The final classification scores of the two streams are added to predict the action label.

4   Experiments

Two large-scale action recognition datasets, namely, NTU RGB+D [11] and Kinetics-Skeleton [3,22], are used to evaluate the performance of the proposed architecture. The top-1 and top-5 classification accuracies are used to evaluate the recognition performance.

Table 1. Comparisons of the validation accuracy with different temporal kernel sizes in the topology-learnable ST-GCN component.

Configurations	Top-1	Top-5
Temporal Kernel size = 3	93.56	99.26
Temporal Kernel size = 5	**93.95**	**99.31**
Temporal Kernel size = 7	93.33	99.30
Temporal Kernel size = 9	93.70	99.27

4.1   Datasets

NTU RGB+D. NTU RGB+D [11] is currently the largest human action dataset with 3D joint annotations. The dataset contains 56,000 action clips in 60 action classes. A total of 40 volunteers perform the actions captured by three cameras at the same height but from different horizontal angles: -45°, 0°, and 45°. Each subject has 25 joints, and each video has no more than two subjects. This dataset has two benchmarks: The first one is the cross-subject (X-Sub), wherein 40,320 and 16,560 videos are used for training and validation, respectively, and the participants in the two subsets are different. The second benchmark is the cross-view (X-View), wherein 37,920 videos are captured by Cameras 2 and 3 for training and 18,960 videos are captured by Camera 1 for validation.

Kinetics-Skeleton. The DeepMind Kinetics human action dataset [3] contains approximately 300,000 video clips, which cover 400 human action classes. The Kinetics dataset has only raw video clips without skeleton data. Yan et al. [22] built the Kinetics-Skeleton dataset by using the publicly available OpenPose toolbox [1]. The skeleton data contain 18 human joints for each frame. The skeleton data of 240,000 clips are used for training, whereas those of 20,000 clips are used for validation.

4.2   Training Details

Our implementation is based on the released code of 2s-AGCN [13], and the same training details are used. In particular, the PyTorch deep learning framework is adopted. Stochastic gradient descent with Nesterov momentum (0.9) is applied for network optimization. The batch size is 64 and the weight decay is 0.0001. Softmax function is used as the final classifier.

The used data preprocessing operations are the same as those in 2s-AGCN [13]. A total of 50 epochs are performed when training on the NTU RGB+D dataset. The initial learning rate is set to 0.1 and divided by 10 at the 30th and 40th epochs. A total of 65 epochs are performed when training on the Kinetics-Skeleton dataset. The initial learning rate is set to 0.1 and divided by 10 at the 45th and 55th epochs.

Table 2. Comparisons of the validation accuracies with different configurations of the GT-LSTM component.

Configurations	Top-1	Top-5
Two layers, bidirectional	**93.95**	**99.31**
Two layers, unidirectional	92.83	99.20
One layers, unidirectional	92.59	99.08

4.3   Ablation Study

The X-View subset of NTU RGB+D is used for the ablation study.

Temporal Kernel Size. ST-GCN [22] and 2s-AGCN [13] set the temporal kernel size of all the layers to 9. Large temporal kernel sizes are necessary to achieve a large temporal receptive field for these two networks. However, the proposed topology-learnable ST-GCN component is designed to learn local spatio-temporal features, whereas the GT-LSTM component is designed to fuse features with the temporal dimension. Therefore, a large temporal receptive field is not crucial for the topology-learnable ST-GCN component. Moreover, increased temporal kernel sizes will result in additional parameters. Different temporal kernel sizes are evaluated, and Table 1 presents the results. The results demonstrate that large temporal kernel sizes do not signify good performances. The default temporal kernel size in this work is set to 5.

Depth of GC-LSTM. Different configurations of the GT-LSTM component are evaluated, and Table 2 presents the results. The configuration "Two layers, bidirectional" means that the Graph-LSTM and Temporal-LSTM components each have two bidirectional LSTM layers. All inputs and hidden states are designed in a fixed-size representation (i.e., 256 channels). The results show that two bidirectional layers achieve excellent performance. In two LSTM layers, the former LSTM layer outputs a sequence of vectors which will be used as an input to a subsequent LSTM layer. This hierarchy of hidden layers enables more complex representation of node sequence, capturing information at different scales. And in Bidirectional LSTM layers, using the two hidden states the cell can combine the information from both past and future. The topology-learnable ST-GCN component is designed to be simple and shallow; therefore, the GT-LSTM component should own a certain complexity to guarantee the performance of the entire architecture.

Node Shuffling. The Graph-LSTM fuses features with the node sequence. However, the nodes are not sorted on the basis of the graph topologies. The evolved graph topologies are excluded in the GT-LSTM process. To verify whether the orders of the nodes will affect the GT-LSTM's performance, node

Table 3. Comparisons of the validation accuracies when node shuffling is performed before feeding graph features into the GT-LSTM component.

Configurations	Top-1	Top-5
w/o Node shuffling, w/o Fine-tuning	**93.95**	**99.31**
w/ Node shuffling, w/o Fine-tuning	93.19	99.19
w/o Node shuffling, w/ Fine-tuning	94.69	**99.32**
w/ Node shuffling, w/ Fine-tuning	**94.72**	99.25

shuffling is performed before feeding the local spatio-temporal features into the GT-LSTM during training. This step means that the node orders may be different at each training and testing iteration. We train the architecture with node shuffling from scratch or fine-tune on the basis of the trained model without node shuffling.We also try to improve the architecture without node shuffling by fine-tuning on the basis of the trained model of the bone-stream. The comparison results in Table 3 show that accuracy loss is not always present when shuffling the nodes. Node shuffling may increase the level of difficulty of training; fine-tuning operations, however, can overcome this difficulty.

4.4   Comparison with State-of-the-Art Methods

Cross-modality fine-tuning can improve network performance. The original NTU RGB+D and Kinetics-Skeleton datasets contain human joint information. Human bone information is a second-order information represented as a vector that points to the target joint from the source joint. We fine-tune the networks of the joint-stream on the basis of the trained models of the bone-stream, and vice versa.

Tables 4 and 5 illustrate the comparison results of the proposed architecture with the state-of-the-art methods on the NTU RGB+D and Kinetics-Skeleton datasets, respectively. Many LSTM-based [5,7,8,11,15,23] and GCN-based [6,10,13,14,17,22] methods for skeleton-based action recognition have been proposed. The proposed architecture achieves state-of-the-art performances. Firstly, the proposed architecture outperforms the stardand spatio-temporal graph convolution networks, such as ST-GCN [22] and 2s-AGCN [13]. Compared to 2s-AGCN [13], our proposed method utilizes Topology-Learnable ST-GCN and GT-LSTM to learn local and global spatio-temporal features, respectively. Secondly, AGC-LSTM [14] is the first network that embeds graph convolution into convolutional LSTM for skeleton-based action recognition, and an attention mechanism is also embedded to enhance the performance of the graph convolutional LSTM. The comparison results indicate that the proposed architecture achieves comparable or better performances, compared with AGC-LSTM. This study provides a new method that combines spatio-temporal graph convolution and LSTM to perform the node information aggregation, graph topology evolution and spatio-temporal graph feature fusion, for skeleton-based action recognition.

Table 4. Comparisons of the validation accuracies of the proposed architecture with the state-of-the-art methods on the NTU RGB+D dataset.

Methods	X-Sub	X-View
Deep LSTM [11]	60.7	67.3
ST-LSTM [8]	69.2	77.7
STA-LSTM [15]	73.4	81.2
VA-LSTM [23]	79.2	87.8
ARRN-LSTM [5]	80.7	88.8
Ind-RNN [7]	81.8	88.0
PA-GCN [10]	80.4	82.7
ST-GCN [22]	81.5	88.3
DPRL+GCNN [17]	83.5	89.8
BPLHM [24]	85.4	91.1
3s RA-GCN (Song et al. [16])	85.9	93.5
AS-GCN [6]	86.8	94.2
PB-GCN (Thakkar et al. [18])	87.5	93.2
2s-AGCN [13]	88.5	95.1
AGC-LSTM [14]	89.2	95.0
GT-LSTM (Joint)	88.4	94.7
GT-LSTM (Bone)	88.0	94.5
GT-LSTM (Both)	**89.2**	**95.2**

Table 5. Comparisons of the validation accuracies of the proposed architecture with the state-of-the-art methods on the Kinetics-Skeleton dataset.

Methods	Top-1	Top-5
Feature Enc. [2]	14.9	25.8
Deep LSTM [11]	16.4	35.3
TCN [4]	20.3	40.0
ST-GCN [22]	30.7	52.8
BPLHM [24]	33.4	56.2
AS-GCN [6]	34.8	56.5
2s-AGCN [13]	36.1	58.7
GT-LSTM (Joint)	35.2	58.0
GT-LSTM (Bone)	34.9	57.8
GT-LSTM (Both)	**36.6**	**59.5**

5   Conclusion

In this paper, we propose a deep architecture that combines spatio-temporal graph convolution and graph-temporal LSTM for skeleton-based action recognition. The topology-learnable spatio-temporal graph convolution network is designed to learn the local spatio-temporal features and adaptively evolve the graph topologies. The graph-temporal LSTM component successively fuses the learned graph features with the node sequence and the temporal dimension. The proposed graph-temporal LSTM effectively aggregates the graph nodes and does not need to operate on each node when fusing with the temporal dimension. The proposed architecture can effectively perform graph node information aggregation, graph topology evolution, and spatio-temporal graph feature fusion.

References

1. Cao, Z., Hidalgo, G., Simon, T., Wei, S.E., Sheikh, Y.: OpenPose: realtime multi-person 2D pose estimation using Part Affinity Fields. arXiv preprint arXiv:1812.08008 (2018)
2. Fernando, B., Gavves, E., Oramas, M.J., Ghodrati, A., Tuytelaars, T.: Modeling video evolution for action recognition. In: CVPR, pp. 5378–5387 (2015)
3. Kay, W., et al.: The kinetics human action video dataset. arXiv preprint arXiv:1705.06950 (2017)
4. Kim, T.S., Reiter, A.: Interpretable 3d human action analysis with temporal convolutional networks. In: CVPRW, pp. 1623–1631 (2017)
5. Li, L., Zheng, W., Zhang, Z., Huang, Y., Wang, L.: Skeleton-based relational modeling for action recognition. arXiv preprint arXiv:1805.02556 (2018)
6. Li, M., Chen, S., Chen, X., Zhang, Y., Wang, Y., Tian, Q.: Actional-structural graph convolutional networks for skeleton-based action recognition. arXiv preprint arXiv:1904.12659 (2019)
7. Li, S., Li, W., Cook, C., Zhu, C., Gao, Y.: Independently recurrent neural network (INDRNN): building a longer and deeper RNN. In: CVPR, pp. 5457–5466 (2018)
8. Liu, J., Shahroudy, A., Xu, D., Wang, G.: Spatio-temporal LSTM with trust gates for 3D human action recognition. In: ECCV, pp. 816–833 (2016)
9. Liu, J., Wang, G., Duan, L.Y., Abdiyeva, K., Kot, A.C.: Skeleton-based human action recognition with global context-aware attention LSTM networks. IEEE Trans. Image Process. **27**(4), 1586–1599 (2017)
10. Qin, Y., Mo, L., Li, C., Luo, J.: Skeleton-based action recognition by part-aware graph convolutional networks. Visual Comput. **36**(3), 621–631 (2019). https://doi.org/10.1007/s00371-019-01644-3
11. Shahroudy, A., Liu, J., Ng, T.T., Wang, G.: NTU RGB+D: a large scale dataset for 3D human activity analysis. In: CVPR, pp. 1010–1019 (2016)
12. Shi, L., Zhang, Y., Cheng, J., Lu, H.: Skeleton-based action recognition with directed graph neural networks. In: CVPR, pp. 7912–7921 (2019)
13. Shi, L., Zhang, Y., Cheng, J., Lu, H.: Two-stream adaptive graph convolutional networks for skeleton-based action recognition. In: CVPR, pp. 12026–12035 (2019)
14. Si, C., Chen, W., Wang, W., Wang, L., Tan, T.: An attention enhanced graph convolutional LSTM network for skeleton-based action recognition. In: CVPR, pp. 1227–1236 (2019)
15. Song, S., Lan, C., Xing, J., Zeng, W., Liu, J.: An end-to-end spatio-temporal attention model for human action recognition from skeleton data. In: AAAI, pp. 4263–4270 (2017)
16. Song, Y.F., Zhang, Z., Wang, L.: Richly activated graph convolutional network for action recognition with incomplete skeletons. In: ICIP (2019)

17. Tang, Y., Tian, Y., Lu, J., Li, P., Zhou, J.: Deep progressive reinforcement learning for skeleton-based action recognition. In: CVPR, pp. 5323–5332 (2018)
18. Thakkar, K.C., Narayanan, P.J.: Part-based graph convolutional network for action recognition. In: BMVC, pp. 1–13 (2018)
19. Tran, D., Wang, H., Torresani, L., Ray, J., LeCun, Y., Paluri, M.: A closer look at spatiotemporal convolutions for action recognition. In: CVPR, pp. 6450–6459 (2018)
20. Tu, Z., Li, H., Zhang, D., Dauwels, J., Li, B., Yuan, J.: Action-stage emphasized spatiotemporal VLAD for video action recognition. IEEE Trans. Image Process. **28**(6), 2799–2812 (2019)
21. Wang, L., et al.: Temporal segment networks for action recognition in videos. IEEE Trans. Pattern Anal. Mach. Intell. **41**(11), 2740–2755 (2018)
22. Yan, S., Xiong, Y., Lin, D., Tang, X.: Spatial temporal graph convolutional networks for skeleton-based action recognition. In: AAAI, pp. 7444–7452 (2018)
23. Zhang, P., Lan, C., Xing, J., Zeng, W., Xue, J., Zheng, N.: View adaptive recurrent neural networks for high performance human action recognition from skeleton data. In: ICCV, pp. 2136–2145 (2017)
24. Zhang, X., Xu, C., Tian, X., Tao, D.: Graph edge convolutional neural networks for skeleton based action recognition. arXiv preprint arXiv:1805.06184 (2018)

Detection of High-Risk Depression Groups Based on Eye-Tracking Data

Simeng Lu[1], Shen Huang[2,7], Yun Zhang[3], Xiujuan Zheng[4], Danmin Miao[2], Jiajun Wang[1(✉)], and Zheru Chi[5,6]

[1] School of Electronic and Information Engineering, Soochow University, Suzhou 215006, People's Republic of China
20185228001@stu.suda.edu.cn, jjwang@suda.edu.cn
[2] Department of Military Medical Psychology, Air Force Medial University, Xi'an 710038, Shaanxi, People's Republic of China
pier711@163.com, miaodanmin@126.com
[3] School of Electronic and Information Engineering, Xi'an Jiaotong University, Xi'an 710049, Shaanxi, People's Republic of China
[4] College of Electrical Engineering, Sichuan University, Chengdu 610065, Sichuan, People's Republic of China
xiujuanzheng@scu.edu.cn
[5] Department of Electronic and Information Engineering, The Hong Kong Polytechnic University, Hung Hom, Kowloon, Hong Kong
enzheru@polyu.edu.hk
[6] PolyU Shenzhen Research Institute, Shenzhen, People's Republic of China
[7] Xi'an Research Institute of Hi-Tech, Xi'an, China

Abstract. Depression is the most common psychiatric disorder in the general population. An effective treatment of depression requires early detection. In reschedule this paper, a novel algorithm is presented based on eye-tracking and a self-rating high-risk depression screening scale (S-hr-DS) for early depression screening. In this algorithm, a subject scan path is encoded by semantic areas of interest (AOIs). AOIs are dynamically generated by the POS (part-of-speech) tagging of Chinese words in the S-hr-DS items. The proposed method considers both temporal and spatial information of the eye-tracking data and encodes the subject scan path with semantic features of items. The support vector machine recursive feature elimination (SVM-RFE) algorithm is employed for feature selection and model training. Experimental results on a data set including 69 subjects show that our proposed algorithm can achieve an accuracy of 81% with 76% in sensitivity and 79% in F1-score, demonstrating a potential application in high-risk depression detection.

Keywords: High-risk depression groups · Word-fixation driven feature coding · Feature selection · Eye-tracking

1 Introduction

Depression is a kind of mental disorder, which is the fourth leading cause of disability and death [24]. More than 300 million people worldwide are suffering

© Springer Nature Switzerland AG 2020
Y. Peng et al. (Eds.): PRCV 2020, LNCS 12306, pp. 492–503, 2020.
https://doi.org/10.1007/978-3-030-60639-8_41

from depression, which equals about 4.4% of the global population [22]. The World Health Organization (WHO) experts point out that, like other diseases, the earlier the depression is detected and treated, the better it will be [19]. High-risk depression groups refer to people who are with personality vulnerable to depression [27] and have developed at least one symptom but haven't reached the DSM-5 [2] diagnosis standards. These people are the majority and have subjective discomfort in social communication, career, and life. They are easier to suffer depression upon stress or pressure. Therefore, this study focuses on developing a novel method effectively identifying high-risk depression groups based on high-risk depression screening scale (S-hr-DS) and the eye-tracking data.

Traditionally, depression is diagnosed with performing psychological self-assessment on scales and inventories. However, these self-assessment results cannot provide the subject's test-taking process data that is objective evidence related to the subject's mental state. Different ways have been proposed to tackle this problem. It can either be explored from the electroencephalogram (EEG) signal [1,8] or the eye-tracking data [6], both acquired during the procedure of self-assessment tests. The eye-tracking technology can both conveniently and non-intrusively record the subject's vision process with high time and special resolution, so we employed it to track the scan path of the subjects when finishing the S-hr-DS and thus provide clues related to the subject's mental state.

The application of eye-tracking technology in mental disorder is an updated attentive [7,23] study area, we utilize it in a novel way. Instead of focusing on conventional features such as the fixation time [25], number of fixations and the spatial distributions of the fixations (heat map) [5], we propose to construct features by encoding the subject scan path when the subject happens to scan different areas of interest (AOIs) in the items of the S-hr-DS. The AOIs in the items of the S-hr-DS are partitioned according to the POS (part-of-speech) tags of Chinese words dynamically. In order to reduce the dimension of the feature and improve the performance of the classifier, we implement the wrapper type feature selection procedure. In our case, the support vector machine (SVM) classifier [17] is chosen for our classification task. Therefore, the recursive feature elimination (RFE) scheme developed specifically for the SVM [21] is employed. Experiment results in a data set containing eye-tracking data of 69 subjects show that our proposed method can achieve an accuracy of 81% with 76% in sensitivity and 79% in F1-score.

In summary, the key contributions of our work are:

1. Specially designed eye-tracking system is utilized to self-rating high-risk depression screening scale for early depression early screening.
2. We dynamically partition the S-hr-D item to different AOIs areas of interest by words or phrases with different POS tags.
3. We propose a word-fixation driven feature encoding method to extract features on the subject scan path and POS based AOI.

2   Theoretical Background

Depression screening mostly relies on self-reported subjective inventories and scores. There have already been many depression measurement scales, such as the center for epidemiological studies depression scale (CES-D) [20], the Beck depression inventory (BDI) [4], the self-rating depression scale (SDS) [26], and so on. However, the major goal of these scales is to assess whether individuals meet the morbidity standards of depression and degree of depression. Unfortunately, most depression scales are inapplicable to screening high-risk depression groups due to the ambiguous definition of state depression and trait depression [14]. Therefore, this study developed a High-Risk Depression Screening Scale (S-hr-DS) based on division and recombination of state and traits of items in some classical scales.

S-hr-DS is a novel self-reported subjective inventory to assess whether individuals who do not meet the morbidity standards of depression and degree of depression but already have relative state and traits of depression. Our research group is composed of psychologists, clinicians and information scientists. We aim to discover the characteristics of the subconscious activities aroused by specific situations under synchronous multimodal behavioral and neurocognitive data.

Although high-risk depression groups can be detected upon performing S-hr-DS, self-assessment tests have obvious disadvantages because these self-assessment tests cannot provide the subject's test-taking process data that is objective evidence related to its mental state. Different ways have been proposed to tackle this problem. As a well-developed cognition research technology, eye-tracking has been one of the common methods to screen dysphrenia and brain cognitive explorations [3]. Recent research has shown that the eye-tracking paradigm of depression has anti-saccade tasks [12], free view tasks [13], staring stability tasks [16] and negative vocabularies or "self" attentions. This technic can track the subject scan path when finishing the S-hr-DS and thus provide clues related to the subject's mental state. The application of eye-tracking technology in mental disorder is an updated attentive study area, we utilize it in a novel way. Instead of focusing on conventional features such as the fixation time [25], number of fixations and the spatial distributions of the fixations (heat map) [5], we propose to extract features by word-fixation driven feature encoding.

In order to reduce the dimension of the feature and improve the performance of the classifier, we implement the wrapper type feature selection procedure. Support vector machine recursive feature elimination (SVM-RFE) can filter out redundant features and insignificant feature components to achieve higher classification performance [9]. The research findings of Harikrishna et al. have shown that this method can not only simplify the computational procedure but also effectively improve classification accuracy in datasets [11]. Therefore, the recursive feature elimination (RFE) scheme developed specifically for the SVM is employed in our case.

3   Experimental Methods

3.1   Data Acquisition

In this study, the randomized control method was applied and subjects were divided into high-risk depression groups and healthy groups according to the S-hr-DS. Items in the S-hr-DS were tested twice on a computer one by one after 10 months to prevent practice effect. The S-hr-DS consisting of 62 items was first imported into the software Tobii Pro Lab. These items are presented on an item-by-item basis to the subjects. The Tobii Pro X3-120 Eye Tracker [13] will record the subjects' eye-tracking data and answers to the S-hr-DS.

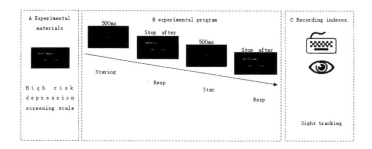

Fig. 1. Procedure of data collection

An item of the S-hr-DS is shown in part A of Fig. 1. The specific process with data collection is shown in part B of Fig. 1. After the eye movement calibration, the subjects were shown a sign of "+ in the middle of the screen in 500 ms. Items of the S-hr-DS are presented on an item-by-item basis to the subjects. Each item in the S-hr-DS should be answered with "Yes" or "No" by pressing the keyboard according to its contents. The eye-tracker will record the subjects' eye-tracking data and answers (the part C of Fig. 1) to the S-hr-DS.

High-risk groups were selected from a total of 697 subjects. All of them were screened according to the S-hr-DS. 39 subjects met the screening standards of the S-hr-DS but didn't reach the diagnosis standards of the Self-Rating Depression Scale (SDS) [26]. They were labeled high-risk depression groups. Healthy groups were also selected from the above mentioned 697 subjects. Healthy groups were those who didn't reach the screening standards of the S-hr-DS and the SDS. The total number of healthy groups are amount to 39 people. We exclude 9 subjects of two groups whose gaze samples are under 75% in the experiment. There is a total of 69 subjects (33 subjects labeled high-risk depression groups and 36 subjects labeled healthy groups) in the data set. Such labels will be used as the ground truth for training the classifier, while features will be extracted from the eye-tracking data.

3.2   Feature Extraction

Traditionally, the clinician makes decisions regarding the mental state of the subject according to a comprehensive assessment of the answers to the item in the scores and inventories. However, these answers cannot provide objective evidence related to the subjects mental state because the subject's test-taking process cannot be recorded. On the other hand, assessment of the answers by psychologists is a laboursome problem, especially for large scale screening. Therefore, we try to develop a computer-assisted screening method based on features extracted from the eye-tracking data acquired during the self-assessment test process.

In order to construct features from the eye-tracking data, each item is decomposed into question and answers part. Our preliminary study shows that high-risk depression groups and the healthy groups show different preferences to Chinese words or phrases with different POS tags. Therefore, we propose to partition the item to different AOIs, including words or phrases with different POS tags, as shown in Fig. 2. In this manner, the time the gaze stays in any area of interest (AOI) reflects the preference level of the subject to the words or phrases with different POS tags in the item and hence can be used as an indicator of the mental state. On the other hand, the order the gaze scans different AOIs (the scan path) provides clues regarding the decision process when the subject is trying to give an answer to the question in the item of S-hr-DS. Based on this analysis, we propose to extract the feature by the way of word-fixation driven feature encoding that the time the gaze stays in different AOIs can be considered simultaneously. A detailed description of word-fixation driven feature encoding process can be found as follows:

Step 1: Mark the segmentation and parsing result of all items in the S-hr-DS and encode each POS tag with an integer in a manner as shown in Table 1.

Step 2: Use the software Tobii Pro Lab to partition the S-hr-DS into different AOIs according to the POS of the words or phrases and assigns a code from Table 1 to each AOI. Figure 2(a) and Fig. 2 (b) give a sample item of the S-hr-DS and the result of AOI partitioning and encoding.

Step 3: Export all eye-tracking data from the software Tobii Pro Lab. Use the software Microsoft Excel to sort out the fixation time in the AOIs of each item. Merge the eye-tracking data for AOIs of the POS tags.

Step 4: Take 60 ms as the basic unit on fixation feature quantization and convert the fixation time on each AOI to that in a time unit of 60 ms.

Step 5: Encode the subject scan path according to the fixation time (in 60 ms unit) and the corresponding POS tag codes of the AOIs. For example, if the gaze scans over an AOI of a verb and stays on this AOI 120 ms, the resulted code for this scan is "11", with "1" denoting the code of the AOI of this POS tag while two "1" implying that the gaze fixes two time units on this AOI.

Table 1. Mapping table of different part-of-speech (POS) tags..

Area of interest	Meaning	Code
v	Verb	1
r	Pronoun	2
n	Noun	3
d	Adverb	4
a	Adjective	5
u	Auxiliary	6
p	Preposition	7
c	Conjunction	8
m	Numeral	9
2	Yes	10
3	No	11
Black	Area of blank	12

   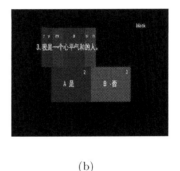

(a)   (b)

Fig. 2. (a) A sample slide of the S-hr-DS. (b)The result of areas of interest partitioning and encoding.

Due to the difference in the time for different subjects spending on answering the questions in the items of the S-hr-DS, the length of the encoded features corresponding to different subjects will be different. The feature lengths corresponding to different subjects will be normalized to that of the longest one (The length of the longest feature in the data set is 10715 features) by padding zeros to tackle such a problem.

3.3   Feature Selection and Model Training

As mentioned above, our feature vector contains 10715 components that is much larger than the number of samples for training in our study. This is a typical under determined problem. On the other hand, there may exist correlations between different feature components [10]. Furthermore, different feature components may contribute quite differently to the classification. Therefore, a wrapper type feature selection procedure (recursive feature elimination, RFE) [9] will be performed specifically for the SVM classifier with respect to the feature vector to remove irrelevant and redundant features.

The SVM-RFE based feature selection method starts with all feature components and recursively removes the feature component with the least importance for classification in a backward elimination manner. The measure for the importance of a feature component is computed from the weighting vector of the SVM [18]:

$$\mathbf{w} = \sum_l \alpha_l \times y_l \times \mathbf{x}^{(l)} \tag{1}$$

where α_l is the Lagrange multipliers, y_l is the class label of the l-th sample, and $\mathbf{x}^{(l)}$ is the feature vector resulted from feature extraction for the l-th sample. With such a weighting vector, the importance of the i-th component c_i can be determined as w_i^2, and hence the feature components can be selected according to their importance as determined above. A detailed description of the feature selection and model training procedure can be found in Algorithm 1.

Feature components are removed recursively one-by-one according to their corresponding weights in the SVM classifier trained in the training data set. The feature selection results are evaluated by the performance of the SVM classifier in the test data set while retrained in the training set based on the reduced feature vector. In this way, the optimal classifier which perform the best in terms of the accuracy can be obtained.

4   Experimental Results and Discussions

4.1   Description of the Data Set

In our study, with the assessment mentioned in Sect. 3.1, 33 of the subjects are labeled high-risk depression groups while the 36 subjects are labeled healthy groups. All subjects have no history of taking psychotropic drugs and also have no history of alcohol and drug abuse. In the test, each subject is asked to sit down in front of a computer where 62 items are displayed sequentially on the screen. For each item, the subject is asked to choose an answer by keyboard, and the eye-tracking data is recorded synchronously.

4.2   Experimental Results and Discussions

We calculate the frequencies of the nine POS tags in extracted features for each of the two groups (healthy groups and those are high-risk depression groups) of subjects. The frequencies of nine POS tags that appeared in items of the S-hr-DS are also calculated as features. Figure 3 (a) and Fig. 3 (b) give the bar plots of these two kinds of frequencies in the S-hr-DS and in the extracted features, respectively.

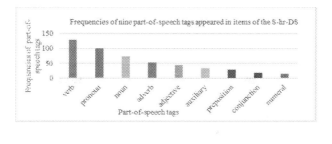

(a)

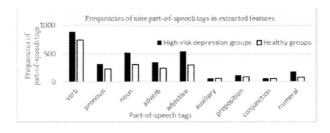

(b)

Fig. 3. (a) The frequencies of nine part-of-speech (POS) tags appeared in items of the self-rating high-risk depression screening scale. Different color bars indicate different POS tags of words. (b) The frequencies of POS tags of words after feature extraction. Black bars indicate the frequencies of POS tags of words in the feature set of high-risk depression groups. White bars indicate the frequencies of POS tags of words in the feature set of healthy groups.

In general, the two groups pay different attention to AOI in different parts of speech. As shown in Fig. 3. (b), both groups pay more attention to nouns, adverbs, and adjectives than to pronouns, although the pronouns appear more frequently than the nouns, the adverbs, and the adjectives in the S-hr-DS (See Fig. 3. (a)). We can also see from Fig. 3. (b) that both groups pay more attention to numbers than to auxiliary words, prepositions, and conjunctions, although Fig. 3. (a) demonstrates that the numbers appear less frequently than the auxiliary words, prepositions and conjunctions in the S-hr-DS. These observations may lend the psychologists a way for improving the S-hr-DS by including more nouns, adjectives, and numbers in it. In addition, it can be observed from Fig. 3. (b) that high-risk depression groups pay more attention to words of each POS tag than the healthy groups do. This population difference is particularly obvious for verbs and adjectives. Such a difference motivates us to develop the classifier based on these eye-tracking features.

In order to demonstrate the robustness of our algorithm 10-fold cross-validation [15] procedure is implemented during the process of feature selection and model training. In this process, the data set is randomly divided into 10 copies for 10 round feature selection and model training. During each round, one copy is randomly selected as the test set, and the remaining data set is used as the training set. This cross-validation procedure results in 10 different classifiers and 10 different subsets of features.

The performance of the classifiers trained in one round of the 10-fold cross-validation procedure in terms of the accuracy under different numbers of feature components is illustrated in Fig. 4. From the curve in this figure, we can see that different numbers of feature components involved in the classification result in different accuracies. We achieve an optimal accuracy of 85% when an optimal subset of 1000 feature components is selected out from the total 10715 feature components. The accuracy saturates at about 71% when more than 2000 components are selected. This is mainly due to the existence of redundant and related features in the feature set. From these observations, we can conclude that upon the proper selection of a subset of feature components, the performance of the corresponding classifier can be significantly improved.

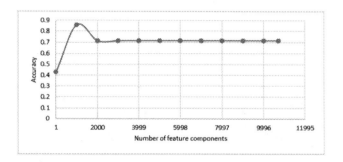

Fig. 4. Performance of the classifiers trained in one round of the 10-fold cross validation procedure in terms of the accuracy under different numbers of feature components

Table 2. Experimental results before & after feature selection.

Feature selection	Accuracy CI(95%)	Sensitivity CI(95%)	F1-score CI(95%)	p-value
Before	0.59	0.54	0.54	
After	0.81	0.76	0.79	$p < 0.05$

The results of statistical assessment of our proposed algorithm in the aforementioned data set in terms of the sensitivity, the F1-score, and the accuracy are shown in Table 2 for cases with and without feature selection procedure. From this table we can see that upon performing the feature selection procedure the performance of the trained model in terms of all the above three parameters is significantly improved.

Algorithm 1: Feature selection and model training algorithm

Input: Pool of features for all subjects $P_{\mathbf{x}} = \{\mathbf{X}^{(1)}, \mathbf{X}^{(2)}, \ldots, \mathbf{X}^{(n_s)}\}$, pool of class labels $P_{lb} \doteq \{y_1, y_2, \ldots, y_{n_s}\}$, the number n_s of subjects, the number L for components of features in the pool.

Output: The accuracy Acc for the optimal classifier.

Initialize: Feature rank list $r = [\ ]$, feature index $I = [1, 2, 3 \ldots L]$, accuracy set of the classifier $R = [\ ]$, the accuracy for the classifier $acc = 0$, the accuracy for the optical classifier $Acc = 0$;

Split: split $P_{\mathbf{x}}$ into training set $X_{tr} = \left[\mathbf{X}^{(1)}, \mathbf{X}^{(2)}, \ldots, \mathbf{X}^{(n_{tr})}\right]^T$ and test set $X_{te} = \left[\mathbf{X}^{(1)}, \mathbf{X}^{(2)}, \ldots, \mathbf{X}^{(n_{te})}\right]^T$, split P_{lb} into class labels for samples in the training set $Y_{tr} = [y_1, y_2, \ldots, y_{n_{tr}}]^T$ and class labels for samples in the test set $Y_{te} = [y_1, y_2, \ldots, y_{n_{te}}]^T$, where n_{tr} is the number of samples in the training set, n_{te} is the number of samples in the test set;

Begin

while $(I \neq \phi)$ **do**

    Train the SVM classifier in the training set with the selected features;

    Perform classification and compute the accuracy acc with the test set ;

    $R \leftarrow [acc, R]$;

    Compute the weighting vector w from Eq.(1);

    Compute the ranking criterion $c_i = (w_i)^2$ for all feature components;

    Find the feature component: $\hat{i} = \arg\min_{i \in I} c_i$;

    $r \leftarrow [\hat{i}, \quad r], I \leftarrow I - \{\hat{i}\}$;

    Set $X_{tr} = X_{tr}(:, I), X_{te} = X_{te}(:, I)$;

end

set $Acc = \max(R)$;

Output: Acc;

End

---

5   Conclusions and Future Work

In this paper, a novel method has been proposed based on the processing of the eye-tracking data recoded when answering S-hr-DS to identify high-risk depression groups. In the method, the eye-tracking features are extracted by a novel way of word-fixation driven feature encoding that the time the gaze stays in different AOIs can be considered simultaneously. The support vector machine recursive feature elimination (SVM-RFE) algorithm is employed for feature selection and model train-ing. We achieve an accuracy of 81% with 76% in sensitivity and 79% in F1-score on a data set including 69 subjects, which demonstrates a potential application of this algorithm in detecting high-risk depression groups.

However, there exist some drawbacks in our method. Firstly, the performance of the trained model is quite different from run to run of 10-fold cross-validation. This may be due to the loss of diversity in each of the 10 partitions of the data set because our data set only contains 69 subjects in total. Therefore, one of our future work will be to acquire more data for both model training and performance evaluations. Secondly, other data sources, including EEG data and the active unit (AU) data of the faces, will be considered in our future work to further improve the performance of the model. Finally, iterative innovation of the S-hr-DS will be performed in the future according to the experimental results of this work.

Acknowledgement. The work reported in this paper was supported by the Awareness and Cognitive Neuroscience based Multimodal Co-sensing Technology Project (ACNMCT) in charge of professor Danmin Miao of Department of military medical psychology, Air Force Medial University, the Natural Science Foundation of China No. 61473243, and the Natural Science Foundation of Jiangsu Province No. BK20171249. The authors thank the anonymous reviewers for their constructive comments and valuable suggestions.

References

1. Armitage, R., Hoffmann, R.F.: Sleep EEG, depression and gender. Sleep Med. Rev. **5**(3), 237–246 (2001)
2. Association, A.P., et al.: DSM-5. diagnostic and statistical manual of mental disorders. Washington, DC: Author (2013)
3. Bartzokis, G., et al.: In vivo evaluation of brain iron in alzheimer disease using magnetic resonance imaging. Arch. Gen. Psychiatry **57**(1), 47–53 (2000)
4. Beck, A.T., Steer, R.A., Brown, G.K., et al.: Beck depression inventory-ii. San Antonio, vol. 78, No. 2, pp. 490–498 (1996)
5. Bianchi, R., Laurent, E.: Emotional information processing in depression and burnout: an eye-tracking study. Eur. Arch. Psychiatry Clin. Neurosci. **265**(1), 27–34 (2015)
6. Ding, X., Yue, X., Zheng, R., Bi, C., Li, D., Yao, G.: Classifying major depression patients and healthy controls using EEG, eye tracking and galvanic skin response data. J. Affect. Disord. **251**, 156–161 (2019)
7. Fan, D.P., Wang, W., Cheng, M.M., Shen, J.: Shifting more attention to video salient object detection. In: Proceedings of the IEEE Conference on Computer Vision and Pattern Recognition, pp. 8554–8564 (2019)
8. Gotlib, I.H.: Eeg alpha asymmetry, depression, and cognitive functioning. Cognition Emotion **12**(3), 449–478 (1998)
9. Guyon, I., Weston, J., Barnhill, S., Vapnik, V.: Gene selection for cancer classification using support vector machines. Mach. Learn. **46**(1–3), 389–422 (2002)
10. Han, D., Kim, J.: Unified simultaneous clustering and feature selection for unlabeled and labeled data. IEEE Trans. Neural Netw. Learn. Syst. **29**(12), 6083–6098 (2018)
11. Harikrishna, S., Farquad, M., et al.: Credit scoring using support vector machine: a comparative analysis. In: Advanced Materials Research, vol. 433, pp. 6527–6533. Trans Tech Publ (2012)

12. Harris, M.S., Reilly, J.L., Thase, M.E., Keshavan, M.S., Sweeney, J.A.: Response suppression deficits in treatment-naive first-episode patients with schizophrenia, psychotic bipolar disorder and psychotic major depression. Psychiatry Res. **170**(2–3), 150–156 (2009)

13. Kellough, J.L., Beevers, C.G., Ellis, A.J., Wells, T.T.: Time course of selective attention in clinically depressed young adults: an eye tracking study. Behav. Res. Ther. **46**(11), 1238–1243 (2008)

14. Koster, E.H., De Raedt, R., Goeleven, E., Franck, E., Crombez, G.: Mood-congruent attentional bias in dysphoria: maintained attention to and impaired disengagement from negative information. Emotion **5**(4), 446 (2005)

15. Ling, H., Qian, C., Kang, W., Liang, C., Chen, H.: Combination of support vector machine and k-fold cross validation to predict compressive strength of concrete in marine environment. Constr. Build. Mater. **206**, 355–363 (2019)

16. Liu, X., et al.: Eye movement pattern and mental retardation in depression. In: 2017 IEEE International Conference on Bioinformatics and Biomedicine (BIBM), pp. 970–974. IEEE (2017)

17. Ma, Y., et al.: Easysvm: a visual analysis approach for open-box support vector machines. Comput. Visual Media **3**(2), 161–175 (2017)

18. Mundra, P.A., Rajapakse, J.C.: SVM-RFE with MRMR filter for gene selection. IEEE Trans. Nanobiosci. **9**(1), 31–37 (2009)

19. Organization, W.H., et al.: Depression and other common mental disorders: global health estimates. World Health Organization, Technical report (2017)

20. Radloff, L.S.: The CES-D scale: a self-report depression scale for research in the general population. Appl. Psychol. Meas. **1**(3), 385–401 (1977)

21. Sahran, S., Albashish, D., Abdullah, A., Shukor, N.A., Pauzi, S.H.M.: Absolute cosine-based SVM-RFE feature selection method for prostate histopathological grading. Artif. Intell. Med. **87**, 78–90 (2018)

22. Trotzek, M., Koitka, S., Friedrich, C.M.: Utilizing neural networks and linguistic metadata for early detection of depression indications in text sequences. arXiv preprint arXiv:1804.07000 (2018)

23. Wang, W., Shen, J., Yang, R., Porikli, F.: Saliency-aware video object segmentation. IEEE Trans. Pattern Anal. Mach. Intell. **40**(1), 20–33 (2017)

24. Wulsin, L.R., Vaillant, G.E., Wells, V.E.: A systematic review of the mortality of depression. Psychosom. Med. **61**(1), 6–17 (1999)

25. Zeng, S., Niu, J., Zhu, J., Li, X.: A study on depression detection using eye tracking. In: Tang, Y., Zu, Q., Rodríguez García, J.G. (eds.) HCC 2018. LNCS, vol. 11354, pp. 516–523. Springer, Cham (2019). https://doi.org/10.1007/978-3-030-15127-0_52

26. Zung, W.W.: A self-rating depression scale. Arch. Gen. Psychiatry **12**(1), 63–70 (1965)

27. Zuroff, D.C., Mongrain, M., Santor, D.A.: Conceptualizing and measuring personality vulnerability to depression: comment on coyne and whiffen (1995) (2004)

Infrared Small Target Detection Based on Prior Constraint Network and Efficient Patch-Tensor Model

Chang Nie[1], Huan Wang[1(✉)], and Ali Lu[2]

[1] Nanjing University of Science and Technology, 200, Xiaolingwei, Nanjing, China
{changnie,wanghuanphd}@njust.edu.cn
[2] Laboratory of Multimedia, Department of Computer Engineering,
Nanjing Institute of Technology, Nanjing, China
luali@njit.edu.cn

Abstract. Infrared small target detection (ISTD) is a key technology in the field of infrared detection and has been widely used in infrared search and tracking systems. In this paper, a novel ISTD approach based on a prior constraint network (PCN) and an efficient patch-tensor model (EPT) is proposed. Firstly, the PCN trained by numerous synthetic image patches is employed to obtain the preliminary segmentation result of small targets, which is later used as a prior constraint. Then the EPT model deals with the target detection problem by solving an optimization problem of recovering low-rank and sparse tensor. Next, the prior constraint is further applied to the target component of the EPT model as a regularization. Finally, the joint PCN-EPT model can be solved efficiently by the Alternating Direction Multiplier Method, and the targets are obtained by applying a simple adaptive threshold segmentation to the obtained target component from the PCN-EPT. Experimental results on multiple real datasets show that the proposed model outperforms the state-of-the-art.

Keywords: Infrared small target detection · Prior constraint network · Robust principal component analysis · Low-rank and sparse decomposition.

1 Introduction

Infrared small and dim target detection (ISTD) are broadly applied in missile detection and guidance, search and rescue in the sea, industrial flaw inspection, wild animal protection. However, great challenges [6,10] exist in the ISTD task: (1) the size of a target is very small due to the long distance between the target and infrared sensors. (2) the signal-to-noise ratio of an infrared small target image is quite low owing to the weakness of the target signal, interference from the background, and sensor noises.

Student paper.

© Springer Nature Switzerland AG 2020
Y. Peng et al. (Eds.): PRCV 2020, LNCS 12306, pp. 504–517, 2020.
https://doi.org/10.1007/978-3-030-60639-8_42

In the literature, two lines of methods are developed to deal with the ISTD task: the sequence-based and single-frame based methods. The sequence-based ones use features such as continuity and smoothness of target motion for detection, which aims to make better use of spatial-temporal information and Representative methods include three-dimensional matching filter [20]. The sequence-based methods are rarely used in real-time applications due to high time-consumption.

On the contrary, The single-frame based methods extract targets from a single image. They are broadly used in real applications and can be roughly categorized into background filtering-based, human vision system (HVS)-inspired and low-rank and sparse decomposition (LSD)-based methods.

The background filtering-based methods use various filtering operations to suppress background and enhance target, say Top-Hat transformation [25], Max-median filtering [11], and two-dimensional adaptive filtering (TDLMS) [12], etc. These methods are efficient and perform well in simple scene backgrounds, but they often fail under complex backgrounds.

Numerous ISTD methods [5,9,15,24] are inspired by the HVS. The Local contrast measure (LCM) [5] uses a ground-effect local contrast calculation method that can effectively enhance the target while suppressing background clutter. Improved local contrast measure (ILCM) [15] is proposed to alleviate the problem of noise enhancement in LCM. The multi-scale patch-based contrast measure (MPCM) [15] and weighted local difference measure (WLDM) [9] are proposed which can further improve the detection performance of small targets. These methods obtain the detection results only based on the gray relationship in a local area, neglecting using the global context information.

The LSD-based methods exploit the non-local self-correlation property of the infrared background image and the sparsity property of small targets. The background and targets are then regarded as sparse components and low-rank components, respectively. Therefore, the ISTD is transformed into the problem of detaching the low-rank and sparse components from a data matrix. The most representative work of LSD is the infrared patch-image (IPI) model [14], in which the data matrix is constructed by vectorizing all image patches of an infrared image and the low-rank and sparse decomposition problem is viewed as a robust principal component analysis (RPCA) [4] problem. The IPI model uses the l_1 norm to measure the sparseness of the target, which could produce some background residuals in the target image. To deal with this issue, the weighted IPI model [8] is proposed to suppress strong edges and noise. The subsequent weighted infrared patch-tensor model (RIPT) [7] tries to use both local and non-local prior information. Sun et al. proposed an improved re-weighted infrared patch-tensor model based on the weighted tensor nuclear norm (WNRIPT) [22], which leverages the advantage of the continuity between consecutive frames. Besides, the total variation regularization is employed to regularize the background component so that the smoothness of the background component can be enhanced. Wang et al. proposed a method based on total variation regularization and principal component pursuit (TV-PCP) [23]. The reference [21] proposed a method via spatial-temporal total variation regularization and weighted tensor nuclear norm. Although these detection methods have good performance in complex scenes, they are lack of efficiency in solving respective optimization problems.

In recent years, deep neural networks have developed rapidly. Due to excellent feature extraction capabilities, they have made great achievements in the fields of general object detection and image segmentation. Convolutional neural network (CNN) for general object detection does not perform well in the detection of dim and small targets, thereby some scholars propose concrete deep methods for ISTD. Fan et al. [13] used modified convolutional neural networks to enhance the contrast of small targets in infrared images. Lin et al. [18] design a seven-layer fully convolutional network to detect small targets and supervised training on the synthetic data generated by oversampling. Zhao et al. [26] regard the target detection problem as a semantic segmentation problem. They use the target extraction module to segment the target and then constrain the segmentation result by using a semantic constraint module. The main bottlenecks of deep methods for ISTD lie in their relying more on a large number of training samples and limited generalization ability to unseen scenes.

To improve the detection accuracy and efficiency of infrared small targets in diverse backgrounds, we propose a new infrared small target method based on prior constraint network and efficient infrared patch-tensor model. This method can combine the advantages of deep neural networks and LSD to achieve better performance in detection accuracy and efficiency. The contributions of this paper are fourfold:

1) A fully convolutional neural network called PCN for ISTD is proposed. It is trained on totally synthetic image patches.
2) We propose a block-wise detection scheme and construct an efficient infrared patch-tensor model (EPT) for the ISTD task.
3) In further, we combine the PCN and EPT to form the PCN-EPT model, where the detection result from PCN is used as a prior constraint to regularize the target component in the EPT. A modified alternating direction multiplier method (ADMM) is designed to solve the PCN-EPT model.
4) Experiments on a single-frame image dataset and five real image sequences show that our proposed method can significantly improve the detection performance and outperform the compared methods.

2   Prior Constraint Network (PCN)

The architecture of PCN is shown in Fig. 1. Note that the color arrows show different layers and the blocks represent the feature maps. In Fig. 1, the lower part is an encode-decode structure, which includes three down-sampling layers and three up-sampling layers. The upper part of the PCN firstly uses bicubic interpolation to increase the resolution of the image. Then it is followed by a residual block and a max-pooling layer, which can increase the detailed characteristics and better distinguish small targets from the background. Each down-sampling process in the encoder process contains a residual block (Upper right dashed block), which has two convolutional layers and each of them is followed by a batch normalization layer and an activation layer. We use a skip-layer to add the feature maps from the first convolutional layer to the final output. Such residual connection can enhance the learning effect of the network and accelerate the network convergence [16]. The network decoding process is completed by the up-sample block (Lower right dashed block) and the intermediate skip connections.

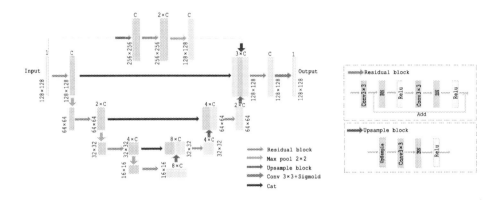

Fig. 1. The architecture of PCN.

In Fig. 1, the sizes are shown around each feature maps, where C denotes the number of channels and we set $C = 8$. The PCN outputs the segmentation image f_P indicating the targets existing in the input image $f_D \in R^{m \times n}$. The PCN model performs supervised learning on a large number of infrared synthetic image patches of size 128×128. The generation procedure of the synthetic image patches will be introduced in the experiment section. The loss function of PCN consists of the target extraction loss and the sparse loss, which are defined as:

$$L_{PCN} = \frac{1}{N} \sum_{i=1}^{N} ||f_{P_i} - f_{T_i}||_F^2 + \lambda ||f_{P_i}||_1, \tag{1}$$

where f_T is the ground truth of infrared image f_D, N is the number of training samples, and the λ is used to trade-off the relationship between the two loss items. The target extraction loss uses the Frobenius norm to measure the difference between the PCN output of f_P and f_D. The target sparse loss is l_1 norm of f_P, which is used to suppress potent noise and reduce the false alarm rate.

Through PCN's learning from numerous training samples, we can obtain a preliminary infrared small detection result f_P with respect to the infrared image f_D. However, this result may not sufficiently capture the background and target information in the real image. In order to achieve high detection performance in an unseen environment, we should utilize more information from the test image. To this end, we consider taking f_P as a prior and building a joint model, which is capable of significantly utilizing target-related information both from a large number of training samples and from the test image.

3   Efficient Patch-Tensor Model

In this section, we will introduce a new patch-tensor model based on a block detection scheme and IPI model. It is efficient in extracting targets only from one test image, so we name it efficient patch-tensor model (EPT).

3.1   Block Detection

For an input infrared image f_D, we adopt a block-wise detection scheme to extract small targets. The motivation behind is that parallel dealing with all sub-images can be more efficient than processing the original image. According to a basic block size $B \times B$, we divide the original image f_D into k detection blocks $(f_{D_1}, f_{D_2}, ..., f_{D_k})$, and the overlap of any two adjacent detection blocks is fixed as 20 pixels. For an infrared image $f_D \in R^{m \times n}$, the number of detection blocks k is $k = \lceil \frac{m}{B} \rceil \times \lceil \frac{n}{B} \rceil$, where $[\cdot]$ is rounding operation. Therefore, using the detection block decomposition method, we formulate an infrared image model to the following form:

$$f_{D_i} = f_{B_i} + f_{T_i} + f_{N_i}, i = 1, 2, ..., k. \tag{2}$$

where f_{D_i}, f_{B_i}, f_{T_i} and f_{N_i} are the original sub-image, background sub-image, small target and noise sub-image of the i-th detection block. Decomposing the original infrared image into detection blocks can significantly improve the detection efficiency, and we will explain in detail in the experiment section.

3.2   Patch-Tensor Model with Block Detection

In fact, we can use tensor structure to rewrite the Eq. (2):

$$\mathcal{D} = \mathcal{B} + \mathcal{T} + \mathcal{N}, \tag{3}$$

where $\mathcal{D}, \mathcal{B}, \mathcal{T}, \mathcal{N} \in R^{M \times N \times k}$ and $\mathcal{D}_{(i)}$ is the infrared patch-image corresponding to the ith detection block f_{D_i}. Due to the lower proportion of small target occupying the pixel area to the infrared image, the corresponding target patch-tensor can be regarded as a sparse component, which is restricted by $||\mathcal{T}||_0 \leqslant l$. Here l is a scalar and determined by the number and size of targets. Generally, the infrared background image have the characteristics of low contrast and slightly blurred. The local and non-local patches in the infrared image have certain correlation properties, so we consider that the background patch-image corresponding to each detection block is low-rank, which can be described as: $||rank(\mathcal{B}_{(i)})||_0 \leqslant r_i, i = 1, 2, ..., k$, where r_i is constant and determined by the background complexity of the ith detection block f_{D_i}. In addition, since the obtained infrared image will be interfered by some noises it can be assumed that the noise is i.i.d. So noise patch-tensor $\mathcal{N}$ for some $\delta > 0$ satisfies the condition $||\mathcal{N}||_F^2 \leqslant \delta$, where the constant δ reflects the degree of image interference by noise. At this point, we can formulate the problem of detecting small targets from an image f_D into the following low-rank and sparse tensor factorization task:

$$\min_{\mathcal{B},\mathcal{T}} \sum_{i=1}^{k} ||\mathcal{B}_{(i)}||_* + \lambda||\mathcal{T}||_1, \quad s.t. \quad \mathcal{D} = \mathcal{B} + \mathcal{T}, \qquad (4)$$

where λ is the parameter that balances target patch-tensor and the background patch-tensor. There we use the tractable nuclear norm and l_1 norm to convex approximations of the rank function and l_0 norm, respectively. We can obtain the detection results of small targets in f_D by solving the problem (4).

4   PCN-EPT Model and Solution

4.1   PCN-EPT Model

To fully utilizing the target information from training samples and the test image, we combine PCN and EPT and obtain the PCN-EPT model. It uses the output of PCN as a prior to constrain the target component in Eq. (4). We first obtain the output f_P from the PCN for the input image f_D, and then we convert f_P to the corresponding patch-tensor $\mathcal{P}$. In order to further reduce the interference of strong edges and noise on the detection results, the target local structure weighted image f_W is also calculated [7]. Likewise, f_W is transformed to the weighted patch-tensor $\mathcal{W}_\mathcal{T}$. We add $\mathcal{P}$ and $\mathcal{W}_\mathcal{T}$ as the prior conditions to regularize the target component in (4). Therefore, we have:

$$\min_{\mathcal{B},\mathcal{T}} \sum_{i=1}^{k} ||\mathcal{B}_{(i)}||_* + \lambda_1||\mathcal{W}_\mathcal{T} \odot \mathcal{T}||_1 + \lambda_2||\mathcal{T} - \mathcal{P}||_F^2, \quad s.t. \quad \mathcal{D} = \mathcal{B} + \mathcal{T}, \qquad (5)$$

where $\odot$ is the notation of Hadamard product and λ_1, λ_2 are the weight parameter. The λ_2 represents the degree of constraint on the target component, which can reflect the reliability of PCN detection. The optimal solution of model (5) can be considered as a good fusion of the detection results of PCN and EPT models, which can incorporate the advantages of both.

4.2   Solution of the PCN-EPT Model

We use the ADMM algorithm [2] to solve our model (5), which helps accelerate the convergence of our model. According to the ADMM, the model (5) can be converted to minimizing the following convex optimization problem (6).

$$
\mathcal{L}(\mathcal{B}, \mathcal{T}, \mathcal{Y}, \beta) = \sum_{i=1}^{k} ||\mathcal{B}_{(i)}||_* + \lambda_1 ||\mathcal{W}_{\mathcal{T}} \odot \mathcal{T}||_1 + \lambda_2 ||\mathcal{T} - \mathcal{P}||_F^2
$$
$$
+ \frac{\beta}{2}||\mathcal{D} - \mathcal{B} - \mathcal{T}||_F^2 + < \mathcal{Y}, \mathcal{D} - \mathcal{B} - \mathcal{T} >, \tag{6}
$$

where $\mathcal{Y} \in R^{M \times N \times k}$ and β are the Lagrange multiplier tensors and positive penalty parameter, respectively. During each iterative process, one variable will individually be considered the decision variable and the corresponding sub-problem is solved while the others are fixed. This process are repeated several times until it meets the convergence condition. The related sub-problems are listed as follows:

1) The sub-problem of the variable $\mathcal{B}$ is:

$$
\min_{\mathcal{B}} \sum_{i=1}^{k} ||\mathcal{B}_{(i)}||_* + \frac{\beta}{2}||\mathcal{D} - \mathcal{B} - \mathcal{T}||_F^2 + < \mathcal{Y}, \mathcal{D} - \mathcal{B} - \mathcal{T} > . \tag{7}
$$

The above problem (7) can be solved by singular value thresholding (SVT) [3]:

$$
\mathcal{B}_{(i)} = SVT_{1/\beta}(\mathcal{D}_{(i)} - \mathcal{T}_{(i)} - \beta^{-1}\mathcal{Y}_{(i)}), i = 1, 2, .., k, \tag{8}
$$

where $SVT_\mu(\cdot)$ is the singular thresholding operator and defined as follows:

$$
SVT_\mu(X) = U\mathcal{S}_\mu(\sum)V^T, \quad where \quad \mathcal{S}_\mu(x) = max(|x| - \mu, 0)sign(x). \tag{9}
$$

Through singular value decomposition we have $X = U\sum V^T$, and extend $\mathcal{S}_\mu(\cdot)$ operation to tensor by applying it to every entry. The function $sign(\cdot)$ is a symbolic function.

2) The sub-problem of the variable $\mathcal{T}$ can be converted to:

$$
\min_{\mathcal{T}} \lambda_1||\mathcal{W}_{\mathcal{T}} \odot \mathcal{T}||_1 + \lambda_2||\mathcal{T} - \mathcal{P}||_F^2 + \frac{\beta}{2}||\mathcal{D} - \mathcal{B} - \mathcal{T}||_F^2 + < \mathcal{Y}, \mathcal{D} - \mathcal{B} - \mathcal{T} >, \tag{10}
$$

which can be solved by soft thresholding operator [1]:

$$
\mathcal{T} = \mathcal{S}_{\frac{\lambda_1}{2\lambda_2 + \beta}\mathcal{W}_{\mathcal{T}}}(\frac{\beta(\mathcal{D} - \mathcal{B}) - \mathcal{Y} + 2\lambda_2\mathcal{P}}{2\lambda_2 + \beta}). \tag{11}
$$

3) For Lagrange multiplier $\mathcal{Y}$ and scale β, the update operation is:

$$
\mathcal{Y} = \mathcal{Y} + \beta(\mathcal{D} - \mathcal{B} - \mathcal{T}), \quad where \quad \beta = \min(\rho\beta, \beta_{max}). \tag{12}
$$

4.3   Overall Detection Procedure

Taking an infrared image f_D as input, the pipeline of the PCN-EPT model is shown in Fig. 2. The main steps are also listed as follows:

1) The PCN trained on numerous synthetic dataset is used to obtain prior detection results f_P, then calculate the target local structure weighted image f_W according to [7].
2) Apply EPT model to image f_D, f_P, f_W and get the corresponding infrared patch-tensor $\mathcal{D}, \mathcal{P}, \mathcal{W}$, respectively. Then we can have the corresponding background patch-tensor $\mathcal{B}$, target patch-tensor $\mathcal{T}$ and noise patch-tensor $\mathcal{N}$.
3) Reconstruct background patch-tensor $\mathcal{B}$ and target patch-tensor $\mathcal{T}$ into corresponding background image f_B and target images f_T.
4) Finally, we use adaptive threshold [14] to extract the target.

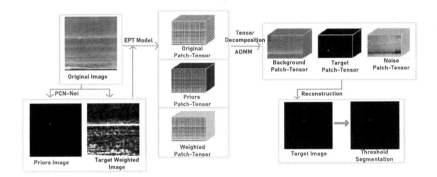

Fig. 2. Infrared small target detection procedure based on PCN-EPT model.

4.4   Complexity Analysis

The computational complexity of the PCN-EPT model mainly consists of two parts, namely the PCN and the EPT model. The complexity of the PCN can be measured by the parameter storage (M_{PCN}) and the number of multiply-add operations (OPs). We only considered convolutional layers for PCN OPs, because the proportion of other layers such as up-sampling layers and activation layers is negligible. During the experiment, we set the parameter $c = 8$, $\lambda = 0.2$, so $M_{PCN} = 1.3 \times 10^5$, $OPs = 2.9 \times 10^8$. For a single infrared image of size 256×256, it can be processed on the device GeForce RTX 2080 Ti in real-time.

Now we analyze the computational complexity of the EPT model. Assuming that the size of the input infrared image is $m \times n$, the IPI model is directly constructed by the patch of size $p \times p$ to obtain a patch-image of size $M \times N$, where $M = p^2, N \sim mn$ and we assume $M > N$. The total cost of each iteration of IPI model is $\mathcal{O}(MN^2)$ operations and our proposed EPT model can be seen as a combination of constructing the IPI model on each detection block. The number of detection blocks is $k = mn/B^2$ when basic block size B be divided with no remainder by m and n. So the complexity of the PET model is $\mathcal{O}(\frac{mn}{B^2} M (\frac{NB^2}{mn})^2)$, which can be simplified to $\mathcal{O}(\frac{MN^2}{k})$. Since the number of iterations of matrix decomposition is related to its size, the actual detection takes less time.

5   Experimental Results

This section presents the experiments on several real datasets to demonstrate the performance of our PCN-EPT model. We first introduce the experimental settings, including evaluation metrics, datasets, and baseline methods used for comparison. Then we discuss the impacts of the key parameters of our model. Finally, the proposed method is tested on the datasets and compared with the baseline methods.

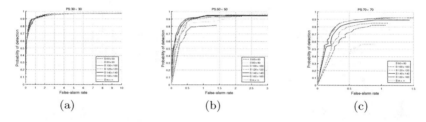

Fig. 3. ROC curves for different size of blocks and patches on Single-frame dataset.

5.1   Metrics, Datasets and Baseline Methods

Metrics. To evaluate our model, we apply the detection probability P_d and false alarm rate F_a [14] and the ROC curve over the two, which are popularly used in the ISTD literature. The detection result obtained after the adaptive threshold is considered to be a target if it meets: there are overlapping pixels between the detection and a real target and their center distance is less than 4 pixels.

Datasets. Our experimental datasets consists of a single-frame dataset containing 100 test images (Single dataset for short) and five real infrared image sequences. All images in Single are obtained from Google Image Search. The real infrared image sequences are captured by infrared cameras. The details of the experimental datasets are shown in the supplementary materials.

The training samples for PCN are obtained by randomly embedding a target T patch of size $a \times b$ into a infrared background image f_B as follows:

$$f_D(x,y) = \begin{cases} max(rT(x - x_0, y - y_0), f_B(x,y)), & x \in (1 + x_0, a + x_0), y \in (1 + y_0, b + y_0), \\ f_B(x,y), & otherwise, \end{cases}$$

(13)

where (x_0, y_0) is randomly generated position of the left upper corner of the target in the background image. The parameter r is the grayscale change factor of the target image, which is a random integer within the range $[h, 255]$, and h is the maximum value of the background image f_B. Each synthetic image randomly contains one to three targets, and the size of the target is changed by bicubic interpolation. The synthetic dataset obtained by the above method can simulate the real environment well [14].

Methods for Comparison. We compared the proposed method with six baseline detection methods, including Top-Hat [25], Max-Median [11], LCM [5], DPS-MGRG [17], FKRW [19], and IPI method [14]. The detailed parameter settings of these detection methods are given in the supplementary materials.

5.2   Parameter Setting and Analysis

We first analyze the effect of basic blocks size B and patch size PS on the detection probability (P_d), false alarm rate (F_a), and detection efficiency on the synthetic dataset. We let the size of the basic block vary between $(60 \times 60,\ 80 \times 80,\ 100 \times 100,\ 120 \times 120,\ 140 \times 140,\ 160 \times 160,\ m \times n)$ and the corresponding patch sizes are $(20 \times 20,\ 30 \times 30,\ 40 \times 40,\ 50 \times 50,\ 60 \times 60,\ 70 \times 70)$. The average detection time of a single infrared image on the synthetic dataset is shown in Table 1, and the corresponding receiver operating characteristic (ROC) curves of simulation experiments is shown in Fig. 3. Due to the patch size PS must be smaller than the basic blocks size B, then part of the data in Table 1 is empty.

Table 1. Average processing time (s) of PCN-EPT model over the Single dataset.

PS	B						
	60×60	80×80	100×100	120×120	140×140	160×160	$m \times n$
20×20	**0.1343**	0.1458	0.1543	0.1718	0.1760	0.1818	0.1884
30×30	**0.1488**	0.1710	0.1675	0.1895	0.2025	0.2154	0.2896
40×40	0.1738	0.1803	**0.1645**	0.2026	0.2488	0.2796	0.4538
50×50	**0.1467**	0.1823	0.1931	0.2086	0.2630	0.3012	0.5876
60×60	*	**0.1758**	0.2210	0.2472	0.3073	0.3420	0.6144
70×70	*	**0.1497**	0.1891	0.2508	0.3317	0.3508	0.6981

It can be seen from Table 1 that the average detection time of a single infrared image will increase with the lager of the basic block. The detection time reaches the maximum when the basic block and original image as the same size. When the size of B is fixed, the average detection time increases first and then decreases as PS becomes larger, and the false alarm rate (F_a) and detection probability (P_d) are decreasing (Fig. 3). Whenever PS is increased by 10, the F_a will decrease by an average of one third, and the small basic block detection probability drop is more obvious. At the same time, when the sizes of B and PS are close, the P_d and detection time will be drop significantly (Fig. 3(c)), but smaller PS can adapt to all size detection block (Fig. 3(a)). The main reason is the aspect ratio of the IPI model constructed by a single detection block is seriously unbalanced, which will cause the lower detection probability. Our experiment found when B meeting condition of $B > 2PS$, then our proposed block detection method will not affect the detection accuracy, and smaller B can improve the detection efficiency. Through the above analysis, in order to trade off the detection efficiency, P_d and F_a, we choose basic block size B is 80×80 and the patch size PS is 30×30.

The parameter sliding step will also affect detection performance. An excessively large sliding step will usually reduce the self-correlation of the patch-image, but too small a sliding step decreases the detection efficiency, we choose the sliding step to 10 according to [14]. We set the parameter $\lambda_1 = 1/\sqrt{max(m,n)}$. The parameter $\lambda_2 = \frac{\gamma}{1-\gamma}$ can be regarded as a weighted combination of the detection results of the PCN and EPT models, where the weight coefficients of the PCN and EPT models are γ and $1 - \gamma$, respectively. When λ_2 is infinity or zero, the PCN-EPT model will degenerate into a single PCN or EPT model. Here we set $\lambda_2 = 0.5$.

5.3   Performance Comparison

The detection results of six representative infrared images from the experimental datasets are shown in Fig. 4. It can be seen that the dim and small targets contained in the image are successfully detected. For the small objects which are obviously

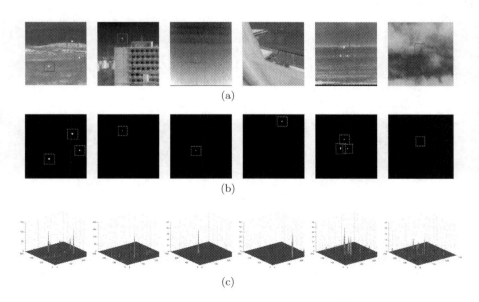

Fig. 4. The detection results of the 6 representative infrared images of with PCN-EPT model. (a) Original image. (b) Detection results. (c) Surface map of target image.

Table 2. The average calculation time of the test methods on the experimental datasets.

Datasets	Top-Hat	Max-Median	LCM	DPS-MGRG	FKRW	IPI	PCN-EPT
Single dataset	0.0108	0.0244	15.0527	0.7333	0.3289	19.3785	0.1710
Sequence 1	0.0145	0.0213	28.2869	1.5713	0.6533	38.9210	0.2673
Sequence 2	0.0149	0.0211	25.4815	1.3957	0.6378	25.8114	0.2844
Sequence 3	0.0144	0.0137	28.5842	0.9491	0.4115	11.4445	0.1768
Sequence 4	0.0154	0.0183	42.7432	1.1721	0.1676	30.5828	0.2601
Sequence 5	0.0181	0.0448	101.5943	2.7116	1.2316	128.8257	0.6963

relative to the surrounding area in the general scene Fig. 4(1,3,5), our model can easily localize the targets by our proposed method. When the low contrast and dim target is located in the complex scene Fig. 4(2,4) or submerged by cloud and noise Fig. 4(6), the PCN-EPT model still has a good detection performance.

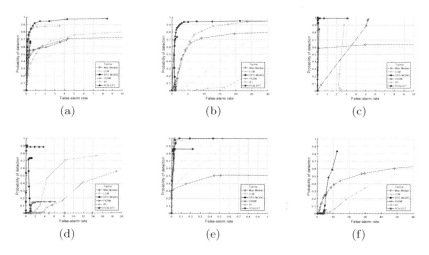

Fig. 5. ROC curves of detection results on experimental datasets: (a) Single-frame dataset; (b) sequence 1; (c) sequence 2; (c) sequence 3; (d) sequence 4; (e) sequence 5; (f) sequence 6.

The ROC curves of PCN-EPT model and six methods for comparison over the datasets are shown in Fig. 5. It can be seen that our proposed method has the best detection performance on the Single dataset and all real image sequences. An obvious advantage of the PCN-EPT model is that it can achieve a high detection probability at a low false alarm rate, rather than other methods (such as IPI and LCM) growing slowly with the increase of false alarm rate. This shows that PCN-EPT can better enhance the target and suppress background. Besides, the PCN-EPT model can also have a good detection performance on 5 real image sequences, while the detection probability of the others methods appears too low (Fig. 5(d)). This is mainly due to the fact that these methods cannot adapt to various complex scenarios. However, our proposed method can correctly and robustly detect small targets thanks to the combination of both PCN and EPT.

The computational time of different methods on the experimental datasets is shown in Table 2. The Top-Hat and Max-median have high detection efficiency but lower detection probability. The IPI model and LCM outperform Top-Hat and Max-median but their time-cost increase too much. FKRW and DPS-MGRG are fast than IPI and LCM. However, they are less efficient than our PCN-EPT model, which is especially efficient for large-scale infrared images.

6 Conclusion

In this paper, a PCN-EPT model is proposed to improve the detection accuracy and efficiency of ISTD, Firstly, a prior constraint network is designed to learn the preliminary segmentation result of small targets from numerous synthetic data. Secondly, a patch-tensor model is proposed to improve the efficiency of single-frame detection method of LSD. The output of PCN is used as a prior constraint to the EPT model to take the advantages of both PCN and EPT. Experiments demonstrate the effectiveness and efficiency of the proposed method. In the future, we will further improve the model by exploiting spatial-temporal context information.

Acknowledgement. This paper is supported by National Science Foundation of China (Grant No. 61703209 and 61773215).

References

1. Beck, A., Teboulle, M.: A fast iterative shrinkage-thresholding algorithm for linear inverse problems. SIAM J. Imaging Sci. **2**(1), 183–202 (2009)
2. Boyd, S., Parikh, N., Chu, E., Peleato, B., Eckstein, J.: Distributed Optimization and Statistical Learning via the Alternating Direction Method of Multipliers. Now Publishers Inc., Zuid-Holland (2010). https://doi.org/10.1561/2200000016
3. Cai, J.F., Candès, E.J., Shen, Z.: A singular value thresholding algorithm for matrix completion. SIAM J. Optim. **20**(4), 1956–1982 (2010)
4. Candès, E.J., Li, X., Ma, Y., Wright, J.: Robust principal component analysis? J. ACM (JACM) **58**(3), 1–37 (2011)
5. Chen, C.P., Li, H., Wei, Y., Xia, T., Tang, Y.Y.: A local contrast method for small infrared target detection. IEEE Trans. Geosci. Remote Sens. **52**(1), 574–581 (2013)
6. Chen, Y., Xin, Y.: An efficient infrared small target detection method based on visual contrast mechanism. IEEE Geosci. Remote Sens. Lett. **13**(7), 962–966 (2016)
7. Dai, Y., Wu, Y.: Reweighted infrared patch-tensor model with both nonlocal and local priors for single-frame small target detection. IEEE J. Sel. Top. App. Earth Obs. Remote Sens. **10**(8), 3752–3767 (2017)
8. Dai, Y., Wu, Y., Song, Y.: Infrared small target and background separation via column-wise weighted robust principal component analysis. Infrared Phys. Technol. **77**, 421–430 (2016)

9. Deng, H., Sun, X., Liu, M., Ye, C., Zhou, X.: Small infrared target detection based on weighted local difference measure. IEEE Trans. Geosci. Remote Sens. **54**(7), 4204–4214 (2016)
10. Deng, H., Sun, X., Liu, M., Ye, C., Zhou, X.: Entropy-based window selection for detecting dim and small infrared targets. Pattern Recogn. **61**, 66–77 (2017)
11. Deshpande, S.D., Er, M.H., Venkateswarlu, R., Chan, P.: Max-mean and max-median filters for detection of small targets. In: Signal and Data Processing of Small Targets, vol. 3809, pp. 74–83. International Society for Optics and Photonics (1999)
12. Fan, H., Wen, C.: Two-dimensional adaptive filtering based on projection algorithm. IEEE Trans. Sig. Proc. **52**(3), 832–838 (2004)
13. Fan, Z., Bi, D., Xiong, L., Ma, S., He, L., Ding, W.: Dim infrared image enhancement based on convolutional neural network. Neurocomputing **272**, 396–404 (2018)
14. Gao, C., Meng, D., Yang, Y., Wang, Y., Zhou, X., Hauptmann, A.G.: Infrared patch-image model for small target detection in a single image. IEEE Trans. Image Proc. **22**(12), 4996–5009 (2013)
15. Han, J., Ma, Y., Zhou, B., Fan, F., Liang, K., Fang, Y.: A robust infrared small target detection algorithm based on human visual system. IEEE Geosci. Remote Sens. Lett. **11**(12), 2168–2172 (2014)
16. He, K., Zhang, X., Ren, S., Sun, J.: Deep residual learning for image recognition. In: Proceedings of the IEEE conference on computer vision and pattern recognition, pp. 770–778 (2016)
17. Huang, S., Peng, Z., Wang, Z., Wang, X., Li, M.: Infrared small target detection by density peaks searching and maximum-gray region growing. IEEE Geosci. Remote Sens. Lett. **16**(12), 1919–1923 (2019)
18. Liangkui, L., Shaoyou, W., Zhongxing, T.: Using deep learning to detect small targets in infrared oversampling images. J. Syst. Eng. Electron. **29**(5), 947–952 (2018)
19. Qin, Y., Bruzzone, L., Gao, C., Li, B.: Infrared small target detection based on facet kernel and random walker. IEEE Trans. Geosci. Remote Sens. **57**(9), 7104–7118 (2019)
20. Reed, I.S., Gagliardi, R.M., Stotts, L.B.: Optical moving target detection with 3-D matched filtering. IEEE Trans. Aerosp. Electron. Syst. **24**(4), 327–336 (1988)
21. Sun, Y., Yang, J., Long, Y., An, W.: Infrared small target detection via spatial-temporal total variation regularization and weighted tensor nuclear norm. IEEE Access **7**, 56667–56682 (2019)
22. Sun, Y., Yang, J., Long, Y., Shang, Z., An, W.: Infrared patch-tensor model with weighted tensor nuclear norm for small target detection in a single frame. IEEE Access **6**, 76140–76152 (2018)
23. Wang, X., Peng, Z., Kong, D., Zhang, P., He, Y.: Infrared dim target detection based on total variation regularization and principal component pursuit. Image Vis. Comput. **63**, 1–9 (2017)
24. Wei, Y., You, X., Li, H.: Multiscale patch-based contrast measure for small infrared target detection. Pattern Recogn. **58**, 216–226 (2016)
25. Zeng, M., Li, J., Peng, Z.: The design of top-hat morphological filter and application to infrared target detection. Infrared Phys. Technol. **48**(1), 67–76 (2006)
26. Zhao, M., Cheng, L., Yang, X., Feng, P., Liu, L., Wu, N.: TBC-Net: A real-time detector for infrared small target detection using semantic constraint (2019). arXiv preprint arXiv:2001.05852

A Novel Unsupervised Hashing Method for Image Retrieval Based on K-Reciprocal Nearest Neighbors

Jinping Wang and Zhichao Lian[✉]

Nanjing University of Science and Technology, Nanjing 210094, China
wjp@njust.edu.cn, lzcts@163.com

Abstract. Hashing methods play an important role in large-scale image retrieval. Unsupervised hashing is popular in practical applications because it does not require labels for supervised training. However, as one of the important steps in unsupervised hashing, construction of semantic relationships between data by k-nearest neighbors has limitations due to the existence of false neighbors among the first k neighbors. In this paper, we propose a novel unsupervised deep hashing method for image retrieval. We firstly construct a semantic similarity matrix which utilizes deep features and the expanded k-reciprocal nearest neighbors to guide the learning of hash codes. After that, we design a deep neural network to preserve the structural information of original images. In addition, a weighted pairwise loss function generated by the positive pairs and negative pairs is employed to solve data imbalance problem. Extensive experiments on CIFAR-10, MIRFLICKR and NUS-WIDE datasets show that our method significantly outperforms the state-of-the-art unsupervised hashing methods.

Keywords: Unsupervised deep hashing · K-reciprocal nearest neighbors · Image retrieval

1 Introduction

With the rapid growth of image and video data on the internet, hashing-based methods for large-scale datasets have received more and more attention. Due to high computational efficiency and low storage burden, hashing-based methods have been applied to many fields [1–4].

Most data in real-world scenarios do not have semantic labels and it takes time to manually make labels for large-scale datasets. Therefore, learning in the unsupervised way is crucial, which is the focus of our work. In recent years, a number of works exploited the similarity relationships between images by the k-nearest neighbors [5, 6]. Deep Discrete Hashing (DDH) [7] constructed a pairwise similarity matrix by k-nearest neighbors and used the similarity matrix to guide the learning of hash codes. However, these methods exist some drawbacks. Firstly, they do not consider that there may be

---

The first author is a student.

© Springer Nature Switzerland AG 2020
Y. Peng et al. (Eds.): PRCV 2020, LNCS 12306, pp. 518–530, 2020.
https://doi.org/10.1007/978-3-030-60639-8_43

some non-neighbor items among the first k neighbors. For example, in Fig. 1, P1, P2, P3 and P4 are four true neighbors to the query image, but some of them are not included in the top-4 ranks such as P4. From the figure, we can also observe some false neighbors (N1−N6) receive high ranks. As a result, directly using the top-k ranked images will introduce noise and influence the final result seriously. Inspired by recent studies [19, 20], we construct the similarity matrix by expanded k-reciprocal nearest neighbors. In addition, the aforementioned methods do not consider the impact of data imbalance problem which is mentioned in the supervised hashing [8]. Data imbalance refers to the issue that the number of similar image pairs is much smaller than that of dissimilar image pairs in a batch of training data.

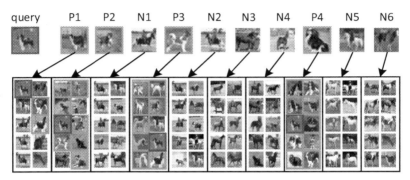

Fig. 1. Illustration of the nearest neighborhoods of a query image in CIFAR-10 dataset. Top: The query image and its 10-nearest neighbors, where P1-P4 are positives, N1-N6 are negatives. Bottom: Each two columns shows 10-nearest neighbors of the corresponding image. Red and orange box correspond to the query image and positives, respectively. (Color figure online)

In order to solve these problems, we propose a novel unsupervised hashing method based on k-reciprocal nearest neighbors and the contributions can be summarized as follows:

1) We employ the expanded k-reciprocal nearest neighbors to build our semantic similarity matrix from deep features, which improves the accuracy of similar matrix.
2) We propose a deep learning framework to learn discrete hashing codes in an unsupervised way to improve the effectiveness of hashing methods, in which a novel weighted loss function for pairwise hashing learning from imbalance data.
3) Extensive experimental results on three popular datasets show that the proposed method outperforms the state-of-the-art unsupervised hashing methods.

2   Related Work

2.1   Hashing Methods

Existing hashing methods include data-independent and data dependent. Locality sensitive hashing (LSH) [9] is a typical data-independent hashing method, the basic idea of LSH is that after two adjacent data points in the original data space are transformed by using random linear projections, the probability that the two data points are still adjacent in the new data space is large. Data-dependent hashing methods can generate more accurate hash codes, so they have attracted more attentions compared to data-independent hashing methods. These methods mainly include two categories: supervised and unsupervised hashing methods.

Supervised hashing methods usually make full use of supervised information such as semantic label to help them obtain superior retrieval performance. Supervised hashing with kernels (KSH) [10] uses the relationship between hamming distance and inner product to obtain an efficient objective function that minimizes the distance between similar pairs and maximizes the distance between dissimilar pairs. Deep hashing network (DHN) [11] minimizes both pairwise cross-entropy loss and pairwise quantization loss to guarantee pairwise similarity and improve the quality of hash codes. Deep learning to hash by continuation (HashNet) [8] improves DHN by balancing the positive and negative pairs in training data, and by continuation technique for lower quantization error. To accelerate the learning process, GAH [12] proposes a new deep hashing model integrated with a novel gradient attention mechanism.

In unsupervised hashing methods, Spectral hashing (SH) [13] provides a relaxed solution to the graph segmentation problem by analyzing the Laplacian matrix eigenvalues and eigenvectors of similar graphs. Density sensitive hashing (DSH) [14] extends LSH by exploring the geometry of the data and avoids the use of projections that are most consistent with the data distribution to randomly select projections. ITQ [15] utilizes an efficient alternating minimization strategy to find a rotation of zero-centered data to minimize the quantization error by mapping these data to the vertexes of a zero-centered binary hypercube. With the appearance of deep learning, deep hashing methods have also been proposed. Deep Hashing (DH) [16] develops a deep neural network to seek multiple hierarchical non-linear transformations to learn hash codes. DeepBit [17] uses original images and their corresponding rotated images as input, and minimizes the distance between their binary descriptors when updating network parameters. Deep discrete hashing (DDH) [7] constructs the similarity matrix which exploits the neighborhood structure through the images in a feature space to guide the training of network. Semantic structure-based unsupervised deep hashing (SSDH) [18] constructs the semantic structure by considering the distribution of data with distances that are estimated by two half Gaussian distributions and designs a pair-wise loss function to preserve the semantic information.

2.2 Stacked Denoising Autoencoder

Autoencoder [29] was proposed by Rumelhart in 1986 and can be used for high-dimensional data processing such as dimensionality reduction and feature learning. Denoising autoencoders can extract more robust characterization, where noise is added to the input data of the model, and then it is expected to recover the original image without noise. SAE [30] is composed of multiple layers of autoencoder. SAE initializes the parameters of the deep network by pretraining of layer-by-layer unsupervised learning. In this paper, we initial our network with a stacked denoising autoencoder (SDAE) [21] which can facilitate more robust hash codes.

3   Methodology

Firstly we introduce the notations $I = \{I_1, I_2, \ldots, I_n\}$ represents the training data of n images without labels, and $X = \{x_1, x_2, \ldots, x_n\} \in R^{d \times n}$ represents the global features of n samples, each of which uses a d-dimensional feature vector. Our goal is to learn their compact binary codes $B = \{b_1, b_2, \ldots, b_n\} \in \{-1, +1\}^{k \times n}$ which can preserve semantic similarities in the original feature space. The training procedure of the proposed framework is shown in Fig. 2. The whole framework has four components. In stage 1, a pretrained ResNet50 [22] network is used to exact 2048 dimensional features from each image. In stage 2, we construct a semantic similarity matrix as information source steering the process of hash learning. In stage 3, we initialize our deep neural network with a stacked denoising autoencoder network. In stage 4, we use all encoder layers of SDAE [21] to learn features and hash function simultaneously. In our method, a pairwise weighted cross-entropy loss preserves the pairwise similarity in deep features and a binary quantization loss function controls the quantization error of binarizing continuous real-values to binary codes.

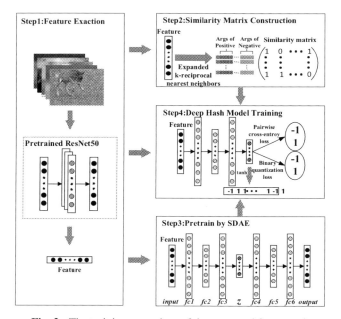

Fig. 2. The training procedure of the proposed framework

3.1   Construction of Semantic Similarity Matrix

To obtain neighborhood structure of the images for unlabeled datasets, we exact the 2048-dimensional features from the pretrained ResNet50 model [22]. Then we use cosine similarity of feature vectors to compare the relationship between image pairs and employ k-nearest neighbors (KNN) method to construct neighbors. For each image, we select k_1 images with highest cosine similarity as its neighbors and get an initial similarity matrix S_1 as:

$$(S_1)_{ij} = \begin{cases} 1, & \textit{if } x_j \textit{ is } k_1 \textit{ NN of } x_i \\ 0, & \textit{otherwise} \end{cases} \tag{1}$$

In fact, the top-k_1 results of the original ranking usually contain some negative samples. In order to find a better approach, we borrow the ideas from k-reciprocal nearest neighbors [19, 20]. The basic idea of k-reciprocal neighbors is that if an image u is one of the k nearest neighbors for an image v and if v is also one of the k nearest neighbors for u, then we can infer u and v are k-reciprocal nearest neighbors. Based on the above description, we generate the second similarity matrix S_2 as:

$$(S_2)_{ij} = \begin{cases} 1, & \textit{if } x_j \textit{ is } k_1 \textit{ NN of } x_i \textit{ and } x_i \textit{ is } k_1 \textit{ NN of } x_j \\ 0, & \textit{otherwise} \end{cases} \tag{2}$$

Here we use $\{G_1, G_2, \ldots, G_n\}$ to denote the ranking lists of points $\{x_1, x_2, \ldots, x_n\}$ by k_1-reciprocal nearest neighbors, G_i is defined as follows:

$$G_i = \{j | (S_2)_{ij} = 1\} \tag{3}$$

However, in practice some images have very few k-reciprocal nearest neighbors, even for a very large k. In order to tackle this problem, more nearest neighbor set G'_d of each candidate $d \in G_i$ is added to form a set R_i. Here, G'_d is defined as follows:

$$G'_d = \{z | (S_3)_{dz} = 1 \land d \in G_i\} \tag{4}$$

And S_3 is defined as follows:

$$(S_3)_{dz} = \begin{cases} 1, & \textit{if } x_z \textit{ is } \frac{1}{2}k_1 \textit{ NN of } x_d \textit{ and } x_d \textit{ is } \frac{1}{2}k_1 NN \textit{ of } x_z \\ 0, & \textit{otherwise} \end{cases} \tag{5}$$

$\frac{1}{2}k_1 NN$ means the first half of k_1 nearest neighbors. Finally, a more robust k_1-reciprocal nearest neighbor set R_i is defined according to following condition:

$$R_i \leftarrow G_i \cup G'_d \tag{6}$$

The above condition ensures that the added neighbors are very likely to be relevant to the original set G_i. Thus, we can obtain final similarity matrix S by:

$$(S)_{ij} = \begin{cases} 1, & \textit{if } j \in R_i \\ 0, & \textit{otherwise} \end{cases} \tag{7}$$

3.2 Architecture of Deep Neural Network

We initialize our deep neural network with a stacked denoising autoencoder [21]. And the aim of initialization is to enhance the discriminability between features. Here we set network dimensions to 2048-5000-3000-5000-k-5000-3000-5000-2048, where k is the length of hash layer z. After that we fine-tune the whole network by minimizing reconstruction loss. $z = f_w(x)$ and $x' = g_{w'}(z)$ respectively represent the encode function and the decode function, where w and w' is the weights of the network. In our method, the reconstruction loss is written as:

$$\mathcal{L} = \sum_{i=1}^{N} \left\| x_i - x_i' \right\|_2^2 \tag{8}$$

After pretraining, we discard the decoder layers and reserve the encoder layers as our deep hash model. We optimize the encoder layers by minimizing the weighted cross-entropy loss and quantization loss. In the hash layer, we use $tanh(\cdot)$ as activation function which aims to promote hash codes at 1 or -1. Eventually we obtain k-bit binary codes by the sign function.

3.3 Loss Function

In order to obtain more accurate hash codes, an appropriate loss function is important which reduce the difference between the current hash codes and the learning goal. As mentioned earlier, data imbalance problem is often ignored in most deep hashing learning training. For example, in CIFAR-10 dataset, supposing that the batch size is 100. In such a batch, the number of similar image pairs is $10 \times (10 \times 9 \div 2) = 450$, and the number of dissimilar image pairs is $100 \times 99 \div 2 - 450 = 4500$. Therefore, the proportion of positive pairs and negative pairs reaches 1: 10. However, too few positive pairs may leads to insufficient learning of positive pairs and reduce the overall retrieval performance. To perform deep hashing from imbalanced data, we jointly preserve similarity information of pairwise images and generate hash codes by maximum likelihood [28]. Given a pair of hash codes b_i and b_j, their hamming distance $dist_B(b_i, b_j)$ and inner product $\langle b_i, b_j \rangle$ have a good relationship. It is defined as:

$$dist_B(b_i, b_j) = \frac{1}{2}\left(k - \langle b_i, b_j \rangle \right) \tag{9}$$

Given the pairwise semantic similarity matrix $S = \{s_{ij}\}$ and hash codes $B = \{b_i\}_{i=1}^n$, the weighted maximum likelihood (WML) estimation of the hash codes $B = \{b_1, b_2, \ldots, b_n\}$ can be formulated as:

$$\log P(S|B) = \sum_{s_{ij} \in S} c_{ij} \log p(s_{ij}|b_i, b_j) \tag{10}$$

c_{ij} is used to balance the loss of similar and dissimilar pairs. The value of c_{ij} does not set a fixed value according to different training sets, but changes dynamically according to similar and dissimilar pairs in a batch of data. If there are many similar pairs in training data, the weight of dissimilar pairs will be increased; otherwise, the weight of similar pairs will be increased. Here we release it as,

$$c_{ij} = \begin{cases} |S_0|/|S_1|, & s_{ij} = 1 \\ |S_1|/|S_0|, & s_{ij} = 0 \end{cases} \tag{11}$$

where $S_1 = \{s_{ij} \in S : s_{ij} = 1\}$ is the set of similar pairs and $S_0 = \{s_{ij} \in S : s_{ij} = 0\}$ is the set of dissimilar pairs. For each pair of hash codes, $p(s_{ij}|b_i, b_j)$ is the conditional probability which can be expressed by the Bernoulli distribution:

$$p(s_{ij}|b_i, b_j) = \begin{cases} \sigma(\langle b_i, b_j \rangle), & s_{ij} = 1 \\ 1 - \sigma(\langle b_i, b_j \rangle), & s_{ij} = 0 \end{cases}$$
$$= \sigma(\langle b_i, b_j \rangle)^{s_{ij}} (1 - \sigma(\langle b_i, b_j \rangle))^{1-s_{ij}} \tag{12}$$

where $\sigma(x) = 1/(1 + e^{-x})$ is the sigmoid function, and $\langle b_i, b_j \rangle$ is the inner product between a pair of hash codes b_i and b_j. By taking Eq. 12 into Eq. 10, we can get the negative log likelihood function as follows:

$$J_1 = \sum_{s_{ij} \in S} c_{ij} \left(log\left(1 + e^{\langle b_i, b_j \rangle}\right) - s_{ij}\langle b_i, b_j \rangle \right) \tag{13}$$

Specifically, in order to facilitate optimization, we introduce a continuous variable z_i instead of the discrete variable b_i. z_i is the output of the last fully connected layer of the encoder and $b_i = sgn(z_i)$, so the negative log likelihood function can be rewritten as:

$$J_1 = \sum_{s_{ij} \in S} c_{ij} \left(log\left(1 + e^{\langle z_i, z_j \rangle}\right) - s_{ij}\langle z_i, z_j \rangle \right) \tag{14}$$

In order to get a more accurate binary code, we add the quantization loss function. $||\cdot||_1$ is the element-wise absolute operation. The loss function is adopted as follows:

$$J_2 = \sum_{i=1}^{N} \sum_{j=1}^{k} \left| ||z_i^j| - 1| \right|_1 \tag{15}$$

The overall loss function can be written as:

$$J = J_1 + \alpha J_2 \tag{16}$$

where α is the hyper-parameter. Equation 16 is not discrete, so we can easily optimize it with Stochastic Gradient Descent (SGD).

4 Experiments

4.1 Datasets

We evaluate the proposed method on three popular datasets: Firstly, CIFAR-10 [23] has 60,000 color images, whose dimension is $32 \times 32 \times 3$. In our experiments, we pick 100 images from each category as the test set, and the left 59,000 images are taken as the retrieval set. We randomly select 10,000 images from the retrieval set as a training set. Secondly, MIRFLICKR [25] includes 25,000 images under 38 concepts. Each image is associated with one or multiple labels. 1,000 images are randomly selected from the dataset as the test set, and the remaining images are left as the retrieval set. We randomly select 10,000 images from the retrieval set as a training set. Finally, NUSWIDE [24] includes 269,648 images assigned with one or multiple labels under totally 81 concepts. We choose the most frequent 21 concepts, and finally obtain 160,236 images. 2,100 images are randomly selected from the dataset as the test set, and the remaining images are left as the retrieval set. We randomly select 21,000 images from the retrieval set as a training set.

4.2 Experimental Settings

We compare our method with some unsupervised hashing methods including LSH [9], PCAH [26], SH [13], ITQ [15], DSH [14], DDH [7] and SSDH [18]. Among them, LSH, PCAH, SH, ITQ and DSH are traditional shallow hashing methods, DDH and SSDH are deep hashing methods. For shallow hashing methods, we exact 2048-dimensional deep features from the pool5-layer of the ResNet50 that is pretrained on ImageNet. For deep hashing methods, we use the same features to generate similarity matrix, others use the same settings in their original papers.

We implement our method based on the open-source framework TensorFlow [27]. Each layer is pretrained with a dropout rate of 20% and the entire network is finetuned without dropout. For pretraining of SDAE, the initial learning rate is set to 0.1, the batch size is 128. For our deep hashing model, the learning rate is set to 0.01, the batch size is 128, the coefficient of quantization loss is 0.01 and the value of k_1 is set to 700, 1000, 1000 for CIFAR-10, MIRFLICKR and NUSWIDE respectively.

Three standard evaluation metrics are adopted to evaluate the performance, where they are mean average precision (MAP), precision-recall(P-R) curves and precision curves with respect to different numbers of top retrieved samples (P@N). We adopt MAR@1000 for CIFAR-10, and MAP@5000 for MIRFLICKR and NUSWIDE.

4.3 Experimental Results

Table 1 shows the MAP results for our method and all baseline methods on three benchmark datasets with hash code lengths to be 16, 32, 64 bits respectively. From the table, we can see that the proposed method obtains the best results in all cases. In particular, compared to the best baseline method, our method achieves relative improvements of 5.4%, 7.7% and 13.6% for different bits on CIFAR-10. On MIRFLICKR, our method achieves relative improvements over the best baseline methods of 6%, 6% and 5.7% for 16 bits, 32 bits and 64 bits respectively. The value of MAP is also higher than other baseline methods on NUSWIDE, and it achieves improvements 8.7%, 6.8% and 8.9% for different bits.

From the table we can also get the following observations. Firstly, comparing to traditional linear methods, although they get good performance using deep features, they

Table 1. Comparison with baselines in terms of MAP@1000, 5000, 5000 on CIFAR-10, MIRFLICKR and NUSWIDE. The best result is shown in bold.

Method	CIFAR-10			MIRFLICKR			NUSWIDE		
	16bits	32bits	64bits	16bits	32bits	64bits	16bits	32bits	64bits
LSH	0.259	0.311	0.324	0.573	0.586	0.606	0.407	0.433	0.465
PCAH	0.484	0.410	0.344	0.645	0.640	0.628	0.526	0.505	0.494
SH	0.428	0.424	0.372	0.603	0.609	0.597	0.484	0.485	0.450
ITQ	0.548	0.581	0.591	0.704	0.696	0.715	0.627	0.665	0.673
DSH	0.453	0.452	0.483	0.618	0.677	0.699	0.565	0.572	0.633
DDH	0.479	0.522	0.561	0.713	0.731	0.755	0.638	0.663	0.689
SSDH	0.473	0.448	0.380	0.725	0.698	0.681	0.571	0.597	0.621
Ours	**0.602**	**0.658**	**0.727**	**0.785**	**0.791**	**0.812**	**0.725**	**0.733**	**0.778**

are still worse than our model which uses deep network to preserve the structural information of the original images. Secondly, we can see that SSDH and DDH obtain better results than traditional linear methods in most cases, which demonstrate the effectiveness of deep architecture. In addition, our method is significantly better than the results of SSDH and DDH. SSDH does not consider the data imbalance problem, so its performance is worse. DDH constructs similarity matrix without considering the non-neighbor items among the first k neighbors, so it can not generate more accurate hash codes.

To further demonstrate the effectiveness of our method, we draw Precision-Recall (P-R) curves and Precision curves @ N retrieved images (P@N) on CIFAR-10 and MIRFLICKR datasets for 64 bit lengths in Fig. 3 and Fig. 4. From the figure, we can observe that the performance of our method is better than several methods.

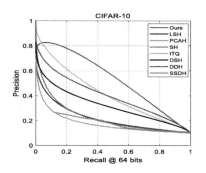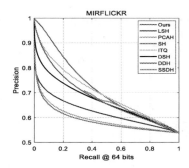

Fig. 3. Precision-recall curves with code length 64.

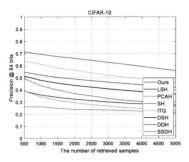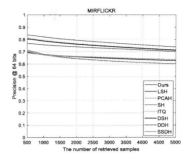

Fig. 4. Precision curves @ N retrieved images (P@N) with code length 64.

4.4 Detailed Analysis

To further evaluate our proposed model, we give three aspects of parameter analysis results. We firstly conduct experiments to analyze the value of k_1, then analyze the weight of solving the data imbalance, finally we analyze the effect of different similarity matrices. In our experiment, we conduct experiments to learn the influence of k_1. We select k_1 from the set $\{100,200,300,400,500,600,700,800,900,1000\}$. We fix other parameters and only change the k_1 value. When the value of k_1 changes, we record the results as shown in Fig. 5. In this figure, the performance increases along with the value of k_1 gradually increasing at the beginning, and finally stable.

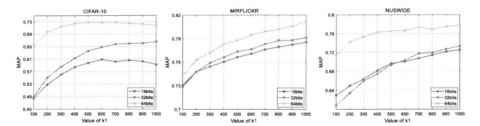

Fig. 5. MAP with respect to the value of k_1

To demonstrate the effectiveness of the sensitive to weight of solving the data imbalance, we conduct experiments on CIFAR-10, MIRFLICKR and NUSWIDE datasets. Both the experiments are conducted with code length 16, 32, 64, and the results are shown in Table 2. From this table, we can observe that the performance with the weight item is better than that without it. This reflects that the weight can effectively solve the data imbalance problem.

In addition, we compare different methods to construct similarity matrix and the results are presented in Fig. 6. From the figure, we can observe that the performance of expanded k-reciprocal matrix is superior to the other two methods. As mentioned before, some images have very few k-reciprocal nearest neighbors, so the performance of non-expanded k-reciprocal matrix is worse than the KNN-based matrix. We can also observe

Table 2. Performance comparison in terms of MAP by training with pairwise loss and with weighted pairwise loss.

Method	Hash bits	CIFAR-10	MIRFLICKR	NUSWIDE
With pairwise loss	16	0.566	0.773	0.699
	32	0.620	0.784	0.704
	64	0.705	0.797	0.746
With weighted pairwise loss	16	0.602	0.785	0.725
	32	0.658	0.791	0.733
	64	0.727	0.812	0.778

that the result of our method is better than the result of KNN, which demonstrates the effectiveness of proposed semantic similarity matrix.

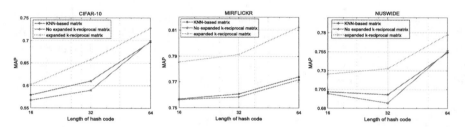

Fig. 6. Performance comparison by different methods to construct similarity matrix.

5   Conclusion

In this paper, we present a novel deep unsupervised hashing method. We construct a deep neural network to learn hash codes. Pairwise similarity matrix is constructed by expanded k-reciprocal nearest neighbors to preserve the semantic relationships between data. Extensive studies on three datasets show that our method significantly outperforms the state-of-the-art methods.

References

1. Strecha, C., Bronstein, A., Bronstein, M., Fua, P.: LDAHash: improved matching with smaller descriptors. IEEE Trans. Pattern Anal. Mach. Intell. **32**(5), 815–830 (2010)
2. Tao, D., Li, X., Wu, X., Maybank, S.J.: General tensor discriminant analysis and gabor features for gait recognition. IEEE Trans. Pattern Anal. Mach. Intell. **29**, 1700–1715 (2007)
3. Mu, Y., Hua, G., Fan, W., Chang, S.-F.: Hash-SVM: scalable kernel machines for large-scale visual classification. In: Proceedings IEEE Conference Computing Vision Pattern Recognition, pp. 979–986 (2014)

4. Zhang, H., Shen, F., Liu, W., He, X., Luan, H., Chua, T.-S.: Discrete collaborative filtering. In: Proceedings 33rd International ACM SIGIR Conference Research Development Information Retrieval, pp. 325–334 (2016)
5. Jegou, H., Harzallah, H., Schmid, C.: A contextual dissimilarity measure for accurate and efficient image search. In: CVPR (2007)
6. Shen, X., Lin, Z., Brandt, J., Avidan, S., Wu, Y.: Object retrieval and localization with spatially-constrained similarity measure and k-nn re-ranking. In: CVPR (2012)
7. Song, J., He, T., Fan, H., Gao, L.: Deep discrete hashing with self-supervised pairwise labels. In: Joint European Conference on Machine Learning and Knowledge Discovery in Databases, pp. 223–238 (2017)
8. Cao, Z., Long, M., Wang, J., Yu, P.S.: Hashnet: Deep learning to hash by continuation. In: Proceedings of the IEEE International Conference on Computer Vision, pp. 5608–5617 (2017)
9. Gionis, A., Indyk, P., Motwani, R.: Similarity search in high dimensions via hashing. In: Proceedings 30th International Conference Very Large Data Bases, pp. 518–529 (1999)
10. Liu, W., Wang, J., Ji, R., Jiang, Y.-G., Chang, Y.-F.: Supervised hashing with kernels. In: CVPR, pp. 2074–2081 (2012)
11. Zhu, H., Long, M., Wang, J., Cao, Y.: Deep hashing network for efficient similarity retrieval. In: Proceedings AAAI Conference Artificial Intelligence, pp. 2415–2421 (2016)
12. Huang, L., Chen, J., Pan, S.: Accelerate learning of deep hashing with gradient attention. In: ICCV (2019)
13. Weiss, Y., Torralba, A., Fergus, R.: Spectral hashing. In: NIPS, pp. 1753–1760 (2009)
14. Jin, Z., Li, C., Lin, Y., Cai, D.: Density sensitive hashing. IEEE Trans. Cybern. 44, 1362–1371 (2014)
15. Gong, Y., Lazebnik, S., Gordo, A., Perronnin, F.: Iterative quantization: a procrustean approach to learning binary codes for large-scale image retrieval. IEEE Trans. Pattern Anal. Mach. Intell. 35, 2916–2929 (2013)
16. Liong, V.E., Lu, J., Wang, G., Moulin, P., Zhou, J.: Deep hashing for compact binary codes learning. In: Proceedings CVPR, pp. 2475–2483 (2015)
17. Lin, K., Lu, C.-S., Chen, J., Zhou, J.: Learning compact binary descriptors with unsupervised deep neural networks. In: Proceedings CVPR, pp. 1183–1192 (2016)
18. Yang, E., Deng, C., Liu, T., Liu, W., Tao, D.: Semantic structure-based unsupervised deep hashing. In: Proceedings IJCAI, pp. 1064–1070 (2018)
19. Zhong, Z., Zheng, L., Cao, D., Li, S.: Re-ranking person re-identification with k-reciprocal encoding. In: Proceedings IEEE Conference Computing Vision Pattern Recognition, pp. 1318–1327 (2017)
20. Qin, D., Gammeter, S., Bossard, L., Quack, T., Van Gool, L.: Hello neighbor: Accurate object retrieval with k-reciprocal nearest neighbors. In: IEEE Conference on Computer Vision and Pattern Recognition, p. 5 (2011)
21. Vincent, P., Larochelle, H., Lajoie, I., Bengio, Y., Manzagol, P.-A.: Stacked denoising autoencoders: learning useful representations in a deep network with a local denoising criterion. J. Mach. Learn. Res. 11, 3371–3408 (2010)
22. He, K., Zhang, X., Ren, S., Sun, J.: Deep residual learning for image recognition (2015). CoRR, abs/1512.03385
23. Krizhevsky, A., Hinton, G.: Learning multiple layers of features from tiny images. Technical report, University of Toronto (2009)
24. Chua, T., Tang, J., Hong, R., Li, H., Luo, Z., Zheng, Y.: Nus-wide: a real-world web image database from national university of Singapore. In: ACM Multimedia, pp. 48–56 (2009)
25. Huiskes, MJ., Thomee, B., Lew, M.S: New trends and ideas in visual concept detection: the mirflickr retrieval evaluation initiative. In: Proceedings of the International Conference on Multimedia Information Retrieval, pp. 527–536 (2010)

J. Wang and Z. Lian

26. Wang, J., Kumar, S., Chang, S.-F.: Semi-supervised hashing for large scale search. IEEE Trans. Pattern Anal. Mach. Intell. **34**, 2393–2406 (2012)
27. Abadi, M., et al.: TensorFlow: Large-scale machine learning on heterogeneous distributed systems (2016). CoRR, abs/1603.04467
28. Dmochowski, J.P., Sajda, P., Parra, L.C.: Maximum likelihood in cost-sensitive learning: Model specification, approximations, and upper bounds. J. Mach. Learn. Res. (JMLR) **11**, 3313–3332 (2010)
29. Rumelhart, D., Hintont, G., Williams, R.: Learning representations by back-propagating errors. Nature **323**, 533–536 (1986)
30. Bengio, Y., Lamblin, P., Popovici, D., Larochelle, H.: Greedy layer-wise training of deep networks. In: Advances in Neural Information Processing Systems, pp. 153–160 (2007)

Micro-Expression Recognition Using Micro-Variation Boosted Heat Areas

Mingyue Zhang[1,2(✉)], Zhaoxin Huan[1,2], and Lin Shang[1,2]

[1] State Key Laboratory for Novel Software Technology, Nanjing University, Nanjing 210023, China
{MF1933117,huanzhx}@samil.nju.edu.cn
[2] Department of Computer Science and Technology, Nanjing University, Nanjing 210023, China
shanglin@nju.edu.cn

Abstract. Micro-Expression Recognition has been a challenging task as transitory micro-expressions only appear in a few of facial areas. In this work, we aim to improve the recognition performance by boosting the micro-variations in the learned areas. This paper proposes an architecture, deriving facial micro-variation heat areas and then integrating in conjunction with the micro-expression recognition network, to learn the micro-expression features in an end-to-end manner. The method is constructed by the Heat Areas Estimator from Heatmap (HAEH), which is to produce micro-variation heat areas as facial geometric structure, and the Temporal Facial Micro-Variation Network (TFMVN) for learning the fusion features. The method can define and capture facial heat areas significantly contributed to the micro-expressions. Our approach activates or deactivates corresponding feature maps from the heat areas to guide feature learning. We perform experiments on CASME II dataset and SAMM dataset. The experimental results show that we achieve state-of-the-art accuracy, and the method demonstrates good generalization ability for cross-dataset. Moreover, we validate three pivotal components' effectiveness within our architecture.

Keywords: Micro-expression recognition · Heatmap · Heat areas

1 Introduction

Micro-Expression Recognition, which refers to the slight facial parts moved facial expression recognition, serves as a key step for lie-detect and other psychological activities.

Different from the traditional Facial Expression Recognition (FER), a micro-expression lasts between 1/5 to 1/25 of a second and usually only involves in several facial key areas [8]. The movement of these facial parts can hardly be recognized even by human eyes. Therefore, it is challenging to distinguish the moved facial key

This work is supported by the National Natural Science Foundation of China (No. 61672276, No. 51975294).

areas between consecutive frames. The objective of this work is to improve the micro-expression recognition performance by focusing on micro-variation.

Recently, [9] proposed a concept called expression-state and used Convolutional Neural Network (CNN) to model the dynamic evolution of expression states. After that, the LSTM takes the features extracted by CNN as input and learns the temporal information between different frames. [8] proposed an Enriched Long-term Recurrent Convolutional Network that encodes each micro-expression frame into a feature vector through CNN modules, then predicts the micro-expression by passing the feature vector through a Long Short-term Memory (LSTM) module. Compared to traditional facial expression recognition, the micro-expression recognition requires different visual features.

Facial landmarks have been proved to effective for recognizing facial expression [6]. However, in micro-expression recognition, the dynamic geometrical variations based on facial landmarks can hardly be caught partially due to the slight movement of facial key areas (usually 1~2 pixels). Besides, the facial landmarks based methods mostly depend on the quality of face alignment algorithms. Even subtle misalignment may weak the performance of micro-expression recognition.

In this paper, we propose a new model to integrate micro-variation information in conjunction with micro-expression recognition. Therefore, the micro-expression recognition can utilize the invariable facial structure information, for the facial geometrical features are essential to locate the facial key areas. We aim to guide the feature extractor automatically focusing on facial key areas which contribute to micro-expression recognition.

In contrast with directly using landmarks to capture the dynamic geometrical variations, our approach uses Fully Convolutional Neural Network (FCN) [2] to generate facial micro-variation heat areas. As noticed in Fig. 1, the heatmaps through FCN can precisely denote the facial key areas such as eyes, mouth and nose. In Sect. 4.1, we experiment and observe that the model guided by ground truth facial micro-variation heat areas can achieve 66.39% accuracy on CASME II dataset [22]. It suggests the richness contained in heat areas.

After generating facial micro-variation heat areas, we integrate heat areas into microexpression recognition. The heat areas serve as structure cue to guide feature learning for CNN. To fully utilize the structure information, we apply facial micro-variation heat areas at multiple stages in the Convolutional Neural Network.

We evaluate the proposed method on popular micro-expression benchmarks CASME II. Our approach outperforms previous state-of-the-art methods. We also conduct the cross-dataset experiment and ablation analysis to demonstrate the generalization and effectiveness of our model.

2   Related Work

In the literature of Micro-Expression recognition, besides classic methods (LBP-TOP [20], DMDSP [14] and Bi-WOOF [11]), recently, state-of-the-art performance has been achieved with Deep Convolutional Neural Networks (DCNNs). These methods mainly fall into two categories, i.e., handcrafted features and Deep Neural Networks.

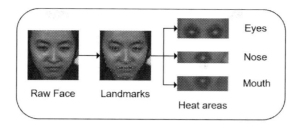

Fig. 1. Given a face image I, we divide the image into three parts: eyes (including left and right eyes), nose and mouth. We transform the landmarks to facial micro-variation heat areas' heatmap according to different parts by initializing the gauss distribution according to the center.

Handcrafted Features. Handcrafted features have mainly chosen Local Binary Pattern with Three Orthogonal Planes (LBP-TOP) [20] as their primary baseline feature extractor. Optical flow has also been widely studied in computer vision for decades. [21] proposed the Facial Dynamics Map (FDM) based on optical flow estimation. [15] learned the temporal characteristics and the corresponding discriminative filters. [10] combined several handcrafted features and classified them by SVM. The handcrafted features based methods have two steps: feature extraction and classification, whereas deep neural networks can automatically extract features and it is end-to-end trainable

Deep Neural Networks. Peng et al. [16] proposed a two-stream 3-D CNN model called Dual Temporal Scale Convolutional Neural Network (DTSCNN). Different streams of the framework were used to adapt to different frame rates of ME video clips. [9] proposed a concept called expression-state. The author used CNN to model the dynamic evolution of expression states. (i.e., onset, onset to apex, apex, apex to offset and offset). After that, the LSTM took the features extracted by CNN as input and learns the temporal information between different frames. This approach provides the motivation towards the design of our proposed method. In this work, we adopt LSTM to extract the temporal appearance features. Recently, [8] proposed an Enriched Long-term Recurrent Convolutional Network that encodes each micro-expression frame into a feature vector through CNN modules, then predicts the micro-expression by passing the feature vector through a Long Short-term Memory (LSTM) module. Our approach is inspired from these deep learning methods, however differs from them in two specific aspects: (1) the heat areas are learned by the heatmap focusing on the micro-variations. (2) The approach is guided by the integrations with the boosted micro-variation heat areas in an end-to-end manner.

3 Proposed Method

As mentioned in the introduction, the movement of facial key areas can hardly be recognized and the dynamic geometrical variations based on facial landmarks can also hardly be caught due to the slight movement. We propose a model to integrate facial micro-variation heat areas into micro-expression recognition. The detailed configuration of our proposed model is illustrated in Fig. 2. It is composed of two closely related

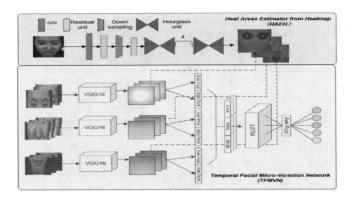

Fig. 2. The model of our approach. The upper box is Heat Areas Estimator from Heatmap (HAEH) and the lower box is Temporal Facial Micro-Variation Network (TFMVN). First we trained the Heatmap Estimator on face alignment dataset. After getting the trained estimator, the micro-expression images are input to the estimator to get the facial heat areas. The heatmaps serves as structure cue to guide feature learning. Based on these heatmaps, TFMVN with three branches were built.

components: Heat Areas Estimator from Heatmap (HAEH) and Temporal Facial Micro-Variation Network (TFMVN). The HAEH produces micro-variation heat areas as facial geometric structure. TFMVN incorporates facial micro-variation information according to heatmaps and uses LSTM to model the temporal features.

We have noticed other choices available for geometric structure representations. [12] has adopted facial parts to aid facial tasks. However, facial parts are too coarse for focusing on facial micro-expression key areas. [4] proposed a mask-cnn to locate the position of objects parts, but the generation approach of mask can not be used on face. On the contrary, facial micro-variation heat areas pay attention to areas which contribute to micro-expression. And the effectiveness of HG network has been proved in other face alignment and pose estimation tasks [13]. Experiments in Sec 4 have shown that heat areas can best boost the performance of micro-expression recognition.

3.1 Heat Areas Estimator from Heatmap

In order to incorporate facial micro-variation heat areas into feature learning, we transform landmarks to facial micro-variation heat areas to aid the learning of feature. The responses of each pixel in heatmap are decided by the distance to the corresponding facial micro-variation heat areas. The details of facial micro-variation heat areas are defined as follows.

Label Transform. The process of label transform is shown in Algorithm 1. Specifically, given a face image I, we divide the image into three parts: eyes (including left and right eyes), nose and mouth. Since the facial boundary of cheek has no contribution to micro-expression, we ignore the facial areas which belong to cheek. We use the face alignment dataset named 300W-LP [17] to train the Stacked-Hourglass network, the

---

Algorithm 1 Process of label transform.

Input:
    A face image, I;
    The corresponding landmarks, $[x_1, y_1, ..., x_{68}, y_{68}]$;
Output:
    Facial micro-variation heat areas, $M_i(i = 1, 2, 3)$;
 1: Initializing the $i - th$ heat areas center:
    compute the midpoint of the 2 boundary landmarks(i.e. lower eyelid and eyebrow);
 2: Setting the radius to the distance between the midpoint and boundary landmarks;
 3: Making gauss distribution according to the center and the radius on M_i
 4: **return** M_i;

---

labels in dataset are 68 landmarks. Therefore, after dividing the face image into three parts, we transform the landmarks to facial micro-variation heat area according to different parts. Specifically, we choose the middle landmarks of lower eyelid and eyebrow (the 41th and 19th landmarks), and compute the midpoint of these two landmarks as the gauss distribution center. Then we initialize the gauss distribution according to the center, and set the radius to the distance between the midpoint and lower eyelid. As for nose and mouth, we take the similar method to initialize these two areas. The heat areas of different parts are shown in Fig. 1.

Hourglass Network. Following the instruction of [13], we adopt four-stacked hourglass network as our backbone network of estimator. The structure of hourglass network is shown in Fig. 2. While [13] uses the bottleneck block of [19] as the main building block for the HG, we go one step further and replace the bottleneck block with the recently introduced hierarchical, parallel and multi-scale block of [1]. This block outperforms the original bottleneck of [19] when the same number of network parameter was used.

Training and Evaluation. We train the heatmap estimator on 300W-LP dataset [17], and evaluate on 300-W test set. Traditionally, the metric used for face alignment is the point-to-point Euclidean distance normalized by the interocular distance. We follow this evaluation metric, in particular, we use the Normalized Mean Error defined as:

$$NME = \frac{1}{N} \sum_{k=1}^{N} \frac{\|x_k - y_k\|_2}{d} \tag{1}$$

, where x denotes the ground truth gauss distribution center for each facial parts ($k = 1, 2, 3$), y is the corresponding prediction and d is the interocular distance of the ground truth centers.

3.2 Temporal Facial Micro-variation Network

In Sect. 3.1, we mentioned that facial micro-variation heat areas can be obtained through heatmap estimator. Based on these heatmaps, a Temporal Facial Micro-Variation Network (TFMVN) with three branches can be built. The TFMVN structure has been shown in Fig. 2. We introduce the operation of each network by taking the eyes part as an example.

Extract Facial Micro-variation Feature. We resize the input image of the eyes branch to 224*224 resolution. The next component is feature learning component. We use VGG-16 [18] as our backbone network to extract the appearance features of each frame. It is worth mentioning that in TFMVN we remove the fully connected layer of the traditional convolutional neural network to extract its convolutional layer features as feature descriptors. Besides, unlike other image classification tasks [19], the classes and frame sequences in micro-facial expression databases are insufficient compared with other video classification problems. If directly using some deeper CNN models, it can easily fall into overfitting when training. Therefore, we retain all convolution operations before the pool3 layer (including the pool3 layer) and remove all convolutional layer after the pool3 layer. For an input image of 224*224, a 28*28*512 convolution feature can be obtained under this condition, which can correspond to a 512-dimensional feature descriptor of 28*28 spatial positions. Then we reorganized the eyes heat areas obtained in the heatmap estimator by bilinear interpolation to a size of 28*28 to select and retain valuable convolution descriptors.

As shown in Fig. 2, the convolution descriptors in the image are reserved in the different semantic branches with the weights corresponding to the heat areas. Meanwhile, if a convolution descriptor is identified as representing an unrelated region, it will be dropped. Specifically, the heatmap is a two-dimensional real-value matrix whose value satisfies the Gaussian distribution. We define a threshold of 0.7 (selected through experiment). If the pixel value of heatmaps is greater than the threshold, it is reserved, otherwise, it will be dropped. The convolution descriptors in the TFMVN "reserve" and "drop" operations are implemented in experiments with element-level multiplication between convolutional features and heatmaps. Therefore, for a descriptor that is determined to represent a facial micro-variation heat areas (a region where the heatmap is larger than the threshold), these descriptors will be reserved according to the weight. Otherwise, the descriptor determined to be the unrelated region will become an all-zero vector.

After getting the convolution descriptors reserved according to heatmaps from three branches (eyes, nose and mouth), we use the maximum and average pooling and separately perform L2 normalization. Then cascade them as the final feature representation. Similarly, the nose and mouth branches have the same operational steps as the eye branches described above.

Extract Temporal Feature. Seeing that we can view the variation of facial expression as an image sequence from neutral expression to peak expression, LSTM can be modeled in accordance with people's understanding behavior of a facial expression. Therefore, after CNN extracts appearance features from each frame, instead of using 3D filters to extract temporal information like [7], we use LSTM to model the temporal relations between expression frames. Moreover, in [5] a CRF module was added to extract the temporal information. But the model is a two-step network. In contrast, by applying the LSTM for modeling consecutive frames in vision problems, our approach can jointly train convolutional and recurrent networks in order to make the FAN end-to-end trainable.

The LSTM structure is shown in Fig. 2. The inputs of LSTM are attention appearance features of each frame. In order to average the predictions of each time step for final classification, we propose an average pooling layer with a weight function W(t):

$$z_i = \frac{1}{N} * \sum_{t=1}^{N} h_{ti} * W(t) \tag{2}$$

, where $H_i = (h_{t1}, h_{t2}, ..., h_{tn})$ is the output of LSTM at time t, $Z = (z_1, z_2, ..., z_n)$ is the result of average pooling. $W(t)$ is a weight function which linearly increases from $0...1$ over frames $t = 0...T$. By applying $W(t)$ in pooling layer, we emphasize the importance of prediction at later frames in which hidden units capture more variation.

Training Method. TFMVN is an end-to-end trainable Convolutional Recurrent Neural Networks and therefore it is trained with Backpropagation Through Time. During the training phase, all model parameters are randomly initialized except for the biases which are initialized with zeros. After a whole facial expression frame sequence has been propagated forward through the TFMVN, the weight parameters begin to update. This end-to-end optimization can update the visual (CNN) and sequential (LSTM) model parameters at the same time.

4   Experiments

In this section, we first introduce the datasets, evaluation methods and experiments configuration. Then, we compare the empirical results of our method with other state-of-the-arts on single domain and cross-dataset evaluation. The experimental results show the effectiveness and potential of our approach. Furthermore, our model consists of several pivotal components, and we also validate their effectiveness on the CASME II dataset.

Datasets. We conduct evaluations on two challenging datasets including CASME II [22] and SAMM [3].

CASME II dataset: CASME II is currently the most widely used benchmark dataset. It consists of 247 video samples, elicited from 26 Asian participants with an average age of 22.03 years old. Each CASME II participants originally has one of five categories of micro-expressions: *Surprise, Disgust, Happiness, Repression* and *Others*.

SAMM dataset: SAMM is a newer challenging dataset. Full set contains 159 micro-movements (one video for each) with a mean age of 33.24 years from 32 participants. SAMM reported 7 basic emotions: *happiness, anger, surprise, fear, disgust, sadness* and *contempt*.

Evaluation Metric. For single domain experiment, we evaluate our algorithm using F1-Score, Unweighted Average Accuracy Recall (UAR), and Accuracy on CASME II. In order to show the generalization ability and verify the capacity of handling cross-dataset micro-expression recognition of our method, we use CASME II to train our model and test it on SAMM dataset. In addition, because our model consists of several pivotal components, i.e., Heatmap Estimator, micro-variation features and average pooling layer with W(t). We validate their effectiveness within our model on the CASME II.

4.1   Comparison with State-of-the-Arts

Evaluation on CASME II. In single domain experiment, we compare our approach with the state-of-the-art methods on CASME II dataset. For all experiments, we use the Leave-One-Subject-Out (LOSO) cross validation protocol to prevents subject bias. The results are shown in Table 1.

The methods compared in Table 1 can be classified into two categories: handcrafted features [8,10,15,21] and deep neural networks [8,9]. For handcrafted features, [21] proposed the Facial Dynamics Map (FDM) based on optical flow estimation. [8] chose LBP-TOP as baseline which utilize only six intersection points in the 3D plane. [10] combined several handcrafted features and classified by SVM. [15] learned the temporal characteristics and the corresponding discriminative filters. As for deep neural networks, [9] used CNN to extract appearance features of each frames and modeled the temporal relations by LSTM. [8] proposed an Enriched Long-term Recurrent Convolutional Network which integrates hierarchical spatial features. Our method outperforms most previous methods on accuracies. Note that, our method achieves 65.37% accuracy on the CASME II dataset which reflects the effectiveness for micro-expression recognition.

We use ground truth facial micro-variation heat areas in the proposed method to verify the effectiveness and potential of heat areas. "TFMVN+gtheatarea" significantly outperforms the TFMVN. The results increased by 1.02% on accuracy, 2.25% on UAR and 3.08% on F1-Score, which demonstrate the effectiveness of heat areas information and show great potential performance gain if the heat areas information can be well captured.

While [10] employed Eulerian video magnification method and combined Local Binary Pattern (LBP), Histograms of Oriented Gradients (HOG) and Histograms of Image Gradient Orientation (HIGO) on three orthogonal planes. The number of subjects and frame sequences in CASME II dataset is insufficient compared with other image classification tasks, which means that it is especially not competent for deep learning methods. Therefore, the handcrafted features like [10] could achieve better accuracies. On the other hand, the handcrafted features based methods have two steps: feature selection and classification, whereas our TFMVN can automatically extract features and the whole network is end-to-end trainable.

Cross-Dataset Evaluation. In cross-dataset evaluation, followed the configuration of [8], we use both CASME II and SAMM dataset. Specifically, we hold out one database at each time: trained on CASME II and tested on SAMM, and vice versa. Table 2. comparing the performance of our TFMVN against the state-of-the-art methods on the cross-dataset evaluation protocol, our model outperforms previous results. We achieve 45.36% accuracy with 36.75% UAR. The accuracy is significantly increased by 2.18%, which indicates the robustness of our method to handle cross-dataset recognition. Whereas the F1-Score is 1.92% lower than [8].

It is worth mentioning that In order to verify the capacity of handling cross-dataset micro-expression recognition of our method, we use heatmaps estimator trained on 300W-LP which has no overlap with CASME II and SAMM dataset and compare the performance with and without using facial micro-variation heat areas (TFMVN without

Table 1. Performance compared with other method on CASME II.

Method	F1-Score	UAR	Acc
FDM [21]	0.4053	N/A	0.4593
LBP-SIP [8]	0.4480	N/A	0.4656
LBP-TOP [8]	0.2941	0.3094	0.4595
EVM+HIGO [10]	N/A	N/A	**0.6721**
MM+LBP-TOP [15]	N/A	N/A	0.5191
ELRCN-TE [8]	0.5000	0.4396	0.5244
CNN-LSTM [9]	N/A	N/A	0.6098
TFMVN	0.4923	0.4587	0.6537
TFMVN+gtheatarea	**0.5231**	**0.4812**	0.6639

heatmap) information. The results are reported in Table 2. There is 4.8% accuracy boost between our method without and with using facial micro-variation heat areas information.

Table 2. Experimental results for cross-dataset evaluation.

Method	F1-Score	UAR	Acc
ELRCN-SE [8]	**0.3411**	0.3522	0.4345
LBP-TOP [8]	0.2162	0.2179	0.3891
HOG-3D [8]	N/A	0.228	0.363
HOOF [11]	N/A	0.348	0.353
TFMVN	0.3219	**0.3675**	**0.4563**
TFMVN without heatareas	0.2814	0.3122	0.4083

Confusion Matrix. To better understand what goes on under the model, we provide the confusion matrix for TFMVN on CASME II dataset. Figure 3 shows the resulting confusion matrices of our model on CASME II. It can be seen that our model achieved relatively high recognition accuracies of each emotion. In particular, our model performs well on *happiness* and *surprise* possibly due to larger amount of training samples. The high confusion in *disgust* expression can be caused by the few number of sequences in dataset. Further analysis is showed that the appearance variations of *disgust* expressions are extremely slight which creates more confusions for our model. Hence, the sample size remains a challenging problem for deep learning based approaches.

4.2 Ablation Study

Our model consists of several pivotal components, i.e., heatmap estimator for heat areas, facial micro-variation features and average pooling layer with W(t). In this section,

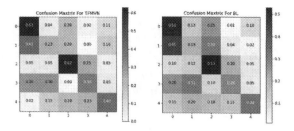

Fig. 3. Confusion matrix for TFMVN and BL on CASME II dataset. *0-Surprise, 1-Disgust, 2-Happiness, 3-Repression, 4-Others.*

we validate their effectiveness within our model on the CASMEII dataset. Based on the baseline CNN-LSTM (BL), we analyze each proposed component, i.e., with the baseline Heatmap estimator (BL+Heatmap) and average pooling with weight function W(t)(APW), by comparing their accuracies on CASEM II.

Facial Micro-Variation Heat Areas. Heatmaps are chosen as geometric structure which represent facial micro-variation heat areas in our model. We verify the potential of other structure information as well, i.e., mask for stress facial key areas. We show the micro-expression recognition accuracies using different structure information in Table 3. It can be observed easily that heat areas map (BL+Heatmap) is the most effective one with the accuracy of 65.37%, increased by 4.39% compared to BL and 3.02% compared to BL+Mask.

Moreover, we use CAM algorithm [20] to visualize the features learned by our model. Figure 4 shows the CAM heatmaps without using heat areas. Many regions which are unrelated to micro-expression have a high response and the model output wrong classification results. Figure 5 shows the CAM heatmaps generated by our model. It can be clearly seen that our model focuses on some facial key areas which are important for recognition. As a result, our model output the ground-truth classification results on these examples.

Pooling Methods. We also conduct experiments to evaluate our pooling methods, including average pooling with a weight function W(t)(APW), max pooling(MP) and directly using the last output of RNN without pooling(WP). The result is shown in Table 4. It can be observed that the performance is increased by 34.75% compared with MP and 12.5% compared to the WP, which indicated that by applying $W(t)$ in pooling layer, TFMVN emphasizes the importance of prediction at later frames in which hidden units capture more information.

Table 3. Accuracies on CASME II for evaluation the potential of facial micro-variation heat areas as the facial structure information.

Methods	BL	BL+Mask	BL+Heatmap
Accuracy	0.6098	0.6253	0.6537

Table 4. Accuracies on CASME II for evaluation the potential of average pooling with a weight function W(t) as the pooling methods.

Methods	MP	WP	APW
Accuracy	0.3062	0.5278	0.6537

Fig. 4. The heatmaps generated by CAM algorithm without using heat areas. The model focuses on many unrelated regions and output wrong classification results.

Fig. 5. The heatmaps generated by CAM algorithm with heat areas. The model focuses on some facial key ares which have an important impact on the classification results.

5   Conclusion

Geometric structure is important for facial expression recognition and landmarks have been proven well performed, however in the Micro-Expression Recognition, it is still challenging due to the micro-variations. In this work, we proposed the method of Heat Areas Estimator from Heatmap (HAEH) and Temporal Facial Micro-Variation Network (TFMVN), which integrates micro-variations heat areas into micro-expression recognition learning. Our approach activate or deactivate corresponding feature maps according to the facial micro-variation heat areas to achieve the micro-variation boosted recognition. We test our approach on CASME II and SAMM datasets. The experimental results show that we achieve state-of-the-art accuracy and the well-performed accuracy on cross-dataset evaluation. We also conduct ablation study to show the effectiveness of our model.

References

1. A. Bulat and G. Tzimiropoulos. Binarized convolutional landmark localizers for human pose estimation and face alignment with limited resources. In: IEEE International Conference on Computer Vision, ICCV 2017, Venice, Italy, 22–29 October 2017, pp. 3726–3734 (2017)
2. Bulat, A., Tzimiropoulos, G.: How far are we from solving the 2D & 3D face alignment problem? (and a dataset of 230, 000 3d facial landmarks). In: IEEE International Conference on Computer Vision, ICCV 2017, Venice, Italy, 22–29 October 2017, pp. 1021–1030 (2017)
3. Davison, A.K., Lansley, C., Costen, N., Tan, K., Yap, M.H.: SAMM: a spontaneous micro-facial movement dataset. IEEE Trans. Affective Comput. **9**(1), 116–129 (2018)
4. Ghiasi, G., Fowlkes, C.C.: Occlusion coherence: localizing occluded faces with a hierarchical deformable part model. In: 2014 IEEE Conference on Computer Vision and Pattern Recognition, CVPR 2014, Columbus, OH, USA, 23–28 June 2014, pp. 1899–1906 (2014)

5. Hassani, B., Mahoor, M.H.: Spatio-temporal facial expression recognition using convolutional neural networks and conditional random fields. In: 12th IEEE International Conference on Automatic Face & Gesture Recognition, FG 2017, Washington, DC, USA, May 30 - June 3, 2017, pp. 790–795 (2017)

6. Huan, Z., Shang, L.: Model the dynamic evolution of facial expression from image sequences. In: Advances in Knowledge Discovery and Data Mining - 22nd Pacific-Asia Conference, PAKDD 2018, Melbourne, VIC, Australia, 3–6 June 2018, Proceedings, Part II, pp. 546–557 (2018)

7. Jung, H., Lee, S., Yim, J., Park, S., Kim, J.: Joint fine-tuning in deep neural networks for facial expression recognition. In: 2015 IEEE International Conference on Computer Vision, ICCV 2015, Santiago, Chile, 7–13 December 2015, pp. 2983–2991 (2015)

8. Khor, H., See, J., Phan, R.C., Lin, W.: Enriched long-term recurrent convolutional network for facial micro-expression recognition. In: 13th IEEE International Conference on Automatic Face & Gesture Recognition, FG 2018, Xi'an, China, 15–19 May 2018, pp. 667–674 (2018)

9. Kim, D.H., Baddar, W.J., Ro, Y.M.: Micro-expression recognition with expression-state constrained spatio-temporal feature representations. In: Proceedings of the 2016 ACM Conference on Multimedia Conference, MM 2016, Amsterdam, The Netherlands, 15–19 October 2016, pp. 382–386 (2016)

10. Li, X., et al.: Towards reading hidden emotions: a comparative study of spontaneous micro-expression spotting and recognition methods. IEEE Trans. Affective Comput. **9**(4), 563–577 (2018)

11. Liong, S., See, J., Wong, K., Phan, R.C.: Less is more: micro-expression recognition from video using apex frame. Sig. Proc. Image Comm. **62**, 82–92 (2018)

12. J. Lv, X. Shao, J. Xing, C. Cheng, and X. Zhou. A deep regression architecture with two-stage re-initialization for high performance facial landmark detection. In 2017 IEEE Conference on Computer Vision and Pattern Recognition, CVPR 2017, Honolulu, HI, USA, 21–26 July 2017, pp. 3691–3700 (2017)

13. Newell, A., Yang, K., Deng, J.: Stacked hourglass networks for human pose estimation. In: Computer Vision - ECCV 2016–14th European Conference, Amsterdam, The Netherlands, October 11–14 (2016), Proceedings, Part VIII, pp. 483–499 (2016)

14. Ngo, A.C.L., See, J., Phan, R.C.: Sparsity in dynamics of spontaneous subtle emotions: analysis and application. IEEE Trans. Affective Comput. **8**(3), 396–411 (2017)

15. Park, S.Y., Lee, S., Ro, Y.M.: Subtle facial expression recognition using adaptive magnification of discriminative facial motion. In: Proceedings of the 23rd Annual ACM Conference on Multimedia Conference, MM 2015, Brisbane, Australia, 26–30 October 2015, pp. 911–914 (2015)

16. Peng, M., Wang, C., Chen, T., Liu, G., Fu, X.: Dual temporal scale convolutional neural network for micro-expression recognition. Front. Psychol. **8**, 1745 (2017)

17. Sagonas, C., Tzimiropoulos, G., Zafeiriou, S., Pantic, M.: 300 faces in-the-wild challenge: the first facial landmark localization challenge. In: 2013 IEEE International Conference on Computer Vision Workshops, ICCV Workshops 2013, Sydney, Australia, December 1–8, 2013, pp. 397–403 (2013)

18. Simonyan, K., Zisserman, A.: Very deep convolutional networks for large-scale image recognition. CoRR, abs/1409.1556 (2014)

19. Verma, A., Qassim, H., Feinzimer, D.: Residual squeeze CNDS deep learning CNN model for very large scale places image recognition. In: 8th IEEE Annual Ubiquitous Computing, Electronics and Mobile Communication Conference, UEMCON 2017, New York City, NY, USA, 19–21 October 2017, pp. 463–469 (2017)

20. Wang, Y., See, J., Phan, R.C., Oh, Y.: LBP with six intersection points: reducing redundant information in LBP-TOP for micro-expression recognition. In: Computer Vision - ACCV 2014–12th Asian Conference on Computer Vision, Singapore, Singapore, 1–5 November 2014, Revised Selected Papers, Part I, pp. 525–537 (2014)
21. Xu, F., Zhang, J., Wang, J.Z.: Microexpression identification and categorization using a facial dynamics map. IEEE Trans. Affective Comput. **8**(2), 254–267 (2017)
22. Yan, W.J., et al.: Casme ii: an improved spontaneous micro-expression database and the baseline evaluation. Plos One **9**(1), e86041 (2014)

Fashion-Sketcher: A Model for Producing Fashion Sketches of Multiple Categories

Junkai Fang[(⊠)], Xiaoling Gu, and Min Tan

Key Laboratory of Complex Systems Modeling and Simulation, School of Computer
Science and Technology, Hangzhou Dianzi University, Hangzhou 310018, China
{kmaeii,guxl,tanmin}@hdu.edu.cn

Abstract. Fashion sketches play a critical role in the initial stages of
fashion product design. This situation has motivated the development of
artificial intelligence (AI) techniques for automatically generating fashion
sketches. We present Fashion-Sketcher, a hybrid deep generative model
able to generate multi-class fashion sketches through two stages. At the
first stage, we design a Contour Generation Network to synthesize contour images with a given categorical vector. At the second stage, we
design a Sketch Translation Network to translate the contour images to
the sketch images by extending the StyleGAN2 model to the conditional
version. The quantitative and qualitative analysis demonstrates that our
method is capable of synthesizing high-quality fashion sketches.

Keywords: Fashion sketch · Deep generative model · Generative
adversarial network · Variational auto-encoder

1 Introdution

Sketch drawings play an important role in assisting humans in communication
and creative design since the ancient period. In ancient times, our ancestors
carved strokes on rocks to record events. Nowadays, sketching is the fundamental
first step for expressing artistic ideas. For instance, in the fashion world, new
designs are presented by fashion experts in the form of hand-drawn sketches
before they're cut and sewn. The fashion sketches usually include key information
such as category and patterns of fashion products, which are the foundation for
fashion style designs.

Besides, AI technologies are transforming the fashion industry in every element of its value chains such as designing at a faster pace than ever. For example, Google has already tested the waters of user-driven AI fashion design with
Project Muze, which creates designs based on users' interests and alignment with

---

This work was supported by the National Science Foundation of China (GrantNo.
61802100, 61972119, 61971172).

J. Fang—Student as the first author.

Y. Peng et al. (Eds.): PRCV 2020, LNCS 12306, pp. 544–556, 2020.
https://doi.org/10.1007/978-3-030-60639-8_45

the style preferences recognized by a pre-trained neural network[1]. Since fashion sketches play a critical role in the initial stages of fashion product design, allowing a fashion designer to quickly draft and visualize their thoughts, which motivates us to generate fashion sketches via AI techniques.

In this work, we aim to automatically generate high-quality multi-class fashion sketches by modeling pixel images of fashion sketches. The diversity of sketches and their abstract structures make fashion sketch synthesis a challenging task. Firstly, the fashion sketches of the same category may be drawn in diverse styles due to the nature of freehand sketches. Secondly, hand-drawn fashion sketches lack texture and necessary contextual information and are generally known to be more ambiguous than natural images.

Recently, there has been rapid improvement in generative modeling of images using neural networks as a generative tool in very recent years, among which Generative Adversarial Networks (GANs) [6] and Variational Auto-Encoder (VAEs) [10,15] have become popular tools in this fast-growing area. However, the majority of those research on image generation mainly deal with natural images and cannot be directly applied for sketch image generation. On the other hand, some works have tried to generate sketch drawings of common objects based on human-drawn inputs [3,8]. However, these works developed generative models of vector images rather than pixel images, which are also not suitable for our problem.

To achieve our goal we design a hybrid deep generative model for fashion sketch synthesis, called Fashion-Sketcher. Our key idea is to automatically generate multi-class fashion sketches by synthesizing contour images as an intermediate step. Such a process of sketch synthesis mimics human painters that first draw the outlines of objects and then finished the detailed structures and contents of the objects. Specifically, our Fashion-Sketcher generates multi-class fashion sketches through two stages. At the first stage, a Contour Generation Network (CGN) is designed to synthesize contour images with given categorical information, which combines a variational auto-encoder with cascade generative adversarial networks. At the second stage, a Sketch Translation Network (STN) is designed to transform the contour images to the sketch images, which extends StyleGAN2 [9] to the conditional version by adding an embedding network for encoding contour images.

Another challenge for fashion sketch synthesis is the lack of datasets for the training of the generative models. Although there have been a few sketch datasets which have been built as benchmarks for sketch-based image retrieval [5,18] and sketch classification [4], those existing sketch datasets cannot be used for the task of fashion sketch synthesis since the categories of objects of those existing sketch datasets are confined to general classes or a few specific classes. Thus, it motivates us to create a new sketch dataset for fashion sketch synthesis. An ideal way to collect the new dataset is to employ fashion experts to draw high-quality fashion sketches. Considering such a process is too expensive and

---

[1] https://blog.google/around-the-globe/google-europe/project-muze-fashion-inspired-by-you/.

labor-intensive, instead, we utilize the advanced image processing techniques to automatically extract sketches of fashion product images from online stores.

To summarize, our contributions are as follows: (1) We propose the first AI technique that enables a computer to automatically generate fashion sketches of multiple categories. (2) We propose a novel hybrid deep generative model called Fashion-Sketcher for automatically generating fashion sketches by synthesizing fashion contours as an intermediate step. (3) We build a new sketch dataset for fashion sketch synthesis by using advanced image processing techniques for extracting sketches of fashion products. (4) Our experiments demonstrate the effectiveness of our proposed model, which produces higher-quality synthesized results compared to other methods.

2   Related Work

2.1   Deep Generative Models

The state of generative image modeling has developed dramatically in recent years. Among various deep generative models, GANs and VAEs have become the most prominent techniques. A Generative Adversarial Net employs a two-player min-max game with two models, a generator and a discriminator. A Variational Auto-Encoder learns a parametric latent variable model by maximizing the marginal log-likelihood of the training data. GANs and VAEs both have their own significant strengths and limitations. Much effort has been devoted to combining the strengths of VAEs and GANs, e.g., VAE/GAN [11] and AAE [14]. In our work, we design a hybrid deep generative model for fashion sketch synthesis by leveraging the strengths of GANs and VAEs.

2.2   Sketch Generation

Very recently, Ha et al. [8] introduced a sequence-to-sequence VAE model called sketch-rnn that is able to generate simple and cursive stroke-based drawings of common objects based on human-drawn inputs. Since its invention, several models have been developed based on Sketch-RNN. Song et al. [16] introduced a stroke-level photo-to-sketch synthesis model based on Sketch-RNN to extract sketches from images. Cao et al. [3] proposed a deep generative model for learning sequential and spatial information from a set of training sketches to automatically generate multi-class sketch drawings with higher quality. The aforementioned methods for sketch drawings are not suitable for our problem because they deal with vector images of sketches, rather than develop generative models of pixel images of sketches.

2.3   Sketch-Based Datasets

Currently, there are a few datasets of human-drawn sketches for sketch-based image retrieval [5,18], sketch-based image classification [4] and sketch synthesis

[8]. As for sketch synthesis, the newly published QuickDraw dataset [8] built by Google has an impressive 50 million vector drawings spanning hundreds of classes of common objects. However, those existing sketch datasets cannot be used for the task of fashion sketch synthesis, because the categories of objects of those existing sketch datasets are confined to general classes or a few specific classes.

3 Dataset

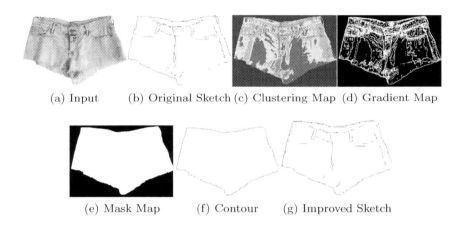

(a) Input   (b) Original Sketch (c) Clustering Map (d) Gradient Map

(e) Mask Map   (f) Contour   (g) Improved Sketch

Fig. 1. The pipeline of fashion sketch creation.

We collect high-resolution (650*650) fashion product images from an online store (e.g., Tradesy) and manually select high-quality images that contain pure products. In total, 12,500 high-quality fashion product images of five representative categories (e.g., dress, pants, shorts, skirts and tops) are collected. We use edge detection and several post-processing steps to extract fashion sketches. We first detect edges of fashion product images with Holistically-nested edge detection (HED). After binarizing the output and thinning all edges, we clean isolated pixels and remove small connected components. Then, we remove the remaining spurs.

However, the originally extracted sketches should be refined because the thinning operation and threshold selection cause the discontinuity of contours. Thus, we take the following steps to further improve the extracted fashion sketches by getting the continuous contours: (1) We cluster the pixels of the original fashion product image by using K-means algorithm. (2) After computing the gradients of the Clustering Matrix[2] and performing a flood-fill operation on the gradient

---

[2] Each entry in the Clustering Matrix corresponds to a pixel in an original fashion product image and the value of each entry is set with the class label of the corresponding pixel.

Table 1. The detailed data split for training and evaluation.

Dataset	Train Set	Test Set	Total
Dress	2,125	375	2,500
Pants	2,125	375	2,500
Shorts	2,125	375	2,500
Skirts	2,125	375	2,500
Tops	2,125	375	2,500
Total	10,625	1,875	12,500

map, we obtain a mask map that separates the foreground and background of the product image. (3) We generate a continuous contour image by computing the gradients of the mask map and binarizing it. (4) Once we get the contour image, the sketch image is then refined by simply overlaying the original sketch image and the contour image. Figure 1 illustrates the pipeline of fashion sketch creation. Finally, we split the 12,500 extracted sketch into the training set and test set with a ratio of 8.5:1.5, where we try to keep the category labels in each set to be evenly distributed. Table 1 displays the detailed data split for training and evaluation.

4   Proposed Method

In this section, we introduce the detailed design and implementation of Fashion-Sketcher. Our Fashion-Sketcher mainly consists of two modules, namely, Contour Generation Network (CGN) and Sketch Translation Network (STN). CGN combines a variational auto-encoder with cascade generative adversarial networks for synthesizing contour images in fine-grained categories. STN extends StyleGAN2 [9] to the conditional version for translating contour images to fashion sketches.

4.1   Contour Generation Network

Network Architecture. To keep the diversity of generated contour images in fine-grained categories during inference, CGN combines a conditional variational auto-encoder with cascade generative adversarial networks. Specifically, CGN consists of following components as shown in Fig. 2:

1. An encoder E, which maps a contour image x to a latent representation z through a learned distribution $P(z|x,c)$, where c is a k-dimensional one-hot conditional vector. The k-dimensional one-hot conditional vector encodes the categorial information of the input contour image. Here, k indicates the number of classes of contour images.

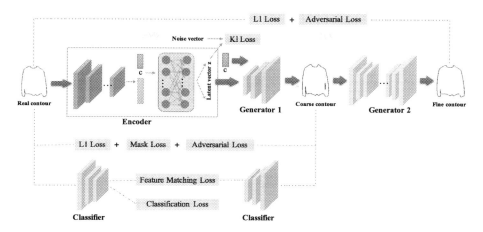

Fig. 2. CGN combines a variational auto-encoder with cascade generative adversarial networks for keeping the diversity of generated contour images.

2. A classifier C, which measures the class probability of a real coutour image x, $P(c|x)$.
3. A generator G_1, which generates a coarse contour image x' given a latent vector z and a conditional vector c by sampling from the learned distribution $P(x|z,c)$.
4. A discriminator D_1, which distinguishes real/fake contours (where x is a real contour image and x' is a fake coarse contour image).
5. A generator G_2, which generates a fine contour image x'' conditioned on a coarse contour image x'. We are motivated to add a cascade generative adversarial network in CGN for refining the contour images since the background of contour images generated by G_1 are usually not clear enough.
6. A discriminator D_2, which distinguishes real/fake contours (where x is a real contour image and x'' is a fake fine contour image).

Overall, the training procedure of CGN follows two steps. In the first step, CGN generates a coarse contour image x' by simultaneously training E, C, G_1, and D_1. In the second step, CGN generate a fine contour image x'' by adding a cascade module (G_2, D_2).

Generating Coarse Contours. Similar to VAE, the encoder E maps a contour image x to a latent representation z. To ensure that random sampling can be used during inference time, the latent distribution $z \sim E(x)$ is regularized using KL-divergence to be close to Gassian distribution $N(0, I)$ during training:

$$\mathcal{L}_{kl} = E_{x \sim p(x)}[D_{KL}(E(x)||N(0, I))] \tag{1}$$

The classifier C accepts a contour image x and outputs a k-dimensional conditional vector, which turns into class probabilities using a softmax function. The classifier C tries to minimize the softmax loss during training:

$$\mathcal{L}_{cla} = -E_{x \sim p(x)} \log p(c|x) \tag{2}$$

For discriminator D_1, it tries to distinguish real contours from synthesized ones. Concretely, the discriminator D_1 tries to minimize the loss function:

$$\mathcal{L}_{dis} = -E_{x \sim p(x)} D_1(x) + E_{c \sim p(c), z \sim E(x)}(D_1(G_1(c,z))) + \lambda_{gp} L_{gp} \tag{3}$$

where the λ_{gp} is a hyperparameter for controlling the gradient penalty. We adopt WGAN-GP for the adversarial learning of D_1 and G_1. WGAN-GP [7] improves WGAN [1] on the implementation of Lipschitz constraint by imposing a gradient penalty on the discriminator instead of weight clipping.

For generator G_1, we improve the GAN loss by incorporating a reconstruction loss, feature matching loss based on the classifier and a mask loss:

$$\begin{aligned}
\mathcal{L}_{gen} = &-\lambda_{gan} E_{c \sim p(c), z \sim E(x)}(D_1(G_1(c,z))) + \\
&\lambda_{rec} E_{c \sim p(c), z \sim E(x)} \|x - G_1(c,z)\|_1 + \\
\lambda_f E_{x \sim p(x), c \sim p(c), z \sim E(x)} \|f(x) &- f(G_1(c,z))\|_1 + \\
\lambda_m E_{x \sim p(x), c \sim p(c), z \sim E(x)} \|m \otimes x &- m \otimes G_1(c,z)\|_1
\end{aligned} \tag{4}$$

Here the first loss term is used to fool the discriminator as the traditional GANs do. The L_1 reconstruction loss (second term) is added to encourage less blurring. The feature matching loss (third term) is used to deal with the unstable gradient of the generator, where $f(x)$ represents features on an intermediate layer of the classifier C. The mask loss (fourth term) is added to refine the contours, where m represents a binary mask map. λ_{gan}, λ_{rec}, λ_f and λ_m are weight parameters that balance different loss terms.

Overall, the goal of the first step of CGN is to minimize the following loss function:

$$\mathcal{L}_{CGN_1} = \lambda_{kl} \mathcal{L}_{kl} + \lambda_{cla} \mathcal{L}_{cla} + \lambda_{dis} \mathcal{L}_{dis} + \lambda_{gen} \mathcal{L}_{gen} \tag{5}$$

where λ_{kl}, λ_{cla}, λ_{dis} and λ_{gen} are weight parameters that balance different loss terms. All models (including E, C, D_1 and G_1) are updated iteratively as all these objectives are complementary to each other, and ultimately enable the model to obtain satisfactory results.

Generating Fine Contours. Once the encoder E, classifier C, discriminator D_1 and generator G_1 have been well trained, a cascade module that consists of G_2 and D_2 is co-trained for refining the generated contours. It is worth noting that, only G_2 and D_2 are updated during training in the second step while all the weights of the models in the first step are fixed. The generator G_2 and discriminator D_2 are essentially a conditional GAN, where G_2 generates a fine contour image $x^{''}$ conditioned on a coarse contour image $x^{'}$ generated by G_1 and D_2 distinguishes a real contour x from a fake contour x''.

We use LSGANs [12] for stable adversarial learning of G_2 and D_2. The goal of the D_2 is to minimize the following loss function:

$$\mathcal{L}_{dis2} = \frac{1}{2} E_{x',x \sim p(x',x)}[(D_2(x^{'},x)-1)^2] + \\ \frac{1}{2} E_{x' \sim p(x')}[D_2(x^{'},G_2(x^{'}))^2] \qquad (6)$$

In contrast, the goal of the G_2 is to minimize the following loss function:

$$\mathcal{L}_{gen2} = \frac{1}{2} E_{x' \sim p(x')}[(D_2(x^{'},G_2(x^{'}))^2-1)^2] \qquad (7)$$

We further utilize the L1 reconstruction loss and feature loss to mix with the GAN objective to encourage less blurring of generated contour images:

$$\mathcal{L}_{gen2} = \frac{1}{2} E_{x' \sim p(x')}[(D_2(x^{'},G_2(x^{'}))^2-1)^2] + \\ \lambda_{rec2} E_{x',x \sim p(x',x)}\|x - G_2(x^{'})\|_1 + \\ \lambda_{f2} E_{x',x \sim p(x',x)}\|f(x) - f(G_2(x^{'}))\|_1 \qquad (8)$$

where λ_{rec2} and λ_{f2} are weight parameters that balance the adversarial loss term, reconstruction loss and feature loss.

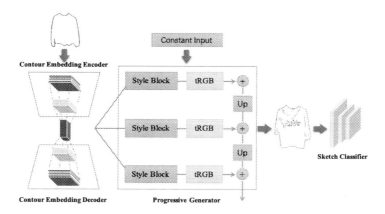

Fig. 3. STN maps a contour image to a target sketch image by extending the Style-GAN2 model to the conditional version.

4.2 Sketch Translation Network

Network Architecture. The task of STN is to map a contour image to a target sketch image. Our STN is built on the state-of-the-art network, StyleGAN2 [9] Concretely, we extend StyleGAN2 to the conditional version by adding an contour embedding autoencoder for encoding contour images and a sketch classifier for improving the quality of synthetic images. As illustrated in Fig. 3, our STN consists of following main components:

1. A contour embedding autoencoder, which consists of the contour embedding encoder and the contour embedding decoder. The encoder encodes a contour image into a 512-dimensional embedding vector for producing the style vector, where the style vector controls the layers of the synthesis network.
2. A progressive generator, which generates the target sketch image with input style vector. Similar with StyleGAN2, the style is fed into each style block. At each style block, weight modulation and demodulation are employed in all convolution layers, except for the output layers leaving out the demodulation.
3. A discriminator, which distinguishes real or fake sketch images. Note that the network structures of the progressive generator and the discriminator are identical to those in StyleGAN2.
4. A sketch classifier, which classifies the synthetic sketch images into five categories.

To train STN, firstly, the contour embedding autoencoder is pretained on the contour images of the training set using the reconstruction loss. Secondly, a VGG16 sketch classifier is pretained on the sketch images of the training set using cross entropy loss. Finally, the training procedure of STN is a minmax two-player game between the progressive generator and discriminator.

The goal of the progressive generator is to minimize the following loss function:

$$\mathcal{L}_{PG} = \lambda_{Adv_g}\mathcal{L}_{Adv_g} + \lambda_{Path}\mathcal{L}_{Path} + \\ \lambda_{Rec}\mathcal{L}_{Rec} + \lambda_{Class}\mathcal{L}_{Class} \tag{9}$$

Here $\mathcal{L}_{Adv_g}$ and $\mathcal{L}_{Path}$ are consistent with the loss terms in StyleGAN2. The $\mathcal{L}_{Rec}$ is the L1 reconstruction loss and $\mathcal{L}_{Class}$ is the classification loss. λ_{Adv_g}, λ_{Path}, λ_{Rec} and λ_{Class} are weight parameters that balance different loss terms.

The goal of the discriminator is to minimize the following loss function:

$$\mathcal{L}_{Dis} = \lambda_{Adv_d}\mathcal{L}_{Adv_d} + \lambda_{R1}\mathcal{L}_{R1} \tag{10}$$

Here $\mathcal{L}_{Adv_d}$ and $\mathcal{L}_{R1}$ are consistent with the loss terms in StyleGAN2. λ_{Adv_d} and λ_{R1} are weight parameters that the adversarial loss term and R1 regularization term.

5 Experiment

5.1 Implementation Details

Training Details. All experiments in this work were conducted on NVIDIA TITAN Xp GPUs. During training CGN, we used adam optimizer with beta1

= 0.5 and beta2 = 0.999. And we trained E, C, D_2 and G_2 with a learning rate of 0.0002 and trained G_1 and D_1 with a learning rate of 0.0001. The weight parameters were set as $\lambda_{kl} = 1$, $\lambda_{cla} = 1$, $\lambda_{dis} = 1$, $\lambda_{gen} = 1$, $\lambda_{gp} = 10$, $\lambda_{gan} = 0.25$, $\lambda_{rec} = 8$, $\lambda_f = 15$, $\lambda_m = 15$, $\lambda_{rec2} = 15$ and $\lambda_{f2} = 10$. In the first stage of CGN, we used a batch size of 22 and trained the network for 90 epochs. In the second stage of CGN, we used a batch size of 30 and trained the network for 100 epochs. During training STN, both the generator and discriminator used the adam optimizer with beta1 = 0, beta2 = 0.99. The weight parameters were set as $\lambda_{Adv_g} = 1$, $\lambda_{Path} = 2$, $\lambda_{Rec} = 3.5$ and $\lambda_{Class=0.01}$, $\lambda_{Adv_d} = 1$ and $\lambda_{R1} = 10$. We used a batch size of 6 and trained STN with a learning rate of 0.0002 for 250 epochs.

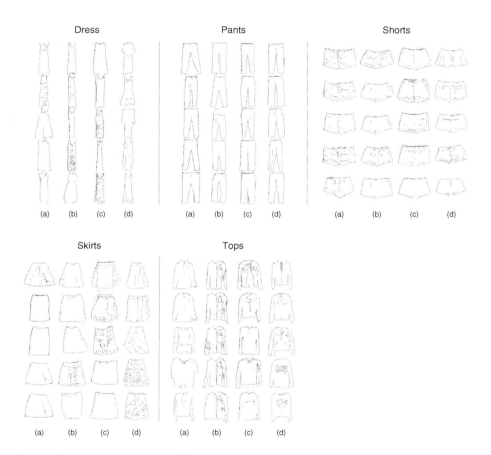

Fig. 4. Comparison results of different approches on multi-class fashion sketch synthesis. Here, (a), (b), (c) and (d) represents ACWGAN, CVAE-GAN, CVAE-WGAN and our method, respectively.

5.2   Evaluation

Baselines. To validate the effectiveness of our proposed method, we compare with the following baseline methods. Unlike our method, the three baselines directly generate fashion sketches conditioned on the conditional vector which encodes the categorial information of sketch images.

- **ACWGAN**: ACWGAN is an improved model of ACGAN [13], where wgan-gp loss [7] replaces the traditional gan loss.
- **CVAE-GAN**: CVAE-GAN combines the conditional variational auto-encoder and the generative adversarial network [2]. Specifically, it consists of four modules, an encoder, a classifier, a generator and a discriminator. Note that the network structures of the encoder, the generator and the discriminator are identical with those of E, G_1 and D_1 in our CGN, respectively. The classifier uses VGG16 model and is fine-tuned on our sketch dataset.
- **CVAE-WGAN**: CVAE-WGAN adopts the same architecture with CVAE-GAN but uses wgan-gp loss instead of the traditional gan loss.

Qualitative Results. Figure 4 presents a visual comparison of the evaluated methods. We compare these methods with the synthetic fashion sketches including dress, pants, shorts, skirts and tops. We observed that the generated fashion sketches by ACWGAN and CVAE-WGAN are blurry and of low quality. Although CVAE-GAN generates more clear fashion sketches than ACWGAN and CVAE-WGAN, it is inferior to our in diversity of synthetic sketches. Overall, comparing with other baselines, our method renders more realistic and diverse textures, especially for skirts and tops.

Quantitative Results. To quantitatively evaluate the quality of the generated images by different models, we adopt four widely used evaluation metrics [17] which include the Inception Score, Fréchet Inception Distance, Kernel MMD and the Wasserstein Distance. The *Inception Score* is the most widely adopted score for GAN evaluation, which measures the quality and diversity of the generated images using a pre-trained neural network. The *Fréchet Inception Distance* is a measure of similarity between two datasets of images, which is then used for evaluating the quality of generated samples. The *Kernel MMD* measures the dissimilarity between P_r and P_g for a fixed kernel function k. A lower MMD means that P_g is closer to P_r. The *Wasserstein Distance* is often referred to as the Earth Mover's Distance, and corresponds to the solution to the optimal transport problem. Similar to the Kernal MMD, the Wasserstein Distance is lower when two distributions are more similar.

Table 2 reports the evaluation results for our method and the other baselines. As we can see, our method consistently outperforms the other baselines in terms of four evaluation metrics, which demonstrates the advatanges of our proposed method that synthesizes contour images as an intermediate step for generating multi-class fashion sketches. By comparing the results of CVAE-GAN and

Table 2. Quantitative comparison of four evaluation metrics.

Model	IS ↑	FID ↓	Kernel MMD ↓	Wasserstein Distance ↓
ACWGAN	1.237	0.037	0.196	7.622
CVAE-GAN	1.232	0.024	0.122	6.919
CVAE-WGAN	1.240	0.035	0.116	7.489
Our Method	**1.268**	**0.018**	**0.069**	**6.721**

CVAE-WGAN, we found that replacing the traditional gan loss with wgan-gp loss dose not necessarily improve the synthetic results. In contrast, wgan-gp loss is crucial for ACWGAN since we found that the model easily suffered from mode collapse using the traditional gan loss in the preliminary experiments.

6   Conclude

We have presented a novel approach for generating high-quality multi-class fashion sketches through two stages. At the first stage, we design CGN for synthesizing contour images with given categorical information by combining a variational auto-encoder with cascade generative adversarial networks. At the second stage, we design STN for transforming the contour images to the sketch images by extending the StyleGAN2 model to the conditional version. We conducted experiments on our fashion sketch dataset, and promising results are achieved both quantitatively and qualitatively, which shows the effectiveness of our proposed approach.

References

1. Arjovsky, M., Chintala, S., Bottou, L.: Wasserstein gan. CoRR abs/1701.07875 (2017)
2. Bao, J., Chen, D., Wen, F., Li, H., Hua, G.: CVAE-GAN: fine-grained image generation through asymmetric training. In: ICCV, pp. 2764–2773 (2017)
3. Cao, N., Yan, X., Shi, Y., Chen, C.: AI-sketcher: a deep generative model for producing high-quality sketches. In: AAAI, pp. 2564–2571 (2019)
4. Eitz, M., Hays, J., Alexa, M.: How do humans sketch objects? ACM Trans. Graph. **31**(4), 44:1–44:10 (2012)
5. Eitz, M., Hildebrand, K., Boubekeur, T., Alexa, M.: Sketch-based image retrieval: benchmark and bag-of-features descriptors. IEEE Trans. Vis. Comput. Graph. **17**(11), 1624–1636 (2011)
6. Goodfellow, I.J., et al.: Generative adversarial nets. In: NIPS, pp. 2672–2680 (2014)
7. Gulrajani, I., Ahmed, F., Arjovsky, M., Dumoulin, V., Courville, A.C.: Improved training of wasserstein gans. In: NIPS, pp. 5767–5777 (2017)
8. Ha, D., Eck, D.: A neural representation of sketch drawings. CoRR abs/1704.03477 (2017)
9. Karras, T., Laine, S., Aittala, M., Hellsten, J., Lehtinen, J., Aila, T.: Analyzing and improving the image quality of stylegan. CoRR abs/1912.04958 (2019)

10. Kingma, D.P., Welling, M.: Auto-encoding variational bayes. In: ICLR (2014)
11. Larsen, A.B.L., Sønderby, S.K., Larochelle, H., Winther, O.: Autoencoding beyond pixels using a learned similarity metric. In: ICML, vol. 48, pp. 1558–1566 (2016)
12. Mao, X., Li, Q., Xie, H., Lau, R.Y.K., Wang, Z., Smolley, S.P.: Least squares generative adversarial networks. In: ICCV, pp. 2813–2821 (2017)
13. Odena, A., Olah, C., Shlens, J.: Conditional image synthesis with auxiliary classifier gans. In: ICML, vol. 70, pp. 2642–2651 (2017)
14. Pidhorskyi, S., Almohsen, R., Doretto, G.: Generative probabilistic novelty detection with adversarial autoencoders. In: NeurIPS, pp. 6823–6834 (2018)
15. Rezende, D.J., Mohamed, S., Wierstra, D.: Stochastic backpropagation and approximate inference in deep generative models. In: ICML, vol. 32, pp. 1278–1286 (2014)
16. Song, J., Pang, K., Song, Y.Z., Xiang, T., Hospedales, T.M.: Learning to sketch with shortcut cycle consistency. In: CVPR, pp. 801–810. IEEE Computer Society (2018)
17. Xu, Q., et al.: An empirical study on evaluation metrics of generative adversarial networks. CoRR abs/1806.07755 (2018)
18. Yu, Q., Liu, F., Song, Y.Z., Xiang, T., Hospedales, T.M., Loy, C.C.: Sketch me that shoe. In: CVPR, pp. 799–807 (2016)

Cross-Modality Person ReID with Maximum Intra-class Triplet Loss

Xiaojiang Hu and Yue Zhou[✉]

Institute of Image Processing and Pattern Recognition,
Shanghai Jiao Tong University, Shanghai, China
{jungle_hu,zhouyue}@sjtu.edu.cn

Abstract. Compared with conventional ReID task, RGB-IR person re-identification is more significant and challenging owing to enormous cross-modality variations between RGB and infrared images. Most existing cross-modality person re-identification approaches in the infrared and visible domains use the joint training of traditional triplet loss and CE loss to reduce the discrepancy between two modalities. However, as the model gradually converges, the number of positive and negative sample pairs that can be optimized by traditional triplet loss also decreases. In this paper, we adopt a dual-stream CNN structure and propose a maximum intra-class distance to triplet loss (MICT loss) for RGB-IR ReID task. The proposed method enjoys several advantages. First, it can learn specific modal information and shared information better through the CNN structure of the dual-stream branch. Second, it can add the optimization of intra-class cross-modality distance to the traditional triplet loss, which can effectively reduce the intra-class distance and increase the training difficulty of the model. Experimental results demonstrate that our proposed algorithm performs favorably against the state-of-the-art methods that rank-1 accuracy is 61.71% and mAP is 58.06% on SYSU-MM01 dataset.

Keywords: Person Re-identification · Triplet loss · Cross modality · Maximum intra-class distance

1 Introduction

Person re-identification is regarded as an image retrieval problem, aiming at matching the same person image of interest between different non-overlapping cameras[25]. With the continuous improvement of smart cities and video surveillance, person re-identification task has attracted more and more attention. It is very challenging owing to the inter-class and intra-class discrepancies resulted from various viewpoints, poses, illuminations. Currently, most of person re-identification methods mainly focus on matching images between visible cameras[4, 8, 9, 12, 16, 24].

X. Hu—Student.

© Springer Nature Switzerland AG 2020
Y. Peng et al. (Eds.): PRCV 2020, LNCS 12306, pp. 557–568, 2020.
https://doi.org/10.1007/978-3-030-60639-8_46

However, only RGB person re-identification is not enough, because conventional visible cameras are often difficult to obtain rich image information in poor light or night environments, causing a large retrieval discrepancy between day and night. Therefore, the visible camera is often used to obtain images during the day, and the infrared camera is used to obtain near-infrared images at night. Compared with RGB person re-identification, IR-RGB person re-identification still needs to consider the difference between two different modalities, making it more difficult to retrieve the same person across disjoint cameras. As shown in Fig. 1, the IR and RGB images have different spectra, and the two types of information are heterogeneous. Therefore, it is difficult to directly use the person re-identification methods for processing RGB images in the RGB-IR cross-modality person re-identification task.

RGB images
In the day

Thermal images
During the night

Fig. 1. Instances of RGB images and thermal images in SYSU-MM01 dataset

To solve this problem, some methods have been proposed in this field. Wu et al. [18] first proposed a large-scale cross-modality re-identification dataset, and used CE loss to train a single-stream deep network structure called zero-padding. Ye et al. [20] proposed to use CE loss and contrastive loss to train the dual-stream network structure named TONE. Due to the lack of flexibility of contrastive loss in learning shared subspaces, Ye et al. [22] proposed to use top-ranking loss to train a dual-stream network model called BDTR. In addition, Dai et al. [3] approached the problem from different angles and used triplet loss and CE loss to train a generative adversarial network called cmGAN to allow the model to learn cross-modal feature representations. Zhang et al. [23] proposed a two-channel public space network that preserves spatial structure and contrast-related networks. The former embeds cross-modal images into a common three-dimensional tensor space, and the latter extracts contrast features by dynamically comparing input images.

However, most of the methods mentioned above are aiming at adjusting the network or trying different loss functions which are existing. As the number of training increases, the effect of inter-class distance on network training is gradually smaller than the effect of intra-class distance, and for cross-modal tasks, the optimization of intra-class distance becomes more important. In this paper, a maximum intra-class distance is added to the traditional triplet loss, ensuring that the network can optimize the intra-class distance in the training process. In addition, we use cascade supervised learning of CE loss and maximum intra-class triplet (MICT) loss to train this network. Both loss functions can be minimized by standard optimization algorithms, such as stochastic gradient descent (SGD) [1].

At the same time, we used a dual-stream network structure. In order to ensure that CE loss and MICT loss are more effective for network training, we first introduced the BNNeck [15] structure to dual-branch network model for RGB-IR ReID, reducing the network optimization differences caused by these loss functions. The network has two parts, the first part is used to extract the feature information of different modalities, and the second part is used to find the shared subspace of the two modalities. A large number of experimental results show that the dual-branch network structure using MICT loss and CE loss can achieve SOTA performance in this field.

The main contributions of this paper can be summarized as follows. (1) We improved the classic triplet loss by adding the maximum intra-class distance, ensuring the network can pay attention to the optimization of intra-class discrepancy during training. Especially for cross-modal problems, the discrepancy between different modalities of the same person is large. (2) We used a simple network structure (ResNet50) and introduced the BNNeck structure to the RGB-IR task to verify its effectiveness. The whole structure is not only simple and effective, but the final result is also very good. (3) In the all-search mode of the SYSU-MM01 dataset, our proposed method achieves 61.71% and 58.06% for rank-1 and mAP, respectively.

2   Related Work

RGB Reid. The main challenge of conventional Reid task is the large intra-class differences caused by different viewpoints, poses and occlusions. The task has two basic contents, namely feature expression and metric learning. In terms of feature expression, some recent researches mainly focus on designing more powerful network structures in order to better extract some features of the person. In metric learning, it mainly focuses on the design of loss functions. [5] first proposed Triplet loss. By selecting three pictures to form a triple, and making the network during the training process, the distance between the positive samples is the smallest, and the distance between the negative samples is increased, which can solve the problems of inter-class similarity and intra-class difference. [7] adds the idea of hard sample mining to traditional triplet loss, and selects the most difficult images from same ID and different ID in a mini-batch to form a triple. [2]proposes the Quadruplet Loss method. Compared with the traditional triplet loss, it mainly considers the absolute distance between negative samples. [19]

proposes the Margin Sample Mining Loss method. Compared with Quadruplet Loss, this method selects only the most difficult positive sample and the most difficult negative sample pair to calculate the loss. However, with the continuous traning of the model, there are many triplets in the late stage of training with a loss of 0. This paper is based on the maximum intra-class triplet loss improved by the tri-hard method, which effectively guarantees that the triplet loss in the mini-batch is non-zero during the entire training process.

Multi-modal Person Re-Identification. Recently, a large number of multi-modal person re-identification models have been proposed. Nguyen et al. [17] used the person re-identification model for the first time on RGB-IR images. Wu et al. [18] designed a deep shape descriptor for noise and rotation and robustness. At the same time, Lin et al. [14] combined attribute information in visible light images. These works usually use multi-modal information to improve the performance of ReID, but we mainly focus on cross-modal person re-identification. For cross-modal person re-identification, [10,11,21] proposed a series of person retrieval methods for text pictures. However, these methods cannot be directly used in RGB-IR Reid tasks.

RGB-IR Reid. In RGB-IR Reid, [22] proposed a two-stage structure that includes feature learning and metric learning. In addition, [18] introduced a deep zero-padding network to learn the shared features of different modalities. [6] proposed an end-to-end dual-stream hyper spherical manifold embedded network with classification and identification constraints. At the same time, a two-stage training scheme to obtain decorrelated features is also designed. [13] used the HPILN framework and proposed a specially designed hard pentaplet that can effectively handle cross-modal and internal modal changes in RGB-IR ReID. [23] proposed a two-path cross-modal feature learning framework based on the characteristic that humans usually notice the difference between two similar objects. Among them, DSCSN embeds the cross-modal image into a common three-dimensional tensor space without losing the spatial structure, and CCN extracts the contrast features by dynamically comparing the input image pairs. Although the above research works have achieved very good results, their model structures are relatively complex and difficult to use efficiently. The model structure proposed in this paper is based on ResNet50 and introduces the BNNeck structure. The whole model is very simple, and the experimental results show that the final effect is also good.

3   Our Proposed Method

In this paper, we use an end-to-end dual-stream convolutional neural network, as shown in Fig. 2. This architecture consists of two parts. The first part is used to extract the unique features of the modal, and the second part uses a shared weight strategy to allow the network to extract common features. With the

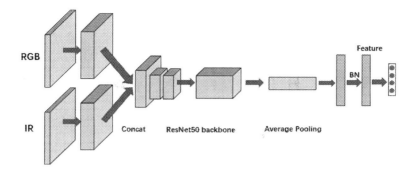

Fig. 2. The Dual-stream Network with BNNeck

joint supervision of CE loss and MICT loss, the model learns the discriminative features. Specifically, the dual-stream network can well optimize the distance between different modalities, and make the model learn a shared space. In addition, in this paper, the maximum intra-class distance is added to the traditional triple loss, so that the intra-class distance can be optimized when the model converges.

3.1 Dual-Stream Network

We used a dual-stream network structure to extract features in the near infrared and visible domains. Specifically, the dual-stream structure contains two parts, feature extraction and feature embedding. The former is mainly to obtain specific information of different image modalities. The latter focuses on learning a multi-modal shared space to bridge the gap between two heterogeneous modalities.

Feature Extraction. We use ResNet50 as the backbone network, which uses the zeroth layer (res-layer0) in ResNet50 to extract shallow features of different modalities. The main reason is that the shallow convolutional layer can obtain low-level information, such as texture, corner. This may be shared among all pictures. Although the structure of extracting shallow features of different modalities is the same, the parameters of the two branches are independent of each other to ensure that unique feature information of different modalities is obtained. In addition, due to the small size of the data set, the network model we use needs to be initialized with the help of pre-trained parameters on ImageNet, which accelerates the convergence of training.

Feature Shared Subspace. In order to learn the embedding space of two heterogeneous modalities, we introduce a shared layer after the feature extraction part. Note that the shared layer contains all the structures after layer1 in ResNet50, and all the parameters are shared. At the same time, we added the BN structure before the FC layer. If the shared layer is not added for feature embedding, the feature space learned in the visible domain and the feature space learned in the near-infrared domain are completely different subspaces. In the following experimental section, we will explain through the experimental results that for RGB-IR Reid, the shared layer can well capture the two different modal shared subspaces, because its function is similar to a projection function.

3.2   Maximum Intra-class Triplet Loss

When embedding the features of visible images and near-infrared images into a shared subspace, we optimized the conventional triplet loss, ensuring that the model can maximize the distance between classes and within classes during the training process. First of all, we review the classic triplet loss and its related deformation forms.

Triplet Loss. There are many image matching works such as face recognition and person re-identification which widely used triplet loss. In person re-identification, we can choose P persons in a mini-batch, and each person has N visible images and N near infrared images, respectively. Using the visible image x_i as the anchor in candidate triplet set, the corresponding label is y_i, we hope that the distance between x_i and the positive sample z_j should be smaller than the distance between x_i and the negative sample z_k, where m is a margin value. Using the convolutional neural network as the feature extractor, all the features used for the triplet loss are normalized by L2. In our method, the Euclidean distance between feature embedding is used as the similarity measure function of two images. Therefore, we further obtain the formula for the triplet loss:

$$L_{triplet} = \sum_{\forall y_i \neq y_k, \forall y_i = y_j} max[m + D(x_i, z_j) - D(x_i, z_k), 0] \tag{1}$$

where Euclidean distance $D(x_i, z_j = ||f(x_i) - f(z_j)||_2$, and the subscripts i and j represent the same identity, while i and k are different identities.

Soft Margin Loss. The loss function is obtained by the classic triplet loss deformation. In traditional triplet loss, the margin value is set artificially to ensure that different classes can be separated by a certain distance during convergence. However, artificially setting the margin value often requires continuous trial and error before it is possible to find a relatively suitable value. Therefore, [7] proposed the concept of soft margin and removed the artificially set margin value. The formula is as follows:

$$L_{softmargin} = \sum_{\forall y_i \neq y_k, \forall y_i = y_j} log[1 + exp(D(x_i, z_j) - D(x_i, z_k))] \tag{2}$$

Maximum Intra-class Triplet Loss. When the first item of max in the traditional triplet loss is zero, the network model will not perform backpropagation to optimize this part, resulting in waste of resources and unbalanced training. Therefore, we propose to replace the second term of max with the maximum intra-class distance to ensure that the network can optimize all samples every time. At the same time, for the RGB-IR Reid task, the distance between different modes of the same category can also be reduced. The improved formula is as follows:

$$L_{mdtri} = \sum_{\forall y_i \neq y_k, \forall y_i = y_j} max[m + D(x_i, z_j) - D(x_i, z_k), D(x_i, z_j)] \qquad (3)$$

In the same way, we can get the maximum intra-class Soft Margin loss:

$$L_{softmargin} = \sum_{\forall y_i \neq y_k, \forall y_i = y_j} log[1 + exp(D(x_i, z_j) - D(x_i, z_k)) + D(x_i, z_j)] \quad (4)$$

Batch Sampling. In order to ensure that the constraint effect in the mini-batch is the same, we have made some adjustments based on mini-batch sampling method. Specifically, P persons are randomly selected for training in each iteration. Then, we randomly select M pictures from different modalities to form a mini-batch, in which a total of $2 \times P \times M$ pictures is fed into the network. In this way, we will be able to select M from the visible image as the anchor and use it to calculate and correspond to the triplet loss of the near infrared image, and vice versa. Due to the randomness of the sampling mechanism, as the number of training increases, it will traverse all possible situations to obtain a global optimal.

4   Experimental Results

4.1   Experimental Settings

SYSU-MM01 is a large-scale cross-modal person re-identification dataset, in which the data is collected by 6 cameras, including 4 visible cameras and 2 thermal cameras. Because the person images in this dataset are collected in two different environments, indoor and outdoor, it is more challenging. It contains a total of 29003 visible images and 15712 near infrared images of 491 persons, and each person collects images from at least two different cameras. Among them, 22258 visible images of 395 people and 11909 near-infrared images were used for training, and 96 images were used for testing. During the test, 3803 near-infrared images were used as the query set, and the gallery set was a sample formed by randomly selecting each person image from different cameras. This is the most difficult situation and the evaluation standard for Single-shot all search, which is widely used by researchers.

Evaluation Metrics. In order to show the performance of our method, standard cumulative matching characteristics (CMC) and mean average accuracy (mAP) were introduced.

Implementation Details. We use the pytorch to implement the algorithm in this paper. Pedestrian pictures are transformed to 288×144. Use random cropping and random horizontal flip as data enhancement. The batch size is set to 64. In order to implement our proposed sampling strategy, there are 8 pedestrian ids in a batch, each of which consists of 4 visible light images and 4 near infrared images. The final feature dimension after pooling is 2048. After passing through the BN layer and the FC layer, the dimension reduction is 395 dimensions. The optimizer used is SGD, where the momentum is set to 0.9. The learning rate was set to 0.2, the learning rate became 0.02 after 20 epochs, and 0.002 after 40 epochs.

4.2 Ablation Study

This section evaluates the effect of our proposed end-to-end dual-stream network structure under different changes, as shown in the following table. In Table 1, we use Triplet loss as the metric function of the network, and compare the effectiveness of adding the BNNeck structure to the network. As can be seen from the data in Table 1, after adding the BNNeck structure, the network performance has been greatly improved, Rank-1 and mAP increased by 17.8% and 14.77%, respectively. In Table 2, we extracted the results of the two modal features at different locations and concat them. From the experimental results, we can know that the effect of concat after res-layer0 is the best. Therefore, in our subsequent experiments, we use the best results in Table 1 and Table 2, that is, use the BNNeck structure and concat after layer0 as the experimental baseline.

Table 1. The effect with BNNeck Structure on SYSU-MM01

	SYSU-MM01	
	R=1	mAP
Triplet loss/no BN	38.23	39.61
Triplet loss/with BN	56.03	54.38

Table 2. Which layer to concat for baseline

	SYSU-MM01	
	R=1	mAP
Res-layer0	60.14	57.05
Res-layer1	59.19	56.73
Res-layer2	58.64	56.39
Res-layer3	55.51	54.28
Res-layer4	51.43	52.04

Table 3 shows the experimental results after adding the method in this article. The classic triplet loss and MICT loss are used in the first two lines, and the soft margin loss and maximum soft margin loss are used in the last two lines. It can be seen from the experimental results that after adding the maximum intra-class distance, rank-1 and mAP in the case of triplet loss are increased by 5.68% and 3.68%, respectively. In the case of soft margin loss, rank-1 and mAP increased by 2.4% and 2.63%, respectively. In addition to verifying the effectiveness of the proposed method in the RGB-IR Reid task, we also conducted related experiments in the RGB Reid task. The baseline in Table 4 uses the code written by [15]. The loss functions include Cross Entropy loss and triplet loss. Using the method of this article on the Market-1501 data set, rank-1 and mAP have been increased by 0.7% and 1.5%. Using the method of this article on the Duke-reid dataset, rank-1 and mAP were improved by 1.0% and 1.7%, respectively. It shows that the method in this paper is to optimize the triplet loss function, and the Reid task using this function has a good performance improvement.

Table 3. Comparison triplet loss and soft margin loss on SYSU-MM01 with baseline

	SYSU-MM01	
	R=1	mAP
Soft margin loss/Baseline	57.74	54.42
Soft margin loss/ours	60.14	57.05
Triplet loss/Baseline	56.03	54.38
Triplet loss/ours	61.58	57.52

Table 4. Using proposed method on RGB-Reid datasets

	Market-1501		Duke-Reid	
	R=1	mAP	R=1	mAP
Baseline	94.3	85.5	86.5	79.5
ours	95.0	87.0	87.5	77.6

4.3 Comparison with State-of-the-art Methods

We compare the method in this paper with the traditional manual feature method and the method based on deep learning. The methods of manual features include the use of HoG and LOMO with different metrics, among which are KISSME, LFDA, CCA, CDFE, GMA. Methods based on deep learning include Zero-padding, TONE + HCML, BDTR, cmGAN, eBDTR, D2RL, DPMBN, HPILN, AlignGAN, TSLFN + HC, JSIA-Reid, Hi-CMD. For these methods that need to be compared, we directly copied from the original paper.

The results on Rank-1 accuracy and mAP are shown in Table 5. The result of the last line is the result of using the dual-stream structure in this article. From Table 5 we can clearly find that the method proposed in this article has a very good effect. Specifically, in the most difficult all-search single-shot mode, the proposed method achieves the state-of-the-art effect, which is 61.71% and 58.06% on Rank-1 and mAP, respectively.

Table 5. Comparison with the state-of-the-arts on SYSU-MM01 datasets. The R=1 and mAP denote Rank-1 accuracy (%) and mean average precision score (%), respectively.

	SYSU-MM01	
	R=1	mAP
HOG	2.76	4.24
LOMO	1.75	3.48
Zero-Padding	14.8	15.95
TONE+HCML	14.32	16.16
BDTR(ResNet50)	17.01	19.66
cmGAN(ResNet50)	26.97	27.8
eBDTR(ResNet50)	27.82	28.42
D2RL(ResNet50)	28.9	29.2
DPMBN(ResNet50)	37.02	40.28
HPILN(ResNet50)	41.36	42.95
AlignGAN	42.4	40.7
TSLFN+HC	56.96	54.95
JSIA-Reid	38.1	36.9
Hi-CMD	34.94	35.94
ours	**61.71**	**58.06**

5   Conclusion

In this paper, for the RGB-IR person re-identification task, we added the BNNeck structure to ResNet50 and used the maximum intra-class triplet loss function to enable the network to learn more distinguishing features. By combining CE loss and maximum intra-class triplet loss, the model can learn the feature representation well, thereby reducing intra-class cross-modal differences and inter-class differences. The whole frame structure is simple and achieves very good performance, confirming the potential of being a good baseline in the future. A large number of experiments show the effectiveness of the proposed method, and the effect exceeds state-of-the-art.

References

1. Bottou, L.: Large-scale machine learning with stochastic gradient descent. In: Lechevallier, Y., Saporta, G. (eds.) Proceedings of COMPSTAT 2010, pp. 177–186. Springer, Heidelberg (2010). https://doi.org/10.1007/978-3-7908-2604-3_16
2. Chen, W., Chen, X., Zhang, J., Huang, K.: Beyond triplet loss: a deep quadruplet network for person re-identification. In: Proceedings of the IEEE Conference on Computer Vision and Pattern Recognition, pp. 403–412 (2017)
3. Dai, P., Ji, R., Wang, H., Wu, Q., Huang, Y.: Cross-modality person re-identification with generative adversarial training. In: IJCAI, vol. 1, p. 2 (2018)
4. Das, A., Chakraborty, A., Roy-Chowdhury, A.K.: Consistent re-identification in a camera network. In: Fleet, D., Pajdla, T., Schiele, B., Tuytelaars, T. (eds.) ECCV 2014. LNCS, vol. 8690, pp. 330–345. Springer, Cham (2014). https://doi.org/10.1007/978-3-319-10605-2_22
5. Ding, S., Lin, L., Wang, G., Chao, H.: Deep feature learning with relative distance comparison for person re-identification. Pattern Recogn. **48**(10), 2993–3003 (2015)
6. Hao, Y., Wang, N., Li, J., Gao, X.: HSME: hypersphere manifold embedding for visible thermal person re-identification. In: Proceedings of the AAAI Conference on Artificial Intelligence, vol. 33, pp. 8385–8392 (2019)
7. Hermans, A., Beyer, L., Leibe, B.: In defense of the triplet loss for person re-identification. arXiv preprint arXiv:1703.07737 (2017)
8. Hirzer, M., Roth, P.M., Köstinger, M., Bischof, H.: Relaxed pairwise learned metric for person re-identification. In: Fitzgibbon, A., Lazebnik, S., Perona, P., Sato, Y., Schmid, C. (eds.) ECCV 2012. LNCS, vol. 7577, pp. 780–793. Springer, Heidelberg (2012). https://doi.org/10.1007/978-3-642-33783-3_56
9. Koestinger, M., Hirzer, M., Wohlhart, P., Roth, P.M., Bischof, H.: Large scale metric learning from equivalence constraints. In: 2012 IEEE Conference on Computer Vision and Pattern Recognition, pp. 2288–2295. IEEE (2012)
10. Li, S., Xiao, T., Li, H., Yang, W., Wang, X.: Identity-aware textual-visual matching with latent co-attention. In: Proceedings of the IEEE International Conference on Computer Vision, pp. 1890–1899 (2017)
11. Li, S., Xiao, T., Li, H., Zhou, B., Yue, D., Wang, X.: Person search with natural language description. In: Proceedings of the IEEE Conference on Computer Vision and Pattern Recognition, pp. 1970–1979 (2017)
12. Liao, S., Hu, Y., Zhu, X., Li, S.Z.: Person re-identification by local maximal occurrence representation and metric learning. In: Proceedings of the IEEE Conference on Computer Vision and Pattern Recognition, pp. 2197–2206 (2015)
13. Lin, J.W., Li, H.: Hpiln: a feature learning framework for cross-modality person re-identification. arXiv preprint arXiv:1906.03142 (2019)
14. Lin, Y., et al.: Improving person re-identification by attribute and identity learning. Pattern Recogn. **95**, 151–161 (2019)
15. Luo, H., Gu, Y., Liao, X., Lai, S., Jiang, W.: Bag of tricks and a strong baseline for deep person re-identification. In: Proceedings of the IEEE Conference on Computer Vision and Pattern Recognition Workshops (2019)
16. Lv, J., Chen, W., Li, Q., Yang, C.: Unsupervised cross-dataset person re-identification by transfer learning of spatial-temporal patterns. In: Proceedings of the IEEE Conference on Computer Vision and Pattern Recognition, pp. 7948–7956 (2018)
17. Nguyen, D.T., Hong, H.G., Kim, K.W., Park, K.R.: Person recognition system based on a combination of body images from visible light and thermal cameras. Sensors **17**(3), 605 (2017)

18. Wu, A., Zheng, W.S., Yu, H.X., Gong, S., Lai, J.: RGB-infrared cross-modality person re-identification. In: Proceedings of the IEEE International Conference on Computer Vision, pp. 5380–5389 (2017)
19. Xiao, Q., Luo, H., Zhang, C.: Margin sample mining loss: a deep learning based method for person re-identification. arXiv preprint arXiv:1710.00478 (2017)
20. Ye, M., Lan, X., Li, J., Yuen, P.C.: Hierarchical discriminative learning for visible thermal person re-identification. In: Thirty-Second AAAI Conference on Artificial Intelligence (2018)
21. Ye, M., Liang, C., Wang, Z., Leng, Q., Chen, J., Liu, J.: Specific person retrieval via incomplete text description. In: Proceedings of the 5th ACM on International Conference on Multimedia Retrieval, pp. 547–550 (2015)
22. Ye, M., Wang, Z., Lan, X., Yuen, P.C.: Visible thermal person re-identification via dual-constrained top-ranking. In: IJCAI, vol. 1, p. 2 (2018)
23. Zhang, S., Yang, Y., Wang, P., Zhang, X., Zhang, Y.: Attend to the difference: cross-modality person re-identification via contrastive correlation. arXiv preprint arXiv:1910.11656 (2019)
24. Zheng, L., Shen, L., Tian, L., Wang, S., Wang, J., Tian, Q.: Scalable person re-identification: a benchmark. In: Proceedings of the IEEE International Conference on Computer Vision, pp. 1116–1124 (2015)
25. Zheng, L., Yang, Y., Hauptmann, A.G.: Person re-identification: past, present and future. arXiv preprint arXiv:1610.02984 (2016)

Locally Consistent Constrained Concept Factorization with L_p Smoothness for Image Representation

Zonghui Weng[1], Zhenqiu Shu[1,2(✉)], Congzhe You[1], Yunmeng Zhang[1],
and Xiao-jun Wu[2]

[1] School of Computer Engineering, Jiangsu University of Technology, Changzhou 231001,
China
shuzhenqiu@126.com
[2] Jiangsu Provincial Engineering Laboratory of Pattern Recognition and Computational
Intelligence, Jiangnan University, Wuxi 231001, China

Abstract. Matrix factorization based on image representation algorithms have been widely used to deal with high-dimensional data. Previous studies have shown that matrix factorization methods can achieve remarkable performances in clustering. In this paper, we propose a novel method for image representation, called Locally Consistent Constrained Concept Factorization with L_p Smoothness (LCCCF-LS). The main contributions of our proposed LCCCF-LS method mainly include as follows: Firstly, the local geometric structure of the data is effectively explored using a graph regularizer. Secondly, the label information of the concepts is consistent with known label information without additional parameters. Finally, we add the L_p smoothness constraint to produce a smooth and more accurate solution, and thus ensure the smoothness of the coefficient matrix. Comprehensive experiments on several image datasets manifest the superiority of the proposed LCCCF-LS method.

Keywords: Matrix factorization · Image representation · Graph · Label information · L_p smoothness

1 Introduction

Data representation is a fundamental problem in machine learning and computer vision. In recent years, the demand for high-dimensional data representation has been increasing in the field of face recognition, document clustering and target tracking [1, 3, 4]. A good representation can effectively discover potential semantic structures, and simultaneously

This work was supported by the National Natural Science Foundation of China [Grant No. 61603159, 61806088, 61902160], Excellent Key Teachers of QingLan Project in Jiangsu Province, Graduate Student Practice Innovation Foundation of Jiangsu Province [Grant No. 20_0930], China Postdoctoral Science Foundation [Grant No. 2017M611695] , Jiangsu Postdoctoral Science Foundation [Grant No. 1701094B].

© Springer Nature Switzerland AG 2020
Y. Peng et al. (Eds.): PRCV 2020, LNCS 12306, pp. 569–580, 2020.
https://doi.org/10.1007/978-3-030-60639-8_47

improve the performances of clustering or classification methods. Therefore, many studies aim to seek an effective data representation method to find a suitable low-dimensional representation for high-dimensional data.

The matrix factorization based on image representation has attracted wide attention due to its effectiveness and efficiency. The main idea of matrix factorization is to find the product of two or more factor matrices to approximate the original matrix. Among the matrix factorization methods, non-negative matrix factorization (NMF) [2] is a well-known method, whose elements in each factor matrix are non-negative. Therefore, NMF only performs additional operations and not subtraction. Thus, it is a parts-based representation method, and have been widely applied in many applications, such as image analysis [3–8], face recognition [9], date clustering [10–12].

In some cases, the nonnegative constraint is too strict to deal with the real data mixed noise or outlier. To solve this issue, Xu *et al.* [3] proposed the Concept Factorization (CF) method for document clustering. The advantages of CF not only deal with high-dimensional data, but also can be easily kernelized. In order to consider the local geometric structure, Cai *et al.* [13] proposed a locally consistent concept factorization (LCCF) for document clustering. It adopts a graph model as an regularizer to extract the consistency concept according to the local consistency theory. Since CF is an unsupervised algorithm, and thus neglects the label information among the data. In order to alleviate this problem, Liu *et al.* [1] proposed a constrained concept factorization (CCF) method, which considers the known label information. It ensures that data points sharing the same label are mapped to the same concept in the low-dimensional space. However, the above-mentioned methods neglect the smoothness of the coefficient matrix. Shu *et al.* [14] proposed a local learning regularized concept factorization (LLRCF) method that considers both the manifold structure and discriminative structure of data using a local learning regularizer. Many studies have shown that the smoothness assumption plays an important role in data representation [15, 16]. To solve this issue, Leng *et al.* [17] proposed to constrain the coefficient matrix with the L_p smoothness. Therefore, it can generate a smoother and more accurate solution for the model.

In this paper, we propose a novel method, called locally consistent constrained concept factorization with L_p smoothness (LCCCF-LS) for image representation. LCCCF-LS not only takes the smoothness of the solution and the geometric manifold structure embedded in data into account using the regularization technology, but also utilizes the label information of the labeled samples with parameter-free. We present an efficient optimization algorithm based on the multiplicative updating algorithm to solve the proposed model. Experiments on several benchmark datasets show that our proposed LCCCF-LS method is superior to related state-of-the-art methods.

The remainder of this paper is organized as follows: Sect. 2 briefly reviews the related works. Section 3 introduces the proposed LCCCF-LS algorithm. Section 3.1 gives the proof of convergence. Section 4 conducts the experiment results, and Sect. 5 concludes the paper.

2   The Related Works

In this section, we briefly review some related works to our proposed method.

2.1   CF

Consider a case that there is a data matrix $X = \left[x_{ij} \right] \in \mathbb{R}^{M \times N}$ from c categories samples. NMF aims to find two non-negative matrixes $U = \left[u_{jk} \right] \in \mathbb{R}^{M \times K}$ and $V = \left[v_{jk} \right] \in \mathbb{R}^{N \times K}$ such that their product approximates to the original data matrix X. Different from NMF, CF seeks to represent each base vector u_j by a linear combination of data samples, i.e., $u_j = \sum_i W_{ij} X_i \left(W_{ij} \geq 0 \right)$, where $W = \left[w_{ij} \right] \in \mathbb{R}^{N \times K}$. The goal of CF is to calculate the following approximate problem as

$$X \approx XWV^T. \tag{1}$$

Using the Frobenius norm as metrics, the objective function of CF can be expressed as follows:

$$O_{CF} = \left\| X - XWV^T \right\|_F^2. \tag{2}$$
$$s.t. \, W > 0, V > 0$$

The multiplicative updating rules of CF in Eq. (3) and Eq. (4) can be derived as follows:

$$W_{ij}^{t+1} \leftarrow W_{ij}^t \frac{(KV)_{jk}}{(KWV^T V)_{jk}}, \tag{3}$$

$$V_{ij}^{t+1} \leftarrow V_{ij}^t \frac{(KW)_{ij}}{(VW^T KW)_{ij}}, \tag{4}$$

where $K = X^T X$.

2.2   LCCF

According to local consistency assumption, LCCF models the local geometric structure of data using a nearest neighbor graph.

By imposing the graph regularization constraint on the model of the original CF, the objective function of LCCF is given as follows:

$$O_{LCCF} = \left\| X - XWV^T \right\|_F^2 + \lambda Tr(V^T LV). \tag{5}$$
$$s.t. \, U \geq 0, V \geq 0$$

where $Tr(\cdot)$ is the trace of the matrix. s is the affine matrix of the nearest neighbor graph, and L denotes a Laplacian matrix, where D is a diagonal matrix, $D_{jj} = \sum_s S_{js}$, $L = D - S$.

By adopting similar optimization scheme, the updating rules of problem (6) is derived as follows:

$$w_{ij} \leftarrow w_{ij} \frac{(KV)_{ij}}{(KWV^T V)_{ij}}, \quad v_{ij} \leftarrow v_{ij} \frac{(KW + \lambda SV)_{ij}}{(VW^T KW + \lambda DV)_{ij}}. \tag{6}$$

3   The Proposed Method

In this section, we introduce the proposed LCCCF-LS method in details.

3.1   Construction of Auxiliary Matrix

In order to make full use of known label information, we construct an auxiliary matrix with parameter-free. Assume that there is a non-negative matrix $\{x_i\}_{i=1}^{n}$ from c categories samples, the first l samples are labeled and the remaining samples are unlabeled. If x_i is from the j-th class, then $m_{ij} = 1$, otherwise $m_{ij} = 0$. Then the indicator matrix A is constructed as follows:

$$A = \begin{pmatrix} M_{l \times c} & 0 \\ 0 & I_{n-l} \end{pmatrix}, \tag{7}$$

where I is an identity matrix. To fully utilize the label information, the label constraint is imposed by introducing an auxiliary matrix Z. The coefficient matrix can be rewritten as follows:

$$V = AZ. \tag{8}$$

3.2   Objective Function of LCCCF-LS

To take advantage of the prior knowledge of data, such as the local geometric information, label information and the smoothness of solution, we propose a novel method, namely LCCCF-LS, for image representation. LCCCF-LS considers more prior knowledge as much as possible, and thus has more representation ability compared with traditional CF methods. The objective function of the proposed LCCCF-LS method is given as follows:

$$O_{LCCCF-LS} = \|X - XWZ^T A^T\|_F^2 + \lambda Tr(Z^T A^T LAZ) + 2\mu \|W\|^p. \tag{9}$$

3.3   Optimization

It is obvious that the proposed model in Eq. (9) is non-convex, and thus cannot find the global optimal solution. Fortunately, we can achieve a local minimum of model (9) using the multiplicative iterative algorithm. Then Eq. (9) can be further rewritten as follow:

$$\begin{aligned} O_{LCCCF-LS} &= \|X - XWZ^T A^T\|_F^2 + \lambda Tr(Z^T A^T LAZ) + 2\mu \|W\|^p \\ &= Tr((X - XWZ^T A^T)^T (X - WZ^T A^T)) + \lambda Tr(Z^T A^T LAZ) + 2\mu \|W\|^p \\ &= Tr(K) - 2Tr(WT^T KAZ) + Tr(W^T KWZ^T A^T AZ) + \lambda Tr(Z^T A^T LAZ) \\ &\quad + 2\mu \|W\|^p \end{aligned} \tag{10}$$

where λ and μ are two nonnegative parameters, respectively, and $K = X^T X$. Let φ_{ij} and ϕ_{ij} be the Lagrange multiplier for constraints $w_{ij} \geq 0$, $z_{ij} \geq 0$ and $\Psi = [\varphi_{ij}]$, $\Phi = [\phi_{ij}]$, the Lagrange function l is given as follows:

$$\ell = O_{LCCCF-LS} + Tr(\Psi W^T) + Tr(\Phi Z^T). \tag{11}$$

Taking the partial derivatives of U and Z for ℓ, we have:

$$\frac{\partial \ell}{\partial W} = -2KAZ + 2KWZ^T A^T AZ + 2\mu PW^{P-1} + \Psi. \tag{12}$$

$$\frac{\partial \ell}{\partial Z} = -2A^T KW + 2A^T AZW^T KW + 2\lambda A^T LAZ + \Phi. \tag{13}$$

According to KKT conditions $\psi_{ij} w_{ij} = 0$ and $\phi_{ij} z_{ij} = 0$, Eqs. (12) and (13) can be further rewritten as follows:

$$w_{ij}^{t+1} \leftarrow w_{ij}^t \frac{(KAZ)_{ij}}{(KWZ^T A^T AZ + \mu PW^{P-1})_{ij}}, \tag{14}$$

$$z_{ij}^{t+1} \leftarrow z_{ij}^t \frac{(A^T KW + \lambda A^T SAZ)_{ij}}{(A^T AZW^T KW + \lambda A^T DAZ)_{ij}}, \tag{15}$$

4 Experiments

To evaluate effectiveness of the proposed LCCCF_LS method, we carried out some clustering experiments on PIE, ORL and YaleB datasets. We compared our proposed method with other methods including k-means (KM), CF, CCF, LCCF, LCCCF and NMF. In our experiments, two popular metrics including accuracy (AC) and normalized mutual information (NMI) are used to evaluate the performances of all methods.

4.1 PIE Image Database

PIE database includes 41368 multi-posture, light, and expression facial images from 68 individuals. Each person was taken 42 images from different light and illumination conditions. The size of all images is 32×32, and each image can be represented as a 1024-dimensional vector. Part of samples from the PIE database are shown in Fig. 1.

Fig. 1. Sample images from PIE database.

In our experiment, P categories samples were randomly picked out as the data subset to evaluate LCCCF-LS and its competitors. The experiments were repeated ten times and then their average results were recorded. Table 1 and Table 2 show the AC and NMI of all methods with different values of P on the PIE database. It is obvious that our proposed LCCCF-LS method is significantly improved over NMF and CF. The main reason is that both NMF ang CF are unsupervised learning algorithms, they take no account of label information. Our proposed LCCCF-LS method not only makes full use of label information among data, but also explores the manifold structure of the data and the smoothness of the solution. It is easily found that the proposed LCCCF-LS method is superior to other state-of-the-art methods.

Table 1. AC on PIE database.

K	KM	NMF	CF	CCF	LCCF	LCCCF	LCCCF-LS
2	0.526	**0.900**	0.612	0.812	0.835	0.647	0.876
4	0.445	0.635	0.421	0.592	0.663	0.477	**0.749**
6	0.348	0.545	0.329	0.500	0.587	0.407	**0.619**
8	0.324	0.465	0.254	0.475	0.500	0.365	**0.682**
10	0.278	0.398	0.217	0.421	0.423	0.341	**0.673**
12	0.264	0.374	0.194	0.382	0.413	0.328	**0.574**
14	0.245	0.317	0.164	0.375	0.356	0.322	**0.579**
AVG	0.347	0.519	0.313	0.508	0.539	0.412	**0.678**

Table 2. NMI on PIE database.

K	KM	NMF	CF	CCF	LCCF	LCCCF	LCCCF-LS
2	0.194	0.767	0.134	0.454	0.550	0.157	**0.654**
4	0.264	0.526	0.248	0.469	0.577	0.271	**0.718**
6	0.249	0.479	0.177	0.445	0.532	0.319	**0.673**
8	0.268	0.453	0.127	0.453	0.515	0.337	**0.743**
10	0.267	0.426	0.119	0.427	0.478	0.345	**0.754**
12	0.274	0.421	0.116	0.429	0.465	0.357	**0.700**
AVG	0.252	0.512	0.154	0.446	0.519	0.297	**0.707**

4.2 YaleB Database

YaleB face database is an extension of Yale face database. It includes in total 16128 images from 38 human subjects under 9 poses and 64 illumination conditions. Figure 2 shows some samples of the YaleB face database.

Fig. 2. Sample images from YaleB database.

We constructed an experimental data subset by randomly choosing K (= 26, ..., 38) categories and 30 samples from each category. We run all methods ten times independently for different values of K, and recorded their average performances as the final results. Table 3 and Table 4 show the AC as well as the NMI of seven algorithms on the

YaleB database. It can be found that LCCCF-LS outperforms other state-of-the-art methods. This is because that LCCCF-LS effectively discovers the local geometric structure embedded in high-dimensional data, and simultaneously considers the label information of labeled samples and the smoothness of the solution.

Table 3. AC on YaleB database.

K	KM	NMF	CF	CCF	LCCF	LCCCF	LCCCF-LS
26	0.497	0.366	0.176	0.217	0.415	0.262	**0.519**
28	0.460	0.373	0.173	0.212	0.420	0.252	**0.531**
30	0.465	0.340	0.164	0.212	0.442	0.250	**0.522**
32	0.445	0.346	0.166	0.216	0.397	0.243	**0.527**
34	0.468	0.341	0.163	0.204	0.388	0.237	**0.468**
36	0.442	0.315	0.160	0.206	0.367	0.237	**0.456**
38	**0.442**	0.337	0.163	0.200	0.353	0.245	**0.431**
AVG	0.460	0.345	0.166	0.209	0.397	0.246	**0.493**

Table 4. NMI on YaleB database.

K	KM	NMF	CF	CCF	LCCF	LCCCF	LCCCF-LS
26	0.643	0.550	0.333	0.404	0.585	0.447	**0.675**
28	0.641	0.574	0.337	0.400	0.560	0.442	**0.690**
30	0.639	0.542	0.345	0.415	0.615	0.452	**0.690**
32	0.634	0.555	0.354	0.428	0.582	0.457	**0.703**
34	0.652	0.565	0.361	0.428	0.580	0.449	**0.666**
36	0.643	0.545	0.370	0.437	0.572	0.466	**0.668**
38	0.659	0.550	0.385	0.443	0.563	0.465	**0.655**
AVG	0.644	0.554	0.355	0.422	0.579	0.454	**0.678**

4.3 ORL Database

The ORL database contains different images of each of 40 distinct subjects. All the images were taken at different times, varying the lighting and facial expressions. Figure 3 shows some sample from ORL database.

Fig. 3. Sample images from ORL database.

Similar to the above experimental setting, we randomly selected samples of K (= 28, …, 40) categories, All methods were run for ten times, and their average performances were reported. Table 5 and Table 6 show the performances of all methods on ORL database. We can see that LCCCF-LS outperforms other state-of-the-art methods in terms of the average AC and NMI. The main reason is that LCCCF-LS takes more prior knowledge of data into account and thus have more discriminative power than other competitors in clustering.

Table 5. AC on ORL database.

K	KM	NMF	CF	CCF	LCCF	LCCCF	LCCCF-LS
28	0.533	0.475	0.197	0.340	**0.606**	0.373	0.600
30	0.545	0.465	0.186	0.332	0.607	0.392	**0.607**
32	0.526	0.433	0.186	0.348	0.591	0.384	**0.605**
34	0.514	0.444	0.180	0.329	**0.605**	0.371	0.595
36	0.537	0.436	0.180	0.341	0.595	0.370	**0.604**
38	0.523	0.418	0.174	0.328	0.599	0.378	**0.607**
40	0.535	0.412	0.165	0.325	0.58	0.367	**0.582**
AVG	0.530	0.440	0.181	0.335	0.597	0.376	**0.600**

Table 6. NMI on ORL database.

K	KM	NMF	CF	CCF	LCCF	LCCCF	LCCCF-LS
28	0.704	0.655	0.371	0.536	0.739	0.589	**0.774**
30	0.710	0.650	0.375	0.520	0.747	0.602	**0.744**
32	0.700	0.636	0.379	0.532	0.735	0.603	**0.775**
34	0.700	0.648	0.384	0.528	0.747	0.604	**0.777**
36	0.715	0.645	0.391	0.557	0.743	0.605	**0.781**
38	0.709	0.638	0.395	0.552	0.747	0.619	**0.788**
40	0.714	0.639	0.388	0.572	0.749	0.609	**0.782**
AVG	0.707	0.644	0.383	0.542	0.744	0.604	**0.774**

4.4   Parameters Selection

The proposed LCCCF-LS model contains three parameters α, μ and P. We randomly selected 30, 20, 28 categories samples from PIE, YaleB and ORL datasets as subset to investigate the parameter sensitivity of the proposed LCCCF-LS method. Specifically, one parameter is varied when other parameters are fixed. Due to the limitation space, Figs. 4, 5 and 6 only show the clustering accuracy of the proposed LCCCF-LS method varied with the parameters α, μ and P, respectively. From Figs. 4, 5 and 6, it can be seen that LCCCF-LS achieves relative stable performance with different values of the parameters α, μ and P.

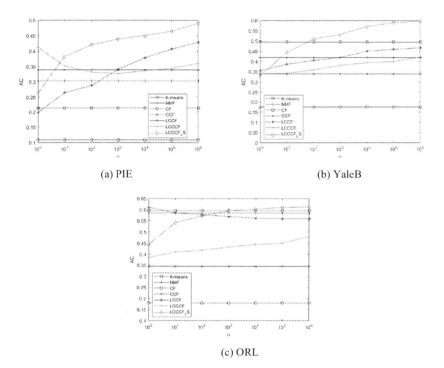

(a) PIE

(b) YaleB

(c) ORL

Fig. 4. The performance of LCCCF-LS varied the parameter α.

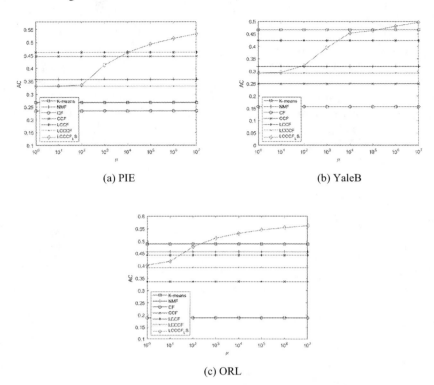

(a) PIE

(b) YaleB

(c) ORL

Fig. 5. The performance of LCCCF-LS varied the parameter μ.

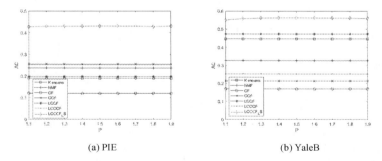

(a) PIE

(b) YaleB

Fig. 6. The performance of LCCCF-LS varied the parameter P.

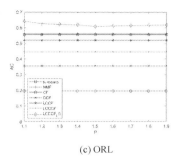

(c) ORL

Fig. 6. (*continued*)

5 Conclusions

In this paper, we propose a novel method named locally consistent constrained concept factorization with L_p smoothness (LCCCF-LS) for image representation. Compared with traditional methods, the advantage of the proposed LCCCF-LS method effectively simultaneously explores more prior knowledge of data including the known label information, the manifold structure and the smoothness of the solution. In addition, an efficient optimization strategy is provided to solve the model of LCCCF-LS. Experimental results on benchmark datasets demonstrate the proposed LCCCF-LS method outperforms related state-of-the-art methods in clustering.

References

1. Shu, Z., Wu, X., You, C., et al.: Rank-constrained nonnegative matrix factorization algorithm for data representation. Inf. Sci. **528**, 133–146 (2020)
2. Lee D.D., Seung H.S.: Algorithms for non-negative matrix factorization. International Conference on Neural Information Processing Systems. MIT Press, Boston (2000)
3. Xu, W., Gong, Y.: Document clustering by concept factorization. In: International ACM SIGIR Conference on Research & Development in Information Retrieval ACM, pp. 202–209 (2004)
4. Tian, Y., Li, X., Wang, K., Wang, F.: Training and testing object detectors with virtual images. IEEE/CAA J. Autom. Sinica **5**(2), 539–546 (2018)
5. He, W., Zhang, H., Zhang, L.: Total variation regularized reweighted sparse nonnegative matrix factorization for hyperspectral unmixing. IEEE Trans. Geosci. Remote Sens. **55**(7), 3909–3921 (2017)
6. Leng, C., Cai, G., Yu, D., Wang, Z.: Adaptive total-variation for non-negative matrix factorization on manifold. Pattern Recogn. Lett. **98**, 68–74 (2017)
7. Gu, J., Hu, H., Li, H.: Local robust sparse representation for face recognition with single sample per person. IEEE/CAA J. Autom. Sinica **5**(2), 547–554 (2018)
8. Liu, H., Wu, Z., Li, X., Cai, D., Huang, T.: Constrained nonnegative matrix factorization for image representation. IEEE Trans. Pattern Anal. Mach. Intell. **34**(7), 1299–1311 (2012)
9. Tan, X., Triggs, B.: Enhanced local texture feature sets for face recognition under difficult lighting conditions. IEEE Trans. Image Process. **19**(6), 1635–1650 (2010)

10. Shu, Z., Wu, X., Fan, H., Ye, F.: Parameter-less auto-weighted multiple graph regularized nonnegative matrix factorization for data representation. Knowl.-Based Syst. **131**, 105–112 (2017)
11. Shu, Z., Wu, X., Huang, P., Fan, H., Liu, Z., Ye, F.: Multiple graph regularized concept factorization with adaptive weights. IEEE Access **6**, 64938–64945 (2018)
12. Shahnaza, F., Berrya, M., Paucab, V., Plemmonsb, R.: Document clustering using nonnegative matrix factorization. Inf. Process. Manag. **642**(2), 373–386 (2006)
13. Cai, D., He, X., Han, J.: Locally consistent concept factorization for document clustering. IEEE Trans. Knowl. Data Eng. **23**(6), 902–913 (2011)
14. Shu, Z., Zhao, C., Huang, P.: Local regularization concept factorization and its semi-supervised extension for image representation. Neurocomputing **152**(22), 1–12 (2015)
15. Lyons, J.: Differentiation of solutions of nonlocal boundary value problems with respect to boundary data. Electron. J. Qual. Theory Differ. Equ. (51), 11 (2011)
16. Xu, L.: Data smoothing regularization, multi-sets-learning, and problem-solving strategies. Neural Netw. **16**(5), 817–825 (2003)
17. Leng, C., Zhang, H., Cai, G.: A novel data clustering method based on smooth non-negative matrix factorization. In: Basu, A., Berretti, S. (eds.) ICSM 2018. LNCS, vol. 11010, pp. 406–414. Springer, Cham (2018). https://doi.org/10.1007/978-3-030-04375-9_35

Direction-Sensitivity Features Ensemble Network for Rotation-Invariant Face Detection

Li-Fang Zhou[1,2,3(✉)] ⓘ, Yu Gu[3] ⓘ, Shan Liang[1] ⓘ, Bang-Jun Lei[4] ⓘ, and Jie Liu[2] ⓘ

[1] College of Automation, Chongqing University, Chongqing 400065, China
zhoulf@cqupt.edu.cn
[2] College of Software Engineering, Chongqing University of Posts and Telecommunications, Chongqing 400065, China
[3] College of Computer Science and Technology, Chongqing University of Posts and Telecommunications, Chongqing 400065, China
[4] China Three Gorges University, Yichang 443002, Hubei, China

Abstract. Recent deep learning based Rotation-Invariant Face Detection (RIFD) algorithms make efforts to explore a mapping function from face appearance to the rotation-in-plane (RIP) orientation. Most methods propose to predict RIP angles in a coarse-to-fine cascade regression style and improve the overall RIFD performance. However, the problem of suboptimal between the models of training phase and testing phase cannot be solved because of its cascaded nature. The weakness of ambiguous mapping between face appearance and its real orientation would also degrade the performance considerably. In this paper, we propose a novel Direction-Sensitivity Features Ensemble Network for rotation-invariant face detection (DFE-Net) which learns an end-to-end convolutional model for RIFD from coarse to fine. Specifically, the incline bounding box regression is implemented by introducing angle prediction based on improved SSD. A Direction-Sensitivity Features Ensemble Module (DFEM) is adopted in the network to progressively focus on the awareness of face angle information, which can learn and accurately extract features of rotated regions and locate rotated faces precisely. Finally, we add multi-task loss to guide the learning process to captures consistent face appearance-orientation relationships. Extensive experiments on two challenging benchmarks demonstrate that the proposed framework achieves favorable performance and consistently outperforms the state-of-the-art algorithms.

Keywords: Rotation-invariant face detection · Rotation convolution · SSD · Deep learning

1 Introduction

Rotation-invariant face detection, aiming at locating and detecting the human faces at the very beginning, has been achieved more and more attention in computer vision because of its wide-range of real-world applications, such as face alignment [3], face recognition [1] and face reenactment [2].

In the rotation-invariant face detection scenario, it is crucial to tackling two important issues: face detection and RIP angle estimation. Traditional rotation-invariant face

© Springer Nature Switzerland AG 2020
Y. Peng et al. (Eds.): PRCV 2020, LNCS 12306, pp. 581–590, 2020.
https://doi.org/10.1007/978-3-030-60639-8_48

detection methods naturally learn the large variations of face appearances from huge training sets and extract representative global features to characterize rotated faces [4–6]. However, these methods are clearly unsatisfactory in the case of practical use which consider both accuracy and processing speed. What is more, the representation ability of deep features is limited to capture rich structure and semantic information of target objects. Therefore, it is a challenge to design discriminative rotation-invariant features which are powerful enough to separate rotated faces from cluttered image background.

Recent rotation-invariant face detectors achieve favorable performance. In order to capture global structural information, Multi-task Convolutional Neural Networks (MTC-NNs) employ a series of calibration processes, e.g., rotating face candidates according to the predicted RIP angles, to extend the face detection model [3, 4, 6–8]. Then the range of RIP angles is gradually decreasing, which helps distinguish faces from non-faces. The result is suboptimal between the model of training phase and testing phase due to its cascaded nature. On the other hand, directly regressing RIP angles of rotated faces introduces cluttered information, which leads to ambiguous mapping between face appearance and its real orientation.

According to the observation above, it is crucial to focus on a detailed representation with richer information about the target object, and design elaborative strategy to integrate rotation-invariant feature maps. To this end, we present a novel Direction-Sensitivity Features Ensemble Network for rotation-invariant face detection (DFE-Net), which explicitly exploits and integrates global structural information by learning the incline bounding box regression. The DFE-Net is an end-to-end convolutional neural network and consists of three parts: (1) we employ the VGG-16 [9] as the backbone of the proposed network, which consists of five convolution blocks Conv-1, Conv-2, Conv-3, Conv-4 and Conv-5; (2) in order to preserve the resolution of feature maps learned in higher-level layers without sacrificing the size of the receptive field, the FCN is appended at the end of the backbone; (3) the Direction-Sensitivity Features Ensemble Module, referring as DFEM, which integrates rotation-invariant feature maps generated in rotation convolution layer and outputs the final detection result with higher quality.

Particularly, the DFEM is appended at the end of the feature extractor. Three DFEMs are learned from the hierarchically integrated feature maps. Considering that feature maps in deeper layers have a global insight into the input image, a rotation convolution operation is applied on such feature maps according to the different directions. As a result, the generated feature maps are equipped with discrimination ability and are denoted as DSF. Based on the DSF, an initial RIP angle is generated. The DSF is subsequently integrated by adding a rotation activation function, from which rotation-invariant features are explored in the DFEM. Then final RIP angle predictions are refined according to initial result from coarse to fine, greatly increasing the model accuracy.

In summary, the main contributions of this work are: 1) A novel DFE network is proposed for rotation-invariant face detection, which explicitly explores and integrates global structural information. 2) A direction-sensitivity features ensemble module is adopted to progressively compensate RIP angle predictions and endows discriminative ability to rotation-invariant feature maps. 3) A multi-task loss is employed to guide the learning process to captures consistent face appearance-orientation relationships.

2   Related Work

Rotation-invariant face detection is inherently involves two different tasks: face detection and pose estimation [3, 12]. Unifying or separating these two tasks will lead to different approaches. In the unified framework [10], the detector is designed to model human faces with diverse RIP angles, while the detector in separated framework addressed rotation variation by multi-class classification approaches, which categorized the entire RIP range into several distinct groups according to their RIP angles [6, 12]. Though a single detectors [19] can achieve accuracy of 87% on the challenging WIDER FACE, learning a large neural networks for powerful image representations in a data augmentation fashion usually leads to low computational efficiency which is unpractical in many applications.

There are alternative strategies for improving rotation-invariant face detection in separated framework apart from data augmentation. In [12], the part-level response signal can be generated by deformable part models (DPM) technique for inferring human faces. [6] divided the full RIP range into several groups and then developed multiple detectors in order to cater to RIFD. Unfortunately, these methods suffer from significant limitations: 1) it causes quantization issue which result from multi-class classification, and 2) it is computationally inefficient.

Currently, the most state-of-the-art RIFD method [4] casts RIFD as a multi-class classification problem and turns the classification results into regression by calculating the expected value as the RIP orientation of each face candidate. The detector, which is used to generate the RIP angle of each face candidate in a coarse-to-fine manner, is easier to achieve fast and accurate calibration by flipping original image few times.

3   Method

This paper proposes an improved SSD model [11] and achieves inclined bounding boxes regression by introducing angle prediction process, where both the coarse-to-fine strategy and the soft classification are incorporated into the formulation of angle prediction, rather than directly performing multi-class classification. By introducing the angle prediction process, the tilted bounding box regression is achieved. Both the refined strategy and the soft classification strategy are incorporated into the learning process of angle prediction to avoid the problem of learning bias caused by the direct execution of multi-classification tasks. Correspondingly, the algorithm adds a new angle offset loss to the general face detection loss function which can be used to supervise the generation of face regions. In order to extract effective rotation features, this paper uses a rotation convolution layer to implement a direction-sensitive feature integration module.

3.1   General Architecture

SSD is a classic single-stage target detection framework. The detection algorithm performs both coordinate regression and classification tasks. Figure 1 shows the architecture of the target detection model proposed in this paper. The model consists of the basic network and the detection function. This paper improves the model detection function based on the SSD model, changes the output result to a slanted bounding box representation method, adds a rotating convolution integration module, and improves the sensitivity of the output feature to the target rotation transformation through convolution kernel rotation and fusion. The loss function is improved so that the model can effectively learn the angle deviation.

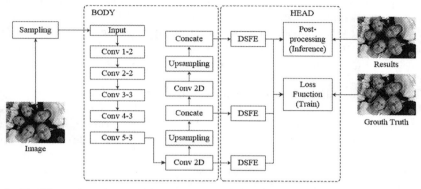

Fig. 1. The illustration of our architecture. The basic network uses the VGG-16 network model which is composed of convolutional layers. The detection head adopts the FPN multi-scale feature fusion structure, and the down-sampled 32-fold, 16-fold and 8-fold feature maps are multi-scale fusion, after each layer of feature maps in FPN. Then a Direction-Sensitivity Features Ensemble Module (DFEM) is implemented to detect the inclined bounding box. Particularly, the DFEM is appended at the end of the feature extractor. Three DFEMs are learned from the hierarchically integrated feature maps.

The basic network uses the VGG-16 network model which is composed of convolutional layers. Considering advantage of the convolutional layer in extracting target features, the convolutional layer with a stride of 2 replaces the pooling layer to complete the feature map. The detection head adopts the FPN [15] multi-scale feature fusion structure, and the down-sampled 32-fold, 16-fold and 8-fold feature maps are multi-scale fusion, after each layer of feature maps in FPN. Then a Direction-Sensitivity Features Ensemble Module (DFEM) is implemented to detect the inclined bounding box.

3.2   Inclined Bounding Box Representation

We have analyzed that the existing method of target representation based on rectangular bounding box which still has bottlenecks. The main problem is that there is a lot of background noise based on the feature extraction of the region. This representation method is difficult to accurately describe the structural information of the rotating face, which interferes with the subsequent classification and positioning of the rotating face. This is the scenario portrayed by FDDB benchmark, which represents the human face with a rotated bounding box [13]. Inspired by this, we simplified the oblique ellipse bounding box as a five-dimensional vector (x, y, w, h, θ), as shown in Fig. 2. It contains the coordinates (x, y) of the center point, and the direction angle θ represents the angle between the side of the slope $k \leq 0$ in the bounding box and the positive direction of the x-axis; w represents width, h represents height.

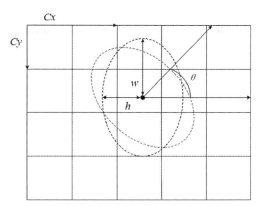

Fig. 2. The incline bounding box representation.

3.3   Direction-Sensitivity Features Ensemble Module

Local feature description is the core of many basic problems in the field of computer vision. In recent years, the emergence of deep learning technology has brought historic breakthroughs to many problems in the field of computer vision. Artificial neural networks with multiple hidden layers have excellent feature learning capabilities, and the learned features can better reflect the essential characteristics of data. Recently, Spatial Transformer Networks (STN) [14] has been introduced into the study of feature expression and learning to obtain ideal features with constant translation, scaling, and rotation. Chen et al.: Used a similar strategy to embed the spatial transformation layer in the face detection network to alleviate the pose change problem [16]. Although the experimental results fully prove the robustness of these methods, the STN module can only process one object at a time, and the calculation complexity is high, which is difficult to meet the real-time requirements in practical applications.

Recent studies have shown that a typical DCNN network, even if it does not change its network structure, can be directly trained on a large-scale multi-directional data set, so that some features including essential attributes of the image can be automatically learned [17]. In order to extract a truly effective rotation-invariant feature from the four features output by the rotation convolution layer, we introduce new effective operations and layers into the depth model, which can further improve the performance of the learned feature representation. First, copy the original convolution kernel in 4 copies and rotate it to four directions. Use convolution to extract the four orientation features in the original image. Second, connect the cross-channel pooling layer behind the rotation convolution layer [18], the output direction is sensitive feature.

Before describing the direction-sensitive feature integration module (see Fig. 3.), we first briefly introduce the angle soft classification strategy. The angle prediction

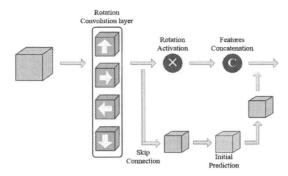

Fig. 3. Detailed structure of Direction-Sensitivity Features Ensemble Module.

task is implemented in DFEN in the form of multiple classifications, defining $\theta \in \{0, \pi/24, 2\pi/24, 3\pi/24, \ldots, 23\pi/24\}$, and dividing the full RIP angle into 24 finite sets. Considering the research and application of rotation-invariant face detection methods, in order to maintain the speed of model detection without sacrificing too much accuracy, we use a strategy from coarse to fine to perform secondary prediction. The first prediction is implemented in the form of four categories, defining $\theta_1 \in \{0, \pi/4, \pi/2, 3\pi/4\}$. The preliminary prediction θ_1 is generated before the direction-sensitive feature extraction by adopting a skip connection structure. The primary prediction generated by the direction-sensitive feature integration module and the final prediction result generated by the prediction module form an angle soft classification, to avoid the inaccurate classification caused by directly performing the multi-classification task.

Subsequently, θ_1 is used as the input parameter of the direction-sensitive feature integration module to realize the direction-sensitive feature integration in the form of an activation function. The activation function is:

$$f = \sum_{i=1}^{4} \frac{\theta_1^i \cdot z^i}{4} \tag{1}$$

in which z denotes the four direction feature map generated by the cyclic convolution layer.

3.4 Loss Function

The training of DFEN model uses a combination of three loss functions, including face classification, bounding box regression and angle classification. These loss functions act on the final convolution layer of the direction-sensitive feature integration module. The face classification loss is mainly used to distinguish the face area and the background. The regression loss is mainly used to learn the relative distance of the face area based on the preset frame. The MSE loss is used to represent face classification and regression errors; the angle classification loss is mainly to fit the person. The global structure of the face uses cross-entropy loss.

$$L(\{p_i\}, \{t_i\}, \{\theta_i\}) =$$

$$\frac{\lambda_{cls}}{N_{cls}} \sum_i L_{cls}(p_i, p_i^*) + \frac{\lambda_{reg}}{N_{reg}} \sum_i L_{reg}(t_i, t_i^*) + \frac{\lambda_{ang}}{N_{ang}} \sum_i L_{ang}(\theta_i, \theta_i^*)$$

$$(2)$$

For each training sample i, p_i represents the corresponding network output face classification score; if it contains a face, $p_i^* = 1$; otherwise, $p_i^* = 0$; t_i represents the bounding box regression vector, $t_i^* = [x_i^*, y_i^*, w_i^*, h_i^*]$ represents the true center coordinates and height and width of the slanted bounding box; θ_i represents the angle classification score output by the network, and θ_i^* represents the real category label of the sample. N represents the number of training samples, set $\lambda_{cls} = 3$, $\lambda_{reg} = 1$, $\lambda_{ang} = 1$.

4 Experiments

4.1 Implementation Details

As for parameter initialization, the basic network of the DFEN method uses the VGG16 network model, and their parameters are initialized from the pre-trained VGG16, and other additional layers are initialized randomly by the "xavier" method. We use the Tensorflow framework and adopt Adam technology for optimization. The weight decay is 0.0005, the momentum is 0.9, the learning rate is fixed at 0.001, and the training round is 100 Epochs. The entire network is trained on Nvidia GTX 1080Ti GPU.

4.2 Methods for Comparison

We compare with the following three representative methods:

Divide-and-Conquer: The first kind addressed the RIFD using multi-class classification approaches by quantizing the entire face ranges into groups. In our experi-ments, we implement an upright face detector based on Cascade CNN [20] and run this detector four times. Then the multi-oriented detection results on test images are combined to output the final decision.

Data Augmentation: These models directly learn a mapping function from the whole training images with great variations of face appearances to the face regions. For more extensive comparison, we employ the most state-of-the-art models, such as Faster R-CNN [19], SSD300 [11] and Cascade CNN.

PCN: The designation of PCN represents rotation variation in a label-learning man-ner and provides a coarse-to-fine way to perform RIFD, which have demonstrated the robustness to faces with full 360 RIP angles.

4.3 Evaluation Results

Results on Multi-oriented FDDB. The FDDB dataset contains 2845 images and 5171 annotated faces. Since the original version of the FDDB dataset contains only a few rotated faces, it is also not suitable for evaluating RIFD, we use its extended version provided by [4] in the experiments. Compared to PCN, Faster R-CNN (VGG16), SSD300 (VGG16) and Cascade CNN, our method achieves the best accuracy on the FDDB dataset. The specific results (Recall rate at 100 False Positive on Multi-Oriented FDDB) are presented in the Table 1. All the results demonstrate the effectiveness of multi-class classification. The speed results are shown in Table 2.

Results on Rotation WIDER FACE. The WIDER FACE dataset [21], remains dominated by faces with small RIP variations. For fair comparisons, we follow the same experimental settings and evaluation metrics as adopted in [4]. To this end, we manually select some images that contain rotated faces from the WIDER FACE test set, denoted as Rotation WIDER FACE, contains 400 images and 1053 rotated faces in the wild. Some detection results can be viewed in Fig. 4.

Table 1. The FDDB recall rate (%) is at 100 false positives.

Method	Recall rate at 100 FP on FDDB				
	Up	Down	Left	Right	Ave
MTCNN	89.6	–	–	–	–
MTCNN-aug	86.2	84.1	83.6	83.5	84.4
Divide-and-Conquer	85.5	85.2	85.5	85.6	85.5
Cascade CNN-aug	85.0	84.2	84.7	85.8	84.9
SSD500 (VGG16)	86.3	86.5	85.5	86.1	86.1
Faster R-CNN (VGG16)	87.0	86.5	85.2	86.1	86.2
PCN	87.8	87.5	87.1	87.3	87.4
Ours	**88.5**	**88.3**	**87.5**	**87.6**	**87.9**

Table 2. Speed comparison between different methods.

Method	Faster R-CNN	SSD500	Ours
CPU	0.5	1	0.5
GPU	10	20	15

Fig. 4. Our DFE-Net's detection results on rotation WIDER FACE.

5   Conclusion

In this work, we propose a novel Direction-Sensitivity Features Ensemble Network for rotation-invariant face detection. To promote the discrimination ability of coarse feature maps, a Direction-Sensitivity Features Ensemble Module is embedded into the detection network. As a result, each directional position of the feature maps can carry over to appearance variations accounted for in our design. Then the fine detailed geometric cues are progressively recovered in the prediction results by learning direction-sensitivity features ensemble module. Future work includes conducting more extensive experiments which would be help further demonstrate the efficacy of our design.

Acknowledgment. This work was supported by the Science and Technology Research Program of Chongqing Municipal Education Commission (Grant No. KJZD-K201900601) and by the National Natural Science Foundation of Chongqing (Grant No. cstc2019jcyj-msxmX0461).

References

1. Zhou, L., Du, Y., Li, W., et al.: Pose-robust face recognition with Huffman-LBP enhanced by divide-and-rule strategy. Pattern Recogn. **78**, 43–55 (2018)
2. Thies, J., Zollhofer, M., Stamminger, M., et al.: Face2face: real-time face capture and reenactment of RGB videos. In: CVPR, pp. 2387–2395. IEEE Press, Las Vegas (2016)
3. Ranjan, R., Patel, V.M., Chellappa, R.: HyperFace: a deep multi-task learning framework for face detection, landmark localization, pose estimation, and gender recognition. IEEE Trans. Pattern Anal. Mach. Intell. **41**(1), 121–135 (2016)
4. Shi, X., Shan, S., Kan, M., et al.: Real-time rotation-invariant face detection with progressive calibration networks. In: CVPR, pp. 2295–2303. IEEE Press, Salt Lake City (2018)
5. Rowley, H.A., Baluja, S., Kanade, T.: Rotation invariant neural network-based face detection. In: CVPR, pp. 38–44. IEEE Press, Santa Barbara (1998)

6. Huang, C., Ai, H., Li, Y., et al.: High-performance rotation invariant multiview face detection. IEEE Trans. Pattern Anal. Mach. Intell. **29**(4), 671–686 (2007)

7. Zhang, K., Zhang, Z., Li, Z., et al.: Joint face detection and alignment using multitask cascaded convolutional networks. IEEE Signal Process. Lett. **23**(10), 1499–1503 (2016)

8. Yang, B., Yang, C., Liu, Q., et al.: Joint rotation-invariance face detection and alignment with angle-sensitivity cascaded networks. In: ACM' MM, pp. 1473–1480. ACM Press, Istanbul (2019)

9. Simonyan, K., Zisserman, A.: Very deep convolutional networks for large-scale image recognition. arXiv preprint arXiv:1409.1556 (2014)

10. Farfade, S.S., Saberian, M.J., Li, L.: Multi-view face detection using deep convolutional neural networks. In: ICMR, pp. 643–650. ACM Press, New York (2015)

11. Liu, W., et al.: SSD: single shot multibox detector. In: Leibe, B., Matas, J., Sebe, N., Welling, M. (eds.) ECCV 2016. LNCS, vol. 9905, pp. 21–37. Springer, Cham (2016). https://doi.org/10.1007/978-3-319-46448-0_2

12. Zhu, X., Ramanan, D.: Face detection, pose estimation, and landmark localization in the wild. In: CVPR, pp. 2879–2886. IEEE Press, Providence (2012)

13. Jain, V., Learned-Miller, E.: FDDB: a benchmark for face detection in unconstrained settings. UMass Amherst Technical report (2010)

14. Jaderberg, M., Simonyan, K., Zisserman, A.: Spatial transformer networks. In: NIPS, pp. 2017–2025. MIT Press, Montreal (2015)

15. Lin, T., Dollár, P., Girshick, R., et al.: Feature pyramid networks for object detection. In: CVPR, pp. 2117–2125. IEEE Press, Honolulu (2017)

16. Chen, D., Hua, G., Wen, F., Sun, J.: Supervised transformer network for efficient face detection. In: Leibe, B., Matas, J., Sebe, N., Welling, M. (eds.) ECCV 2016. LNCS, vol. 9909, pp. 122–138. Springer, Cham (2016). https://doi.org/10.1007/978-3-319-46454-1_8

17. Liu, Y., Shen, Z., Lin, Z., et al,: GIFT: learning transformation-invariant dense visual descriptors via group CNNs. In: NIPS, pp. 6992–7003. MIT Press, Vancouver (2019)

18. Marcos, D., Volpi, M., Tuia, D.: Learning rotation invariant convolutional filters for texture classification. In: ICPR, pp. 2012–2017. IEEE Press (2016)

19. Ren, S., He, K., Girshick, R., et al.: Faster R-CNN: towards real-time object detection with region proposal networks. In: NIPS, pp. 91–99. MIT Press, Montreal (2015)

20. Li, H., Lin, Z., Shen, X., et al.: A convolutional neural network cascade for face detection. In: CVPR, pp. 5325–5334. IEEE Press, Boston (2015)

21. Yang, S., Luo, P., Loy, C., et al.: Wider face: a face detection benchmark. In: CVPR, pp. 5525–5533. IEEE Press, Las Vegas (2016)

Branch Information Correction Network for Human Pose Estimation

Qingzhan Ni, Chenxing Wang$^{(\boxtimes)}$, and Feipeng Da

Southeast University, Nanjing 210018, China
cxwang@seu.edu.cn

Abstract. The main task of human keypoint detection is to detect the position of human bone joints in pictures or videos. In the branch-based network, key points are classified according to different properties, and each branch of the network is responsible for the prediction of a certain set of keypoints. Compared with the direct prediction of the network of all the key points, the advantage of this method is that it considers the structural constraints on the human body and the internal relationship between the key points. Based on the branch network, this paper studies the information-sharing relationship and the negative transfer relationship between branches. It proposes a new branch information correction network to make full use of the complementary information on branches. The experiment proves that the method proposed in this paper can further improve the accuracy of the keypoint prediction of the human body, and can correct some key points which are easily affected by the environment.

Keywords: Human pose estimation · Convolutional neural network · Human skeleton joints

1 Introduction

Human Pose Estimation (HPE) is a foundational problem in the field of computer vision. It is also known as the location of Human skeleton key points, which is to use robust algorithms to estimate the position of human skeleton key points in pictures or videos.

The traditional algorithm [1–6] for HPE in the early stage is based on the graph model of parts. However, due to the inconsistency of the human scale in the picture and the complex distribution of distance and angle of key points, the generalization performance of these models is usually insufficient. In recent years, due to the outstanding performance of the convolutional neural network in image classification and feature extraction, many HPE works use deep netural network to realize feature extraction and spatial position constraint. DeepPose [7] for the first time proposed the use of deep convolution neural network for coordinate regression of key points. The Iterative Error Feedback model [8] proposed by

---

C. Wang—Student Paper.

© Springer Nature Switzerland AG 2020
Y. Peng et al. (Eds.): PRCV 2020, LNCS 12306, pp. 591–601, 2020.
https://doi.org/10.1007/978-3-030-60639-8_49

Joao Carreira et al. gradually changes the initial hypothetical position of points through the feedback error.

Some deep learning methods explicitly add tree structure [9] and graph structure [10] to the network, which usually requires high computation overhead and a clear description of the distribution. So recently, many methods use heatmap regression to learn structural information implicitly at the same time to avoid the above problems. CPM [11] (Convolutional Pose Machines) proposed by Shih-En Wei et al. has become a classical method in the field of human pose estimation, which uses large convolution kenel sequence to expand the receptive field and implicitly learn structural information. The Stacks Hourglass Networks of Newell [12] can learn high-level semantic information through multi-scale receptive field mechanism. At the same time, the encoding and decoding parts with the same resolution of the network are added to get enhanced pose estimation information. The network [13–15] based on Hourglass, changes the upsampling method of feature graph or introduces the network structure such as attention mechanism. HRNet [16] has changed the traditional method of recovering high-resolution representation from low resolution and has always maintained reliable high-resolution representation, so it has better performance.

Pose estimation is essentially homogeneous multi-task learning, with each task responsible for the prediction of one or a series of key points. Some recent work has used the multi-task sharing mechanism to locate key points, where the branch network is a representation of the network structure. Compared with the common network that directly estimates all the key points, the advantage of the branch network is that it takes into account the structural constraints of the human body and the internal relationship between the key points. CrossInfoNet [17] divided hand posture estimation into two subtasks, palm and finger, and adopted two branch cross-linking structures to share beneficial information. CPN [18] (Cascaded Pyramid Network) divided the human posture estimation task into two subtasks: coarse critical point detection and fine critical point detection. The fine detector was built on multiple feature layers of the coarse detector to improve the accuracy of the difficult key points.

Part-based branching Network (PBN) [19] believes that traditional end-to-end learning, all key points share the same rich pose representation, and ignores the negative transfer between some key points which can lead to adverse effect. Therefore, based on the nature of mutual information, the author divided the pose estimation task into five subtasks and adopted a multi-branch structure to avoid the interaction of negative correlation feature of some key points.

This paper studies the roles of the backbone and the branch of PBN. Based on the information-sharing relationship between branches and the correlation of key points, the corresponding network structure is designed to optimize the branch structure and improve the independence of branches. The branch information correction network is proposed, which makes full use of the complementary information between the branches to select the needed features from the rich coupling features, and at the same time enhances the independence of the branches and realizes the enhancement of the features. The experiment shows that it can correct some key points which are easily affected by the environment.

2 Related Work

PBN [19] treats the joint position to be predicted as a random variable $l_m \in L$, $m \in \{1, \cdots, M\}$, where L is spatial domain, m represents any key point and M is the number of joints. Therefore, the correlation between the two variables is measured by calculating the mutual information $I(l_m, l_n)$. The higher the value of $I(l_m, l_n)$ means that the features closely related to the joint m also provide some clues to the joint n and vice versa. Finally, they treat $\{I(l_m, l_n)\}$, with $m, n \in \{1, \cdots, M\}$ as an affinity matrix and use spectral clustering to group related parts. On the basis of the experiment, the key points were divided into five groups.

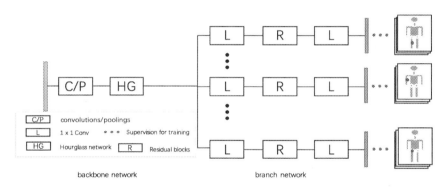

Fig. 1. A part-based branching network (PBN)

PBN uses the structure of a backbone network cascading branch network, and the superposition of multiple modules is to continually refine the accuracy of pose estimation. The branch network is composed of five branches, and the five branches separately predict the key points which are divided into five groups according to the mutual information. PBN's backbone is used to extract the semantic features of human pose, and the branch network independently estimates the distribution of a certain set of key points (See Fig. 1). Some simple common feature representations learned from the backbone are provided to each branch, and then the individual feature is expected to be learned from each branch, which thereby mitigates the impact of negative migration between key point groups. However, the PBN's backbone contains at least twelve residual blocks and much more convolutional layers, which is thus much deeper than each branch that contains only one residual block. As a result, the backbone provides abundant global high-level pose semantic feature, and this feature is coupled with the information of each key point, while the branch network is only a simple feature selector and has such a limited decoupling ability that cannot fully extract or independently learn the features of this group of key points. That is why increasing the number of residuals in a branch network does not improve network performance. Besides, if the backbone network only learns the general features, and each branch network independently learns the features of each group of key points, there should be some repetitive structures, which increase redundant network parameters.

In response to the questions raised above, this paper proposes a branch information correction network (BICNet). It uses HRNet as the backbone to extract high-level semantic features that retain reliable high resolution. The branch network is improved so that it can extract features from the rich coupling features of all key point in an optimized way.

3 Our Approach

The BICNet proposed in this paper is based on the backbone network cascading branch network structure, and its main improvement is the branch feature enhancement module in the branch network. Our work is inspired by CrossInfoNet [17] and applies it to human pose estimation, extending it to the case of the five branches, and analyzing its ability to enhance its independence. The information extraction capability of the branch network is optimized by using the branch structure and the information between branches. At the same time, the mutually beneficial information between the branches is used to enhance the feature and the independence of the branch network. The proposed branch information correction network is shown in Fig. 2.

The grey rectangle in the figure represents the input image. The backbone uses HRNet network to learn the rich pose feature (See Fig. 2). Each branch learns the preliminary features of the keypoints by residual block and then obtains the prediction feature map of each branch through our branch feature enhancement module (red dotted box in Fig. 2). A concatenation of feature maps from each branch is used for intermediate supervision (See Fig. 3).

Based on PBN, we divide the key points of human body into five groups. Each branch is set as $b_i, i \in [1, 5]$, and the feature obtained by backbone network is f. In the branch information enhancement module, fm_i (the ith branch) in the figure represents the initial features obtained by convolution of the residual block $R(\cdot)$ with f.

$$fm_i = R(f) \tag{1}$$

For the b_n, f_{bn} is obtained by subtracting the features of b_i, $i \in [1, 5]$, $i \neq n$ from the common feature f. Then, f_{bn} is concatenating with the initial feature of b_n, and the result fm_i' can be used as the branch feature after enhancement. That is:

$$fm_i' = \left(f - \Sigma_{j=1, j \neq i}^5 fm_j\right) \odot fm_i \tag{2}$$

where $\odot$ represents the concatenation of feature map.

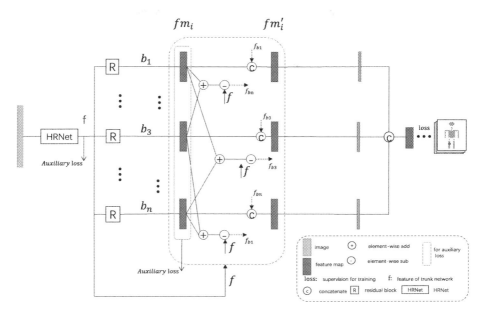

Fig. 2. Branch information correction network. Inside the red dotted rectangle is the branch feature enhancement module (Color figure online)

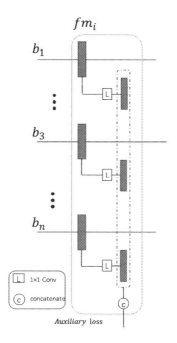

Fig. 3. The intermediate supervision of branch feature enhancement module

596   Q. Ni et al.

If we only have the structure of the network with no specific constraints, it is not correct to rely solely on the network to learn the functions of the modules we expect. Therefore, a fundamental premise of this module is that fm_i must be a feature of some of the key points that the b_i wants to predict. Therefore, this article introduces an auxiliary loss, the green box part of the picture (See Fig. 3). The intermediate supervision is added to the network so that the initial estimation can be optimized continuously. At the same time, the intermediate supervision can also prevent the gradient from disappearing and promote convergence.

This is shown in Fig. 3. Since the number of channels of the initial feature map fm_i in the box is the same, new feature map with different channels is obtained by 1×1 convolution, and the number of channels of the new feature map is the same as the number of key points to be predicted by this branch. In this way, the newly obtained feature map can be concatenated to obtain the feature maps of 16 channels (MPII [20] dataset has 16 joint points) for intermediate supervision. That is:

$$J_B = L\left(fm_1\right) \odot L\left(fm_2\right) \ldots \ldots \odot L\left(fm_5\right) \tag{3}$$

The auxiliary loss function of the intermediate supervision is:

$$LossB = \frac{1}{2n} \sum_x \|J_B - Y\|^2 \tag{4}$$

where x is each element in the feature map, and Y is the ground-truth label. At the end of the backbone, J_A obtained from the backbone network with 1×1 convolution with f. The auxiliary loss function of it is:

$$LossA = \frac{1}{2n} \sum_x \|J_A - Y\|^2 \tag{5}$$

Assuming that the output of the whole branch network is J, the loss function of the network can be defined as:

$$LossO = \frac{1}{2n} \sum_x \|J - Y\|^2 \tag{6}$$

Therefore, the total loss function of the whole network is:

$$Loss = LossA + LossB + LossO \tag{7}$$

The point of subtraction of feature map is that each branch either excludes the direct information from other branches or the information coupling with other branches. The resulting feature maps are then cascaded with the initial feature maps fm_i learned through supervision, and the resulting new feature maps are enhanced compared to the previous ones. Importantly, the network has a specific decoupling capability. More importantly, since extraction and cascading are both directed at this branch, the independence between branches is also increased.

Intermediate supervision only plays a role in training and does not need to work in network reasoning, so the network improves its performance without improving the parameters of PBN's branch network.

4   Experiments and Results

The performance of the pose estimation was evaluated on the MPII dataset to evaluate the effectiveness of the proposed method. The images in the MPII dataset are all drawn from YouTube videos and cover a variety of human activities. It contains 25K images, including 40,000 people with richly annotated human joints. MPII's annotation in the single-person pose estimation includes 16 key points of the human body correspond to the coordinates of the original image, among which "visible" indicates whether the key points are visible, and the coordinates of some invisible points beyond the image boundary are set as (0, 0). "Scale" refers to the ratio of the height of the human body in the original figure to 200, which represents the size of the human body. "Center" marks the central position of the human body, which is obtained from the average of the left and right buttocks and left and right shoulders.

Data preprocessing includes random scale enhancement, rotation enhancement and shear enhancement, and image mirroring with a 50% probability. The image is cropped to 256 × 256 with the "center" annotation as the centre. HRNet is used as the backbone network for pose feature extraction, and HRNet's pretraining model was used for training. PCKh is used as the evaluation measure for the prediction results. In this paper, two 2080Ti GPUs are used for the experiment under Pytorch framework.

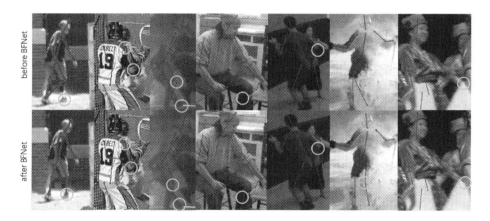

Fig. 4. Upgrade the 3 stacks PBN with the branch feature enhancement moudle. The first row is the result of 3 stacks PBN, and the second row is the result of 3 stacks PBN with branch feature enhancement moudle.

The experiments were conducted ablation studies on branch information correction network. Compare the performance of a network that uses HRNet as its backbone and does not include branches. At the same time, we also compared the PBN with the backbone of HRNet with our BICNet.

598   Q. Ni et al.

Table 1. The experimental results on the MPII verification set.

Method	Head	Shoulder	Elbow	Wrist	Hip	Knee	Ankle	Mean
HRNet	97.1	95.3	90.5	85.4	89.2	85.8	82.2	90.1
Baseline	97.2	96.5	91	86.2	88.8	86.3	83.3	90.3
BICNet*	97.2	95.9	90.7	86.1	88.6	87.2	83.8	90.3
BICNet	97.2	96.6	91.1	86.7	89.6	86.4	83.5	90.6

The Baseline in Table 1 is a network structure with HRNet as the backbone network and five branches of PBN. BICNet is the experimental result of the branch information correction network proposed in this paper on the MPII verification set. The experimental results show that the branch feature enhancement module can improve the accuracy on the basis of the reference network. With an average accuracy of 90.1% for HRNet only, our BICNet improves accuracy by 0.5%. Our branch network is also 0.3% more accurate than the baseline. BICNet* means no intermediate supervision was used in the experiment. We tried to adjust the parameters in the hope of improving the results, but after multiple experiments, the results remained at 90.3%. It is explained that intermediate supervision plays a decisive role in our approach because it ensures that the structure we design has specific physical meaning. This is only significant when a small number of parameters are introduced in the intermediate supervision section. However, intermediate supervision only plays a role in training and does not need to work in network reasoning, so the network improves its performance without improving the parameters of PBN's branch network.

Fig. 5. The partial results obtained by BICNet on the MPII dataset.

The network training parameters of the above three experiments are consistent. Because each branch of the feature enhancement module shares its learned features in a mutually beneficial way, the feature enhancement is realized. In the actual prediction, its reasoning effect in the occlusion of the key points is more prominent.

PBN's idea is to design branch networks by grouping key points through mutual information. Comparisons with PBN (the backbone network is Hourglass) need only be different in branching networks, all else being equal. Based on the same HRNet as the backbone, compare the branches of PBN and ours can not only verify the effectiveness of the PBN branch, but also verify the effectiveness of our method. Because the results of PBN with the backbone of HRNet are higher than that of HRNet, the results of our approach are higher than those of PBN with the backbone of HRNet.

Our experimental results are not obtained by tuning parameters, but by using the parameters of HRNet. Perhaps its parameters do not fit our network, but our experimental results are improved. We have also tried to design more complex network structural relationships, but the results have not improved.

Figure 4 shows some of the results from the PBN revamped using BICNet's branches compared to the original 3stacks PBN. The results showed that BICNet could effectively correct the position of vulnerable joints such as wrists, ankles and knees. In this paper, the predefined information sharing structure encodes part of the information and learns the corresponding information through the network. Columns 2, 5, and 6 indicate that BICNet effectively corrects some of the key points where PBN locates it to someone nearby (See Fig. 4). Figure 5 is a partial result of the branch information correction network on the MPII dataset.

5   Conclusion

Through the benchmark experiment and some ablation comparison experiments, the branch information correction network proposed by us is valid and has a good effect on the prediction of the key points that are difficult to reason and easy to block, such as wrist and ankle. The branch feature enhancement module can correct some key points which are easily affected by the environment. It combines the reciprocal information between branches to help other branches learn better with the features learned by this branch. The module also optimizes the feature extraction method to exclude the information of other branches from the rich backbone network features, and then concatenate the information of this branch to isolate the features between the branches. That is to say, the branch feature enhancement module enhances the independence of the branches while enhancing the features.

Acknowledgements. This work is supported by the National Natural Science Foundation of P. R. China (61828501) and the Fundamental Research Funds for the Central Universities.

References

1. Felzenszwalb, P., McAllester, D., Ramanan, D.: A discriminatively trained, multi-scale, deformable part model. In: 2008 IEEE Conference on Computer Vision and Pattern Recognition, pp. 1–8. IEEE (2008)
2. Fischler, M.A., Elschlager, R.A.: The representation and matching of pictorial structures. IEEE Trans. Comput. **100**(1), 67–92 (1973)
3. Andriluka, M., Roth, S., Schiele, B.: Pictorial structures revisited: People detection and articulated pose estimation. In: 2009 IEEE Conference on Computer Vision and Pattern Recognition, pp. 1014–1021. IEEE (2009)
4. Felzenszwalb, P.F., Huttenlocher, D.P.: Pictorial structures for object recognition. Int. J. Comput. Vision **61**(1), 55–79 (2005)
5. Yang, Y., Ramanan, D.: Articulated pose estimation with flexible mixtures-of-parts. In: CVPR 2011, pp. 1385–1392. IEEE (2011)
6. Ye, M., Yang, R.: Real-time simultaneous pose and shape estimation for articulated objects using a single depth camera. In: Proceedings of the IEEE Conference on Computer Vision and Pattern Recognition, pp. 2345–2352 (2014)
7. Toshev, A., Szegedy, C.: DeepPose: human pose estimation via deep neural networks. In: Proceedings of the IEEE Conference on Computer Vision and Pattern Recognition, pp. 1653–1660 (2014)
8. Carreira, J., Agrawal, P., Fragkiadaki, K., Malik, J.: Human pose estimation with iterative error feedback. In: Proceedings of the IEEE Conference on Computer Vision and Pattern Recognition, pp. 4733–4742 (2016)
9. Tompson, J.J., Jain, A., LeCun, Y., Bregler, C.: Joint training of a convolutional network and a graphical model for human pose estimation. In: Advances in Neural Information Processing Systems, pp. 1799–1807 (2014)
10. Yang, W., Ouyang, W., Li, H., Wang, X.: End-to-end learning of deformable mixture of parts and deep convolutional neural networks for human pose estimation. In: Proceedings of the IEEE Conference on Computer Vision and Pattern Recognition, pp. 3073–3082 (2016)
11. Wei, S.E., Ramakrishna, V., Kanade, T., Sheikh, Y.: Convolutional pose machines. In: Proceedings of the IEEE Conference on Computer Vision and Pattern Recognition, pp. 4724–4732 (2016)
12. Newell, A., Yang, K., Deng, J.: Stacked hourglass networks for human pose estimation. In: Leibe, B., Matas, J., Sebe, N., Welling, M. (eds.) ECCV 2016. LNCS, vol. 9912, pp. 483–499. Springer, Cham (2016). https://doi.org/10.1007/978-3-319-46484-8_29
13. Chen, Y., Shen, C., Wei, X.S., Liu, L., Yang, J.: Adversarial PoseNet: a structure-aware convolutional network for human pose estimation. In: Proceedings of the IEEE International Conference on Computer Vision, pp. 1212–1221 (2017)
14. Xiao, B., Wu, H., Wei, Y.: Simple baselines for human pose estimation and tracking. In: Proceedings of the European Conference on Computer Vision (ECCV), pp. 466–481 (2018)
15. Insafutdinov, E., Pishchulin, L., Andres, B., Andriluka, M., Schiele, B.: DeeperCut: a deeper, stronger, and faster multi-person pose estimation model. In: Leibe, B., Matas, J., Sebe, N., Welling, M. (eds.) ECCV 2016. LNCS, vol. 9910, pp. 34–50. Springer, Cham (2016). https://doi.org/10.1007/978-3-319-46466-4_3
16. Sun, K., Xiao, B., Liu, D., Wang, J.: Deep high-resolution representation learning for human pose estimation. In: Proceedings of the IEEE Conference on Computer Vision and Pattern Recognition, pp. 5693–5703 (2019)

17. Du, K., Lin, X., Sun, Y., Ma, X.: CrossInfoNet: multi-task information sharing based hand pose estimation. In: Proceedings of the IEEE Conference on Computer Vision and Pattern Recognition, pp. 9896–9905 (2019)
18. Huang, S., Gong, M., Tao, D.: A coarse-fine network for keypoint localization. In: Proceedings of the IEEE International Conference on Computer Vision, pp. 3028–3037 (2017)
19. Tang, W., Wu, Y.: Does learning specific features for related parts help human pose estimation? In: Proceedings of the IEEE Conference on Computer Vision and Pattern Recognition, pp. 1107–1116 (2019)
20. Andriluka, M., Pishchulin, L., Gehler, P., Schiele, B.: 2D human pose estimation: new benchmark and state of the art analysis. In: Proceedings of the IEEE Conference on Computer Vision and Pattern Recognition, pp. 3686–3693 (2014)

Automatic Detection of Cervical Cells Using Dense-Cascade R-CNN

Lin Yi[1], Yajie Lei[2], Zhichen Fan[2], Yingting Zhou[2], Dan Chen[2], and Ran Liu[2(✉)]

[1] Chongqing University Cancer Hospital, Chongqing 400030, China
[2] College of Computer Science, Chongqing University, Chongqing 400044, China
ran.liu_cqu@qq.com

Abstract. It is necessary to realize the automatic detection of cervical cells in Pap Smears. We present an automatic cervical cell detection approach based on the so-called Dense-Cascade R-CNN (Dense-Cascade Region-based Convolutional Neural Networks). The approach consists of three modules: data augmentation, training set balancing (TSB), and the Dense-Cascade R-CNN. The data augmentation module carries out operations such as rotation, scale, flip, etc. on input images to increase the samples. The TSB module is used to balance the number of cervical cell samples of various classes in the training set after data augmentation. As for the Dense-Cascade R-CNN module, the residual neural network (ResNet) with 101 layers in a Cascade R-CNN is replaced by a dense connected convolution neural network (DenseNet) with 121 layers so as to improve the detection performance of cervical cells. We evaluated the proposed method on the Herlev dataset. The results show that our approach can improve both mean average precision (mAP) and mean average recall (mAR) for Cascade R-CNN. Our cervical cell automatic detection approach can be used as an auxiliary diagnostic tool for cervical cancer screening.

Keywords: Cervical cancer · Cervical cells · Automatic detection · Pap smear · Deep learning

1 Introduction

With the rapid development of social economy, the natural and social factors that endanger women's health are increasing, and the incidence of gynecological cancer is also increasing year by year. Cervical cancer is the most common gynecological malignant tumor. In China, cervical cancer is the second most common gynecological malignant tumor. As cervical cancer is the only gynecological malignant tumor that can be successfully cured by early diagnosis and treatment, screening for precancerous lesions of the cervix is particularly important. In the 1940s, the Pap Smear method was established by Papanicolaou who used cervical exfoliated cells directly for cervical cancer screening. It was simple for operation and low in price. The widespread of this technique has reduced

---

L. Yi and Y. Lei—These authors contributed equally to this work and should be considered co-first authors.

Y. Peng et al. (Eds.): PRCV 2020, LNCS 12306, pp. 602–613, 2020.
https://doi.org/10.1007/978-3-030-60639-8_50

the incidence and mortality of cervical cancer significantly. Now Pap Smear method is a widely used method for early screening of cervical cancer, which is of great significance for the early diagnosis and treatment of cervical cancer [1, 2]. However, this method relies on manual microscopy, the doctor's workload is large, the screening cost is high, and the accuracy of the screening results depends on the doctor's experience. In addition, the cumbersome operation in the high-powered environment is likely to cause the doctor to fatigue and thus affect the critical result. Therefore, it is necessary to achieve automatic detection of cervical cell in the Pap Smear.

In recent years, the automatic detection of cervical cell images by image processing technology has become the norm [1–3]. Neghina et al. introduced a method of cervical cell segmentation and classification based on a polar transformation. One aspect of this method is that a number of parameters can be observed conveniently and evaluated as fuzzy memberships to the non-cell class in the segmented polar representation, out of which the final decision can be determined [4]. Chankong et al. used the fuzzy C-means (FCM) clustering technique to segment a single-cell image into the nucleus, cytoplasm, and background; hence, the 2-class problem can be achieved [4].

The above research mainly uses the traditional image classification method. The problems of these methods are as follows: (1) Manual selection of features is required, so technicians need to master professional knowledge of cytology and pathology; (2) Different classifiers for different types of cervical cells need to be designed. These classifiers are not universal and are difficult to apply to different datasets. As the number of cell types increases, the tasks of the classifier design will increase dramatically; (3) These methods are only good at classification, and the cervical cells cannot be accurately marked in the image. In recent years, deep learning is widely used in the automatic detection of medical images [5]. Some scholars have used a deep convolutional neural network (DCNN) to perform feature extraction and classification on gynecological examination smears such as thinprep cytologic test (TCT), and cross-validation is used to verify its validity [4]. The results show that the performance of the deep learning method significantly outperforms that of the traditional method. However, deep learning is still less used in the automatic detection of cervical cells in Pap Smears, which contains both classification and segmentation tasks. The purpose of this paper is to explore the application of deep learning in the automatic detection of cervical cells. Since the detection performance of DCNN depends heavily on the size of the sample, the balance of the samples of different classes in the input image, and the structure of the DCNN, this paper focuses on the investigation of data augmentation, training set balancing (TSB), and the structure of region-based CNN (a kind of DCNN) to find ways to improve the automatic detection performance of cervical cells. The main contributions of this paper are as follows: (1) A data augmentation method that combing 7 carefully-selected operations is presented to expand the sample size of the data set; (2) A training set balancing (TSB) algorithm is used to balance the number of samples in the data set; (3) A improved network, named Dense-Cascade R-CNN, is given based on Cascade R-CNN: the 101-layer ResNets in the Cascade R-CNN are replaced with 121 layers of DenseNets, and a segmentation branch is added to each detection branch of the Cascade R-CNN to implement segmentation task. All these tricks help to improve the performance of our network significantly in contrast to other networks.

2   Materials and Methods

2.1   Dataset

We use the Herlev dataset for experiments. The Herlev dataset was produced in collaboration with the Technical University of Denmark and the Herlev University Hospital [4]. The images in this dataset are saved as BMP format. Each image is a single intact cell image with an average resolution of 150×140 that has been divided into normal class and abnormal class by cytologists. In this paper, 917 samples (675 abnormal cells and 242 normal cells) in the Herlev dataset are used for experiments [6]. All cells can be further divided into 7 classes (C1–C7, see Table 1). The entire dataset is divided into a training set and a test set. Furthermore, a part of the samples from the training set (about 10%) is taken as the verification set. Table 1 lists the number of training and test samples for the original Herlev dataset.

Table 1. Number of training images and test images

Type	Class	Training set	Test set
Normal cell	Superficial squamous epithelial (C1)	58	16
	Intermediate squamous epithelial (C2)	54	16
	Columnar epithelial (C3)	82	16
Abnormal cell	Mild squamous non-keratinizing dysplasia (C4)	166	16
	oderate squamous non-keratinizing dysplasia (C5)	130	16
	Severe squamous non-keratinizing dysplasia (C6)	181	16
	Squamous cell carcinoma in situ intermediate (C7)	134	16
	Total	805	112

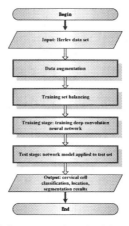

Fig. 1. Flowchart of the proposed method for cervical cell detection

2.2 Detection Method

Data Augmentation

Since the number of samples in the Herlev dataset is too small, we use data augmentation methods to enrich the dataset by creating new similar samples before training our proposed DCNN. This operation improves the accuracy of DCNNs and reduces overfitting [6]. Random translation [7], rotation [8], scale [8], ZCA whitening [9], feature standardization [10], horizontal flip [7], and vertical flip [7] are used to increase the number of samples. After data augmentation, the number of samples is increased from 805 to 7000.

Training Set Balancing

As the number of samples in different classes of the Herlev dataset is unbalanced, we balance the number of samples in each class by training set balancing (TSB). In this paper, we first generate a large number of new samples by various methods in the data augmentation module, and then select the required number of samples randomly from generated samples together with the original ones to form the training set, so that the number of cell images in each class is equal.

Dense-Cascade R-CNN

Currently, there are many object detection networks based on DCNN, such as R-CNN [4], Fast R-CNN [4], Faster R-CNN [4], Mask R-CNN [11], and Cascade R-CNN [4]. They are networks based on regional recommendations. Note that the Mask R-CNN can simultaneously perform classification, location, and pixel-level segmentation [4], hence it is always chosen as a benchmark for performance evaluation. In our experiments, we also chose it as a benchmark for comparison purpose.

Different from Mask R-CNN or Faster R-CNN, Cascade R-CNN is composed of a sequence of detectors trained with increasing IoU thresholds, and the detectors are trained sequentially, using the output of a detector as the training set for the next. Actually, Cascade R-CNN is a multi-stage extension of the Faster R-CNN. In the two-stage architecture of the Faster R-CNN, The first stage is a proposal sub-network, which is applied to produce preliminary detection hypotheses. In the second stage, these hypotheses are processed, a final classification score and a bounding box are assigned per hypothesis. Cascade R-CNN contains four stages, i.e. an RPN network and three detectors. In these three detectors, the input of each detector is the result of the regression of the boundary box of the previous detector, and the IoU thresholds of the three detectors are different and increasing.

Cascade R-CNN using ResNet backbone has achieved excellent results in the segmentation of many large datasets [11]. It outperforms most state-of-the-art detectors. However, ResNet is not an optimal choice for Cascade R-CNN for feature extraction for small datasets. Some research argues that for a relatively small dataset, the network called DenseNet (Densely Connected Convolutional Neural Network) always shows better performance [12], because it can solve the problem of overfitting better [12].

The basic idea of DenseNet is similar to ResNet, but it establishes a dense connection between all the front and back layers. DenseNet is mainly divided into two modules:

Dense block and Transition layer. The Dense block consists of two convolution layers with a convolution kernel size of 1 × 1 and 3 × 3, respectively. The Transition layer consists of a convolution layer of 1 × 1 convolution kernel and an average pooled layer. Similar to ResNet, DenseNet is initially a 7 × 7 convolution layer and a 3 × 3 maximum pooling layer, followed by the Dense block and Transition layer, and finally the 7 × 7 global pooling layer and the fully connected layer. The feature map size of the corresponding layer of DenseNet can match the feature map size of the 5 stages output in ResNet. Because of this architecture, we can replace the ResNet in Cascade R-CNN with DenseNet. Another feature of DenseNet is that it can reuse feature through the connection of features on Channel [12]. These features allow DenseNet to achieve better performance than ResNet in the case of fewer parameters and computational costs. Studies have shown that the feature extraction performance of DenseNet is superior to ResNet [12].

Inspired by this conclusion, we attempt to replace the 101-layer ResNet in Cascade R-CNN with the 121-layer DenseNet. Moreover, we add a segmentation branch to each detection branch of Cascade R-CNN to implement segmentation tasks. Figure 2(b) shows the architecture of our proposed DCNN, which uses the 121-layer DenseNet instead. We call this network Dense-Cascade R-CNN. We also manage to load the weights of DenseNet and Cascade R-CNN pre-trained models to initialize the proposed network. This operation speeds up the convergence rate of the gradient. We also replace ResNet in Mask R-CNN for comparison purpose. We call it Dense R-CNN.

3   Results and Analysis

In this paper, we use the samples listed in Table 1 for the experiment. We first evaluate the performance of the data augmentation module and TSB module, showing how they can contribute to detection; then, we evaluate the whole approach and present the cervical cell detection results.

3.1   Evaluation of Data Augmentation

In order to evaluate the effect of the data augmentation module in this paper, we design the ablation experiment shown in Table 2, in which the first row is the number of each experiment, and the first column lists the methods available in the data augmentation module. The "$\sqrt{}$" in the table indicates that this method is used in the experiment of this column. We evaluate the effect of each method on cervical cell performance by combining the methods in the module. Table 3 shows the detection performance scores on the test set when using different data augmentation methods. In order to make a fair comparison, E1–E7 does not use the TSB method, but uses each data augmentation method separately to expand the original training set in Table 1 by one time to get 805 × 3 training samples (original included), and then randomly selects 24 samples from each class (hence 192 samples) to form the verification set. E8 merges the training sets of E1–E7 together to obtain 805 × 8 training samples, and then randomly selects 96 samples from each class to form the verification set. The test set is always the same, and the default hyperparameters of the network are applied.

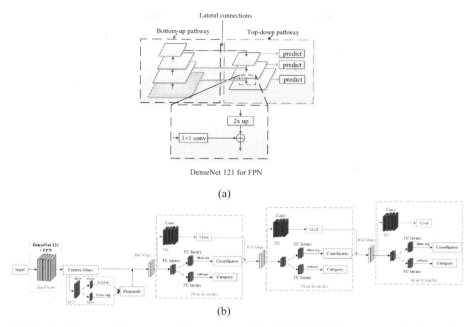

Fig. 2. Architecture of the proposed Dense-Cascade R-CNN. (a) FPN is constructed by DenseNet 121. (b) Overall architecture.

Table 2. Results of ablation experiments on the data augmentation module. Methods used in the module were combined to evaluate the impact of each method on the detection performance. The "$\checkmark$" in the table indicates that the method was used in the experiment listed in the same column

No. Method	E1	E2	E3	E4	E5	E6	E7	E8
Random translation	$\checkmark$							$\checkmark$
Rotation		$\checkmark$						$\checkmark$
Scale			$\checkmark$					$\checkmark$
ZCA whitening				$\checkmark$				$\checkmark$
Feature standardization					$\checkmark$			$\checkmark$
Horizontal flip						$\checkmark$		$\checkmark$
Vertical flip							$\checkmark$	$\checkmark$

In Table 2, E1–E7 are experiments using a single data augmentation method to evaluate the impact of the method on the detection performance of cervical cells. The purpose of E8 is designed to assess the impact of the combination of all data augmentation methods on the detection performance of cervical cells. It can be seen from Table 3 that when a single data augmentation method is applied, the best performance of mean

Table 3. Detection results on the test set using different data augmentation methods

No. Metric	E1	E2	E3	E4	E5	E6	E7	E8
mAP	0.539	0.483	0.505	0.464	0.478	0.539	0.525	0.787
mAR	0.620	0.577	0.588	0.547	0.560	0.611	0.589	0.833

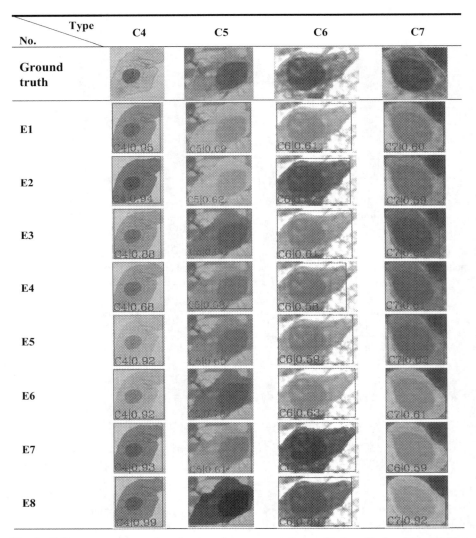

Fig. 3. Cell segmentation results of various data augmentation methods on Herlev dataset, taking abnormal cells as an example

average precision (mAP) is E1 (53.9%), the worst is E4 (46.4%), the best performance of mean average recall (mAR) is E1 (62.0%), and the worst is E4 (54.7%). This shows that for cervical cell detection, different data augmentation methods have different impacts on the detection results; combined with various data augmentation methods, mAP and mAR have been significantly improved compared to the best results using single method: mAP increased by 24.8% point, and mAR increased by 21.3% point. Figure 3 shows the segmentation results of abnormal cells with different data augmentation methods (E1–E8).

3.2 Evaluation of Training Set Balancing

In order to verify whether the TSB is beneficial to the cervical cell automatic detection performance of the network, we conduct a comparative experiment of the network cell segmentation performance without and with TSB. Default hyperparameters of the network are applied. In the experiment without TSB, the original Herlev dataset shown in Table 1 is used, with a total of 805 training set samples. For the experiment with TSB, the new training set (the number of cells in each class is 1000) and data augmentation method E8 in Table 2 are used. The results show that, when there is no TSB, the mAP of the network is 29.6%, and the mAR is 39.1%, which are much smaller than the values when TSB is applied (mAP: 79.2%, mAR: 89.0%).

Figure 4 compares the detection performance of the network without and with TSB for each class in the test set. It can be seen from Fig. 4 that TSB improves the performance of the network in all classes, and it has the greatest influence on class C2 (intermediate squamous epithelial).

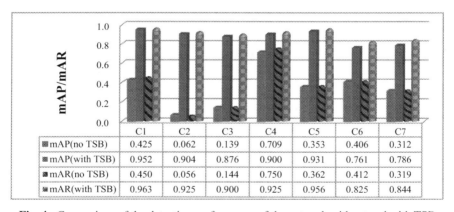

	C1	C2	C3	C4	C5	C6	C7
mAP(no TSB)	0.425	0.062	0.139	0.709	0.353	0.406	0.312
mAP(with TSB)	0.952	0.904	0.876	0.900	0.931	0.761	0.786
mAR(no TSB)	0.450	0.056	0.144	0.750	0.362	0.412	0.319
mAR(with TSB)	0.963	0.925	0.900	0.925	0.956	0.825	0.844

Fig. 4. Comparison of the detection performance of the network without and with TSB

3.3 Evaluation of Dense-Cascade R-CNN

In this section, we apply the proposed approach to the test set to evaluate its performance. The experiments are implemented based on PyTorch and MMDetection.

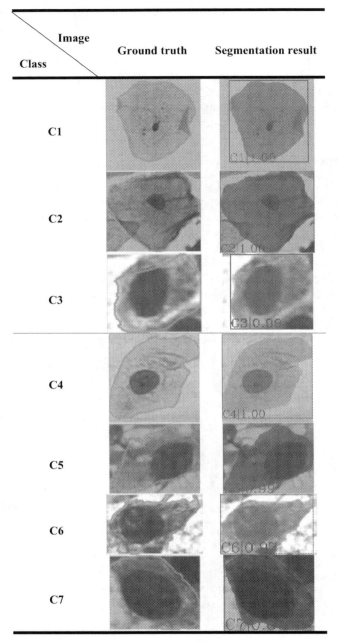

Fig. 5. Cell segmentation results of the proposed methods on the Herlev dataset. C1–C3: normal cell; C4–C7: abnormal cell

We fine-tune the hyperparameters of the Dense-Cascade R-CNN to get an excellent result. Figure 5 shows the detection results. As can be seen from Fig. 5, all the cells are classified, located, and segmented accurately.

We compare the performance of several state-of-the-art approaches (Mask R-CNN, Dense R-CNN, Cascade R-CNN [13], and Dense Cascade-R-CNN) on the test set. Table 4 shows the comparison results.

Table 4. Performances of different approaches for cervical cell detection. The data underlined indicate the best performance metrics among these approaches. (Epoch = 200)

Metric	Mask R-CNN	Dense R-CNN	Cascade R-CNN	Dense-Cascade R-CNN
mAP	0.922	0.899	0.964	**0.979**
mAR	0.968	0.959	0.973	**0.988**

As can be seen from Table 4, Dense R-CNN achieves both the lowest mAP and mAR, indicating that DenseNet replacing ResNet is the cause of performance degradation, hence the framework of Mask R-CNN might not suitable for DenseNet. Please note both mAP and mAR are increased after ResNets are replaced by DensNets in Cascade R-CNN: the mAP of Dense-Cascade R-CNN is 1.5% points higher than that of Cascade R-CNN; for the mAR, it also increases 1.5 percentage points. Both mAP and mAR of Dense-Cascade R-CNN are highest among all approaches. This indicates our method does work better for the cervical cell detection task, and it is served as the detector according to clinical needs.

In addition, we also compare the performance of the above approaches in dealing with abnormal cells and normal cells. As shown in Fig. 6, the detection performance

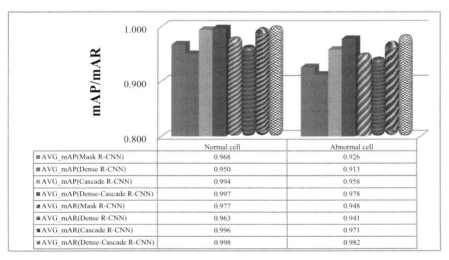

Fig. 6. Performance comparison of different methods for segmenting abnormal and normal cells

of our approach outperforms that of other methods for both abnormal and normal cells. This may be because our network can better learn the shape and the contour details of cells.

4   Conclusion

Currently, most hospitals depend on the doctors to examine Pap smears for the screening of precancerous lesions in the cervix. This method is time-consuming and labor-intensive, and the accuracy of the screening results depends on the experience of the doctor. In addition, it can easily lead to fatigue of the doctor. It is of great significance to implement the automatic detection of cervical cells in Pap Smear. In this paper, a new approach for the automatic detection of Pap Smear cervical cells is proposed. Three modules, data augmentation, training set balancing, and Dense-Cascade R-CNN, are contained in our approach. One important aspect of the proposed Dense-Cascade R-CNN is that the DenseNet is used to extract features instead of ResNet in the networks so as to improve the detection performance of cervical cells. Experimental results show that compared with other state-of-the-art approaches, our proposed system can improve the automatic detection performance of cervical cells. Therefore, the proposed approach can be used as an auxiliary diagnostic tool for cervical cancer screening.

As the performance of the network relies heavily on hyperparameters, our method needs more experiments to verify the parameter sensitivity. We plan to design and carry out these experiments to investigate the effects of hyperparameters in the next step.

Acknowledgment. This work was supported in part by the Fundamental Research Funds for the Central Universities (2018CDXYJSJ0026, 2019CDYGZD004); the Chongqing Foundation and Advanced Research Project (cstc2019jcyj-msxmX0622); the Science and Technology Research Program of Chongqing Municipal Education Commission (KJQN201800111), the Sichuan Science and Technology Program (2019YFSY0026), and the Entrepreneurship and Innovation Program for Chongqing Overseas Returned Scholars (No. cx2017094).

Appendix

The supplementary materials (including the source code of the proposed approach) for this paper can be downloaded from https://github.com/threedteam/cell_detection.

References

1. Wang, P., Wang, L., Li, Y., Song, Q., Lv, S., Hu, X.: Automatic cell nuclei segmentation and classification of cervical pap smear images. Biomed. Sig. Process. Control **48**, 93–103 (2019)
2. Kurnianingsih, et al.: Segmentation and classification of cervical cells using deep learning. IEEE Access **7**, 116925–116941 (2019)
3. Sompawong, N., et al.: Automated pap smear cervical cancer screening using deep learning. In: 41st Annual International Conference of the IEEE Engineering in Medicine and Biology Society (EMBC), pp. 7044–7048. Institute of Electrical and Electronics Engineers Inc., Berlin (2019)

4. Lin, T.Y., Dollar, P., Girshick, R., He, K., Hariharan, B., Belongie, S.: Feature pyramid networks for object detection. In: 30th IEEE Conference on Computer Vision and Pattern Recognition (CVPR), pp. 936–944. Institute of Electrical and Electronics Engineers Inc., Honolulu (2017)

5. Kermany, D.S., et al.: Identifying medical diagnoses and treatable diseases by image-based deep learning. Cell **172**(5), 1122–1131 (2018)

6. Zhang, L., Lu, L., Nogues, I., Summers, R.M., Liu, S., Yao, J.: DeepPap: deep convolutional networks for cervical cell classification. IEEE J. Biomed. Health Inform. **21**(6), 1633–1643 (2017)

7. Liu, R., Zhang, Y., Zheng, Y., Liu, Y., Zhao, Y., Yi, L.: Automated detection of vulnerable plaque for intravascular optical coherence tomography images. Cardiovasc. Eng. Technol. **10**(4), 590–603 (2019). https://doi.org/10.1007/s13239-019-00425-2

8. Shorten, C., Khoshgoftaar, T.M.: A survey on image data augmentation for deep learning. J. Big Data **6**(1), 1–48 (2019)

9. Chiu, T.Y.: Understanding generalized whitening and coloring transform for universal style transfer. In: 17th IEEE/CVF International Conference on Computer Vision (ICCV), pp. 4451–4459. Institute of Electrical and Electronics Engineers Inc., Seoul (2019)

10. Reiche, B., Moody, A.R., Khademi, A.: Effect of image standardization on FLAIR MRI for brain extraction. SIViP **9**(1 supplement), 11–16 (2015)

11. He, K., Gkioxari, G., Dollar, P., Girshick, R.: Mask R-CNN. IEEE Trans. Pattern Anal. Mach. Intell. **42**(2), 386–397 (2018)

12. Gao, H., Zhuang, L., Maaten, L.V.D., Weinberger, K.Q.: Densely connected convolutional networks. In: 30th IEEE Conference on Computer Vision and Pattern Recognition (CVPR), pp. 2261–2269. Institute of Electrical and Electronics Engineers Inc., Honolulu (2017)

13. Singh, K.R., Chaudhury, S.: A cascade network for the classification of rice grain based on single rice Kernel. Complex Intell. Syst. **6**, 321–334 (2020)

Feature Selection and Classification of Texture Images Based on Local Structure and Low-Rank Constraints

Rihong Li, Hongxia Gao[✉], Jiaxiang Luo, Haiming Liu, Weipeng Yang, and An Chen

School of Automation Science and Engineering, South China University of Technology,
510640 Guangzhou, Guangdong, People's Republic of China
hxgao@scut.edu.cn

Abstract. Texture description is a challenging problem in computer vision and pattern recognition. The task of texture classification is to classify texture into the class it belongs to, which is influenced by variations in scale, illumination, and changes in perspective. There are many texture descriptors in the literature. In this paper, we combine five texture descriptors for texture classification, which obtained better performance than the single descriptor at the price of high dimensionality. To solve this problem, we proposed a novel unsupervised feature selection method based on local structure and low-rank constraints, which can not only reduce the dimensions but also further improve the classification accuracy. To evaluate the performance of combing multiple descriptors and the proposed feature selection method, we design a variety of experiments in two typical texture datasets, namely KTH-TIPS-2a and CURET. Finally, the result shows the proposed method outperforms the state-of-the-art methods.

Keywords: Texture classification · Feature selection · Local structure learning

1 Introduction

Texture classification is still a difficult problem in computer vision and pattern recognition. The task of it is to assign a class label to the texture category it belongs to. Recently, a lot of texture feature descriptors have been proposed in the literature [1–6]. One of the most famous texture descriptors is the Local Binary Patterns (LBP) [1], which describes the neighborhood of an image pixel by comparing its gray value with the neighborhood pixels near it and finally forming a binary code. Except for texture classification, the texture descriptors have been employed to solve other vision tasks, such as object detection, face recognition, and defect detection.

Texture description may suffer the impact of multiple factors, such as variations in scale, illumination, and changes in perspective, which makes the single texture descriptor not fit all the situations. People try to combine multiple features to solve this problem [7].

R. Li—The first author is a student.

© Springer Nature Switzerland AG 2020
Y. Peng et al. (Eds.): PRCV 2020, LNCS 12306, pp. 614–625, 2020.
https://doi.org/10.1007/978-3-030-60639-8_51

In this paper, we proposed to use a set of complementary feature descriptors to extract features.

Combing multiple texture descriptors may cause the dimension to be too high. To solve this problem, we proposed a novel feature selection method. Inspired by the recent development of low-rank constraint representation [8], we design low-rank constraints to learn the global structure of feature space and remove the noise. Additionally, local structure plays an important role in feature selection [9]. We design local structure learning by combing the Euclidean distance with the KNN graph. As our experiments show, the proposed method can select more informative features.

In this paper, we proposed to combine multiple texture descriptors to improve the performance compared to single texture descriptors. Moreover, we designed a novel feature selection method to reduce the dimension without loss of accuracy. Those improve the accuracy in texture classification.

The rest of the paper is organized as follows. In Sect. 2, we present related work focusing on texture descriptors and feature selection. Section 3 describes the multiple texture descriptors. In Sect. 4, we introduce the feature selection method we proposed. The experimental results are shown in Sect. 5. Finally, we provide the conclusion in Sect. 6.

2 Related Work

Our texture classification framework involves texture feature description and feature selection. In this section, we will discuss them separately.

2.1 Texture Descriptors

Texture descriptors reflect the spatial distribution of image pixels. Recently, a variety of texture descriptors have been proposed. In [1], a multiresolution approach based on LBP was also proposed for rotation invariants texture classification. Because of LBP's simplicity and efficiency, lots of variants have been proposed, such as CLBP [2]. In [3], a method based on a deep convolution network consisting of computing successive wavelet transforms and modulus nonlinearity was proposed for invariance to scaling. Moreover, people also introduced a method that uses vector quantization based on the lookup table for texture description [10].

It is an interesting problem to fuse multiple texture descriptors for robust classification. For example, Li [11] uses the combination of HOG, LBP, and Gabor features for gender classification. We also use the combination of descriptors for feature extracting.

2.2 Feature Selection

Feature selection is a method that reduces the dimensionality by selecting a subset of most informative features, which may improve the efficiency and accuracy of classification.

In terms of label availability, feature selection methods can be classified into supervised methods and unsupervised methods. The supervised methods can effectively select discriminative features to distinguish samples from different classes. However, with the absence of a class label, it is difficult for the unsupervised methods to define feature relevance. To solve this problem, one of the criteria is to select the features which can preserve the data similarity or manifold structure in the original feature space [12]. These methods always generate cluster labels via clustering algorithms to guide the feature selection, such as MCFS [13] and UFSwithOL [14].

However, these methods may have a common drawback, that is, they ignore the effect of noise on the estimation of data's underlying structures. To solve this problem, we proposed a novel unsupervised method to learn global and local structures simultaneously, and remove the noise from data.

3 Multiple Texture Descriptors

In this selection, we show the multiple texture descriptors and the representation constructed by them. Similarity to the paper [15], we combine five texture descriptors, which are completed local binary patterns (CLBP) [2], wavelet scattering coefficient (SCAT) [3], binary Gabor pattern (BGP) [4], local phase quantization (LPQ) [5] and binarized statistical features (BSIF) [6]. They are briefly described below.

- **CLBP:** The completed local binary patterns (CLBP) extends the conventional LBP operator, which incorporates local difference and sign-magnitude transform information (LDSMT). The LDSMT further consists of two components, the difference sign and difference magnitude, which is encoded by a binary code. Similar to the conventional LBP, a region is also represented by its center pixel encoded by a binary code after global thresholding. Finally, the image is represented by the concatenation of three binary codes, which form a single histogram.
- **SCAT:** The wavelet scattering coefficient is a joint translation and invariant representation of image patches. It is implemented with a deep convolution network, which computes successive wavelet transforms and modulus nonlinearity. Invariants to scaling, shearing, and small deformations are calculated with linear operators in the scattering domain. SCAT obtains excellent results on texture databases with uncontrolled view conditions.
- **BGP:** The binary Gabor pattern is an efficient and effective multi-resolution approach to gray-scale and rotation invariant texture classification. Unlike MR8 filters [16], BGP uses predefined rotation invariant binary patterns without the pre-training phase. To counter the noise sensitivity, BGP adopts the difference of regions instead of the difference between two single pixels.
- **LPQ:** The local phase quantization is based on quantizing the phase information of the local Fourier transform. It is a powerful image descriptor and robust against the most common image blurs. LPQ is showed to provide excellent results for texture and face recognition tasks.
- **BSIF:** The binarized statistical features computes a binary code for each pixel by linearly projecting local image patches onto a subspace, whose basis vectors are learned from natural images via independent component analysis, and by binarizing the coordinates in this basis via thresholding. The number of basis vectors determines the length of the pixel binary codes which are used to construct the final histogram of an image.

In our paper, we slightly modified the scope of the CLBP descriptor, calculating the coded values of the inner and outer circles, which makes it have multi-scale effects. We called it CLBP_ext, and others remain the same. Then, each image is represented by the five texture description methods. The final representation is obtained by concatenating all five representations into a single histogram, H = [$h1$, $h2$, $h3$, $h4$, $h5$]. This histogram is the original multi-textured description, namely x_i, and then we optimize it.

4 Unsupervised Feature Selection of Local Structures and Low-Rank Constraints

In this selection, we use an unsupervised feature selection method to optimize the original texture features. Compared with the supervised method, the unsupervised feature selection method pays more effort to find the most informative features. With the absence of a class label, the selected features should maintain the internal structure as the data presented by the feature before selected.

To solve this problem, we propose to use low-rank constraints to preserve the global structure and adjust the local structures with the heat kernel function calculated by the k-nn graph.

Let $\mathbf{X} = \{x_1, x_2, \ldots, x_n\} \in R^{d \times n}$ be the data matrix with each column correspond to the data instance x_i and row to feature. Then we summarize some notation and norms used in the following selections. The bold uppercase characters are used to denote matrices, and the bold lowercase characters to denote vectors. For an arbitrary matrix $\mathbf{M} \in R^{m \times n}$, M_{ij} means the (i, j)-th entry of $\mathbf{M}$, m_i means the i-th column vector of $\mathbf{M}$ and m_j^T denotes the j-th row vector of $\mathbf{M}$. The $l_{2,1}$-norms of matrix $\|\mathbf{M}\|_{2,1}$ is defined as $\sum_{i=1}^{m} \sqrt{\sum_{j=1}^{n} M_{i,j}^2}$

4.1 Global Low-Rank Constraints

In the last few decades, people proposed lots of algorithms to analyze the global structure of data, such as PCA. Recently, the similarity preserving feature selection framework has demonstrated promising performance, which selects a feature subset with the pairwise similarity between high-dimensional samples. However, much redundant information and noise exist in the original high dimensional space.

Inspired by the recent development of low-rank constraint representation [8], we use the low-rank reconstruction to extract the global structure of data and remove noise. According to the theory of latent low-rank representation [17], we can get the below function:

$$\min_{Z,L,E} \|Z\|_* + \|L\|_* + \lambda \|E\|_{2,1} \tag{1}$$

$$s.t. X = XZ + LX + E$$

where $\mathbf{Z} \in R^{n \times n}$ is the low-rank matrix, L is used to extract salient features, and E is the noise component, λ is used to balance the noise component. Compared with pairwise similarity, the low-rank representation can remove the noise in the samples and represent the principal feature in the data. Finally, the optimal solution can be obtained by the iterative method.

To preserve the global and low-rank reconstruction structure, we propose a row sparse feature selection and transformation matrix $\mathbf{W} \in R^{d \times c}$ to reconstruct it, and get

$$\min_{W} \left\| W^T X - W^T XZ \right\|^2 + \beta \| W \|_{2,1} \tag{2}$$

$$s.t. W^T XX^T W = I$$

where β is a regularization parameter to ensure that matrix $\mathbf{W}$ is row sparse, and $W^T X$ denotes low dimensional space after dimension reduction. From Eq. (2), the global structure captured by $\mathbf{Z}$ can lead to finding the principle feature. Without noise, global structure estimation can be more accurate.

4.2 Local Structure Learning

Recently, the importance of preserving local geometric data structures in feature dimensionality reduction has been well recognized [9, 18], especial when transforming high-dimensional data to a low-dimensional space for analysis. What's more, the local geometric structure of data can be considered as a data-dependent regularization of the transformation matrix, which leads to maintaining the local manifold structure.

In this paper, we first build a KNN graph with Heat kernel weight. Then, we can get the weight matrix $\mathbf{P} \in R^{n \times n}$. For each data sample x_i, only k nearest points $\{x_j\}_{j=1}^{k}$ are considered its neighborhood with weight P_{ij}. In the original feature space, the following equation can obtain minimum value:

$$\sum_{i,j} \| x_i - x_j \|_2^2 P_{ij} \tag{3}$$

With the weight matrix $\mathbf{P}$, the induced Laplacian $L_P = D_P - (P + P^T)/2$ can be used for local manifold characterization, where D_P is a diagonal matrix whose i-th diagonal element is $\sum_j (P_{ij} + P_{ji})/2$.

To maintain the local structure after dimension reduction, we propose to recognize Eq. (3) as a regularization with transformation matrix $\mathbf{W}$, and we get

$$\min_{W} \sum_{i,j}^{n} \left\| W^T x_i - W^T x_j \right\|_2^2 P_{ij} \tag{4}$$

Thus, the optimization problem of Eq. (4) can be considered as a local structure learning.

Based on the low-rank constraints and local structure learning presented in Eq. (2) and Eq. (4), we propose a novel unsupervised feature selection method by solving the following optimization problem:

$$\min_{W} \left\| W^T X - W^T XZ \right\|^2 + \alpha \sum_{i,j}^{n} \left\| W^T x_i - W^T x_j \right\|_2^2 P_{ij} + \beta \| W \|_{2,1} \tag{5}$$

$$\text{s.t.} \mathbf{W}^T XX^T \mathbf{W} = I$$

where α and β are regularization parameters balancing the fitting error of local structure learning and sparsity of transformation matrix $\mathbf{W}$. It can be seen that our method removes the noise by low-rank constraints and learns global and local structure simultaneously.

4.3 Optimization Algorithm

With the only variable to be solved, it is easy to derive the approximate optimum solution in an iterative way. Let $\mathbf{L}_Z = (I - Z)(I - Z)^T$, $\mathbf{L}_P = \mathbf{D}_P - (P + P^T)/2$ and $\mathbf{L} = \mathbf{L}_Z + \alpha \mathbf{L}_P$, the Eq. (5) can be rewritten as:

$$\min_{\mathbf{W}} Tr(\mathbf{W}^T X \mathbf{L} X^T \mathbf{W}) + \beta \| \mathbf{W} \|_{2,1} \tag{6}$$

$$\text{s.t.} \mathbf{W}^T XX^T \mathbf{W} = I$$

With the t-th estimation $\mathbf{W}^t$, and denote $\mathbf{D}_{\mathbf{W}^t}$ be a diagonal matrix whose i-th diagonal element is $\frac{1}{2\|w_i^t\|^2}$, the Eq. (6) can be rewritten as:

$$\min_{\mathbf{W}} Tr(\mathbf{W}^T X (\mathbf{L} + \beta \mathbf{D}_{\mathbf{W}^t}) X^T \mathbf{W}) \tag{7}$$

$$\text{s.t.} \mathbf{W}^T XX^T \mathbf{W} = I$$

The optimal solution of $\mathbf{W}$ are the eigenvectors corresponding to c smallest eigenvalues of generalized eigenproblem:

$$X(\mathbf{L} + \beta \mathbf{D}_{\mathbf{W}^t}) X^T \mathbf{W} = \Lambda XX^T \mathbf{W} \tag{8}$$

Where Λ is a diagonal matrix whose diagonal elements are eigenvalues.

The complete algorithm of the feature selection method is summarized in algorithm 1.

Algorithm 1 This paper's feature selection method

Input: The data matrix $\mathbf{X} \epsilon R^{d \times n}$, the regularization parameters λ, α, β, the dimension of the transformed data c, the parameter k of KNN.
  1) Compute $\mathbf{Z}$ by Eq. (1);
  2) Compute $\mathbf{P}$ by KNN graph with Heat kernel weight;
  3) **repeat**
        Update W by Eq. (8);
     until Converges
Output: Sort all the d features according to $\|\mathbf{W}_i\|_2$ $(i = 1, ..., d)$ in descending order and select the top m ranked features

---

5 Experiments

In this section, we conduct extensive experiments to evaluate the performance of the proposed multiple descriptors combination and feature selection method in texture classification.

5.1 Data Sets

To validate the proposed method, we use the following two texture datasets, namely KTH-TIPS-2a [19] and CURET [20]. The KTH-TIPS-2a dataset consists of 11 texture categories with images at 9 different scales, 3 poses, and 4 different illumination conditions. According to the standard protocol [21], we randomly select 1 sample for the test and the remaining 3 samples for training. The CURET dataset consists of 61 texture categories, 92 images per class. We randomly select 46 images for training and the remaining for the test. The example images are shown in Fig. 1.

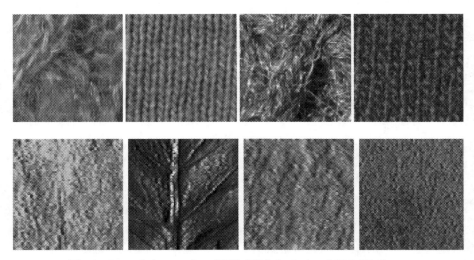

Fig. 1. Example images from KTH_TIPS_2a (up) and CURET (down)

Throughout our experiments, we use one-versus-all SVM using the RBF kernel [22].

5.2 Combining Multiple Texture Features

We start by showing the results for multi-texture representations. They are presented in Table 1. Compared to other single texture features, BGP provides the best performance. And the combination of five texture features significantly improves the classification accuracy. Although BSIF provides the worst performance, it still improves the accuracy of the combination. The results suggest that different texture representations possess complementary information, we should make good use of it.

5.3 The Effect of Feature Selection Method

As shown above, the combination of features improves the accuracy at the price of high dimensionality. We use the proposed unsupervised feature selection method to remove the redundancy features and reduce dimensions. How many dimensions are appropriate is an open question. In this paper, we reduce the final dimension to 300. We set the

Table 1. Classification accuracy (%) of different texture representations and their combinations

Method	Dimension	KTH-TIPS-2a	CURET
CLBP_ext	432	69.25	97.01
SCAT	391	76.48	98.64
BGP	216	82.13	98.82
LPQ	256	73.42	96.65
BSIF	256	58.42	96.29
CLBP_ext + SCAT	823	72.52	98.61
CLBP_ext + SCAT + BGP	1039	75.74	98.97
CLBP_ext + SCAT + BGP + LPQ	1295	76.19	99.14
CLBP_ext + SCAT + BGP + LPQ + BSIF	1551	84.72	99.47

Table 2. Classification accuracy (%) obtained with and without feature selection method

Method	Dimension	KTH-TIPS-2a	CURET
Original texture feature	1551	84.72	99.47
Feature after selection	300	87.69	99.29

parameters $k = 5, \lambda = 0.1, \alpha = 0.1, \beta = 0.5$, and c is set to be the number of classes. The result is shown in Table 2.

The result shows our selection method reduces the dimensions without any significant loss in accuracy. Especially, on the KTH-TIPS-2a, our feature selection method improves the performance by 2.97% compared to the original representation.

To evaluate the effect of the local structure learning, we design the experiments with and without local structure learning. The result is shown in Table 3. The result shows that local structure learning improves performance, especially in KTH-TIPS-2a.

Table 3. Classification accuracy (%) obtained with and without local structure learning

Method	KTH-TIPS-2a	CURET
Without local structure learning	85.37	99.25
With local structure learning	87.69	99.29

Additionally, we also compare our feature selection method with other classical feature selection methods, such as CFS [23], MCFS [13], UDFS [24] and UFSwithOL [14]. We provide a brief introduction to the above methods:

- CFS: a correlation-based feature selection method.

- MCFS: it selects the features by adopting spectral regression with l_1-norm regularization.
- UDFS: it exploits local discriminative information and feature correlations simultaneously.
- UFSwithOL: it uses a triplet-based loss function to enforce the selected feature groups to preserve the ordinal locality of original data.

Table 4 shows all 4 feature selection methods can reduce texture features' dimensions to 300, but they may cause varying degrees of decline in accuracy, especially MCFS. Our feature selection method significantly outperforms other methods.

Table 4. Classification accuracy (%) of different feature select methods

Method	Dimension	KTH-TIPS-2a	CURET
CFS	300	81.57	99.03
MCFS	300	79.63	97.90
UDFS	300	85.46	99.00
UFSwithOL	300	84.63	98.79
The proposed	300	87.69	99.29

5.4 Computation Cost and Parameter Sensitivity

The experiment was running on a machine with Window10, Matlab R2018a, NVIDIA GeForce GTX 1080, Intel (R) Core (TM) i5-9400 CPU @ 2.90 GHz, 8 GB RAM. We recorded the time to solve Eq. (5) in the two data sets. It took 16.26 s in KTH_TIPS_2a and 14.57 s in CURET. Moreover, the classification accuracy is not very sensitive to λ, α and β in wide ranges.

5.5 Comparison with State-of-the-Art

Table 5 shows the classification accuracy of various methods on two databases, which come from either original or related publications. It can be seen that our texture classification method outperforms typical and state-of-the-art methods.

Table 5. Classification accuracy (%) of various methods, and "-" means the lack of related original or publication

Method	KTH-TIPS-2a	CURET
MR8 [16]	-	93.5
RP [25]	-	98.5
SIFT + IFV [26]	76.6	98.1
DMD + IFV [26]	80.3	98.4
LHS [27]	73	-
COV-KLBPD [28]	74.9	-
scLBP [29]	78.4	-
NDV [30]	77.1	-
LZMHPP [31]	-	98.38
The proposed	87.69	99.29

6 Conclusion

In this paper, we introduce a novel idea of fusing complementary texture features, which significantly improves the accuracy of texture classification. To reduce the dimensions of fusing texture features without loss of accuracy, we proposed a novel unsupervised feature selection method. We use low-rank constraints to learn global structures, and design a regularization to learn local structure simultaneously. Finally, our experimental results demonstrate that the framework combining multiple texture features and feature selection outperforms the state-of-the-art in texture classification.

In the future, we plan to design a feature complimentary evaluation method, which helps us to find more complementary features and further improves classification accuracy. Moreover, we plan to validate the performance of our feature selection method in a wider dataset.

Acknowledgments. This work was supported by the Natural Science Foundation of Guangdong Province, China under Grant number 2019A1515011041, Xijiang Innovation Team Project, and Natural Science Foundation of China under Grant number 61603105.

References

1. Ojala, T., Pietikainen, M., Maenpaa, T.: Multiresolution gray-scale and rotation invariant texture classification with local binary patterns. IEEE Trans. Pattern Anal. Mach. Intell. **24**(7), 971–987 (2002)
2. Guo, Z., Zhang, L., Zhang, D.: A completed modeling of local binary pattern operator for texture classification. IEEE Trans. Image Process. **19**(6), 1657–1663 (2010)
3. Sifre, L., Mallat, S.: Rotation, scaling and deformation invariant scattering for texture discrimination. In: Proceedings of the IEEE Conference on Computer Vision and Pattern Recognition, pp. 1233–1240 (2013)
4. Zhang, L., Zhou, Z., Li, H.: Binary gabor pattern: an efficient and robust descriptor for texture classification. In: 19th IEEE International Conference on Image Processing, pp. 81–84 (2012)

5. Rahtu, E., Heikkilä, J., Ojansivu, V., et al.: Local phase quantization for blur-insensitive image analysis. Image Vis. Comput. **30**(8), 501–512 (2012)
6. Kannala, J., Rahtu, E.: Bsif: binarized statistical image features. In: Proceedings of the 21st International Conference on Pattern Recognition, pp. 1363–1366 (2012)
7. Tan, X., Triggs, B.: Fusing gabor and LBP feature sets for kernel-based face recognition. In: Zhou, S.K., Zhao, W., Tang, X., Gong, S. (eds.) AMFG 2007. LNCS, vol. 4778, pp. 235–249. Springer, Heidelberg (2007). https://doi.org/10.1007/978-3-540-75690-3_18
8. Liu, G., Lin, Z., Yu, Y.: Robust subspace segmentation by low-rank representation. In: Proceedings of the 27th International Conference on Machine Learning, pp. 663–670 (2010)
9. Liu, X., Wang, L., Zhang, J., et al.: Global and local structure preservation for feature selection. IEEE Trans. Neural Networks Learn. Syst. **25**(6), 1083–1095 (2013)
10. ul Hussain, S., Triggs, B.: Visual recognition using local quantized patterns. In: Fitzgibbon, A., Lazebnik, S., Perona, P., Sato, Y., Schmid, C. (eds.) ECCV 2012. LNCS, vol. 7573, pp. 716–729. Springer, Heidelberg (2012). https://doi.org/10.1007/978-3-642-33709-3_51
11. Li, M., et al.: Head-shoulder based gender recognition. In: 2013 IEEE International Conference on Image Processing, pp. 2753–2756 (2013)
12. He, X., Cai, D., Niyogi, P.: Laplacian score for feature selection. In: Advances in Neural Information Processing Systems, pp. 507–514 (2006)
13. Cai, D., Zhang, C., He, X.: Unsupervised feature selection for multi-cluster data. In: Proceedings of the 16th ACM SIGKDD International Conference on Knowledge Discovery and Data Mining, pp. 333–342 (2010)
14. Guo, J., et al.: Unsupervised feature selection with ordinal locality. In: 2017 IEEE International Conference on Multimedia and Expo, pp. 1213–1218 (2017)
15. Khan, F.S., Anwer, R.M., van de Weijer, J., et al.: Compact color–texture description for texture classification. Pattern Recogn. Lett. **51**, 16–22 (2015)
16. Varma, M., Zisserman, A.: A statistical approach to texture classification from single images. Int. J. Comput. Vis. **62**(1–2), 61–81 (2005)
17. Liu, G., Yan, S.: Latent low-rank representation for subspace segmentation and feature extraction. In: 2011 International Conference on Computer Vision, pp. 1615–1622 (2011)
18. Du, L., Shen, Y.D.: Unsupervised feature selection with adaptive structure learning. In: Proceedings of the 21th ACM SIGKDD International Conference on Knowledge Discovery and Data Mining, pp. 209–218 (2015)
19. Mallikarjuna, P., et al.: The kth-tips2 database. KTH Royal Institute of Technology (2006)
20. Hayman, E., Caputo, B., Fritz, M., Eklundh, J.-O.: On the significance of real-world conditions for material classification. In: Pajdla, T., Matas, J. (eds.) ECCV 2004. LNCS, vol. 3024, pp. 253–266. Springer, Heidelberg (2004). https://doi.org/10.1007/978-3-540-24673-2_21
21. Caputo, B., Hayman, E., Mallikarjuna, P.: Class-specific material categorization. In: Tenth IEEE International Conference on Computer Vision, pp. 1597–1604 (2005)
22. Chang, C.C., Lin, C.J.: LIBSVM: a library for support vector machines. ACM Trans. Intell. Syst. Technol. **2**(3), 1–27 (2011)
23. Hall, M.A.: Correlation-based feature selection of discrete and numeric class machine learning (2000)
24. Yang, Y., et al.: L2, 1-norm regularized discriminative feature selection for unsupervised. In: Twenty-Second International Joint Conference on Artificial Intelligence (2011)
25. Liu, L., Fieguth, P.: Texture classification from random features. IEEE Trans. Pattern Anal. Mach. Intell. **34**(3), 574–586 (2012)
26. Mehta, R., Eguiazarian, K.E.: Texture classification using dense micro-block difference. IEEE Trans. Image Process. **25**(4), 1604–1616 (2016)

27. Sharma, G., ul Hussain, S., Jurie, F.: Local Higher-Order Statistics (LHS) for texture categorization and facial analysis. In: Fitzgibbon, A., Lazebnik, S., Perona, P., Sato, Y., Schmid, C. (eds.) ECCV 2012. LNCS, vol. 7578, pp. 1–12. Springer, Heidelberg (2012). https://doi.org/10.1007/978-3-642-33786-4_1

28. Hong, X., Zhao, G., Pietikäinen, M., et al.: Combining LBP difference and feature correlation for texture description. IEEE Trans. Image Process. 23(6), 2557–2568 (2014)

29. Ryu, J., Hong, S., Yang, H.S.: Sorted consecutive local binary pattern for texture classification. IEEE Trans. Image Process. 24(7), 2254–2265 (2015)

30. Zhang, W., Zhang, W., Liu, K., et al.: A feature descriptor based on local normalized difference for real-world texture classification. IEEE Trans. Multimedia 20(4), 880–888 (2017)

31. Roy, S.K., Dubey, S.R., Chaudhuri, B.B.: Local ZigZag Max histograms of pooling pattern for texture classification. Electron. Lett. 55(7), 382–384 (2019)

Efficient Human Pose Estimation with Depthwise Separable Convolution and Person Centroid Guided Joint Grouping

Jie Ou[ID] and Hong Wu[(✉)][ID]

School of Computer Science and Engineering,
University of Electronic Science and Technology of China, Chengdu 611731, China
oujieww6@gmail.com, hwu@uestc.edu.cn

abstract
Abstract. In this paper, we propose efficient and effective methods for 2D human pose estimation. A new ResBlock is proposed based on depthwise separable convolution and is utilized instead of the original one in Hourglass network. It can be further enhanced by replacing the vanilla depthwise convolution with a mixed depthwise convolution. Based on it, we propose a bottom-up multi-person pose estimation method. A rooted tree is used to represent human pose by introducing person centroid as the root which connects to all body joints directly or hierarchically. Two branches of sub-networks are used to predict the centroids, body joints and their offsets to their parent nodes. Joints are grouped by tracing along their offsets to the closest centroids. Experimental results on the MPII human dataset and the LSP dataset show that both our single-person and multi-person pose estimation methods can achieve competitive accuracies with low computational costs.

Keywords: Human pose estimation · Depthwise separable convolution · Hourglass network · Joint grouping

1 Introduction

Human pose estimation aims to locate human body joints from a single monocular image. It is a challenge and fundamental task in many visual applications, e.g. surveillance, autonomous driving, human-computer interaction, etc. In the last a few years, considerable progress on human pose estimation has been achieved by deep learning based approaches [17–19, 29, 34, 35].

Most existing research works on human pose estimation focus on improving the accuracy and develop deep networks with large model size and low computational efficiency, which prohibits their practical application. To adopt deep networks in real-time applications and/or on limited resource devices, the model

The first author is student. This work was supported in part by the Sichuan Science and Technology Program, China, under grants No.2020YFS0057, and the Fundamental Research Funds for the Central Universities under Project ZYGX2019Z015.

© Springer Nature Switzerland AG 2020
Y. Peng et al. (Eds.): PRCV 2020, LNCS 12306, pp. 626–638, 2020.
https://doi.org/10.1007/978-3-030-60639-8_52

should be compact and computational efficient. Inception module [30] is used to build deeper networks without increase model size and computational cost. Depthwise separable convolution [5,10,31], has been utilized as the key building block in many successful efficient CNNs. In this paper, we follow these successful design principles to develop efficient deep networks for human pose estimation.

For multi-person pose estimation, it is needed to distinguishing poses of different persons. The approaches can mainly be divided into two categories: top-down strategy and bottom-up strategy. The top-down approaches [6,13,22,29,34] employ detectors to localize person instances and then apply joint detector to each person instance. Each step of top-down approaches requires a very large amount of calculations, and the run-time of the second step is proportional to the number of person. In contrast, the bottom-up approaches [3,11,12,15,21,24] detect all the body joints for only once and then group/allocate them into different persons. However, they suffer from very high complexity of joint grouping step, which usually involves solving a NP-hard graph partition problem. Different methods have been proposed to reduce the grouping time. Recently, some one-stage multi-person pose estimation approaches [19,27,32] have been proposed, but their performance lag behind the two-stage ones. In this paper, we also focus on the bottom-up strategy.

In this paper, we propose efficient and effective methods for 2D human pose estimation. A new ResBlock is proposed with two depthwise separable convolutions and a squeeze-and-excitation (SE) module and utilized in place of the original ResBlock in Hourglass network. Its representation capability is further enhanced by replacing the vanilla depthwise convolution with a mixed depthwise convolution. The new Hourglass networks is very light-weighted and can be directly applied to single-person pose estimation. Base on this backbone network, we further propose a new bottom-up multi-person pose estimation method. A rooted tree is used to represent human pose by introducing person centroid as the root which connecting to all the joints directly or hierarchically. Two branches of sub-networks are used to predict the centroids, body joints and their offsets to their parent nodes. Joints are grouped by tracing along their offsets to the closest centroids. Our single-person pose estimation method is evaluated on MPII Human Pose dataset [1] and Leeds Sports Pose dataset [14]. It achieves competitive accuracy with only 4.7 GFLOPs. Our multi-person pose estimation method is evaluated on MPII Human Pose Multi-Person dataset [1], and achieves competitive accuracy with only 13.6 GFLOPs.

2 Related Works

2.1 Efficient Neural Networks

To adopt deep neural networks in real-time applications and/or on resource-constrained devices, many research works have been devoted to build efficient neural networks with acceptable performance. Depthwise separable convolution was originally presented in [28]. It can achieve a good balance between the representation capability and computational efficiency, and has been utilized as the key building block in many successful efficient CNNs, such as Xception [5], MobileNets [10, 26] and ENAS [23]. MixConv [31] extends vanilla depthwise convolution by partitioning channels into multiple groups and apply different kernel sizes to each of them, and achieves better representation capability.

2.2 Multi-person Pose Estimation

Top-Down Methods. Top-down multi-person pose estimation methods first detect people by a human detector (*e.g.* Faster-RCNN [25]), then run a single-person pose estimator on the cropped image of each person to get the final pose predictions. Representative top-down methods include PoseNet [22], RMPE [6], Mask R-CNN [8], CPN [4] and MSRA [34]. However, top-down methods depend heavily on the human detector, and their inference time will significantly increase if many people appear together.

Bottom-Up Methods. Bottom-up methods detect the human joints of all persons at once, and then allocate these joints to each person based on various joint grouping methods. However, they suffer from very high complexity of joint grouping step, which usually involves solving a NP-hard graph partition problem. DeepCut [24] and DeeperCut [12] solve the joint grouping with an integer linear program which results in the order of hours to process a single image. Later works drastically reduce prediction time by using greedy decoders in combination with additional tools. Cao *et al.* [3] proposed part affinity fields to encode location and orientation of limbs. Newell and Deng [16] presented the associative embedding for grouping joint candidates. PPN [18] performs dense regressions from global joint candidates within a embedding space of person centroids to generate person detection and joint grouping. But it need to adopt the Agglomerative Clustering algorithm [2] to determine the person centroids. In this paper, we avoid the time-consuming clustering by regressing the person centroids together with body joints and using them to guide the joint grouping.

Recently, Nie *et al.* [19] proposed a one-stage multi-person pose estimation method (SPM) which predicts root joints (person centroids) and joint displacements directly. Although both SPM and our method predict person centroid, they use centroid plus displacements to recover joints and we use centroid to guide joint grouping. We argue that joints can be predicted more precisely than its displacements.

3 The Proposed Light-Weight Hourglass Network

3.1 Hourglass Network

Although Hourglass network has been utilized in many human pose estimation methods [18–20, 35], it is hard to been adapted in practical applications due to its large model size. Original Hourglass network consists of eight stacked hourglass modules, whose structure is illustrated in Fig. 1. The ResBlock used in original Hourglass network has a bottleneck structure (Fig. 2(a)). In this paper, we try to improve the efficiency of Hourglass network by replacing the original ResBlocks with the light-weight ones (Fig. 1). The proposed Hourglass network is called DS-Hourglass network. More details are given as follows.

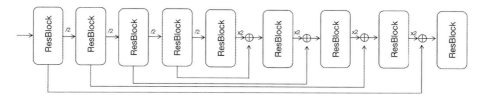

Fig. 1. An illustration of a Hourglass module, "/2" means downsampling operation and "x2" means upsampling operation.

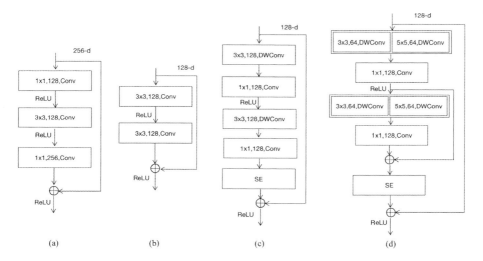

Fig. 2. Architecture of the ResBlocks. (a) The original ResBlock in Hourglass [17]. (b) another kind of ResBlock proposed in ResNet [9]. (c) DS-ResBlock corresponding to (b), (d) replace the DWConv in (c) with Mix-Cov.

3.2 Depthwise Separable Convolutions

A depthwise separable convolution decomposes a standard convolutional operation into a depthwise convolution (capture the spatial correlation) followed by a pointwise convolution (capture the cross-channel correlation).

A standard convolution operation needs $c_1 \times c_2 \times k \times k$ parameters and about $h \times w \times c_1 \times c_2 \times k \times k$ computational cost, where $h \times w$, c_1/c_2 and

$k \times k$ are the spatial size of input and output feature maps, the number of input/output feature channels and the convolutional kernel size, respectively. While a depthwise separable convolution operation only needs $c_1 \times k \times k + c_1 \times c_2$ parameters and about $h \times w \times c_1(k^2 + c_2)$ computational cost. For example, if we set k to 3 and c_2 to 128, the number of parameters and the computational cost of the depthwise separable convolution is only about 1/9 of the corresponding standard convolution.

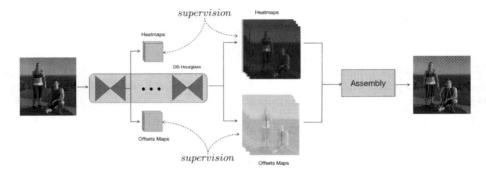

Fig. 3. Our pipeline. Our method takes color image as input, the DS-Hourglass build with 8 stage and predict 2 sets, heatmaps and offsetmaps, finally return joints for each person in image at once by our efficient joint grouping algorithm.

3.3   Light-Weight ResBlock

To develop an efficient ResBlock, we first reduce the number of its input/output feature channels from 256 to 128, and use two stacked 3×3 convolutions (Fig. 2(b)). Then, we replace the two standard 3×3 convolutions with two depthwise separable convolutions followed with a squeeze-and-excitation (SE) block to get a light-weight ResBlock (Fig. 2(c)). The SE block is very efficient and used to relocate features and strengthen features. To capture the information of different scales, we further replace the depthwise convolution with a mixed depthwise convolution (MixConv [31]) to get another version of light-weight ResBlock (Fig. 2(d)). In MixConv, the input feature channels are first split into groups, then depthwise convolutions with different kernel sizes are applied to different groups, finally, the output of each depthwise convolution are concatenated. In this paper, we apply kernels of 3×3 and 5×5 to two groups of channels respectively to trade-off the representation capability and the computational costs. In our study, we found that adding a skip connection around the second depthwise separable convolution can make the training of this ResBlock (Fig. 2(d)) more stable.

4   Multi-person Pose Estimation

Figure 3 illustrates the overall pipeline of our network. Our multi-person pose estimation method first predicts joints of all person at once, then the joint candidates are grouped into different persons. To improve the efficiency of joint grouping, a rooted tree is used to represent human pose by using person centroid as the root which connecting to all the joints directly or hierarchically. Another network branch is used to predict the offset from each joint to its parent node. The person centroid is treated as a pseudo joint and predicted together with body joints. After that, body joints are grouped by tracing along their offsets to the closed centroids.

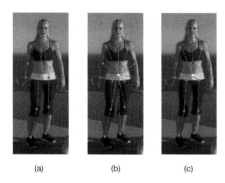

(a)   (b)   (c)

Fig. 4. Pose representations: (a) kinematic structure; (b) centroid-rooted tree; (c) hierarchical centroid-rooted tree

4.1   Centroid-Rooted Tree Structure to Define the Offsets

The centroid-rooted tree structure is illustrated in Fig. 4(b), where the person centroid (root node) directly connect to all body joints (leaf nodes). The drawback of this representation is that it leads to some long-range offsets which are hard to be precisely predicted, e.g. from ankle, knee and wrist to the centroid. To alleviate this problem, we further proposed a hierarchical centroid-rooted tree (Fig. 4(c)) based on the kinematic structure (Fig. 4(a)), where the long-range offsets are decomposed into short-range or middle-range offsets.

4.2   Joint and Offset Prediction

Our network has two branches of sub-networks for joint and offset prediction. Both sub-networks have only one 1×1 convolution and share the same feature maps from hourglass module.

Joint Prediction. Our ground-truth heatmap is generated according to following equation,

$$H_j(x,y) = min(\sum_{i=1}^{N} exp(-\frac{((x,y) - (x_j^i, y_j^i))^2}{2\sigma^2}), 1), \tag{1}$$

where (x_j^i, y_j^i) is the coordinate of joint j of person i, N is the number of person in the image and σ controls the spread of the peak, and minimum function is used to guarantee the value not greater than 1.

Offset Prediction. we construct a dense offset map for each body joint as the ground-truth for offset prediction. We first construct a offset map O_j^i for joint j of person i as:

$$O_j^i(x,y) = \begin{cases} \frac{1}{Z}((x_c^i, y_c^i) - (x,y)) & \text{if } (x,y) \in N_j^i \\ 0 & \text{otherwise} \end{cases} \tag{2}$$

where $N_j^i = \left\{ (x,y) | \sqrt{(x,y) - (x_j^i, y_j^i)} \leq \tau \right\}$ denotes the area of neighbors of joint j of person i, (x_c^i, y_c^i) is the coordinate of the centroid of the person i, $Z = \frac{1}{2}min(W, H)$ is a normalization coefficient, W and H are the width and height of the input image. If a location belongs to multiple people, these vectors are averaged. If the hierarchical centroid-rooted tree is used to represent human pose, we only need to replace the centroid in Eq. (2) with the parent node of the joint j of the person i.

The MSE Loss is used for joint prediction, and Smooth L1-Loss is used for offset prediction.

4.3   Centroid-Guided Joint Grouping

Based on the predicted person centroids, we develop a greedy method for joint grouping. First, we apply NMS to the heatmaps of the last stage of DS-Hourglass to get the coordinates of all candidate joints, and sort them in descending order of their score. For the centroid-rooted tree representation, joint allocation is performed independently for each body joint. Given a candidate of joint j, its centroid's coordinate can be predicated as

$$(\hat{x}, \hat{y})_c = (x,y) + Z \times O_j(x,y), \tag{3}$$

where (x,y) is the coordinate of the candidate joint, $(\hat{x}, \hat{y})_c$ is the predicated coordinate of its person centroid, and O_j is the offset map of joint j. Then the predicated centroid is compared to each person centroid generated from heatmap and allocated to the nearest one under the constrain that one person has only one instance for each joint category.

For hierarchical centroid-rooted tree representation, the joints are also grouped hierarchically. Base on the intuition that the joints close to the torso can be predicted more reliably. We classify joints into three levels, the first level contains *shoulder*, *hip*, and *neck*, the second level contains *head*, *elbow* and *knee*, and third level contains *wirst* and *ankle*. The joint allocation is performed from the first level to the third level, each joint candidate is associated to its parent node in the tree structure according to Eq. (3) and the nearest-neighbour rule.

5   Experiments

5.1   Experiment Setup

Datasets. MPII Single-Person Dataset consists of around 25k images with annotations for multiple people providing 40k annotated samples (28k training, 11k testing) for single-person pose estimation. The MPII Multi-Person dataset consists of 3,844 and 1,758 groups of multiple interacting persons for training and testing. The LSP dataset has 11K training samples and 1K test samples, with 14 annotated joints for a person.

Table 1. Comparison of ResBlocks on MPII single-person validation set

Methods	Mean	Stages	Param	FLOPs
Hourglass [17]	90.52	8	26M	26.2G
SBN [34]	89.6	–	68.6M	21G
HRNet [29]	90.3	–	28.5M	9.5G
FPD [35]	89.04	4	3M	3.6G
DS-Hourglass*	88.71	8	2.9M	3.3G
DS-Hourglass w/o SE	89.47	8	2.9M	4.7G
DS-Hourglass	89.87	8	4.2M	4.7G
DS-Hourglass (mix)	89.94	8	4.6M	4.8G

Training Details. We randomly augmented the samples with rotation degrees in [−40, 40], scaling factors in [0.7, 1.3], translation offset in [−40, 40] and horizontally mirror, adopt 256 × 256 as training size. The initial learning rate is 0.0025, learning rate decay at step 150, 170, 200 and 230 with total 250 epochs by 0.5.

5.2   Ablation Study

Ablation for ResBlocks. In Table 1, DS-Hourglass* uses Fig. 2(c) with only one depthwise separable convolution. DS-Hourglass w/o SE uses Fig. 2(c) without SE module. DS-Hourglass and DS-Hourglass (mix) use Fig. 2(c) and Fig. 2(d) respectively. ResBlock with two depthwise separable convolution can improve 1.1% in PCKh than ResBlock with only one, but with 1.4 GFLOPs computational cost increased. SE model can improve 0.4% in PCKh; Mixed depthwise convolution only brings very little improvement for single-person pose estimation. Compared with the excellent methods, our DS-Hourglass(mix) is 0.9% higher than FPD. And 0.36% lower than HRNet [29], However, the GFLOPs is only half of it and the parameters are only 16% of HRNet.

Ablation for Assembly Methods. Table 2 indicated that our centroid-guided method improves 0.4% over PPN, and our hierarchical centroid-guided method improves 1.8% over PPN. DS-Hourglass (mix) improves 1% over DS-Hourglass with hierarchical centroid-guided assembly method, as it can handles multi-scale problem better, even better than PPN [18] based on original Hourglass one (79.8% vs 79.4%), we can save 50GFLOPs and only need 21% parameters of PPN.

5.3   Comparisons to State-Of-The-Art Methods

MPII Single-person Dataset. From Table 3, we can find that our method is very lightweight and efficient. Our model has greatly reduced the deployment cost, while still achieving a high PCKh of 91.5%. Compared our method with the best performer, PIL [20], the DS-Hourglass needs only 16% of its computational cost but has only 0.9% drop in PCKh. Our method outperforms FPD [35] (91.5% vs 91.1% AP) which needs knowledge distillation and pretrained weights.

LSP Dataset. Our method also achieve the 90.8% PCK@0.2 accuracy on LSP dataset which is same as FPD [35], without using extra dataset. Because space is limited, the comparison is not listed in the form of a table.

Table 2. Comparison of assembly methods on MPII multi-person validation set

Model	Stage	Method	Mean	Param	FLOPs
Hourglass	8	PPN [18]	79.4	22M	62.9G
Hourglass	1	PPN	74.4	3.0M	10.8G
Hourglass	1	Center	75.8	3.0M	10.9G
Hourglass	1	Cent.Hier.	76.2	3.0M	10.9G
DS-Hourglass	8	Cent.Hier.	78.8	4.4M	13.3G
DS-Hourglass (mix)	8	Cent.Hier.	79.8	4.6M	13.6G

MPII Multi-person Dataset. In Table 4, we compare our method with the leading methods in recent years. It should be noted that our method does not use single-person pose estimation to refine the results. Research works [3,18] have reported that the single-pose refinement can improve the result by about 2.6%. However, the refinement is always time-consuming, so we did not use it. We get very competitive result 77.4% achieve the state-of-the-art among the methods without refinement. Our model has only 4.6M parameters and needs 13.6 GFLOPs when using input size of 384 × 384. To best of our knowledge,

Table 3. Result on MPII single-person test set. ∗ means the use of extend dataset.

Method	PCKh	**Auxiliaries**	Stage	**Pre.**	Input size	Out. size	FLOPs	#Param
DeeperCut [12]	88.5	–	–	Yes	344 × 344	43 × 43	37G	66M
CPM* [33]	88.5	–	6	Yes	368 × 368	46 × 46	175G	31M
SHG [17]	90.9	–	8	No	256 × 256	64 × 64	26.2G	26M
PIL [20]	**92.4**	Segment	8	No	256 × 256	64 × 64	29.2G	26.4M
Sekii [27]	88.1	–	–	Yes	384 × 384	12 × 12	6G	16M
FPD [35]	91.1	Know. dist.	4	Yes	256 × 256	64 × 64	**3.6G**	**3.2M**
HRNet [29]	92.3	–	–	Yes	256 × 256	64 × 64	9.5G	28.5M
DS-Hourglass	91.5	-	8	No	256 × 256	64 × 64	4.7G	4.2M

PPN [18] is the state-of-the-art on MPII Multi-Person dataset, from our ablation Table 2, we can find that our method is better than PPN, and our method can reduced the computational cost by approximately 50GFLOPs.

Table 5 lists the results on MPII 288 test set, and our method gets 81.0%, only 0.3% lower than the best one [7] which uses refinement. It can be found that our method has great advantages in the distal part of the body (*e.g.* wrist, knee, ankle, *etc.*), as we use hierarchical centroid-rooted tree to avoid long-range offset prediction.

Table 4. Results on the Full MPII Multi-person test set.

Methods	Head	Shoulder	Elbow	Wrist	Hip	Knee	Ankle	Total	**Refine.**
CMU-Pose [3]	91.2	87.6	77.7	66.8	75.4	68.9	61.7	75.6	Yes
AE-Pose [16]	92.1	89.3	78.9	69.8	76.2	71.6	64.7	77.5	Yes
RefinePose [7]	91.8	89.5	80.4	69.6	77.3	71.7	65.5	78.0	Yes
PPN [18]	92.2	89.7	82.1	74.4	78.6	76.4	69.3	80.4	Yes
SPM [19]	89.7	87.4	80.4	72.4	76.7	74.9	68.3	78.5	Yes
DeepCut [24]	73.4	71.8	57.9	39.9	56.7	44.0	32.0	54.1	No
DeeperCut [12]	89.4	84.5	70.4	59.3	68.9	62.7	54.6	70.0	No
Levinkov *et al.* [15]	89.8	85.2	71.8	59.6	71.1	63.0	53.5	70.6	No
ArtTrack [11]	88.8	87.0	75.9	64.9	74.2	68.8	60.5	74.3	No
RMPE [6]	88.4	86.5	78.6	70.4	74.4	73.0	**65.8**	76.7	No
Sekii [27]	**93.9**	**90.2**	**79.0**	**68.7**	74.8	68.7	60.5	76.6	No
DS-hourglass (mix)	91.6	88.3	78.0	68.3	**77.9**	**72.5**	65.1	**77.4**	No

Table 5. Results on a set of 288 images from MPII Multi-person test set.

Methods	Head	Shoulder	Elbow	Wrist	Hip	Knee	Ankle	Total	**Refine.**
CMU-Pose [3]	92.9	91.3	82.3	72.6	76.0	70.9	66.8	79.0	Yes
AE-Pose [16]	91.5	87.2	75.9	65.4	72.2	67.0	62.1	74.5	Yes
RefinePose [7]	93.8	91.6	83.9	75.2	78.0	75.6	70.7	81.3	Yes
DeepCut [24]	73.1	71.7	58.0	39.9	56.1	43.5	31.9	53.5	No
Iqbal and Gall [13]	70.0	65.2	56.4	46.1	52.7	47.9	44.5	54.7	No
DeeperCut [12]	92.1	88.5	76.4	67.8	73.6	68.7	62.3	75.6	No
ArtTrack [11]	92.2	91.3	80.8	71.4	79.1	72.6	67.8	79.3	No
RMPE [6]	89.4	88.5	81.0	75.4	73.7	75.4	66.5	78.6	No
Sekii [27]	**95.2**	**92.2**	**83.2**	73.8	74.8	71.3	63.4	79.1	No
DS-Hourglass (mix)	93.2	91.1	81.5	**74.7**	**81.0**	**75.5**	**70.2**	**81.0**	No

6 Conclusions

In this paper, we develop a light-weight Hourglass network by applying depthwise separable convolution and mixed depthwise convolution. The new network can be directly applied to single-person pose estimation. Based on this backbone network, we further proposed an efficient multi-person pose estimation method. Both our single-person and multi-person pose estimation methods can achieve competitive accuracies on public datasets with low computational costs.

References

1. Andriluka, M., Pishchulin, L., Gehler, P., Schiele, B.: 2D human pose estimation: new benchmark and state of the art analysis. In: Proceedings of the IEEE Conference on Computer Vision and Pattern Recognition, pp. 3686–3693 (2014)
2. Bourdev, L., Malik, J.: Poselets: body part detectors trained using 3D human pose annotations. In: International Conference on Computer Vision, pp. 1365–1372 (2009)
3. Cao, Z., Simon, T., Wei, S.E., Sheikh, Y.: Realtime multi-person 2D pose estimation using part affinity fields. In: Proceedings of the IEEE Conference on Computer Vision and Pattern Recognition, pp. 7291–7299 (2017)
4. Chen, Y., Wang, Z., Peng, Y., Zhang, Z., Yu, G., Sun, J.: Cascaded pyramid network for multi-person pose estimation. In: Proceedings of the IEEE Conference on Computer Vision and Pattern Recognition, pp. 7103–7112 (2018)
5. Chollet, F.: Xception: deep learning with depthwise separable convolutions. In: Proceedings of the IEEE Conference on Computer Vision and Pattern Recognition, pp. 1251–1258 (2017)
6. Fang, H.S., Xie, S., Tai, Y.W., Lu, C.: RMPE: regional multi-person pose estimation. In: Proceedings of the IEEE International Conference on Computer Vision, pp. 2334–2343 (2017)
7. Fieraru, M., Khoreva, A., Pishchulin, L., Schiele, B.: Learning to refine human pose estimation. In: Proceedings of the IEEE Conference on Computer Vision and Pattern Recognition Workshops, pp. 205–214 (2018)

8. He, K., Gkioxari, G., Dollár, P., Girshick, R.: Mask R-CNN. In: Proceedings of the IEEE International Conference on Computer Vision, pp. 2961–2969 (2017)
9. He, K., Zhang, X., Ren, S., Sun, J.: Deep residual learning for image recognition. In: Proceedings of the IEEE Conference on Computer Vision and Pattern Recognition, pp. 770–778 (2016)
10. Howard, A.G., et al.: Mobilenets: efficient convolutional neural networks for mobile vision applications. arXiv preprint arXiv:1704.04861 (2017)
11. Insafutdinov, E., et al.: Arttrack: articulated multi-person tracking in the wild. In: Proceedings of the IEEE Conference on Computer Vision and Pattern Recognition, pp. 6457–6465 (2017)
12. Insafutdinov, E., Pishchulin, L., Andres, B., Andriluka, M., Schiele, B.: DeeperCut: a deeper, stronger, and faster multi-person pose estimation model. In: Leibe, B., Matas, J., Sebe, N., Welling, M. (eds.) ECCV 2016. LNCS, vol. 9910, pp. 34–50. Springer, Cham (2016). https://doi.org/10.1007/978-3-319-46466-4_3
13. Iqbal, U., Gall, J.: Multi-person pose estimation with local joint-to-person associations. In: Hua, G., Jégou, H. (eds.) ECCV 2016. LNCS, vol. 9914, pp. 627–642. Springer, Cham (2016). https://doi.org/10.1007/978-3-319-48881-3_44
14. Johnson, S., Everingham, M.: Clustered pose and nonlinear appearance models for human pose estimation. In: British Machine Vision Conference, vol. 2, p. 5 (2010)
15. Levinkov, E., et al.: Joint graph decomposition & node labeling: problem, algorithms, applications. In: Proceedings of the IEEE Conference on Computer Vision and Pattern Recognition, pp. 6012–6020 (2017)
16. Newell, A., Huang, Z., Deng, J.: Associative embedding: end-to-end learning for joint detection and grouping. In: Advances in Neural Information Processing Systems, pp. 2277–2287 (2017)
17. Newell, A., Yang, K., Deng, J.: Stacked hourglass networks for human pose estimation. In: Leibe, B., Matas, J., Sebe, N., Welling, M. (eds.) ECCV 2016. LNCS, vol. 9912, pp. 483–499. Springer, Cham (2016). https://doi.org/10.1007/978-3-319-46484-8_29
18. Nie, X., Feng, J., Xing, J., Yan, S.: Pose partition networks for multi-person pose estimation. In: Ferrari, V., Hebert, M., Sminchisescu, C., Weiss, Y. (eds.) ECCV 2018. LNCS, vol. 11209, pp. 705–720. Springer, Cham (2018). https://doi.org/10.1007/978-3-030-01228-1_42
19. Nie, X., Feng, J., Zhang, J., Yan, S.: Single-stage multi-person pose machines. In: Proceedings of the IEEE International Conference on Computer Vision, pp. 6951–6960 (2019)
20. Nie, X., Feng, J., Zuo, Y., Yan, S.: Human pose estimation with parsing induced learner. In: Proceedings of the IEEE Conference on Computer Vision and Pattern Recognition, pp. 2100–2108 (2018)
21. Papandreou, G., Zhu, T., Chen, L.-C., Gidaris, S., Tompson, J., Murphy, K.: PersonLab: person pose estimation and instance segmentation with a bottom-up, part-based, geometric embedding model. In: Ferrari, V., Hebert, M., Sminchisescu, C., Weiss, Y. (eds.) Computer Vision – ECCV 2018. LNCS, vol. 11218, pp. 282–299. Springer, Cham (2018). https://doi.org/10.1007/978-3-030-01264-9_17
22. Papandreou, G., et al.: Towards accurate multi-person pose estimation in the wild. In: Proceedings of the IEEE Conference on Computer Vision and Pattern Recognition, pp. 4903–4911 (2017)
23. Pham, H., Guan, M.Y., Zoph, B., Le, Q.V., Dean, J.: Efficient neural architecture search via parameter sharing. arXiv preprint arXiv:1802.03268 (2018)

24. Pishchulin, L., et al.: Deepcut: joint subset partition and labeling for multi person pose estimation. In: Proceedings of the IEEE Conference on Computer Vision and Pattern Recognition, pp. 4929–4937 (2016)
25. Ren, S., He, K., Girshick, R., Sun, J.: Faster R-CNN: towards real-time object detection with region proposal networks. In: Proceedings of the International Conference on Neural Information Processing Systems, pp. 91–99 (2015)
26. Sandler, M., Howard, A., Zhu, M., Zhmoginov, A., Chen, L.C.: Mobilenetv 2: inverted residuals and linear bottlenecks. In: Proceedings of the IEEE Conference on Computer Vision and Pattern Recognition, pp. 4510–4520 (2018)
27. Sekii, T.: Pose proposal networks. In: Ferrari, V., Hebert, M., Sminchisescu, C., Weiss, Y. (eds.) ECCV 2018. LNCS, vol. 11217, pp. 350–366. Springer, Cham (2018). https://doi.org/10.1007/978-3-030-01261-8_21
28. Sifre, L., Mallat, S.: Rigid-motion scattering for image classification. Ph.D. thesis (2014)
29. Sun, K., Xiao, B., Liu, D., Wang, J.: Deep high-resolution representation learning for human pose estimation. In: Proceedings of the IEEE Conference on Computer Vision and Pattern Recognition, pp. 5693–5703 (2019)
30. Szegedy, C., Vanhoucke, V., Ioffe, S., Shlens, J., Wojna, Z.: Rethinking the inception architecture for computer vision. In: Proceedings of the IEEE Conference on Computer Vision and Pattern Recognition, pp. 2818–2826 (2016)
31. Tan, M., Le, Q.V.: Mixconv: Mixed depthwise convolutional kernels. arXiv preprint arXiv:1907.09595 (2019)
32. Tian, Z., Chen, H., Shen, C.: Directpose: Direct end-to-end multi-person pose estimation. arXiv preprint arXiv:1911.07451 (2019)
33. Wei, S.E., Ramakrishna, V., Kanade, T., Sheikh, Y.: Convolutional pose machines. In: Proceedings of the IEEE conference on Computer Vision and Pattern Recognition, pp. 4724–4732 (2016)
34. Xiao, B., Wu, H., Wei, Y.: Simple baselines for human pose estimation and tracking. In: Ferrari, V., Hebert, M., Sminchisescu, C., Weiss, Y. (eds.) ECCV 2018. LNCS, vol. 11210, pp. 472–487. Springer, Cham (2018). https://doi.org/10.1007/978-3-030-01231-1_29
35. Zhang, F., Zhu, X., Ye, M.: Fast human pose estimation. In: Proceedings of the IEEE Conference on Computer Vision and Pattern Recognition, pp. 3517–3526 (2019)

Discriminative Regions Erasing Strategy for Weakly-Supervised Temporal Action Localization

Huanbin Zeng, Suguo Zhu$^{(\boxtimes)}$, and Jun Yu

Key Laboratory of Complex Systems Modeling and Simulation, School of Computer
Science and Technology, Hangzhou Dianzi University, Hangzhou 310018, China
zsg2016@hdu.edu.cn

Abstract. Weakly-supervised temporal action localization (WTAL) has
recently attracted attentions. Many of the state-of-the-art methods usually utilize temporal class activation map (T-CAM) to obtain target
action temporal regions. However, class-specific T-CAM tends to cover
only the most discriminative part of the actions, not the entire action.
In this paper, we propose an erasing strategy for mining discriminative regions in weakly-supervised temporal action localization (DRES).
DRES achieves better performance with action localization, which can
be attribute to two aspects. First, we employ the salient detection module, which suppresses the background to obtain the most discriminative regions. Second, we design the eraser module to discover the missed
action regions by the salient detection module, which complements action
regions. Based on experiments, we demonstrate that DRES improve the
state-of-the-art performance on THUMOS'14.

Keywords: Weakly-supervised temporal action localization ·
Temporal class activation map · Erasing strategy

1 Introduction

Temporal action localization, as a crucial and challenging task in the field of
video content analysis, has received more and more attention. It not only requires
classifying the categories of actions correctly, but also needs to precisely locate
the temporal boundaries of each action in an untrimmed video. Most methods
with excellent performance are based on full supervision (e.g. [3,23] etc.), that
is, they usually require temporal annotations of action intervals. However, the
diversity of actions and the complexity of the background in real scenes make the
large scale temporal annotations of actions are prohibitively expensive and time-
consuming. Therefore, weakly-supervised temporal action localization (WTAL),

---

H. Zeng—Student as the first author.
This work was supported by the National Natural Science Foundation of China, with
grant numbers 61902101 and 61806063 respectively.

© Springer Nature Switzerland AG 2020
Y. Peng et al. (Eds.): PRCV 2020, LNCS 12306, pp. 639–651, 2020.
https://doi.org/10.1007/978-3-030-60639-8_53

which required only video-level labels (i.e. whether each video contains action frames of interest) is increasingly favored by the research community.

Drawing on the method of generating Class Activation Map (CAM) for object localization in the field of weakly-supervised object Localization (e.g. [19,26,29] etc.), some previous methods utilize temporal Class Activation Map (T-CAM) to obtain target action temporal regions (e.g. [14,21,28] etc.). However, due to the lack of negative examples of the background, these methods often did not explicitly consider the background class, but only use the action categories contained in the current video sequence as a positive example for training the classification network. Ignoring the background class is equivalent to adding a lot of noise to the data, which will cause the background frames to be incorrectly classified into action classes during the training process, and cause more interference to the action classifier. To tackle this problem, Lee et al.[8] proposed a background suppression network (BaSNet) to better exploit background class. By introducing an auxiliary class representing the background, and design a two-branch architecture with weights sharing, meanwhile, elaborately design a filtering module in the suppression branch, and introduce L1 norm as the sparsity supervision, the interference of background frames to the localization results is effectively eliminated.

Although BaSNet [8] has achieved good results after modeling the background class, it still exist following defects: Using L1 norm as the sparsity supervision when training the classification network may cause the model to cover only the most discriminative part of the action, and ultimately make the localization results a sparse subset of a series of key segments rather than the entire action. To alleviate this defect, we proposed an erasing strategy, which first makes the most discriminative part of the action instances obtain a high response through salient detection module, and a preset threshold is used on the existing detection results to generate a binary mask to erase the corresponding high response regions in initial feature maps, such erasing strategy is equivalent to hiding the most discriminative part of the initial features. Based on erased feature, an eraser module, which the structure is the same as the salient detection module, is used to classify the erased feature. Since the most discriminative regions of the feature have been hidden, the eraser module can only tap the remaining relatively insignificant areas in the same action instances as much as possible to obtain the same classification result in the common forward propagation process. Finally, by fusing the localization results of the salient detection module and the eraser module, a more complete action localization result is obtained.

To sum up, the main contributions of our work are as follows:

We propose an erasing strategy, which force the eraser model to mining relatively insignificant regions in the same action instance by erasing the most discriminative part of action instance, and through fusion to make the final detected action instance more complete.

2   Related Work

Fully-Supervised Temporal Action Localization (TAL): During the past few years, driven by the development of deep learning, TAL has made great progress. Inspired by SSD [11], Lin et al. [9] proposed an anchor-based one-stage architecture detector which directly localizes the instances by temporal convolution with predetermined anchors, but its performance is not satisfactory enough due to limit of fixed temporal scales. R-C3D [23] generate candidate regions using the anchor segment method, and then classify the categories and regression the temporal boundaries of these candidate proposals through a classification subnet, but temporal boundaries of candidate proposals which generated by the anchor segment method only are not fine enough. On the basis of R-C3D, PCAD [5] draws on the BSN [10] and proposes an auxiliary boundary proposal network (BPN), which complements the candidate proposal generation network based on anchor segment method to generate more elaborate candidate proposals, which further improves the detection performance of the model.

Weakly-Supervised Temporal Action Localization (WTAL): Unlike TAL whose training labels contains both frame-wise labels and video-level labels, WTAL solves the same problem by using only video-level labels. Drawing on the method of generating Class Activation Map (CAM) [29] for object localization in the field of weakly-supervised object detection, some previous methods utilize temporal class activation map (T-CAM) to obtain target action temporal regions. [14] first proposed the use of T-CAM for WTAL, and introduced additional L1 norm as sparseness supervision, so that the designed attention module can obtain the class-agnostic attention weight to represent the frame-level importance of the segment, and ensure that the final localization results are more complete. Unlike other methods that use convolutional networks for modeling, [24] uses long short-term memory networks to model the relationship between different actions, and a permissive coverage mechanism is proposed to solve the problem of overlapping attention weights when action segments are close to or coincide in temporal dimension. [25] proposed marginalized average attention network (MAAN) to alleviate the problem that attention weights are easily dominated by some subsets of segments that have a large contribution to the action classification. To better model the background class, [8] proposed a background suppression network (BaSNet) to better exploit background class.

3   Our Approach

We proposed an end-to-end trainable model for weakly-supervised temporal action localization (WTAL). The overall architecture of our method is shown in Fig. 1. The model consists of three sub-modules: a feature extraction module, a salient detection module (SDM), and an eraser module (EM). In order to more effectively use the appearance and motion characteristics of the video, we simultaneously encode the RGB and optical flow features of the input video streams with pre-trained feature extractor, and stack the features obtained by each in temporal dimension. For better model the background class, we designed a segment-level activator, which is two-branch structure with weights sharing, to better suppresses background frames while mining foreground actions. The SDM effectively mines the most discriminative regions of the action instances, but the obtained action interval prediction is not complete. In view of this, we generate a binary mask, to erase the most discriminative regions in the initial feature map of the corresponding action instances by a preset threshold, and designed an eraser module (EM) with the same structure as the SDM to mine the lower response part of the action instances. Finally, a more complete action localization prediction is obtained after fusion the prediction results of the SDM and EM. We describe our proposed method in detail with the following sections.

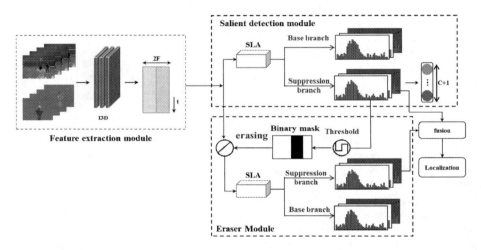

Fig. 1. Overall architecture of the proposed method. It consists of three modules: (a) Feature extraction Module, (b) Salient Detection Module (SDM) and (c) Eraser Module (EM), where the SLA represents Segment-level Activator whose detailed structure is shown in Fig. 2

3.1 Feature Extraction

Suppose $V = \{V_n\}_{n=1}^N$ denote as N training videos we are given, and their video-level labels $Y = \{y_n\}_{n=1}^N$, where y_n is C-dimensional binary vector with $y_{n;c} = 1$ if n-th video contains c-th action category otherwise 0 for C classes. For each input video v_n , we first divide it in a 16-frame non-overlapping manner, and each divided video can be expressed as $V_n = \{S_{n,i}\}_{i=1}^I$ which contain I segments, due to its video lengths are large variation, we sample a fixed number of T segments $\{S_{n,i}\}_{i=1}^T$ from each video. Afterwards, we use pre-trained I3D network to extract the RGB and optical flow features of the input video streams, respectively. Then stack those two features in temporal dimension to integrate the appearance and motion characteristics of the input video, and the initial feature obtained can be denoted as $F_{init} \in R^{(T,2F)}$, where T represents the size of the temporal dimension of the extracted feature (i.e. the number of fixed sampling segments for each video), and F represents the dimension of each segment output after passing through the I3D network .

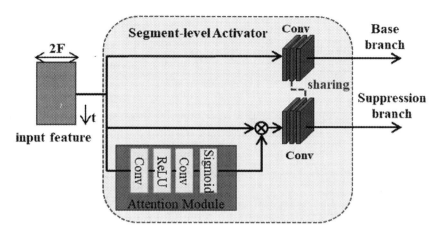

Fig. 2. The detailed structure of the segment-level activator, which is two-branch structure with weights sharing.

3.2 Salient Detection Module (SDM)

To effectively model the background class, and reduce the negative impact of the background frames on final detection result, in SDM, we first introduced an auxiliary class representing the background, and then designed a segment-level activator, which is a two-branch structure with weights sharing, whose structure is shown in Fig. 2. Among them, base branch takes the original feature as input and treat the background as a positive sample, while suppression branch regards the background as a negative sample. Then, an additional attention module, is designed to predict the probability of whether each segment contains foreground actions. The greater the corresponding attention weight, the higher probability that the current segment contains foreground actions.

Specifically, the base branch in segment-level activator take the F_{init} as input, then temporal class activation map (T-CAM) corresponding to each class is generated. The overall process can be summarized as:

$$M_{base} = Activator(F_{init}; \Phi) \tag{1}$$

where Φ represents the trainable parameter of the corresponding convolutional layer of the base branch in the segment-level activator, $M_{base} \in R^{T \times (C+1)}$ donate the T-CAM generated for each class, where $C + 1$ represent that we use C action classes and one auxiliary class for the background.

In order to generate single video-level class score, following previous work (Wang et al. [21] ; Paul, Roy, and RoyChowdhury [15])), we use the top-k mean technique to aggregate segment-level class scores, and then a softmax function is applied to generate the probability score for single class. The overall process can be described as:

$$P_{base} = Softmax(topk(M_{base}, k)) \tag{2}$$

where $k = \lfloor \frac{T}{r} \rfloor$, and r is a hyperparameter to control the ratio of selected segments in a video.

Different from base branch, an additionally attention module, which is used to predict the probability of whether each segment contains foreground actions, is designed in the suppression branch, and uses L1 norm as the sparsity supervision for attention module. The output of attention module can be denoted as $W_{att} \in R^T$, which range from 0 to 1. Then multiply the obtained attention weight W_{att} by the F_{init} to obtain the background suppression feature, which can be described as:

$$F_{supp} = F_{init} \bigotimes W_{att} \tag{3}$$

where $F_{supp} \in R^{2F \times T}$ and $\bigotimes$ denotes element-wise multiplication over temporal dimension. Following the steps (1) and (2) in the base branch, we correspondingly generate T-CAM and single probability scores of suppression branch, which can be denote as M_{supp} and P_{supp}.

The base branch and the suppression branch share the weights of the Segment-level activator, so that the attention module can effectively learn the weights of each frame as the foreground, and effectively suppress the background frames.

3.3 Eraser Module (EM)

Due to the use of L1 norm as sparseness supervision, the salient detection module can only cover the most discriminative part in the action instances, not the entire action, which can be seen from Fig. 3. In view of this, we propose a complementary eraser module for mining relatively insignificant regions in action instances, and the visualization process is shown in Fig. 4.

Specifically, we first obtain the temporal class activation map (T-CAM) M_{supp} from the suppression branch of the SDM. Since the corresponding video-level labels are available during training, we select the activation sequence corresponding to the class labels in M_{supp}. Note that a video segments may contain more than one class of video-level labels, to facilitate erasing, we selected the activation sequence with the highest response. The process above can be described as:

$$\begin{cases} A_s = M_{supp}[index(y)] \\ \overline{A}_s = \max_{a \in A_s} a \end{cases} \tag{4}$$

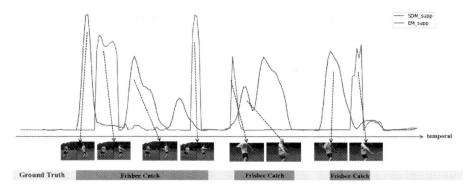

Fig. 3. Visualization of action localization by using activation sequence generated from suppression branch of different modules, where the blue polyline represents the activation sequence generated from Salient detection module (SDM) with current action category (i.e. Frisbee Catch), while the orange polyline represents the activation sequence generated from Eraser module (EM) with the same action category. It can be seen that the high-response part of the blue polyline covers most of the area of the current action, but some relatively insignificant areas, such as edge area at the beginning or end of the action, are easily ignored. In contrast, the orange polyline has a higher response to these relatively insignificant areas. By fusion the results of two modules, the final action localization results are more complete. (Color figure online)

where y is C-dimensional binary vector with $y_c = 1$ if current video contains c-th action category otherwise 0 for C classes ,note that a video may contain multiple action classes, i.e. $\sum_{c=1}^{C} y_c \geq 1$, index(y) refers to the index of the position equal to 1 in the y vector, A_s donates the corresponding activation sequence selected from M_{supp} through the video-level labels y, and $\overline{A}_s \in R^{T \times 1}$ represents the activation sequence with the highest response selected from A_s.

To better guide the model erase the most discriminative regions in the action instances, we use sigmoid function to map the activation sequence $\overline{A}_s$ to the range of 0 to 1. Then a preset threshold ε is used to generate binary mask, whose dimension is the same as $\overline{A}_s$, if the value of the corresponding segment of the $\overline{A}_s$ after sigmoid mapping is greater than the threshold ε, the value of the corresponding position in the binary mask is 0, otherwise it is 1. The whole process can be described as:

$$\begin{cases} dim(mask) = dim(\overline{A}_s) \in R^{T \times 1} \\ mask[sigmoid(\overline{A}_s) \geq \varepsilon] = 0 \\ mask[sigmoid(\overline{A}_s) < \varepsilon] = 1 \end{cases} \tag{5}$$

After obtaining the binary mask, we multiply it with the initial feature F_{init}, and obtain the erased feature:

$$F_{erased} = mask \bigotimes F_{init} \tag{6}$$

where $F_{erased} \in R^{T \times 2F}$ represents the erased feature obtained after erasing the most discriminative part in the action instances.

After obtaining the erased feature, we use eraser module, whose structure is the same as SDM, to further mine the remaining part of the action instances in the erased feature. The specific process is exactly the same as SDM, except that the input of the model has changed: in SDM, the input used is F_{init}, while the input used by EM is F_{erased}.

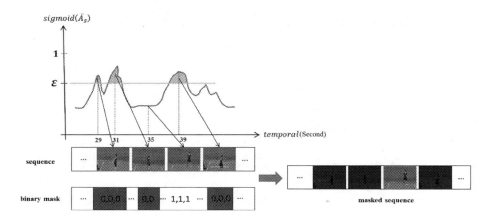

Fig. 4. Visualization of generating binary mask corresponding to a specific category of activation sequence, where $sigmoid\,(\overline{A}_s)$ represents the mapping value of the activation sequence corresponding to the current action category (JavelinThrow in this example) after being mapped by the sigmoid function, and ε represents the preset threshold. Frames in the frame sequence corresponding to the feature map of the current segment sequence will be masked when the corresponding mapping value is higher than the threshold ε, and the value of the corresponding position of the binary mask will be set to 0, otherwise it is 1.

3.4   Result Fusion

In inference time, we use the predicted probability greater than the class-threshold Θ_{class} as the video-level labels of the current testing video. In order to obtain a more complete localization result, we use video-level labels to select the activation sequence from T-CAM generated by the suppression branches of the SDM and EM, which the process can be described by Eq. 4 above. Let $\overline{A}_s$ and $\overline{A}_e$ be the activation sequence selected from SDM and EM, respectively. The activation sequence after fusion can be described as:

$$\overline{A}_f = \max([\overline{A}_s, \overline{A}_e], dim = t) \tag{7}$$

where $\overline{A}_f$ represents the fusion activation sequence obtained after fusion operation, and $dim = t$ means that the max operation is performed in the temporal dimension.

After fusion activation sequence $\overline{A}_f$ is obtained, frame-wise action intervals are inferred by thresholding $\overline{A}_f$. Since $\overline{A}_f$ contains not only the most discriminative regions, but also the relatively insignificant regions in the action instances, the final action localization results are more complete.

4   Experiments and Analysis

4.1   Dataset

To verify the effectiveness of our method, we evaluate our model on a large-scale action detection dataset THUMOS'14 [7]. THUMOS'14 contains a total of 20 different categories of sport activities, which include three subsets: training set, validation set and test set. However, training set only contains trimmed videos which are not suitable for our tasks. So actually we use validation set and test set to match our tasks. Among this, validation set contains a total of 200 untrimmed videos with 3007 action instances, which we use to train our model, test set contains 213 untrimmed videos with 3358 action instances, which we use to evaluate our trained model.

4.2   Evaluation Metrics

In order to make a fair comparison with other works in the same task, and proves the superiority of our model. We evaluate our results in terms of mean Average Precision – mAP@α where α denotes different Intersection over Union (IoU) thresholds. It is also a common metrics to demonstrate model performance in this task.

4.3   Implementation Details

We use I3D [2] networks which pre-trained on ImageNet [4] and Kinetics [2] respectively to extract features for video segments. To better use the motion information of the input videos, we use TVL1 algorithm [22] for generating optical flow of segments. Since the video lengths are large variation, we sample a fixed number of 750 segments for each video, and adopted different sampling strategies during the training and testing phases: stratified random perturbation during training and uniform sampling during test. We set the threshold $\varepsilon = 0.615$ to generate binary mask both in training and testing. The network is trained using Adam optimizer with learning rate 10^{-4}. At testing time, we set the class-threshold $\Theta_{class} = 0.25$, and reject classes whose video-level probabilities are lower than Θ_{class}. To delete proposals with high overlap rates, we perform non-maximum suppression (NMS) with threshold 0.7.

Table 1. Comparison results of using different strategies in terms of mAP at IoU threshold 0.1-0.9 under the weakly-supervised temporal action localization (WTAL) task on THUMOS'14 testing set. Among them, SDM only means that only Salient detection module is used for WTAL, while SDM + EM means that Eraser module is introduced through the erasing strategy based on result of SDM only.

mAP@IoU	0.1	0.2	0.3	0.4	0.5	0.6	0.7	0.8	0.9
SDM only	57.01	51.2	43.8	35.2	26.06	17.8	9.92	3.77	0.39
SDM + EM	58.87	52.43	44.33	36.04	27.33	18.77	10.22	3.84	0.46

Table 2. Comparison results of our work with some representative works in terms of mAP at IoU thresholds 0.1–0.9 (denoted as α) under the weakly-supervised temporal action localization task on THUMOS'14 testing set.

Strong subversion	α								
	0.1	0.2	0.3	0.4	0.5	0.6	0.7	0.8	0.9
SSN [27]	66	59.4	51.9	41	29.8	–	–	–	–
CBR [6]	60.1	56.7	50.1	41.3	31	19.1	9.9	–	–
BSN [10]	–	–	53.5	45	36.9	28.4	20	–	–
S-CNN [18]	47.7	43.5	36.3	29.7	19	10.3	5.3	–	–
CDC [16]	–	–	40.1	29.4	23.3	13.1	7.9	–	–
R-C3D [23]	54.5	51.5	44.8	35.6	28.9	–	–	–	–
Weak subversion	α								
	0.1	0.2	0.3	0.4	0.5	0.6	0.7	0.8	0.9
UntrimmedNet [21]	44.4	37.7	28.2	21.1	13.7	–	–	–	–
Hide-and-seek [20]	36.4	27.8	19.5	12.7	6.8	–	–	–	–
STPN(I3D) [14]	52.0	44.7	35.5	25.8	16.9	9.9	4.3	1.2	0.1
W-TACL(I3D) [15]	55.2	49.6	40.1	31.1	22.8	–	7.6	–	–
AutoLoc [17]	–	–	35.8	29.0	21.2	13.4	5.8	–	–
MAAN [25]	59.8	50.8	41.1	30.6	20.3	12.0	6.9	2.6	0.2
3C-Net [13]	59.1	53.5	44.2	34.1	26.6	–	8.1	–	–
RefineLoc [1]	–	–	33.9	–	22.1	–	6.1	–	–
Clean-Net [12]	–	–	37.0	–	30.9	13.9	7.1	–	–
BaSNet(I3D) [8]	58.2	52.3	44.6	36.0	27.0	18.6	10.4	3.9	0.5
Ours(I3D)	58.87	52.43	44.33	36.04	27.33	18.77	10.22	3.84	0.46

4.4 Ablation Studies

To thoroughly evaluate the effectiveness of our erasing strategy for the weakly-supervised temporal action localization, we performed extensive ablation experiments on the THUMOS'14 dataset. We first use only Salient detection module for weakly-supervised temporal action detection, based on this basis, we then take erasing strategy to hide the most discriminative part of the action instances, and add an additional eraser module to further mine the remaining actions in the erased feature. The comparison results are shown in Table 1. We observe that by adopting the erasing strategy in the existing Salient detection module, the action localization performance of our method has been effectively improved.

4.5 Results Comparison

As shown in Table 2, we compare our result with some representative works in terms of mAP at IoU thresholds 0.1–0.9 (denoted as α) under the same task to evaluate the effectiveness of our methods.

From Table 2, we can see that under the same supervision conditions, our method exhibits excellent performance. Even compared to some fully-supervised methods with more annotated information, our method also demonstrates superiority.

5   Conclusion

In this paper, we introduce an erasing strategy for weakly-supervised temporal action localization. In order to better explore the relatively insignificant regions of the action instances, a preset threshold is used to generate a binary mask to erase the most discriminative part of the action instances, and force the model to find those regions that contribute less to the classification task in the action instances, and ultimately improve the action localization performance. In the future, we will continue to explore weakly-supervised temporal action localization methods and further improve and upgrade our method.

References

1. Alwassel, H., Heilbron, C.F., Thabet, K.A., Ghanem, B.: RefineLoc: iterative refinement for weakly-supervised action localization. arXiv Computer Vision and Pattern Recognition (2019)
2. Carreira, J., Zisserman, A.: Quo Vadis, Action Recognition? A New Model and the Kinetics Dataset (2017)
3. Chao, Y.W., Vijayanarasimhan, S., Seybold, B., Ross, D.A., Deng, J., Sukthankar, R.: Rethinking the Faster R-CNN Architecture for Temporal Action Localization (2018)
4. Deng, J., Dong, W., Socher, R., Li, L.j., Li, K., Li, F.f.: ImageNet: a large-scale hierarchical image database. In: CVPR, pp. 248–255 (2009)
5. Fang, Z., Zhu, S., Yu, J., Tian, Q.: PCPCAD: proposal complementary action detector. In: 2019 IEEE International Conference on Multimedia and Expo (ICME), pp. 424–429, July 2019. https://doi.org/10.1109/ICME.2019.00080
6. Gao, J., Yang, Z., Nevatia, R.: Cascaded boundary regression for temporal action detection. In: BMVC (2017)
7. Jiang, Y.G., et al.: Thumos Challenge: Action Recognition with a Large Number of Classes (2014)

8. Lee, P., Uh, Y., Byun, H.: Background suppression network for weakly-supervised temporal action localization. In: AAAI (2020)

9. Lin, T., Zhao, X., Shou, Z.: Single shot temporal action detection. In: Proceedings of the 25th ACM International Conference on Multimedia, pp. 988–996. ACM (2017)

10. Lin, T., Zhao, X., Su, H., Wang, C., Yang, M.: BSN: Boundary sensitive network for temporal action proposal generation (2018)

11. Liu, W., Anguelov, D., Erhan, D., Szegedy, C., Reed, S., Fu, C.-Y., Berg, A.C.: SSD: single shot multibox detector. In: Leibe, B., Matas, J., Sebe, N., Welling, M. (eds.) ECCV 2016. LNCS, vol. 9905, pp. 21–37. Springer, Cham (2016). https://doi.org/10.1007/978-3-319-46448-0_2

12. Liu, Z., et al.: Weakly Supervised Temporal Action Localization Through Contrast Based Evaluation Networks, pp. 3899–3908 (2019)

13. Narayan, S., Cholakkal, H., Khan, S.F., Shao, L.: 3C-net - category count and center loss for weakly-supervised action localization. In: ICCV, pp. 8678–8686 (2019)

14. Nguyen, P., Han, B., Liu, T., Prasad, G.: Weakly supervised action localization by sparse temporal pooling network. In: 2018 IEEE/CVF Conference on Computer Vision and Pattern Recognition, pp. 6752–6761 (2018)

15. Paul, S., Roy, S., Roy-Chowdhury, A.K.: W-TALC: weakly-supervised temporal activity localization and classification. In: Ferrari, V., Hebert, M., Sminchisescu, C., Weiss, Y. (eds.) ECCV 2018. LNCS, vol. 11208, pp. 588–607. Springer, Cham (2018). https://doi.org/10.1007/978-3-030-01225-0_35

16. Shou, Z., Chan, J., Zareian, A., Miyazawa, K., Chang, S.F.: CDC: convolutional-de-convolutional networks for precise temporal action localization in untrimmed videos (2017)

17. Shou, Z., Gao, H., Zhang, L., Miyazawa, K., Chang, S.-F.: AutoLoc: weakly-supervised temporal action localization in untrimmed videos. In: Ferrari, V., Hebert, M., Sminchisescu, C., Weiss, Y. (eds.) ECCV 2018. LNCS, vol. 11220, pp. 162–179. Springer, Cham (2018). https://doi.org/10.1007/978-3-030-01270-0_10

18. Shou, Z., Wang, D., Chang, S.F.: Temporal action localization in untrimmed videos via multi-stage CNNs (2016)

19. Singh, K.K., Lee, Y.J.: Hide-and-seek: forcing a network to be meticulous for weakly-supervised object and action localization. In: International Conference on Computer Vision (ICCV) (2017)

20. Singh, K.K., Lee, J.Y.: Hide-and-seek: forcing a network to be meticulous for weakly-supervised object and action localization. In: ICCV, pp. 3544–3553 (2017)

21. Wang, L., Xiong, Y., Lin, D., Gool, V.L.: UntrimmedNets for weakly supervised action recognition and detection. In: CVPR (2017)

22. Wedel, A., Pock, T., Zach, C., Bischof, H., Cremers, D.: An improved algorithm for TV-L^1 optical flow. In: Cremers, D., Rosenhahn, B., Yuille, A.L., Schmidt, F.R. (eds.) Statistical and Geometrical Approaches to Visual Motion Analysis. LNCS, vol. 5604, pp. 23–45. Springer, Heidelberg (2009). https://doi.org/10.1007/978-3-642-03061-1_2

23. Xu, H., Das, A., Saenko, K.: R-C3D: region convolutional 3D network for temporal activity detection. In: Proceedings of the International Conference on Computer Vision (ICCV) (2017)

24. Xu, Y., et al.: Segregated temporal assembly recurrent networks for weakly supervised multiple action detection. In: National Conference on Artificial Intelligence (2019)

25. Yuan, Y., Lyu, Y., Shen, X., Tsang, I., Yeung, D.Y.: Marginalized average attentional network for weakly-supervised learning. In: International Conference on Learning Representations (2019)
26. Zhang, X., Wei, Y., Feng, J., Yang, Y., Huang, T.: Adversarial complementary learning for weakly supervised object localization. In: IEEE CVPR (2018)
27. Zhao, Y., Xiong, Y., Wang, L., Wu, Z., Tang, X., Lin, D.: Temporal action detection with structured segment networks (2017)
28. Zhong, J.X., Li, N., Kong, W., Zhang, T., Li, H.T., Li, G.: Step-by-step erasion, one-by-one collection: a weakly supervised temporal action detector. In: MM 2018: ACM Multimedia Conference Seoul Republic of Korea October 2018, pp. 35–44 (2018)
29. Zhou, B., Khosla, A., Lapedriza, A., Oliva, A., Torralba, A.: Learning deep features for discriminative localization. In: Computer Vision and Pattern Recognition (2016)

Joint Feature Learning Network for Visible-Infrared Person Re-identification

Kunfeng Chen, Zhisong Pan$^{(\boxtimes)}$, Jiabao Wang$^{(\boxtimes)}$, Shanshan Jiao,
Zhicheng Zeng, and Zhuang Miao

Army Engineering University of PLA, Nanjing 210007, China
`hotpzs@hotmail.com`, `jiabao_1108@163.com`

Abstract. Visible-infrared person re-identification (VI-ReID) is a significant technology in night-time surveillance applications. Compared to traditional person re-identification that focuses on only visible imaging system, the modality discrepancy between infrared and visible images brings an additional challenge to VI-ReID. Taking the shortcomings of existing VI-ReID works in terms of feature representation into account, in this paper, we propose an end-to-end network named Joint Feature Learning Network (JFLN) to jointly utilize global and local features, high-level and middle-level features, as well as shared and specific features. In addition, based on these extracted features, we introduce a multi-loss supervision strategy in the training process. Extensive experiments conducted on public SYSU-MM01 dataset demonstrate that the proposed method outperforms the state-of-the-arts, with relatively less computational cost.

Keywords: Person re-identification · Visible-infrared · Joint feature · Multi-loss supervision

1 Introduction

The purpose of person re-identification (ReID) is to find a specific person across different cameras. This technology can make the analysis of surveillance videos more intelligent, which is of great significance to public security. Since day-and-night video surveillance is more suitable for actual scenarios, visible-infrared ReID (VI-ReID) has attracted increasing attention recently. As we can see in Fig. 1, the purpose of VI-ReID is to retrieve person images captured by different spectrum cameras. Therefore, in addition to the traditional problems like pedestrian pose, viewpoint or illumination variations of ReID, the modality discrepancy between infrared and visible images resulting from different imaging systems makes VI-ReID more challenging.

Since Wu et al. [13] released the first large-scale visible-infrared cross-modality dataset named SYSU-MM01, analyzing three different network structures and proposed a deep zero-padding one-stream network, many works have

---

K. Chen—Student.

© Springer Nature Switzerland AG 2020
Y. Peng et al. (Eds.): PRCV 2020, LNCS 12306, pp. 652–663, 2020.
https://doi.org/10.1007/978-3-030-60639-8_54

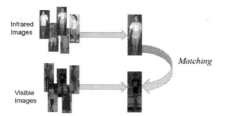

Fig. 1. An example of visible-infrared person re-identification (VI-ReID). The person images captured by different spectrum cameras can be divided into four classes: 1) images of same modality and same identity, 2) images of same modality and different identities, 3) images of different modalities and same identity, 4) images of different modalities and different identities. In VI-ReID, given a infrared image, we want to retrieve the same identity image from a large visible gallery.

been done in this field. Ye et al. [14] proposed a two-stream CNN network to learn modality-shared feature representation. Subsequently, more two-stream networks [3] have been explored to improve VI-ReID performance. In addition, cmGAN [2] is the first work which applied generative adversarial network technology to VI-ReID. It achieved better performance than before and provided a new idea [10,12] for solving VI-ReID problem. Zhu et al. [18] started the first attempt to apply local feature to VI-ReID, and got impressive accuracy. As shown in Fig. 1, the aim of VI-ReID is to retrieve images with same identity and different modalities from a gallery dataset. Some works [13] map the visible features and the infrared features into a same feature space to extract modality-shared feature, fully exploiting the cross-modality invariable characteristics. Hetero-Center (HC) loss [18] effectively narrows the discrepancy of intra-class cross-modality images, which greatly improves the performance of VI-ReID. Although above works make great progress in VI-ReID, they have the following shortcomings: 1) they mostly only utilized global high-level features, which has limited discriminant ability. And the simple partition strategy likes [8] only produced local regional features, without using the global feature. 2) the widely used high-level feature can only represent the high-level semantic information. The low-level and middle-level features, which has massive detail information, are neglected by above methods. 3) the additional computation cost caused by GAN becomes additional challenge. 4) only using modality-shared feature can make the features of the same identities compressed, and HC loss is hard to distinguish different identities due to the loss of modality-specific feature.

In this paper, we propose a novel architecture named Joint Feature Learning Network (JFLN), which fully exploit the joint utilization of global and local features, high-level and middle-level features, as well as shared and specific features. In addition, a multi-loss supervision strategy is introduced in the training process. The main contributions of our work can be concluded as follows:

(i) The joint feature strategies are proposed to extract more discriminative feature representation. It is the first attempt to utilize the combination of joint global and local features, joint high-level and middle-level features and joint shared and specific features. Besides, different kinds of losses are used to supervise the training process.

(ii) A novel Joint Feature learning Network is proposed. Extensive experiments on SYSU-MM01 dataset show that the proposed JFLN achieves 66.11% in Rank-1 and 64.93% in mAP, which outperforms the state-of-art methods in VI-ReID.

2   Proposed Approach

We proposed a Joint Feature Learning Network (JFLN), which utilized Joint Global and Local (JGL) features, Joint High-level and Middle-level (JHM) features, and Joint Shared and Specific (JSS) features to improve VI-ReID. In addition, a multi-loss Supervision Strategy (MSS) is proposed for training. Figure 2 shows the architecture. We utilize pretrained ResNet50 with four stages as the backbone network, and the convolution stride of Stage 4 is changed from 2 to 1. The red solid and black dashed boxes in the figure represent implementations of JGL and JHM, respectively. Above JGL and JHM are modality-shared features, which are mapped to a same feature space from different modalities by a weight-shared fully-connected layer. The modality-specific feature is obtained using a fully connected layer without shared weight.

According to [18], Hetero-Center (HC) loss and Cross Entropy (CE) loss are used for the modality-shared features. To learn better modality-specific feature, we used mixed-modality Triplet loss [11] and CE loss on modality-specific feature to reduce the intra-class cross-modality discrepancy and enlarge the inter-class cross-modality dissimilarity simultaneously.

More details are presented in the following each subsection.

2.1   Joint Global and Local (JGL) Features

Most works of VI-ReID extract global feature to learn the whole representation of a person image [1,13,14]. However, global features only focus on whole information of pedestrian, ignoring some fine-grained features which play an important role to VI-ReID. Recently, local features were proved effective for VI-ReID [18]. Because of the different details between pedestrians contained in local feature, it is possible to distinguish various identities. However, learning accurate local features is still a difficult task, due to the difference in actual scenes (such as diversified pedestrian poses, the distance of the camera view, and the block features are not aligned and calibrated, etc.). Then, we can conclude that it is not ideal to learning global features or local features alone.

Considering the advantages and disadvantages of global features and local features, and inspired by visible ReID, we propose a multi-scale horizontal partitioning strategy in VI-ReID tasks. As shown in Fig. 2, for the features extracted by Stage 4, we use a multi-level block strategy for pooling to obtain global feature, halving horizontally feature and trisecting horizontally feature. Global feature can be viewed as the macroscopic characteristic of the pedestrian's entire

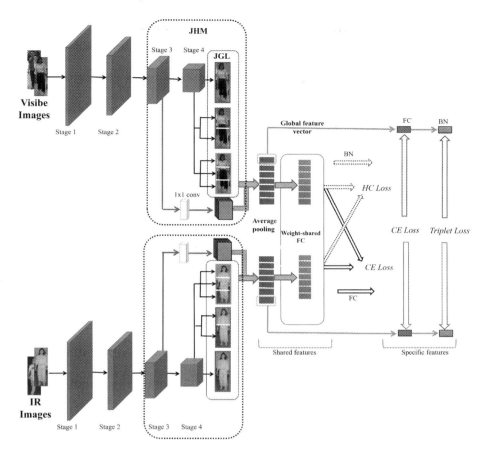

Fig. 2. Structure of JFLN. There are two branches for visible and infrared images. For each branch, the input images goes through each stage of ResNet50 to form a 3D tensor described low-level, middle-level or high-level features. To get JHM features, we convolved the feature map extracted from Stage 3 with 1×1 kernel, then combining it with the feature map extracted from Stage 4. Besides, JGL feature representation is obtained by putting three partition schemes on feature map extracted from Stage 4. Concatenating these 7 tensors, we can get 7 feature vectors after a conventional average pooling layer. Then, the dimension of each vector is reduced by a weight-sharing FC layer, and we combine them to form shared features. Afterward, each strip of shared features goes though a BN layer and a FC layer, used to calculate HC loss and CE loss. In addition, the global feature vector going though a FC layer represents specific feature. Finally, we use a BN layer to get two feature spaces for CE loss and Triplet loss on each modality.

body, while the two-part and three-part features represent different scales of pedestrian detail. The combining of local feature and global feature can get a more informative representation of pedestrian. To the best of our knowledge, our work is the first attempt to comprehensively consider global and local features in VI-ReID.

2.2   Joint High-Level and Middle-Level (JHM) Features

In our ReID task, researchers often use a deep network as a backbone to extract advanced semantic features of images to represent pedestrians. However, it is inappropriate to only use the high-level features, because VI-ReID requires fine-grained features (clothing texture, hairstyle characteristics, etc.) of image to distinguish different identities. Additionally, the poor semantic expression in infrared image and the significant modality discrepancy between Visible and IR images also make it hard to only use high-level features to retrieve pedestrians in VI-ReID. Middle-level features mainly focus on fine-grained features, which is complementary for high-level features to achieve VI-ReID. Therefore, middle-level features should not be completely abandoned.

In our work, to avoid making the model complicated and the feature dimension too large, only a branch is added after Stage 3 to get middle-level feature representations. 1×1 convolution was used to make the middle-level features increase from 1024 dimensions to 2048 dimensions. Then, middle-level features are concatenated with high-level features in the last stage. The fusion of high-level features and middle-level features makes the network effectively learn complex and abstract visual concepts, and the more informative feature representation benefits VI-ReID.

2.3   Joint Shared and Specific (JSS) Features

It is a huge challenge to keep as much as possible discriminative information when learning feature representation from cross-modality images. Many excellent methods were proposed to extract modality-shared features for VI-ReID and achieved good performance. However, existing research mainly focuses on learning modality-shared features, abandoning modality-specific features, which means a huge loss of useful identity information. Therefore, it is inappropriate to consider modality-shared features alone. Recently, [6] proposed a shared-specific transfer network (SSTN) in which the modality-shared and modality-specific features are utilized in VI-ReID. Their work achieves good performance and proves the great importance to jointly use of modality-shared and modality-specific features in VI-ReID.

Inspired by traditional two-stream network, we use the weight-shared fully connected layer to extract the modality-shared feature. Besides, the fully-connected layer without shared-weight was proposed to extract the modality-specific features. Then supervised learning was performed on the modality-shared features and modality-specific features in the training process. Experiments show that although we only add a simple branch for specific feature extraction, the using of modality-specific features has significantly improved the overall performance.

2.4   Multi-loss Supervision Strategy (MSS)

Early VI-ReID methods [13,14] often used CE loss and Triplet loss, where Triplet loss can reduce intra-class distance and enlarge inter-class distance simultaneously. Recently, Zhu et al. [18] proposed HC loss, which is combined with CE loss to achieve a new milestone and effectively improve the performance for VI-ReID. Different from Triplet loss, HC loss is designed specially for reducing the discrepancy between intra-class samples in different modalities. As a result, based on CE loss and HC loss, we introduced Triplet loss to improve feature distribution. Luo et al. [7] analyzed that CE loss and Triplet loss face convergence problem when the two losses acted on the same feature. They introduced a Batch Normalization (BN) layer to alleviate the confliction. Similarly, there exist conflicts between CE loss and HC loss, Triplet loss and HC loss, which have been proved in experiments. Therefore, the BN layers were applied to eliminate the conflicts.

For modality-shared feature, similar to [18], we use the combination of CE loss and HC loss to supervise the modality-shared feature learning. The formula is

$$L_{SH} = L_{sh-CE} + \lambda L_{sh-HC}$$
$$= -\sum_{i=1}^{K} \log \frac{e^{W_{yi}^{\top} x_i + b_{yi}}}{\sum_{j=1}^{n} e^{W_j^T x_i + b_j}} + \lambda \sum_{i=1}^{U} \left[\|c_{i,1} - c_{i,2}\|_2^2 \right], \tag{1}$$

For CE loss in Eq. 1, the batch size is K. The ith sample belonging to the y_i class, we extracted feature x_i. Besides, W_j denotes the weight and b is the bias term. For HC loss, U denotes the number of classes, as well as $\|c_{i,1} - c_{i,2}\|_2^2$ means center-sample feature distance of images with different modalities and same identity. $c_{i,1} = \frac{1}{M} \sum_{j=1}^{M} x_{i,1,j}$, $c_{i,2} = \frac{1}{N} \sum_{j=1}^{N} x_{i,2,j}$ denotes the centers of feature distribution of visible modality and infrared modality in the ith class. M and N are the numbers of visible images and infrared images in the ith class. $x_{i,1,j}$ and $x_{i,2,j}$ denotes the jth visible image and infrared image in the ith class. And λ in Eq. 1 is the weight between HC loss and CE loss, to balance the two losses.

To supervise the modality-specific feature learning, the mixed-modality Triplet loss [11] is introduced to reduce the intra-class distance and enlarge the inter-class distance. This mixed-modality Triplet loss is expressed as

$$L_{sp-mTri}(V) = \sum_{i=1}^{P} \sum_{a=1}^{K} [\alpha + \max_{p=1,\dots,K} d\left(f_{vt}^a(i), f_{vt}^p(i)\right)$$
$$- \min_{j=1,\dots,P, n=1,\dots,K, j \neq i} d\left(f_{vt}^a(i), f_{vt}^n(j)\right)]_+ \tag{2}$$

Above mixed-modality Triplet loss is computed from the hard triplet examples. To construct triplet examples, at each iteration, P identities are randomly selected, and then K visible images and K infrared images of each selected identity are randomly selected to form a mini-batch. There are $2PK$ images in total. Where $f_{vt}^a(i)$, $f_{vt}^p(i)$ and $f_{vt}^n(j)$ refer to the features extracted from the anchor sample, positive sample, and negative sample selected from the sample batch of mixed visible and infrared images, respectively. In addition, the distance here is measured by Euclidean distance, and + means that when the value in [] is bigger than zero, the value is taken as loss, while the loss is zero if it is less than zero.

So the loss function on the modality-specific feature can be expressed as

$$L_{SP} = L_{sp-CE} + L_{sp-mTri} \tag{3}$$

The overall joint loss can be expressed as

$$L = \sum_{i=1}^{7} L_{SH_i} + L_{SP} \tag{4}$$

Because we extracted 7 part of local features by modality-shared branches, the total loss is the sum of 7 modality-shared loss and 1 modality-specific loss.

3   Experiments and Results

3.1   Dataset and Evaluation Metrics

We conducted the experiments to evaluate the proposed JFLN -Net on the large-scale VI-ReID dataset, SYSU-MM01 [13]. The images in SYSU-MM01 are captured from 6 cameras: four visible cameras (CAM 1,2,4,5) and two thermal cameras (CAM 3,6). There are 29003 visible images and 15712 infrared images belonging to 491 pedestrians in the dataset, among which 22258 visible images and 11909 infrared images from 395 identities are used for training and the images of the remaining 96 identities are used for testing. In testing set, 3803 thermal images consist of query set, and the gallery set is formed by random sampling one visible sample from each identity. The above method is the single-shot setting in the widely used all-search mode, which is also the most difficult test protocol.

Each given query image is matched by calculating the Euclidean distance to these images in gallery set. Then, a similarity ranking list is formed by sorting the distance in descending order. Cam 2 and Cam 3 are in the same position, which means easier match between images captured by them. Consequently, if the infrared image in the query set is obtained from camera 3, the visible image from camera 2 is skipped directly in the gallery set. The widely used Rank-x (x=1, 10, 20) of Cumulative Match Characteristic (CMC) and mean average precision (mAP) are reported for comparison. Considering the randomness of the gallery set, during the test, we repeated the above evaluation method 100 times to get the average performance.

3.2   Implementation Details

Our experiments were conducted on a NVIDIA GeForce 1080Ti GPU, using the PyTorch framework. The pedestrian images are resized to 384×128. During the training process, we randomly selected 4 identities, and then randomly selected

8 visible images and 8 infrared images for each identity. As a result, the batch size in an epoch was set to 64. To balance HC loss and CE loss, the weight of HC loss in Eq. 1 was set to 0.5. The margin of Triplet loss was set to 0.3. The SGD algorithm is used as the optimizer with a momentum of 0.9. The network was trained 80 epochs with warm up strategy for first 10 epochs. The learning rate $lr(t)$ at epoch t is compute as Eq. 5. In addition, we utilized modality-shared and modality-specific features to optimize the network in the training process. In testing, only the modality-shared features were used to evaluate the similarity between query images and gallery images. The reason is that, under the influence of modality-specific features, the modality-shared features extracted have been able to effectively describe the image through end-to-end collaborative learning, which was proved in our experiment. Another reason is that using shared features alone can speed up the computation in testing.

$$\text{lr}(t) = \begin{cases} 0.1 \times \frac{t}{10} & \text{if } t \leq 10 \\ 0.1 & \text{if } 10 < t \leq 20 \\ 0.01 & \text{if } 20 < t \leq 50 \\ 0.001 & \text{if } 50 < t \leq 80 \end{cases} \tag{5}$$

Table 1. Comparison with the state-of-art methods.

Methods	All-search & Single-shot			
	Rank-1(%)	Rank-10(%)	Rank-20(%)	mAP(%)
Deep Zero-Padding(ICCV2017)	14.80	54.12	71.33	15.95
BDTR(IJCAI2018)	17.01	55.43	71.96	19.66
HSME(AAAI2019)	20.68	58.31	74.43	23.12
IPVT-1 and MSR(Access2019)	23.18	–	–	22.49
cmGAN(IJCAI2018)	26.97	67.51	80.56	27.80
D2RL(CVPR2019)	28.90	70.60	82.40	29.20
Hi-CMD(CVPR2020)	34.94	77.58	–	35.94
DSCSN+CCN(ArXiv2019)	35.10	77.60	88.90	37.40
EDFL(ArXiv2019)	36.94	84.52	93.22	40.77
JSIA(AAAI2020)	38.10	80.70	89.90	36.90
HPILN(IET IP2019)	41.36	84.78	94.51	42.95
AlignGAN(ICCV2019)	42.40	85.00	93.70	40.70
BDTR(IJCAI2018 with AWG)	47.50	–	–	47.65
TSLFN+HC(ArXiv2019)	56.96	91.50	96.82	54.95
cm-SSFT(CVPR2020)	61.60	89.20	93.90	63.20
JFLN (ours)	66.11	95.69	98.79	64.93

3.3   Comparison with the State-of-the-Art Methods

We compared JFLN -Net with the state-of-the-art (SOTA) methods, including Deep Zero-Padding [13], cmGAN [2], BDTR [15], D2RL [12], HSME [3],

(IPVT-1 and MSR) [4], EDFL [5], AlignGAN [10], HPILN [17], TSLFN+HC
[18], DSCSN+CCN [16], Hi-CMD [1], cm-SSFT [6], JSIA [9]. The experiments
are conducted under all-search and single-shot test mode for the above methods.
The results are shown in the Table 1.

Among these methods, our JFLN is similar with TSLFN+HC, but exceeds it
by +9.15% in Rank-1 and +9.98% in mAP. Besides, we note that cm-SSFT is the
best one in all compared methods. Although cm-SSFT achieves 61.60% in Rank-
1 and 63.20% in mAP, our results exceed it by +4.51% in Rank-1 and +1.73% in
mAP. Furthermore, cm-SSFT has a more complicated network structure, which
leads to more parameters and computations.

3.4   Ablation Study

In order to verify the effectiveness of each component of our proposed JFLN,
we take TSLFN+HC for comparison as it is similar with our method. We
replaced the part-based feature of TSLFN+HC by Joint Global and Local
(JGL) feature, and named it as Baseline(JGL). Then, we added Joint High-
level and Middle-level (JHM) feature to the Baseline and named it as Base-
line(JGL)+JHM. Besides, we added Jointly Shared and Specific (JSS) feature
to Baseline(JGL)+JHM and named it as Baseline(JGL)+JHM+JSS, which is
our proposed JFLN. The results are shown in the Table 2.

Table 2. Components analysis. We take TSLFN + HC for comparison. JGL, JHM
and JSS represent the components of the proposed JFLN.

Methods	Rank-1(%)	Rank-10(%)	Rank-20(%)	mAP(%)
TSLFN+HC	56.96	91.50	96.82	54.95
Baseline(JGL)	61.23	94.10	97.98	59.95
Baseline(JGL)+JHM	64.69	94.08	98.00	62.38
Baseline(JGL)+JHM+JSS(JFLN)	**66.11**	**95.69**	**98.79**	**64.93**

From the results of Table 2, we can find that our Baseline(JGL) achieves
61.23% in Rank-1 and 59.95% in mAP, which exceeds TSLFN+HC by +4.27%
in Rank-1 and +5.00% in mAP. This means the Joint Global and Local (JGL)
feature is more effective then part-based feature. After adding JHM and JSS to
the Baseline(JGL), the results are improved respectively. As the result, we concluded that Joint Global and Local (JGL) feature, Joint High-level and Middle-
level (JHM) feature and Jointly Shared and Specific (JSS) feature are effective
and can be combined together to achieve the best performance.

In order to determine the best partitioning level for Joint Global and Local
(JGL) feature, we designed the following experiment. We used different partitioning strategies to get different scale features, and added them with the global
feature. The combination of global feature and 2 partitioning is named as Scale 1.
Then, Scale 2 contains global feature, 2 partitioning features and 3 partitioning

features. It is similar for Scale 3, which contains global feature, 2 partitioning features, 3 partitioning features and 4 partitioning features. The results are reported in Table 3, where we also report the results of TSLFN+HC.

Table 3. Analysis of multi-scale feature. Here, Scale 1, Scale 2 and Scale 3 represent different partitioning strategies.

Methods	Rank-1(%)	Rank-10(%)	Rank-20(%)	mAP(%)
TSLFN+HC	56.96	91.50	96.82	54.95
Scale 1	55.78	92.38	97.52	54.67
Scale 2	**61.23**	**94.10**	**97.98**	**59.95**
Scale 3	60.60	87.09	97.65	59.89

From Table 3, we can find that the best Joint Global and Local (JGL) feature is Scale 2, which has global feature, 2 partitioning features and 3 partitioning features. Less scale and more scale drop the performance. According to our knowledge, a person has three parts (head, body and legs) in structure, which has different appearances, so Scale 3 is appropriate for person representation.

Furthermore, we also did experiments to find the best Joint High-level and Middle-level (JHM) feature. Based on the best Joint Global and Local (JGL) feature, we extracted different level feature to form multi-level feature. The feature extracted form Level x (x = 2, 3) are combined with JGL features. The results are shown in Table 4.

Table 4. Analysis of multi-level feature

Methods	Rank-1(%)	Rank-10(%)	Rank-20(%)	mAP(%)
Baseline(JGL)	61.23	94.10	97.98	59.95
Baseline(JGL) + Level 2	60.66	93.75	97.87	59.76
Baseline(JGL) + Level 3	**64.69**	**94.08**	**98.00**	**62.38**
Baseline(JGL) + Level 2 + Level 3	63.77	94.79	98.05	61.58

From Table 4, we can find that the best Multi-Level (ML) feature is Level 3. It achieved 64.69% in Rank-1 and 62.38% in mAP, which exceeds JGL by +3.46% in Rank-1 and +2.43% in mAP. Note that Baseline(JGL)+Level 2 has lower performance than Baseline(JGL), and Baseline(JGL) + Level 2 + Level 3 also has lower performance than Baseline(JGL)+Level 3. So we can conclude that the feature extracted from Level 2 always drops the performance. The reason may be the feature extracted from Level 2 has too low-level information, which has no contribution for semantic classification.

There exists a converging problem when CE loss and Triplet loss are applied to the same feature simultaneously. Inspired by [7] we introduce a BN layer on the modality-specific feature, and then calculate CE loss and Triplet loss. We used the same strategy as [18] to calculate HC and CE for modality-shared features. Two settings are tested: (1)CE-HC-on-SH&Tri-CE-on-SP: Triplet loss is applied before BN while CE loss is applied after BN; (2)CE-HC-on-SH&CE-Tri-on-SP: CE loss is applied before BN while Triplet loss is applied after BN. From Table 5, we find that using Triplet loss after BN is better.

Table 5. Analysis of loss-settings

Methods	Rank-1(%)	Rank-10(%)	Rank-20(%)	mAP(%)
CE-HC-on-SH&Tri-CE-on-SP	63.32	95.85	98.87	61.77
CE-HC-on-SH&CE-Tri-on-SP	**66.11**	**95.69**	**98.79**	**64.93**
ALL-on-SH	58.71	93.94	98.33	58.73

We also added Triplet loss to the modality-shared feature as HC loss, however both mAP and Rank-1 drop greatly. The reason is that HC loss and Triplet loss may have a conflict due to their different optimization objectives.

4   Conclusion

In this paper, we propose a novel network named Joint Feature Learning Network for visible and infrared cross-modality person re-identification. Combining joint global and local features, joint high-level and middle-level features, and joint shared and specific features, the network can fully exploit most information contained in input and learns better discriminative feature. Besides, our proposed multi-loss supervision strategy plays an important role in training. Finally comprehensive experiments demonstrate that the proposed method outperforms the state-of-the-arts.

References

1. Choi, S., Lee, S., Kim, Y., Kim, T., Kim, C.: Hi-CMD: hierarchical cross-modality disentanglement for visible-infrared person re-identification. arXiv preprint arXiv:1912.01230 (2019)
2. Dai, P., Ji, R., Wang, H., Wu, Q., Huang, Y.: Cross-modality person re-identification with generative adversarial training. In: IJCAI, vol. 1, p. 2 (2018)
3. Hao, Y., Wang, N., Li, J., Gao, X.: Hsme: hypersphere manifold embedding for visible thermal person re-identification. In: Proceedings of the AAAI Conference on Artificial Intelligence, vol. 33, pp. 8385–8392 (2019)
4. Kang, J.K., Hoang, T.M., Park, K.R.: Person re-identification between visible and thermal camera images based on deep residual CNN using single input. IEEE Access **7**, 57972–57984 (2019)
5. Liu, H., Cheng, J., Wang, W., Su, Y., Bai, H.: Enhancing the discriminative feature learning for visible-thermal cross-modality person re-identification. Neurocomputing **398**, 11–19 (2020)

6. Lu, Y., et al.: Cross-modality person re-identification with shared-specific feature transfer. arXiv preprint arXiv:2002.12489 (2020)
7. Luo, H., et al.: A strong baseline and batch normalization neck for deep person re-identification. IEEE Trans. Multimedia (2019)
8. Sun, Y., Zheng, L., Yang, Y., Tian, Q., Wang, S.: Beyond part models: person retrieval with refined part pooling (and a strong convolutional baseline). In: Ferrari, V., Hebert, M., Sminchisescu, C., Weiss, Y. (eds.) ECCV 2018. LNCS, vol. 11208, pp. 501–518. Springer, Cham (2018). https://doi.org/10.1007/978-3-030-01225-0_30
9. Wang, G.A., Yang, T.Z., Cheng, J., Chang, J., Liang, X., Hou, Z., et al.: Cross-modality paired-images generation for RGB-infrared person re-identification. arXiv preprint arXiv:2002.04114 (2020)
10. Wang, G., Zhang, T., Cheng, J., Liu, S., Yang, Y., Hou, Z.: RGB-infrared cross-modality person re-identification via joint pixel and feature alignment. In: Proceedings of the IEEE International Conference on Computer Vision, pp. 3623–3632 (2019)
11. Wang, J., Jiao, S., Li, Y., Miao, Z.: Two-stage metric learning for cross-modality person re-identification. In: Proceedings of the 5th International Conference on Multimedia and Image Processing, pp. 28–32 (2020)
12. Wang, Z., Wang, Z., Zheng, Y., Chuang, Y.Y., Satoh, S.: Learning to reduce dual-level discrepancy for infrared-visible person re-identification. In: Proceedings of the IEEE Conference on Computer Vision and Pattern Recognition, pp. 618–626 (2019)
13. Wu, A., Zheng, W.S., Yu, H.X., Gong, S., Lai, J.: RGB-infrared cross-modality person re-identification. In: Proceedings of the IEEE International Conference on Computer Vision, pp. 5380–5389 (2017)
14. Ye, M., Lan, X., Li, J., Yuen, P.C.: Hierarchical discriminative learning for visible thermal person re-identification. In: Thirty-Second AAAI Conference on Artificial Intelligence (2018)
15. Ye, M., Wang, Z., Lan, X., Yuen, P.C.: Visible thermal person re-identification via dual-constrained top-ranking. In: IJCAI, vol. 1, p. 2 (2018)
16. Zhang, S., Yang, Y., Wang, P., Zhang, X., Zhang, Y.: Attend to the difference: cross-modality person re-identification via contrastive correlation. arXiv preprint arXiv:1910.11656 (2019)
17. Zhao, Y.B., Lin, J.W., Xuan, Q., Xi, X.: HPILN: a feature learning framework for cross-modality person re-identification. IET Image Process. 13(14), 2897–2904 (2019)
18. Zhu, Y., Yang, Z., Wang, L., Zhao, S., Hu, X., Tao, D.: Hetero-center loss for cross-modality person re-identification. Neurocomputing 386, 97–109 (2019)

Pavement Crack Detection Using Attention U-Net with Multiple Sources

Junfeng Wang[1,2], Fan Liu[1,2(✉)], Wenjie Yang[2], Guoyan Xu[2], and Zhang Tao[3]

[1] Key Laboratory of Ministry of Education for Coastal Disaster and Protection, and College of Computer Information, Hohai University, Nanjing 210098, China
fanliu@hhu.edu.cn
[2] College of Computer and Information, Hohai University, Nanjing 210098, China
[3] School of Information Engineering, YangZhou University, Yangzhou 225009, China

Abstract. The detection of road cracks is the main basis of highway maintenance, and the noise, shadows, and irregularities of road images will bring great challenges to traditional detection. Therefore, we propose a multi-source attention U-net network, which can effectively avoid these interferences and get satisfactory results. In this method, we use transfer learning to make up for the lack of data, then use the U-net add attention mechanism to increase the weights of the cracks, and finally get more accurate results through model fusion. To prove the effectiveness of the method, we verify it by comparative experiments, and the experimental results show that the proposed approach is superior to the state of the art method in crack detection task.

Keywords: U-net network · Transfer learning · Attention mechanism · Model fusion

1 Introduction

Highway is one of the most basic transportation facilities, which plays an irreplaceable role in national construction and development. As pavement cracks are the main hidden danger of highway safety, the detection of cracks plays a vital role in the maintenance of pavement.

Traditional crack detection methods [1–5] are based on the typical features of the crack in the image, such as the transformation of crack features in different directions, the continuity feature crack, the edge feature of the crack, and so on. These methods can resist noise, light, shadow, and other interference in the image. But their ability to resist interference is limited, so we need to look for other ways.

The emergence of deep neural network helps us open a new door. In [6], CNN overcomes the interference brought by noise and shadow. CrackSegnet [7] improves the ability of neural network to segment cracks. In [8], U-net solves the problem caused by less dataset samples. These techniques significantly improve our detection results.

© Springer Nature Switzerland AG 2020
Y. Peng et al. (Eds.): PRCV 2020, LNCS 12306, pp. 664–672, 2020.
https://doi.org/10.1007/978-3-030-60639-8_55

In order to obtain more accurate crack detection results, a new crack detection method is proposed in this paper. The main structure of this method is the U-net network, which is more robust and effective than the DCNN network of concrete crack detection in [9] and has a better effect on the small sample problem. Considering the small number of road samples and the inaccuracy of crack feature extraction, we use transfer learning based on the U-net network. To avoid the interference caused by noise and small crack width in the image, we introduce the attention mechanism of increasing the weight of the crack area. Finally, in order to avoid the differences in the results, which caused by the transfer learning of different datasets, we use model fusion to improve the accuracy of results.

The paper is organized as the following. In the next section, we review some existing work. The third section describes the implementing details of the neural network, including the U-net network, transfer learning, attention mechanism, and model fusion. Section four gives data preparation, evaluation methods, and experimental results. The fifth section provides the conclusion of this paper.

2 Related Work

In recent years, some mainstream target detection methods have been used to detect cracks. For example, in [10], presents an approach for detecting cracks in infrared thermal imaging steel sheets using Convolutional Neural Networks (CNN), which overcome the drawbacks of vision-based methods. Tang et al. [11] propose a Multitask Enhanced dam crack image detection method based on Faster R-CNN (ME-Faster R-CNN) to adapt the detection of dam cracks in different lighting environments and lengths. Aiming at the problems of poor real-time performance and low accuracy of traditional pavement crack detection, Nie et al. [12] designed a method based on Yolo v3 for pavement crack detection. Although the target detection can identify the crack, the detection result is only in the rough range, so we need to adopt other methods to conduct more accurate positioning. In [13], Mask R-CNN was used to localize cracks on concrete surfaces, and this method lowers the time consumption in the process of detection, reduces costs, and increases the personnel's safety. Wang et al. [14] use a MAV equipped with an HD camera to capture images of the concrete dam, and reconstructing the 3D point cloud model of these images; then the FCN is used to train the data to obtain the crack detection model iteratively. In [15], a new crack detection and segmentation method is proposed, which can lower time-consuming, and suppress false alarms. Chen et al. [16] propose an encoder-decoder structural model with a fully convolutional neural network, namely, PCS, which significantly improve the test results.

For crack detection, we often face the difficulty of fewer data sets, so we need to take other ways to solve this problem. The emergence of transfer learning [17, 18] can help us to solve this problem. Through transfer learning, the network can not only acquire the characteristics of the same kind of data but also accelerate the convergence of the model and complete the training faster. The attention mechanism [19, 20] increases the weight of the required parts of the image, and the other parts reduce the weight to highlight the areas we need, making the segmentation of the image more efficient and accurate.

3   The Proposed Approach

The network structure of multi-source attention U-net is shown in Fig. 1, which including four parts: U-net network, transfer learning, attention mechanism, and model fusion. The functionality of each section is described in detail below.

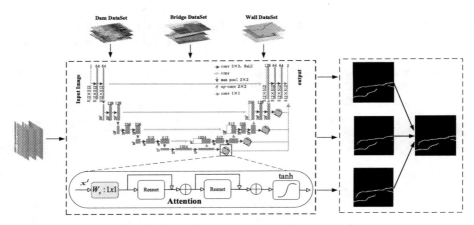

Fig. 1. The multi-source attention U-net network.

3.1   U-Net Network

U-net network consists of two parts, and one is used for image feature extraction; the other is used for the precise location of the segmentation region. For the datasets input into the network, firstly, the feature information in the image is collected by convolution, and then the image size is reduced by pooling. In this way, the features in the image are extracted after repeated four times. After that, the crack is located by deconvolution. In the process of deconvolution, the local feature extracted previously is combined with the new feature mapping to get a more accurate location.

3.2   Transfer Learning with Multiple Sources

Due to the small number of road data samples, it will have a negative impact on the results of the experiment. Therefore, we introduce the transfer learning method. In this paper, we use the dam, wall, and bridge data sets for transfer learning. Firstly, three sets of data sets are trained by U-net network, and three different network models are generated. Then the weights of nodes in each network model are transferred to the new U-net network, and the road data sets are input into these three new networks for training.

The use of transfer learning accelerates the convergence of the model and helps to identify the characteristics of road cracks more accurately and improve the detection accuracy through the feature transfer of dam cracks, wall cracks, and bridge cracks.

3.3 Attention U-Net for Pavement Crack

Although migration learning improves the accuracy of detection, it still has some inter-ference information. In order to minimize this interference information, after transfer learning, we add attention mechanism in the U-net network.

In the process of locating road cracks, after each deconvolution layer, we transform the feature map into an attention map: convolute the feature map, transform the size, then use the residual block to reduce the error, and finally use the Hyperbolic Tangent activation function. Four attention maps are generated, marked as W_1, W_2, W_3, W_4 respectively, and then these four attention modules are fused with the detection results to get more accurate detection results, as shown in Fig. 2.

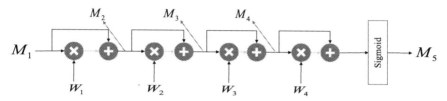

Fig. 2. The fusion process of attention map and detection results.

The formula of image fusion is as follows:

$$m_5 = \{1 + \exp[-((((m_1 w_1 + m_1)w_2 + m_2)w_3 + m_3)w_4 + m_4)]\}^{-1} \qquad (1)$$

where: m_1, m_2, m_3, m_4, m_5 is the response value of the same position of the shadow masks M_1, M_2, M_3, M_4, M_5, and w_1, w_2, w_3, w_4 is the weight value of the corresponding position of the weight maps W_1, W_2, W_3, W_4.

3.4 Model Fusion

Because it is the result of three different datasets after transfer learning, there may be some differences between them. To eliminate the contingency of the results, we fuse the three groups of results, which are respectively from the training results of U-net network with attention mechanism after using three different datasets for transfer learning.

Firstly, we performed binarization processing on the three training results, which separate the crack from the background more obviously:

$$f(pi) = \begin{cases} pi = 255, pi \geq x; \\ pi = 0, pi < x. \end{cases} \qquad (2)$$

where i is the location of the pixel, pi is the pixel value, and x is the threshold value. When the threshold value of the pixel point exceeds x, it is the crack area; otherwise, it is the background area.

Then, for the processed results, we make decisions based on the pixel values of the same location of these images, which are obtained from the same road image training:

$$g(pi) = \begin{cases} pi = 255, \frac{\sum_k^3 pi}{255} \geq y; \\ pi = 0, \frac{\sum_k^3 pi}{255} < y. \end{cases} \qquad (3)$$

Where k is the prediction graph of group kth, i is the position of pixel points in the image, and pi is the pixel value, and y is the decision indicator. When more than the indicator, we see the pixels as part of the crack part, or as the background.

4   Experiments

The network is implemented in Python with Tensorflow and Keras. All experiments are performed using an Intel Core i5-9400F CPU with 2.90 GHz, 1080. The batch size for each iteration is one, and the number of epochs is 30 for both datasets.

4.1   Dataset

In this article, we use the four groups of datasets: pavement dataset (including 114 training sets and 81 test sets), 288 dam training images, 64 Bridge training images, and 167 wall training images, with the size of 512 * 512 per image. The parameter setting in data augmentation is rotation range, width shift range, height shift range, shear range, zoom range, horizontal flip.

4.2   Evaluation

For the comparison of results, Precision (P_r), Recall (R_e) and F-Measure (F_1) [21] were used to measure the accuracy of the method:

$$P_r = \frac{TP}{TP + FP} \tag{4}$$

$$R_e = \frac{TP}{TP + FN} \tag{5}$$

$$F1 = \frac{\left(1 + \beta^2\right) \times P_r \times R_e}{\beta^2 \times P_r + R_e} \tag{6}$$

where TP, FP, FN are the numbers of true positive, false positive, and false negative respectively. In this experiment, the value of β^2 is 0.3, which makes the weight of the accuracy rate higher than the recall rate.

4.3   Loss Function

The experiment in this paper is about the segmentation of road cracks, a pixel-level binary classification problem. Therefore, we choose binary_crossentropy as the loss function of the experiment:

$$loss = -\sum\nolimits_{i=1}^{n} \left(y_i \log \hat{y}_i + (1 - y_i)\log\left(1 - \hat{y}_i\right)\right) \tag{7}$$

Where y_i is the classification label of pixels in the ground truth, $\hat{y}_i$ is the classification label of pixels in the prediction results.

When y_i and $\hat{y}_i$ are equal, the loss is 0; otherwise, the loss is a positive number. And when the probability difference is larger, the loss will be larger.

4.4 Result

In this section, to test the effectiveness of the proposed method, we conducted several comparative experiments. First of all, we compared several results in the experimental process, as shown in Fig. 3. It can be seen that in different experimental stages, the results we get are quite different, and the method proposed in this paper is more accurate than other results. Table 1 is the comparison data of these results. From the perspective of accuracy, all the results in the experimental process have a significant improvement compared with U-net. For recall, different results have different degrees of decline, but most remain in a relatively stable range. As for the final evaluation index, it can be seen that compared with the results of U-net, the results of other methods have significantly improved. Among them, the effects of adding the attention mechanism are enhanced by 1.74%, and the improvement of the three transfer learning results is between 5% and 7%. Our proposed method is improved more, with an increase of 8.28%, which is much higher than other results.

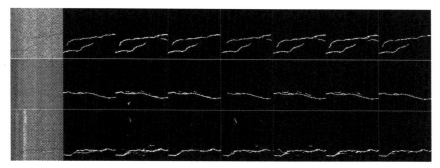

Fig. 3. Parts of the road crack detection results (from left to right: original image, ground truth, U-net, U-net + Attention, bridge transfer learning, dam transfer learning, wall transfer learning and the proposed method)

Table 1. Results at different stages.

Stages	P_r	R_e	F_1
U-net	0.7479	0.8122	0.7519
U-net + Attention	0.7881	0.7699	0.7693
Bridge transfer learning	0.9075	0.6361	0.8069
Dam transfer learning	0.8345	0.7695	0.8095
Wall transfer learning	0.8761	0.7381	0.8149
The Proposed Method	0.8940	0.7302	0.8347

Then the canny algorithm, Frequency tuning, Image salience, and vgg19 network are used to compare with the method in this paper. The results are shown in Fig. 4. It can be found that the Canny algorithm and Frequency tuning have higher requirements for the image; otherwise, it is easy to be interfered. Image saliency and vgg19 network have good results, but they are also easily interfered with by the white line in the image, leading to a decrease in accuracy. Although there are some defects in our method, the interference of external factors is largely eliminated, so that the results can be better improved. Table 2 is the comparison data of these algorithms, from which we can see that the data of the Canny algorithm and Frequency tuning are both minimal, indicating that these two algorithms are not suitable for this experiment data. The recall rate of image significance is high, but its accuracy is low, making the F-Measure poor. The accuracy of the vgg19 network is low, but the recall rate is high, which improves the F-Measure. The method proposed in this paper has high accuracy and recall rate, so the F-Measure is the best, which also shows the method's effectiveness.

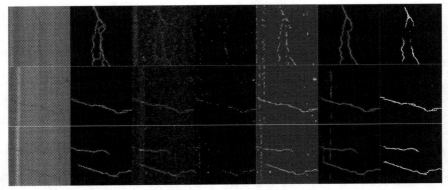

Fig. 4. Results of the comparative experiments (from left to right: original image, ground truth, canny algorithm, gray histogram statistical comparison method, image significance method, vgg19 and the proposed method)

Table 2. Results of different methods.

Methods	P_r	R_e	F_1
Canny	0.2255	0.4154	0.2359
Frequency tuning	0.5630	0.1192	0.2875
Image salience	0.4364	0.7062	0.4703
Vgg19	0.6420	0.8347	0.6665
The proposed method	0.8940	0.7302	0.8347

5 Conclusion

In this paper, we made some improvements based on the U-net network. We added the attention mechanism on its expansion path, to increase the weight of the crack region in the image, to facilitate segmentation. Besides, the idea of transfer learning is adopted to improve the model's shrinkage speed and introduce the characteristics of other cracks, on the other hand, to facilitate the identification of cracks. Finally, the method of model fusion is used, which not only helps to extract more accurate fracture areas but also helps to exclude other external interference, making the results more accurate and better. We compared various parts of the method, and the results showed that the method had significant improvement compared with the individual parts. Compared with different algorithms, the results show that the method has a higher detection efficiency than the traditional algorithm, and it has a more accurate effect than other neural networks. The method proposed in this paper is only an improvement of the U-net network, and it has more potential, so we will further explore the U-net network in the following research.

Acknowledgement. This work was partially funded by Natural Science Foundation of Jiangsu Province under grant No. BK20191298, National Key R & D Program of China under grant no. 2018YFC0407106, Key Laboratory of Coastal Disaster and Protection of Ministry of Euducation, Hohai University under grant no. 201905 and Fundamental Research Funds for the Central Universities under Gran No. B200202175.

References

1. Nguyen, T.S., Begot, S., Duculty, F., Bardet, J.-C., Avila, M.: Pavement cracking detection using an anisotropy measurement. In: 11th IASTED International Conference on Computer Graphics and Imaging, CGIM, Innsbruck, Austria (2010)
2. Li, Q.Q., Liu, X.L.: A model for segmentation and distress statistic of massive pavement image based on muli-sacle strategies. Int. Arch. Photogrammetr. Remote Sens. Spatial Inf. Sci. **37**(B5), 63–68 (2008)
3. Yan, M.D., Bo, S.B., He, Y.Y.: A method of image detection and analysis for pavement crack based on morphology. J. Eng. Graph. **29**(2), 142–147 (2008)
4. Gangadhar, H., Srinivasan, E.: Performance comparison of ROAD statistic based nonlinear filters for image denoising. In: Industrial and Information Systems, 2008, ICIIS, December 2008
5. Chen, X., Yan, X., Chu, X., Lv, Z.: Recognition of pavement image with shadow based on image decomposition. In: First International Conference on Transportation Engineering, July 2007
6. Li, S., Zhao, X.: Image-based concrete crack detection using convolutional neural network and exhaustive search technique. In: Advances in Civil Engineering, vol. 2019 (2019)
7. Ren, Y., et al.: Image-based concrete crack detection in tunnels using deep fully convolutional networks. In: Construction and Building Materials, vol.234, February 2020
8. Ronneberger, O., Fischer, P., Brox, T.: U-Net: Convolutional Networks for Biomedical Image Segmentation. arXiv, May 2015
9. Liua, Z., Caoa, Y., Wanga, Y., Wangb, W.: Computer vision-based concrete crack detection using U-net fully convolutional networks. In: Automation in Construction, pp. 129–139, April 2019

10. Yang, J., Wang, W., Lin, G., Li, Q., Sun, Y., Sun, Y.: Infrared thermal imaging-based crack detection using deep learning. IEEE Access **7**, 182060–182077 (2019)
11. Tang, J., Mao, Y., Wang, J., Wang, L.: Multi-task enhanced dam crack image detection based on faster R-CNN. In: 2019 IEEE 4th International Conference on Image, Vision and Computing (ICIVC), pp. 336–340 (2019)
12. Nie, M., Wang, C.: Pavement crack detection based on yolo v3. In: 2019 2nd International Conference on Safety Produce Informatization (IICSPI), pp. 327–330 (2019)
13. Attard, L., Debono, C.J., Valentino, G., Di Castro, M., Masi, A., Scibile, L.: Automatic crack detection using mask R-CNN. In: 2019 11th International Symposium on Image and Signal Processing and Analysis (ISPA), pp. 152–157 (2019)
14. Wang, S., Zhang, H., Wang, H., Chen, B., Li, Y., Chen, C.: Combination of point-cloud model and FCN for dam crack detection and scale calculation. In: 2019 Chinese Automation Congress (CAC), pp. 5859–5862 (2019)
15. Fang, F., Li, L., Rice, M., Lim, J.-H.: Towards real-time crack detection using a deep neural network with a Bayesian fusion algorithm. In: 2019 IEEE International Conference on Image Processing (ICIP), pp. 2976–2980 (2019)
16. Chen, T., et al.: Pavement crack detection and recognition using the architecture of segNet. J. Ind. Inf. Integr. **18** (2020)
17. Sousa, M.J., Moutinho, A., Almeida, M.: Wildfire detection using transfer learning on augmented datasets. Expert Syst. Appl. **142** (2020)
18. Li, X., Hu, Y., Li, M. Zheng, J.: Fault diagnostics between different type of components: a transfer learning approach. Appl. Soft Comput. **86** (2020)
19. Song, S., Lan, C., Xing, J., Zeng, W., Liu, J.: An end-to-end spatio-temporal attention model for human action recognition from skeleton data. In: The Thirty-First AAAI Conference on Artificial Intelligence AAAI, pp. 4263–4270 (2019)
20. Woo, S., Park, J., Lee, J.-Y., Kweon, I.S.: CBAM: convolutional block attention module. In: Ferrari, V., Hebert, M., Sminchisescu, C., Weiss, Y. (eds.) ECCV 2018. LNCS, vol. 11211, pp. 3–19. Springer, Cham (2018). https://doi.org/10.1007/978-3-030-01234-2_1
21. Brank, J., Mladenić, D., Grobelnik, M.: F-Measure: Encyclopedia of Machine Learning. Springer, Boston (2011). https://doi.org/10.1007/978-0-387-30164-8_315

A Cooperative Tracker by Fusing Correlation Filter and Siamese Network

Bin Zhou[1], Xin Liu[2], and Bineng Zhong[3]([✉]) [iD]

[1] Beijing Aerospace Automatic Control Institute, Beijing, China
giggsnet@163.com
[2] Beijing Seetatech Technology Co., Beijing, China
xin.liu@seetatech.com
[3] Huaqiao University, Xiamen, China
bnzhong@hqu.edu.cn

Abstract. The robustness of model-free trackers is always supported by a model updater and a motion model. However, most state-of-the-art trackers (e.g. correlation-filter or Siamese-network based trackers) are unbalanced in both aspects. Consequently, they drift easily when encountering challenging scenarios such as fast motion, occlusion or background clutter. Inspired by the complementarity of different tracking mechanisms, we propose an adaptive cooperation tracker, where correlation filter and Siamese networks complement each other in their shortcomings. Specifically, our tracker consists of three components: a context-aware correlation filter network (termed as CaCFNet), a Siamese network and a tracking failure estimator. In the online tracking, the Siamese network component locates the target coarsely in a larger search region, and then CaCFNet refines the coarse position for higher accuracy. The Siamese network component is activated adaptively according to the result of failure estimator, which keeps the tracker in real time and avoids interference between two different mechanisms. Moreover, context-aware correlation filter network and Siamese network are trained offline for better feature representation for visual tracking task. Comprehensive experiments are performed on three popular benchmark: OTB2013, OTB2015, VOT2017 to demonstrate the effectiveness of the proposed tracker, and the proposed tracker achieves state-of-the-art results on these benchmark.

Keywords: Visual tracking · Correlation filter · Siamese network

1 Introduction

Visual tracking is a fundamental problem in computer vision, which tracks a specified target given in the first frame in a changing video sequence automatically. Although much progress [2,4,14,15,24–26] has been made in recent years, it remains very challenging to track the target at a high speed while robust for changing scenarios such as occlusions, fast motion, illumination variations, background clutter and deformation and so forth.

© Springer Nature Switzerland AG 2020
Y. Peng et al. (Eds.): PRCV 2020, LNCS 12306, pp. 673–685, 2020.
https://doi.org/10.1007/978-3-030-60639-8_56

Recent works [2,4,6,12] demonstrating significant performance improvement on several benchmarks [10,13,22,23], can be fallen roughly into two broad categories: correlation-filter (CF) based trackers and Siamese-network based trackers. They are two different tracking mechanisms.

In this work, we propose an adaptive cooperation tracker via cooperating correlation filter with Siamese network. Specifically, our tracker consists of three basic components: a correlation filter component, a Siamese component, and a tracking failure estimator.

The correlation filter component is a context-aware correlation filter network (CaCFNet), which is designed to gain better feature extractor for correlation filter task. We equip the CaCFNet with light weight convolution network, which is trained offline in an end-to-end way using positive and negative samples for larger discriminative power. The Siamese network is different from traditional SiamFC [2], which uses the shallow Alexnet [11]. In this paper, we choose the more robust modified VGG-16 [17] for larger discriminative power (Fig. 1).

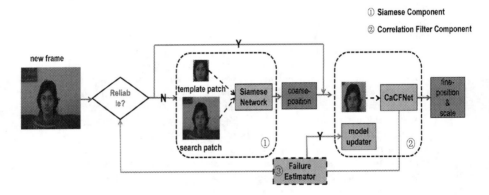

Fig. 1. The adaptive cooperation tracking architecture.

The main contributions of this work are as follow:

1) We propose a cooperative tracker that takes advantage of complementary of correlation filter and Siamese network for robust tracking.
2) We propose a context-aware correlation filter network to improve the performance of traditional correlation filter. The network can be trained offline using positive and negative samples to obtain better and more discriminative feature presentation for correlation filter task.
3) We propose an adaptive strategy to automatically to guide the cooperation of CaCFNet and Siamese network. This strategy keeps the tracker in real time and avoids interference between two different mechanisms.

2   Related Works

In this section, we give a brief related works closely related to this work.

Correlation Filter for Visual Tracking. Recently, correlation filter based trackers have achieve great success. However, there are also two major issues that limit performance improvements: boundary effect and feature representation. Different from using ineffective and inefficient hand-craft and raw convolution feature, inspired by [19,20], we transform context-aware correlation filter [15] into a differentiable layer coupled with a lightweight convolution network to obtain more discriminative and effective feature. To address boundary effect, [6,9] introduce a spatial regularization component to ensure tracker can work on a large search region effectively, which are at the price of optimizing the regularized objective function costly, even in the Fourier domain.

Siamese Network for Visual Tracking. The Siamese-network based trackers have drawn much attention in recent years. These trackers not only track the object in a larger search region with high accuracy but also be real time. SiamFC [2] and SiamRPN [12] are two typical algorithms. Many trackers [3,21,27] have been done to make further improvement based on them. In this work, we apply the Siamese network to search the target in larger region to alleviate the boundary effect of correlation filter.

3 The Proposed Tracker

3.1 Overview

An adaptive cooperation tracker utilizing the complementarity advantage of correlation filter and Siamese network is proposed. Specifically, our tracker consists of three core components: the proposed context-aware correlation network (termed as CaCFNet), Siam-VGG and a tracking failure estimator.

In online tracking, once the tracking failure estimator finds that CaCFNet misses the target encountering difficult scenarios such as out of view because of the limited search region, Siamese network is activated adaptively to tracking coarsely the target in a larger search region, and then CaCFNet tracks the target finely based the position provided by Siamese Network. The adaptive cooperation tracking architecture is shown as in Fig. 3.

3.2 Context-Aware Correlation Filter Network

Inspired by the context aware correlation filter [15], our correlation filter is regularized by distracted patches for larger discrimination power. The distracted patches can be viewed as hard negative samples which contain various distractors and diverse background. Then, the correlation filter is learned that has a high response for the target patch and close to zero response for distracted patches:

$$min_w||X_0w - y||_2^2 + \lambda_1||w||_2^2 + \lambda_2\sum_{i=1}^{k}||X_iw||_2^2 \qquad (1)$$

where X_i and X_0 represent circulant feature matrices from negative samples and positive samples respectively. w represents the learned correlation filter. λ_1

and λ_2 is the regularization coefficient. And the regression objective y is a 2D Gaussian. The closed-form solution in the Fourier domain for our CF is:

$$\widehat{w} = \frac{\widehat{x_0}^* \odot \widehat{y}}{\widehat{x_0}^* \odot \widehat{x_0} + \lambda 1 + \lambda 2 \sum_{i=1}^{k} \widehat{x_i}^* \odot \widehat{x_i}}, \tag{2}$$

where x_i and x_0 represent the feature patches of negative samples and positive sample respectively. $\wedge$ denotes the discrete Fourier transform and $*$ represents the complex conjugate. $\odot$ denotes the Hadamard product.

Different from most CF based trackers using raw convolution or hand-craft features, we propose to actively learn robust representations fitting to a CF by transforming context-aware correlation filter into a differentiable layer. We equip the differentiable CF layer with a lightweight convolution network for real-time speed. For more larger discriminative power, we train this network offline using two positive samples and four distracted negative samples from other classes as shown in the Fig. 2. This network is termed as context-aware correlation filter network (CaCFNet). And the CF differentiable layer is termed as CaCF layer.

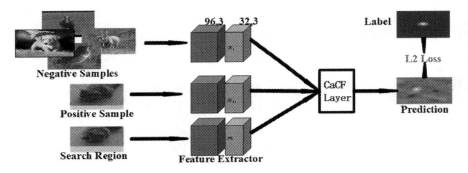

Fig. 2. Architecture of context-aware correlation filter network. The numbers above the convolution layers represent the number of channels and the size of kernel respectively.

3.3   Fully Convolution Siamese Network

The goal of fully convolution Siamese network is to learn a metric of similarity in the embedding space. We can easily calculate the similarity to the exemplar patch at all translation of a large candidate patch using cross convolution. The candidate patch x is much larger than the exemplar patch z. The similarity at all translation of the candidate patch can be calculated using:

$$f(z, x) = \varphi(z) * \varphi(x) + b_1 \tag{3}$$

where $*$ is the general convolution operation, b_1 is the bias which takes value in every location. The $f(z, x)$ is a score map that represents the similarity of each translation. The network can be trained using the logistic loss:

$$l(y, v) = \log(1 + exp(-yv)) \tag{4}$$

Where y is the ground-truth label and v is the similarity score of a single exemplar-candidate pair. The output of the network is a score map $f(z, x)$ and therefore we define the loss of the whole network as the mean of the individual losses,

$$L(y, f) = \frac{1}{|D|} \sum_{u \in D} l(y[u], f[u]), \tag{5}$$

where $y[u]$ represents the ground-truth label for each position $u \in D$ in the score map and $f[u]$ represents the similarity score for each position in the score map.

In this work, the Siamese network complements the correlation filter to alleviate boundary effect. Compared with correlation filter trackers, the advantage of Siamese network provide as input to the network a much larger search image and it will compute the similarity at all translated sub-windows on a dense grid in a single evaluation using cross convolution. This larger search image equips a tracker with a larger file of vision to cope with abrupt motion and heavy occlusion. In this work, we choose the more robust backbone VGG-16, which is modified for balance of speed and accuracy, as shown in Fig. 3. The padding of this backbone is removed. The first seven layers are initialized using pre-trained VGG-16 model on ImageNet [16].

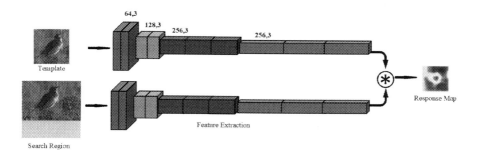

Fig. 3. Architecture of full convolution Siamese network. We use a modified VGG architecture as the feature extractor. The numbers above the convolution layers represent the number of channels and the size of kernel respectively.

3.4 Adaptive Model Update and Target Search

Most frames in a sequence are easy cases, where targets with distinct and move slowly. In those cases, our CaCFNet can track a target well and the model should be updated to adapt to target appearance changes. However, when the target is occluded or out of the view, the target drifts and may never be found again. In these cases, model update is not needed and the prediction of the potential target regions in larger search region is needed for finding the target again. Instead of tracking with model update and Siamese network in each frame, model update and Siamese network are activated automatically by evaluating the tracking results of CaCFNet.

For correlation filter, idea response map should have one peak value in target position and damps quietly from peak to the rest region. However, when encountering challenging scenarios, such as heavy occlusion and out of view, the response map fluctuates intensely and damps slowly as shown in Fig. 4. The distribution of response map can reveal the confidence degree of tracking results to some extent. Based on above observation, we design a novel criterion called Region-Peak-Sidelobe-Ratio (RPSR) to reveal distribution of the response peak and thus evaluate the tracking status.

Specially, we divide the response map F into two region: peak region F_p and side-lobe region F_s. F_p represents the area near the maximum value, F_s represents the rest region of response map except the F_p, called side-lobe region. The $RPSR$ is defined as

$$RPSR = \frac{mean(F_p - f_{min})^2}{mean(F_s - f_{min})^2} \tag{6}$$

where

$$f_{min} = min(F) \tag{7}$$

where f_{min} is corresponding minimum value of response map F. Meanwhile, the maximum value in response map is also a important criteria to reveal the status of tracking to some extent.

$$f_{max} = max(F) \tag{8}$$

When the response map is ideal, $RPSR$ and f_{max} are larger, and fall into a small value when response map fluctuating intensely.

Then, we calculate historical average values of $RPSR$ and f_{max} within the first m frame of the current frame, termed as $RPSR_t^m$ and $f_{max_t^m}$. In online tracking, When these two criteria $RPSR$ and f_{max} of the current frame are greater than $RPSR_t^m$ and $f_{max_t^m}$ with certain ratios β_1, β_2, the tracking result in the current frame is considered to be high-confidence. The proposed tracking model will be updated online with a learning rate parameter η as,

$$w_t = \eta w + (1 - \eta)w_{t-1} \tag{9}$$

Further, When encountering occlusion, CaCFNet drifts easily and out of the view because of the limited search region. As shown in Fig. 5, when encountering challenging scenarios, such as occlusion, CaCFNet drifts and cannot find the target because of the limited search region, in contrast, Siamese network based trackers can find the target. Based on the above observation, the Siamese network is activated in these cases.

Fig. 4. The shots of sequence Jogging-1 from OTB2015, where red box represents the ground-truth. (Color figure online)

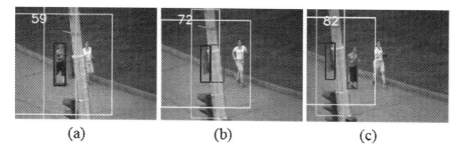

Fig. 5. The comparison of search region. Red, yellow green, blue boxes represent the ground-truth, the search region from Siamese network, the search region from correlation filter and the results of CaCFNet, respectively. (Color figure online)

4 Experiments

In this section, firstly, we introduce the implementation details of our tracker. Secondly, we analyze the contribution to the performance of each component of our tracker through the results of the ablation experiments. Finally, the results on OTB2013, OTB2015 and VOT2017 demonstrate the effectiveness and robustness of our approach.

4.1 Implementation Details

Training Parameters. We apply stochastic gradient descent (SGD) with momentum of 0.9 to train both correlation network and Siamese network. For correlation filter, the learning rate is annealed geometrically at each epoch from $1e-2$ to $1e-5$. The training is performed for 50 epochs with a mini-batch size of 32. For Siamese network, the learning rate is annealed geometrically at each epoch from $1e-3$ to $1e-6$. The training is performed for 50 epochs with a mini-batch size of 16.

The Training Dataset. To avoid the over-fitting, the training and testing are selected from different datasets. We use the data from video of ImageNet Large Scale Visual Recognition Challenge (ILSVRC) [16] as our training dataset for both CaCFNet and Siam-VGG. For CaCFNet, the inputs include two positive samplers and four negative samplers. The two positive samplers are collected within the nearest 10 frames in the same sequence, and the negative samplers are collected in different sequence. The all samplers are cropped with a padding size of 2, and then resized to the input size of $125 \times 125 \times 3$. For Siamese network, we collect each pair of frames within the nearest 100 frames and the processing of training samples can be referred to SiamFC [2].

Tracking Parameters. For correlation filter, λ_1 and λ_2 in (2) is set as 1e-4 and 0.1 respectively. For $RPSR$, we set the area of peak region as the area which distance from the center position is less than 16. The β_1 and β_2 is set 0.65 and 0.43 respectively. m is set 7. The η in (9) is set 0.01. To make the tracker adapt to the scale variations of target, scale pyramid is used in this work. The scale interval in CaCFNet is set as S = 1.031 and 3 scale are exploited. And only one search patch is used in Simaese network, which is cropped based on the position and scale of previous frame.

The proposed tracker is implemented with the PyTorch wrapper with python and all the experiments are executed on a PC with an Intel i5-3470 CPU, 22 GB RAM and Nvidia GTX 2080Ti GPU.

Tracking Benchmarks. Tracking performance evaluations of our tracker and state-of-the-art trackers are performed on the OTB2013 [22], OTB2015 [23] and VOT2017 [10]. On OTB benchmarks, precision plot and success plot are exploited as are evaluation metrics. The precision plot show the relative number of frames in the sequence where the center location error is smaller than the fixed threshold of 20 pixels. The success plot shows the percentage of successful frames where the bounding box overlap exceeds the given threshold ranging from 0 to 1. The area under curve (AUC) of each success plot is used to rank trackers. On VOT2017, the expected average overlap (EAO) is exploited to quantitatively analyze the tracking performance. The EAO considering both bounding box overlap ratio (accuracy) and the re-initialization times (robustness).

4.2   Evaluation on OTB2013

OTB2013 [22] contains 50 fully annotated sequences that are collected from commonly used tracking sequences. The proposed tracker is compared with recent state-of-the-art trackers as shown in Fig. 6. These trackers can be roughly categorized into two types: (I) correlation filter based trackers: C-COT [8], ECO-hc [4], SRDCF-deep [7], SRDCF [6], Staple [1], DSST [5]; (II) Siamese network and deep learning based tracker: Siamfc [2], CFNet [19], CREST [18]. Our tracker achieves precision of 89.4% and AUC of 67.6% with real-time speed of 35 fps. The results demonstrate that robust and accurate target tracking can be achieved by the cooperation of different tracking mechanisms, namely the proposed context-aware correlation filter network and the Siamese network.

Ablation study. The basic idea in this work is how to take advantage of their complementary strengths from correlation filters and Siamese networks. So our tracker uses two basic component to construct a robust tracker: CaCFNet and

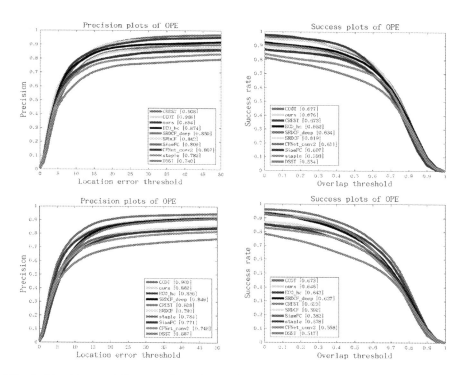

Fig. 6. Precision and Success plots of OPE(one pass evaluation) on OTB2013 [22] and OTB2015 [23] respectively. The numbers in the legend indicate the precision at 20 pixels for precision plots and the AUC score for success plots respectively.

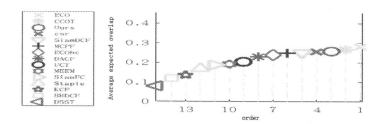

Fig. 7. Expected average overlap(EAO) ranking on VOT2017. We only show a part of trackers in this figure for clarity.

682   B. Zhou et al.

Siamese network. Besides, a tracking failure estimator is introduced for efficiency in this work. We evaluate their contributions by checking the performance of the tracker when the examined component is missing. "CaCFNet+RPSR" represents the CaCFNet with RPSR. "CaCFNet+Siam" represents without RPSR. "Ours+Alexnet" represents using the AlexNet as the backbone for Siamese network. Table 1 shows the effectiveness of the cooperation of correlation filter and Siamese network.

Table 1. The results of ablation experiments on OTB2013

	AUC	Precision	Speed (fps)
CaCFNet	0.657	0.871	120
CaCFNet+RPSR	0.664	0.879	**143**
CaCFNet+Siam	0.632	0.846	27
Ours+Alexnet	0.665	0.887	50
Ours	**0.676**	**0.894**	35

Table 1 shows the effectiveness of the cooperation of correlation filter and Siamese network. $RSPR$ enhances the performance of CaCFNet via measuring tracking state and can combines effectively the correlation filter and Siamese network. The degraded performance of "CaCFNet+Siam" demonstrates Siamese network distracts CaCFNet in easy case during tracking. The comparison of "Ours+Alexnet" and "Ours" represent a more discriminant Siamese network is better for our tracker.

4.3   Evaluation on OTB2015

The OTB2015 [23] benchmark is the extension of OTB2013. The evaluation criteria of two benchmarks are identical. Compared to the state-of-the-art trackers, as shown in Fig. 6, our tracker achieves a precision of 86.2% and AUC of 64.6%. Considering both speed and accuracy, our tracker achieves a very competitive result.

4.4   Evaluation on VOT2017

VOT2017 contains 60 challenging sequences and trackers are re-initialized when missing the target. As indicated in the VOT2017 [10], the strict state-of-the-art bound is about 0.21 under the EAO metric. That means trackers are considered as state-of-the-art one when their EAO values exceed 0.21.

The proposed tracker is evaluated on VOT2017 with state-of-the-art trackers as shown in Fig. 7. Among these trackers, ECO [4] and CCOT [8] has a higher EAO than our tracker but has low speed. And our tracker can run at 14 fps and with state-of-the-art performance. According to the analysis of VOT2017 [10] and the definition of the strict state-of-the-art bound, our tracker can be regarded as a state-of-the-art tracker.

5   Conclusion

In this work, we propose an adaptive cooperation tracker inspired by the complementary of CF based tracker and Siamese network based trackers. These two basic components with different tracking mechanism complement each other in motion model and model updater. Instead of using off-the-shelf components directly, we improve the basic components. Specifically, a context-aware correlation filter network and a Siamese network with more robust backbone are designed, which can be trained offline to obtain more better feature representation for visual tracking task. Further, an automatic activating strategy is designed to guide their cooperation and reliable model update. Experiments on four popular benchmarks demonstrate state-of-the-art performance with high speed frame-rates. We believe that considerably higher performance could be obtained by substituting the two base trackers with more advanced trackers.

Acknowledgements. This work is supported by the Nature Science Foundation of China (No. 61972167, No.61802135).

References

1. Bertinetto, L., Valmadre, J., Golodetz, S., Miksik, O., Torr, P.H.S.: Staple: complementary learners for real-time tracking. In: Proceedings of the IEEE Conference on Computer Vision and Pattern Recognition, pp. 1401–1409 (2016)
2. Bertinetto, L., Valmadre, J., Henriques, J.F., Vedaldi, A., Torr, P.H.S.: Fully-convolutional Siamese networks for object tracking. In: Hua, G., Jégou, H. (eds.) ECCV 2016. LNCS, vol. 9914, pp. 850–865. Springer, Cham (2016). https://doi.org/10.1007/978-3-319-48881-3_56
3. Chen, Z., Zhong, B., Li, G., Zhang, S., Ji, R.: Siamese box adaptive network for visual tracking. In: Proceedings of the IEEE/CVF Conference on Computer Vision and Pattern Recognition, pp. 6668–6677 (2020)
4. Danelljan, M., Bhat, G., Khan, F.S., Felsberg, M.: Eco: efficient convolution operators for tracking. In: Proceedings of the IEEE Conference on Computer Vision and Pattern Recognition, pp. 6638–6646 (2017)
5. Danelljan, M., Häger, G., Khan, F., Felsberg, M.: Accurate scale estimation for robust visual tracking. In: British Machine Vision Conference, Nottingham, 1–5 September 2014. BMVA Press (2014)
6. M. Danelljan, G. Hager, F. S. Khan, M. Felsberg. Learning spatially regularized correlation filters for visual tracking. In: Proceedings of the IEEE International Conference on Computer Vision, pp. 4310–4318 (2015)
7. Danelljan, M., Hager, G., Khan, F.S., Felsberg, M.: Adaptive decontamination of the training set: a unified formulation for discriminative visual tracking. In: Proceedings of the IEEE Conference on Computer Vision and Pattern Recognition, pp. 1430–1438 (2016)
8. Danelljan, M., Robinson, A., Shahbaz Khan, F., Felsberg, M.: Beyond correlation filters: learning continuous convolution operators for visual tracking. In: Leibe, B., Matas, J., Sebe, N., Welling, M. (eds.) ECCV 2016. LNCS, vol. 9909, pp. 472–488. Springer, Cham (2016). https://doi.org/10.1007/978-3-319-46454-1_29

9. Galoogahi, H.K., Fagg, A., Lucey, S.: Learning background-aware correlation filters for visual tracking. In: Proceedings of the IEEE International Conference on Computer Vision, pp. 1135–1143 (2017)

10. Kristan, M., et al.: The visual object tracking vot2017 challenge results. In: IEEE International Conference on Computer Vision Workshop (2017)

11. Krizhevsky, A., Sutskever, I., Hinton, G.E.: Imagenet classification with deep convolutional neural networks. In: Advances in Neural Information Processing Systems, pp. 1097–1105 (2012)

12. Li, B., Yan, J., Wu, W., Zhu, Z., Hu, X.: High performance visual tracking with Siamese region proposal network. In: Proceedings of the IEEE Conference on Computer Vision and Pattern Recognition, pp. 8971–8980 (2018)

13. Liang, P., Blasch, E., Ling, H.: Encoding color information for visual tracking: algorithms and benchmark. IEEE Trans. Image Process. 24(12), 5630–5644 (2015)

14. Lin, Y., Zhong, B., Li, G., Zhao, S., Chen, Z., Fan, W.: Localization-aware meta tracker guided with adversarial features. IEEE Access 7, 99441–99450 (2019)

15. Mueller, M., Smith, N., Ghanem, B.: Context-aware correlation filter tracking. In: Proceedings of the IEEE Conference on Computer Vision and Pattern Recognition, pp. 1396–1404 (2017)

16. Russakovsky, O., et al.: Imagenet large scale visual recognition challenge. Int. J. Comput. Vis. 115(3), 211–252 (2015)

17. Simonyan, K., Zisserman, A.: Very deep convolutional networks for large-scale image recognition. arXiv preprint arXiv:1409.1556 (2014)

18. Song, Y., Ma, C., Gong, L., Zhang, J., Lau, R.W.H., Yang, M.-H.: Crest: convolutional residual learning for visual tracking. In: Proceedings of the IEEE International Conference on Computer Vision, pp. 2555–2564 (2017)

19. Valmadre, J., Bertinetto, L., Henriques, J., Vedaldi, A., Torr, P.H.S.: End-to-end representation learning for correlation filter based tracking. In: Proceedings of the IEEE Conference on Computer Vision and Pattern Recognition, pp. 2805–2813 (2017)

20. Wang, Q., Gao, J., Xing, J., Zhang, M., Hu, W.: Dcfnet: discriminant correlation filters network for visual tracking. arXiv preprint arXiv:1704.04057 (2017)

21. Wang, Q., Teng, Z., Xing, J., Gao, J., Hu, W., Maybank, S.: Learning attentions: residual attentional Siamese network for high performance online visual tracking. In: Proceedings of the IEEE Conference on Computer Vision and Pattern Recognition, pp. 4854–4863 (2018)

22. Wu, Y., Lim, J., Yang, M.-H.: Online object tracking: a benchmark. In: Proceedings of the IEEE Conference on Computer Vision and Pattern Recognition, pp. 2411–2418 (2013)

23. Yi, W., Lim, J., Yang, M.-H.: Object tracking benchmark. IEEE Trans. Pattern Anal. Mach. Intell. 37(9), 1834–1848 (2015)

24. Zhong, B., Bai, B., Li, J., Zhang, Y., Yun, F.: Hierarchical tracking by reinforcement learning-based searching and coarse-to-fine verifying. IEEE Trans. Image Process. 28(5), 2331–2341 (2018)

25. Zhong, B., Yao, H., Chen, S., Ji, R., Chin, T.-J., Wang, H.: Visual tracking via weakly supervised learning from multiple imperfect oracles. Pattern Recogn. 47(3), 1395–1410 (2014)

26. Zhou, Q., Zhong, B., Zhang, Y., Li, J., Yun, F.: Deep alignment network based multi-person tracking with occlusion and motion reasoning. IEEE Trans. Multimedia **21**(5), 1183–1194 (2018)
27. Zhu, Z., Wang, Q., Li, B., Wu, W., Yan, J., Hu, W.: Distractor-aware Siamese networks for visual object tracking. In: Ferrari, V., Hebert, M., Sminchisescu, C., Weiss, Y. (eds.) ECCV 2018. LNCS, vol. 11213, pp. 103–119. Springer, Cham (2018). https://doi.org/10.1007/978-3-030-01240-3_7

Author Index

692   Author Index

Printed in the United States
By Bookmasters